W. Braune   O. Fischer

The Human Gait

Translators:

P. Maquet   R. Furlong

With 101 Figures in 170 Separate Illustrations,
Some in Colour, and 120 Tables

Springer-Verlag
Berlin Heidelberg New York
London Paris Tokyo

Wilhelm Braune †   Otto Fischer †

Translators:

Dr. Paul Maquet
25, Thier Bosset, B-4070 Aywaille, Belgium

Ronald Furlong, M.B., B.S., F.R.C.S.
149, Harley Street, GB-London W1N2DE, United Kingdom

Title of the original German edition:
Der Gang des Menschen. Published by B.G. Teubner, 1895–1904

ISBN-13:978-3-642-70328-7   e-ISBN-13:978-3-642-70326-3
DOI: 10.1007/978-3-642-70326-3

Library of Congress Cataloging in Publication Data. Main entry under title:
Braune, W. (Wilhelm) The Human Gait. Translation of: Der Gang des
Menschen. Includes index. 1. Gait in humans. 2. Walking. 3. Human
locomotion. 4. Human mechanics. 5. Body, Human.   I. Fischer, O. (Otto),
1861–1916. II. Title.   QP310.W3B7313 1987 612'.76

2124/3140-543210

Foreword

The different chapters of the present book were published separately each as a complete entity in the Proceedings of the Royal Saxon Society for Sciences. Chapter 1 appeared in 1895 under the names of Wilhelm Braune and Otto Fischer although Braune died immediately after the initial experiments, before the recordings had been interpreted. Chapters 2–6 were signed by Fischer only and appeared in 1899, 1900, 1901, 1903 and 1904.

Basic data needed for this investigation of the human gait had been provided previously. A research on the centre of gravity of the human body and its different segments by both authors was published in 1889, determination of the moments of inertia of the human body and its segments in 1892. So far only the first of these two works has been published in English. The other has been translated and awaits publication.

Springer-Verlag must be congratulated for the quality of this edition and for the care they took in reproducing the original figures. This was certainly no easy task. We thank them for the patience they displayed towards the translators.

Publication of the present book was made possible financially by Prof. M. Müller, Bern. We are grateful to him for his generosity and so will be the scientific community.

Aywaille and London, June 1987   *P. Maquet*
                                                  R. Furlong

Preface

The present work represents a further step in the series of studies which I had the honour of conducting with my distinguished teacher, Prof. Wilhelm Braune, medical advisor. It was our opinion that following the investigations of Weber and Weber, further understanding of the mechanics of walking could only be achieved empirically. We therefore investigated some mechanical properties of the human body in movement in a number of experiments. In the cadaver we measured the sizes and weights of the different parts of the body; we determined the position of their centres of gravity and the magnitude of their moments of inertia. In a study on the knee we made preliminary investigations on how the joints move the adjacent parts of the limb. To complete our investigations we undertook to ascertain, as accurately as possible, the movements involved in walking using photography and all other means at our disposal. We began this last piece of research because we were convinced that the existing recordings of the movements involved in walking, although they have broadened our knowledge of the human gait, are not really sufficient to shed light on the laws of movement in full detail and accuracy.

Only after many tedious and sometimes disappointing preliminary experiments, in which Prof. Braune was tireless as always when a new method of research had to be elaborated, did we believe we had found a pattern in the experiments from which we could draw sufficienty accurate results. The experiments themselves were very time-consuming and fatiguing. From ten to twelve hours of uninterrupted activity were often necessary since preparing the experimental subject required the utmost care, as did the accurate arranging and insulating of the Geissler tubes. Decisive experiments had to be carried out at night because there was no means of darkening the room in which we performed the studies. It is due to the enormous energy of Prof. Braune that every hindrance to the work was finally overcome and the experiments could be carried out despite all obstacles.

The data resulting from the experiments permitted the transposition of the process of movement into a system of tridimensional rectangular co-ordinates. The calculations involved were voluminous and required continuous work for several months. The High Ministry, which was most generous in providing us with exceptional means to plan the experiments, also assisted us in carrying out the calculations according to the equations we had established. I am greatly indebted to the High Ministry for its tremendous help in our endeavours.

Unfortunately, Prof. Braune was unable to enjoy the results of the research to which he had devoted all his energies and could not harvest the fruits of his labours. Death took him away in the middle of his work, even before the measurements of the co-ordinates on all the photographic plates were finished. Therefore, I had to carry out the further studies alone. I have attempted to perform this task as Braune would have done.

Leipzig, October 1894   *Otto Fischer*

Contents

Chapter 1   Experiments on Man, Loaded and Unloaded   1

Introduction   1
Description of the Experimental Method   12
Deduction of the Tridimensional Rectangular Co-ordinates
from the Serial Pictures   32
Finding the Tridimensional Co-ordinates of the Joint Centres .   59
Trajectories of the Joint Centres, Vertex, Centres of Gravity
of the Feet and Tips of the Feet   80
Rotations and Deformations of the Trunk   92
Rotations of the Hip Line   94
Rotations of the Shoulder Line   103
Rotations of the Trunk Line   107
Rotations of the Head   110
Gait of the Loaded Man   112
Conclusions   116

**Chapter 2   The Movement of the Total Centre of Gravity
            and the External Forces**   117

Introduction   117
Methods of Determining the Trajectory of the Centre of
Gravity   120
Trajectory of the Total Centre of Gravity of the
Human Body   137
Velocities and Accelerations of the Total Centre of Gravity . .   168
On the External Forces   198
Conclusions   203

**Chapter 3   Considerations on the Further Goals of the Research
            and Overall View of the Movements
            of the Lower Extremities**   205

Considerations on the Further Process and Final Goals
of the Investigation   205
General Behaviour of the Legs   217
Critical Analysis of the Leg Movements According
to Weber and Weber   231
Positional Angles of the Segments of the Lower Extremities .   233
The Angles of the Knee and Ankle Joints   252
Summary   253

**Chapter 4   On the Movement of the Foot and the Forces
            Acting on It** 255

Introduction 255
On the Forces Acting on the Foot and Their Torques 257
The Resultant Force Couples of the Muscles, of Gravity
and of the Effective Forces 269
The Movement Equations for the Feet 272
Overall View of the Activity of the Muscles of the Foot
During Walking 273
Velocities and Accelerations of the Centre of Gravity
of the Foot 276
Effective Forces and Momenta of Both Feet 300
Summary 301

Chapter 5   Kinematics of the Swing of the Leg 315

Introduction 315
On the Typical Walker's Step 318
On the Forces Involved During the Swinging Period and
How They Act on the Segments of the Leg 328
Angular Velocities and Angular Accelerations of the Rotation of
the Three Segments of the Leg During the Swinging Period .. 338
Velocities and Accelerations of the Centres of Gravity of the
Three Segments of the Leg During the Swinging Period 357
Summary 382

**Chapter 6   On the Influence of Gravity and the Muscles
            on the Swinging Movement of the Leg** 385

Introduction 385
The Components of the Effective Forces 387
On the Effect of Movement on Joint Compression 401
The Torques Exerted by Gravity 403
The Torques Exerted by the Effective Forces 421
The Torques Exerted by the Internal Forces 428
On the Activity of the Muscles During the Swinging
of the Leg 431
Summary 436

Name and Subject Index 439

Readers are advised to consult the Glossary on the overleaf page
preceding the Name and Subject Index, where they will find
English translations of the German terms appearing in the
diagrams.

Experiments on Man, Loaded and Unloaded

Introduction

The classical researches of Weber and Weber[1] on the mechanics of human gait have demonstrated that the movements of man, particularly in walking and running, are amenable to mechanical processing. This is based on the possibility of obtaining accurate knowledge of a particular movement by direct measurement. The results, empirically obtained, constitute the foundation for subsequent studies. One has to deal with a mechanical problem: from the movements carried out by the different parts of the human body the forces necessary to cause those movements have to be deduced. This problem can in principle always be solved if one understands the process of movement and if one knows the magnitude of the mass, the position of the centre of gravity, the magnitude of the moments of inertia of each part of the body and in what way the movements of the different parts of the body are conditioned by articulation and external forces such as friction against the ground. Conversely, difficulties very often oppose the precise solution of the reverse problem, that is to deduce the movements from a knowledge of the forces.

The conclusions that are drawn concerning the forces from the movements become more reliable with a deeper knowledge of the process of movement. Determining the range and type of movement, as accurately as possible, is of particular importance here. Natural phenomena cannot be completely understood because of the limitation of our senses and measuring instruments. Thus, there is a limit to our knowledge of the movements involved in walking and running and of the forces in action. However, this knowledge can be increased if greater accuracy can be attained in measuring.

From numerous measurements, Weber and Weber accumulated observations concerning the inclination and vertical oscillations of the trunk, the length of the leg in different degrees of extension, the periodicity of the free swinging leg with and without clothing, the velocity of the fastest gait, the relationship between the duration and the length of the step and other magnitudes and relationships regarding the forward movement of man. These observations provided Weber and Weber with the foundations for a theory of walking and running. Hitherto, they have not been outdated by any measurements of a similar kind, and they remain valid. Weber and Weber had indeed obtained the maximum possible information with the means at their disposal, and the conclusions drawn by the two authors from their direct measurements are very reliable. For instance, it is now accepted as a fact that the trunk comes closer to the ground during walking than in standing, and the faster one moves the shorter the distance. Or, to cite another example, the measurements of Weber and Weber have established beyond any doubt that the double limb support period of walking is shorter the faster a person moves.

However, Weber and Weber took the liberty of drawing conclusions which did not result directly from their measurements and for which they could not provide more accurate

---

[1]   Weber W, Weber E (1836) Mechanik der menschlichen Gehwerkzeuge. Göttingen. Also in: Weber W, Gesammelte Werke, vol 6

observations with the means then available; these conclusions are of course much less reliable. For example, Weber and Weber had no way of determining empirically, with sufficient accuracy, the movement and changes in the shape of the leg during a double step. They could only measure the alternate shortening and lengthening of the leg, the rising and falling of the trunk and the length of the step. They deduced the positions and changes in the shape of the leg from these measurements. The same applies to the arm. Only approximate results can be obtained in this manner, because a slight inaccuracy in the direct measurements entails a great deviation in the calculation of the position of the segments of the leg or arm.

A new series of measurements with new and better methods is, therefore, required to correct the results deduced by Weber and Weber and to provide new insights into the movements of the limbs and the external and internal forces involved in walking and running. However, any further results can only represent extensions of the fundamental work carried out by Weber and Weber in their classic studies.

The mechanics of the locomotor apparatus as conceived by Weber and Weber aroused an interest in human gait. Earlier, only individual anatomists, mathematicians and physicians, such as Gassendi, Borelli, Haller, Barthez, Magendie, Gerdy and Poisson, had turned their attention to a study of walking and running, and then they mostly considered one aspect of the movement process. After the publication of Weber and Weber's researches, discussions on the mechanism of walking became common in almost all important works of anatomy and physiology. In many cases these discussions reproduced the results of Weber and Weber and the conclusions they had drawn. Only a few authors expressed new points of view and new ideas opposing the theories of Weber and Weber. Still fewer authors presented new facts discovered empirically concerning walking and running. Doubt has been expressed about the validity of several conclusions and concepts of Weber and Weber. However, in most cases it is not possible to present empirical data which irrefutably demonstrate their falsity. For instance, much has been written against Weber and Weber's concept that during walking the leg not on the ground swings forwards from behind, like a pendulum moved by its own weight. This, however, is only an opinion; this and similar points do not force one to abandon the theories of Weber and Weber. In most cases it is objected that although the main swinging effect of the leg results from gravity, at the same time muscular action accompanies that of gravity. However, to assume that Weber and Weber deny any accompanying muscular contraction is to seize too literally the concept of pendulum movement of the leg. Indeed, they write (op. cit. §17):

"The leg alters its shape when hanging from the trunk and swinging like a pendulum. If it remained extended as it is the instant it is raised from the ground, it would strike the ground and could not swing freely beneath the trunk. Consequently, it is flexed at knee level and thus shortened. . . .
The shape of the swinging leg is altered in the opposite way at the end of its swing when it is again placed on the ground. It is lengthened by extending the knee until the heel strikes the ground. . . . (§104) When swinging forwards during walking, the leg must be shortened so as not to come into contact with the ground, so much so that the hip joint is closer to the ground when walking than when standing."

One could perhaps assume that flexion and extension of the knee accompanying the swinging of the leg should be attributed to the same action of gravity. Clearly, an articulated pendulum, as represented by the leg, does not behave in the same way in swinging as a rigid pendulum. The two articulated parts move in relation to each other when swinging. For example, when a large bell is rung it represents an articulated or double pendulum. The bell swings about an axis. The clapper in turn is articulated with the bell. When the bell is rung it is made to swing like the pendulum of a clock; it is moved from its position of rest and is then subjected to the action of gravity. If the articulated system of the bell and its clapper behaved

CHAPTER 1

as a rigid body not even the most extreme swinging would produce any sound, since the clapper would always maintain the same position relative to the bell. The clapper strikes the bell only because it moves at its joint with the bell when the whole system swings.

The leg behaves similarly when it is moved from its position of rest under the fixed pelvis and subjected solely to the action of gravity. The thigh can be compared with the bell, and the lower leg and foot with the clapper. The thigh and lower leg generally carry out rotations of different magnitudes resulting in flexion or extension of the knee. It is possible to imagine that, if the posterior leg is put into the position from which the swing begins, this movement of the knee will proceed as it does in human gait. The initial shortening and subsequent lengthening of the whole leg observed by Weber and Weber could then be attributed solely to gravity and not to the action of muscles. If this had been the conviction of the two authors they would have certainly expressed it in the detailed explanation of their ideas and demonstrated it again. This, however, they did not do; on the contrary, they said that the leg is extended at the knee. Thus, they did not expressly deny voluntary contraction of the muscles in swinging the leg, as assumed by many researchers. They left open the question of whether muscular action is necessary for the shortening and lengthening of the swinging leg since they could not arrive, from the means at their disposal, at a definite conclusion. They did not have a sufficient basis to discuss this question as they lacked precise knowledge of the successive alterations in the shape of the leg during swinging, and experiments and theories of the swinging of a double pendulum did not exist.

It would be objected that the empirical foundations were insufficient for Weber and Weber's assumption to be considered an irrefutable truth, according to which the swinging leg behaves as a pendulum moved only by gravity, except for its shortening and lengthening. This objection is justified and the authors would have been the first to recognize its truth. In their work the pendulum theory appears only as a hypothesis which explains all the processes in walking they directly observed. Measuring the velocity during the fastest gait (Weber and Weber, op. cit. §101) they always found the same value whenever they repeated the experiment and irrespective of whether their subject was rested or tired. Therefore, they concluded:

"As long as the muscles retained some strength to carry out the movement, the velocity depended not on the strength of the muscles but on the length of the legs and the external force acting on them. ... From many experiments (§102), concerning not only the duration of a single swing of the leg but also the duration of a step when walking fast, it appeared that the proportion between these two lengths of time was almost exactly 2:1. In order to convince themselves that the simplicity of this proposition was not something individual, they repeated the experiments on many subjects.... From this correspondence between the duration of the step in walking fast and the half length of time of the pendulum swing of the leg, they concluded that (§104) the former is determined by the latter: the leg raised from the ground behind swings forward, moved only by gravity (§17). ... When the leg strikes the ground before ending its swing, the exclusive action of gravity which has brought it there thus makes the precise repetition of the steps, with the same duration, extraordinarily easy. Then, without being conscious of the leg, one is aware that the leg will repeat each part of its swing course in the same time and that, in a certain time after the beginning of its swing, it will have taken up a particular position in relation to the rest of the body. The mechanism whereby the leg hanging from the trunk swings forward from behind, like a pendulum, is very useful in that the steps follow each other with a certain similarity of length and duration."

The hypothesis of the pendulum swing of the leg explained, above all, to Weber and Weber why successive steps were carried out with such similarity of length and duration. Vierordt[2]

---

[2] Vierordt (1881) Über das Gehen des Menschen in gesunden und kranken Zuständen. Tübingen, p 42

was against the theory of pure swinging. He wrote: "I must confess that the theory of pure swinging, which excludes any participation by the muscles during the swinging of the leg, does not seem very plausible to me. Could so precise a movement depend only on the result of gravity?"

This hypothesis of the pendulum swing of the leg and the whole of Weber and Weber's theory of walking fit very well with the then current knowledge of the process of movement. This naturally does not preclude the possibility that the hypothesis may be replaced by another hypothesis and perhaps by another theory should facts be discovered which directly contradict it. This can only result from more accurate knowledge of the process of movement arrived at by measurements obtained using more sophisticated methods. Among the more recent researches on human gait, only those which empirically establish new data go beyond the mechanics of the locomotor apparatus of Weber and Weber and represent progress.

The present work initially deals only with the experimental determination of the process of movement, without considering the cause. At first, we shall only consider experiments that precisely record movement. Works which aim at determining the causes of locomotion within and outside the human body, such as those of H. von Meyer, Henke, Pettigrew, A. Fick, Strassen and Duchenne, will be considered in later studies.

Carlet[3] previously recorded the movements in walking using new methods in the laboratory of Marey. Firstly, he endeavoured to consider the magnitudes of the length and duration of the steps, the duration of the stance and swing period of the legs, oscillations and inclinations of the trunk, etc. These were automatically recorded on the drum of a kymograph. His researches thus constitute progress in that, for the first time, they involved an experimental determination of the trajectory in space which a point of the human body follows during walking. An ingenious mechanism enabled Carlet to record simultaneously on a kymograph the vertical as well as the horizontal oscillations of a point of the body. In this way he found, for instance, that a point on the pubic symphysis describes a curve which can be likened to a furrow with its convexity pointing downwards. The minima are at the bottom of the furrow and the maxima at its edges. The generatrix of this half cylinder is parallel to the direction of walking. The minima correspond to the middle of the double-support periods and the maxima to the middle of the single-support periods.

Vierordt attempted to record the parameters of walking in space and time in a more direct way through a series of elaborate experiments. These involved a special mechanism on the shoe of the walking subject which printed the position and long axis of the foot on the floor or on paper when the foot struck the ground. As a result he could measure with great accuracy the length of the different steps and the average and longest step for each leg, the width of the track, the angle formed by the long axis of the foot and the direction of gait when the foot struck the ground, the average sideways deviation of the heel of the homolateral foot during the double step and other data relevant to the knowledge of the locomotor apparatus. In this way he arrived at very interesting data about the behaviour of one leg and both legs during the course of several steps. Weber and Weber could not have arrived at these data by their method, which consisted of deducing the step length by dividing the length of the whole distance by the number of steps.

Further, Vierordt undertook to record by direct graphic means the movements of the legs and arms during walking. This involved small nozzles being fixed to different parts of the body. They were connected by thin rubber tubes to a glass tank full of coloured fluid. The tank was carried on the back of the experimental subject at shoulder height. During walking, the coloured fluid squirted out of the nozzles with great velocity in a thin stream and marked out curves on paper spread on the ground or held vertically at the side. The nozzles were attached to the foot, the arm and at two places each on the upper and lower leg and on the

---

[3] Carlet (1872) Essai expérimental sur la locomotion humaine: Etude de la marche, vol 15. Annales des Sciences Naturelles: Zoologie

trunk. On the foot, the two nozzles were at right angles to each other, one vertical, the other horizontal, as the subject stood erect. The other nozzles were horizontal and perpendicular to the median plane of the body, like the horizontal foot nozzles. Only one horizontal nozzle lay in the median plane on the trunk itself. The vertically orientated nozzle on the foot gave, during walking, a horizontal projection, and the horizontally positioned nozzle a vertical projection of the trajectories of the corresponding parts of the body. The horizontal trunk nozzle in the median plane demonstrated the sideways displacements of the trunk in relation to the floor.

It is obvious that this "projection" method presents considerable sources of error, and this was not denied by Vierordt. The vertical nozzle was subjected to great variations of directions since it was attached to the foot; the lateral oscillations of the body caused excessive variations in the distance between the horizontal nozzles, which all except for one sprayed onto the vertical surface of the paper, and the plane of projection. A reliable picture of the curves of movement thus could not be provided. The same applies to the median trunk nozzle. Furthermore, the nozzles were not positioned on parts of the body at rest or which moved evenly but on parts affected by simultaneous different movements.

Though it was later replaced by a much more accurate technique of projecting on planes the trajectories of different parts of the body during walking, the method of Vierordt cannot be denied a certain historical value. It remains the first experiment to record simultaneously the trajectories of different parts of the body. It could, even today, provide the medical practitioner with a means of obtaining an approximate overall view of abnormalities of walking and running occasioned by particular pathological conditions, though photography is an incomparably better method. The research worker in the field of physiological mechanics would, however, find the projection methods to be of little use. Quick and easily obtained results are not so essential here, and the research worker can spend time and trouble to achieve as accurate a knowledge as possible of the laws operating during walking and hence gain a deep insight into the forces involved.

Finally, Vierordt also recorded the "time-related conditions of the movements in walking". By means of an appropriate mechanism, he was able to record the duration of the steps, the length of time of the swing period and the time of the support of the whole foot, the heel and the forefoot alone on the drum of a kymograph. Whereas Carlet, following his teacher Marey, used air as the transmission medium, Vierordt used electricity for his measurements of time. With this very accurate experimental method he again obtained very interesting data concerning the variable behaviour of one leg and of both legs in the course of several steps. Weber and Weber had only deduced the average duration of one step by dividing the length of time of several steps by the number of steps. They also measured directly the duration of the single-support period. From these data they then calculated the other relevant times, such as the duration of the swing of a leg and the duration of the double-support period. They assumed not only that each leg repeated its movement exactly during successive steps, but also that both legs behaved in exactly the same way in walking. Thus they appear to have ignored the dissimilarities which Vierordt considered as normal in walking.

The physiologist Marey of Paris must be credited with a great advance in the knowledge of the mechanics of the locomotor apparatus. He devised a completely new method of research, making practical use of chronophotography for the direct measurement of the process of movement. This allowed not only the projection of movements of parts of the body, with the utmost accuracy on a plane, but also permitted the time relationships to be taken into account.

Photography as a means of determining the phases of movement was first employed by the American photographer Muybridge of San Francisco. He was the first to photograph a series of successive movement phases of a horse. To do this, he set up a series of cameras close to each other, which were briefly opened one after another at short intervals. The accomplishment of Muybridge is remarkable since photographic plates at the time did not have the high degree of sensitivity they now possess, which enables instantaneous pictures to be taken. Photographing the successive movement phases during the short exposure time necessitated by the fast movement of the horse was made possible by a white sheet in front of which the horse moved. This sheet was positioned so as to reflect the rays of the sun towards the camera. Consequently, the horse appeared as a dark body sharply outlined against a very light background. These first instantaneous pictures are still imperfect since they show the silhouette of the animal but almost no details. However, they permitted the clear recognition of the different positions of the legs and head during the successive phases of movement. This first series of instantaneous pictures was described by Willmann[4]. The publication was sponsored by Stanford, Governor of California, whom Muybridge invited to the experiments.

Encouraged by the success of his first experiments, Muybridge then undertook to take instantaneous pictures of the phases of movement of other animals and, most importantly, of man. After the discovery of the extremely sensitive bromine-silver-gelatine plate, he soon obtained instantaneous pictures on which both the outlines and details appeared with the greatest accuracy. He then took a series of excellent pictures of man and different animals during all kinds of movement. Some of these pictures were taken simultaneously from two different sides.

After the lead of Muybridge, the photographer Anschütz, previously in Lissa, now in Berlin, and Londe, Directeur du Service Photographique de la Salpêtrière[5], photographed movement phases of man and various animals from one side. Both achieved a high degree of perfection. The achievements of Muybridge, Anschütz and Londe are very important for artists, particularly those who depict people and animals in motion.

The use of photography as a scientific research tool and the improvement of cameras to this end are due, above all, to Marey as mentioned above.

To permit comparison between the successive instantaneous pictures of the same movement and to achieve reliable results, it was necessary that the cameras were directed exactly parallel to one another, and that the pictures were of the same magnification. The distances between the axes of the cameras, standing side by side, had to correspond to the phases of movement and the different cameras had to be optically similar. The latter prerequisite could to a certain degree be fulfilled but not the former. Indeed, the distance between the different cameras depends on the velocity of the moving body, which generally is not known beforehand and must be different for each body and each type of movement of the same body. In the experiments of Muybridge, Anschütz and Londe, the cameras standing side by side were separated by a predetermined distance that was not related to the velocity of the photographed body. If the pictures of the phases were to be compared, it was necessary to take into account the different directions from which these pictures were taken. This is significantly easier if all the directions intersect at one point, i.e. if the exposures are taken from the same point. This also corresponds much better to what we see when we follow a moving object with our eyes.

Marey[6] thus took a series of pictures with a single camera, and this represented considerable progress in the use of photography to analyse the movements of humans and

[4] Willmann (1882) The horse in motion, as shown by instantaneous photography. Turner, London

[5] Londe (1893) La photographie médicale. Gauthier, Villars et fils, Paris; (1894) La photo-chronographie appliquée aux études médicales. Internationale Medizinische-photographische Monatsschrift, vol 1, p 9

[6] Marey (1882) Sur la reproduction, par la photographie, des diverses phases du vol des oiseaux. Comptes rendus, vol 94. Naples, p 683; (1882) Photographies instantanées d'oiseaux au vol. Comptes rendus, vol 94, Naples, p 823; (1882) Emploi de la photographie instantanée pour l'analyse des mouvements chez les animaux. Comptes rendus, vol 94. Naples, p 1013

animals. His first endeavour was to record a series of successive phases of the movement of a bird in flight whereas Muybridge had obtained only separate instantaneous pictures of such a flight. Marey built a camera which resembled a shotgun; thus, he was able to aim the camera at the bird and follow its flight. The sensitive plate was larger than the back wall of the camera and it could be rotated quickly about an axis outside the camera and parallel to the optical axis. In this way, fresh areas of the plate always formed the back wall of the camera prior to exposure. A special device kept interrupting the rotation of the plate briefly whilst an instantaneous picture was taken. The exposure was 1/700 s since Marey could make use of sensitive bromine-silver-gelatine plates. The plate rotated about its axis in 1 s, and in so doing was stopped 12 times while the camera admitted light. In this way, Marey obtained a dozen pictures of a flying bird in one second. The pictures were regularly distributed round the periphery of the sensitive plate.

Marey[7] achieved a further improvement in experimental photographic methods by recording the different phases of the movement of a man on the still plate of an ordinary camera. A man illuminated by sunlight walked against a dark background (a cave). The part of the plate where no picture was yet printed retained its sensitivity longer because of the dark background. The various phases appeared through a fenestrated disc rotating in front of the lens. Later, Marey[8] replaced this disc by a wheel with ten spokes which rotated 10 times/s. In this way, 1/100 s elapsed between the beginning of two successive phases. With this accumulation of the phases, the pictures are superimposed in many areas in such a way that it is difficult and sometimes impossible to disentangle them. This difficulty can be overcome either by moving the plate or by photographing only a very small part of the moving body. Marey applied and developed both methods.

For the photographic rifle, as he called his first camera with the moving plate, described above, the sensitive plate was as usual made of glass. However, the weight of the glass plate did not allow the frequent movements and stoppings that would be required if the number of phases were increased beyond 12/s. Therefore, Marey[9] used a thin film covered by a sensitive layer. The film was stretched between two rollers and was wound quickly from one roller to the other in such a way that the area of film between the two rollers formed the back of the camera. With this new device Marey achieved a series of consecutive phase pictures of very high quality. Although these clearly represent changes in the position of the moving body, they do not immediately show the velocity of the movement at different times. This is because the phase pictures are separated in one direction by the distance travelled by the film between two exposures and the camera cannot be kept absolutely immobile in this setup. In order to measure the velocity accurately an absolutely constant movement of the film is necessary as well as complete absence of vibration of the camera.

For accurate results, the sensitive plate had to be kept immobile and only those parts of the body that were essential to the attitude were to be photographed. Marey fulfilled these prerequisites by dressing a man in black and attaching white stripes to the suit between the main joints of the extremities[10]. With this arrangement he could photograph on the same plate a considerable number of movement phases of a man or animal moving in any way

---

[7] Marey (1882) Analyse du mécanisme de la locomotion au moyen d'une série d'images photographiques recueillies sur une même plaque et représentant les phases successives du mouvement. Comptes rendus, vol 95. Naples, p 14

[8] Marey (1882) Emploi de la photographie pour déterminer la trajectoire des corps en mouvement, avec leurs vitesses à chaque instant et leurs positions relatives. Application à la mécanique animale. Comptes rendus, vol 95. Naples, p 267

[9] Marey (1890) Appareil photochronographique applicable à l'analyse de toutes sortes de mouvements. Comptes rendus, vol 111. Naples, p 626

[10] Marey (1883) Emploi des photographies partielles pour étudier la locomotion de l'homme et des animaux. Comptes rendus, vol 96. Naples, p 1827

without mutual disturbance of the pictures. Since the very short time intervals between the phases were equal, the series of exposures made by Marey provides a true projection of the whole process of movement in relation to space and time.

If all the points of a human or animal body moved in one plane during walking or running, these series of pictures would represent not only a one-sided projection but also a true picture of the whole process of movement. However, movement in space has to be taken into account, whereby the centres of all joints describe double curves. Thus, the projection achieved by Marey's method is insufficient to describe completely the movement in space. For instance, when pictures are taken perpendicular to the direction of gait they give an idea of the movement in the direction of gait and of the simultaneous rising and lowering of the individual parts of the body, but one has no insight into the sideways oscillations of the body. More precisely, these lateral exposures do not even provide an accurate means of determining the movement forwards and upwards. This would only be the case if one had parallel projections on the plane of gait. Photography, however, does not deliver parallel projections, rather a central projection of the body on a plane perpendicular to the optical axis of the camera. This is easily explained. The central system of lenses contained in the objective of the camera breaks up the light rays in such a way that a ray directed to point K of the system of lenses is, after refraction, orientated in the camera as if it came from another point K' (Fig. 1). These two particular points lie on the axis of the lens system and are called the nodal points of the dioptical system. In cameras they are usually close to each other in such a way that, for many measurements, they can be replaced by a single point located between them without adding any significant error to the other sources of errors unavoidable in measurement. The point which replaces the two nodal points is called the optical middle point of a single lens, if the thickness of the latter is discounted. If one assumes one single optical middle point O, all the light rays directed through O emerge from the lens system apparently unrefracted and undisplaced.

If in a measurement all sources of error are reduced to the minimum possible, it can happen that the error due to the replacement of two nodal points by an optical middle point becomes so great compared with the other unavoidable errors as to impair the accuracy. In such cases,

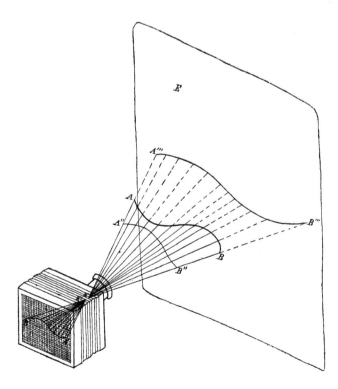

Fig. 1. Projection of a curve on a photographic plate

one must of course retain the two nodal points. Nevertheless, one can easily imagine the course of the light rays and the reproduction of the picture on the photographic plate by assuming that the beam of refracted light comes from the incident beam and is displaced towards the optical axis by the distance separating the two nodal points. This is how the two beams are drawn in Fig. 1. The rays of the incident light are formed by the points of the double-curved line AB. The image $A'B'$ imprinted on the photographic plate by the rays of refracted light is in general also a curve. However, this curve $A'B'$ is not usually similar to the curve in space AB. This results from the fact that any other curve $A''B''$ situated in the area formed by the incident light must give the same image, and also that the curved image $A'B'$ lies in one plane. Even if AB were a plane curve, the curve $A'B'$ could not have the same shape. For instance, a circle generally appears as an ellipse on a photographic plate. The curve on the picture can present the same shape as the curve AB only if the latter lies in a plane perpendicular to the optical axis. If the beams of the incident light are prolonged beyond the curve AB until they meet the plane E perpendicular to the optical axis, they imprint on this plane the curve $A'''B'''$ which is similar to the curve $A'B'$. The distance between this plane and the camera is irrelevant. The picture curve $A'B'$ remains similar to the intersection curve $A'''B'''$ whether the plane is moved away from or closer to the camera. Only the proportional relationship between the two curves $A'''B'''$ and $A'B'$ changes.

Curve $A'''B'''$ is nothing but a central projection of the curve in space AB on the plane E, in which the first nodal point K acts as the centre of projection. Thus, photography gives a true, usually reduced, picture of the central projection of the body on a plane perpendicular to the optical axis of the camera. Consequently, the picture of the movement curves, such as those obtained by Marey on one plate, will be subject to the same defects associated with central projection.

If, for instance, two points A_1 and A_2 move with the same velocity in the same direction parallel to the plane of projection E, the equal velocity can be recognized on the central projection only if the two points are equidistant from the projection plane. If one of the points, for instance A_2, is further away from E than the other A_1, A_2 will traverse a greater distance in the same time on the projection than A_1. This is shown in Fig. 2, in which the projection plane is perpendicular to the plane of the drawing and intersects the latter at EE. K is the centre of projection, $\overline{A_1B_1}$ and $\overline{A_2B_2}$ are the equal displacements of A_1 and A_2. The projection of the individual points is indicated by the horizontal bars. It is immediately obvious that $\overline{A_2'B_2'}$ is longer than $\overline{A_1'B_1'}$.

The photography of a movement thus cannot give a precise comparison between the velocities of two different regions of the body, even when it is known that both movements are perpendicular to the optical axis of the camera. If, for instance, the lateral view of human gait shows that at a certain time the hip and wrist move forwards with the same speed one should not assume that at that moment both joints actually move forward with equal velocity. On the contrary, from the previous argument it would appear that in this case the wrist has moved a little more slowly. The error here depends both on the evaluation of the

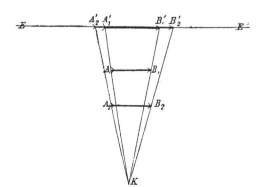

Fig. 2. Projection on the plane EE of two points moving from A_1 and A_2 respectively to B_1 and B_2 with the same velocity

distance between two moving points in the direction of the optical axis of the camera and on the distance between the camera and the photographed object. The error is greater the longer the former distance and the shorter the latter, and vice-versa. The error changes almost proportionally to the relationship between the two distances. If the camera is very distant and the two points relatively close to each other, the error can be so small that it does not exert any significant influence on the accuracy of the measurement. However, as a result of the great distance between object and camera the picture becomes very small, which diminishes the precision of the measurements. If very accurate measurements are required the camera must not be too distant. But then a source of error arises from the different distances of the different points from the camera, and this becomes significant. One must always strive for great accuracy if one wishes to determine by direct measurement not only the different positions of a moving point but also the velocities of the movement of this point.

Not only do unverifiable errors occur when comparing the velocities of various points, but false conclusions can be drawn with regard to the velocity of a point in its different positions from photographs of the trajectory of this point.

In Fig. 3 the curve AB represents the trajectory of a point moving in a horizontal plane. The small segments $\overline{A_1 B_1}$, $\overline{A_2 B_2}$, $\overline{A_3 B_3}$ are traversed in the same time. The figure shows that in A_1 and A_3 the point has about the same velocity. In A_2 its velocity is much less. From the central projection one would have concluded that the velocity is the same in A_1 and A_2, and that it would be greater in A_3.

The central projection does not even allow conclusions to be drawn from the reciprocal position of the points of a figure at rest. Figure 4 represents the projection of a quadrangle situated in a vertical plane. Unlike Figs. 2 and 3 which view the process from above, this figure is a lateral view. Although in reality the corner R lies higher than S, the latter appears higher in the projection on plane E; the two corners P and Q actually lie close to each other in the same horizontal plane, though P appears to be closer to the floor.

A central projection thus generally appears to be insufficient for analysing processes of movement if one limits oneself to examining the movement in one determined direction or parallel to one determined plane. The chronophotographic series of pictures obtained by Marey are likewise insufficient to determine exactly the laws of movement despite their value and the progress they represent for the analysis of human gait.

In order to record completely a movement in space, at least two simultaneous projections are necessary from directions as different as possible. It is irrelevant whether these projections are central or parallel or in which direction they are obtained. For discussion and further evaluation, rectangular parallel projections in two perpendicular planes are particularly appropriate since such projections immediately describe the movement in a fixed tridimensional rectangular system of co-ordinates. The fact that photography gives only central projections makes the transposition into a system of co-ordinates less easy, but not impossible.

Two simultaneous photographic exposures of the same movement are sufficient to determine the movement in any direction. Based on this point of view, we have for years used "two-sided chronophotography", as it could be named, to measure the movement of the lower leg in relation to the thigh in the living man[11]. The method we used was a follows. Two-sided chronophotography requires the cameras to be opened and shut at short intervals at precisely the same time. This requirement can only be achieved using a highly complicated mechanism. Therefore, to interrupt the exposure, we relied not on shutting the camera but on altering the photographic object itself, so that it was possible to dispense with a particular mechanism for shutting the camera.

---

[11] Braune W, Fischer 0 (1891) Die Bewegungen des Kniegelenkes nach einer neuen Methode am lebenden Menschen gemessen. Abhandlungen der mathematisch-physischen Klasse der Königl. Sächsischen Gesellschaft der Wissenschaften, vol 17, no 2. Leipzig

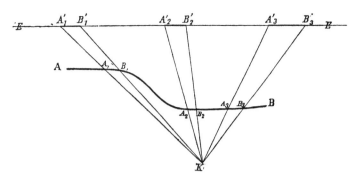

Fig. 3. Projection on the plane *EE* of the trajectory of a point moving from *A* to *B*. The times for moving from A_1 to B_1, from A_2 to B_2 and from A_3 to B_3 are equal

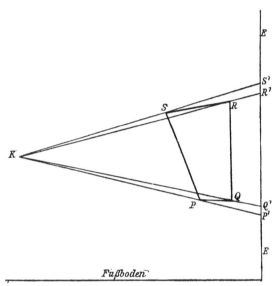

Fig. 4. Projection of a quadrangle *PQRS* on a vertical plane *EE*

We proceeded as follows. In order to photograph, say, the movement of a point on the lower leg, we used an electric current to produce intermittent lighting. To this end, the secondary current from a Ruhmkorff coil was conducted, using appropriate equipment, to the region to be studied, where we interrupted the circuit. The wires ended as fine points, close to and opposite each other. When the induction coil was activated, sparks of light sprang between the two points. Despite the very short duration this gave a very clear and sharply delineated picture on the photographic plate, not blurred by the movement of the lower leg. The duration of the electric sparks was less than $1/1\,152\,000$ s, as measured by a Wheatstone bridge. Therefore, even during the quickest movements carried out by a solid body on the surface of the earth, this body does not significantly alter its position during the electric spark.

The electric spark thus possesses two properties which make it invaluable for taking instantaneous pictures: its great luminosity and its short duration. Furthermore, it appears particularly appropriate for two-sided chronophotography because it emits rays of equal intensity in all directions. When Marey carried out his series of pictures of human gait, he illuminated the white stripes of a black tight-fitting suit worn by the subject by utilizing sunlight in such a way that the greatest amount of reflected rays were directed towards the camera. His object was thus illuminated, not self-illuminated. However, a body lit by the sun does not reflect rays evenly in all directions. If, therefore, Marey positioned his camera in a similar way to Muybridge to capture as many of the reflected rays of light as possible, any other direction became unfavourable for photographic exposure. As a consequence, in order to obtain other equally good pictures, a longer exposure was required. A longer exposure brings with it a disadvantage, namely that a moving object is less well defined. In general, therefore, a series of simultaneous pictures taken from different sides and lit by the sun is unevenly delineated and does not provide values evenly suitable for measurement. This drawback could perhaps be overcome by the use of mirrors. But then another source of error could remain: the white stripes would not appear to be the same width from all sides.

A secondary but practical advantage of our method, not to be underrated, is that using an electric spark one does not depend on the sun or weather but can work indoors.

We are not mistaken in considering the electric spark as a useful tool for the analysis of joint movement in the living subject. The results obtained from two-sided chronophotography of movement of the lower limb in four experiments carried out at different times in two different individuals show a surprising concordance. These results encouraged us to record human gait by two-sided chronophotography and thus to transpose the whole process of movement into a system of tridimensional co-ordinates. This could not be achieved using any of the previous series of pictures. Muybridge had taken pictures simultaneously from different sides. His serial pictures, however, as previously stated, were taken from different standpoints. Furthermore, they resolved the step into too few phases to enable one to deduce from them all the peculiarities of the movement. In particular, it is not possible to deduce from them the velocities with sufficient accuracy. The laws of movement of a body can only be completely known if one knows not only the trajectories of different points of the body but also the velocity of these points everywhere along their trajectories and the acceleration of the body during the movement.

In order reliably to deduce the velocities and accelerations of the different parts of the walking subject from the co-ordinate data, these data must be very accurate indeed. Every conceivable precaution must be taken in planning the experiment and every measurement must be improved as much as the dimensions will allow and the available instruments permit. Every effort should be made to avoid any source of error, especially with regard to simplifying assumptions, necessary in all such researches on living subjects. One cannot deduce results with unsatisfactory tools. The results can only be arrived at by using precise methods and the means with which mathematics provides us.

The following method of research is based on these beliefs.

Description of the Experimental Method

In order to analyse walking we modified the method which we had used to measure movement of the knee: we illuminated different parts of the human body by Geissler tubes rather than electric sparks. These tubes were filled with rarefied nitrogen since, when incandescent, this gas emits many chemically active rays. In order to record the successive positions of the human body in walking without taking into account the contingent deformations of the hand and foot, it would have been sufficient to attach small tubes to all the big joints. However, we decided to use long, thin, straight tubes, so as to provide a better overall view of the changes in the position of the body and a more reliable determination of the successive phases of the different curves of displacement. The tubes were attached to, e. g. the upper arm and thigh (between the two adjacent joints), the forearm, lower leg and foot (from one of the two joints to beyond the middle of the limb).

The experimental subject wore a black jersey suit (Fig. 5), similar to the one used by Marey. The suit provided a dark background for the tubes and permitted better attachment of the tubes to the body. The tubes could not be fixed directly to the clothing as the subject would have been at risk from the electric current. Therefore, we cut long strips of thick gutta-percha thickened at the ends with several layers of discs of the same material. We sewed these strips onto the jersey in the direction of the long axes of the various parts of the body, just as Marey did with his white stripes. To ensure that they were immobile with respect to the adjacent part of the body, we fixed them additionally with rubber straps as can be seen in

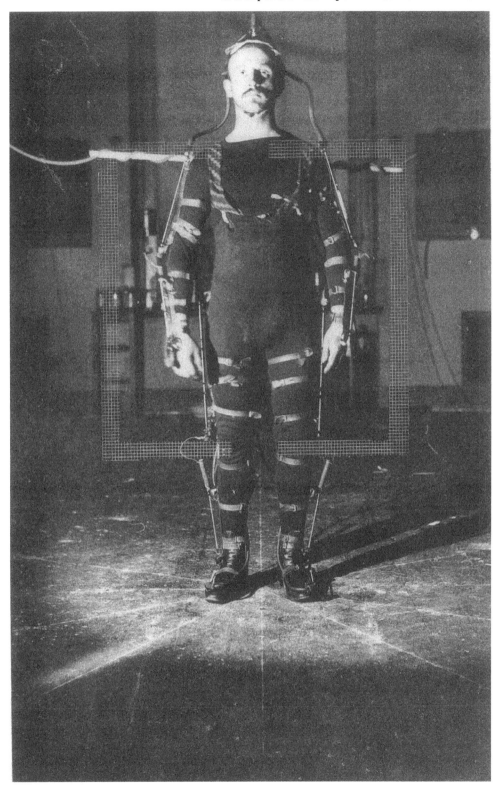

Fig. 5. Subject wearing the experimental suit

Fig. 5. We took particular care in calculating the dimensions of the gutta-percha strips and in fixing them to the body so that they would not hinder or influence movement in any joint not intended to be recorded, as for example the wrist.

Each straight Geissler tube rested on the thickened parts of the insulating strip at its two ends and thus could not come into contact with the body. This ensured complete insulation of the tube and also reduced the risk of breakage. Thick rubber sleeves were used to cover the somewhat widened ends of the Geissler tubes. These rubber sleeves were also intended to insulate the fixation points of the fine wires which connected the tubes of the various parts of the body; at the joints they forced these wires into a wide loop spanning the joints sideways. This appears clearly at the right knee in Fig. 5. Therefore, it was not necessary to insulate these short connecting wires better, which may have hindered joint movement. The connecting wires which did not span a joint but ran over a whole segment of the body were particularly well insulated in thick rubber sleeves. Perhaps we were overconcerned with regard to the insulation and could have saved ourselves a great deal of work since it usually took us between 6 and 8 hours to dress the experimental subject. However, we thought that the subject would walk naturally if he knew that the electric current, of which so many people are afraid, would not come into contact with his body. Naturally, before the experiment we allowed him to walk for a long time with the inductor working in order that he should feel quite safe. This gave us the opportunity to observe with the unaided eye the whole process of movement in its different phases. Similarly, in the Gassiot star experiment, one sees the successive phases of movement with a Geissler tube rotating about an axis. With this arrangement one has excellent means of demonstrating before an audience the course of a movement which otherwise would be too fast to be followed by the unaided eye.

Eleven tubes were used, one for the head and one for each thigh, lower leg, foot, upper arm and forearm. When walking, each hand and the forearm of the subject moved as one. The eleven tubes were connected in series to the secondary winding of a large Ruhmkorff induction coil. In order that the wire from the coil should not hinder the free mobility of the subject, a light wooden rod was fixed across his shoulders; at either end of the rod the incoming wires were connected to the circuit on the body. These wires connected to the Ruhmkorff coil were hanging from the ceiling on both sides of the gangway; they reached the ends of the wooden rod by forming a large loop on either side. Thus, the subject could walk about 10 m, the length of the room necessary for the experiment, without being impeded in his movements by electric wires.

In some places the Geissler tubes were surrounded by narrow rings of black Japan varnish; these places thus appeared as short interruptions in the line of light in the pictures. They were located near the ends of the tubes and at the same level as the centre of the joint. They thus marked the corresponding positions of the joints as isolated points of light in the photographs, as is shown in Fig. 5. Similar black rings were also located at the places corresponding to the centre of gravity of each segment of the body; the different centres of gravity appeared, then, as black dots on the white lines in the photograph. In the same way, a point was marked on the small head tube to be used as a guide mark for measuring the movement of the vertex. Finally, the two points dividing the thigh tubes into three equal parts were each marked by three rings close to one another, the middle one being very thin. This was done so as to facilitate the determination, by construction or calculation, of the location of the centre of the hip joint in the positions of the body in which the centre of the hip joint is hidden by the forearm.

The points at the ends of the Geissler tubes of course cannot immediately give the location of the corresponding joint centre for the different phases of movement. Neither could the position of the extremities of the white stripes of Marey give this location. Photography can never indicate directly the position of a joint centre. It gives the projection of the joint centre somewhere on the surface of the body. The points on the Geissler tubes were only to be considered as guides, enabling us to determine the true position of the joint centres. It follows, therefore, that they had to be orientated in a precisely known way in relation to the joints.

For the hips and shoulders we placed the points on the tubes, on both sides, on the line connecting the two hip or the two shoulder joints as accurately as palpation allowed on the living subject. For the hips, the orientation was made easier by the fact that with the body erect, the axis of the hips, i. e. the line connecting the centres of the two hip joints, if extended, reaches the skin on both sides at the point where the tip of the greater trochanter is palpable. Neither is it difficult to determine a projection on the skin of a line connecting the centres of the two shoulder joints. After placing the points on the tubes we measured the distance to the corresponding joint centres. This measurement of the joint centres can be carried out with sufficient accuracy in the sagittal plane by palpation and comparison with an anatomical specimen of the same size. These indications are sufficient to deduce, by construction or calculation the co-ordinates of the centres of the hips and shoulders from the tridimensional co-ordinates of the points on the Geissler tubes of the hips and shoulders. Geometrically this consists of joining by a straight line two points the position of which is known in space, and subtracting a known length from the two ends of this line.

So as not to overcomplicate the study, we placed the points of the Geissler tubes corresponding to the elbows, knees and ankles in such a way that, with the body erect, they were on horizontal lines perpendicular to the plane of gait passing through the centre of the joint. We then measured the distance between each point and the centre of the joint. Strictly speaking, this method is not perfect for determining the position of the centre of the joint from the position of the point on the tube. This would be the case if the line connecting the point on the tube and the centre of the joint was always at right angles to the direction of gait and moved only parallel to itself. This is only approximately true during human gait. The deviation of this line from its direction in the erect position is no greater than, for example, the change in direction of the axis of the elbow in relation to the two articulated bones during flexion. In analysing gait one must neglect these oscillations of the axis and assume a fixed axis in relation to the two bones, if only to make the problem tangible. It is also permitted, to begin with, to neglect the oscillations of the lines connecting the points on the tubes and the joint centres. However, for analysing movements of the human body which, contrary to walking, are not carried out following a well-determined direction, another arrangement must be found which makes precise determination of the centre of the joint possible. To this end, a second Geissler tube would have to be fixed to the medial side of the extremity and the end of this tube situated on the medial prolongation of the joint axis. One could then obtain the co-ordinates of two points, as for the hips and shoulders. The joint centre would lie on the line connecting these two points at a particular distance from one of them. If the Geissler tube on the medial side were to disturb movement of the limb, the two tubes could be placed in such a way that they would be on a line passing through the centre of the joint in another direction, for instance from anterior to posterior. One would certainly need to take advantage of such an arrangement for particular analyses of the movement of an individual joint. For analysing human gait, the advantage achieved by this arrangement would be minimal compared with other hitherto unavoidable sources of error.

The tube on the head was placed in the midline somewhat in front of the vertex. The foot tubes were on the axis of the feet and reached almost to the ends of the feet. The centre of gravity of the foot was also marked on its tube.

In order that the different phases of movement could be recorded at equal intervals, the primary circuit of the Ruhmkorff coil was interrupted by a large tuning fork. In recording movement of the knee photographically, we had used a normal contact breaker since in this piece of research we dealt with the geometric process of movement of the lower leg in relation to the thigh; we did not consider the velocity of this movement. In analysing human gait, however, we had a different aim: we wanted to determine not only the spatial but also the temporal patterns in such an accurate way that one could deduce from our measurements the tridimensional trajectories of the joint centres and also the velocities and accelerations of these points along their trajectories. We, therefore, had to devise an accurate means of recording time.

The number of vibrations emitted by the tuning fork when switched into the primary circuit was of course lower than when it vibrated freely. We determined the frequency by recording the vibrations on the rotating drum of a kymograph on which we simultaneously recorded time with a seconds pendulum. This measurement was carried out shortly after the experiments, whilst the tuning fork was still switched on to the primary circuit, and the Geissler tubes to the secondary circuit. It appeared that the tuning fork vibrated 260.9 times in 10 s. The frequency was thus 26.09. Consequently, between two successive phases recorded photographically a time of $1/26.09$ or 0.0383 s elapsed; generally, between the first and nth of a series of successive phases the time was thus $(n-1)/26.09$ s.

To enable us to draw the trajectories in a system of tridimensional co-ordinates, after photographing the phases of movement we photographed on the same plate a network of 1-cm squares printed on a glass plate covered with Japan varnish. In order to improve accuracy we built a large wooden frame on which a 1-m square table of co-ordinates could rotate about a vertical axis. This frame rested on four small screws for which four metal recesses were prepared in the floor. The frame and the table of co-ordinates could thus be brought into exactly the same position at any time.

The axis of the table of co-ordinates coincided with the middle vertical line of the network of co-ordinates. Consequently, the axis was moved until the vertical middle line remained in exactly the same position when the table was rotated. This was checked by the use of a telescope: the vertical cross-wire of the latter was aimed to coincide with the middle vertical line. When this position of the table in the frame was attained, it was only necessary to ensure that the axis of rotation of the table was vertical by adjusting the screws in the floor. This was indicated by a plumb line hanging in front of the middle vertical line of the network of co-ordinates.

The recesses in the floor for the screws of the frame indicated where the table was to be placed in the centre of the room. The subject had to walk over this area during the experiment.

It was almost impossible to see all the Geissler tubes on both sides from any one direction. The phases of movement were, therefore, photographed from both sides. We used two sets of two cameras, i.e. four in all. If it had appeared necessary or even expedient to take pictures from a fifth direction, perhaps from above, we could have used a fifth, sixth, or any number of standard cameras, without having to make any special arrangements. This is a great advantage of the use of electrical sparks or Geissler tubes: one is limited neither in the direction nor in the number of simultaneous exposures that can be taken.

For our experiments, two cameras on each side proved sufficient. One of each pair of cameras had its optical axis perpendicular to the plane of gait. To calculate the tridimensional co-ordinates from the direct measurements, it would have been advantageous to have the optical axis of the second camera on each side in a horizontal direction in the plane of gait. In addition to the two lateral cameras one would then have needed only a third camera which could have taken a picture from the front or back. The direction of gait was the only one from which one could see simultaneously the Geissler tubes on both sides. We had to reject this arrangement, however, because the different phases of movement of the tubes partly overlapped each other and together they gave a picture which was not sufficiently clear. Therefore, we placed a second camera on each side; its optical axis formed an angle of $60°$ with that of the first camera and it gave an oblique view from the front.

Figure 6 shows the position of the four cameras clearly. DE is the path of the subject during the experiment; the path was 9 m long. The four points L_1, L_2, L_3 and L_4 show the location of the positioning screws for the table of co-ordinates. Point O lies between them; it is 90 cm above the floor and marks the centre of the table of co-ordinates. This point was considered as the origin of the system of co-ordinates in which the whole movement would be drawn. The horizontal line in the direction of gait and passing through O was designated as the positive X axis. The Y axis was horizontal and perpendicular to the X axis. The Y axis to the right of the subject represented the positive Y axis. The third axis of the system, the Z axis, was formed by the vertical line through O. The upper part from O represented the positive Z axis. The optical axes of the four cameras were in the horizontal plane through O and intersected each other at O, the origin of the system of co-ordinates. The centres of the photographic plates, designated as A_1, B_1, A_2, B_2, were situated 90 cm above the level of the floor. The cameras were so arranged that a clear picture of the origin O, well marked on the table of co-ordinates, appeared in the centre of the photographic plates. The distances between the four points A, B and the origin of the co-ordinates were:

$$\overline{OA_1} = 438 \text{ cm}, \ \overline{OB_1} = 446 \text{ cm}, \ \overline{OA_2} = 634.5 \text{ cm}, \ \overline{OB_2} = 588.5 \text{ cm}.$$

The four cameras and their four simultaneous exposures of the same process of movement were designated as 1^a, 1^b, 2^a, 2^b, as indicated in Fig. 6. The cameras 1^a and 2^a provided the two central projections for the points on the right side of the body, and 1^b and 2^b the two central projections for the points on the left. A projection of the single point of the head tube was obtained from both sides. This was used later as an invaluable means of checking the accuracy of the measurements.

The photographer Schleicher of Leipzig arranged the cameras and developed the pictures. We are most grateful to him for his expert assistance in planning and performing the experiments which were carried out at night.

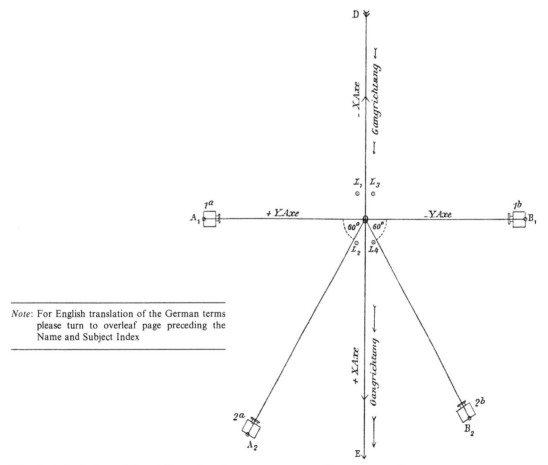

Note: For English translation of the German terms please turn to overleaf page preceding the Name and Subject Index

Fig. 6. Calculation of the tridimensional rectangular co-ordinates x, y, z from the projections of an object on the planes at right angles to the optical axes of cameras

Figure 7 shows the four pictures of the subject at rest equipped with the Geissler tubes. These pictures were taken simultaneously by the four cameras in a semi-darkened room. The subject stood in the centre of the four recesses for the positioning screws. In this way, his longitudinal axis coincided approximately with the Z axis of the system of co-ordinates, and he faced the direction of gait. After photographing the standing subject, the table of co-ordinates was put into place, illuminated with a magnesium flare from behind and photographed successively on the four plates in such a way that this table of co-ordinates was each time perpendicular to the optical axis of the camera. In order not to hide too much of the network of co-ordinates, most of the table was covered with black paper leaving only a small square area with the origin of the co-ordinates in the centre and a peripheral strip 5–6 cm wide, as had been done in the knee investigation. Whereas the pictures of the network of co-ordinates 1^b, 2^a and 2^b are well delineated, the picture 1^a appears blurred. This is not due to a less accurate arrangement of the camera 1^a but to an unfortunate oversight which occurred during the first exposure and could not be rectified later: we used a plate which had been exposed previously during one of the many preliminary experiments to record a series of phases of movement. Thus one can see printed on this plate the series of a gait experiment and a table of co-ordinates superimposed, if somewhat faintly. This unfortunate occurrence may have had a good side: it helps in understanding the sequence of pictures better than any description! Consequently, only the exposures of the standing man pictured in Fig. 7 are used. They will help to clarify the serial pictures especially those designated 2^a and 2^b on which the tubes of the opposite side also partly appear. They were not, however, used further in the actual measurements.

The decisive experiments were carried out at night since we had no way of darkening the whole room. As it was midsummer it was not so dark that the subject who had often walked along the experimental path, could not find his way as precisely as he had done in the daytime. We illuminated the room with a gas flame as necessary. In the dark we were able to open the four cameras without exposing the sensitive plates.

After a final check of the position of the Geissler tubes, the subject began walking at a distance of about 5 m from the origin of the co-ordinates in a natural, not too slow manner. After approximately five steps, the Ruhmkorff coil was switched on, and after an additional three or four steps it was switched off. The subject continued walking a few steps further. The starting point was chosen such that the subject passed through the Z axis of the system of co-ordinates in the middle of the photographed steps. The steps to be measured were photo-graphically recorded in this manner. The four cameras were then shut and the light was turned on. The frame with the table of co-ordinates – which of course was not used during the experiment – was brought to the predetermined position by means of the four recesses in the floor. The table was then set at right angles to one of the four optical axes. The corresponding camera was again opened, and the table was photographed using a magnesium flash placed behind it. The camera then was shut. The procedure was repeated with the other cameras until the network of co-ordinates was printed on the four plates.

From the many series taken on different nights we selected three for further discussion. These were obtained on the night of the 24th–25th July 1891. The experiments from which the first two series originated were carried out in an identical manner, without loading the subject. In order to investigate the influence of a relatively heavy load on gait, in the third experiment the subject carried an army regulation knapsack, three full cartridge pouches and an 88 rifle in the "shoulder-arms" position. These pieces of equipment were provided by the Royal 8th Infantry Regiment, Prince Johann Georg, Nr. 107. The helmet had to be discarded because it would have required removal of the head tube. The series of pictures and the corresponding tables of co-ordinates, the perpendicular projections, etc. are all differen-tiated thus: experiment I, experiment II and experiment III (with loading). The serial pictures of experiment I are given in Fig. 8, those of experiment II in Fig. 9, and those of experiment III in Fig. 10.

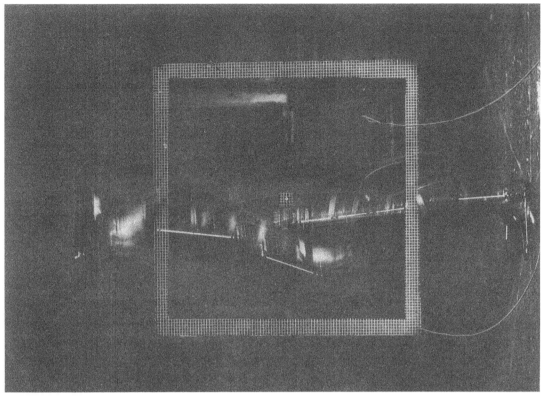

Fig. 7a–d. Four views of the subject at rest while wearing the suit with the Geissler tubes: **a** from right; **b** from left; **c** from front right; **d** from front left

Fig. 7d

Fig. 7c

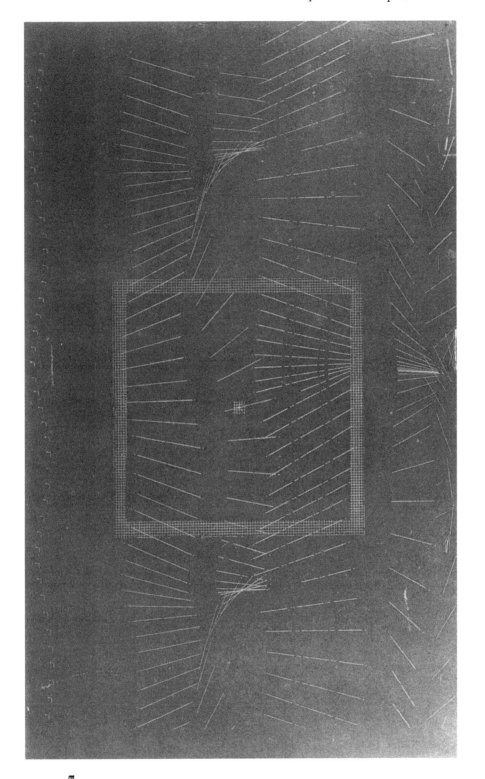

Fig. 8a

Fig. 8a–d. Exposures showing the positions of the subject's limbs during experiment I (without loading): **a** from the right; **b** from front right; **c** from the left; **d** from front left

Fig. 8b

Fig. 8c

Fig. 8d

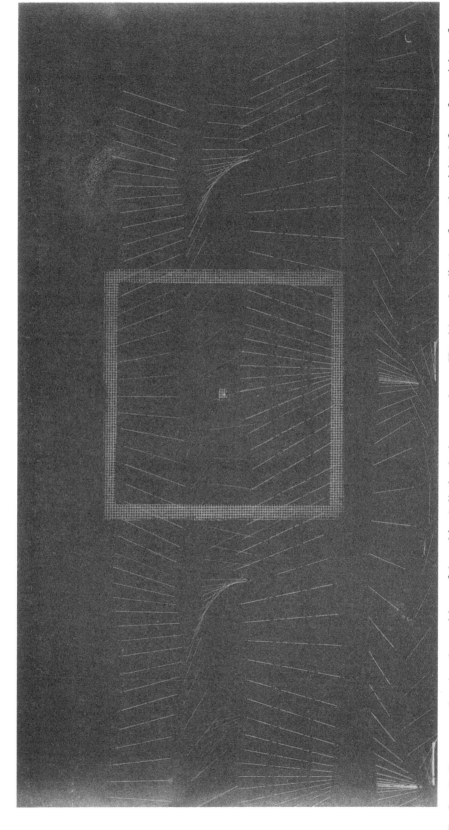

Fig. 9a

Fig. 9a–d. Exposures showing the positions of the subject's limbs during experiment II (without loading): **a** from the right; **b** from front right; **c** from the left; **d** from front left

Fig. 9b

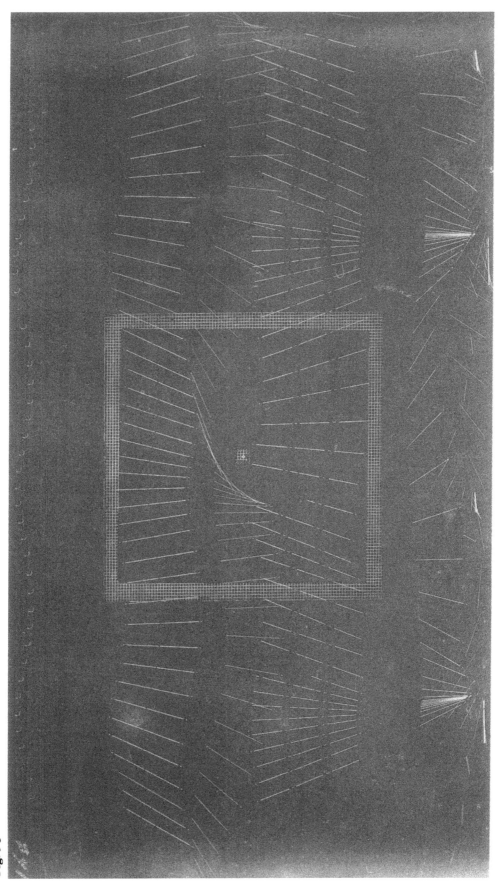

Fig. 9c

Fig. 9d

Fig. 10a

Fig. 10a–d. Exposures showing the positions of the subject's limbs during experiment III (with loading): **a** from the right; **b** from front right; **c** from the left; **d** from front left

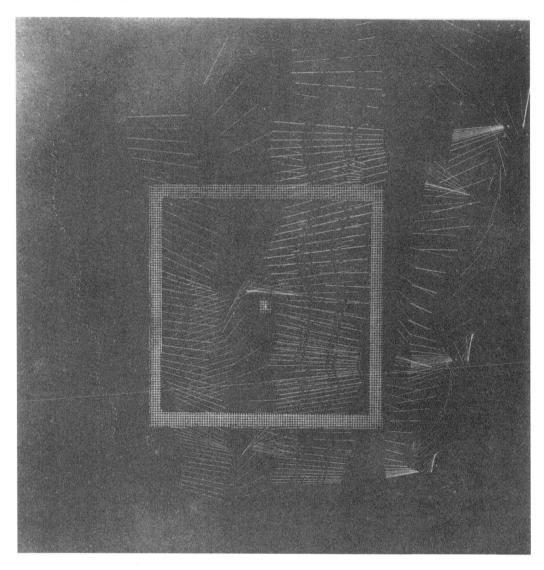

Fig. 10b

Fig. 10c

Fig. 10d

Deduction of the Tridimensional Rectangular Co-ordinates from the Serial Pictures

In two-sided chronophotography the trajectories of the points on the Geissler tubes corresponding to the different human joints are projected on two vertical planes passing through the origin of the co-ordinates. These planes are perpendicular to the optical axes of the two cameras. They were determined by the table of co-ordinates in the experiment. The projection planes corresponding to cameras 1^a and 2^a will be designated as $\mathfrak{A}_1$ and $\mathfrak{A}_2$, those corresponding to cameras 1^b and 2^b as $\mathfrak{B}_1$ and $\mathfrak{B}_2$. Because of the orientation of the cameras the plane $\mathfrak{B}_1$ formed the reverse side of $\mathfrak{A}_1$ whereas the planes $\mathfrak{A}_2$ and $\mathfrak{B}_2$ formed an angle of $60°$. The four simultaneous exposures of each experiment provide reduced pictures of these projection planes with the projection curves.

In order to determine exactly the different points on these projection curves, a system of co-ordinates has to be introduced into each projection plane. A rectangular system of co-ordinates was chosen the origin of which coincided with the origin O of the tridimensional system of co-ordinates in the four planes of projection. One axis was horizontal, the other vertical. The horizontal co-ordinates in planes $\mathfrak{A}_1$ and $\mathfrak{B}_1$ were designated as ξ_1, and the vertical as ζ_1. In both planes the ξ_1-co-ordinates had the same axis as the x-co-ordinates in the tridimensional system and were positive in the direction of gait; the ζ_1-co-ordinates had the same axis as the z-co-ordinates in the tridimensional system and were positive upwards. In the opposite direction they were negative. The horizontal co-ordinates in the planes $\mathfrak{A}_2$ and $\mathfrak{B}_2$ were designated as η_2; the vertical co-ordinates, completely different from ζ_1, were designated as ζ_2. The η_2-co-ordinates were positive to the left in $\mathfrak{A}_2$, and to the right in $\mathfrak{B}_2$. In both planes the ζ_2-co-ordinates were positive upwards.

The tridimensional rectangular co-ordinates x, y, z of a point P on the right side of the body must be calculated from the bidimensional pairs of co-ordinates ξ_1, ζ_1, and η_2, ζ_2 of the two projections P_1 and P_2 on the planes $\mathfrak{A}_1$ and $\mathfrak{A}_2$.

In Fig. 11, K_1 and K_2 represent the nodal points of the optical systems of the two cameras 1ᵃ and 2ᵃ closest to the projection planes $\mathfrak{A}_1$ and $\mathfrak{A}_2$. The projection points P_1 and P_2 of a point P are the intersections of the rays passing through P from K_1 and K_2 with the planes $\mathfrak{A}_1$ and $\mathfrak{A}_2$. The bidimensional co-ordinates ξ_1, ζ_1 of P_1 are represented by the segments $\overline{OQ_1}$ and $\overline{Q_1P_1}$, and the bidimensional co-ordinates η_2, ζ_2 of P_2 by the segments $\overline{OQ_2}$ and $\overline{Q_2P_2}$. The horizontal plane passes through O (not indicated in the figure); this plane represents the plane XY in the tridimensional system of co-ordinates. The perpendicular PQ from P to this plane represents the z-co-ordinate (positive upwards). The perpendicular QR_1 from Q to the

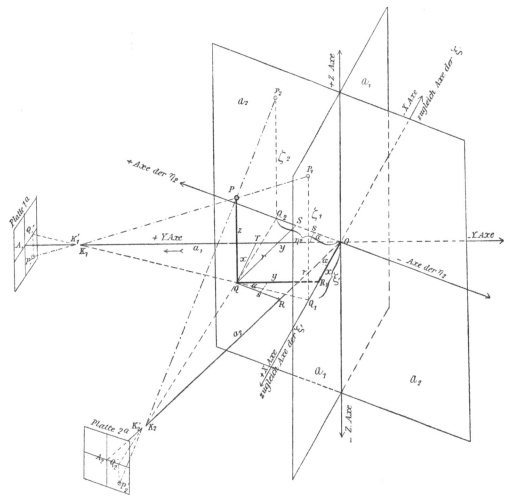

Fig. 11. Determination of the co-ordinates of a point P from the projections of P on two planes a_1 and a_2 as seen from cameras 1ᵃ and 2ᵃ

X axis in $\mathfrak{A}_1$ represents the y-co-ordinate (positive in the direction R_1Q). Finally the line OR_1 drawn from R_1 to the X-axis represents the x-co-ordinate of point P. Figure 12 represents the horizontal plane through O, as seen from above. The first nodal points K_1, K_2 and the optical axes OK_1 and OK_2 of the two cameras lie in this plane. The optical axis OK_1 coincides with the Y axis of the tridimensional system of co-ordinates, positive in the direction OK_1, whereas the optical axis OK_2 forms with the X axis an angle α which was 30° as our cameras were set up. The axis of the ξ_1-co-ordinates of the plane $\mathfrak{A}_1$ coincides with the axis X, whereas the horizontal axis of the η_2-co-ordinates in $\mathfrak{A}_2$ forms an angle α (30°) with the Y axis. The rays K_1QQ_1 and K_2QQ_2 represent the rectangular projections of the two rays K_1PP_1 and K_2PP_2 on the horizontal plane. They reach the axes of the ξ_1- and η_2-co-ordinates at the points Q_1 and Q_2 which delineate the ξ_1- and η_2-co-ordinates on these axes. Therefore, Fig. 12 gives not only the tridimensional x- and y-co-ordinates, but also the bidimensional ξ_1- and η_2-co-ordinates, unshortened. If one draws perpendiculars from Q to the optical axis OK_2 and to the axis of the η_2-co-ordinates, the perpendiculars delineate segments $\overline{OR}$ and $\overline{OS}$ on the axes. These segments will be designated as r and s. Since $QROS$ is a rectangle, QS has also the length r, and QR the length s. Since $QR \perp OK_2$ and $QR_1 \perp OQ_1$, the straight lines QR and QR_1 form the angle α. If one draws the perpendicular QT from Q to OK_1, one has a rectangle QR_1OT and, consequently, $\overline{TQ} = x$ and $\overline{OT} = y$.

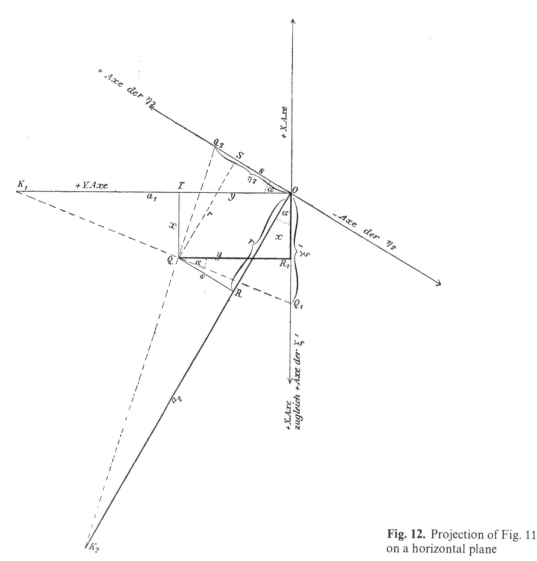

Fig. 12. Projection of Fig. 11 on a horizontal plane

In Fig. 12 it can be seen that:

$$\overline{OQ_1} : \overline{TQ} = \overline{OK_1} : \overline{TK_1} \quad \text{and}$$

$$\overline{OQ_2} : \overline{RQ} = \overline{OK_2} : \overline{RK_2} .$$

If we designate as a_1 and as a_2 the distances $\overline{OK_1}$ and $\overline{OK_2}$ between O and the nodal points K_1, K_2, we can write these two equations as follows:

$$\xi_1 : x = a_1 : (a_1 - y) \tag{1}$$

$$\eta_2 : s = a_2 : (a_2 - r) \tag{2}$$

The following equations result from Fig. 11:

$$\overline{Q_1 P_1} : \overline{QP} = \overline{Q_1 K_1} : \overline{QK_1}$$

$$\overline{Q_2 P_2} : \overline{QP} = \overline{Q_2 K_2} : \overline{QK_2} .$$

Since according to Fig. 12:

$$\overline{Q_1 K_1} : \overline{QK_1} = \overline{OK_1} : \overline{TK_1}$$

$$\overline{Q_2 K_2} : \overline{QK_2} = \overline{OK_2} : \overline{RK_2}$$

one can deduce from the above equations the new ones:

$$\overline{Q_1 P_1} : \overline{QP} = \overline{OK_1} : \overline{TK_1}$$

$$\overline{Q_2 P_2} : \overline{QP} = \overline{O_2 K_2} : \overline{RK_2} .$$

which can also be written:

$$\zeta_1 : z = a_1 : (a_1 - y) \tag{3}$$

$$\zeta_2 : z = a_2 : (a_2 - r) \tag{4}$$

If, in Fig. 12, the broken line $OR_1 Q$ formed by x and y is projected on the optical axis OK_2 and on the axis of the η_2-co-ordinates, one obtains r and s expressed by x, y and the angle α. r and s are the projections of the resultant OQ of the aforementioned broken line on the two axes. Then:

$$r = x \cos \alpha + y \sin \alpha \quad \text{and}$$

$$s = y \cos \alpha - x \sin \alpha .$$

If we introduce these values into Eqs. (2) and (4), we have:

$$\eta_2 : (y \cos \alpha - x \sin \alpha) = a_2 : (a_2 - x \cos \alpha - y \sin \alpha) \tag{5}$$

$$\zeta_2 : z = a_2 : (a_2 - x \cos \alpha - y \sin \alpha) \tag{6}$$

Equations (1) and (5) are used to determine x and y and the Eqs. (3) and (6) give two values for the z-co-ordinate, which act as mutual checks.

Resolving the equations concerned gives:

$$x = \frac{a_1 a_2 \cos\alpha \cdot \xi_1 - (a_2 - a_1 \sin\alpha) \cdot \xi_1 \eta_2}{a_1 a_2 \cos\alpha + a_2 \sin\alpha \cdot \xi_1 + a_1 \sin\alpha \cdot \eta_2 - \cos\alpha \cdot \xi_1 \eta_2}, \tag{7}$$

$$y = \frac{a_1 a_2 \sin\alpha \cdot \xi_1 + a_1 a_2 \cdot \eta_2 - a_1 \cos\alpha \cdot \xi_1 \eta_2}{a_1 a_2 \cos\alpha + a_2 \sin\alpha \cdot \xi_1 + a_1 \sin\alpha \cdot \eta_2 - \cos\alpha \cdot \xi_1 \eta_2}, \tag{8}$$

$$z = \frac{(a_1 - y) \cdot \zeta_1}{a_1} \quad \text{and} \quad z = \frac{(a_2 - x\cos\alpha - y\sin\alpha) \cdot \zeta_2}{a_2}. \tag{9}$$

The two x- and y-co-ordinates are involved in the equations for z. In order to find z from the bidimensional co-ordinates ξ_1, η_2, ζ_1 and ζ_2, one must introduce the values of x and y into the equations for z. It is more convenient, however, not to do so and to calculate z only after the values of x and y have been found by the first equations. All the other magnitudes in the equations, a_1, a_2 and the angle α, are constants which must be determined definitively before calculating the tridimensional co-ordinates x, y, z.

These equations are valid for the points on the right side of the body. They give the tridimensional co-ordinates x, y, z from the pairs of bi-dimensional co-ordinates ξ_1, ζ_2 and η_2, ζ_2, which can be directly measured on the photographs 1ª and 2ª. The same equations can be used for the points on the left side of the body. By convention, however, the ξ_1-co-ordinate on photograph 1[b] is positive to the left, and the η_2-co-ordinate on photograph 2[b] is positive to the right. The opposite is true for the coordinates ξ_1 on photograph 1ª and η_2 on photograph 2ª. Taking this into account, we obtain the positive value of x from Eq. (7) and the positive value of z from Eq. (9) for the left side of the body. The positive value of y is given by Eq. (8) after its sign has been changed.

The angle α is given by the arrangement of the cameras; it is 30°. In order to determine the two other constant magnitudes a_1 and a_2, which represent the distances between the nodal points K_1, K_2 closest to the planes of projection and the origin O of the co-ordinates, one must consider more closely the optical properties of the system of lenses.

If, in an optical system consisting of three or more spherically limited lenses, the first and last lenses are similar, the nodal points coincide with two other important points, the principal points. The nodal points have then the same properties as the principal points. Furthermore, the two focal lengths, i.e. the distances between the foci and their principal points are equal. This is always the case for the system of lenses of a photographic camera because the initial and final optical environment is air.

In Fig. 13, K and K' represent the two nodal points, and F, F' the two foci of the optical system of one of the cameras. O is the origin of the co-ordinates, and A its image in the middle of the sensitive plate. If one relates the position of the foci and the two points O and A to the nodal points, the relationship between their distances becomes simple because the nodal points and the principal points coincide. If we designate the distance KO as a, the distance $K'A$ as a', the two equal focal distances KF and $K'F'$ as f, we find the following relationship between the three magnitudes

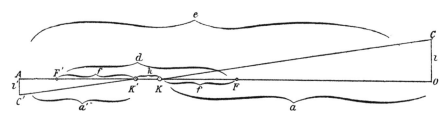

Fig. 13. Image AC' of an object OC on a photographic plate. K, K', nodal points; F, F', foci

$$\frac{1}{a} + \frac{1}{a'} = \frac{1}{f} \tag{10}$$

If C is a point on the table of co-ordinates and C' its image on the photographic plate, the ray KC will run parallel to the ray $K'C'$ because of the above-mentioned property of the nodal points. If the segment $\overline{OC}$ has the length l and its image $\overline{AC'}$ the length l', the following relationship exists between these two lengths and the magnitudes a, a':

$$\frac{l}{a} = \frac{l'}{a'} \tag{11}$$

If e represents the distance $\overline{OA}$ between the photographic plate and the table of co-ordinates, k the distance between two nodal points and d the distance between the two foci, the following relationships exist between these three distances and the magnitudes a, a' and f. They can be seen in Fig. 13.

$$a + a' + k = e \tag{12}$$

$$2f + k = d \tag{13}$$

Of the eight magnitudes involved in Eqs. (10)–(13), we could only determine four by direct measurement: l, l', e and d.

We measured the distance d between the two foci for each of the four cameras in the following manner. The objective was removed and exposed to the sun's rays in such a way that these were incident on one side, following the direction of the optical axis. The rays thus converged at their exit and were focused at the opposite side. The position and the distance of the focus from a mark on the objective were determined fairly exactly by means of a paper screen perpendicular to the optical axis and movable along this axis. The objective was then turned round so that the sun's rays fell on the opposite side and converged at the other focal point. The distance of this second focal point from the mark on the objective was also measured. The actual distances between the focal points and the mark were not measured, rather their projections on the optical axis. The sum of the two distances is d. Measurement of the four objectives gave: for camera 1^a, $d = 52$ cm; 2^a, $d = 69$ cm; 1^b, $d = 55,5$ cm; 2^d, $d = 67$ cm.

As stated above, the distance e between the photographic plates and the table of co-ordinates was measured immediately after arranging the cameras. As shown in Eq. (17) for a, we are interested in the relation l/l' rather than in the actual magnitudes of l and l'. The ratio l/l' is that of a length on the table of co-ordinates to its image on the photographic plate. Therefore, in order to determine l and l' it is not necessary to compare a segment from the origin of the co-ordinate O with its image. It is possible to measure any length on the table of co-ordinates and its image. It is thus possible to choose as long a segment l as possible on the table. This improves the accuracy of the determination.

In our studies only the distance a between the nodal point K and the origin of the co-ordinates O was important. It can be expressed starting from the four magnitudes l, l', e, and d. Eliminating a' from the Eqs. (10) and (11) gives:

$$\frac{l+l'}{l'} \cdot \frac{1}{a} = \frac{1}{f} \tag{14}$$

Eliminating a' and k from Eqs. (11)–(13) gives:

$$\frac{l+l'}{l} a - 2f = e - d \tag{15}$$

Eliminating the focal distance f from Eqs. (14) and (15) gives:

$$\frac{l+l'}{l} a - \frac{2l'}{l+l'} a = e - d \qquad (16)$$

and thus:

$$a = \frac{l(l+l')}{l^2+l'^2} (e-d)$$

The expression of a depends only on the ratio of the two magnitudes l and l'. This is easily shown by dividing numerator and denominator by l'^2. If the ratio of l/l' is designated λ, one has:

$$a = \frac{\lambda(\lambda+1)}{\lambda+1} (e-d) \qquad (17)$$

For our work on the knee, we had accepted a simplification in determining a. We neglected the distance between the two nodal points and assumed that the two points coincided at the optical middle point. The error in determining a is here very small. If, in Fig. 13, one imagines the two nodal points K and K' brought together, besides Eq. (11)

$$\frac{l}{a} = \frac{l'}{a'}$$

one obtains the equation which is only approximate:

$$a + a' = e.$$

Eliminating a' from both equations, one arrives at the approximate equations used in the study on the knee:

$$a = \frac{l}{l+l'} \cdot e \quad \text{or} \quad a = \frac{\lambda}{\lambda+1} \cdot e$$

For experiment I, for instance, we found:

$$l = 86 \text{ cm}, \ l' = 5.8979 \text{ cm}[12], \ e = 438 \text{ cm}, \ d = 52 \text{ cm}$$

If these values are introduced into the exact equation (16), the value of a is 410.54 cm. If the approximate equation is used, the value becomes 409.89 cm for a. The difference is thus 0.65 cm. Since a is a length of over 4 m, one is compelled to recognize that it is justifiable to use the approximate equation instead of the exact equation without increasing the sources of error significantly. One must remember that direct measurement of such a length as e can in itself entail an error of some millimetres if one does not use particularly accurate measuring instruments.

Though there might be some doubt as to whether we could have improved the accuracy of the determination of a by taking into account the nodal points, we spared no efforts in ensuring that the results were as accurate as possible.

Besides depending on the angle $\alpha = 30°$ and the distance a between the nodal points, the calculation of which has been explained, the three tridimensional co-ordinates x, y, z of the

---

[12] Average of ten measurements

points on the Geissler tubes corresponding to the different joints depended only on the bidimensional pairs of co-ordinates ξ_1, ζ_1, and η_2, ζ_2. The latter had to be measured on the photographs. In our work on the knee, with this in view, we had completed the network of co-ordinates on the pictures by drawing the parallels to the two axes of co-ordinates, partly indicated on the photograph. Since these parallels ran at an interval of 1 cm, by evaluating one-tenth of this distance we were able to read the co-ordinates with an accuracy of about 1 mm. The accuracy thus attained corresponded approximately to the magnitude of the picture obtained by the spark and was then sufficient to determine the geometric pattern of the movement.

In the present study we strove to greater accuracy from the very beginning since our aims were much higher. Using Geissler capillary tubes marked off into very small segments we produced points on the photographic plates much smaller than the spark images in our work on the knee joint. Because of the smaller size, these points enabled a much more accurate determination of the co-ordinates than could be achieved by drawing the network of co-ordinates. This circumstance had to be taken into account since, given the care taken in reducing all sources of error to a minimum, the accuracy of the final results was to improve proportionally to the accuracy in measuring the co-ordinates on the serial pictures.

The reliability of a measurement depends above all on the measuring instrument. Prints made from a (glass) photographic plate are never identical to one another or to the image on the plate itself because the paper becomes distorted as a result of the effects of the chemical fluids. Consequently, it was necessary to measure the co-ordinates on the photographic plates themselves. However, we had no instrument and we knew of none that could serve our purpose. Therefore, we devised our own instrument which would enable us to measure the rectangular co-ordinates of tiny points on photographic plates, 18 × 24 cm, with an accuracy of 0.001 mm. This instrument was built by Mr. E. Zimmermann, a precision mechanic in Leipzig, and fulfilled our requirements exactly. Figures 14 and 15 show this measuring instrument.

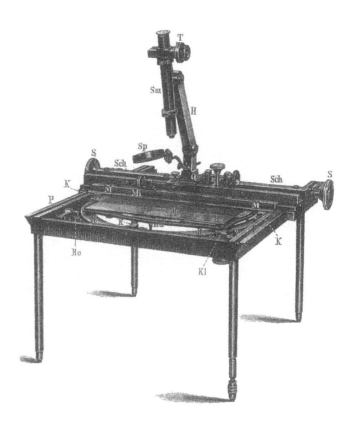

Fig. 14. Instrument used for the measurement of the co-ordinates

Fig. 15. Top view of the instrument used for the measurement of the co-ordinates

The essential constituents of this instrument for measuring the co-ordinates are as follows: a mobile ring R on which the photographic plate Pl is fixed; a rule M, which can be displaced directly over the plate Pl; and a microscope S_m which can be moved parallel to itself so that any point on the plate Pl can be brought into view.

The ring R is held by four rollers Ro, which are located in a groove along the edge of the ring in such a way that the ring can be rotated about its centre. On the ring, a frame Ra can firmly grasp the photographic plate with the sensitive layer uppermost. The frame Ra is not permanently fixed to the ring; four screws allow it to be raised above the ring to a certain extent. It is thus possible to bring the sensitive layer of the photographic plate to a certain predetermined level in the apparatus, despite the different thicknesses of the plates. The position of the plate is ensured by two clamps Kl which fix the ring. These clamps are equipped with micrometer screws to allow for very slight displacements of the ring.

The rule M can move on a slide Sch. The latter carries out a perfectly straight movement by means of a prism mechanism P, P. This movement is perpendicular to the long axis of the rule. The rule itself can be moved longitudinally a little in relation to the slide Sch by a micrometer screw Mi which is made of silver and has a millimetre scale visible only under the microscope. The zero point of the scale is in the middle; on each side the scale is marked in millimetres.

There is a groove in the slide Sch for a prism Q, perpendicular to the movement of the slide. The prism constitutes the base of a stand H sufficiently strong to carry the microscope Sm. In this manner the microscope can move anywhere in two perpendicular directions. It can move with the slide Sch or move along this slide with the prism Q. Thus the microscope can be moved parallel to itself anywhere over the photographic plate and provide a sharp view of the sensitive layer. Because of the importance of the position of the microscope in the direction of the rule M, the prism Q is equipped with a micrometer screw.

The microscope Sm gives a slightly magnified picture of the scale of the rule and of the image on the photographic plate. In the plane of the picture there are two parallel cross hairs stretched over a small frame. This frame can be moved sideways in the plane of the picture, perpendicular to the direction of the hairs by a very accurate micrometer screw with a limited range. Ten revolutions of the micrometer screw move the cross hairs from one stroke on the millimetre scale of the rule to the next when the microscope is stationary. One revolution thus corresponds to a sideways displacement of the cross hairs by 0.1 mm. The number of

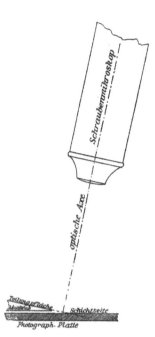

Fig. 16. Arrangement required for simultaneous clear view of the point under scrutiny and of the scale

screw revolutions, i.e. the number of tenths of a millimetre, can be read on a part similar to a toothed rack fixed in the plane of the picture. The top of the micrometer screw carries a drum *T* on the edge of which there is a division into 100 units. Thanks to a mark on the external side of the head of the microscope, one-hundreth of a screw revolution can be read, i.e. 0.001 mm of the sideways displacement of the cross hairs. (If the height of the micrometer screw was such that after five revolutions the cross hairs appeared in the ocular as displaced sideways by two 1-mm strokes, the drum should then be divided into 200 parts to allow for a direct reading of 0.001 mm.)

The stand *H* of the microscope carries a concave mirror *Sp* which reflects the light of a lamp on the scale of the rule. The photographic plate is brightly illuminated from underneath by the light, natural or artificial, reflected by an oblique plane mirror.

The main advantage of this measuring instrument is that it is able to show simultaneously in the same field of the microscope the point the co-ordinates of which are to be measured and the scale of the rule. This is achieved by the fact that the surface of the rule with the scale is oblique to the photographic plate. It is, therefore, possible to place the point to be measured in the plane of the scale (Fig. 16). Consequently, the microscope must also be oblique in such a way that its optical axis is perpendicular to the plane of the scale. It is thus possible to obtain simultaneously a sharp picture of the scale and of the point the co-ordinates of which are to be measured.

The measurement is carried out in the following manner. The photographic plate is orientated by turning the ring *R* until the vertical ζ-axis (middle vertical line on the table of co-ordinates) lies exactly parallel to the direction in which the slide *Sch* can be moved. This position is easily checked with the microscope: when the slide is moved, the central vertical line on the table of co-ordinates should remain between the two cross hairs. The rule is then moved sideways using the micrometer screw *Mi* until the 0 of the scale coincides with the ζ-axis in a vertical plane: the 0 of the scale should always appear between the two cross hairs together with the part of the ζ-axis visible under the microscope.

When the photographic plate and the rule are positioned with sufficient accuracy, it is possible to measure to within 0.001 mm the co-ordinate perpendicular to the ζ-axis of any point on the photographic plate (i.e. the ξ-coordinates on plates 1[a] and 1[b] and the η-co-ordinates on plates 2[a] and 2[b]). It is prerequisite that the magnitude and shape of the point allow such an accuracy in the positioning of the cross hairs and that the person performing the measuring is sufficiently trained in this exercise. The slide *Sch* and the microscope on the latter are moved simultaneously using the prism Q until the point the co-ordinates of which are to be measured appears in the field of the microscope. With the micrometer screw which allows the fine positioning of the microscope, the cross hairs and the point in the field of the microscope are brought into the position indicated in Fig. 17. Even if the diameter of the point is somewhat greater than the distance between the two hairs, this positioning can still be reliably achieved since it can easily be judged whether the point symmetrically overlaps the two cross hairs. As a rule, the cross hair will not be exactly superimposed on a millimetre line. If it lies, for instance, between the lines marking 12 and 13 mm, as in Fig. 17, 12 expresses the co-ordinate in full millimetres. In order to measure the fraction of millimetre beyond 12, the cross hair is moved to the right by turning the drum T, the microscope being fixed, until the cross hair is exactly superimposed on the 12 line. The distance can then be read directly to within 0.001 mm on the drum, according to the number of screw revolutions required. In our example it is 0.317 mm. The co-ordinate of the point thus considered is 12.317 mm. By repeating measurements of the same point at different times, we were convinced that the position of most of the points on our photographic plates could be determined with an accuracy of some thousandths of a millimetre. We could, therefore, rely absolutely on an accuracy of 0.01 mm.

Whether the value of the co-ordinates actually reached this accuracy depended then on the quality of the micrometer screw in the microscope and on the precision and thinness of the scale of the rule. To be completely sure here, the microscope and the scaling of the silver rule were made by Mr. Wanschaff, precision mechanic in Berlin, to whom the physikalisch-technische Reichanstalt is indebted for a series of very precise instruments, such as accurate machines for measuring lengths and angles.

After one co-ordinate has been measured for all the points on the photographic plate, the ring R is rotated through 90°. The other co-ordinate (ζ-co-ordinate) can then be measured in a similar manner. In order to enable the rotation of the ring R with sufficient accuracy, we had it equipped with four silver chocks K at equal distances from each other. Mr. Wanschaff carved on each chock a fine line directed towards the centre of the ring. These lines were radii of the ring and perpendicular to each other. Before rotation, the microscope was brought into a position in which the line lay in the cross hair. The ring was then rotated until the next line came to occupy the same place in the microscope. Before measurement of the co-ordinates could begin in the new position of the plate, the rule was again moved sideways with the micrometer screw Mi until the 0 of the scale was brought exactly to the plumb line of the origin of the co-ordinates, especially marked in the middle of the table of co-ordinates. The measurement was then carried out exactly as before.

The measurement on the photographic plates does not directly give the true values of ξ_1, ζ_1, η_2 and ζ_2. The latter are the co-ordinates of the projections on the table of co-ordinates perpendicular to the optical axis of the camera and passing through the origin O of the

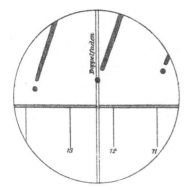

Fig. 17. Measurement of a co-ordinate

co-ordinates. The photographs give only a reduced picture of this table of co-ordinates with the projections. If we designate the co-ordinates measured on the photographic plates as ξ'_1, ζ'_1, η'_2 and ζ'_2 these must be multiplied in each individual case by the factor of magnification. This is the ratio of the length on the table of co-ordinates to its image in the photograph. This factor is the magnitude l/l' or λ mentioned previously.

The unit of measurement of the machine is the millimetre. If the co-ordinates directly measured are multiplied by λ, one obtains the bidimensional co-ordinates on the table of co-ordinates, also in millimetres. However, it seems more appropriate to use the centimetre as the unit of length for the human gait so as to avoid too large figures. If we wish to express the co-ordinates ξ_1, ζ_1, η_2, ζ_2 in centimetres from the co-ordinates ξ'_1, ζ'_1, η'_2, ζ'_2 directly measured in millimetres, we must multiply by λ and divide by 10. If we replace λ by its value l/l' and provide these magnitudes with the index 1 when they relate to cameras 1^a or 1^b and with the index 2 when they relate to cameras 2^a or 2^b, we obtain:

$$\xi_1 = \frac{l_1}{10\,l'_1}\,\xi'_1 \quad \zeta_1 = \frac{l}{10\,l'_1}\,\zeta'_1 \quad \eta_2 = \frac{l_2}{10\,l'_2}\,\eta'_2 \quad \zeta_2 = \frac{l_2}{10\,l'_2}\,\zeta'_2 \tag{18}$$

It must not be forgotten that the magnitudes l and l' have different values in the equations: in the first two equations they relate to a different camera than in the last two. Further the magnitudes l and l' in the first two equations are different for the right and left sides of the body because they relate to two different cameras, 1^a and 1^b. Correspondingly, the magnitudes l and l' in the last two equations are to be deduced from different cameras for the right side (1^a) of the body and for the left (1^b).

Tables 1–6 give the values of the four magnitudes ξ'_1, ζ'_1, η'_2 and ζ'_2 arrived at by direct measurement in the three experiments. Since the constant magnitudes are different for the right and left sides of the body, at the end of each table related to one side of the body we give the values of the constants and the resultant equations used in the calculation. Strictly speaking, the constants corresponding to one side of the body should be the same for the three experiments. However, small variations in the magnitude of l', corresponding to the chosen length $l = 86$ cm, appeared in the three experiments. These slight discrepancies are due to differences in the way the cassette was inserted, whereby the sensitive layer of the photographic plate was not always exactly in the same place. This also resulted in somewhat different values for the distance e between the table of co-ordinates and the photographic plate. However, these differences are so small in relation to the magnitude of e that they can be ignored.

For each experiment we obtained 31 successive phases of movement symmetrically grouped on both sides of the middle of the table of co-ordinates. To avoid an accumulation of data, we measured the co-ordinates of the centres of the joints, the vertex, the centre of gravity and the tip of the foot. The centres of gravity of the different limbs were initially neglected. The co-ordinates of the vertex were measured from both sides. Thus, $9 \times 4 \times 31 = 1116$ measurements had to be carried out for each side, $2 \times 1116 = 2232$ for a whole experiment and 6696 for the three experiments. Some of these measurements were not used. In several places the points were so unclear that a measurement could not be made with sufficient accuracy. For instance in experiment III, this was the case for the positions of the centre of gravity of the foot, the tip of the foot and most of the left hand. In some phases, the hip joint was hidden by the forearm, reducing the number of measurements. However, in experiment III, the co-ordinates of a point on the medial side of the Geissler tube attached to the thigh were additionally determined. The results of these direct measurements are given in Tables 1–6.

The constants of the first camera 1^a in experiment I were $e = 438$ cm, $d = 52$ cm, $l_1 = 86$ cm and $l'_1 = 5.8979$ cm. The distance between the nodal point and the origin of the co-ordinates was calculated using Eq. (16): $a_1 = 410.54$ cm.

Table 1. Experiment I, right

No.	Right shoulder				Right elbow				Right wrist				No.
	ξ'_1	η'_2	ζ'_1	ζ'_2	ξ'_1	η'_2	ζ'_1	ζ'_2	ξ'_1	η'_2	ζ'_1	ζ'_2	
1	−59,437	+32,529	+28,626	+22,872	−56,797	+35,340	+10,342	+ 8,100	−43,553	+31,854	− 6,695	−5,394	1
2	−55,366	+31,127	+29,735	+23,950	−53,831	+34,005	+11,068	+ 8,760	−44,704	+32,872	− 7,892	−6,310	2
3	−51,271	+29,727	+30,731	+	−51,852	+33,186	+11,824	+ 9,409	−46,125	+34,139	− 7,938	−6,356	3
4	−47,211	+28,347	+31,370	+25,601	−50,211	+32,697	+12,615	+10,050	−47,333	+35,290	− 7,517	−6,060	4
5	−43,109	+27,003	+31,604	+25,961	−48,255	+32,176	+13,242	+10,590	−47,573	+35,960	− 7,014	−5,440	5
6	−39,078	+25,634	+31,391	+25,981	−45,744	+31,465	+13,512	+10,833	−46,714	+36,010	− 6,721	−5,460	6
7	−34,966	+24,237	+30,762	+25,664	−42,562	+30,552	+13,235	+10,640	−45,075	+36,050	− 6,683	−5,530	7
8	−30,496	+22,750	+29,753	+24,982	−38,456	+29,291	+12,373	+10,000	−42,358	+35,767	− 7,104	−5,730	8
9	−26,116	+21,272	+28,763	+24,360	−33,941	+27,800	+11,304	+ 9,240	−38,502	+34,963	− 7,776	−6,340	9
10	−21,821	+19,926	+28,143	+24,000	−29,156	+26,348	+10,454	+ 8,630	−33,318	+33,482	− 8,573	−7,100	10
11	−16,889	+18,423	+27,870	+23,960	−23,512	+24,555	+ 9,893	+ 8,240	−26,076	+31,050	− 9,629	−8,100	11
12	−11,887	+16,812	+27,979	+24,290	−17,331	+22,617	+ 9,614	+ 8,090	−17,626	+27,970	−10,413		12
13	− 7,014	+15,114	+28,421	+24,920	−10,611	+20,419	+ 9,695	+ 8,230	− 7,767	+24,180	−10,545	−8,990	13
14	− 2,825	+13,528	+29,291	+25,900	− 4,302	+18,250	+10,340	+ 8,960	+ 2,224	+19,940	− 9,511	−8,300	14
15	+ 1,671	+11,739	+30,678	+27,430	+ 2,047	+16,060	+11,802	+10,200	+12,934	+14,943	− 6,683	−5,912	15
16	+ 5,603	+10,146	+31,721	+28,230	+ 7,189	+14,369	+13,037	+11,560	+21,470	+10,625	− 3,417	−3,040	16
17	+10,146	+ 8,283	+32,350	+29,450	+13,251	+12,465	+14,084	+12,680	+30,669	+ 5,660	+ 0,645	+0,750	17
18	+14,185	+ 6,550	+32,260	+29,490	+18,833	+10,680	+14,577	+13,300	+38,134	+ 1,140	+ 4,050	+4,030	18
19	+18,451	+ 4,570	+31,678	+29,320	+24,819	+ 8,500	+14,682	+13,600	+44,986	− 2,935	+ 6,852	+6,790	19
20	+22,882	+ 2,490	+30,880	+28,880	+30,521	+ 6,150	+14,376	+13,330	+50,939	− 6,850	+ 8,432	+8,300	20
21	+27,469	+ 0,250	+29,979	+28,330	+35,742	+ 3,570	+13,662	+12,800	+56,269	− 9,540	+ 8,662	+8,650	21
22	+31,957	− 2,100	+29,154	+27,860	+40,366	+ 0,950	+12,730	+11,620	+60,916	−11,870	+ 7,588	+8,030	22
23	+36,362	− 4,420	+28,653	+27,560	+44,281	− 1,300	+11,871	+11,200	+64,701	−13,700	+ 5,392	+5,660	23
24	+40,804	− 6,820	+28,409	+27,750	+47,786	− 3,540	+11,168	+10,800	+67,614	−15,270	+ 2,321	+2,270	24
25	+45,205	− 9,270	+28,614	+28,240	+50,812	− 5,470	+10,794	+10,600	+69,399	−15,620	− 1,123	−1,430	25
26	+49,964	−11,800	+28,997	+28,855	+54,230	− 7,842	+10,739	+10,700	+70,102	−15,100	− 4,851	−4,750	26
27	+53,915	−13,800	+29,550	+29,500	+57,424	−10,060	+10,961	+10,840	+69,701	−14,000	− 7,395	−7,370	27
28	+57,843	−15,870	+30,441	+30,830	+60,015	−12,090	+11,453	+11,464	+68,752	−12,560	− 8,146	−8,050	28
29	+61,360	−17,640	+31,197	+31,830	+61,615	−13,300	+12,031	+12,100	+67,824	−11,300	− 7,962	−8,020	29
30	+65,119	−19,520	+31,691	+32,460	+63,177	−14,270	+12,633	+12,740	+67,153	−10,470	− 7,532	−7,160	30
31	+68,888	−21,500	+31,812	+32,960	+65,047	−15,230	+13,068	+13,165	+67,484	−10,000	− 7,153	−6,770	31

No.	Right hip				Right knee				Right ankle				No.
	ξ'_1	η'_2	ζ'_1	ζ'_2	ξ'_1	η'_2	ζ'_1	ζ'_2	ξ'_1	η'_2	ζ'_1	ζ'_2	
1	−58,332		−5,793		−58,330	+31,672	−33,384	−26,885	−79,137	+35,487	−46,956	−36,855	1
2	−53,410	+29,490	−4,680	−3,876	−50,024	+28,637	−32,029	−26,200	−69,702	+32,342	−47,499	−37,943	2
3		+27,990			−42,324	+25,787	−30,467	−25,243	−59,243	+28,689	−48,942	−39,861	3
4	−43,987	+26,285	−2,947	−2,457	−35,445	+23,072	−29,164	−24,500	−48,220	+24,713	−50,766	−42,211	4
5	−39,643	+24,805	−2,875	−2,393	−29,249	+20,487	−28,387	−24,142	−36,577	+20,306	−52,430	−44,558	5
6	−35,723	+23,370	−3,471	−2,915	−24,130	+18,329	−28,434	−24,416	−24,963	+15,854	−53,401	−46,391	6
7	−31,979	+22,020	−4,744	−3,980	−19,955	+16,327	−29,424	−25,530	−14,163	+11,734	−53,328	−47,260	7
8	−27,801	+20,630	−6,289	−5,331	−15,677	+14,130	−30,800	−27,000	− 4,681	+ 8,162	−52,447	−47,351	8
9	−23,197	+19,083	−7,101	−6,072	−10,611	+12,100	−31,310	−27,770	+ 1,020	+ 5,617	−52,595	−48,037	9
10	−18,461	+17,360	−7,219	−6,240	− 4,792	+10,165	−30,954	−27,770	+ 3,695	+ 4,008	−54,006	−49,697	10
11	−12,891	+15,560	−7,071	−6,168	− 0,121	+ 8,640	−31,324	−28,240	+ 6,567	+ 2,771	−54,577	−50,521	11
12	− 8,077	+14,116	−6,345	−5,530	+ 4,702	+ 7,615	−30,703	−27,890	+ 7,652	+ 2,336	−54,966	−50,984	12
13	− 3,927	+12,827	−5,792	−5,095	+ 8,291	+ 6,310	−30,530	−27,880	+ 8,111	+ 2,049	−55,060	−51,209	13
14		+11,576		−4,413	+ 9,472	+ 5,870	−30,708	−28,110	+ 8,202	+ 1,985	−55,049	−51,127	14
15	+ 3,751	+ 9,756	−3,972	−3,600	+10,043	+ 5,410	−30,834	−28,210	+ 8,213	+ 1,920	−54,945	−51,050	15
16	+ 6,663		−3,622		+10,474	+ 4,770	−30,845	−28,290	+ 8,242	+ 1,850	−54,863	−51,000	16
17	+10,257	+ 7,139	−3,532	−3,229	+11,134	+ 4,120	−30,890	−28,470	+ 8,276	+ 1,780	−54,809	−50,970	17
18	+13,749	+ 5,500	−3,631	−3,374	+11,927	+ 3,500	−30,907	−28,620	+ 8,297	+ 1,700	−54,804	−50,980	18
19	+17,670	+ 3,700	−4,069	−3,840	+13,214	+ 2,820	−31,025	−28,860	+ 8,394	+ 1,623	−54,798	−50,990	19
20	+21,543	+ 1,930	−4,797	−4,523	+15,106	+ 1,720	−31,243	−29,270	+ 8,581	+ 1,511	−54,733	−50,997	20
21	+25,418	+ 0,140	−5,342	−5,110	+17,216	+ 0,748	−31,355	−29,470	+ 8,978	+ 1,303	−54,409	−50,764	21
22	+29,570	− 1,992	−5,608	−5,420	+20,273	− 0,500	−31,350	−29,650	+ 9,743	+ 1,071	−53,763	−50,202	22
23	+34,173	− 4,401	−5,808	−5,640	+24,418	− 2,130	−31,574	−30,120	+11,056	+ 0,635	−52,734	−49,366	23
24	+39,118	− 7,100	−5,873	−5,750	+29,717	− 4,120	−31,900	−30,840	+13,492	− 0,444	−51,210	−48,223	24
25	+44,080	−10,010	−5,912	−5,840	+36,888	− 6,840	−32,733	−31,920	+17,377	− 2,153	−49,349	−46,843	25
26	+49,451	−13,100	−6,133		+46,004	−11,120	−33,721	−33,482	+23,822	− 4,990	−47,364	−45,587	26
27	+54,342		−5,674		+54,664	−15,890	−33,457	−33,834	+31,850	− 8,869	−47,067	−46,086	27
28	+59,098	−18,470	−4,635	−4,730	+62,802	−20,410	−32,188	−33,154	+41,315	−13,895	−47,826	−47,832	28
29		−20,680		−3,750	+69,805	−24,470	−30,799	−32,180	+50,892	−19,126	−49,270	−50,366	29
30	+68,082	−23,030	−3,076	−3,120	+76,433	−28,250	−29,588	−31,400	+61,630	−25,091	−51,183	−53,584	30
31	+72,028	−25,000	−3,080	−3,170	+82,137	−31,610	−28,964	−31,107	+72,616	−31,234	−52,901	−56,771	31

Table 1 (continued)

No.	Centre of gravity of right foot				Tip of right foot				Point on head, from right				No.
	ξ'_1	η'_2	ζ'_1	ζ'_2	ξ'_1	η'_2	ζ'_1	ζ'_2	ξ'_1	η'_2	ζ'_1	ζ'_2	
1	−80,828	+35,705	−51,246	−40,057	−75,818	+32,010	−57,623	−45,695	−52,678	+20,183	+52,166	+43,537	1
2	−70,761	+32,410	−51,950	−41,341	−64,849	+28,300	−57,602	−46,617	−48,195	+18,417	+53,035	+44,653	2
3	−59,356	+28,553	−53,488	−43,509	−52,224	+24,084	−57,932	−48,021	−43,891	+16,750	+53,806	+45,684	3
4	−47,307	+28,287	−55,195	−45,939	−39,314	+19,479	−58,120	−49,333	−39,844	+15,235	+54,215	+46,383	4
5	−34,568	+19,531	−56,498	−48,206	−26,061	+14,459	−57,425	−50,053	−35,766	+13,721	+54,159	+46,713	5
6	−22,019	+14,679	−56,886	−49,798	−13,536	+ 9,180	−55,746	−49,930	−31,739	+12,210	+53,733	+46,704	6
7	−10,352	+10,238	−56,147	−50,251	− 2,269	+ 5,170	−53,097	−48,656	−27,670	+10,683	+52,975	+46,389	7
8	− 0,307	+ 6,604	−54,450	−49,722	+ 7,097	+ 2,130	−49,808	−46,490	−23,326	+ 9,047	+52,074	+45,996	8
9	+ 5,396	+ 4,212	−54,434	−50,275	+12,694	+ 0,380	−49,554	−46,800	−19,173	+ 7,491	+51,406	+45,767	9
10	+ 7,799	+ 2,646	−56,231	−52,266	+15,534	− 1,487	−52,137	−49,573	−15,174	+ 6,050	+51,049	+45,787	10
11	+10,248	+ 1,267	−57,734	−54,000	+18,602	− 3,825	−55,703	−53,257	−10,990	+ 4,479	+51,084	+46,189	11
12	+10,858	+ 0,846	−58,421	−54,680	+19,343	− 4,210	−57,666	−55,108	− 7,048	+ 3,031	+51,514	+46,947	12
13	+11,018	+ 0,824	−58,497	−54,760	+19,542	− 4,444	−58,171	−55,600	− 3,109	+ 1,556	+52,245	+47,953	13
14	+11,025	+ 0,801	−58,477		+19,551	− 4,490	−58,206	−55,660	+ 0,738	+ 0,006	+53,192	+49,181	14
15	+11,032	+ 0,770	−58,460		+19,559	− 4,520	−58,240	−55,720	+ 4,898	− 1,821	+54,232	+50,605	15
16	+11,039	+ 0,750	−58,439		+19,569	− 4,560	−58,275	−55,780	+ 8,384	− 3,431	+54,882	+51,572	16
17	+11,046	+ 0,720	−58,418		+19,578	− 4,610	−58,310	−55,840	+12,518	− 5,403	+55,152	+52,287	17
18	+11,053	+ 0,680	−58,396		+19,587	− 4,660	−58,343	−55,900	+16,397	− 7,267	+54,931	+52,523	18
19	+11,062	+ 0,650	−58,378	−54,580	+19,598	− 4,700	−58,377	−55,960	+20,599	− 9,346	+54,330	+52,431	19
20	+11,081	+ 0,610	−58,357	−54,500	+19,622	− 4,750	−58,485	−56,030	+24,863	−11,447	+53,507	+52,116	20
21	+11,283	+ 0,532	−58,149	−54,420	+19,768	− 4,800	−58,655	−56,100	+29,122	−13,666	+52,680	+51,813	21
22	+11,670	+ 0,445	−57,760	−54,100	+20,043	− 4,900	−58,898	−56,360	+33,388	−16,035	+52,027	+51,665	22
23	+12,322	+ 0,257	−57,153	−53,580	+20,418	− 5,070	−59,230	−56,660	+37,625	−18,509	+51,584	+51,782	23
24	+13,614	− 0,419	−55,832	−52,460	+21,050	− 5,380	−59,675	−57,200	+41,885	−21,175	+51,502	+52,244	24
25	+16,205	− 1,058	−53,933	−50,970	+21,815	− 5,667	−60,092	−57,580	+45,824	−23,664	+51,746	+53,014	25
26	+21,944	− 4,394	−51,693	−49,430	+25,688	− 8,150	−58,791	−56,914	+50,141	−26,375	+52,284	+54,151	26
27	+30,017	− 8,569	−51,410	−50,060	+33,789	−13,770	−57,465	−56,877	+54,376	−28,892	+53,043	+55,506	27
28	+40,156	−13,885	−52,302	−52,070	+44,845	−19,880	−57,532	−58,442	+58,963	−31,570	+53,901	+57,004	28
29	+50,706	−19,460	−53,816	−54,957	+56,655	−25,930	−57,992	−60,457	+63,086	−33,940	+54,557	+58,261	29
30	+62,560	−25,848	−55,646	−58,320	+69,595	−32,770	−58,364	−62,666	+67,229	−36,183	+54,896	+59,139	30
31	+74,642	−32,468	−57,064	−61,517	+82,294	−39,700	−58,079	−64,210	+71,207	−38,343	+54,787	+59,557	31

The constants of the second camera 2^a in experiment I were $e = 634.5$ cm, $d = 69$ cm, $l_2 = 86$ cm and $l'_2 = 5.4618$ cm. The distance from the corresponding nodal point was found by using Eq. (16): $a_2 = 599.00$ cm.

From these constants one obtains, using Eq. (18):

$$\xi_1 = 1.45815 \cdot \xi'_1 \quad \zeta_1 = 1.45815 \cdot \zeta'_1$$

$$\eta_2 = 1.5746 \cdot \eta'_2 \quad \zeta_2 = 1.5746 \cdot \zeta'_2$$

If we introduce these values into Eqs. (7)–(9), considering that $\alpha = 30°$, we obtain the equations necessary to calculate the tridimensional co-ordinates x, y, z from the values of the bidimensional co-ordinates ξ'_1, ζ'_1, η'_2 and ζ'_2 given in Table 1.

We can improve these equations so as to facilitate calculation by removing the coefficient of $\xi'_1 \eta'_2$ in the denominator of the first two equations and the coefficient of y in the numerator of the last two. This gives:

$$x = \frac{156\,175 \cdot \xi'_1 - 454.64 \cdot \xi'_1 \eta'_2}{107\,105 + 219.63 \cdot \xi'_1 + 162.55 \cdot \eta'_2 - \xi'_1 \eta'_2} \tag{19}$$

$$y = \frac{90\,168 \cdot \xi'_1 + 194\,737 \cdot \eta'_2 - 410.54 \cdot \xi'_1 \eta'_2}{107\,105 + 219.63 \cdot \xi'_1 + 162.55 \cdot \eta'_2 - \xi'_1 \eta'_2} \tag{20}$$

$$z_1 = \frac{(410.54 - y) \cdot \zeta_1}{281.55} \quad \text{and} \quad z_2 = \frac{(1198 - 1.732 \cdot x - y) \cdot \zeta'_2}{760.84} \tag{21}$$

in which the two values of z that check each other are distinguished by the indices 1 and 2. These equations were used in this form in the calculation.

Table 2. Experiment I, left

No.	Left shoulder				Left elbow				Left wrist				No.
	ξ'_1	η'_2	ζ'_1	ζ'_2	ξ'_1	η'_2	ζ'_1	ζ'_2	ξ'_1	η'_2	ζ'_1	ζ'_2	
1	−66,794	+36,615	+31,864	+22,633	−67,769	+41,202	+11,637	+ 8,099	−60,143	+44,908	− 9,579	−7,372	1
2	−62,066	+35,265	+33,233	+23,764	−61,296	+39,286	+13,062	+ 9,191	−50,072	+41,080	− 6,968	−6,628	2
3	−57,258	+33,838	+34,522	+24,872	−55,128	+37,460	+14,515	+10,304	−40,441	+36,996	− 3,580	−4,905	3
4	−52,585	+32,436	+35,244	+25,577	−49,396	+35,870	+15,471	+11,096	−31,828	+32,869	− 0,157	−2,547	4
5	−47,722	+30,862	+35,296	+25,811	−43,656	+34,245	+15,838	+11,456	−24,026	+28,943	+ 3,031	−0,041	5
6	−42,817	+29,205	+34,808	+25,633	−37,986	+32,571	+15,653	+11,402	−17,236	+25,208	+ 5,538	+2,334	6
7	−37,819	+27,404	+33,925	+25,218	−32,395	+30,752	+14,992	+11,039	−11,192	+21,943	+ 6,973	+4,257	7
8	−32,463	+25,405	+32,733	+24,566	−26,629	+28,728	+13,913	+10,315	− 5,313	+19,125	+ 7,230	+5,431	8
8	−27,316	+23,388	+31,741	+24,043	−21,416	+26,689	+12,829	+ 9,603	− 0,237	+16,516	+ 6,095	+5,698	9
10	−22,309	+21,311	+31,150	+23,818	−16,691	+24,642	+12,046	+ 9,126	+ 4,093	+14,346	+ 3,941	+4,849	10
11	−17,171	+19,047	+30,853	+23,841	−11,857	+22,445	+11,616	+ 8,857	+ 8,033	+12,774	+ 0,652	+3,151	11
12	−12,256	+16,926	+31,189	+24,230	− 7,431	+20,342	+11,588	+ 8,914	+10,697	+11,552	− 2,931	+0,558	12
13	− 7,375	+14,888	+31,558	+24,849	− 2,858	+18,167	+11,810	+ 9,193	+11,999	+11,129	− 6,060	−2,328	13
14	− 3,127	+13,209	+32,254	+25,614	+ 0,832	+16,383	+12,269	+ 9,593	+12,349	+11,563	− 7,755	−4,941	14
15	+ 1,318	+11,441	+33,388	+26,737	+ 3,691	+15,116	+13,103	+10,291	+12,207	+12,530	− 7,860	−6,264	15
16	+ 5,055	+ 9,950	+34,225	+27,573	+ 5,659	+14,482	+13,906	+10,958	+12,127	+13,817	− 7,420	−6,353	16
17	+ 9,571	+ 8,171	+34,717	+28,179	+ 8,106	+13,841	+14,641	+11,557	+12,531	+14,850	− 7,022	−6,040	17
18	+13,726	+ 6,455	+34,675	+28,338	+10,755	+13,014	+14,899	+11,778	+13,767	+15,667	− 6,889	−5,734	18
19	+18,165	+ 4,526	+34,170	+28,165	+14,267	+11,851	+14,753	+11,704	+15,951	+15,585	− 7,034	−5,754	19
20	+22,722	+ 2,645	+33,324	+27,684	+18,361	+10,354	+14,140	+11,276	+19,041	+14,964	− 7,472	−6,100	20
21	+27,461	+ 0,679	+32,324	+27,103	+23,207	+ 8,601	+13,158	+10,553	+23,101	+13,960	− 8,152	−6,745	21
22	+32,432	− 1,325	+31,465	+26,666	+28,374	+ 6,568	+12,187	+ 9,845	+28,347	+12,192	− 8,946	−7,509	22
23	+37,521	− 3,297	+30,947	+26,490	+34,041	+ 4,371	+11,465	+ 9,367	+34,786	+ 9,625	− 9,854	−8,315	23
24	+43,015	− 5,425	+30,750	+26,553	+40,361	+ 1,951	+11,043	+ 9,109	+42,574	+ 6,162	−10,574	−9,064	24
25	+48,439	− 7,656	+30,930	+27,063	+46,964	− 0,771	+10,995	+ 9,243	+51,012	+ 2,100	−10,832	−9,530	25
26	+54,049	−10,150	+31,466	+27,820	+54,319	− 3,871	+11,318	+ 9,626	+61,027	− 3,119	−10,385	−9,247	26
27	+58,763	−12,157	+32,315	+28,910	+60,985	− 6,820	+12,167	+10,569	+70,441	− 8,723	− 9,001	−8,148	27
28	+63,649	−14,536	+33,665	+30,488	+67,543	− 9,981	+13,631	+12,037	+79,717	−14,625	− 6,222	−5,837	28
29	+68,158	−16,830	+34,790	+31,851	+73,291	−12,736	+14,990	+13,503	+87,704	−20,227	− 3,137	−3,016	29
30	+72,818	−19,289	+35,388	+32,760	+79,115	−15,456	+15,922	+14,559	+95,320	−25,962	+ 0,019	−0,063	30
31	+77,299	−21,776	+35,319	+33,068	+84,542	−18,001	+16,214	+15,029	+101,758	−31,312	+ 2,802	+2,786	31

No.	Left hip				Left knee				Left ankle				No.
	ξ'_1	η'_2	ζ'_1	ζ'_2	ξ'_1	η'_2	ζ'_1	ζ'_2	ξ'_1	η'_2	ζ'_1	ζ'_2	
1		+34,732		−1,932	−54,303	+30,693	−32,017	−23,117	−52,072	+26,263	−61,379	−45,135	1
2	−58,957	+33,428	−1,923	−1,309	−53,078	+30,309	−31,797	−23,014	−52,077	+26,240	−61,306	−45,010	2
3	−55,148		−1,362		−52,027	+29,764	−31,548	−22,881	−52,105	+26,220	−61,190	−44,960	3
4	−51,346	+30,948	−1,063	−0,760	−51,031	+29,194	−31,343	−22,806	−52,073	+26,210	−61,100	−44,890	4
5	−47,320	+29,634	−1,180	−0,853	−49,694	+28,484	−31,224	−22,810	−51,988	+26,135	−60,970	−44,840	5
6	−43,112	+28,259	−1,591	−1,155	−47,947	+27,743	−31,248	−22,912	−51,887	+26,040	−60,890	−44,820	6
7	−38,643	+26,771	−2,303	−1,726	−45,681	+26,861	−31,444	−23,169	−51,712	+25,922	−60,822	−44,790	7
8	−33,741	+25,083	−3,136	−2,373	−42,746	+25,848	−31,681	−23,493	−51,324	+25,798	−60,530	−44,612	8
9	−28,794	+23,296	−3,650	−2,837	−39,113	+24,810	−31,870	−23,772	−50,617	+25,739	−59,886	−44,160	9
10	−23,705	+21,328	−3,995	−3,190	−34,585	+23,593	−32,161	−24,184	−49,331	+25,584	−58,741	−43,422	10
11	−17,892	+18,884	−4,059	−3,231	−28,582	+21,969	−32,527	−24,698	−46,620	+25,038	−56,982	−42,257	11
12	−12,091	+16,457	−4,069	−3,282	−20,759	+19,839	−33,365	−25,670	−42,048	+23,736	−54,610	−40,825	12
13	− 6,502	+13,957	−4,038	−3,267	−11,570	+16,612	−34,162	−26,720	−35,133	+21,481	−51,949	−39,340	13
14	− 1,282		−3,290		− 2,449	+13,053	−33,779	−26,877	−26,380	+18,353	−51,027	−39,296	14
15	+ 3,981	+ 9,352	−1,995	−1,697	+ 6,381	+ 9,279	−32,380	−26,221	−15,371	+14,055	−51,896	−40,815	15
16		+ 7,397		−0,937	+13,611	+ 6,112	−31,034	−25,553	− 5,197	+ 9,829	−53,550	−42,972	16
17	+13,824	+ 5,203	−0,383	−0,357	+21,553	+ 2,455	−29,784	−24,924	+ 7,559	+ 4,366	−55,878	−46,001	17
18	+18,224	+ 3,421	−0,489	−0,461	+27,859	− 0,666	−29,257	−24,794	+19,693	− 0,972	−57,845	−48,775	18
19	+22,583	+ 1,607	−1,249	−1,113	+33,489	− 3,570	−29,473	−25,320	+32,663	− 6,833	−59,230	−51,304	19
20	+27,040	− 0,261	−2,417	−2,102	+37,768	− 5,999	−30,573	−26,542	+45,131	−12,514	−59,366	−52,720	20
21	+31,856	− 2,340	−3,689	−3,208	+41,465	− 8,300	−32,089	−28,104	+55,275	−17,147	−58,499	−53,043	21
22	+37,058	− 4,519	−4,396	−3,849	+46,783	−10,870	−32,661	−28,930	+61,163	−20,199	−58,709	−53,959	22
23	+42,350	− 6,669	−4,527	−3,989	+53,707	−13,620	−32,240	−28,940	+63,548	−21,801	−60,273	−55,785	23
24	+48,445	− 9,105	−4,179	−3,756	+57,902	−15,232	−32,665	−29,494	+66,530	−23,197	−60,756	−56,608	24
25	+53,217	−10,850	−3,654	−3,310	+62,747	−17,106	−32,183	−29,389	+67,825	−23,912	−61,074	−57,118	25
26		−12,855		−2,849	+67,429	−19,035	−31,688	−29,160	+68,370	−24,290	−61,170	−57,263	26
27		−14,970		−2,073	+69,310	−20,054	−31,556	−29,239	+68,463	−24,340	−61,090	−57,270	27
28	+67,030	−17,281	−1,432	−1,437	+70,754	−20,850	−31,361	−29,221	+68,550	−24,420	−61,010	−57,200	28
29	+70,698		−1,127		+71,860	−21,850	−31,220	−29,232	+68,635	−24,500	−60,930	−57,200	29
30	+74,597	−21,360	−1,005	−1,131	+72,932	−22,609	−31,080	−29,223	+68,720	−24,580	−60,850	−57,130	30
31	+78,421	−23,422	−1,209	−1,300	+74,243	−23,391	−31,088	−29,260	+68,800	−24,660	−60,770	−57,060	31

Table 2 (continued)

No.	Centre of gravity of left foot				Tip of left foot				Point on head, from left				No.
	ξ'_1	η'_2	ζ'_1	ζ'_2	ξ'_1	η'_2	ζ'_1	ζ'_2	ξ'_1	η'_2	ζ'_1	ζ'_2	
1	−48,732	+25,223	−65,006	−48,050	−39,170	+21,446	−65,230	−49,190	−58,045	+23,508	+57,386	+42,849	1
2	−48,681	+25,207	−64,991	−48,035	−39,162	+21,447	−65,254	−49,200	−53,167	+21,906	+58,414	+43,933	2
3	−48,679	+25,191	−64,976	−48,020	−39,154	+21,448	−65,278	−49,210	−48,457	+20,284	+59,292	+44,943	3
4	−48,642	+25,174	−64,960	−48,005	−39,146	+21,449	−65,303	−49,220	−43,999	+18,654	+59,735	+45,620	4
5	−48,640	+25,157	−64,944	−47,990	−39,137	+21,450	−65,327	−49,230	−39,492	+16,941	+59,684	+45,944	5
6	−48,630	+25,140	−64,928	−47,975	−39,128	+21,451	−65,348	−49,240	−35,031	+15,203	+59,180	+45,899	6
7	−48,620	+25,123	−64,910	−47,960	−39,119	+21,452	−65,370	−49,250	−30,522	+13,414	+58,308	+45,567	7
8	−48,583	+25,107	−64,763	−47,910	−38,987	+21,386	−65,580	−49,349	−25,727	+11,425	+57,271	+45,179	8
9	−48,224	+25,106	−64,435	−47,687	−38,719	+21,340	−65,850	−49,524	−21,136	+ 9,462	+56,444	+44,898	9
10	−47,516	+25,074	−63,920	−47,293	−38,375	+21,190	−66,245	−49,849	−16,725	+ 7,507	+55,922	+44,892	10
11	−46,055	+24,751	−62,740	−46,511	−37,503	+20,853	−66,760	−50,295	−12,089	+ 5,403	+55,841	+45,210	11
12	−43,318	+23,831	−60,190	−44,812	−36,694	+20,444	−66,640	−50,305	− 7,778	+ 3,390	+56,164	+45,908	12
13	−37,002	+21,910	−57,214	−43,072	−32,564	+19,294	−65,307	−49,642	− 3,463	+ 1,313	+56,789	+46,849	13
14	−28,171	+18,743	−56,265	−43,052	−23,270	+15,740	−64,055	−49,553	+ 0,724	− 0,607	+57,706	+48,043	14
15	−16,221	+14,070	−57,190	−44,783	−10,205	+10,195	−63,851	−50,697	+ 5,226	− 2,636	+58,751	+49,365	15
16	− 5,106	+ 9,500	−58,830	−47,200	+ 2,013	+ 5,040	−64,150	−52,242	+ 8,978	− 4,300	+59,424	+50,331	16
17	+ 8,935	+ 3,672	−60,920	−50,200	+17,255	− 1,290	−64,444	−54,120	+13,472	− 6,272	+59,728	+51,009	17
18	+22,272	− 2,050	−62,434	−52,935	+31,176	− 7,250	−63,976	−55,357	+17,673	− 8,168	+59,472	+51,239	18
19	+36,291	− 8,416	−63,048	−55,096	+45,478	−13,764	−62,146	−55,540	+22,219	−10,268	+58,806	+51,138	19
20	+49,690	−14,626	−62,152	−55,885	+58,320	−20,000	−58,766	−54,210	+26,821	−12,457	+57,916	+50,807	20
21	+60,451	−19,390	−60,253	−55,438	+68,252	−24,510	−55,117	−51,960	+31,421	−14,608	+57,024	+50,527	21
22	+66,447	−22,412	−60,380	−56,310	+74,167	−26,866	−55,144	−52,620	+36,103	−16,742	+56,369	+50,413	22
23	+68,565	−24,000	−62,500	−58,660	+77,020	−28,970	−58,340	−56,070	+40,742	−18,796	+55,972	+50,524	23
24	+70,886	−25,435	−64,090	−60,490	+79,800	−31,070	−62,300	−60,234	+45,452	−20,869	+55,979	+51,008	24
25	+71,710	−26,280	−64,800	−61,240	+80,850	−32,200	−64,520	−62,816	+49,833	−22,831	+56,326	+51,807	25
26	+71,760	−26,195	−64,910	−61,380	+80,660	−32,214	−65,000	−62,880	+54,587	−25,063	+56,973	+52,971	26
27	+71,778	−26,196	−64,860	−61,340	+80,692	−32,228	−65,000	−62,930	+59,242	−27,469	+57,179	+54,303	27
28	+71,796	−26,197	−64,810	−61,300	+80,724	−32,242	−65,000	−62,980	+64,149	−30,208	+58,602	+55,747	28
29	+71,814	−26,198	−64,760	−61,260	+80,756	−32,256	−65,000	−63,030	+68,516	−32,809	+59,162	+56,918	29
30	+71,833	−26,199	−64,710	−61,210	+80,788	−32,270	−65,000	−63,070	+72,841	−35,580	+59,330	+57,712	30
31	+71,853	−26,200	−64,650	−61,160	+80,820	−32,284	−65,000	−63,106	+76,837	−38,256	+58,970	+58,080	31

Measurements on the photographs gave the bidimensional co-ordinates in Table 2 for the points of the left side of the body. The constants involved were $\alpha = 30°$ and for camera 1^b: $e = 446$ cm, $d = 55.5$ cm, $l_1 = 86$ cm, $l'_1 = 6.43823$ cm [13]; for camera 2^b: $e = 588.5$ cm, $d = 67$ cm, $l_2 = 86$ cm, $l'_2 = 5.3310$ cm. Equation (8) gives the negative values for y for the left side. Therefore, this equation must be preceded by a negative sign. After introducing the values of the constants into Eqs. (16) and (7)–(9), we obtain the following equations which are valid for the left side in experiment I:

$$x = \frac{142\,806 \cdot \xi'_1 - 396.09 \cdot \xi'_1 \eta'_2}{106\,909 + 197.53 \cdot \xi'_1 + 180.41 \cdot \eta'_2 - \xi'_1 \eta'_2} \tag{22}$$

$$y = -\frac{82\,449 \cdot \xi'_1 + 199\,068 \cdot \eta'_2 - 417.39 \cdot \xi'_1 \eta'_2}{106\,909 + 197.53 \cdot \xi'_1 + 180.41 \cdot \eta'_2 - \xi'_1 \eta'_2} \tag{23}$$

$$z_1 = (1.33577 - 0.00320 \cdot y) \cdot \zeta'_1 \quad \text{and}$$
$$z_2 = (1.61257 - 0.00253 \cdot x - 0.00146 \cdot y) \cdot \zeta'_2 \tag{24}$$

The last two equations are different from Eq. (21) because they were initially divided by their denominator. This proved useful for the calculation in this and later cases.

---

[13] These and the values of all the following l'_1 and l'_2 are each the average of ten measurements

Table 3. Experiment II, right

No.	Right shoulder ξ'_1	η'_2	ζ'_1	ζ'_2	Right elbow ξ'_1	η'_2	ζ'_1	ζ'_2	Right wrist ξ'_1	η'_2	ζ'_1	ζ'_2	No.
1	−69,805	+37,062	+28,719	+22,686	−67,423	+40,763	+10,921	+ 8,605	−49,279	+35,324	− 1,549	−1,002	1
2	−65,773	+35,684	+29,115	+23,173	−64,134	+39,740	+11,044	+ 8,720	−48,240	+35,010	− 4,305	−3,209	2
3	−61,813	+34,361	+29,900	+23,964	−60,959	+38,127	+11,495	+ 9,202	−48,249	+35,166	− 6,263	−4,846	3
4	−57,742	+33,051	+30,845	+24,896	−58,222	+36,883	+12,116	+ 9,684	−48,927	+35,720	− 7,183	−5,597	4
5	−53,979	+31,855	+31,445	+25,517	−56,211	+36,008	+12,609	+10,108	−49,571	+36,404	− 7,417	−5,734	5
6	−49,981	+30,510	+31,647	+25,800	−54,111	+35,250	+12,976	+10,413	−49,845	+36,900	− 7,505	−5,803	6
7	−45,761	+29,170	+31,411	+25,850	−51,605	+34,463	+13,111	+10,584	−49,327	+36,960	− 7,659	−5,905	7
8	−41,932	+27,896	+30,916	+25,650	−48,928	+33,620	+12,957	+10,507	−48,136	+36,999	− 7,874	−6,220	8
9	−37,303	+26,371	+30,073	+25,137	−45,115	+32,460	+12,436	+10,118	−45,877	+36,636	− 8,319	−6,600	9
10	−32,883	+24,916	+29,155	+24,554	−40,957	+31,167	+11,601	+ 9,528	−42,805	+36,109	− 8,896	−7,075	10
11	−28,159	+23,431	+28,429	+24,135	−35,937	+29,636	+10,722	+ 8,833	−38,379	+35,149	− 9,481	−7,613	11
12	−23,512	+22,060	+28,148	+24,070	−30,700	+28,020	+10,200	+ 8,610	−32,807	+33,656	− 9,866	−8,073	12
13	−18,326	+20,525	+28,232	+24,374	−24,797	+26,280	+ 9,971	+ 8,410	−25,567	+31,492	−10,316	−8,440	13
14	−13,545	+18,981	+28,631	+24,944	−18,828	+24,495	+10,036	+ 8,600	−17,425	+28,660	−10,377	−8,550	14
15	− 8,915	+17,319	+29,415	+25,882	−12,402	+22,569	+10,569	+ 9,144	− 7,820	+25,281	− 9,807	−8,254	15
16	− 4,659	+15,730	+30,594?	+27,132	− 6,313	+20,573	+11,644	+10,156	+ 1,882	+21,310	− 8,008	−6,760	16
17	− 0,433	+14,150	+31,768	+28,370	− 0,542	+18,739	+12,865	+11,458	+11,258	+17,075	− 5,423	−4,602	17
18	+ 3,758	+12,690	+32,463	+28,900	+ 5,258	+16,922	+13,847	+12,303	+20,365	+12,618	− 2,474	−2,073	18
19	+ 8,118	+10,860	+32,452	+29,500	+11,357	+14,980	+14,258	+12,858	+29,040	+ 7,970	+ 0,467	+0,700	19
20	+12,530	+ 8,640	+31,913	+29,350	+17,509	+12,869	+14,271	+13,080	+36,794	+ 3,416	+ 3,047	+3,150	20
21	+16,771	+ 6,840	+31,194	+28,946	+23,075	+10,500	+13,990	+13,200	+43,079	− 0,140	+ 4,591	+4,700	21
22	+21,120	+ 4,840	+30,391	+28,500	+28,327	+ 8,000	+13,404	+13,040	+48,630	− 3,280	+ 5,119	+5,274	22
23	+26,106	+ 2,450	+29,593	+28,060	+33,635	+ 5,310	+12,561	+12,600	+53,989	− 5,900?	+ 4,464	+4,560	23
24	+30,769	+ 0,040	+29,116	+27,920	+37,864	+ 3,120	+11,812	+11,500	+57,939	− 8,100	+ 2,846	+3,060	24
25	+35,387	− 2,270	+28,923	+27,970	+41,615	+ 0,950	+11,222	+11,000	+60,997	− 9,130	+ 0,396	+1,200	25
26	+39,864	− 4,650	+29,069	+28,350	+44,832	− 0,860	+10,928	+10,640	+62,869	− 9,570	− 2,399	−2,147	26
27	+44,500	− 6,940	+29,431	+28,970	+48,097	− 2,880	+10,890	+10,720	+63,772	− 9,600	− 5,244	−5,000?	27
28	+48,507	− 8,740	+30,040	+29,700	+51,151	− 4,880	+11,227	+11,040	+63,791	− 8,520	− 7,223	−7,000?	28
29	+52,273	−10,660	+30,994	+31,000	+53,672	− 6,585	+11,889	+11,840	+63,351	− 8,000	− 7,748	−7,140?	29
30	+55,919	−12,610	+31,842	+32,056	+55,507	− 7,896	+12,619	+12,500	+63,040	− 7,100	− 7,565	−7,000?	30
31	+59,987	−14,700	+32,347	+32,900	+57,542	− 8,920	+13,333	+13,270	+63,043	− 6,500	− 7,255	−7,000?	31

? signifies uncertainty

No.	Right hip ξ'_1	η'_2	ζ'_1	ζ'_2	Right knee ξ'_1	η'_2	ζ'_1	ζ'_2	Right ankle ξ'_1	η'_2	ζ'_1	ζ'_2	No.
1	−70,730	+36,801	−6,297	−4,841	−75,652	+38,464	−33,456	−26,081	−93,617	+40,363	−48,742	−37,360	1
2	−65,720		−6,195		−67,160	+35,887	−33,812	−26,741	−86,895	+38,408	−47,181	−36,561	2
3	−60,880		−5,395		−58,669	+32,990	−32,970	−26,450	−78,316	+35,737	−47,187	−37,160	3
4	−56,081	+31,683	−4,317	−3,366	−50,580	+30,120	−31,419	−25,591	−68,344	+32,407	−48,236	−38,672	4
5		+30,193		−2,660	−43,616	+27,606	−29,936	−24,675	−58,260	+29,016	−49,835	−40,705	5
6	−46,953	+28,710	−3,004	−2,640	−37,139	+25,113	−28,841	−24,067	−47,097	+25,190	−51,740	−43,132	6
7	−42,641	+27,211	−3,329	−2,606	−31,407	+22,781	−28,572	−24,138	−35,158	+20,986	−53,356	−45,473	7
8	−39,063	+25,932	−4,278	−3,400	−27,143	+20,941	−29,197	−24,881	−24,657	+17,233	−53,960	−46,865	8
9	−35,011	+24,564	−5,907	−4,786	−23,036	+18,806	−30,743	−26,403	−13,713	+13,369	−54,454	−47,368	9
10	−30,700	+23,094	−7,044	−5,803	−18,622	+16,789	−31,698	−27,560	− 6,449	+10,419	−52,937	−47,576	10
11	−25,583	+21,406	−7,289	−6,060	−12,154	+14,525	−31,220	−27,504	− 2,496	+ 8,167	−53,890	−48,919	11
12	−20,269	+19,600	−7,380	−6,177	− 6,760	+12,550	−31,248	−27,766	+ 0,249	+ 6,555	−54,608	−49,933	12
13	−15,085	+18,083	−6,766	−5,727	− 2,515	+11,650	−31,359	−27,894	+ 2,003	+ 5,816	−55,104	−50,548	13
14	−10,749	+16,789	−6,061	−5,173	+ 1,763	+10,213	−30,809	−27,650	+ 2,569	+ 5,593	−55,316	−50,834	14
15	− 6,770	+15,474	−5,311	−4,555	+ 3,574	+ 9,540	−30,965	−27,810	+ 2,770	+ 5,452	−55,293	−50,858	15
16		+13,943		−3,837	+ 4,624	+ 8,935	−31,004	−27,910	+ 2,816	+ 5,402	−55,272	−50,838	16
17	+ 0,727	+12,482	−3,893	−3,373	+ 5,350	+ 8,410	−31,029	−28,100	+ 2,854	+ 5,336	−55,243	−50,820	17
18	+ 4,229		−3,733		+ 5,960	+ 7,780	−31,064	−28,320	+ 2,879	+ 5,270	−55,127	−50,750	18
19	+ 7,813	+ 9,608	−3,776	−3,365	+ 6,800	+ 7,200	−31,105	−28,470	+ 2,929	+ 5,210	−55,074	−50,706	19
20	+11,612	+ 7,937	−4,139	−3,691	+ 7,950	+ 6,575	−31,225	−28,770	+ 3,005	+ 5,140	−55,064	−50,710	20
21	+15,253	+ 6,323	−4,706	−4,295	+ 9,465	+ 5,740	−31,487	−28,950	+ 3,166	+ 5,030	−54,994	−50,700	21
22	+18,970	+ 4,680	−5,231	−4,800	+11,540	+ 4,780	−31,567	−29,250	+ 3,544	+ 4,832	−54,720	−50,459	22
23	+23,579	+ 2,467	−5,532	−5,178	+15,000	+ 3,530	−31,631	−29,530	+ 4,382	+ 4,583	−54,044	−49,907	23
24	+28,280	+ 0,218	−5,762	−5,444	+19,397	+ 1,930	−31,873	−30,000	+ 5,763	+ 4,134	−52,990	−49,076	24
25	+33,302	− 2,403	−5,887	−5,586	+24,950	− 0,140	−32,382	−30,770	+ 8,360	+ 3,103	−51,394	−47,881	25
26	+38,485	− 5,229	−5,967	−5,763	+32,550	− 2,788	−33,286	−31,990	+12,642	+ 1,506	−49,423	−46,424	26
27	+43,866	− 7,990	−6,242	−6,200	+41,743	− 6,975	−34,118	−33,440	+19,254	− 1,064	−47,486	−45,214	27
28	+48,795		−5,641		+50,629	−11,550	−33,553	−33,310	+27,671	− 4,708	−47,171	−45,663	28
29	+53,335	−13,050	−4,537	−4,363	+58,547	−15,560	−32,041	−32,590	+36,978	− 9,144	−47,872	−47,244	29
30		−15,304		−3,444	+65,683	−19,380	−30,456	−31,390	+46,885	−14,036	−49,368	−49,786	30
31	+62,943	−17,749	−2,786	−2,783	+72,477	−23,170	−29,163	−30,470	+58,204	−19,892	−51,443	−53,200	31

Table 3 (continued)

No.	Centre of gravity of right foot				Tip of right foot				Point on head, from right				No.
	ξ'_1	η'_2	ζ'_1	ζ'_2	ξ'_1	η'_2	ξ'_1	ξ'_1	ξ'_1	η'_2	ζ'_1	ζ'_2	
1	−95,128	+40,703	−53,090	−40,543	−90,226	+38,203	−60,194	−46,469	−64,233	+25,523	+51,394	+42,158	1
2	−88,545	+38,665	−51,423	−39,700	−83,856	+35,716	−58,359	−45,662	−59,946	+23,811	+52,026	+43,047	2
3	−79,630	+35,712	−51,520	−40,430	−74,188	+31,812	−57,444	−45,880	−55,662	+22,213	+52,951	+44,138	3
4	−68,876	+32,237	−52,684	−42,180	−62,370	+27,880	−57,707	−47,070	−51,250	+20,636	+53,883	+45,286	4
5	−57,863	+28,636	−53,316	−44,398	−50,348	+24,008	−58,117	−48,454	−47,251	+19,244	+54,476	+46,139	5
6	−45,672	+24,603	−56,082	−46,872	−37,378	+19,864	−58,246	−49,696	−43,116	+17,826	+54,649	+46,610	6
7	−32,649	+20,084	−57,220	−49,036	−23,984	+15,268	−57,339	−50,164	−38,825	+16,356	+54,327	+46,714	7
8	−21,258	+15,973	−57,234	−50,132	−12,740	+11,078	−55,368	−49,623	−35,014	+14,991	+53,726	+46,533	8
9	− 9,420	+11,825	−55,960	−50,140	− 1,491	+ 7,322	−52,103	−47,756	−30,469	+13,355	+52,795	+46,129	9
10	− 1,966	+ 9,076	−54,886	−49,840	+ 5,411	+ 5,450	−50,010	−46,377	−26,244	+11,813	+52,038	+45,822	10
11	+ 1,740	+ 6,934	−56,189	−51,540	+ 9,497	+ 3,466	−52,006	−48,684	−21,813	+10,216	+51,573	+45,794	11
12	+ 4,140	+ 5,140	−57,451	−53,040	+12,454	+ 0,633	−54,887	−51,800	−17,532	+ 8,670	+51,436	+46,047	12
13	+ 5,324	+ 4,473	−58,500	−54,150	+13,856	− 0,517	−57,390	−54,270	−13,118	+ 7,071	+51,684	+46,645	13
14	+ 5,619	+ 4,294	−58,740	−54,440	+14,140	− 0,944	−58,267	−55,170	− 9,193	+ 5,617	+52,254	+47,494	14
15	+ 5,626	+ 4,266	−58,720	−54,426	+14,170	− 0,963	−58,365	−55,210	− 4,997	+ 3,995	+53,131	+48,676	15
16	+ 5,640	+ 4,242	−58,714	−54,404	+14,195	− 0,990	−58,407	−55,230	− 0,861	+ 2,246	+54,080	+49,934	16
17	+ 5,647	+ 4,216	−58,690	−54,390	+14,210	− 1,004	−58,430	−55,280	+ 3,109	+ 0,498	+54,802	+51,025	17
18	+ 5,662	+ 4,190	−58,660	−54,360	+14,225	− 1,021	−58,473	−55,330	+ 7,057	− 1,280	+55,134	+51,733	18
19	+ 5,674	+ 4,176	−58,645	−54,340	+14,240	− 1,040	−58,520	−55,354	+11,137	− 3,186	+54,984	+52,053	19
20	+ 5,694	+ 4,150	−58,620	−54,320	+14,254	− 1,085	−58,580	−55,390	+15,318	− 5,169	+54,452	+52,011	20
21	+ 5,725	+ 4,100	−58,594	−54,280	+14,273	− 1,115	−58,640	−55,460	+19,248	− 7,025	+53,740	+51,769	21
22	+ 5,940	+ 4,044	−58,420	−54,180	+14,420	− 1,170	−58,820	−55,560	+23,206	− 8,986	+53,015	+51,512	22
23	+ 6,333	+ 3,884	−58,016	−53,790	+14,744	− 1,296	−59,084	−55,890	+27,753	−11,369	+52,379	+51,426	23
24	+ 6,973	+ 3,722	−57,356	−53,216	+15,150	− 1,476	−59,417	−56,210	+32,010	−13,723	+52,004	+51,566	24
25	+ 8,391	+ 3,023	−55,965	−52,055	+15,790	− 1,772	−59,900	−56,720	+36,341	−16,177	+51,950	+52,035	25
26	+11,287	+ 1,843	−53,995	−50,467	+16,578	− 2,143	−60,387	−57,210	+40,330	−18,537	+52,181	+52,760	26
27	+17,363	− 0,695	−51,866	−49,110	+21,208	− 4,780	−58,705	−56,376	+44,401	−20,964	+52,707	+53,800	27
28	+25,830	− 4,494	−51,607	−49,670	+29,780	− 9,710	−57,570	−56,408	+48,557	−23,269	+53,539	+55,200	28
29	+35,828	− 9,172	−52,428	−51,570	+40,640	−14,922	−57,740	−57,900	+52,812	−25,833	+54,412	+56,656	29
30	+46,761	−14,350	−54,022	−54,375	+52,846	−20,574	−58,340	−60,010	+56,949	−28,142	+55,103	+57,904	30
31	+59,216	−20,593	−56,015	−57,994	+66,410	−27,340	−58,805	−62,296	+61,162	−30,442	+55,439	+58,807	31

In experiment II measurements gave the bidimensional co-ordinates in Table 3 for the points on the right side. The constants of camera 1^a in experiment II were: $e = 438$ cm, $d = 52$ cm, $l_1 = 86$ cm, $l'_1 = 5.88277$ cm; and for camera 2^a: $e = 634.5$ cm, $d = 69$ cm, $l_2 = 86$ cm; $l'_2 = 5.47414$ cm. Consequently, the equations for calculating the tridimensional co-ordinates for the right side in experiment II were:

$$x = \frac{156\,527 \cdot \xi'_1 - 454.75 \cdot \xi'_1 \eta'_2}{107\,071 + 220.16 \cdot \xi'_1 + 162.11 \cdot \eta'_2 - \xi'_1 \eta'_2} \tag{25}$$

$$y = \frac{90\,371 \cdot \xi'_1 + 194\,234 \cdot \eta'_2 - 410.48 \cdot \xi'_1 \eta'_2}{107\,071 + 220.16 \cdot \xi'_1 + 162.11 \cdot \eta'_2 - \xi'_1 \eta'_2} \tag{26}$$

$$z_1 = (1.46190 - 0.00356 \cdot y) \cdot \zeta'_1 \quad \text{and}$$
$$z_2 = (1.57102 - 0.00227 \cdot x - 0.00131 \cdot y) \cdot \zeta'_2 \tag{27}$$

These equations are somewhat different from Eqs. (19)–(21) because of slight differences in the values of l'_1 and l'_2. One could however use Eqs. (19)–(21) instead of these since a difference appeared only in the last decimal place of the co-ordinate values.

The bidimensional co-ordinates of the left side in experiment II are given in Table 4.

The corresponding constants of camera 1^b in experiment II were: $e = 446$ cm, $d = 55.5$ cm, $l_1 = 86$ cm, $l'_1 = 6.42504$ cm; and of camera 2^b: $e = 588.5$ cm, $d = 67$ cm, $l_2 = 86$ cm, $l'_2 = 5.31771$ cm. From these values the equations to calculate the tridimensional co-ordinates of the left side in experiment II were deduced:

Table 4. Experiment II, left

No.	Left shoulder				Left elbow				Left wrist				No.
	ξ'_1	η'_2	ζ'_1	ζ_2	ξ'_1	η'_2	ζ'_1	ζ_2	ξ'_1	η'_2	ζ'_1	ζ_2	
1	−76,053	+38,319	+30,427	+21,342	−79,259	+43,301	+10,247	+ 7,057	−76,833	+45,734	−11,397	− 7,687	1
2	−71,312	+36,987	+31,020	+21,890	−72,636	+41,239	+10,607	+ 7,383	−67,425	+42,110	−10,940	− 7,500	2
3	−66,689	+35,611	+32,125	+22,829	−66,089	+39,091	+11,684	+ 8,218	−57,489	+37,979	− 9,216	− 6,443	3
4	−61,796	+34,112	+33,562	+24,041	−59,571	+37,019	+13,278	+ 9,419	−47,262	+33,401	− 6,241	− 4,435	4
5	−57,217	+32,680	+34,689	+25,015	−53,721	+35,269	+14,680	+10,502	−38,223	+29,565	− 2,885	− 1,760?	5
6	−52,399	+31,152	+35,252	+25,640	−47,723	+33,589	+15,655	+11,374	−29,644	+24,880	+ 0,785	+ 0,649	6
7	−47,464	+29,489	+35,094	+25,752	−41,520	+31,824	+16,054	+11,709	−21,796	+20,882	+ 4,241	+ 3,306	7
8	−42,103	+27,939	+34,457	+25,441	−36,094	+30,159	+15,934	+11,720	−15,598	+17,738	+ 6,593	+ 5,140	8
9	−37,893	+26,011	+33,399	+24,901	−29,867	+28,030	+15,262	+11,378	− 9,113	+14,734	+ 8,231	+ 6,562	9
10	−32,957	+24,142	+32,339	+24,300	−24,494	+25,956	+14,302	+10,743	− 3,639	+12,364	+ 8,461	+ 6,716	10
11	−27,717	+22,064	+31,510	+23,926	−19,443	+23,807	+13,256	+10,075	+ 1,537	+10,481	+ 7,417	+ 5,966	11
12	−22,837	+19,986	+31,096	+23,855	−15,023	+21,792	+12,544	+ 9,622	+ 5,719	+ 9,129	+ 5,216	+ 4,253	12
13	−18,057	+17,965	+31,042	+24,034	−10,972	+19,893	+12,095	+ 9,360	+ 9,246	+ 8,388	+ 2,090	+ 1,749	13
14	−13,581	+16,092	+31,478	+24,578	− 7,285	+18,233	+12,223	+ 9,600	+11,403	+ 8,365	− 1,193	− 0,973	14
15	− 9,175	+14,321	+32,221	+25,359	− 3,243	+16,505	+12,758	+ 9,975	+12,434	+ 9,111	− 4,366	− 3,519	15
16	− 4,912	+12,633	+33,150	+26,327	+ 0,055	+15,069	+13,394	+10,538	+12,594	+10,475	− 5,891	− 4,840	16
17	− 0,677	+10,966	+33,969	+27,060	+ 2,663	+14,171	+13,954	+11,004	+12,357	+11,848	− 6,525	− 5,363	17
18	+ 3,617	+ 9,214	+34,356	+27,634	+ 5,154	+13,493	+14,286	+11,313	+12,335	+13,091	− 6,970	− 5,320	18
19	+ 8,188	+ 7,400	+34,263	+27,776	+ 7,818	+12,822	+14,454	+11,417	+12,790	+14,301	− 7,091	− 5,775	19
20	+12,777	+ 5,506	+33,918	+27,720	+10,937	+11,919	+14,368	+11,392	+14,216	+14,597	− 7,327	− 5,800	20
21	+17,075	+ 3,558	+33,275	+27,510	+14,416	+10,730	+13,967	+11,130	+16,556	+14,597	− 7,539	− 6,227	21
22	+21,539	+ 1,808	+32,464	+27,040	+18,539	+ 9,187	+13,263	+10,640	+19,792	+13,920	− 8,019	− 6,550	22
23	+26,842	− 0,378	+31,590	+26,578	+23,943	+ 7,050	+12,340	+ 9,995	+24,779	+12,400	− 8,656	− 7,160	23
24	+31,888	− 2,376	+31,116	+26,440	+29,445	+ 4,763	+11,646	+ 9,530	+30,814	+ 9,955	− 9,309	− 7,880	24
25	+37,217	− 4,510	+30,854	+26,500	+35,510	+ 2,271	+11,174	+ 9,231	+38,223	+ 6,123	−10,108	− 8,530	25
26	+42,478	− 6,750	+30,923	+26,850	+41,877	− 0,500	+11,031	+ 9,257	+46,475	+ 2,220	−10,418	− 9,000	26
27	+47,577	− 8,990	+31,356	+27,510	+48,589	− 3,561	+11,290	+ 9,692	+55,813	− 3,040	−10,002	−▼8,780	27
28	+52,201	−11,140	+32,176	+28,625	+54,917	− 6,495	+12,145	+10,526	+64,974	− 8,900	− 8,504		28
29	+56,787	−13,380	+33,401	+30,020	+61,038	− 9,377	+13,551	+11,961	+73,636	−14,600	− 5,749	− 5,430?	29
30	+61,278	−15,550	+34,517	+31,374	+66,929	−12,130	+14,935	+13,440	+81,596	−20,700	− 2,553	− 2,430	30
31	+65,969	−17,910	+35,141	+32,237	+73,061	−14,950	+15,981	+14,570	+89,243	−26,750	+ 0,185	+ 0,900?	31

No.	Left hip				Left knee				Left ankle				No.
	ξ'_1	η'_2	ζ'_1	ζ_2	ξ'_1	η'_2	ζ'_1	ζ_2	ξ'_1	η'_2	ζ'_1	ζ_2	
1	−72,614	+36,027	−3,996	−2,673	−61,003	+31,324	−31,897	−22,833	−56,393	+26,670	−60,893	−44,483	1
2		+34,958		−2,120	−58,408	+30,637	−31,742	−22,824	−56,248	+26,589	−60,910	−44,511	2
3		+33,577		−1,459	−56,754	+30,228	−31,554	−22,759	−56,185	+26,560	−60,880	−44,493	3
4	−59,865		−1,593		−55,513	+29,694	−31,437	−22,741	−56,153	+26,530	−60,840	−44,470	4
5	−56,169		−1,240		−54,562	+29,246	−31,301	−22,680	−56,128	+26,520	−60,770	−44,447	5
6	−52,329	+30,068	−1,207	−0,826	−53,578	+28,763	−31,175	−22,655	−56,073	+26,500	−60,710	−44,420	6
7	−48,250	+28,775	−1,513	−1,089	−52,075	+28,166	−31,179	−22,741	−56,030	+26,460	−60,630	−44,393	7
8	−44,406	+27,529	−2,062	−1,523	−50,402	+27,496	−31,289	−22,915	−55,987	+26,420	−60,570	−44,350	8
9	−39,665	+25,892	−2,919	−2,182	−47,980	+26,572	−31,511	−23,172	−55,728	+26,320	−60,404	−44,247	9
10	−35,167	+24,265	−3,523	−2,661	−44,881	+25,553	−31,687	−23,436	−55,203	+26,201	−59,890	−43,959	10
11	−30,118	+22,364	−3,870	−3,005	−40,358	+24,306	−32,021	−23,886	−54,063	+26,045	−58,975	−43,287	11
12	−24,791	+20,202	−4,022	−3,151	−34,965	+22,830	−32,377	−24,373	−51,948	+25,654	−57,430	−42,301	12
13	−18,808	+17,672	−4,040	−3,202	−27,297	+20,744	−33,120	−25,238	−47,687	+24,509	−55,101	−40,895	13
14	−13,496	+15,409	−4,178	−3,346	−18,607	+17,743	−34,038	−26,349	−41,563	+22,601	−52,417	−39,315	14
15	− 8,081	+12,927	−3,627	−2,954	− 8,860	+13,933	−33,874	−26,722	−32,731	+19,514	−50,811	−38,753	15
16	− 3,140		−2,403		− 0,364	+10,449	−32,494	−26,074	−22,493	+15,562	−51,220	−39,834	16
17	+ 1,786	+ 8,764	−1,485	−1,076	+ 7,480	+ 7,126	−30,954	−25,218	−11,454	+11,031	−52,842	−42,012	17
18		+ 6,611		−0,529	+14,935	÷ 3,825	−29,630	−24,527	+ 0,458	+ 5,983	−55,032	−44,814	18
19		+ 4,572		−0,493	+21,721	+ 0,531	−28,970	−24,306	+13,298	+ 0,352	−57,237	−47,835	19
20	+16,506	+ 2,518	−1,324	−1,106	+27,485	− 2,541	−29,227	−24,889	+26,439	− 5,624	−58,754	−50,424	20
21	+20,829	+ 0,680	−2,440	−2,072	+31,463	− 4,953	−30,323	−26,090	+38,088	−11,110	−58,962	−51,874	21
22	+25,381	− 1,331	−3,612	−3,122	+34,777	− 7,152	−31,745	−27,500	+47,829	−15,821	−58,026	−52,098	22
23	+30,825	− 3,635	−4,346	−3,764	+40,395	−10,100	−32,269	−28,100	+54,209	−19,551	−58,020	−52,945	23
24	+35,935	− 5,783	−4,488	−3,919	+47,095	−12,900	−31,851	−28,200	+56,574	−21,411	−59,381	−54,599	24
25	+41,939	− 8,207	−4,137	−3,700	+51,196	−14,630	−32,209	−28,730	+59,501	−22,953	−59,760	−55,341	25
26	+46,705	−10,007	−3,595	−3,256	+56,561	−16,700	−31,531	−28,600	+60,690	−23,654	−60,027	−55,788	26
27		−11,980		−2,900	+60,615	−18,500	−31,266	−28,600	+61,263	−24,014	−60,108	−55,930	27
28	+55,670	−14,049	−2,335	−2,144	+62,653	−19,500	−31,153	−28,640	+61,354	−24,100	−60,100	−55,914	28
29	+59,699		−1,670		+64,014	−20,420	−31,038	−28,650	+61,381	−24,140	−60,085	−55,870	29
30	+63,340	−18,080	−1,276	−1,550	+65,033	−21,260	−30,884	−28,640	+61,422	−24,190	−60,040	−55,827	30
31	+62,232	−20,129	−1,117	−1,092	+66,043	−21,950	−30,756	−28,645	+61,470	−24,240	−59,970	−55,810	31

? signifies uncertainty

Table 4 (continued)

No.	Centre of gravity of left foot				Tip of left foot				Point on head, from left				No.
	ξ_1'	η_2'	ζ_1'	ζ_2'	ξ_1'	η_2'	ζ_1'	ζ_2'	ξ_1'	η_2'	ζ_1'	ζ_2'	
1	−52,787	+25,470	−64,522	−47,505	−43,380	+21,425	−64,520	−48,410	−69,853	+26,368	+56,042	+41,108	1
2	−52,795	+25,470	−64,510	−47,500	−43,360	+21,425	−64,555	−48,430	−65,403	+25,084	+56,885	+41,976	2
3	−52,800	+25,465	−64,503	−47,495	−43,340	+21,425	−64,574	−48,450	−60,755	+23,559	+57,945	+43,052	3
4	−52,817	+25,460	−64,490	−47,490	−43,330	+21,425	−64,600	−48,470	−55,945	+21,869	+58,936	+44,147	4
5	−52,835	+25,450	−64,470	−47,480	−43,320	+21,425	−64,620	−48,490	−51,523	+20,226	+59,553	+44,929	5
6	−52,826	+25,447	−64,450	−47,470	−43,310	+21,425	−64,650	−48,510	−46,936	+18,452	+59,642	+45,356	6
7	−52,810	+25,425	−64,430	−47,450	−43,310	+21,425	−64,670	−48,530	−42,231	+16,574	+59,238	+45,458	7
8	−52,800	+25,416	−64,420	−47,430	−43,310	+21,425	−64,690	−48,560	−38,054	+14,921	+58,539	+45,232	8
9	−52,780	+25,408	−64,400	−47,380	−43,310	+21,410	−64,770	−48,620	−33,073	+12,890	+57,479	+44,833	9
10	−52,440	+25,383	−64,180	−47,280	−43,117	+21,340	−65,035	−48,780	−28,430	+10,972	+56,553	+44,471	10
11	−51,964	+25,338	−63,747	−46,956	−42,660	+21,220	−65,370	−49,017	−23,594	+ 8,891	+55,946	+44,433	11
12	−50,830	+25,144	−62,900	−46,366	−41,990	+20,995	−65,847	−49,405	−18,902	+ 6,810	+55,672	+44,594	12
13	−48,445	+24,374	−60,660	−44,884	−41,050	+20,503	−66,090	−49,626	−14,117	+ 4,632	+55,795	+45,157	13
14	−43,254	+22,883	−57,780	−43,111	−38,483	+19,732	−65,294	−49,358	− 9,848	+ 2,667	+56,255	+45,954	14
15	−34,621	+19,865	−56,082	−42,504	−29,790	+16,733	−63,600	−48,840	− 5,335	+ 0,617	+57,088	+47,045	15
16	−23,630	+15,624	−56,525	−43,780	−17,995	+11,706	−63,242	−49,700	− 0,889	− 1,375	+58,006	+48,265	16
17	−11,517	+10,677	−58,136	−46,140	− 4,709	+ 6,109	−63,500	−51,260	+ 3,362	− 3,253	+58,703	+49,295	17
18	+ 1,588	+ 5,243	−60,175	−49,020	+ 9,603	+ 0,110	−63,810	−53,020	+ 7,588	− 5,128	+59,024	+49,960	18
19	+15,741	− 0,786	−61,913	−51,965	+24,558	− 6,176	−63,516	−54,456	+11,952	− 7,074	+58,846	+50,245	19
20	+30,010	− 7,230	−62,624	−54,230	+39,124	−12,600?	−61,880	−54,810	+16,406	− 9,112	+58,269	+50,222	20
21	+42,603	−13,140	−61,790	−55,020	+51,225	−18,700	−58,700	−53,524	+20,594	−11,095	+57,450	+49,960	21
22	+52,935	−18,160	−59,872	−54,550	+60,726	−24,100?	−54,970	−51,070	+24,832	−13,070	+56,671	+49,672	22
23	+59,423	−21,780	−59,570	−55,180	+67,028	−26,450	−54,260	−51,500	+29,694	−15,297	+56,013	+49,633	23
24	+61,518	−23,680	−61,556	−57,380	+69,663	−28,650	−57,404	−54,950	+34,283	−17,393	+55,675	+49,826	24
25	+63,735	−25,560	−62,780	−58,910	+72,300	−31,240	−60,970	−58,610	+38,930	−19,506	+55,602	+50,202	25
26	+64,660	−25,730	−63,590	−59,580	+73,100	−31,800	−63,080	−60,670	+43,238	−21,481	+55,914	+50,890	26
27	+64,600	−26,000	−63,670	−59,620	+73,285	−32,100	−63,584	−61,230	+47,657	−23,563	+56,525	+52,000	27
28	+64,610	−26,000	−63,640	−59,640	+73,290	−32,130	−63,610	−61,250	+52,093	−25,803	+57,374	+53,320	28
29	+64,620	−26,000	−63,620	−59,680	+73,300	−32,150	−63,620	−61,270	+56,618	−28,217	+58,224	+54,760	29
30	+64,630	−26,000	−63,600	−59,710	+73,300	−32,170	−63,640	−61,300	+60,959	−30,644	+58,865	+55,870	30
31	+64,640	−26,000	−63,580	−59,730	+73,300	−32,190	−63,660	−61,330	+65,329	−33,313	+59,096	+56,790	31

$$x = \frac{142\,356 \cdot \xi_1' - 396.02 \cdot \xi_1' \eta_2'}{106\,354 + 196.93 \cdot \xi_1' + 180.02 \cdot \eta_2' - \xi_1' \eta_2'} \tag{28}$$

$$y = -\frac{82\,189 \cdot \xi_1' + 198\,607 \cdot \eta_2' - 417.34 \cdot \xi_1' \eta_2'}{106\,354 + 196.93 \cdot \xi_1' + 180.02 \cdot \eta_2' - \xi_1' \eta_2'} \tag{29}$$

$$z_1 = (1,33851 - 0.003207 \cdot y) \cdot \zeta_1' \quad \text{and}$$
$$z_2 = (1.61624 - 0.002539 \cdot x - 0.001466 \cdot y) \cdot \zeta_2' \tag{30}$$

The differences between these and the corresponding Eqs. (22)–(24) of the first experiment are again so small that with both groups of equations one obtains values the differences of which would not jeopardize the accuracy which one can attain for the co-ordinates.

In experiment III the subject was heavily loaded. The direct measurements gave the co-ordinates listed in Table 5 for the right side. Besides the points used previously there was one additional point situated on the Geissler tube of the thigh.

The values of the constants for cameras 1^a in experiment III were: $e = 438$ cm, $d = 52$ cm, $l_1 = 86$ cm, $l_1' = 5.86671$; and for camera 2^a: $e = 634.5$ cm, $d = 69$ cm $l_1 = 86$ cm, $l_2' = 5 \cdot 43188$ cm. The equations for the right side in experiment III were thus:

$$x = \frac{155\,233 \cdot \xi_1' - 454.51 \cdot \xi_1' \eta_2'}{105\,896 + 218.37 \cdot \xi_1' + 161.65 \cdot \eta_2' - \xi_1' \eta_2'} \tag{31}$$

$$y = \frac{89\,624 \cdot \xi_1' + 193\,597 \cdot \eta_2' - 410.42 \cdot \xi_1' \eta_2'}{105\,896 + 218.37 \cdot \xi_1' + 161.65 \cdot \eta_2' - \xi_1' \eta_2'} \tag{32}$$

Table 5. Experiment III, right

No.	Right shoulder				Right elbow				Right wrist				No.
	ξ'_1	η'_2	ζ'_1	ζ'_2	ξ'_1	η'_2	ζ'_1	ζ'_2	ξ'_1	η'_2	ζ'_1	ζ'_2	
1	−61,373	+33,423	+30,721	+24,022	−60,327	+36,632	+12,017	+9,202	−50,062	+34,367	−8,976	−7,200	1
2	−57,405	+32,284	+31,492	+24,842	−58,068	+36,259	+12,775	+9,807	−50,613	+35,682	−9,109	−7,307	2
3	−53,492	+31,159	+31,981	+25,400	−55,897	+36,047	+13,438	+10,360	−50,760	+36,773	−9,014	−7,217	3
4	−49,594	+30,024	+32,070	+25,664	−53,734	+35,800	+13,886	+10,742	−50,382	+37,715	−8,837	−7,010	4
5	−45,682	+28,904	+31,743	+25,550	−51,213	+35,374	+13,964	+10,842	−48,319	+38,371	−8,836	−7,064	5
6	−41,748	+27,850	+31,097	+25,250	−48,033	+34,667	+13,572	+10,546	−47,359	+38,597	−9,160	−7,316	6
7	−37,729	+26,680	+30,270	+24,750	−44,209	+33,765	+12,844	+10,054	−44,422	+38,435	−9,759	−7,785	7
8	−33,430	+25,446	+29,470	+24,316	−39,691	+32,635	+11,953	+9,439	−40,247	+37,777	−10,483	−8,445	8
9	−28,987	+24,274	+28,970	+24,109	−34,615	+31,304	+11,250	+8,954	−34,630	+36,365	−11,200	−9,027	9
10	−24,493	+23,008	+28,900	+24,260	−29,135	+29,776	+10,961	+8,830	−27,797	+34,236	−11,554	−9,463	10
11	−20,107	+21,730	+29,370	+24,890	−23,223	+28,084	+11,176	+9,130	−19,737	+31,369	−11,360	−9,360	11
12	−16,033	+20,372	+30,152	+25,760	−17,417	+26,252	+11,814	+9,801	−11,015	+27,840	−10,330?		12
13	−12,084	+19,008	+31,207	+25,920	−11,529	+24,299	+12,942	+10,904	−1,822	+23,710	−8,407	−7,192	13
14	−8,309	+17,660	+32,294		−5,784	+22,291	+14,290	+12,210	+7,032	+19,250	−5,790	−4,885	14
15	−4,525	+16,200	+33,085		+0,100	+20,191	+15,430	+13,480	+15,477	+14,415	−2,696	−2,230	15
16	−0,835	+14,750	+33,327	+29,463	+5,249	+18,141	+16,180	+14,310	+22,924	+9,807	+0,260	+0,468	16
17	+2,955	+13,200	+33,063		+10,465	+16,060	+16,360	+14,440	+29,639	+5,300	+2,860	+2,980	17
18	+6,819	+11,560	+32,396	+29,220	+15,475	+13,840	+16,120	+14,480	+35,472	+1,300	+4,934	+4,980	18
19	+10,908	+9,725	+31,533	+28,720	+20,202	+11,540	+15,544	+14,240	+40,619	−2,160	+6,060	+6,170	19
20	+15,013	+7,750	+30,700	+28,230	+24,744	+9,300	+14,725	+13,720	+45,012	−5,000?	+6,030		20
21	+19,252	+5,660	+29,950	+27,860	+28,859	+7,190	+13,854	−13,120	+48,990	−7,250?	+4,742		21
22	+21,392	+3,510	+29,483	+27,710	+32,557	+5,000	+13,036	−12,580	+52,132	−8,800?	+2,593		22
23	+27,522	+1,220	+29,317	+27,800	+35,704	+3,230	+12,400	−12,050	+54,539	−9,560	−0,248		23
24	+31,638	−0,980	+29,450	+28,200	+38,558	+1,500	+12,030	−11,810	+55,871	−9,220	−3,360		24
25	+35,578	−3,000	+29,834		+41,113	−0,125	+11,940	−11,720	+56,142	−8,100	−6,106		25
26	+39,279	−4,880	+30,561	+29,839?	+43,413	−1,720	+12,200	−12,080	+55,601	−6,600	−7,880		26
27	+42,947	−6,730	+31,348	+30,900	+45,277	−2,853	+12,730	−12,600	+54,902	−4,970?	−8,530		27
28	+56,506	−8,760	+31,900	+31,820	+46,978	−3,600	+13,306	−13,170	+54,495	−3,800	−8,600		28
29	+50,281	−10,820	+32,086	+32,270	+48,952	−4,410	+13,770	−13,715	+54,572	−3,000?	−8,565		29
30	+54,119	−12,900	+31,874	+32,280	+51,336	−5,476	+13,930	−13,920	+55,453	−3,000?	−8,617		30
31	+57,930	−15,030	+31,286	+32,090	+54,231	−6,810	+13,700	−13,780	+57,235	−3,000?	−8,884		31

? signifies uncertainty

No.	Right hip				Right knee				Right ankle				No.
	ξ'_1	η'_2	ζ'_1	ζ'_2	ξ'_1	η'_2	ζ'_1	ζ'_2	ξ'_1	η'_2	ζ'_1	ζ'_2	
1	−63,954		−1,755		−62,441	+31,990	−29,392	−23,422	−79,703	+35,137	−47,908	−37,459	1
2	−60,184		−0,826		−55,869	+30,074	−28,181	−22,694	−69,796	+32,199	−49,542	−39,346	2
3					−49,773	+28,420	−27,172	−22,100	−59,154	+28,998	−61,546	−41,695	3
4		+29,310		−0,272	−44,462	+26,799	−26,638	−21,878	−48,128	+25,772	−53,372	−44,000	4
5		+28,239		−0,662	−40,175	+25,364	−26,930	−22,277	−36,986	+22,597	−54,473	−45,795	5
6	−45,369	+27,090	−1,748	−1,480	−36,626	+24,104	−28,057	−23,382	−26,530	+19,679	−54,482	−46,637	6
7	−41,401	+25,983	−2,836	−2,381	−33,248	+22,739	−29,308	−24,605	−18,264	+17,356	−53,772	−46,655	7
8	−37,238	+24,915	−3,522	−2,942	−28,160	+21,223	−29,760	−25,199	−13,550	+15,833	−54,367	−47,622	8
9	−32,717	+23,645	−3,930	−3,282	−22,475	+19,991	−29,807	−25,458	−11,048	+14,802	−55,591	−48,957	9
10		+22,580		−3,087	−19,235	+19,197	−30,170	−25,910	−8,881	+14,075	−56,253	−49,750	10
11		+21,421		−2,538	−14,530	+18,061	−29,626	−25,583	−7,953	+13,663	−56,517	−50,113	11
12	−19,073	+20,162	−2,363	−2,002	−10,878	+16,968	−29,142	−25,373	−7,680	+13,489	−56,549	−50,171	12
13	−15,232	+19,017	−1,525	−1,268	−9,808	+16,484	−28,969	−25,295	−7,661	+13,432	−56,545	−50,160	13
14	−11,784		−0,820		−9,431	+16,235	−28,703	−25,071	−7,650	+13,430	−56,494	−50,147	14
15	−8,185		−0,410		−8,711	+15,777	−28,330	−24,804	−7,632	+13,366	−56,393	−50,087	15
16	−4,601	+14,979	−0,434	−0,374	−7,432	+15,114	−28,163	−24,753	−7,600	+13,255	−56,335	−50,049	16
17	−0,901	+13,531	−0,724	−0,574	−6,171	+14,531	−28,063	−24,753	−7,560	+13,167	−56,284	−50,020	17
18	+2,619	+12,072	−1,183	−1,037	−4,964	+13,912	−27,926	−24,779	−7,515	+13,097	−56,269	−50,025	18
19	+6,410	+10,501	−1,843	−1,657	−2,968	+12,944	−28,059	−25,008	−7,368	+12,978	−56,177	−49,990	19
20	+10,211	+8,809	−2,424	−2,210	−0,548	+11,854	−28,064	−25,136	−7,010	+12,850	−55,838	−49,720	20
21	+14,308	+6,773	−2,682	−2,426	−2,653	+10,533	−27,915	−25,179	−6,282	+12,663	−55,159	−49,167	21
22	+18,796	+4,422	−2,799	−2,553	+6,997	+8,736	−28,020	−25,494	−4,944	+12,222	−54,005	−48,280	22
23	+23,498	+1,661	−2,743	−2,502	+12,454	+6,226	−28,258	−26,034	−2,398	+11,002	−52,306	−47,054	23
24	+28,137	−1,164	−2,690	−2,405	+19,574	+3,105	−28,973	−27,101	+1,792	+8,958	−49,971	−45,419	24
25	+32,422		−2,550		+27,495	−0,790	−29,823	−28,423	+7,809	+5,939	−47,685	−43,963	25
26	+36,498		−1,869		+35,439	−5,020	−29,507	−28,556	+15,707	+1,651	−47,030	−44,227	26
27	+40,463	−8,457	−0,992	−0,734	+42,314	−8,887	−28,532	−27,928	+24,476	−3,413	−47,583	−45,744	27
28	+44,141	−10,483	−0,417	−0,153	+48,587	−12,379	−27,635	−27,553	+33,590	−8,706	−48,997	−48,231	28
29		−12,457		+0,045	+54,479	−15,758	−26,960	−27,258	+43,773	−14,613	−50,952	−51,463	29
30		−14,438		−0,220	+59,634	−18,777	−26,770	−27,354	+54,578	−20,920	−52,804	−54,807	30
31		−16,397		−0,932	+64,011	−21,620	−27,160	−28,136	+65,699	−27,283	−53,976	−57,546	31

Table 5 (continued)

No.	Centre of gravity of right foot				Tip of right foot				No.
	ξ'_1	η'_2	ζ'_1	ζ'_2	ξ'_1	η'_2	ζ'_1	ζ'_2	
1	−80,300?	+34,952	−52,312?	−40,729	−73,920	+30,846	−57,387	−45,687	1
2	−69,500	+31,769	−53,920	−42,804	−62,020	+27,290	−57,889	−46,942	2
3	−57,800	+28,275	−55,730	−45,193	−49,560	+23,404	−57,947	−48,094	3
4	−45,830	+24,782	−57,170	−47,387	−37,110	+19,630	−57,459	−48,815	4
5	−33,780	+21,301	−57,690	−48,909	−25,010	+16,295	−56,053	−48,662	5
6	−22,550	+18,212	−56,992	−49,302	−14,370	+13,420	−53,411	−47,338	6
7	−13,851	+15,857	−55,494	−48,753	− 6,407	+11,413	−50,605	−45,490	7
8	− 9,175	+14,444	−56,143	−49,720	− 1,700	+10,366	−51,172	−46,376	8
9	− 7,005	+13,371	−57,921	−51,572	+ 1,080	+ 8,733	−54,313	−49,474	9
10	− 5,262	+12,565	−59,546	−53,173	+ 3,244	+ 7,164	−57,648	−52,790	10
11	− 4,860	+12,317	−59,863	−53,510	+ 3,700	+ 6,923	−59,113	−54,104	11
12	− 4,855	+12,273	−59,839	−53,500	+ 3,740	+ 6,847	−59,300	−54,284	12
13	− 4,850	+12,235	−59,815	−53,485	+ 3,748	+ 6,825	−59,320	−54,303	13
14	− 4,845	+12,195	−59,791	−53,465	+ 3,756	+ 6,815	−59,340	−54,322	14
15	− 4,840	+12,180	−59,768	−53,446	+ 3,764	+ 6,810	−59,360	−54,341	15
16	− 4,835	+12,163	−59,745	−53,427	+ 3,772	+ 6,805	−59,380	−54,360	16
17	− 4,830	+12,150	−59,722	−53,408	+ 3,780	+ 6,795	−59,400	−54,380	17
18	− 4,825	+12,130	−58,699	−53,389	+ 3,812	+ 6,790	−59,460	−54,413	18
19	− 4,820	+12,100	−59,676	−53,370	+ 3,838	+ 6,770	−59,520	−54,480	19
20	− 4,617	+12,044	−59,443	−53,200	+ 3,957	+ 6,710	−59,654	−54,620	20
21	− 4,209	+11,970	−59,052	−52,853	+ 4,231	+ 6,626	−59,830	−54,810	21
22	− 3,468	+11,724	−58,386	−52,300	+ 4,708	+ 6,400	−60,088	−55,060	22
23	− 2,063	+10,861	−57,001	−51,223	+ 5,425	+ 5,790	−60,524	−55,645	23
24	+ 0,783	+ 9,366	−54,594	−49,380	+ 6,320	+ 5,430	−66,838	−55,960	24
25	+ 6,323	+ 6,399	−51,989	−47,686	+10,530	+ 2,704	−59,000	−54,890	25
26	+14,311	+ 1,883	−51,376	−48,017	+18,585	− 2,970	−57,470	−54,780	26
27	+23,716	− 3,496	−51,903	−49,764	+28,880	− 9,015	−57,254	−56,040	27
28	+33,645	− 9,132	−53,387	−52,504	+39,970	−15,170	−57,550	−57,980	28
29	+44,813	−15,546	−55,152	−55,843	+51,960	−22,067	−57,810	−60,040	29
30	+56,594	−22,379	−56,612	−59,168	+64,330	−29,277	−57,350	−61,500	30
31	+68,641	−29,312	−57,187	−61,639	+76,486	−36,320	−55,970	−61,900	31

No.	Point on Geissler tube on right thigh				Point on head, from right				No.
	ξ'_1	η'_2	ζ'_1	ζ'_2	ξ'_1	η'_2	ζ'_1	ζ'_2	
1	−62,964	+32,142	−19,943	−15,950	−52,971	+20,570	+53,182	+43,771	1
2	−57,349	+30,484	−18,836	−15,220	−49,065	+19,296	+53,725	+44,621	2
3	−52,107	+29,110	−18,010	−14,630	−45,247	+18,056	+53,980	+45,157	3
4	−47,322	+27,662	−17,598	−14,437	−41,420	+16,813	+53,844	+45,354	4
5	−43,265	+26,343	−17,963	−14,840	−37,465	+15,518	+53,395	+45,299	5
6	−39,608	+25,132	−19,155	−15,853	−33,482	+14,199	+52,618	+45,025	6
7	−36,019	+23,838	−20,267	−16,990	−29,669	+12,970	+51,823	+44,673	7
8	−31,245	+22,499	−20,802	−17,539	−25,798	+11,739	+51,134	+44,448	8
9	−25,946	+21,259	−20,993	−17,816	−21,884	+10,466	+51,740	+44,457	9
10	−22,223	+20,362	−21,129	−18,062	−18,055	+ 9,137	+50,863	+44,917	10
11	−17,493	+19,224	−20,534	−17,676	−14,256	+ 7,752	+51,309	+45,688	11
12	−13,666	+18,064	−19,992	−17,352	−10,523	+ 6,225	+52,000	+46,664	12
13	−11,618	+17,345	−19,642	−17,058	− 6,786	+ 4,683	+52,917	+47,859	13
14	−10,214	+16,756	−19,214	−16,766	− 3,090	+ 3,155	+55,749	+49,023	14
15	− 8,524	+15,955	−18,830	−16,456	+ 0,566	+ 1,604	+54,191	+49,875	15
16	− 6,448	+15,054	−18,710	−16,486	+ 4,106	+ 0,083	+54,250	+50,315	16
17	− 4,376	+14,176	−18,741	−16,594	+ 7,645	− 1,488	+53,940	+50,443	17
18	− 2,367	+13,283	−18,865	−16,780	+11,218	− 3,164	+53,321	+50,284	18
19	+ 0,229	+12,120	−19,183	−16,173	+14,846	− 4,991	+52,524	+50,012	19
20	+ 3,129	+10,822	−19,361	−17,433	+18,479	− 6,877	+51,786	+49,743	20
21	+ 6,624	+ 9,262	−19,346	−17,568	+22,290	− 8,943	+51,233	+44,666	21
22	+11,107	+ 7,286	−19,425	−17,803	+26,092	−11,114	+50,915	+49,884	22
23	+16,238	+ 4,678	−19,558	−17,140	+29,879	−13,370	+51,052	+50,548	23
24	+22,499	+ 1,659	−20,000	−18,807	+33,667	−15,644	+51,436	+51,441	24
25	+29,183	− 1,762	−20,488	−18,536	+37,309	−17,748	+52,066	+52,585	25
26	+35,818	− 5,405	−20,072	−19,393	+41,003	−19,907	+52,858	+53,912	26
27	+41,667	− 8,738	−19,117	−19,739	+44,701	−21,963	+53,577	+55,171	27
28	+47,059	−11,697	−18,329	−18,150	+48,193	−23,894	+54,014	+56,081	28
29	+52,213	−14,618	−17,860	−17,870	+51,803	−25,919	+54,018	+56,652	29
30	+56,853	−17,258	−17,784	−17,974	+55,538	−28,058	+53,663	+56,798	30
31	+60,915	−19,810	−18,294	−17,728	+59,386	−30,413	+52,973	+56,701	31

Table 6. Experiment III, left

No.	Left shoulder				Left elbow				Left wrist				No.
	ξ_1'	η_2'	ζ_1'	ζ_2'	ξ_1'	η_2'	ζ_1'	ζ_2'	ξ_1'	η_2'	ζ_1'	ζ_2'	
1	−65,463	+37,915	+34,618	+26,587	−66,760	+42,708	+14,569	+11,073	−43,712		+ 8,016		1
2	−61,069	+36,352	+35,390	+27,405	−62,408	+41,262	+15,441	+11,800	−39,443		+ 8,743		2
3	−56,728	+34,781	+35,800	+27,832	−58,085	+39,910	+15,928	+12,185	−35,260	+27,385?	+ 9,163?	+7,374	3
4	−52,417	+33,120	+35,625	+27,948	−53,828	+38,410	+15,846	+12,254		+25,797		+7,163	4
5	−48,054	+31,420	+34,980	+27,696	−49,575	+36,751	+15,196	+11,833		+24,081		+6,560	5
6	−43,643	+29,573?	+34,020	+27,000?	−45,217	+34,963	+14,198	+11,141		+22,170		+5,700	6
7	−39,205	+27,705	+32,927	+26,485	−40,723	+33,094	+13,100	+10,343					7
8	−34,650	+25,645		+25,910	−35,994	+31,035	+12,164	+ 9,697	−13,960	+17,960	+ 4,830	+4,092	8
9	−30,130	+23,541	+31,333	+25,617	−31,381	+28,912	+11,524	+ 9,190	− 9,272	+15,670	+ 4,320	+3,687	9
10	−25,586	+21,480	+31,214	+25,727	−26,515	+26,637	+11,310	+ 9,147	− 4,530	+13,375	+ 4,390	+3,780	10
11	−21,273	+19,430	+31,480	+26,192	−21,788	+24,409	+11,546	+ 9,430	+ 0,130	+11,080	+ 5,050	+4,347	11
12	−17,070	+17,498	+32,110	+26,935	−17,431	+22,372	+12,197	+10,000	+ 4,490	+ 8,937	+ 5,840	+5,074	12
13	−13,130	+15,780		+27,948	−13,430	+20,606	+13,132	+10,873	+ 8,450	+ 7,178	+ 6,840	+5,947	13
14	− 9,444	+14,160		+29,000	− 9,655	+18,956	+14,138	+11,774		+ 5,561?		+6,950	14
15	− 5,640	+12,480	+34,710?	+29,710	− 5,861	+17,332	+14,757	+12,366	+16,150		+ 8,540		15
16	− 1,810	+10,820	+34,840?	+30,010	− 2,052	+15,726	+14,838?	+12,540	+19,850		+ 8,630		16
17	+ 2,140	+ 9,077		+29,895	+ 1,981	+14,015	+14,458	+12,279					17
18	+ 6,106	+ 7,348	+33,670?	+29,490	+ 6,155	+12,340	+13,740	+12,791					18
19	+10,215	+ 5,594	+32,830?	+29,050	+10,490	+10,613	+12,905	+11,129					19
20	+14,445	+ 3,896	+32,010	+28,500	+14,925	+ 8,910	+12,040	+10,476					20
21	+18,953	+ 2,144	+31,470?	+28,188	+19,510	+ 7,173	+11,474	+10,077	+41,140		+ 5,440		21
22	+23,522	+ 0,430	+31,210	+28,211	+24,227	+ 5,470	+11,230	+ 9,866	+45,970		+ 5,364		22
23	+28,135	− 1,300	+31,500	+28,745	+29,238	+ 3,700	+11,5?	+10,201	+51,000		+ 5,820		23
24	+32,773	− 3,150	+32,240	+29,694	+34,310	+ 1,934	+12,260	+11,046	+56,070		+ 6,795		24
25	+37,220	− 4,950?	+33,130	+30,672	+39,155	+ 0,216	+13,265	+12,047					25
26	+41,610	− 6,874	+34,240	+32,025	+43,780	− 1,763	+14,346	+13,181	+65,270		+ 8,999		26
27	+46,010	− 8,860	+35,140	+33,272	+48,232	− 3,816	+15,320	+14,147	+69,700		+10,060		27
28	+50,215	−10,920	+35,750	+34,187	+52,530	− 5,740	+16,000	+14,919					28
29	+54,557	−13,070	+35,925	+34,561	+56,980	− 7,644	+16,198	+15,225					29
30	+58,870	−15,265	+35,480	+34,518	+61,244	− 9,704	+15,736	+14,942					30
31	+63,200	−17,630	+34,624	+34,000	+65,366	−11,930	+14,824	+14,193					31

? signifies uncertainty

No.	Left hip				Left knee				Left ankle				No.
	ξ_1'	η_2'	ζ_1'	ζ_2'	ξ_1'	η_2'	ζ_1'	ζ_2'	ξ_1'	η_2'	ζ_1'	ζ_2'	
1	−67,155	+38,722	−3,962	−2,802	−62,030	+34,948	−33,700	−25,462		+32,338		−46,700	1
2	−63,456	+37,288	−3,586	−2,563	−61,060	+34,419	−33,610	−25,446		+32,270		−46,650	2
3	−59,730	+35,846	−3,610	−2,536	−60,010	+33,827	−33,590	−25,460	*)	+32,180	ca.	−46,600	3
4	−55,980	+34,378	−3,710	−2,722	−58,814	+33,208	−33,600	−25,557	ca.	+32,060		−46,550	4
5	−52,160	+32,847	−4,150	−3,131	−57,470	+32,578	−33,680	−25,711	− 63	+31,960	−61,4	−46,500	5
6	−48,310	+31,371	−4,870	−3,733	−55,890	+31,857	−33,780	−25,811		+31,830		−46,430	6
7	−44,300	+29,729?	−5,630		−53,644	+30,970	−33,900	−26,157		+31,700		−46,360	7
8	−39,980	+27,902	−6,135	−4,840	−50,394	+29,864	−34,350	−26,511	−62 ca.	+31,490	?	−46,200	8
9	−35,190	+25,812	−6,500	−5,178	−46,254	+28,553	−34,500	−26,921	−60,5ca.	+31,180	?	−45,250	9
10	−29,950	+23,320	−6,560	−5,312	−40,530	+26,638	−34,880	−27,502	−58,2ca.	+30,380	−57,500	−44,050	10
11	−24,615	+20,709	−6,480	−5,320	−32,130	+23,894		−28,544	−53,720	+28,820	−54,920	−42,600	11
12	−19,730	+18,331	−6,290	−5,222	−23,000	+20,010		−29,580	−46,960	+26,300	−52,000?	−41,000	12
13	−14,932	+16,118	−5,330	−4,473	−13,990	+15,924	−35,400?	−29,418	−38,150	+22,420	−51,080	−41,000	13
14	−10,520	+14,237	−4,090	−3,431	− 6,206	+12,635	−33,845	−28,542	−28,300	+17,850	−51,460	−42,100	14
15	− 6,257	+12,420	−3,110	−2,675	+ 1,193	+ 9,444	−32,255	−27,602	−17,763	+13,070	−53,015	−44,300	15
16	− 2,222	+10,775	−2,700	−2,323	+ 7,625	+ 6,561	−31,044	−26,942	− 6,700	+ 8,310	−55,225	−47,000	16
17	+ 1,820	+ 9,174	−2,790	−2,407	+13,225	+ 4,078	−30,510	−26,796	+ 5,000	+ 3,430	−57,284	−49,730	17
18	+ 5,557	+ 7,633	−3,450	−2,959	+17,653	+ 2,025	−30,834	−27,331	+16,450	− 1,290	−58,675	−52,100	18
19	+ 9,411	+ 6,044	−4,530	−3,954	+21,120	+ 0,158	−32,065	−28,616	+27,570	− 5,720	−59,135	−53,500	19
20	+13,790	+ 4,325	−5,680	−5,009	+24,100?	− 1,746		−30,216	+36,100	− 8,660	−58,600?	−54,100	20
21	+18,600	+ 2,394	−6,340	−5,655	+28,900?	− 3,838		−31,133	+40,900?	−11,130	−59,339	−55,000	21
22	+23,607	+ 0,597	−6,410	−5,780	+34,350?	− 6,006		−31,355	+43,600	−12,600	−60,460	−56,300	22
23	+28,680	− 1,111	−5,880	−5,381	+38,690	− 7,180	−34,060	−31,490	+45,450	−13,630	−60,670	−57,000	23
24	+33,750	− 2,827	−5,345	−4,913	+43,420	− 8,930	−33,640	−31,410	+46,680	−14,140	−60,800	−57,700	24
25	+38,280	− 4,643	−4,685	−4,285	+46,642	−10,306	−33,340	−31,383	+46,920	−14,270	−61,050	−57,600	25
26	+42,644	− 6,486	−3,864	−3,736	+48,345	−11,231	−33,270	−31,404	+46,970	−14,340	−61,000	−57,500	26
27	+46,600	− 8,416	−3,360	−3,325	+49,170	−11,736	−33,150	−31,420	+47,020	−14,380	−60,940	−57,400	27
28	+50,275	−10,243	−3,126	−3,165	+49,910	−12,292	−33,000	−31,375	+47,080	−14,440	−60,880	−57,300	28
29	+54,080	−12,155	−3,245	−3,280	+50,980	−13,032	−32,080	−31,406	+47,140	−14,500	−60,820	−57,200	29
30	+57,720	−14,051	−3,687	−3,777	+52,190	−13,790	−33,040	−31,533	+47,200	−14,580	−60,760	−57,100	30
31	+61,360	−15,980	−4,310	−4,412	+53,748	−14,669	−33,140	−31,852	+47,260	−14,680	−60,700	−57,000	31

* These co-ordinates could not be determined precisely

Table 6 (continued)

No.	Point on head, from left			
	ξ_1'	η_2'	ζ_1'	ζ_2'
1	−57,049	+24,547	+56,820	+45,645
2	−52,849	+22,820	+57,410	+46,450
3	−48,626	+21,070	+57,620	+46,973
4	−44,513	+19,240	+57,350	+47,166
5	−40,220	+17,380	+56,760	+47,040
6	−35,880	+15,470	+55,890	+46,680
7	−31,713	+13,470	+54,910	+46,205
8	−27,500	+11,330	+54,110	+45,900
9	−23,265	+ 9,300	+53,560	+45,830
10	−19,140	+ 7,230	+53,550	+46,220
11	−15,118	+ 5,280	+54,000	+46,940
12	−11,150	+ 3,470	+54,620	+47,950
13	− 7,168	+ 1,630	+55,600	+49,175
14	− 3,326	− 0,160	+56,400	+50,300
15	+ 0,588	− 2,100	+56,890	+51,124
16	+ 4,350	− 3,910	+56,970	+51,575
17	+ 8,090	− 5,750	+56,590	+51,650
18	+11,883	− 7,560	+56,010	+51,461
19	+15,749	− 9,380	+55,180	+51,124
20	+19,582	−11,190	+54,490	+50,837
21	+23,677	−13,030	+53,950	+50,800
22	+27,743	−14,840	+53,755	+51,050
23	+31,836	−16,650	+54,080	+51,655
24	+36,081	−18,630	+54,600	+52,626
25	+39,991	−20,610	+55,260	+53,800
26	+43,976	−22,650	+56,225	+55,177
27	+47,990	−24,870	+57,010	+56,430
28	+51,667	−27,080	+57,425	+57,382
29	+55,538	−29,360	+57,390	+57,910
30	+59,590	−31,720	+56,980	+58,070
31	+63,540	−34,230	+56,260	+57,825

$$z_1 = (1.46590 - 0.00357 \cdot y) \cdot \zeta_1' \quad \text{and}$$
$$z_2 = (1.58325 - 0.00229 \cdot x - 0.00132 \cdot y) \cdot \zeta_2' \tag{33}$$

Finally the bidimensional co-ordinates of the left side as directly measured in experiment III are given in Table 6. The last series of measurements were in many respects less satisfactory than the previous ones. Several points, such as the centre of gravity of the foot, the tip of the foot and partly also the wrist and ankle were not sufficiently clear on the photographic plate to allow for measurement of the co-ordinates. Other points were less sharply delineated than in the other photographic plates and thus the measurement was less accurate.

The constants of the left side for camera 1^b in experiment III were: $e = 446$ cm, $d = 55.5$ cm, $l_1 = 86$ cm, $l_2' = 6.34830$ cm; and for camera 2^b: $e = 588.5$ cm, $d = 67$ cm, $l_1 = 86$ cm, $l_2' = 5.67074$ cm.

Hence the equations to calculate the tridimensional co-ordinates of the left side in experiment III were:

$$x = \frac{152\,206 \cdot \xi_1' - 398.32 \cdot \xi_1' \eta_2'}{112\,354 + 210.71 \cdot \xi_1' + 177.74 \cdot \eta_2' - \xi_1' \eta_2'} \tag{34}$$

$$y = - \frac{87\,876 \cdot \xi_1' + 196\,751 \cdot \eta_2' - 417.05 \cdot \xi_1' \eta_2'}{112\,354 + 210.71 \cdot \xi_1' + 177.74 \cdot \eta_2' - \xi_1' \eta_2'} \tag{35}$$

$$z_1 = (1.35469 - 0.00325 \cdot y) \cdot \zeta_1' \quad \text{and}$$
$$z_2 = (1.51656 - 0.00237 \cdot x - 0.00137 \cdot y) \cdot \zeta_2' \tag{36}$$

Dr. G. Höckner helped us in determining the moments of inertia of the human body and its limbs [14]. He calculated the tridimensional co-ordinates using Eqs. (19)–(36). The calculations were carried out with the utmost care and with the help of a calculating machine. The results were checked so that, despite the huge amount of data produced by the direct measurement, an error would have been almost impossible.

We are highly indebted to Dr. Höckner for having carried out this difficult work tirelessly. Without his help we should not have been able to deduce the tridimensional co-ordinates of the joint centres from the directly measured data for some considerable time.

A description of the intermediate steps of these calculations would be of little interest. The resulting co-ordinates were not the final aim of our research since they corresponded to the points chosen arbitrarily on the Geissler tubes and not to the centres of the joints. Therefore, the intermediate calculations and the preliminary results will not be given so as to avoid unnecessarily increasing the size of this work.

In order to make available at any time the determination of the tridimensional co-ordinates of the points on the tubes and the checking of the calculations, the relevant data have been deposited in the archives of the Königlich Sächsische Gesellschaft der Wissenschaften. In the course of the following discussions it will often be necessary to deal with the co-ordinates of the points on the tubes. These points are listed in special tables for each study, which are not published here. The tables are designated as Tables A, B or C, depending on whether they belong to experiments I, II or III. In each of these tables the value of the z-co-ordinate is the average of the two values z_1 and z_2. The co-ordinates of the point on the head are also the average values of the two series of co-ordinates obtained from different sides.

The reliability of all further conclusions depends on the accuracy of these co-ordinate data. To judge this reliability, it is necessary to consider first the accuracy of the values of the co-ordinates. Our experimental method provides us with two means of doing so. Firstly, we deduced the z-co-ordinate in two different ways using different data obtained by measurement. The more or less high correspondence between the two magnitudes designated as z_1 and z_2 characterizes the accuracy arrived at. Secondly, measuring one point, the point on the head, from two sides was another and more reliable test of accuracy. Whereas both z-values depend on each other in a certain way since the same y-value occurs in Eq. (9), the co-ordinates of the point on the head obtained by cameras 1ᵃ and 2ᵃ are completely independent of the co-ordinates of the same point obtained by cameras 1ᵇ and 2ᵇ.

With regard to the z-co-ordinates, some of the data from the material deposited in the archives will serve as examples. Experiment I gave the z-values for the right hip joint listed in Table 7. The phases for which no value of z is indicated were hidden by the forearm during the experiment. It can be seen that the two z-values coincide exactly in some places and only in a few cases do they present a discrepancy of up to 0.1 cm. It should not be imagined that here we have chosen co-ordinate values showing the closest correspondence. For example, we can consider a series of values for z for the point on the head photographed from the right in experiment II, a series taken at random (Table 8).

There are some instances in which the two z-values show somewhat greater discrepancies. In such cases, the discrepancies show a continuous increase or decrease and thus indicate a systematic error, as for instance in the series from the table of experiment III related to the point on the head, seen from the right (Table 9). These differences decrease too regularly to be due to chance errors in observation. It is possible that during the measurement a photographic plate or even the table of co-ordinates printed on it rotated somewhat from its correct position. These differences would then exert some influence on the orientation but not on the shape of the head point trajectory.

---

[14] cf. Abhandlungen der mathematisch-physischen Klasse der Königlich Sächsischen Gesellschaft der Wissenschaften, vol 18, nr 8, p 46, note

Table 7. z-values for the right hip joint in experiment I

No.	z_1	z_2	No.	z_1	z_2	No.	z_1	z_2
1	—	—	12	− 8,82	− 8,71	23	− 8,06	− 8,12
2	− 6,57	− 6,69	13	− 8,03	− 7,95	24	− 8,15	− 8,19
3	—	—	14	—	—	25	− 8,22	− 8,23
4	− 4,13	− 4,16	15	− 5,51	− 5,53	26	− 8,54	—
5	− 4,03	− 4,02	16	—	—	27	—	—
6	− 4,86	− 4,86	17	− 4,89	− 4,89	28	− 6,46	− 6,45
7	− 6,64	− 6,59	18	− 5,03	− 5,08	29	—	—
8	− 8,80	− 8,75	19	− 5,64	− 5,73	30	− 4,27	− 4,16
9	− 9,92	− 9,87	20	− 6,65	− 6,69	31	− 4,28	− 4,19
10	− 10,08	− 10,05	21	− 7,40	− 7,50			
11	− 9,85	− 9,82	22	− 7,77	− 7,88			

Table 8. z-values for the point on the head, from the right, in experiment II

No.	z_1	z_2	No.	z_1	z_2	No.	z_1	z_2
11	+ 75,21	+ 75,19	16	+ 78,41	+ 78,37	21	+ 77,82	+ 77,79
12	+ 74,91	+ 74,92	17	+ 79,43	+ 79,41	22	+ 76,76	+ 76,73
13	+ 75,16	+ 75,17	18	+ 79,89	+ 79,83	23	+ 75,85	+ 75,84
14	+ 75,90	+ 75,89	19	+ 79,66	+ 79,62	24	+ 75,35	+ 75,34
15	+ 77,08	+ 77,07	20	+ 78,87	+ 78,83		etc.	

Table 9. z-values for the point on the head, from the right, in experiment III

No.	z_1	z_2	Difference	No.	z_1	z_2	Difference
11	+ 74,74	+ 74,35	+ 0,39	18	+ 77,44	+ 77,49	− 0,05
12	+ 75,71	+ 75,35	+ 0,36	19	+ 76,31	+ 76,47	− 0,16
13	+ 77,01	+ 76,72	+ 0,29	20	+ 75,27	+ 75,47	− 0,20
14	+ 78,18	+ 77,91	+ 0,27	21	+ 74,52	+ 74,74	− 0,22
15	+ 78,78	+ 78,65	+ 0,13	22	+ 74,13	+ 74,46	− 0,33
16	+ 78,83	+ 78,73	+ 0,10		etc.		
17	+ 78,35	+ 78,33	+ 0,02				

The two determinations of the co-ordinates of the point on the head, completely independent of each other, absolutely guarantee the accuracy. Better than figures, direct observation allows evaluation of the coincidence of the two trajectories of the head point arrived at from different sides. Figure 18 shows two corresponding rectangular life-size projections (on the vertical plane of gait and on the horizontal floor) of segments of the trajectories of the point on the head in experiments I and II. One projection relates to the part of the step between phases 9 and 13 in experiment I and to the part of the step between phases 14 and 18 in experiment II. The size of the diagram does not allow life-size reproduction of the phases from 1 to 31. However, these short portions are sufficient to appreciate the accuracy of our method. The points obtained from the left with cameras 1[b] and 2[b] are connected by an interrupted line to distinguish them from the corresponding points obtained from the right with cameras 1[a] and 2[a]. In the projection on the plane of gait, there is almost absolute correspondence. All the projections on the horizontal plane arrived at from the left

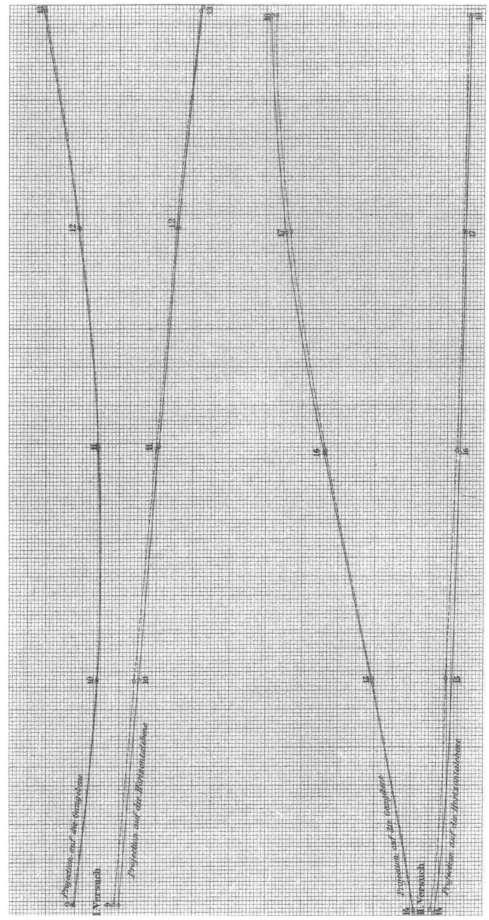

18. Projection of parts of the curves of the vertex determined in two different ways. Curves obtained from the left are indicated by broken lines. Life-size projection deviate a little to the left, compared with the others. This is easily explained by the fact that the illuminated object on the head was not a mathematical point but a small cylinder. The cameras on the right thus provide us with a view of the surface of the cylinder seen from the right and the cameras on the left with a view of the surface of the cylinder seen from the left. The two curves must, therefore, be separated by the thickness of the illuminated part of the Geissler capillary tube. If this is borne in mind, a surprising correspondence can also be seen in the horizontal projections. This correspondence is greater than we expected considering the numerous sources of error which were only recognized after the experiments had been carried out when it was too late to eliminate them. The two segments of the curves of the point of the head drawn in Fig. 18 were in a relatively favourable position for photographic recording and measurement of the co-ordinates, since they did not wander too much to the right or to the left from the middle of the photographic plates. The parts of the curve located closer to the edges cannot be determined with the same accuracy. On one hand the pictures are not so well delineated and do not allow such precise measurement of the co-ordinates as the points in the vicinity of the centre of the plates. On the other hand even with the best lenses the pictures at the edge are a little distorted. The curves of the point on the head near the initial phase 1 and near the final phase 31 thus showed somewhat more deviation than the parts drawn in Fig. 18.

These examples have fully demonstrated that our efforts were not in vain. Transposing the movements of walking into a tridimensional system of co-ordinates provided a reliable foundation for further studies.

Finding the Tridimensional Co-ordinates of the Joint Centres

Our next objective was to determine the tridimensional co-ordinates of the joint centres from the co-ordinates corresponding to the points on the Geissler tubes. Even though the Geissler tubes were firmly attached to the different parts of the body, they did not represent lines along the limbs that were particularly advantageous with regard to studying certain movements. Thus, they were not appropriate to indicate by their position the exact attitude of the whole body directly. In order to gain a direct view of the relative position of the different limbs we had to find out the position of their long axes with the help of photography. Considering the 11 parts into which we divided the body, we designated as long axes the straight lines connecting the centres of the two adjacent joints for the thigh, lower leg and upper arm. For the head, foot and the forearm-hand system, the long axis is the straight line connecting the centre of gravity of the limb to the adjacent joint. For the trunk, the long axis joins the midpoint of the axis of the hips and the centre of the occipito-atlantal joint.

When the position of the long axes of the different parts of the body is known, the attitude of the whole body is determined if one disregards contingent rotation of the head, foot or forearm about their long axes. Since the spine does not actually bend forwards or backwards in walking, the attitude and shape of the trunk are determined sufficiently by the position of the shoulder and hip joints. The upper arms, thighs and lower legs must undergo some rotation about their long axes. This rotation can be determined by comparing the positions of their long axes and the long axes of the forearms and feet. If the long axes of the upper arm and forearm are fixed, it is no longer possible to rotate the upper arm between the trunk and the forearm. Each rotation of the upper arm about its long axis will entail a simultaneous movement of the forearm. This can to a certain extent be used as an indicator of the rotation of the upper arm. Furthermore, if the three long axes of the thigh, lower leg and foot are fixed, it is not possible to rotate voluntarily the thigh alone, the lower leg alone or, in full extension of the knee, the system thigh + lower leg about their long axis. With the knee flexed, any rotation of the thigh about its long axis will entail a movement of the lower leg, and any rotation of the lower leg about its long axis a change in the direction of the long axis of the foot. Therefore, the axis of the lower leg and axis of the foot can be used as indicators of the rotation of the thigh and lower leg, respectively. When the knee is extended the rotation of the leg about its long axis is generally indicated by the position of the long axis of the foot. However, as a consequence of the construction of the joints, it would be possible for example for the leg to be simultaneously rotated by a certain angle about its long axis, and the foot by the same angle in the opposite direction. Then the three long axes of the thigh, lower leg and foot would retain their positions. But a man cannot voluntarily carry out such a movement because the lower limbs are used almost entirely for walking and the muscles and motor nerves of the leg muscles are not trained to perform this movement.

Consequently, determining the position of the long axes during the different phases of walking is sufficient for defining the attitude of the whole body except for contingent rotation of the head, foot and the forearm + hand system about their long axes. Since all these long axes are within body segments, it is impossible to determine directly by photography their tridimensional position for the different phases of the movement. One can deduce their position indirectly, using lines or points located either on the surface of the body or completely outside but rigidly fixed to the body, like our Geissler tubes. Photography can only show these lines or points.

To make possible the deduction of the tridimensional co-ordinates of the joint centres from the co-ordinates of the points on the Geissler tubes, the tubes were positioned on the body segments as described in the previous section. The luminous points corresponding to the two hip joints were chosen such that the line connecting them passed through the centres of these joints. Similarly, the points corresponding to the shoulders lay on the upper arm in such a way that the line connecting them passed as accurately as possible through the centres of the humeral heads. For all the other joints, the luminous points were placed so that, for a normal attitude of the body, the line connecting them with the corresponding joint centre was perpendicular to the median plane through the body. Moreover, the distances between the points on the tubes and the middle of the corresponding joints were measured as accurately as possible, using as a reference a skeleton of the same proportions as the experimental subject. Comparison of the horizontal projections – given by the tables of co-ordinates – of the trajectories of the luminous points corresponding to two opposite joints of the same human body provided a way of checking and, where necessary, adjusting somewhat the distances directly measured.

The distances between the luminous points and the corresponding joint centres are given in Table 10. They appear to coincide very closely with the actual figures.

The inaccuracy which obviously besets these distances does not affect the accuracy obtained by the measurements. Since the movements of the different body segments in walking are all almost parallel to a vertical plane, that of gait, a point on the axis of the knee, for instance, which lies 1 cm away from the centre of the joint will follow almost exactly a course which is congruent with the trajectory of the centre of the knee. This course will only be 1 cm further away from the plane of gait. However, we are above all concerned with the shape of the trajectories and the velocities and accelerations of the points moving on these trajectories.

Thus, the trajectories of the centres of the elbows, wrists, knees and ankles were assumed to be directly congruent with the curves of the corresponding luminous points. These are

Table 10. Distances between luminous points and corresponding joint centres

	Shoulder	Elbow	Wrist	Hip	Knee	Ankle
Right	$4\frac{1}{2}$ cm	$5\frac{1}{4}$ cm	3 cm	$11\frac{3}{4}$ cm	$8\frac{1}{4}$ cm	$5\frac{1}{4}$ cm
Left	$5\frac{1}{2}$ cm	$5\frac{1}{2}$ cm	3 cm	$12\frac{1}{4}$ cm	$9\frac{1}{2}$ cm	$6\frac{1}{2}$ cm

brought closer to the plane of gait by the determined distance. There was no other course we could take since only one luminous point had been attributed to each of these joints.

The situation was more favourable for the shoulders and hips. Here it was possible to determine with the utmost accuracy the true trajectories of the joint centres using the co-ordinates of both shoulder points and both hip points. This was further helped by the fact that the distance between the two hip centres remains absolutely constant and between the two shoulder centres is, at least, very nearly constant during the movements considered.

Before deducing the tridimensional co-ordinates of the joint centres, the system of co-ordinates hitherto used had to be subjected to another small displacement for the following reason. In choosing this system we had neglected to make an axis coincide with the direction of gait (X-axis). Before the experiment this was not possible with sufficient accuracy. It is indeed very difficult to walk precisely in a particular direction. The direction of gait will always deviate somewhat from that prescribed if the experimental subject walks without constraints. This is in fact what occurred in the three experiments.

It is easy to determine the small angle by which the direction of gait has deviated from the previously determined direction of the X-axis using thee XY projection, the view from above of the trajectories of the different luminous points. There are several approaches to the problem. The simplest way would be to establish the direction followed by the successive tracks of the same foot. This assumes that the foot strikes the ground every time in precisely the same way. However, it is possible that the foot is set 1 cm more medially or more laterally than in the previous step without the upper part of the body, and particularly the total centre of gravity, carrying out this small displacement sideways. When walking on uneven ground this often happens. Since the movement of the total centre of gravity depends much more on the movement of the upper part of the body than on that of one leg, it is more reliable to determine the direction of the movement forwards of a point on the trunk or on the head. Thus, one should precisely ascertain the movement of the total centre of gravity itself. This is not possible if only the movement of the luminous points on the surface of the body is known and not the movement of the centres of the joints inside the body. Therefore, in order to determine the direction of the movement, we used the trajectory of the head point in experiment I and the midpoint on the hip axis in experiments II and III.

The point on the head follows an undulating line when projected on the vertical plane of gait (plane XZ), as shown by the photographs. Its projection on the horizontal plane (plane XY) also presents the shape of an undulating line as a consequence of the periodic sideways oscillations of the whole trunk and head. The direction of the two tangents to this horizontal undulating line is the direction of gait. The same applies when the midpoint on the hip axis is used.

Drawing the horizontal projection of the trajectory of the head point life-size on millimetre paper showed that in experiment I the direction of gait deviated by an angle of $1°21'$ clockwise from the prescribed direction of the positive X-axis, as seen from above. This angle can be measured to the minute and even to the fraction of a minute since the line the direction of which is to be determined has a length of about 2 m. The co-ordinates x_1, y_1 and x_2, y_2 of two points P_1 and P_2 (Fig. 19), about 2 cm apart on this line are read and the angle designated as ε in Fig. 19, is calculated using the equation:

$$\tan \varepsilon = \frac{y_2 - y_1}{x_2 - x_1}$$

In experiment I, $\tan \varepsilon$ was 0.02356 from which the angle of $1°21'$ was deduced.

In order to analyse precisely the movements in walking, a projection of the movement curves on the plane of gait must first be obtained. The extent to which the different curves deviate from this plane is then determined. In order to obtain a precise projection on the plane of gait from the tridimensional system of co-ordinates, one plane of co-ordinates

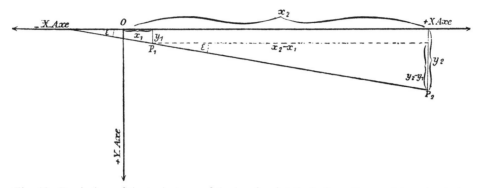

Fig. 19. Deviation of the trajectory of the head point $P_1 P_2$ from the predetermined plane of gait OX

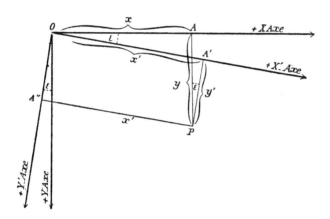

Fig. 20. Rotation of the system of co-ordinates about the axis OZ

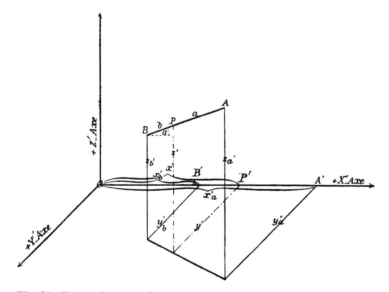

Fig. 21. Determination of the co-ordinates of a point P on the line AB from the co-ordinates of points A and B

(plane XZ) must be exactly parallel to the plane of gait. In other words, one axis (X-axis) must coincide with the direction of gait. Therefore, it is necessary to rotate clockwise the system of co-ordinates thus far used through the vertical Z-axis by the angle ε. The z-co-ordinates remain unchanged, whereas the horizontal x- and y-co-ordinates are modified. The equations corresponding to this rotation are found as follows.

The two horizontal axes of the new system of co-ordinates resulting from the rotation are designated as X'-axis and Y'-axis and their co-ordinates as x' and y' (Fig. 20). The co-ordinate $OA' = x'$ of a point P which in the previous system of co-ordinates had the co-ordinates $OA = x$ and $AP = y$, signifies the projection of the geometrical figure OAP on the X'-axis, as shown in Fig. 20. The projection of OA on the X'-axis is equal to $x \cos \varepsilon$ and the corresponding projection of AP is equal to $y \sin \varepsilon$. The projection of A on the X'-axis falls between O and A'. Therefore, OA' or x' is the sum of both projections.

$$x' = x \cos \varepsilon + y \sin \varepsilon \tag{37}$$

The y'-co-ordinate $A'P = OA''$ of the point P is equal to the projection of the open geometrical figure OAP on the Y'-axis. Since the projection of A on the Y'-axis would fall outside OA'', as appears in Fig. 20, y' is equal to the difference of the projections of AP and OA on the Y'-axis. The former is $y \cos \varepsilon$, the latter is $x \sin \varepsilon$.

$$y' = y \cos \varepsilon - x \sin \varepsilon \tag{38}$$

$$z' = z \tag{39}$$

We have the three equations which lead from the initial tridimensional co-ordinates of any point in space to the co-ordinates of the system resulting from the rotation. Since $\varepsilon = 1°21'$, thus $\cos \varepsilon = 0.99972$ and $\sin \varepsilon = 0.02356$ and the equations become:

$$\left. \begin{aligned} x' &= 0{,}99972 \cdot x + 0{,}02356 \cdot y \\ y' &= 0{,}99972 \cdot y - 0{,}02356 \cdot x \\ z' &= z \end{aligned} \right\} \tag{40}$$

All the tridimensional co-ordinates of the luminous points for experiment I, collected in Table A (not published), were subjected to this transformation. Since they underwent further transformations, the results will not be published here in extenso so as to avoid unnecessarily increasing the volume of this work. The corresponding table will be designated as Table A'. Drawing the horizontal projections $(X'Y')$ of all the trajectories resulting from the co-ordinates confirmed that the direction of gait coincides with the positive direction of the X'-axis. It appeared also that the axis of the undulating line of the head point did not as yet coincide with the X' axis itself but ran parallel to it at a distance, by chance, of exactly 1.00 cm to the negative side of the Y'-axis. To correct this difference, it was thus necessary to subject the whole system of co-ordinates to a parallel displacement of 1.00 cm in the direction of the positive Y'-axis.

Before this further transformation of the co-ordinates was carried out, the tridimensional co-ordinates of the joint centres inside the human body were deduced, as described above, from the co-ordinates x', y', z' of the points on the Geissler tubes in Table A' (unpublished).

To determine the co-ordinates of the centres of the shoulders and hips, the following points were of importance. As a rule, to deduce the co-ordinates x', y', z' of a point P on the line connecting A and B from the tridimensional co-ordinates x'_a, y'_a, z'_a and x'_b, y'_b, z'_b of these two points A and B, it is useful to find the angle formed by the connecting line $\overline{BA}$ and the positive direction of the three axes of co-ordinates. We designate these three angles as α, β, γ and the distance between the points A and B as e. The projection $\overline{B'A'}$ of the segment $\overline{BA}$ on the X'-axis is $e \cdot \cos \alpha$ (Fig. 21). Since $\overline{OA'} = x'_a$ and $\overline{OB'} = x'_b$, thus $B'A' = x'_a - x'_b$ and:

$$e \cdot \cos \alpha = x'_a - x'_b \tag{41}$$

The distance e is correlated to the co-ordinates of A' and B' by the equation:

$$e^2 = (x'_a - x'_b)^2 + (y'_a - y'_b)^2 + (z'_a - z'_b)^2 \tag{42}$$

thus:

$$\cos\alpha = \frac{x'_a - x'_b}{\sqrt{(x'_a - x'_b)^2 + (y'_a - y'_b)^2 + (z'_a - z'_b)^2}}$$

and, correspondingly, for the other angles:

$$\cos\beta = \frac{y'_a - y'_b}{\sqrt{(x'_a - x'_b)^2 + (y'_a - y'_b)^2 + (z'_a - z'_b)^2}} \tag{43}$$

$$\cos\gamma = \frac{z'_a - z'_b}{\sqrt{(x'_a - x'_b)^2 + (y'_a - y'_b)^2 + (z'_a - z'_b)^2}}$$

These three cosines are called the directional cosines of the connecting line $\overline{BA}$. They have been calculated for the line connecting the two shoulder points and for the line connecting the two hip points for the 31 phases. Point A belongs to the right and point B to the left side of the body. These values simultaneously represent the directional cosines of the line connecting the centres of the shoulders and the line connecting the centres of the hips (the shoulder line and hip line). They are given in Table 11 for experiment I. They describe the movements of the shoulder and hip lines during walking, as will be explained.

With these directional cosines, the co-ordinates x', y', z' of a point P on the connecting line $\overline{BA}$ can be easily calculated if either the distance a between P and A or the distance b between P and B is known. If P lies between B and A, as in Fig. 21, either each of the three co-ordinates of point A must be diminished by the projection of the segment $\overline{PA}$ on the corresponding axis of co-ordinates, or each of the three co-ordinates of point B must be increased by the projection of the segment $\overline{BP}$ on the corresponding axis of co-ordinates. If, for instance, P' is the projection of P on the X'-axis:

$$\overline{OP'} = \overline{OA'} - \overline{P'A'} \quad \text{or} \quad \overline{OP'} = \overline{OB'} + \overline{B'P'}.$$

Since the segments $\overline{PA}$ and $\overline{BP}$ have the same directional cosines as $\overline{BA}$, the following equations can be derived:

$$\left.\begin{aligned} x' &= x'_a - a \cdot \cos\alpha \\ y' &= y'_a - a \cdot \cos\beta \\ z' &= z'_a - a \cdot \cos\gamma \end{aligned}\right\} \tag{44}$$

or:

$$\left.\begin{aligned} x' &= x'_b + b \cdot \cos\alpha \\ y' &= y'_b + b \cdot \cos\beta \\ z' &= z'_b + b \cdot \cos\gamma \, . \end{aligned}\right\} \tag{45}$$

The two ways of calculating each co-ordinate provide a means of checking the calculation. The co-ordinates of the centres of the shoulder and hip joints are calculated using these equations.

Table 11. Directional cosines, experiment I

No.	Shoulder line			Hip line			No.
	$\cos\alpha$	$\cos\beta$	$\cos\gamma$	$\cos\alpha$	$\cos\beta$	$\cos\gamma$	
1	− 0,0042	+ 0,9996	+ 0,0325	− 0,1340	+ 0,9870	− 0,0899	1
2	− 0,0125	+ 0,9995	+ 0,0294	− 0,0176	+ 0,9900	− 0,0724	2
3	− 0,0183	+ 0,9998	+ 0,0230	+ 0,0353	+ 0,9918	− 0,0706	3
4	− 0,0279	+ 0,9995	+ 0,0114	+ 0,0353	+ 0,9929	− 0,0759	4
5	− 0,0377	+ 0,9993	− 0,0025	+ 0,0471	+ 0,9953	− 0,0782	5
6	− 0,0475	+ 0,9989	− 0,0141	+ 0,0485	+ 0,9966	− 0,0694	6
7	− 0,0533	+ 0,9984	− 0,0172	+ 0,0701	+ 0,9956	− 0,0635	7
8	− 0,0555	+ 0,9984	− 0,0141	+ 0,0734	+ 0,9944	− 0,0780	8
9	− 0,0563	+ 0,9982	− 0,0124	+ 0,0662	+ 0,9923	− 0,1060	9
10	− 0,0572	+ 0,9982	− 0,0122	+ 0,0758	+ 0,9891	− 0,1254	10
11	− 0,0564	+ 0,9982	− 0,0138	+ 0,0954	+ 0,9882	− 0,1210	11
12	− 0,0476	+ 0,9991	− 0,0143	+ 0,1127	+ 0,9864	− 0,1190	12
13	− 0,0203	+ 1	− 0,0127	+ 0,1057	+ 0,9898	− 0,0977	13
14	+ 0,0011	+ 0,9998	− 0,0147	+ 0,0892	+ 0,9935	− 0,0696	14
15	+ 0,0190	+ 0,9998	− 0,0120	+ 0,0562	+ 0,9966	− 0,1024	15
16	+ 0,0287	+ 0,9996	− 0,0022	+ 0,0294	+ 0,9959	− 0,1029	16
17	+ 0,0380	+ 0,9993	+ 0,0106	− 0,0002	+ 0,9963	− 0,0862	17
18	+ 0,0438	+ 0,9991	+ 0,0172	− 0,0382	+ 0,9959	− 0,0935	18
19	+ 0,0471	+ 0,9987	+ 0,0225	− 0,0718	+ 0,9953	− 0,0953	19
20	+ 0,0521	+ 0,9984	+ 0,0186	− 0,0975	+ 0,9906	− 0,0975	20
21	+ 0,0592	+ 0,9984	+ 0,0123	− 0,1106	+ 0,9903	− 0,0820	21
22	+ 0,0658	+ 0,9980	+ 0,0129	− 0,1288	+ 0,9898	− 0,0619	22
23	+ 0,0677	+ 0,9978	+ 0,0145	− 0,1446	+ 0,9881	− 0,0499	23
24	+ 0,0683	+ 0,9977	+ 0,0152	− 0,1448	+ 0,9881	− 0,0532	24
25	+ 0,0591	+ 0,9982	+ 0,0170	− 0,1602	+ 0,9848	− 0,0671	25
26	+ 0,0484	+ 0,9989	+ 0,0200	− 0,1305	+ 0,9875	− 0,0876	26
27	+ 0,0459	+ 0,9989	+ 0,0190	− 0,1294	+ 0,9865	− 0,1000	27
28	+ 0,0358	+ 0,9993	+ 0,0101	− 0,0765	+ 0,9853	− 0,0765	28
29	+ 0,0206	+ 1	+ 0,0061	− 0,0500	+ 0,9818	− 0,0594	29
30	+ 0,0049	+ 1	− 0,0004	− 0,0088	+ 0,9765	− 0,0565	30
31	− 0,0042	+ 1	− 0,0018	+ 0,0503	+ 0,9743	− 0,0561	31

[a] The numbers printed in italics in this and subsequent tables were not directly measured but interpolated from other values

With regard to the shoulders, Eq. (44) are used to calculate the co-ordinates of the right side, Eq. (45) to calculate those of the left side. The point on the Geissler tube of the right side was designated as point A, the point on the left side as point B. The co-ordinates of the centre of the right shoulder are designated as x_r', y_r', z_r', and those of the centre of the left shoulder as x_l', y_l', z_l'. Directly measured (p. 60), distance $a = 4.5$ cm, and distance $b = 5.5$ cm. The equations for the centre of the right shoulder are thus:

$$
\left.
\begin{aligned}
x_r' &= x_a' - 4.5 \cdot \cos\alpha \\
y_r' &= y_a' - 4.5 \cdot \cos\beta \\
z_r' &= z_a' - 4.5 \cdot \cos\gamma
\end{aligned}
\right\}
\tag{46}
$$

and those for the centre of the left shoulder:

$$\left. \begin{array}{l} x'_l = x'_b + 5.5 \cdot \cos\alpha \\ y'_l = y'_b + 5.5 \cdot \cos\beta \\ z'_l = z'_b + 5.5 \cdot \cos\gamma \end{array} \right\} \tag{47}$$

In these equations, the values of $\cos\alpha$, $\cos\beta$, $\cos\gamma$ for experiment I were taken from the first part of Table 11 and those of x'_a, y'_a, z'_a; x'_b, y'_b, z'_b from the table of co-ordinates A' (not published) obtained by rotating the system of co-ordinates through the angle ε. The results of this calculation will not be given since they are not the final results.

The corresponding equations for the centres of the hip joints could be obtained by replacing a by $11\frac{3}{4}$ cm, b by $12\frac{1}{4}$ cm and x'_a, y'_a, z'_a and x'_b, y'_b, z'_b by the co-ordinates of the two luminous hip points. However, the equations were not both used.

In determining the co-ordinates of the centres of the hip joints, it is important to take into account that the distance between the two centres remains absolutely constant. It is possible to start from the same side for both hips. The trajectory of the left hip point was free of any error which could have resulted from the foot suddenly striking the ground, at least in the middle of the part of the movement that was measured, since the left leg was swinging at this time. This trajectory was, therefore, chosen for the calculation. The distance between the two centres of the hip joints was directly measured as 17 cm, with the approximation involved in such a measurement. The distance between the centre of the left hip and the left luminous point (Table 10) was 12.25 cm and that between the centre of the right hip and the left luminous point was $12.25 + 17 = 20.25$ cm. Consequently, we were able to calculate the co-ordinates of the centres of the right hip:

$$\left. \begin{array}{l} x'_r = x'_b + 29.25 \cdot \cos\alpha \\ y'_r = y'_b + 29.25 \cdot \cos\beta \\ z'_r = z'_b + 29.25 \cdot \cos\gamma \end{array} \right\} \tag{48}$$

and those of the centre of the left hip:

$$\left. \begin{array}{l} x'_l = x'_b + 12.25 \cdot \cos\alpha \\ y'_l = y'_b + 12.25 \cdot \cos\beta \\ z'_l = z'_b + 12.25 \cdot \cos\gamma \end{array} \right\} \tag{49}$$

The values of the directional cosines of experiment I are taken from the second part of Table 11. x'_b, y'_b, z'_b are the co-ordinates of the luminous point of the left hip.

The results thus far obtained cannot as yet be given in their present form.

With regard to the co-ordinates of the centres of the elbows, wrists, knees and ankles, x' and z' coincide with the corresponding co-ordinates of the luminous points. Only the y'-co-ordinates differ from the two corresponding points by the distances given in Table 10. Using the same designations as above we have the following equations:

Centre of the right elbow:	$y'_r = y'_a - 5.5$	(50)
Centre of the left elbow:	$y'_l = y'_b + 5.5$	(51)
Centre of the right wrist:	$y'_r = y'_a - 3$	(52)
Centre of the left wrist:	$y'_l = y'_b + 3$	(53)
Centre of the right knee:	$y'_r = y'_a - 8.5$	(54)

Centre of the left knee:	$y'_l = y'_b + 9.5$	(55)
Centre of the right ankle:	$y'_r = y'_a - 5.5$	(56)
Centre of the left ankle:	$y'_l = y'_b + 6.5$	(57)

y'_a and y'_b are taken from the A' table of co-ordinates (not published) resulting from the rotation of the system of co-ordinates through the angle ε.

Lastly, we had still to deduce the tridimensional co-ordinates of points of the human body that are not the centre of a joint. These were: the vertex, the centres of gravity of the two feet and the tips of the two feet. According to direct measurement, the vertex was 3 cm below and 4 cm behind the point marked on the head tube. If the co-ordinates of this luminous point are designated as x'_k, y'_k, z'_k, and those of the vertex as x'_0, y'_0, z'_0, and if we assume that the head does not bend significantly forwards and backwards during walking, we have the following equations for the vertex:

$$\left. \begin{array}{l} x'_0 = x'_k - 4 \\ y'_0 = y'_k \\ z'_0 = z'_k - 3 \end{array} \right\} \tag{58}$$

If the head is kept straight, its centre of gravity lies beneath the vertex at a distance of 10 cm. This figure was obtained from direct measurement of the dimensions of the head of the experimental subject and compared with earlier finding on the position of the centre of gravity of the head[15]. Thus, assuming that the head was kept straight, which is not absolutely true, and based on the co-ordinates of the luminous point, we found the co-ordinates x'_s, y'_s, z'_s of the centre of gravity of the head using the following equations:

$$\left. \begin{array}{l} x'_s = x'_k - 4 \\ y'_s = y'_k \\ z'_s = z'_k - 13 \end{array} \right\} \tag{59}$$

The point on the Geissler tube corresponding to the centre of gravity of the foot lay 5 cm lateral to the centre of gravity of the right foot, 6 cm lateral to that of the left foot. The anterior extremity of the foot tube was 1 cm outside the long axis of the right foot and 3.5 cm outside that of the left foot. This last point did not correspond to the tip of the foot but to a point on the long axis of the foot situated approximately 3 cm behind the tip. Nevertheless, it will be designated as the "tip of the foot".

One thus obtains the following equations for the y'-co-ordinates; they are only approximate:

Centre of gravity of the right foot:	$y'_r = y'_a - 5$	(60)
Centre of gravity of the left foot:	$y'_l = y'_b + 6$	(61)
Tip of the right foot:	$y'_r = y'_a - 1$	(62)
Tip of the left foot:	$y'_l = y'_b + 3.5$	(63)

[15] Braune W, Fischer O (1889) Über den Schwerpunkt des menschlichen Körpers mit Rücksicht auf die Ausrüstung des deutschen Infanteristen. Abhandlungen der mathematisch-physischen Klasse der Königlichen Sächsischen Gesellschaft der Wissenschaften, vol 15, nr 7. English translation: (1984) On the centre of gravity of the human body. Springer, Berlin Heidelberg New York Tokyo

The x'- and z'-co-ordinates maintain the same values.

All the above-mentioned transformations of the co-ordinates were combined with parallel displacements of the system of co-ordinates in such a way that the co-ordinates could directly provide the answer to a number of questions. Of course, the co-ordinates set into a tridimensional system chosen at random furnished sufficient basis for analysing the whole process of movement. However, this involved as a rule much more calculation to arrive at a precise result. If, from the outset, the whole movement is transposed into an appropriate system of co-ordinates for the particular case, either significant results are obtained without the need for further calculation, or the calculation necessary to determine certains results is reduced to a minimum. It is, therefore, worthwhile to transpose the process of movement which is at first represented in a random system of co-ordinates, into a more appropriate system.

To obtain the movement of the individual joint centres in relation to the plane of gait directly from the co-ordinates, the whole system of co-ordinates had to be shifted by 1 cm parallel to the y'-axis as explained above. This implies an increase of all the y'-co-ordinates by 1.

To represent directly the distance between the joint centres and the horizontal floor, it was necessary to lower the horizontal plane $X'Y'$ to the floor, parallel to its former position. Since the origin of the co-ordinates was initially 90 cm above the floor, this displacement increased all the z'-co-ordinates by 90 cm.

Finally, the calculation is made easier by arranging the system of co-ordinates so that the whole movement takes place on the same side of the vertical plane $Y'Z'$. With this aim in view, the system of co-ordinates is pushed backwards in the direction opposite to gait. This entails an increase of the x'-co-ordinates. In experiment I it was sufficient to push back the system by 120 cm in order to measure 31 phases. Consequently, 120 was added to all co-ordinates x'.

These three displacements are added to the previous successive transformations of the tridimensional co-ordinates of the luminous points given in Table A (unpublished), starting with the rotation of the system of co-ordinates by the angle ε. We thus obtain the following transformations which translate the co-ordinates of the luminous points directly into the co-ordinates of the joint centres, of the centres of gravity of the head and of the feet, of the vertex and of the tips of the feet. To distinguish them, the co-ordinates are in brackets and provided with the indices r, l or o depending on whether their corresponding luminous points lay on the right or on the left side, or in the median plane of the body.

Centre of right shoulder joint:

$$\left.\begin{array}{l} x = [x_r] \cdot \cos\varepsilon + [y_r] \cdot \sin\varepsilon - 4.5 \cdot \cos\alpha + 120 \\ y = [y_r] \cdot \cos\varepsilon - [x_r] \cdot \sin\varepsilon - 4.5 \cdot \cos\beta + 1 \\ z = [z_r] - 4.5 \cdot \cos\gamma + 90 \end{array}\right\} \tag{64}$$

Centre of left shoulder joint:

$$\left.\begin{array}{l} x = [x_l] \cdot \cos\varepsilon + [y_l] \cdot \sin\varepsilon + 5.5 \cdot \cos\alpha + 120 \\ y = [y_l] \cdot \cos\varepsilon - [x_l] \cdot \sin\varepsilon + 5.5 \cdot \cos\beta + 1 \\ z = [z_l] + 5.5 \cdot \cos\gamma + 90 \end{array}\right\} \tag{65}$$

Centre of right elbow joint:

$$\left.\begin{array}{l} x = [x_r] \cdot \cos\varepsilon + [y_r] \cdot \sin\varepsilon + 120 \\ y = [y_r] \cdot \cos\varepsilon - [x_r] \cdot \sin\varepsilon - 4.5 \\ z = [z_r] + 90 \end{array}\right\} \tag{66}$$

Centre of left elbow joint:

$$
\left.\begin{array}{l}
x = [x_l] \cdot \cos \varepsilon + [y_l] \cdot \sin \varepsilon + 120 \\
y = [y_l] \cdot \cos \varepsilon - [x_l] \cdot \sin \varepsilon + 6.5 \\
z = [z_l] + 90
\end{array}\right\} \tag{67}
$$

Centre of right wrist joint:

$$
\left.\begin{array}{l}
x = [x_r] \cdot \cos \varepsilon + [y_r] \cdot \sin \varepsilon + 120 \\
y = [y_r] \cdot \cos \varepsilon - [x_r] \cdot \sin \varepsilon - 2 \\
z = [z_r] + 90
\end{array}\right\} \tag{68}
$$

Centre of left wrist joint:

$$
\left.\begin{array}{l}
x = [x_l] \cdot \cos \varepsilon + [y_l] \cdot \sin \varepsilon + 120 \\
y = [y_l] \cdot \cos \varepsilon - [x_l] \cdot \sin \varepsilon + 4 \\
z = [z_l] + 90
\end{array}\right\} \tag{69}
$$

Centre of right hip joint [16]:

$$
\left.\begin{array}{l}
x = [x_l] \cdot \cos \varepsilon + [y_l] \cdot \sin \varepsilon + 29.25 \cdot \cos \alpha + 120 \\
y = [y_l] \cdot \cos \varepsilon - [x_l] \cdot \sin \varepsilon + 29.25 \cdot \cos \beta + 1 \\
z = [z_l] + 12.25 \cdot \cos \gamma + 90
\end{array}\right\} \tag{70}
$$

Centre of left hip joint:

$$
\left.\begin{array}{l}
x = [x_l] \cdot \cos \varepsilon + [y_l] \cdot \sin \varepsilon + 12.25 \cdot \cos \alpha + 120 \\
y = [y_l] \cdot \cos \varepsilon - [x_l] \cdot \sin \varepsilon + 12.25 \cdot \cos \beta + 1 \\
z = [z_l] + 12.25 \cdot \cos \gamma + 90
\end{array}\right\} \tag{71}
$$

Centre of right knee joint:

$$
\left.\begin{array}{l}
x = [x_r] \cdot \cos \varepsilon + [y_r] \cdot \sin \varepsilon + 120 \\
y = [y_r] \cdot \cos \varepsilon - [x_r] \cdot \sin \varepsilon - 7.5 \\
z = [z_r] + 90
\end{array}\right\} \tag{72}
$$

Centre of left knee joint:

$$
\left.\begin{array}{l}
x = [x_l] \cdot \cos \varepsilon + [y_l] \cdot \sin \varepsilon + 120 \\
y = [y_l] \cdot \cos \varepsilon - [x_l] \cdot \sin \varepsilon + 10.5 \\
z = [z_l] + 90
\end{array}\right\} \tag{73}
$$

Centre of right ankle joint:

$$
\left.\begin{array}{l}
x = [x_r] \cdot \cos \varepsilon + [y_r] \cdot \sin \varepsilon + 120 \\
y = [y_r] \cdot \cos \varepsilon - [x_r] \cdot \sin \varepsilon - 4.5 \\
z = [z_r] + 90
\end{array}\right\} \tag{74}
$$

---

[16] It should be noted that the co-ordinates of the centre of the right hip were deduced from the co-ordinates of the hip point of the left thigh tube

Centre of left ankle joint:

$$\left.\begin{aligned}
x &= [x_l] \cdot \cos\varepsilon + [y_l] \cdot \sin\varepsilon + 120 \\
y &= [y_l] \cdot \cos\varepsilon - [x_l] \cdot \sin\varepsilon + 7.5 \\
z &= [z_l] + 90
\end{aligned}\right\} \tag{75}$$

Centre of gravity of right foot:

$$\left.\begin{aligned}
x &= [x_r] \cdot \cos\varepsilon + [y_r] \cdot \sin\varepsilon + 120 \\
y &= [y_r] \cdot \cos\varepsilon - [x_r] \cdot \sin\varepsilon - 4 \\
z &= [z_r] + 90
\end{aligned}\right\} \tag{76}$$

Centre of gravity of left foot:

$$\left.\begin{aligned}
x &= [x_l] \cdot \cos\varepsilon + [y_l] \cdot \sin\varepsilon + 120 \\
y &= [y_l] \cdot \cos\varepsilon - [x_1] \cdot \sin\varepsilon + 7 \\
z &= [z_l] + 90
\end{aligned}\right\} \tag{77}$$

Tip of right foot [17]:

$$\left.\begin{aligned}
x &= [x_r] \cdot \cos\varepsilon + [y_r] \cdot \sin\varepsilon + 120 \\
y &= [y_r] \cdot \cos\varepsilon - [x_r] \cdot \sin\varepsilon \\
z &= [z_r] + 90
\end{aligned}\right\} \tag{78}$$

Tip of left foot:

$$\left.\begin{aligned}
x &= [x_l] \cdot \cos\varepsilon + [y_l] \cdot \sin\varepsilon + 120 \\
y &= [y_l] \cdot \cos\varepsilon - [x_l] \cdot \sin\varepsilon + 4.5 \\
z &= [z_l] + 90
\end{aligned}\right\} \tag{79}$$

Vertex:

$$\left.\begin{aligned}
x &= [x_o] \cdot \cos\varepsilon + [y_o] \cdot \sin\varepsilon + 116 \\
y &= [y_o] \cdot \cos\varepsilon - [x_o] \cdot \sin\varepsilon + 1 \\
z &= [z_o] + 87
\end{aligned}\right\} \tag{80}$$

Centre of gravity of head:

$$\left.\begin{aligned}
x &= [x_o] \cdot \cos\varepsilon + [y_o] \cdot \sin\varepsilon + 116 \\
y &= [y_o] \cdot \cos\varepsilon - [x_o] \cdot \sin\varepsilon + 1 \\
z &= [z_o] + 77
\end{aligned}\right\} \tag{81}$$

All these equations are valid for experiment I for which $\cos\varepsilon = 0.99972$ and $\sin\varepsilon = 0.02356$. The values of $\cos\alpha$, $\cos\beta$ and $\cos\gamma$ in the equations for the centres of the shoulder and hip joints are given in Table 11 (p. 65). Except for the equations for the centre of the right hip, the co-ordinates in brackets always correspond to the point of the Geissler tube attributed to the centre of the joint centre involved.

These equations are only slightly changed for experiments II and III. In these two experiments the angle ε is of course different.

---

[17] More precisely, point on the long axis of the foot, 3 cm behind the tip

In experiment II, $\tan \varepsilon = 0.03191$

Therefore, $\cos \varepsilon = 0.99949$ and $\sin \varepsilon = 0.03189$,

$$\varepsilon = 1°49'39''.$$

In experiment III, $\tan \varepsilon = 0.02974$

Therefore, $\cos \varepsilon = 0.99956$ and $\sin \varepsilon = 0.02973$,

$$\varepsilon = 1°42'13''.$$

These values of $\cos \varepsilon$ and $\sin \varepsilon$ thus have to be introduced into the equations for the other experiments. After rotation of the system of co-ordinates through the appropriate angle ε and drawing of the projection $X'Y'$, it appeared that the plane of gait was parallel to the direction of the positive X'-axis. However it was still 1.65 cm away from the plane $X'Z'$ on the positive side of the Y'-axis in experiment II and 1.00 cm in experiment III. Whereas in experiment I an addition of 1 to all the y'-co-ordinates was necessary, in experiment II these y'-co-ordinates had to be decreased by 1.65, and in experiment III by 1. Finally, in experiment II the plane $Y'Z'$ was not pushed back by 120 cm, as in experiments I and III, but by 140 cm, in order that the whole process of movement could take place on the same side of this plane of co-ordinates. Otherwise, the transformations of the co-ordinates for the last two experiments were exactly the same as for the first.

Consequently, the equations of experiment I are valid for experiment II only if the values of $\cos \varepsilon$ and $\sin \varepsilon$ are adequately modified and if 20 is added to all equations for the x-co-ordinates and 2.65 subtracted from all equations for the y-co-ordinates. In experiment III, 2 must be subtracted from all equations of experiment I for the y-co-ordinates. The x-co-ordinates are calculated by exactly the same equations as for experiment I. In the last two experiments the directional cosines of the shoulder and hip lines are of course different. They are calculated as in experiment I using Eq. (43). The results of this calculation are given in Tables 12 and 13.

All the transformation equations have been written out in detail to allow the final results to be checked at any time without having to consult the calculations deposited in the archives. The final tridimensional co-ordinates of the joint centres, etc. given in Tables 14–16 were not calculated using these equations. The different successive transformations of the co-ordinates of Tables A–C were carried out in the manner detailed for experiment I. The intermediate results were not listed so as to avoid unnecessarily increasing the volume of the present study. However, because of this, it was important to give the equations which summarize all the successive transformations that a co-ordinate must undergo.

It must be mentioned that in experiment III, when the right hip joint was hidden by the forearm, its co-ordinates were calculated from those of the knee point and those of another point on the thigh tube using equations similar to Eqs. (44) and (45). From the positions of the thigh tube in which three points on the latter could be measured, it appeared that the ratio of the distance between the middle point and the knee point to the distance between the hip point and the knee point was 1 : 2.935. This was taken into account in the calculations.

Tables 14–16 give the tridimensional co-ordinates of the joint centres, the centres of gravity of the feet, the tips of the feet and the vertex for the three experiments. The co-ordinates of the midpoints of the shoulder and hip lines are also given. These represent the arithmetical averages of the co-ordinates of the two shoulder centres or hip centres during the same phase of movement. The co-ordinates of the centre of gravity of the head are not given in the tables since they could not be determined with sufficient accuracy using the method detailed above.

All details of the process of movement are implicitly contained in these tables of co-ordinates. They thus provide the basis for the solution of all problems concerning the laws of movement in human gait as long as the human body can be assumed to be made up of different solid masses. These tables enable us, for instance, to find the tridimensional trajectories not only of the joint centres but also of other mechanically important points such as the centres of gravity of the limb segments or of the whole body. They enable us to determine the change in position of the long axes of the different limbs, the movement of adjacent body segments in relation to each other and hence the movements of the joints. They enable us to deduce the velocities and accelerations of different points of the human body on their trajectories and thus permit conclusions to be drawn with regard to the moving forces inside the body, etc. They can thus widen our knowledge of human gait in many different ways.

These results concern only the gait of our experimental subject. Each person has his own characteristic way of walking. However, the differences in gait between two individuals depend only on slight quantitative variations in the size of the bones, variations in the shape

Table 12. Directional cosines, experiment II

No.	Shoulder line			Hip line			No.
	cos α	cos β	cos γ	cos α	cos β	cos γ	
1	+ 0,0324	+ 0,9998	− 0,0042	− 0,0529[a]	+ 0,9965	− 0,1206	1
2	+ 0,0275	+ 0,9995	− 0,0069	+ 0,0005	+ 0,9949	− 0,1047	2
3	+ 0,0217	+ 0,9995	− 0,0085	+ 0,0500	+ 0,9935	− 0,0829	3
4	+ 0,0187	+ 0,9998	− 0,0129	+ 0,0924	+ 0,9934	− 0,0689	4
5	+ 0,0124	+ 1,0000	− 0,0078	+ 0,1160	+ 0,9912	− 0,0620	5
6	+ 0,0029	+ 0,9998	− 0,0007	+ 0,1240	+ 0,9898	− 0,0699	6
7	− 0,0069	+ 0,9998	+ 0,0036	+ 0,1162	+ 0,9896	− 0,0901	7
8	− 0,0157	+ 0,9998	+ 0,0031	+ 0,1091	+ 0,9874	− 0,1161	8
9	− 0,0221	+ 0,9995	− 0,0009	+ 0,1152	+ 0,9853	− 0,1258	9
10	− 0,0272	+ 0,9995	− 0,0057	+ 0,1223	+ 0,9858	− 0,1187	10
11	− 0,0161	+ 0,9991	− 0,0100	+ 0,1353	+ 0,9843	− 0,1119	11
12	+ 0,0016	+ 0,9998	− 0,0145	+ 0,1215	+ 0,9887	− 0,0853	12
13	+ 0,0152	+ 0,9995	− 0,0164	+ 0,0914	+ 0,9935	− 0,0681	13
14	+ 0,0250	+ 0,9995	− 0,0100	+ 0,0676	+ 0,9947	− 0,0553	14
15	+ 0,0377	+ 0,9993	+ 0,0016	+ 0,0271	+ 0,9970	− 0,0722	15
16	+ 0,0538	+ 0,9986	+ 0,0107	− 0,0265	+ 0,9965	− 0,0971	16
17	+ 0,0656	+ 0,9973	+ 0,0167	− 0,0608	+ 0,9921	− 0,1079	17
18	+ 0,0719	+ 0,9971	+ 0,0147	− 0,0796	+ 0,9908	− 0,1088	18
19	+ 0,0837	+ 0,9975	+ 0,0125	− 0,0828	+ 0,9913	− 0,1004	19
20	+ 0,0904	+ 0,9957	+ 0,0125	− 0,0906	+ 0,9918	− 0,0882	20
21	+ 0,0946	+ 0,9954	+ 0,0140	− 0,1093	+ 0,9916	− 0,0673	21
22	+ 0,0930	+ 0,9954	+ 0,0145	− 0,1300	+ 0,9896	− 0,0546	22
23	+ 0,0862	+ 0,9961	+ 0,0115	− 0,1360	+ 0,9887	− 0,0571	23
24	+ 0,0696	+ 0,9973	+ 0,0164	− 0,1537	+ 0,9856	− 0,0691	24
25	+ 0,0545	+ 0,9982	+ 0,0185	− 0,1273	+ 0,9878	− 0,0873	25
26	+ 0,0440	+ 0,9989	+ 0,0155	− 0,1059	+ 0,9900	− 0,0982	26
27	+ 0,0338	+ 0,9993	+ 0,0083	− 0,0794	+ 0,9924	− 0,1129	27
28	+ 0,0171	+ 0,9998	− 0,0027	− 0,0402	+ 0,9926	− 0,1127	28
29	− 0,0027	+ 0,9998	− 0,0117	− 0,0059	+ 0,9929	− 0,0900	29
30	− 0,0199	+ 0,9993	− 0,0145	+ 0,0265	+ 0,9971	− 0,0692	30
31	− 0,0321	+ 0,9993	− 0,0105	+ 0,0387	+ 0,9968	− 0,0635	31

of the articular surfaces and, above all, variations in the distribution of the body mass. The sequence and type of the simultaneous movements of the different body segments is the same in all humans. For instance, in walking, the trunk carries out certain oscillations and rotations. These may differ quantitatively but not qualitatively among different people. Some people rotate their shoulders and hips considerably so that these movements are noticeable at a distance. There are others who keep their shoulders and hips relatively still when walking so that the rotations can be seen only close to. The type of shoulder and hip movement, however, is the same in both instances. The shoulder and the hip lines rotate about a vertical axis always towards the arm or leg swinging forwards. Even movements which are not absolutely essential to locomotion, such as the swinging of the arms, occur in a particular order in relation to the movements of the other body segments as long as they are not hindered. The arm always swings forwards simultaneously with the opposite leg. Some

Table 13. Directional cosines, experiment III

No.	Shoulder line			Hip line			No.
	cos α	cos β	cos γ	cos α	cos β	cos γ	
1	− 0,0325	+ 0,9986	− 0,0394	− 0,1009	+ 0,9934	+ 0,0578	1
2	− 0,0290	+ 0,9989	− 0,0384	− 0,0812	+ 0,9938	+ 0,0772	2
3	− 0,0269	+ 0,9989	− 0,0347	− 0,0652	+ 0,9934	+ 0,0949	3
4	− 0,0240	+ 0,9993	− 0,0290	− 0,0483	+ 0,9933	+ 0,1061	4
5	− 0,0224	+ 0,9993	− 0,0237	− 0,0379	+ 0,9940	+ 0,1030	5
6	− 0,0212	+ 0,9998	− 0,0173	− 0,0214	+ 0,9955	+ 0,0911	6
7	− 0,0197	+ 0,9998	− 0,0137	− 0,0055	+ 0,9968	+ 0,0780	7
8	− 0,0129	+ 1	− 0,0124	+ 0,0077	+ 0,9975	+ 0,0714	8
9	− 0,0005	+ 1	− 0,0124	+ 0,0168	+ 0,9975	+ 0,0699	9
10	+ 0,0110	+ 1	− 0,0133	+ 0,0194	+ 0,9966	+ 0,0805	10
11	+ 0,0248	+ 0,9993	− 0,0163	+ 0,0190	+ 0,9946	+ 0,1013	11
12	+ 0,0302	+ 0,9995	− 0,0046	+ 0,0081	+ 0,9929	+ 0,1196	12
13	+ 0,0390	+ 0,9991	0,0000	− 0,0123	+ 0,9929	+ 0,1185	13
14	+ 0,0489	+ 0,9989	+ 0,0035	− 0,0334	+ 0,9944	+ 0,1010	14
15	+ 0,0551	+ 0,9982	+ 0,0109	− 0,0484	+ 0,9948	+ 0,0882	15
16	+ 0,0577	+ 0,9981	+ 0,0153	− 0,0556	+ 0,9958	+ 0,0713	16
17	+ 0,0595	+ 0,9981	+ 0,0107	− 0,0584	+ 0,9961	+ 0,0647	17
18	+ 0,0634	+ 0,9981	+ 0,0226	− 0,0582	+ 0,9956	+ 0,0685	18
19	+ 0,0698	+ 0,9972	+ 0,0201	− 0,0521	+ 0,9954	+ 0,0798	19
20	+ 0,0741	+ 0,9972	+ 0,0178	− 0,0646	+ 0,9930	+ 0,0953	20
21	+ 0,0763	+ 0,9970	+ 0,0175	− 0,0750	+ 0,9925	+ 0,1081	21
22				− 0,0790	+ 0,9914	+ 0,1054	22
23	+ 0,0669	+ 0,9977	+ 0,0009	− 0,0768	+ 0,9928	+ 0,0924	23
24	+ 0,0618	+ 0,9979	− 0,0146	− 0,0739	+ 0,9942	+ 0,0772	24
25	+ 0,0564	+ 0,9981	− 0,0249	− 0,0677	+ 0,9957	+ 0,0643	25
26	+ 0,0451	+ 0,9961	− 0,0331	− 0,0615	+ 0,9961	+ 0,0623	26
27	+ 0,0322	+ 0,9986	− 0,0377	− 0,0508	+ 0,9956	+ 0,0785	27
28	+ 0,0223	+ 0,9991	− 0,0361	− 0,0400	+ 0,9948	+ 0,0923	28
29	+ 0,0152	+ 0,9995	− 0,0290	− 0,0330	+ 0,9943	+ 0,1012	29
30	+ 0,0104	+ 0,9995	− 0,0244	− 0,0228	+ 0,9936	+ 0,1105	30
31	+ 0,0035	+ 0,9998	− 0,0153	− 0,0187	+ 0,9938	+ 0,1109	31

Table 14. Final co-ordinates of the centres of the joints, experiment I

No.	Centre of shoulder Right x	y	z	Centre of shoulder Left x	y	z	Midpoint of shoulder line x	y	z	Centre of elbow Right x	y	z	Centre of elbow Left x	y	z	No.
1	37,29	+15,93	129,91	36,17	-18,80	130,06	36,73	-1,44	129,99	42,88	+22,57	104,04	36,54	-27,97	104,26	1
2	43,01	+15,79	131,48	42,05	-18,95	131,72	42,53	-1,58	131,60	46,80	+21,78	105,08	44,39	-27,90	106,01	2
3	48,76	+15,68	133,02	48,01	-19,06	133,31	48,39	-1,69	133,17	49,45	+21,42	106,13	51,88	-27,89	107,78	3
4	54,48	+15,70	133,73	53,82	-19,16	134,18	54,15	-1,73	133,96	51,73	+21,51	107,19	58,90	-28,20	108,94	4
5	60,26	+15,83	133,99	59,83	-19,11	134,26	60,05	-1,64	134,13	54,47	+21,76	108,04	65,91	-28,50	109,37	5
6	65,97	+15,90	133,64	65,87	-18,96	133,67	65,92	-1,53	133,66	57,98	+22,00	108,38	72,82	-28,72	109,11	6
7	71,77	+16,02	132,73	72,01	-18,64	132,61	71,89	-1,31	132,67	62,40	+22,29	107,96	79,63	-28,68	108,31	7
8	78,08	+16,24	131,26	78,62	-18,26	131,15	78,35	-1,01	131,21	68,07	+22,55	106,76	86,64	-28,43	106,97	8
9	84,25	+16,48	129,88	85,01	-17,78	129,92	84,63	-0,65	129,90	74,27	+22,65	105,33	92,97	-27,90	105,62	9
10	90,30	+16,90	129,00	91,23	-17,18	129,19	90,77	-0,14	129,10	80,89	+23,07	104,16	98,74	-27,13	104,76	10
11	97,17	+17,59	128,53	97,72	-16,40	128,86	97,45	+0,60	128,70	88,64	+23,46	103,38	104,64	-26,23	104,23	11
12	104,05	+18,14	128,65	104,00	-15,83	129,14	104,03	+1,16	128,90	97,12	+24,01	102,97	110,09	-25,30	104,23	12
13	110,75	+18,49	129,23	110,23	-15,46	129,79	110,49	+1,52	129,51	106,29	+24,53	103,04	115,74	-24,43	104,55	13
14	116,51	+18,61	130,39	115,66	-15,37	130,72	116,09	+1,62	130,56	114,86	+24,92	103,94	120,32	-23,73	105,11	14
15	122,65	+18,65	132,27	121,37	-15,32	132,22	122,01	+1,67	132,25	123,48	+25,37	105,70	123,88	-23,45	106,14	15
16	128,02	+18,68	133,66	126,19	-15,32	133,30	127,11	+1,68	133,48	130,42	+25,95	107,54	126,31	-23,65	107,12	16
17	134,21	+18,71	134,49	131,97	-15,42	133,92	133,09	+1,65	134,21	138,58	+26,84	108,94	129,32	-24,18	107,98	17
18	139,74	+18,67	134,37	137,28	-15,42	133,87	138,51	+1,63	134,12	146,03	+27,66	109,58	132,59	-24,53	108,26	18
19	145,81	+18,45	133,60	142,98	-15,30	133,18	144,40	+1,58	133,39	154,02	+28,18	109,73	136,91	-24,93	108,03	19
20	151,85	+18,20	132,50	148,80	-15,43	132,08	150,33	+1,39	132,29	161,63	+28,32	109,20	141,96	-25,17	107,25	20
21	158,00	+17,91	131,27	154,82	-15,61	130,81	156,41	+1,15	131,04	168,65	+27,86	108,25	147,76	-25,47	106,02	21
22	164,20	+17,43	130,19	161,08	-15,96	129,71	162,64	+0,74	129,95	174,91	+27,02	106,89	154,30	-25,76	104,83	22
23	170,33	+17,02	129,37	167,43	-16,49	128,99	168,88	+0,27	129,18	180,24	+26,34	105,78	161,27	-26,11	103,92	23
24	176,51	+16,56	129,24	174,21	-17,14	128,68	175,36	-0,29	128,96	185,06	+25,45	105,01	169,02	-26,65	103,37	24
25	182,76	+16,10	129,54	180,92	-17,67	128,91	181,84	-0,79	129,23	189,23	+24,74	104,59	177,12	-27,01	103,27	25
26	189,40	+15,84	130,04	187,91	-18,02	129,52	188,66	-1,09	129,78	194,00	+23,69	104,58	186,15	-27,47	103,70	26
27	194,90	+15,84	130,82	193,74	-18,51	130,54	194,32	-1,34	130,68	198,46	+22,78	104,80	194,32	-27,80	104,76	27
28	200,39	+15,83	132,09	199,80	-18,65	132,19	200,10	-1,41	132,14	202,14	+21,75	105,53	202,43	-27,88	106,55	28
29	205,30	+15,89	133,15	205,40	-18,71	133,56	205,35	-1,41	133,36	204,42	+21,20	106,33	209,49	-28,09	108,26	29
30	210,52	+16,06	133,81	211,21	-18,72	134,31	210,87	-1,33	134,06	206,58	+20,97	107,15	216,60	-28,49	109,40	30
31	215,73	+16,15	133,91	216,85	-18,64	134,27	216,29	-1,25	134,09	209,15	+21,80	107,63	223,19	-28,94	109,76	31

No.	Centre of wrist Right x	y	z	Centre of wrist Left x	y	z	Centre of hip Right x	y	z	Centre of hip Left x	y	z	Midpoint of hip line x	y	z	No.
1	61,38	+26,72	80,93	46,53	-34,79	78,47	39,00	+6,44	83,85	39,90	-10,50	85,40	39,45	-2,03	84,38	1
2	60,04	+28,01	79,36	58,32	-32,55	81,57	45,44	+6,52	84,58	45,44	-10,39	86,36	45,44	-1,94	85,47	2
3	58,41	+29,58	79,30	69,74	-30,28	85,66	51,65	+6,61	85,75	50,80	-10,28	87,16	51,23	-1,84	86,46	3
4	57,05	+31,09	79,85	80,08	-28,06	89,87	57,64	+6,67	86,65	56,07	-10,22	87,83	56,86	-1,78	87,24	4
5	56,96	+32,26	80,53	89,55	-25,95	93,79	63,37	+6,64	86,71	61,40	-10,21	87,78	62,39	-1,79	87,24	5
6	58,24	+32,91	80,93	97,89	-24,27	96,91	68,88	+6,61	85,97	66,77	-10,21	87,15	67,83	-1,80	86,56	6
7	60,65	+34,02	80,93	105,38	-23,10	98,72	74,24	+6,61	84,45	72,26	-10,22	85,98	73,25	-1,81	85,22	7
8	64,49	+35,16	80,52	112,73	-22,33	99,05	80,18	+6,61	82,62	78,33	-10,17	84,60	79,26	-1,78	83,61	8
9	69,77	+36,00	79,63	119,09	-21,91	97,64	86,57	+6,71	81,65	84,61	-10,04	83,79	85,59	-1,67	82,72	9
10	76,73	+36,40	78,54	124,52	-22,10	94,93	93,16	+7,06	81,37	91,08	-9,69	83,39	92,12	-1,32	82,38	10
11	86,35	+36,36	77,12	129,43	-22,69	90,84	100,84	+7,69	81,52	98,54	-9,05	83,42	99,69	-0,68	82,47	11
12	97,54	+36,04	75,92	132,70	-23,78	86,38	107,73	+8,35	82,27	105,67	-8,46	83,73	106,70	-0,06	83,00	12
13	110,40	+35,64	76,00	134,23	-25,41	82,45	113,92	+9,19	82,83	112,37	-7,70	83,99	113,15	+0,75	83,41	13
14	123,81	+34,61	77,38	134,54	-27,29	80,41	119,90	+9,78	83,97	118,75	-7,18	84,91	119,33	+1,33	84,44	14
15	138,06	+33,20	81,15	134,24	-29,36	80,31	125,38	+10,16	85,28	124,92	-6,79	86,51	125,15	+1,69	85,90	15
16	149,47	+31,83	85,49	134,03	-31,05	80,85	130,40	+10,28	85,95	130,85	-6,66	87,60	130,63	+1,81	86,78	16
17	161,88	+30,15	90,98	134,42	-32,68	81,36	135,39	+10,20	86,32	136,41	-6,67	88,16	135,90	+1,77	87,24	17
18	172,07	+28,29	95,58	135,89	-33,48	81,52	140,41	+10,01	86,16	141,77	-6,83	88,01	141,09	+1,59	87,08	18
19	182,22	+26,88	99,36	138,44	-34,75	81,39	145,87	+9,92	85,42	147,28	-6,94	87,13	146,58	+1,49	86,28	19
20	189,70	+25,70	101,44	142,12	-35,70	80,89	151,32	+9,78	84,31	152,86	-7,08	85,81	152,09	+1,35	85,06	20
21	196,91	+25,20	101,77	146,95	-36,67	80,07	156,90	+9,63	83,30	158,76	-7,22	84,45	157,83	+1,21	83,88	21
22	203,19	+25,21	100,54	153,24	-37,20	79,09	162,91	+9,25	82,78	165,12	-7,58	83,71	164,02	+0,84	83,25	22
23	208,25	+25,35	97,44	161,02	-37,30	77,97	169,40	+8,70	82,56	171,71	-8,11	83,53	170,56	+0,30	83,05	23
24	212,18	+25,26	93,00	170,49	-37,03	77,15	176,53	+7,87	82,64	179,00	-8,89	83,81	177,84	-0,51	83,23	24
25	214,41	+26,06	88,00	180,83	-36,50	76,77	183,22	+7,03	82,80	185,38	-9,76	84,28	184,30	-1,37	83,54	25
26	215,09	+27,27	83,60	193,21	-35,60	77,36	189,90	+6,65	83,29	191,70	-10,18	84,96	190,80	-1,77	84,13	26
27	214,28	+28,48	80,18	205,08	-34,12	79,04	196,60	+6,41	83,96	197,95	-10,46	85,88	197,28	-2,03	84,92	27
28	212,73	+29,76	79,23	216,91	-32,52	82,34	203,16	+6,30	84,81	203,85	-10,57	86,73	203,51	-2,14	85,77	28
29	211,26	+30,84	79,38	227,31	-30,77	86,11	209,20	+6,34	85,81	209,80	-10,54	87,34	209,25	-2,10	86,58	29
30	210,23	+31,50	80,01	237,39	-28,91	89,97	214,76	+6,48	86,58	214,30	-10,48	87,75	214,53	-2,00	87,17	30
31	210,49	+32,37	80,80	246,11	-26,95	93,49	219,98	+6,42	86,51	219,32	-10,53	87,59	219,65	-2,06	87,05	31

Table 14   (continued)

No.	Centre of knee						Centre of ankle						No.
	Right			Left			Right			Left			
	x	y	z	x	y	z	x	y	z.	x	y	z	
1	38,77	+10,87	43,26	50,87	−10,39	49,50	7,51	+7,60	23,04	52,56	−5,58	10,70	1
2	50,33	+10,34	45,13	52,41	−10,41	49,78	20,95	+7,29	22,24	52,55	−5,52	10,85	2
3	61,10	+ 9,94	47,33	53,67	− 9,98	50,05	35,83	+6,87	20,17	52,50	−5,46	10,96	3
4	70,67	+ 9,39	49,12	54,85	− 9,46	50,25	51,55	+6,49	17,56	52,55	−5,46	11,08	4
5	79,15	+ 8,75	50,18	56,47	− 8,88	50,32	68,11	+6,05	15,19	52,64	−5,36	11,20	5
6	86,47	+ 8,27	50,10	58,64	− 8,54	50,26	84,68	+5,89	13,86	52,75	−5,23	11,26	6
7	92,27	+ 7,48	48,62	61,47	− 8,25	49,96	100,11	+6,12	14,10	52,96	−5,11	11,33	7
8	98,21	+ 6,48	46,58	65,18	− 8,19	49,64	113,62	+6,67	15,49	53,46	−5,12	11,70	8
9	105,36	+ 6,44	45,84	69,82	− 8,57	49,43	121,71	+6,40	15,28	54,42	−5,49	12,61	9
10	113,60	+ 7,21	46,42	75,62	− 9,25	49,11	125,49	+5,58	13,11	56,14	−6,07	14,16	10
11	120,19	+ 7,91	46,07	83,29	−10,17	48,74	129,57	+5,60	12,31	59,73	−6,83	16,59	11
12	127,00	+ 9,65	47,13	93,25	−11,43	47,79	131,12	+5,66	11,78	65,64	−7,35	19,71	12
13	131,99	+10,02	47,45	104,85	−11,72	46,81	131,77	+5,52	11,56	74,52	−7,64	23,15	13
14	133,64	+10,33	47,22	116,38	−11,57	47,29	131,90	+5,48	11,63	85,74	−7,63	24,31	14
15	134,43	+ 9,76	47,05	127,59	−11,07	49,01	131,92	+5,37	11,76	99,87	−7,23	23,12	15
16	135,05	+ 8,97	46,94	136,78	−10,73	50,65	131,96	+5,27	11,84	112,96	−6,73	20,91	16
17	135,97	+ 8,35	46,76	146,78	−10,24	52,19	132,01	+5,17	11,90	129,45	−6,28	17,82	17
18	137,09	+ 7,89	46,64	155,02	− 9,67	52,83	132,04	+5,05	11,88	145,17	−6,10	15,28	18
19	138,89	+ 7,68	46,42	162,29	− 9,09	52,48	132,18	+4,99	11,88	161,99	−6,19	13,50	19
20	141,56	+ 7,24	46,02	167,87	− 8,38	51,01	132,45	+4,94	11,92	178,11	−6,72	13,45	20
21	144,52	+ 7,18	45,89	172,77	− 7,50	49,00	133,01	+4,89	12,36	191,15	−7,49	14,72	21
22	148,80	+ 7,39	45,93	179,59	− 7,52	48,27	134,09	+5,07	13,34	198,79	−7,55	14,44	22
23	154,58	+ 7,81	45,66	188,31	− 8,58	48,92	135,96	+5,33	14,86	202,00	−7,08	12,31	23
24	161,92	+ 8,56	45,23	193,54	− 9,34	48,50	139,41	+5,35	17,01	205,81	−7,41	11,73	24
25	171,80	+ 9,62	44,32	199,55	−10,28	49,16	144,93	+5,43	19,70	207,49	−7,40	11,29	25
26	184,42	+ 9,90	43,04	205,38	−11,04	49,89	154,06	+5,67	22,58	208,23	−7,29	11,16	26
27	196,50	+ 9,36	43,42	207,79	−11,02	50,01	165,43	+5,60	23,04	208,35	−7,29	11,22	27
28	207,84	+ 9,06	45,20	209,63	−10,99	50,21	178,88	+5,08	21,94	208,47	−7,24	11,27	28
29	217,59	+ 8,76	47,18	211,20	−10,44	50,30	192,48	+4,68	19,88	208,59	−7,19	11,37	29
30	227,13	+ 8,73	48,89	212,63	−10,19	50,44	207,69	+4,50	17,22	208,71	−7,15	11,47	30
31	234,68	+ 8,69	49,79	214,34	−10,11	50,33	223,13	+4,69	14,90	208,84	−7,10	11,58	31

No.	Centre of gravity of foot						Tip of foot						Vertex			No.
	Right			Left			Right			Left						
	x	y	z	x	y	z	x	y	z	x	y	z	x	y	z	
1	4,86	+7,39	16,85	56,87	−6,23	6,07	10,66	+ 6,59	6,85	69,05	− 7,69	5,50	38,81	−1,40	163,55	1
2	19,27	+7,18	15,87	56,94	−6,23	6,10	26,54	+ 6,40	6,89	69,06	− 7,70	5,47	45,32	−1,59	164,84	2
3	35,58	+6,99	13,67	56,93	−6,20	6,11	44,91	+ 6,70	6,50	69,07	− 7,71	5,45	51,59	−1,65	165,98	3
4	52,80	+6,78	11,17	56,98	−6,19	6,14	63,59	+ 6,83	6,36	69,08	− 7,72	5,43	57,53	−1,57	166,56	4
5	70,97	+6,48	9,38	56,99	−6,16	6,16	82,70	+ 6,84	7,39	69,09	− 7,72	5,40	63,50	−1,41	166,48	5
6	88,88	+6,32	8,82	56,99	−6,13	6,18	100,69	+ 6,31	9,71	69,10	− 7,73	5,39	69,41	−1,23	165,81	6
7	105,53	+6,70	10,03	57,00	−6,10	6,20	116,93	+ 7,59	13,75	69,12	− 7,74	5,37	75,39	−1,03	164,64	7
8	119,84	+7,66	12,64	57,05	−6,09	6,34	130,34	+ 9,38	18,76	69,28	− 7,70	5,15	81,76	−0,75	163,29	8
9	127,92	+7,73	12,71	57,54	−6,35	6,81	138,26	+10,65	19,30	69,64	− 7,80	4,86	87,86	−0,42	162,23	9
10	131,33	+6,82	10,00	58,50	−6,78	7,60	142,32	+ 9,64	15,50	70,08	− 7,74	4,34	93,74	−0,01	161,61	10
11	134,81	+6,30	7,76	60,42	−7,13	9,19	146,74	+ 8,06	10,23	71,20	− 7,68	3,68	99,89	+0,41	161,56	11
12	135,68	+6,04	6,77	63,94	−7,15	12,50	147,80	+ 8,00	7,50	72,22	− 7,42	3,78	105,62	+0,86	162,10	12
13	135,91	+6,12	6,68	72,08	−7,68	16,41	148,10	+ 7,76	6,76	77,60	− 8,07	5,62	111,43	+1,33	163,04	13
14	135,92	+6,09	6,71	83,42	−7,63	17,61	148,12	+ 7,69	6,68	89,58	− 7,75	7,17	117,03	+1,58	164,33	14
15	135,93	+6,04	6,74	98,76	−7,15	16,33	148,13	+ 7,64	6,61	106,47	− 6,72	7,23	123,08	+1,66	165,76	15
16	135,93	+6,01	6,77	113,08	−6,68	14,03	148,14	+ 7,58	6,53	122,37	− 6,33	6,72	128,13	+1,63	166,71	16
17	135,94	+5,96	6,80	131,24	−6,55	11,35	148,16	+ 7,51	6,45	142,20	− 6,70	6,38	134,14	+1,53	167,11	17
18	135,95	+5,90	6,83	148,50	−6,72	9,36	148,18	+ 7,43	6,37	160,29	− 7,37	7,04	139,77	+1,45	166,74	18
19	135,97	+5,86	6,87	166,68	−6,91	8,54	148,20	+ 7,37	6,29	178,85	− 8,04	9,42	145,88	+1,37	165,85	19
20	136,00	+5,80	6,94	183,98	−7,49	9,79	148,24	+ 7,31	6,16	195,50	− 8,62	13,75	152,06	+1,32	164,62	20
21	136,28	+5,82	7,17	197,72	−8,69	12,43	148,44	+ 7,34	5,99	208,17	− 9,87	18,65	158,24	+1,15	163,42	21
22	136,83	+5,97	7,74	205,44	−8,97	12,30	148,83	+ 7,39	5,66	215,52	−11,23	18,83	164,46	+0,81	162,48	22
23	137,76	+6,15	8,66	208,35	−8,36	9,43	149,37	+ 7,39	5,24	219,42	−10,56	14,60	170,63	+0,32	161,89	23
24	139,59	+5,98	10,54	211,40	−8,14	7,34	150,27	+ 7,37	4,58	223,23	− 9,90	9,43	176,87	−0,30	161,82	24
25	143,27	+5,87	13,23	212,62	−7,60	6,39	151,35	+ 7,50	4,08	224,81	− 9,20	6,28	182,66	−0,88	162,23	25
26	151,42	+5,71	16,45	212,64	−7,76	6,24	156,95	+ 6,43	5,75	224,60	− 9,02	5,89	188,97	−1,40	163,06	26
27	162,92	+5,15	16,80	212,66	−7,77	6,30	168,79	+ 3,83	7,17	224,64	− 9,03	5,86	195,16	−1,63	164,14	27
28	177,34	+4,66	15,53	212,68	−7,79	6,37	184,73	+ 3,18	7,00	224,68	− 9,03	5,83	201,78	−1,67	165,32	28
29	192,31	+4,53	13,35	212,70	−7,80	6,43	201,58	+ 3,75	6,53	224,72	− 9,04	5,80	207,71	−1,58	166,18	29
30	209,06	+4,66	10,88	212,72	−7,81	6,50	219,91	+ 5,44	6,35	224,76	− 9,04	5,78	213,49	−1,26	166,52	30
31	226,00	+5,14	9,00	212,75	−7,83	6,58	237,70	+ 5,68	6,92	224,80	− 9,06	5,76	219,19	−0,89	166,23	31

Table 15. Final co-ordinates of the centres of the joints, experiment II

No.	Centre of shoulder						Midpoint of shoulder line			Centre of elbow						No.
	Right			Left						Right			Left			
	x	y	z	x	y	z	x	y	z	x	y	z	x	y	z	
1	43,48	+16,41	129,88	43,63	-18,55	128,75	43,56	-1,07	129,32	49,39	+24,83	+104,91	41,63	-27,53	+102,69	1
2	49,11	+16,12	130,47	49,54	-18,73	129,44	49,33	-1,31	129,96	53,86	+24,75	+105,05	49,64	-27,15	+103,08	2
3	54,64	+15,96	131,59	55,28	-18,79	130,79	54,96	-1,42	131,19	57,87	+23,40	+105,76	57,54	-26,65	+104,50	3
4	60,39	+15,93	132,94	61,36	-18,81	132,54	60,88	-1,44	132,74	61,43	+22,56	+106,58	65,50	-26,41	+106,46	4
5	65,72	+15,98	133,78	67,03	-18,81	133,87	66,38	-1,42	133,83	64,08	+22,05	+107,26	72,69	-26,52	+108,19	5
6	71,35	+15,91	134,04	73,01	-18,85	134,53	72,18	-1,47	134,29	66,92	+21,86	+107,74	80,11	-26,92	+109,43	6
7	77,30	+16,05	133,74	79,16	-18,73	134,34	78,23	-1,34	134,04	70,37	+21,87	+107,94	87,78	-27,38	+109,85	7
8	82,67	+16,08	133,04	84,59	-18,52	133,53	83,63	-1,22	133,29	74,05	+21,89	+107,72	94,46	-27,59	+109,68	8
9	89,16	+16,20	131,82	91,10	-18,17	132,24	90,13	-0,99	132,03	79,32	+22,04	+106,96	102,12	-27,50	+108,88	9
10	95,34	+16,35	130,49	97,30	-17,85	130,91	96,32	-0,75	130,70	85,05	+22,18	+105,83	108,72	-27,05	+107,69	10
11	101,96	+16,71	129,43	103,89	-17,39	129,90	102,93	-0,34	129,67	91,96	+22,45	+104,57	114,93	-26,34	+106,44	11
12	108,44	+17,27	128,96	110,07	-16,78	129,45	109,26	+0,25	129,21	99,17	+22,75	+103,95	120,39	-25,54	+105,59	12
13	115,57	+17,93	129,00	116,27	-16,28	129,43	115,92	+0,83	129,22	107,31	+23,30	+103,53	125,41	-24,78	+105,06	13
14	122,12	+18,34	129,51	122,08	-15,92	130,01	122,10	+1,21	129,76	115,47	+23,87	+103,63	130,00	-24,26	+105,30	14
15	128,43	+18,49	130,56	127,78	-15,77	130,97	128,11	+1,36	130,77	124,25	+24,56	+104,31	135,06	-23,91	+105,89	15
16	134,25	+18,56	132,21	133,27	-15,72	132,24	133,76	+1,42	132,23	132,51	+24,97	+105,72	139,19	-23,62	+106,69	16
17	140,03	+18,68	133,61	138,72	-15,75	133,25	139,38	+1,47	133,43	140,31	+25,53	+107,42	142,45	-23,84	+107,35	17
18	145,76	+19,02	134,40	144,24	-15,73	133,80	145,00	+1,65	134,10	148,10	+26,20	+108,59	145,55	-24,37	+107,75	18
19	151,72	+18,89	134,47	150,06	-15,86	133,68	150,89	+1,52	134,08	156,29	+26,92	+109,13	148,84	-25,03	+107,86	19
20	157,76	+18,19	133,84	155,98	-15,93	133,20	156,87	+1,13	133,52	164,51	+27,45	+109,16	152,68	-25,63	+107,72	20
21	163,54	+18,15	132,86	161,54	-15,75	132,45	162,54	+1,20	132,66	171,93	+27,28	+108,95	157,00	-26,01	+107,20	21
22	169,48	+17,90	131,79	167,25	-16,09	131,35	168,37	+0,91	131,57	178,96	+26,90	+108,34	162,12	-26,28	+106,32	22
23	176,33	+17,53	130,70	174,02	-16,38	130,21	175,18	+0,58	130,46	186,10	+26,10	+107,37	168,83	-26,49	+105,17	23
24	182,75	+16,99	130,09	180,44	-16,87	129,57	181,60	+0,06	129,83	191,79	+25,58	+106,02	175,70	-26,61	+104,31	24
25	189,15	+16,64	129,79	187,14	-17,42	129,21	188,15	-0,39	129,50	196,88	+24,82	+105,23	183,23	-26,90	+103,70	25
26	195,38	+16,17	129,97	193,74	-17,83	129,29	194,56	-0,83	129,63	201,24	+24,29	+104,74	191,16	-27,06	+103,54	26
27	201,76	+16,02	130,45	200,19	-18,19	129,80	200,98	-1,09	130,13	205,71	+23,53	+104,73	199,54	-27,16	+103,92	27
28	207,27	+16,21	131,21	206,03	-18,40	130,87	206,65	-1,10	131,04	209,92	+22,70	+105,14	207,45	-27,30	+104,94	28
29	212,50	+16,10	132,60	211,78	-18,53	132,39	212,14	-1,22	132,50	213,41	+21,97	+106,11	215,08	-27,50	+106,71	29
30	217,59	+15,90	133,76	217,42	-18,75	133,78	217,51	-1,43	133,77	215,98	+21,36	+107,03	222,38	-27,83	+108,46	30
31	223,20	+15,86	134,47	223,35	-18,92	134,53	223,28	-1,53	134,50	218,73	+21,31	+108,00	229,95	-28,35	+109,73	31

No.	Centre of wrist						Centre of hip						Midpoint of hip line			No.
	Right			Left			Right			Left						
	x	y	z	x	y	z	x	y	z	x	y	z	x	y	z	
1	74,23	+27,57	88,10	46,03	-36,80	76,23	43,46	+7,63	82,41	45,74	-9,14	83,94	44,60	-0,76	83,18	1
2	75,65	+27,60	84,39	56,93	-34,73	76,75	57,25	+7,54	83,58	51,55	-9,29	84,81	51,40	-0,88	84,20	2
3	75,70	+27,90	81,69	68,56	-32,23	78,74	57,85	+7,44	85,06	57,25	-9,42	86,26	57,55	-0,99	85,66	3
4	74,92	+28,58	80,45	80,67	-29,34	82,34	63,50	+7,31	85,80	62,90	-9,57	87,09	63,20	-1,13	86,45	4
5	74,25	+29,54	80,18	91,62	-26,85	86,97	69,05	+7,18	86,24	68,25	-9,74	87,57	68,65	-1,28	86,91	5
6	74,05	+30,36	80,08	102,02	-24,06	91,04	74,48	+7,03	86,48	73,65	-9,91	87,66	74,07	-1,44	87,07	6
7	74,84	+30,79	79,90	111,75	-21,72	95,41	80,09	+6,91	86,23	79,07	-10,01	87,31	79,58	-1,55	86,77	7
8	76,62	+31,61	79,52	119,54	-20,08	98,38	85,22	+6,81	85,10	83,97	-10,10	86,42	84,60	-1,65	85,76	8
9	79,80	+32,28	78,94	127,76	-19,07	100,54	90,99	+6,81	83,18	89,86	-10,05	84,98	90,43	-1,62	84,08	9
10	84,09	+33,13	78,21	134,72	-18,65	100,76	96,93	+6,88	81,83	95,64	-9,93	83,86	96,29	-1,53	82,90	10
11	90,16	+33,99	77,46	141,31	-18,98	99,45	103,88	+7,10	81,48	102,26	-9,69	83,53	103,07	-1,30	82,51	11
12	97,69	+34,55	76,91	146,61	-19,60	96,65	111,12	+7,54	81,33	109,20	-9,23	83,35	110,16	-0,85	82,34	12
13	107,38	+34,95	76,42	151,02	-20,85	92,69	118,49	+8,26	81,92	116,69	-8,56	83,58	117,59	-0,15	82,75	13
14	118,18	+34,80	76,40	153,66	-22,36	88,49	124,77	+8,87	82,56	123,25	-8,02	83,75	124,01	+0,43	83,16	14
15	130,92	+34,74	77,14	154,83	-24,38	84,54	130,72	+9,68	83,37	129,75	-7,26	84,53	130,24	+1,21	83,95	15
16	143,72	+33,96	79,58	154,85	-26,81	82,60	136,30	+9,96	84,55	135,80	-6,97	85,75	136,05	+1,50	85,15	16
17	156,14	+32,80	82,98	154,40	-28,98	81,83	141,73	+10,06	85,67	141,74	-6,88	87,13	141,74	+1,59	86,40	17
18	168,26	+31,40	86,84	154,22	-31,08	81,61	147,10	+10,07	86,13	147,75	-6,86	87,72	147,43	+1,61	86,93	18
19	179,89	+29,69	90,82	154,61	-33,42	81,23	152,48	+10,03	86,10	153,70	-6,89	87,72	153,09	+1,57	86,91	19
20	190,39	+27,82	94,27	156,25	-34,86	81,10	157,82	+9,94	85,45	159,48	-6,91	87,11	158,65	+1,52	86,28	20
21	198,91	+26,67	96,37	159,00	-36,40	80,75	162,99	+9,83	84,45	164,88	-7,01	85,85	163,94	+1,41	85,15	21
22	206,46	+25,80	97,09	162,86	-37,42	80,20	168,31	+9,75	83,51	170,50	-7,07	84,56	169,41	+1,34	84,04	22
23	213,65	+25,67	96,12	168,85	-38,22	79,43	174,84	+9,39	82,93	177,30	-7,41	83,78	176,07	+0,99	83,36	23
24	219,00	+25,25	93,99	176,19	-38,29	78,57	181,36	+8,99	82,66	183,83	-7,81	83,57	182,60	+0,59	83,12	24
25	222,98	+25,90	91,07	185,41	-37,23	77,67	188,54	+8,17	82,69	191,27	-8,58	83,83	189,91	-0,21	83,26	25
26	225,35	+26,56	86,95	195,59	-36,86	77,24	195,41	+7,41	82,79	197,63	-9,37	84,28	196,52	-0,98	83,54	26
27	226,45	+27,14	83,16	207,36	-35,47	77,74	202,10	+6,97	83,26	203,85	-9,80	84,96	202,98	-1,42	84,11	27
28	226,18	+28,65	80,53	219,17	-33,48	79,50	208,50	+6,68	84,60	209,80	-10,07	85,90	209,15	-1,70	85,25	28
29	225,51	+29,06	80,08	230,42	-31,73	82,79	214,80	+6,47	85,83	215,65	-10,22	86,84	215,23	-1,88	86,34	29
30	224,90	+30,09	80,31	241,12	-29,29	86,80	220,95	+6,30	86,53	221,10	-10,30	87,49	221,03	-2,00	87,01	30
31	224,75	+30,92	80,53	251,49	-27,05	90,70	227,00	+6,27	86,89	226,15	-10,29	87,84	226,58	-2,01	87,37	31

Table 15 (continued)

No.	Centre of knee Right x	y	z	Left x	y	z	Centre of ankle Right x	y	z	Left x	y	z	No.
1	34,99	+11,58	43,34	61,61	-9,34	49,27	6,73	+6,73	20,42	66,44	-5,51	10,79	1
2	46,97	+11,64	42,89	64,95	-9,65	49,49	16,45	+6,80	22,73	66,63	-5,43	10,76	2
3	58,79	+11,18	44,09	67,09	-9,91	49,74	28,77	+6,69	22,73	66,71	-5,41	10,80	3
4	70,04	+10,70	46,23	68,62	-9,63	49,85	42,98	+6,28	21,24	66,74	-5,38	10,84	4
5	79,74	+10,34	48,32	69,79	-9,34	50,00	57,40	+6,07	18,98	66,77	-5,38	10,91	5
6	88,74	+9,82	49,82	70,98	-9,00	50,12	73,38	+6,00	16,34	66,84	-5,37	10,97	6
7	96,69	+9,22	50,15	72,86	-8,79	50,07	90,46	+6,03	14,15	66,89	-5,31	11,04	7
8	102,60	+8,64	49,24	74,96	-8,54	49,89	105,45	+6,28	13,45	66,93	-5,23	11,11	8
9	108,23	+7,45	47,01	78,02	-8,30	49,59	121,05	+6,92	14,37	67,26	-5,23	11,33	9
10	114,36	+6,77	45,56	81,98	-8,37	49,36	131,33	+6,80	15,15	67,96	-5,35	11,97	10
11	123,51	+7,17	46,28	87,82	-8,99	48,99	136,86	+5,64	13,61	69,51	-5,84	13,31	11
12	131,08	+7,48	46,32	94,75	-9,82	48,57	140,73	+4,80	12,44	72,32	-6,53	15,41	12
13	137,09	+8,83	46,45	104,60	-11,10	47,75	143,22	+4,79	11,73	77,89	-7,17	18,51	13
14	143,07	+9,34	47,27	115,66	-11,46	46,60	144,03	+4,81	11,44	85,84	-7,63	22,05	14
15	145,60	+9,45	47,12	128,08	-11,27	46,78	144,31	+4,71	11,44	97,23	-7,74	24,10	15
16	147,06	+9,14	47,05	138,93	-11,05	48,50	144,38	+4,66	11,47	110,44	-7,42	23,50	16
17	148,07	+8,75	46,89	148,98	-10,86	50,45	144,43	+4,58	11,49	124,74	-6,93	21,28	17
18	148,91	+8,10	46,68	158,56	-10,61	52,10	144,47	+4,48	11,62	140,24	-6,56	18,34	18
19	150,08	+7,71	46,55	167,34	-10,06	52,91	144,53	+4,41	11,69	157,00	-6,34	15,41	19
20	151,67	+7,46	46,27	174,85	-9,28	52,47	144,64	+4,34	11,69	174,19	-6,34	13,44	20
21	153,79	+7,12	46,00	180,11	-8,35	50,96	144,87	+4,27	11,75	189,44	-6,54	13,17	21
22	156,69	+6,99	45,82	184,56	-7,32	49,06	145,41	+4,20	12,16	202,19	-6,86	14,45	22
23	161,53	+7,38	45,76	191,92	-6,95	48,50	146,59	+4,39	13,16	210,70	-6,27	14,30	23
24	167,64	+7,91	45,51	200,50	-7,79	49,05	148,56	+4,64	14,70	214,01	-5,38	12,32	24
25	175,32	+8,54	44,91	205,73	-8,35	48,65	152,23	+4,78	16,98	217,86	-5,46	11,81	25
26	185,76	+9,80	43,90	212,47	-9,42	49,49	158,30	+5,23	19,88	219,45	-5,39	11,42	26
27	198,43	+9,98	42,80	217,63	-9,93	49,87	167,63	+5,84	22,75	220,23	-5,33	11,30	27
28	210,71	+9,61	43,61	220,25	-10,07	50,02	179,46	+6,17	23,33	220,36	-5,28	11,32	28
29	221,61	+9,56	45,72	222,08	-9,79	50,13	192,54	+6,12	22,42	220,40	-5,24	11,36	29
30	231,43	+9,46	48,00	223,51	-9,38	50,25	206,43	+6,12	20,38	220,46	-5,20	11,42	30
31	240,78	+9,29	49,84	224,87	-9,18	50,36	222,27	+6,11	17,55	220,53	-5,17	11,48	31

No.	Centre of gravity of foot Right x	y	z	Left x	y	z	Tip of foot Right x	y	z	Left x	y	z	Vertex x	y	z	No.
1	4,45	+6,98	14,22	71,10	-6,02	6,01	10,65	+8,52	3,63	83,10	-6,84	5,65	42,38	-0,91	+162,34	1
2	13,88	+6,77	16,63	71,09	-6,02	6,02	19,47	+7,24	6,02	83,11	-6,85	5,62	48,45	-1,36	+163,35	2
3	26,56	+6,22	16,45	71,08	-6,00	6,03	32,93	+5,28	6,98	83,14	-6,87	5,59	54,68	-1,52	+164,70	3
4	42,00	+6,04	14,81	71,06	-5,98	6,03	50,13	+5,17	6,69	83,16	-6,88	5,56	61,12	-1,48	+166,03	4
5	57,84	+6,04	12,53	71,03	-5,95	6,05	67,71	+5,69	6,29	83,17	-6,89	5,53	67,00	-1,32	+166,86	5
6	75,38	+6,30	10,15	71,04	-5,95	6,07	86,64	+6,70	6,40	83,19	-6,89	5,49	73,09	-1,06	+167,02	6
7	94,03	+6,57	8,65	71,05	-5,91	6,10	106,02	+7,50	7,95	83,19	-6,89	5,46	79,38	-0,75	+166,51	7
8	110,29	+6,84	8,76	71,06	-5,90	6,12	122,16	+7,89	10,83	83,19	-6,89	5,42	84,97	-0,56	+165,58	8
9	127,15	+7,69	10,80	71,09	-5,90	6,18	138,26	+9,38	15,77	83,18	-6,86	5,31	91,63	-0,29	+164,19	9
10	137,70	+8,16	12,49	71,56	-6,09	6,46	148,09	+11,17	19,10	83,43	-6,86	5,01	97,83	-0,06	+162,98	10
11	142,90	+7,06	10,43	72,21	-6,34	7,10	153,85	+10,75	16,25	84,04	-6,95	4,64	104,32	+0,28	+162,23	11
12	146,27	+5,69	8,41	73,73	-6,75	8,31	158,05	+8,11	11,72	84,91	-6,97	4,04	110,60	+0,65	+161,93	12
13	147,94	+5,40	6,88	76,82	-6,85	11,25	160,06	+7,20	8,08	86,10	-6,90	3,71	117,04	+1,05	+162,19	13
14	148,36	+5,31	6,50	83,59	-7,48	15,10	160,48	+6,70	6,73	89,47	-6,95	4,70	122,78	+1,38	+162,90	14
15	148,37	+5,27	6,52	94,73	-7,57	17,29	160,52	+6,69	6,63	100,81	-7,31	6,96	128,88	+1,63	+164,07	15
16	148,39	+5,23	6,54	108,94	-7,19	16,60	160,56	+6,65	6,59	116,16	-6,28	7,22	134,88	+1,70	+165,37	16
17	148,40	+5,19	6,56	124,65	-6,71	14,37	160,58	+6,64	6,53	133,57	-5,82	6,75	140,64	+1,68	+166,37	17
18	148,41	+5,15	6,61	141,72	-6,57	11,65	160,60	+6,62	6,46	152,37	-6,04	6,32	146,34	+1,60	+166,79	18
19	148,43	+5,14	6,63	160,20	-6,75	9,33	160,62	+6,60	6,41	171,99	-6,94	6,78	152,25	+1,49	+166,55	19
20	148,46	+5,11	6,66	178,86	-6,97	8,35	160,64	+6,54	6,54	191,05	-8,97	8,98	158,28	+1,37	+165,77	20
21	148,51	+5,05	6,71	195,31	-7,38	9,43	160,67	+6,50	6,23	206,99	-8,16	13,02	163,96	+1,33	+164,70	21
22	148,81	+5,11	6,93	208,80	-7,85	11,97	160,88	+6,52	6,05	219,64	-7,77	17,98	169,67	+1,17	+163,61	22
23	149,37	+5,12	7,56	217,38	-7,65	12,25	161,35	+6,55	5,64	227,57	-9,49	18,85	176,25	+0,85	+162,71	23
24	150,27	+5,32	8,55	220,39	-6,53	9,45	161,92	+6,56	5,20	231,30	-8,50	14,52	182,42	+0,44	+162,21	24
25	152,28	+5,17	10,54	223,56	-5,58	7,66	162,84	+6,54	4,54	235,20	-6,99	9,77	188,68	-0,03	+162,09	25
26	156,39	+5,30	13,41	224,65	-5,50	6,81	163,96	+6,52	3,90	236,29	-6,87	6,98	194,47	-0,51	+162,44	26
27	164,98	+5,55	16,50	224,69	-5,65	6,67	170,58	+5,65	6,06	236,61	-6,61	6,25	200,39	-0,98	+163,24	27
28	176,93	+5,64	16,99	224,70	-5,66	6,68	182,90	+4,21	7,48	236,63	-6,57	6,21	206,37	-1,27	+164,41	28
29	191,01	+5,71	15,91	224,71	-5,67	6,67	198,37	+4,46	7,39	236,65	-6,55	6,19	212,51	-1,48	+165,63	29
30	206,33	+6,05	13,81	224,72	-5,68	6,66	215,57	+5,47	6,85	236,65	-6,52	6,15	218,44	-1,54	+166,52	30
31	223,72	+6,38	11,14	224,73	-5,68	6,66	234,56	+6,40	6,50	236,66	-6,49	6,11	224,44	-1,39	+166,95	31

Table 16. Final co-ordinates of the centres of the joints, experiment III

No.	Centre of shoulder Right x	y	z	Left x	y	z	Midpoint of shoulder line x	y	z	Centre of elbow Right x	y	z	Left x	y	z	No.
1	34,74	+15,19	132,80	35,83	−18,46	134,12	35,29	−1,64	133,46	37,90	+21,65	106,21	36,01	−27,27	108,29	1
2	40,38	+15,46	133,86	41,36	−18,26	135,16	40,87	−1,40	134,51	41,20	+22,34	107,25	41,37	−27,16	109,37	2
3	45,94	+15,76	134,48	46,85	−18,04	135,65	46,40	−1,14	135,07	44,41	+23,30	108,11	46,73	−27,26	109,91	3
4	51,48	+16,06	134,54	52,29	−17,68	135,52	51,89	−0,81	135,03	47,58	+24,18	108,68	51,98	−27,07	109,86	4
5	57,04	+16,43	133,99	57,80	−17,34	134,79	57,42	−0,46	134,39	51,20	+24,92	108,75	57,19	−26,61	109,06	5
6	62,65	+16,98	133,04	63,36	−16,82	133,62	63,01	+0,08	133,33	55,67	+25,52	108,15	62,54	−26,04	107,83	6
7	68,31	+17,38	131,83	68,97	−16,33	132,29	68,64	+0,53	132,06	61,02	+26,16	107,15	68,08	−25,45	106,45	7
8	74,36	+17,88	130,69	74,79	−15,63	131,10	74,58	+1,13	130,90	67,27	+26,81	105,94	73,93	−24,75	105,31	8
9	80,58	+18,62	129,94	80,00	−14,90	130,35	80,59	+1,86	130,15	74,27	+27,45	104,97	79,65	−23,92	104,46	9
10	86,84	+19,25	129,78	86,47	−14,33	130,23	86,66	+2,46	130,01	81,77	+28,01	104,57	85,70	−23,07	104,26	10
11	92,91	+19,81	130,09	92,08	−13,70	130,64	92,50	+3,06	130,37	89,74	+28,55	104,86	91,63	−22,28	104,59	11
12	98,53	+20,06	131,40	97,53	−13,26	131,55	98,03	+3,40	131,48	97,67	+28,90	105,72	97,12	−21,66	105,39	12
13	103,96	+20,23	132,84	102,66	−13,07	132,84	103,31	+3,58	132,84	105,61	+29,13	107,25	102,19	−21,33	106,61	13
14	109,13	+20,38	134,30	107,50	−12,94	134,19	108,32	+3,72	134,25	113,32	+29,24	109,06	106,97	−21,10	107,87	14
15	114,29	+20,33	135,37	112,46	−12,81	135,01	113,38	+3,76	135,19	121,20	+29,36	110,67	111,79	−20,97	108,65	15
16	119,36	+20,30	135,69	117,45	−12,79	135,18	118,41	+3,76	135,44	128,09	+29,16	111,70	116,63	−20,92	108,77	16
17	124,56	+20,18	135,37	122,60	−12,76	135,02	123,58	+3,71	135,20	135,06	+29,03	111,80	121,77	−20,90	108,26	17
18	129,85	+20,02	134,48	127,77	−12,83	133,73	128,81	+3,60	134,11	141,76	+28,61	111,57	127,08	−21,09	107,34	18
19	135,43	+19,73	133,33	133,13	−13,00	132,67	134,28	+3,37	133,00	148,23	+28,01	110,91	132,57	−21,36	106,28	19
20	141,06	+19,26	132,10	138,63	−13,41	131,52	139,85	+2,93	131,81	154,22	+27,34	109,90	138,20	−21,78	105,18	20
21	146,97	+18,74	131,29	144,46	−13,99	130,72	145,72	+2,38	131,01	159,76	+26,74	108,81	143,99	−22,31	104,45	21
22	*152,80*	*+18,15*	*150,75*	150,32	−14,72	130,28	151,56	+1,72	130,52	164,80	+25,79	107,83	149,92	−23,03	104,06	22
23	158,38	+17,36	130,60	156,18	−15,48	130,56	157,28	+0,94	130,58	169,08	+25,17	106,97	156,18	−23,90	104,36	23
24	164,11	+16,79	130,89	162,07	−16,15	131,37	163,09	+0,32	131,13	172,99	+24,45	106,51	162,49	−24,88	105,34	24
25	169,61	+16,45	131,50	167,73	−16,79	132,33	168,67	−0,17	131,92	176,51	+23,73	106,36	168,49	−25,82	106,56	25
26	174,78	+16,21	132,59	173,27	−17,16	133,70	174,03	−0,48	133,15	179,71	+22,90	106,77	174,28	−26,24	107,91	26
27	179,91	+16,01	133,58	178,82	−17,69	134,85	179,37	−0,84	134,22	182,27	+22,49	107,47	179,86	−26,51	109,08	27
28	184,91	+15,52	134,44	184,16	−17,92	135,65	184,54	−1,20	135,05	184,56	+22,56	108,21	185,23	−26,91	109,92	28
29	190,20	+15,20	134,84	189,69	−18,20	135,81	189,95	−1,50	135,33	187,20	+22,72	108,86	190,75	−27,47	110,13	29
30	195,54	+14,95	134,48	195,20	−18,43	135,30	195,37	−1,74	134,89	190,40	+22,80	109,06	196,08	−27,75	109,55	30
31	200,86	+14,66	133,72	200,75	−18,53	134,23	200,81	−1,94	133,98	194,29	+22,88	108,74	201,29	−27,76	108,41	31

No.	Centre of wrist Right x	y	z	Left x	y	z	Centre of hip Right x	y	z	Left x	y	z	Midpoint of hip line x	y	z	No.
1	52,51	+26,12	77,74				31,02	+ 4,59	86,61	32,64	−12,30	85,83	31,83	−3,86	86,22	1
2	52,22	+28,43	77,60				36,18	+ 5,04	87,83	37,57	−11,86	86,52	36,88	−3,41	87,18	2
3	52,48	+30,53	77,79	74,32	−20,96	101,71	41,30	+ 5,42	88,37	42,41	−11,47	86,75	41,86	−3,03	87,56	3
4	53,44	+32,65	78,14				46,50	+ 5,80	88,48	47,32	−11,08	86,68	46,91	−2,64	87,58	4
5	55,28	+34,61	78,14				51,58	+ 6,25	87,79	52,23	−10,64	86,04	51,91	−2,20	86,92	5
6	58,24	+36,24	77,78				56,93	+ 6,56	86,50	57,30	−10,37	84,96	57,12	−1,91	85,73	6
7	62,44	+37,69	77,06				62,46	+ 7,00	85,13	62,55	− 9,95	83,81	62,51	−1,48	84,47	7
8	68,20	+38,91	76,12	101,43	−18,83	96,24	68,31	+ 7,51	84,26	68,17	− 9,45	83,04	68,24	−0,97	83,65	8
9	75,74	+39,55	75,27	107,46	−18,19	95,58	74,65	+ 8,11	83,73	74,37	− 8,85	82,55	74,51	−0,37	83,14	9
10	84,77	+39,65	74,79	113,57	−17,66	95,67	81,39	+ 9,04	83,93	81,06	− 7,90	82,57	81,23	+0,57	83,25	10
11	95,32	+39,24	75,11	119,60	−17,16	96,51	88,18	+ 9,98	84,60	87,85	− 6,93	82,88	88,02	+1,53	83,74	11
12	106,75	+38,22	76,44	125,26	−16,76	97,53	94,14	+10,71	85,37	94,00	− 6,17	83,34	94,07	+2,27	84,36	12
13	118,80	+36,69	78,93	130,39	−16,79	98,79	99,75	+11,18	86,57	99,96	− 5,70	84,55	99,86	+2,74	85,56	13
14	130,49	+34,66	82,47				104,86	+11,32	87,68	105,43	− 5,59	85,97	105,15	+2,87	86,83	14
15	141,74	+32,06	86,54				109,94	+11,35	88,53	110,77	− 5,56	87,03	110,36	+2,90	87,78	15
16	151,80	+29,45	90,52				114,97	+11,23	88,59	115,92	− 5,70	87,37	115,45	+2,77	87,98	16
17	161,00	+26,80	94,10				120,14	+10,96	88,29	121,13	− 5,98	87,19	120,64	+2,49	87,74	17
18	169,07	+24,54	96,92				125,01	+10,74	87,57	126,00	− 6,18	86,41	125,51	+2,28	86,99	18
19	176,24	+22,80	98,51				130,18	+10,50	86,49	131,06	− 6,43	85,14	130,62	+2,04	85,82	19
20	182,36	+21,60	98,57				135,56	+ 9,96	85,47	136,61	− 6,93	83,85	136,09	+1,52	84,66	20
21	187,85	+21,07	96,74				141,39	+ 9,43	84,99	142,66	− 7,44	83,15	142,03	+1,00	84,07	21
22	192,15	+21,02	93,80				147,68	+ 8,48	84,83	149,02	− 8,38	83,04	148,35	+0,05	83,94	22
23	195,32	+21,61	89,91				154,21	+ 7,37	85,09	155,52	− 9,51	83,52	154,87	−1,07	84,31	23
24	196,89	+23,05	85,68				160,68	+ 6,23	85,41	161,93	−10,67	84,10	161,31	−2,22	84,76	24
25	196,97	+24,85	81,98				166,60	+ 5,59	85,93	167,75	−11,33	84,84	167,18	−2,87	85,39	25
26	195,93	+26,62	79,62				172,34	+ 5,05	86,80	173,39	−11,89	85,74	172,87	−3,42	86,27	26
27	194,68	+28,47	78,80				177,72	+ 4,85	87,90	178,59	−12,07	86,56	178,16	−3,61	87,23	27
28	193,90	+29,87	78,75				182,75	+ 4,62	88,58	183,43	−12,23	87,01	183,09	−3,81	87,80	28
29	193,81	+31,08	78,83				187,84	+ 4,47	88,70	188,41	−12,43	86,98	188,13	−3,98	87,84	29
30	194,89	+31,67	78,78				192,81	+ 4,33	88,38	193,20	−12,56	86,50	193,01	−4,12	87,44	30
31	197,03	+32,84	78,47				197,62	+ 4,21	87,60	197,94	−12,69	85,72	197,78	−4,24	86,66	31

Table 16 (continued)

No.	Centre of knee Right x	y	z	Left x	y	z	Centre of ankle Right x	y	z	Left x	y	z	No.
1	32,11	+ 7,37	48,51	39,73	−10,75	47,02	5,83	+ 5,24	21,24	37,43	−7,73	10,52	1
2	41,58	+ 7,85	50,31	40,89	−10,35	47,09	20,27	+ 5,65	19,08	37,41	−7,59	10,54	2
3	50,40	+ 8,62	51,83	42,16	− 9,87	47,10	35,79	+ 6,24	16,40	37,38	−7,41	10,56	3
4	57,99	+ 8,99	52,62	43,61	− 9,46	47,02	51,84	+ 7,29	14,05	37,34	−7,15	10,56	4
5	64,05	+ 9,07	52,25	45,29	− 9,13	46,87	68,02	+ 8,77	12,82	37,30	−6,96	10,58	5
6	69,06	+ 9,03	50,66	47,26	− 8,77	46,75	83,10	+10,46	13,18	37,25	−6,68	10,60	6
7	73,78	+ 8,69	48,88	50,13	− 8,58	46,45	94,91	+11,90	14,52	37,21	−6,41	10,62	7
8	81,03	+ 9,30	48,33	54,31	− 8,66	46,00	101,58	+12,39	13,77	38,56	−6,70	11,19	8
9	89,21	+10,91	48,45	59,70	− 9,01	45,90	105,08	+12,26	12,04	40,57	−7,13	13,13	9
10	93,82	+11,68	48,03	67,09	− 9,36	45,28	108,13	+12,48	11,16	43,57	−7,16	15,53	10
11	100,49	+12,83	48,96	77,97	−10,13	44,30	109,42	+12,39	10,75	49,40	−7,21	18,70	11
12	105,60	+13,35	49,65	89,69	− 9,51	43,44	109,80	+12,27	10,69	58,18	−7,08	22,32	12
13	107,07	+13,21	49,87	101,33	− 8,70	44,46	109,82	+12,18	10,69	69,56	−5,97	23,32	13
14	107,59	+13,02	50,24	111,42	− 8,66	46,46	109,85	+12,18	10,74	82,38	−4,62	22,62	14
15	108,57	+12,69	50,72	121,04	− 8,68	48,49	109,85	+12,08	10,84	96,22	−3,76	20,42	15
16	110,32	+12,39	50,91	129,41	− 8,64	50,02	109,89	+11,90	10,89	110,85	−3,75	17,62	16
17	112,07	+12,21	51,01	136,69	− 8,77	50,70	109,94	+11,76	10,94	126,33	−4,41	15,11	17
18	113,73	+11,95	51,12	142,46	− 8,79	50,27	109,99	+11,66	10,93	141,41	−5,53	13,45	18
19	116,49	+11,62	50,91	147,01	− 8,45	48,67	110,20	+11,56	11,02	155,94	−7,19	13,22	19
20	119,86	+11,39	50,90	150,99	− 7,72	46,62	110,70	+11,58	11,51	166,91	−9,37	14,10	20
21	124,31	+11,33	51,10	157,21	− 8,14	45,78	111,74	+11,76	12,51	173,21	−9,43	13,38	21
22	130,36	+11,28	50,97	164,23	− 9,02	46,02	113,61	+11,93	14,15	176,78	−9,39	11,97	22
23	137,94	+10,87	50,60	169,66	−10,58	46,28	117,17	+11,56	16,27	179,22	−9,33	11,54	23
24	147,84	+10,75	49,62	175,65	−11,62	46,88	123,04	+10,95	19,63	180,81	−9,57	11,08	24
25	158,88	+10,10	48,38	179,75	−12,08	47,28	131,48	+10,03	22,72	181,12	−9,57	11,00	25
26	169,98	+ 9,17	48,82	181,97	−12,06	47,39	142,61	+ 8,50	23,41	181,20	−9,50	11,10	26
27	179,63	+ 8,24	50,22	183,06	−11,96	47,49	155,04	+ 6,61	22,35	181,27	−9,48	11,21	27
28	188,44	+ 7,63	51,37	184,07	−11,74	47,63	168,05	+ 4,96	20,08	181,35	−9,45	11,32	28
29	196,73	+ 7,03	52,28	185,50	−11,51	47,67	182,62	+ 3,57	17,07	181,44	−9,41	11,46	29
30	203,98	+ 6,55	52,56	187,11	−11,37	47,58	198,12	+ 2,53	14,27	181,53	−9,35	11,53	30
31	210,22	+ 5,82	51,91	189,15	−11,33	47,37	213,93	+ 2,12	12,58	181,63	−9,25	11,63	31

No.	Centre of gravity of foot Right x	y	z	Left x	y	z	Tip of foot Right x	y	z	Left x	y	z	Vertex x	y	z	No.
1	4,70	+ 4,89	14,92				12,55	+ 4,11	6,53				38,51	−2,20	164,42	1
2	20,51	+ 5,41	12,75				30,20	+ 4,93	6,14				44,22	−1,88	165,18	2
3	37,60	+ 6,15	10,34				48,55	+ 5,78	6,27				49,89	−1,57	165,50	3
4	55,06	+ 7,38	8,56				66,86	+ 7,19	7,28				55,49	−1,14	165,23	4
5	72,54	+ 8,96	8,14				84,60	+ 9,50	9,78				61,31	−0,76	164,47	5
6	88,73	+10,97	9,55				100,01	+11,82	13,95				67,18	−0,40	163,31	6
7	101,14	+12,73	12,03				111,44	+13,91	18,31				72,82	+0,13	162,05	7
8	107,76	+13,44	11,29				118,13	+15,42	17,77				78,52	+0,80	161,02	8
9	110,76	+13,03	8,72				122,01	+14,53	13,26				84,27	+1,28	160,36	9
10	113,19	+12,82	6,44				125,04	+13,33	8,36				89,88	+1,76	160,45	10
11	113,75	+12,65	5,97				125,68	+13,24	6,34				95,40	+2,07	161,08	11
12	113,75	+12,58	5,98				125,74	+13,14	6,06				100,83	+2,10	162,03	12
13	113,75	+12,51	6,00				125,74	+13,11	6,02				106,25	+2,14	163,37	13
14	113,75	+12,45	6,03				125,75	+13,10	6,00				111,57	+2,18	164,52	14
15	113,77	+12,42	6,06				125,76	+13,09	5,97				116,91	+2,29	165,19	15
16	113,77	+12,21?	6,05				125,77	+13,09	5,94				122,06	+2,30	165,28	16
17	113,77	+12,37	6,11				125,79	+13,08	5,90				127,19	+2,32	164,81	17
18	113,78	+12,35	6,14				125,83	+13,09	5,84				132,36	+2,20	163,95	18
19	113,79	+12,29	6,16				125,87	+13,08	5,75				137,64	+1,93	162,84	19
20	114,07	+12,34	6,49				126,04	+13,06	5,55				142,89	+1,63	161,82	20
21	114,65	+12,50	7,09				126,42	+13,11	5,35				148,45	+1,17	161,06	21
22	115,69	+12,58	8,07				127,09	+13,07	4,98				153,97	+0,60	160,73	22
23	117,64	+12,05	9,94				128,11	+12,52	4,21				159,51	−0,04	161,03	23
24	121,62	+11,44	13,23				129,37	+12,55	3,84				165,14	−0,62	161,68	24
25	129,39	+10,25	16,68				135,32	+10,95	6,13				170,46	−0,98	162,64	25
26	140,65	+ 8,37	17,29				146,86	+ 7,37	7,59				175,86	−1,34	163,88	26
27	154,02	+ 6,40	16,16				161,68	+ 5,33	7,52				181,29	−1,49	164,94	27
28	168,19	+ 4,83	13,78				177,67	+ 4,22	6,85				186,36	−1,51	165,57	28
29	184,18	+ 3,45	10,99				194,98	+ 3,23	6,30				191,63	−1,58	165,59	29
30	201,05	+ 2,48	8,66				212,77	+ 2,68	6,86				197,10	−1,74	165,06	30
31	218,22	+ 2,12	7,77				230,11	+ 2,79	8,90				202,62	−1,81	164,09	31

people cannot walk swinging forward the arm and leg of the same side simultaneously. Most people can walk in this way for any length of time only after a certain amount of training. Such movement of the limbs is found very tiring by the person walking and looks unnatural to an observer.

Thus, the results reported in the tables of co-ordinates are valid not only for an individual. They also represent general laws of the movements of the limbs in human gait. Conversely, they will probably provide the means to determine quantitatively the slight differences which give each gait its characteristic features.

The tables of co-ordinates immediately provide the answers to many questions. The answer only has to be presented in a clear form. Further transformation or special mechanical interpretation of the data may be necessary to obtain other information. The number of problems that can be solved through the co-ordinates given in the tables is so great that they cannot be dealt with in one study. For the time being, those results will be considered which are either immediately obtainable or do not require long calculation. We shall initially discuss the trajectories described by the different joint centres. Later we shall deal with the movements and deformations of body segments such as the trunk or head. Only the geometrical aspect of movement will be considered. The determination of the velocities and accelerations, the movement of the different centres of gravity and of the total centre of gravity, the analysis of the forces at work inside and outside the human body to produce the movement described in detail in the tables of co-ordinates and other important problems will be the subject of later chapters.

Trajectories of the Joint Centres, Vertex, Centres of Gravity of the Feet and Tips of the Feet

The trajectories of the joint centres, the vertex, the centres of gravity of the feet and the tips of the feet are known from the co-ordinates. They are double-curved and cannot, therefore, be drawn in one plane. In order to construct a good representation of their tridimensional course, it is necessary to project them on at least two non-parallel planes. The co-ordinates immediately give their projections on the three planes of co-ordinates since two of the three tridimensional co-ordinates always represent the bidimensional co-ordinates of a projection. One can thus draw the projections of all the trajectories directly on the planes of co-ordinates. In our system, the x- and z-co-ordinates give the projection on the plane of gait (lateral view), the x- and y-co-ordinates the projection on the horizontal floor (view from above), and the y- and z-co-ordinates the projection on the vertical plane perpendicular to that of gait (coronal view). The last projection, the view from in front or behind, does not offer a very clear picture because the movements parallel to the coronal plane are too small. The projections on the plane of gait and on the horizontal plane better illustrate the process of movement because the successive phases of movement are more distant in the direction of gait.

Figure 22 gives these two projections for experiment I at a scale of 1 : 10, and Fig. 23 the same projections for experiment II. It is thus possible to compare the two recorded processes of movement and see their similarity. The corresponding projections for experiment III which concerns the gait of the loaded subject, are not given here because the deviations caused by loading are too small to appear with sufficient clarity in the small scale of the diagrams. The main differences between the results of experiment III and those of the other two can be seen in special large-scale diagrams (Figs. 28, 36).

Figures 22 and 23 not only indicate the successive positions of the different joint centres, etc., but also give the continuous trajectories by connecting these positions. The vertex, the two shoulder centres and the two hip centres are surrounded by small circles to distinguish them from the other points. Every fifth circle is drawn thicker. The centres of gravity of the feet are marked by small asterisks. Since the trajectories of the wrists and ankles are drawn differently from those of other joints, it is easy to distinguish the trajectories of the three joints of each extremity. The trajectories of the knees and elbows are uninterrupted lines, except at the 31 places of measurement. They are thus different from the trajectories of the ankles and wrists. One can immediately recognize the trajectories of the hips and shoulders since the places of measurement are indicated by small circles.

In order to represent the variations in shape and attitude of the extremities, the long axes of the limb segments are drawn by connecting the adjacent joint centres. The long axis of the foot is not shown, rather the part of its greatest diameter lying between the centre of gravity and a point 3 cm behind the tip. The heel thus lies on the prolongation backwards of this segment, at a distance approximately equal to two-thirds of this segment. The tip of the foot lies in front at about one-quarter of the part of the greatest diameter mentioned above. It must be stressed that the lines drawn in the diagram do not represent the Geissler tubes used for the experiment but the long axes situated inside the limbs. So as to provide a better overall view of the positions of the different limb segments belonging to the same phase of movement, every fifth position is marked by thicker lines, as the hip and shoulder centres are marked by thicker circles. This enables the 31 successive phases to be recognized without having to print the corresponding numbers, thereby making the figure much clearer. Failure of the thicker lines to coincide does not indicate a difference between the results of experiments I and II. It is due to the fact that the initial phase of the part of the movement that was measured was not exactly the same in the two experiments. The second phase of experiment II corresponds roughly to the first phase of experiment I and, more generally, the nth phase of the second experiment to the $n-1^{th}$ phase of the first. Consequently, one could characterize the relationship between the phases of the two experiments by the fact that the thick lines belong to one phase further in the direction of gait in experiment II than in experiment I. Of course this cannot be absolutely exact. If it were it would be pure luck since the two experiments were carried out completely independently of each other. However, we avoided such a demonstration of approximate coincidence between the two series of phases so as not to create a false, too restricted concept of the degree of correspondence. The actual processes of movement correspond, not the arbitrary series of phases used in measuring.

The projection on the plane of gait, the view from the right, needs no further explanation, but the projection on the horizontal plane, the view from above, is less obvious. The successive positions of the lower extremities cover one another in such a way that only a thorough study of the picture provides a clear concept of the movement of the leg as seen from above. To facilitate understanding, the projections of the long axes of the thighs and lower legs of both sides were drawn. The horizontal projection of the foot line was not added to the picture but drawn separately. Using the vertical projection of the foot and the horizontal projection of the ankle, it is not difficult to reinsert the foot in its correct place in the overall picture. No additional means are necessary to clarify the projection of the upper extremities. The situation is sufficiently obvious and a clear picture is given of the movement of the arms as seen from above.

A good deal of practice was required before a clear picture of the tridimensional movement could be obtained from the projections on two planes perpendicular to each other. Therefore, using the co-ordinates of the joint centres, the vertex, the centres of gravity of the feet and the tips of the feet, the author built a tridimensional model directly representing the trajectories of the different points and the attitude of the human body, which changes continuously during walking. The plane of gait was represented by a thin panel of wood. The movement was projected on this panel (Figs. 22 and 23). At each point that represented the projection of a joint centre, etc. a needle was passed perpendicular to the wooden panel. Its length was equal to the magnitude of the corresponding y-co-ordinate. The extremities of all the needles belonging to one and the same joint centre described in space the double-curved trajectory of this joint centre. In order also to illustrate the movement of the long axes of the different segments of limbs and hence the change in the

Ansicht von rechts.

I. Versuch (*⅒ natürlicher Größe*)

Kopfscheitelpunkt.

Linker Schultergelenk
Rechtes Schultergelenk

Linkes Ellbogengelenk
Rechtes Ellbogengelenk

Linkes Handgelenk
Linkes Mittelgelenk
Linkes Zeigefeld
Rechtes Mittelgelenk

Linkes Kniegelenk
Rechtes Kniegelenk

Rechtes Fußgelenk
Linkes Fußgelenk

Längsaxe des rechten Fußes

(*: Schwerpunkt des rechten Fußes, der vordere Punkt der Fußlängsaxe liegt nach 3 cm hinter der Fußspitze; entsprechend links.*) + Notiz

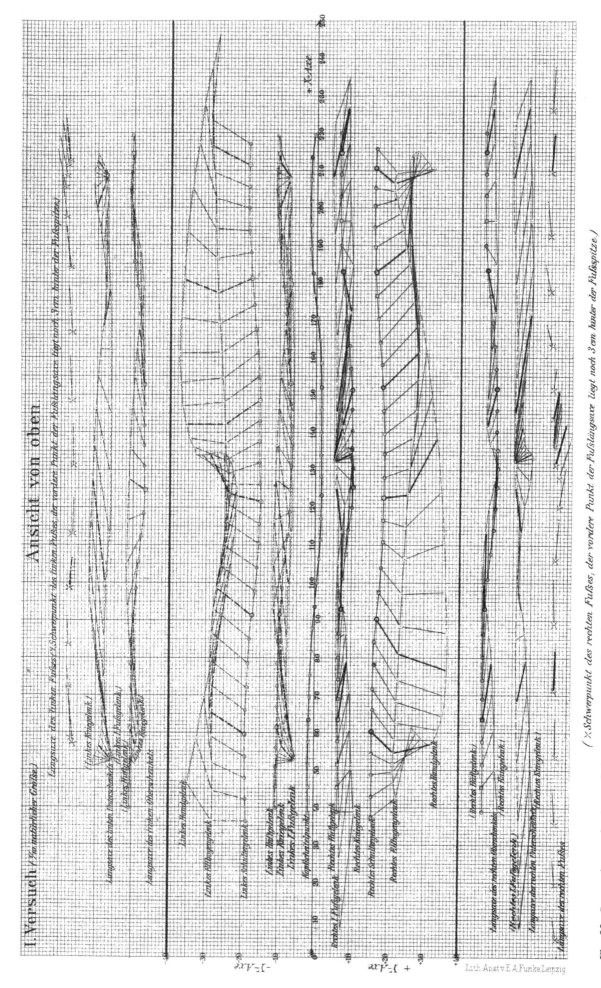

Fig. 22. Successive projections of the vertex, centres of the joints and axes of the body segments in experiment I. *Scale,* $^1/_{10}$

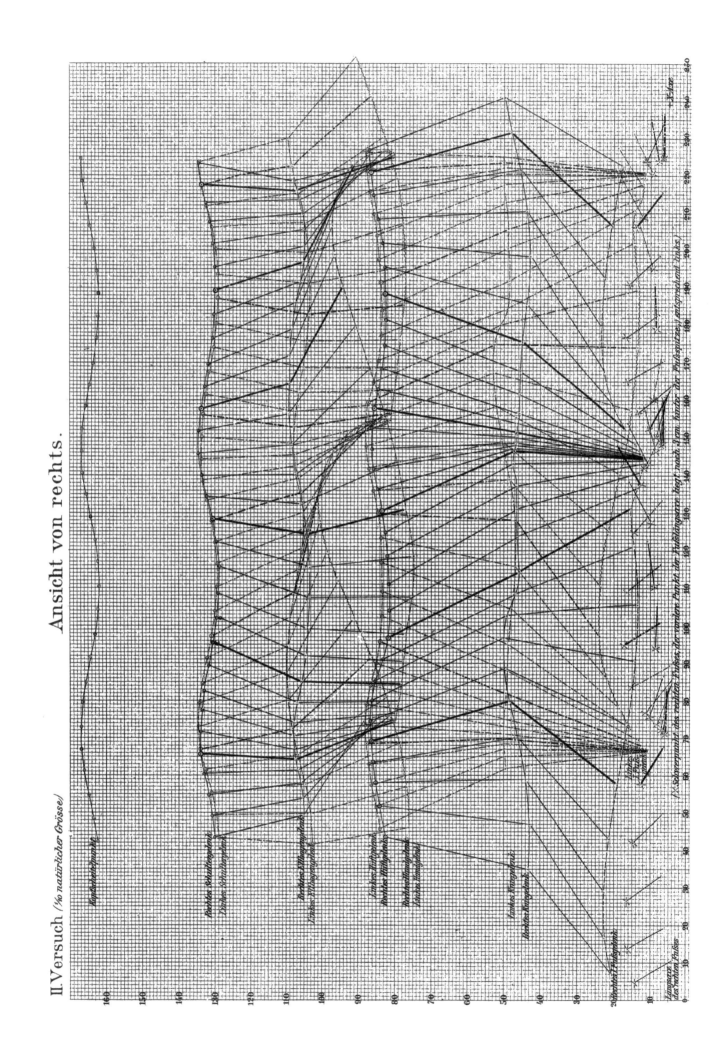

Ansicht von rechts.

II. Versuch (⅒ natürlicher Grösse)

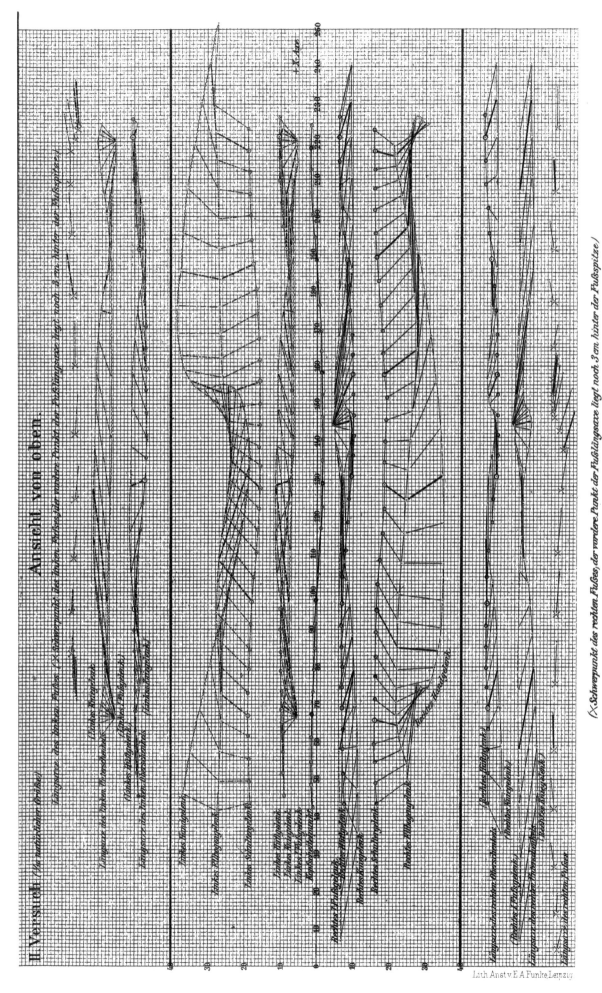

(×Schwerpunkt des rechten Fußes, der vordere Punkt der Fußlängsaxe liegt noch 3 cm hinter der Fußspitze.)

Fig. 23. Successive projections of the vertex, centres of the joints and axes of the body segments in experiment II. *Scale,* 1/10

body attitude, a silk thread was stretched between the two joint centres of a segment for each phase of the movement. The thread belonging to every fifth phase had a special colour. This enabled the position of the limb segments to be easily found for each of the phases of movement. The model was built at a scale of 1 : 10.

From the preliminary model, Mr. E. Zimmermann, mechanic of Leipzig, built a bigger one at a scale of 3 : 10. The panel representing the plane of gait is made of aluminium and interrupted to allow a simultaneous view of the positions of the limbs on both sides. The model consists of five parts: one part shows the trajectory of the vertex; the others represent the movements of the four extremities. Figure 24 shows this model from the left and Fig. 25 from behind. The model records the results of experiment I.

The two projections of the movement in Figs. 22 and 23 as well as the tridimensional model clearly show all the details of the movement within certain limits due to the small scale of the drawing. Further detailed description is thus superfluous. However, some points will be considered which do not appear in the figures with sufficient clarity.

First of all it is necessary to discuss some data concerning time and particularly to determine which phases delineate the different parts of the movement. In experiment I the right heel strikes the floor in the tenth phase. This can be easily recognized if one imagines the line of the right foot in the tenth phase as prolonged backwards to the heel. Of course this and the following data with respect to time can only be accurate to a certain degree. In fact, the heel probably strikes the floor somewhat later, between the 10th and the 11th phase. If the figure does not present any phase for the precise time in question, this time is certainly closer to the lower phase, e.g. closer to the 10th than to the 11th phase. We thus assume that right heel strike occurs in the tenth phase. The error must be less than half the time interval between two successive phases. It must thus be less than 0.038 s : 2, i.e. less than 0.019 s or, more approximately, less than 1/5 s. Left heel strike occurs in the 23rd phase. Consequently, one step in experiment I comprises 13 phase intervals or 0.498 s, approximately 0.5 s. Left toe-off occurs in the 12th phase, right toe-off in the 25th. The period of double support is thus 0.077 s, single support 15 intervals or 0.575 s, and the swinging period 11 intervals or 0.422 s. The step length is roughly 78 cm.

In experiment II the events occur one phase later. Right heel strike occurs in the 11th phase, left heel strike in the 24th. The duration of a step is thus the same, 0.498 s. Left toe-off occurs in the 13th phase, right toe-off in the 26th. Thus, the other data concerning time are the same as in experiment I. The step length is somewhat shorter in experiment II, roughly 77 cm.

In addition to other points, the following was evident from the two experiments. The vertex as well as the centres of the shoulder joints and hip joints describe a double-curved trajectory. This, when projected on the plane of gait as well as on the horizontal plane presents a regular wave line. The two lines will be referred to as "vertical" and "horizontal" wave lines. The vertical wave line possesses a wave-length equal to the length of the step. The wave-length of the horizontal wave line is twice as long, equal to the length of a double step. In the vertical wave line, the segment from the maximum down to the minimum is always traversed more quickly than that from the minimum up to the maximum. This is evident from the number of phases of movement in the intervals. In the horizontal waves the course between two successive sideways maximum deviations is always completed in remarkably the same time. Thus far, the trajectories of the vertex and of the centres of the shoulder and hip joints display the same behaviour.

However, if the five wave lines are analysed more closely essential discrepancies appear. These are obvious in the trajectories of the two centres of the shoulder joints as well as in those of the two centres of the hip joints. It is clear that these discrepancies must be caused by the changes of direction which the shoulder line on the one hand and the hip line on the other undergo in the course of the movement. For instance, the centre of the right hip goes further forwards than that of the left hip in phase 9 of experiment I and in phase 10 of experiment II.

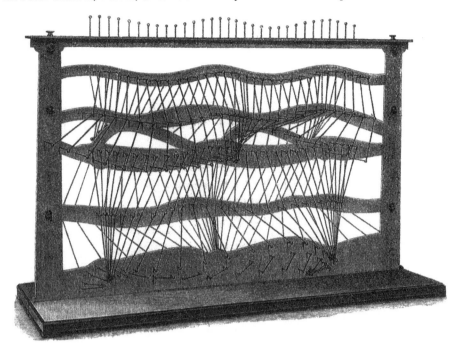

Fig. 24. Tridimensional model representing the attitudes of the human body during walking. (Seen from left)

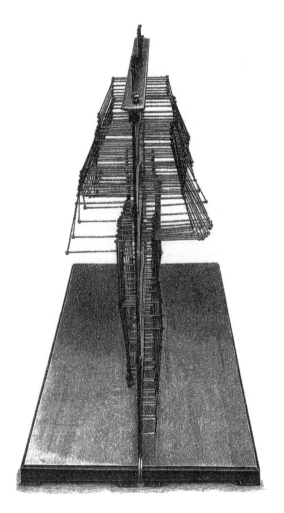

Fig. 25. Back view of tridimensional model representing the attitudes of the human body during walking

The opposite behaviour appears in I, 22 and II, 23 [18]. In the first case, the right leg is extended forwards and the left backwards and the hip line has rotated horizontally counter-clockwise, as seen from above. In the second case, the reverse took place. But this is not the only discrepancy between the two cases under consideration. In I, 9 and II, 10 the vertical distance between the two centres of the hip joints was evaluated to be greater than in I, 22 and II, 23. In the latter case the right hip joint came up closer to the level of the left. Moreover, the hip line must have rotated about an axis parallel to the direction of gait in such a way that the right hip was raised. This rotation is not at all dependent on the fact that the trajectory of the centre of the left hip joint lies overall higher than that of the right hip in both figures. It is difficult to decide whether this behaviour results from a one-sided attitude of the pelvis of the experimental subject or from a slight error in the arrangement of one of the Geissler tubes fixed to the thighs. Even if the latter were the case, the error affecting the shape of the trajectory of the hip and knee joint of one side would have been so slight as to have been imperceptible, despite the accuracy of the measurement. The left hip trajectory and knee trajectory were closely congruent with the corresponding curve of the other side and could be superimposed on the latter after a move backwards by one step and appropriate lowering.

The two phases of movement thus considered show differences in the positions of the hip joints. They also show differences in the positions of the two centres of the shoulder joints. The left shoulder lies further forward at I, 9 and II, 10; the right lies further forward at I, 22 and II, 23. Conversely, in the latter case the right shoulder is raised so that it finally lies higher than the left whereas in the former case the left shoulder was higher. This double behaviour points to a rotation of the shoulder line about a vertical axis and about a horizontal axis parallel to the direction of gait. The consequences of these rotations of the hip and shoulder lines are clearly evident in Figs. 22 and 23. However, it is impossible to follow the overall magnitude and the whole course of these variations with sufficient accuracy in these figures because the scale is too small. This is even more true of the movements of the trunk and head which give rise to differences between the trajectories of the centres of the hip joints and shoulder joints or between the latter and the trajectory of the vertex.

It is thus necessary to make a better analysis of the shape of these trajectories than is possible from the projections at a scale of 1:10. It is above all important to measure accurately the movements perpendicular to the direction of gait since these largely determine the wave shape, the deviations of the trajectories from a straight line parallel to the direction of gait. Since the displacements of the vertex and centres of shoulders or hips parallel to a plane perpendicular to the direction of gait are relatively small, they can be easily reproduced life-size. They actually represent the projection on the third plane of co-ordinates, the plane perpendicular to the direction of gait (plane YZ). We can thus assess the deviations from the straight forward movement without calculation using the y- and z-co-ordinates. This was carried out at life size in Fig. 26 for the two experiments for the vertex and centres of the shoulder and hip joints and for the midpoints of the shoulder and hip lines in such a way that a view in the direction of gait, i.e. from behind, was obtained.

These diagrams represent the repeating curves for a precisely straight forward movement of the experimental subject on a perfectly horizontal floor. They are very appropriate to illustrate the trajectories in question. If the velocity of the points along the curves in the direction of gait were constant and known, this projection alone would be sufficient to give the precise shape of the double-curved trajectories. One would only have to imagine that the points in question on the human body travel along the curves in Fig. 26 in the direction of the arrows, with the variable velocity evident from the varying distance between the successive phases. Thus, the figures would simultaneously recede from the eyes of the observer in the

---

[18] This signifies phase 22 of experiment I and phase 23 of experiment II. In the following, the phases of movement of the two experiments will be referred to in this abbreviated way

direction of gait, i.e. perpendicular to the plane of the drawing, with the known velocity of the forward motion. Since each of these points does not progress in the direction of gait with constant velocity, its trajectory is more contracted at some places and stretched at others than would be expected from the way it is produced. This does not change essentially the shape of the trajectory. In every case the trajectory has to run on a cylinder the generatrix of which is perpendicular to the plane of the drawing and the cross section of which is represented by the closed curve of the ideal case drawn in Fig. 26. This shows also that the concept of Carlet, following his analysis of the shape of the tridimensional curve describing the upper point of the pubic symphysis (p. 4), cannot be completely right. If it were correct, this curve could be inscribed in one furrow with its convexity pointing downwards. The curves of Fig. 26, and above all the curve corresponding to the midpoint of the hip line, would then represent a single segment, passed through twice, on a roughly elliptical curve. However, Carlet's interpretation does not deviate too much from reality.

The curves relating to the head, shoulders and hips present characteristic differences. Comparison of the three middle curves, that of the vertex and those of the midpoints of the shoulder and hip lines, show that the first two curves have almost the same height, about 5.5 cm and the same width, about 3.5 cm, but that the curve of the midpoint of the hip line is lower and wider. The two segments of curves, between phases 5 and 17 and phases 17 and 30 in experiment I, between phases 6 and 18 and phases 18 and 31 in experiment II, are of unequal height in all the curves. The first of the two segments is lower. This discrepancy cannot result from an asymmetrical gait only, otherwise it could not occur in all the middle curves and in the two experiments in such a concordant way. Right heel strike occurs in the middle of the first segment of the curves, left heel strike in the middle of the second segment. The experimental subject may lower his whole body more before right heel strike than before left heel strike. Thus, a slight limp on the right side appears in the curves. This cannot be seen when observing the movement because it is only a matter of a few millimetres. If the two legs were used equally, these two segments of the three middle curves would show complete symmetry.

The segments of the other curves related to the shoulder and hip joint centres are also not symmetrical in this ideal case. As shown in the curves of Fig. 26, the minimum of the first segment is lower for the curves of the right side of the body, that of the second segment is lower for the curves of the left. For the right shoulder the point where the second segment follows the first is higher, whereas on the left side the point where the second segment rejoins the first is higher. The opposite is true for the hips.

The minimum of the first segment of the curve lies at I, $10\frac{1}{2}$; II, 12 for the point on the head, at I, 11; II, $12\frac{1}{2}$ for the three shoulder points and at I, 10; II, 12 for the three hip points. This minimum occurs at the moment of or (e.g. for the shoulder points) shortly after right heel strike, in the double-support period of walking. The second minimum corresponds to the phases I, $23\frac{1}{2}$; II, 25 for the point on the head, I, 24; II, $25\frac{1}{2}$ for the shoulder points and I, 23; II, 24 for the hip points. This means that the second minimum of the hip curves coincides exactly with left heel strike whereas the corresponding minimum of the other four points takes place a short time after this moment, later for the shoulder points than for the point on the head. This minimum thus occurs in the double-support period for the seven curves. The first minimum for the point on the head and for the shoulder points lies to the right of the second; for the hip points it lies to the left of the second minimum. Therefore, the two curve segments are brought closer together in the latter than in the former. Consequently, at right heel strike the upper part of the trunk and the head tilt to the right. The opposite occurs at left heel strike; the trunk is then tilted to the left.

The points at which the two segments of curve meet – I, $4\frac{1}{2}$ [19]; II, $5\frac{1}{2}$ and I, $17\frac{1}{2}$, II, $18\frac{1}{2}$ – represent the maxima of the vertical wave lines. They correspond to the moment when the

---

[19] This moment thus lies between the fourth and fifth phases of movement

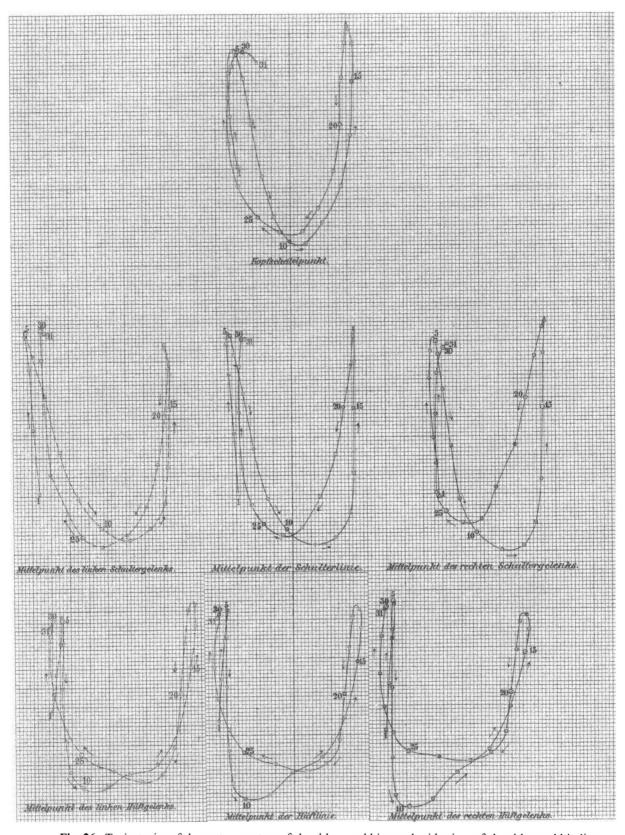

Fig. 26. Trajectories of the vertex, centres of shoulders and hips and midpoints of shoulder and hip lines

II. Versuch.

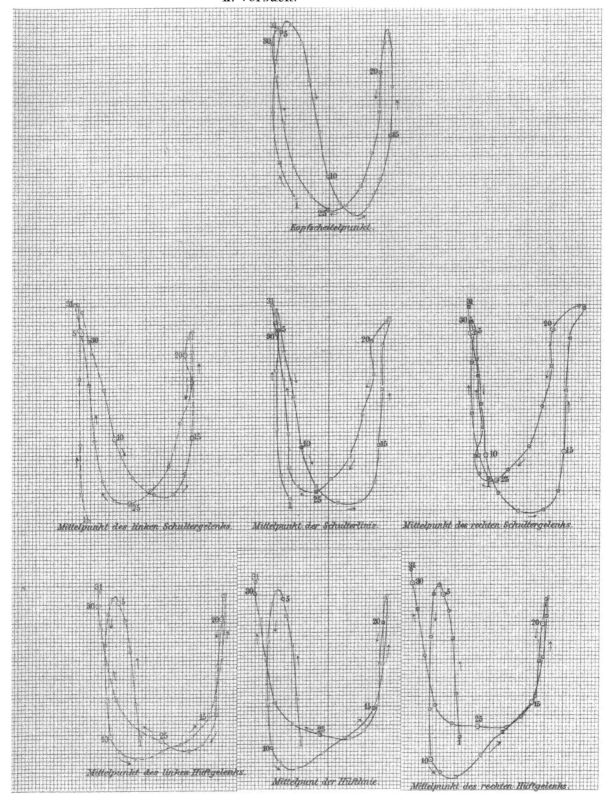

Kopfscheitelpunkt.

Mittelpunkt des linken Schultergelenks.   Mittelpunkt der Schulterlinie.   Mittelpunkt des rechten Schultergelenks.

Mittelpunkt des linken Hüftgelenks.   Mittelpunkt der Hüftlinie.   Mittelpunkt des rechten Hüftgelenks.

centre point of the hip joint of the supporting leg lies above the centre of gravity of the loaded foot (see the vertical projections in Figs. 22, 23). This also occurs almost simultaneously with the centre of gravity of the foot of the swinging leg crossing the plumb line through the centre of the ipsilateral hip joint. These maxima thus correspond to the two legs simultaneously being in the vertical position.

From the above-mentioned normal asymmetry of the segments of curves related to one of the shoulder or hip centres, it appears that, for the shoulders and hips of both sides of the body, the lower minimum of the vertical wave line occurs at the moment of or shortly after the homolateral heel strike. Ideally, the higher maximum of the shoulder curves occurs when the supporting leg of the same side is in the vertical position. That of the hip curves occurs when the homolateral swinging leg is vertical.

A further difference can be seen between the shoulder curves and hip curves. In the shoulder curves, as in the head curve, the two curve segments intersect each other once in the projection; in the hip curves they intersect each other three times.

Figure 26 also shows the greatest sideways deviations, i.e. the maxima and minima of the horizontal wave lines. They correspond to where the curves are furthest away from the vertical middle line. For the head point this occurs at I, 15; II, 16 to the right and at I, 3; II, 3 and I, 28; II, 30 to the left. For a straight forward movement the left minima should be at I, 2 and II, 29 instead of I, 3 and II, 30. The subject did not usually maintain the prescribed direction in a completely precise manner over the duration of the displacement. Therefore, the positions of the maxima of the horizontal wave lines are less constant than those of the vertical wave lines. Whereas in the trajectory of the vertex the horizontal maximum to the right and to the left occurs over two phase intervals, i.e. 0.08 s, before the vertical maximum, for the curves of the shoulder points the horizontal maximum seems to coincide with the vertical maximum. Here, the curve before the maximum presents almost a straight course in the projection in Fig. 26 as well as in the horizontal projection (Figs. 22, 23). For the hip line the horizontal maximum again lies before the vertical maximum but not so much as for the vertex, at only 1 1/2 phase intervals roughly 0.05 s. The horizontal maximum is found to the right at I, 16; II, 17, to the left at I, 1; II, 1 and I, 28; II, 31. The latter data clearly show the disturbing influence on the position of the horizontal maxima of a gait which is not absolutely straight; otherwise, these figures would read I, 3; II, 4 and I, 29; II, 30.

If one disregards the irregularities of the curves resulting from the individual asymmetrical and not precisely straight gait, the seven curves of Fig. 27 show a typical course. On these curves right heel strike is marked by R and left heel strike by L. The type of these curves is probably the same for every person. Differences would consist of variations in height and width and in different proportions of the two dimensions. They would also appear, as with our experimental subject, as asymmetries of the two parts of the curves. These asymmetries would be particularly common in pathological gaits.

The seven types represented in Fig. 27 show characteristic differences from each other. These differences result from particular movements and rotations of the trunk and head which have to be more closely analysed. Related questions can also be answered from our tables of co-ordinates with the assistance of a few calculations.

Rotations and Deformations of the Trunk

Three parameters have to be considered in the movements of the trunk: the changes in orientation of the hip line, the shoulder line and the line connecting the midpoints of the hip and shoulder lines, i.e. the trunk line. Through reciprocal rotations of these three lines the movements and modifications in shape which the trunk undergoes during walking are known as far as they can be determined by our method of research in which no other points of the trunk were measured than those of the shoulder and hip joints.

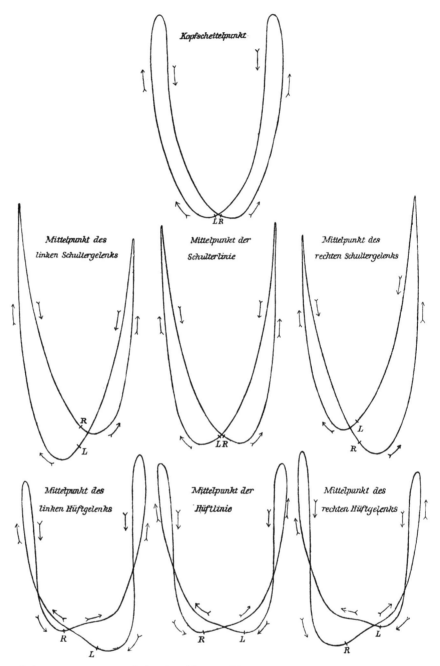

Fig. 27. Trajectories of the vertex, centres of the shoulders and hips, and midpoints of shoulder and hip lines

For clarity, it is convenient to regard the changes in direction of the hip line as rotations about its midpoint, the changes in direction of the shoulder line as rotations about its midpoint and the changes in direction of the trunk line as rotations about the lower extremity of this line, i.e. about the midpoint of the hip line. If we consider also the movement of the midpoint of the hip line on its double-curved trajectory which has been analysed above, we have a clear picture of the whole movement of the trunk. The movement of the two centres of the hip joints is determined by the trajectory of the midpoint of the hip line in conjunction with the rotations of the hip line. The same trajectory of the midpoint of the hip line and the rotations of the trunk line indicate the trajectory of the midpoint of the shoulder line. The latter in conjunction with the rotations of the shoulder line shows the movement of the two centres of the shoulder joints. At the same time the rotation of the trunk line about the midpoint of the hip line offers a clear representation of the tilting forwards, backwards and sideways of the whole trunk during walking. A comparison of the rotations of the hip line and shoulder line about their midpoints reveals the rotation and simultaneous deformation of the trunk.

Finally, the movement of the vertex can be determined in relation to the midpoint of the shoulder line. Then, at least an approximate overall view of the movements of the head on the trunk is obtained, even though the head does not exactly turn about the midpoint of the shoulder line as it would about a fixed articular centre. The movement of the head in relation to the trunk cannot be represented as a rotation about a fixed point of the trunk since the cervical spine undergoes considerable bending in the process. In order to determine the precise movement of the head in relation to the trunk, it would have been necessary to employ more points on the head, such as the ear openings. This was omitted since the movements of the head during walking are not of major importance.

Rotations of the Hip Line

The rotation of the hip line about its midpoint is known if the movement of one of the two hip centres relative to the midpoint of the hip line has been determined. The co-ordinates of this relative movement are the differences between the co-ordinates of one of the hip centres and those of the midpoint of the hip line in the tridimensional system of co-ordinates recorded in the preceding tables. For instance, if A represents the right hip joint centre and B the midpoint of the hip line in Fig. 21, and if B is considered as the origin of the co-ordinates, the system of co-ordinates can be thought of as displaced to this point, parallel to itself. The three tridimensional co-ordinates of the point A are $x'_a - x'_b$, $y'_a - y'_b$ and $z'_a - z'_b$, according to the designations chosen earlier. The co-ordinates of the centre of the right hip relative to the midpoint of the hip line were thus determined from the previous co-ordinates for the two first experiments. They are given in Table 17.

The direction of the hip line deviates only slightly from that of the Y-axis of the co-ordinates during its rotations. This appears in the table from the almost constant value of the y-co-ordinates and from the small values of the other two co-ordinates. Therefore, one obtains an adequate view of the oscillations of the hip line by analysing only the variations in the x- and z-co-ordinates.

The variations in the z-co-ordinates illustrate the vertical rising and lowering of the centres of the hips and also the rotation of the pelvis about a horizontal axis through the midpoint of the hip line. This axis must not necessarily lie in the direction of gait although it does not deviate much from it. The axis forms with the direction of gait the same angle by which the pelvis simultaneously rotates about the vertical axis from its normal position. In this normal position the hip line is perpendicular to the direction of gait. In order to show the elevations and lowerings, it is appropriate to represent them in a diagram in which the magnitudes of the z-co-ordinates are drawn as ordinates at the same distance from each other, perpendicular to a horizontal abscissa line. These equal distances on the abscissa can be chosen at will. However, they must not be disproportionate to the magnitudes of the ordinates. If these distances are too long, the curve connecting the ordinate values will be too flat to show clearly the details of the elevation and lowering. Considering the magnitudes of the vertical rising and lowering, it is appropriate to separate the points along the abscissa by 1 mm. In this manner, the upper curve shows the elevations and lowerings of the centre of the right hip, the lower curve those of the centre of the left hip, in graph 1 of Fig. 28 for experiment I and in graph 9 of the same figure for experiment II. These curves have a similar shape except for a deviation at the beginning and end of graph 9. This can be attributed only to the fact that the co-ordinates of the centres of the hip in experiment II were not found by direct measurements for each of the four successive phases 2–5 and 27–30, but by interpolation. From the previous data, one could expect that the maxima and minima of the curves of experiment II would be displaced 1 mm (one phase) to the right further than those in experiment I. The curves of the centre of the left hip form, naturally, mirror images of those of the right hip in relation to a horizontal plane. When the right hip is raised relative to the midpoint of the hip line, the left must be lowered equally and inversely.

If the curves are projected onto a vertical line at the side, an accurate representation of the relative vertical movement of the centres of the hip joint is obtained, since the movement occurs along the same vertical line. The lateral resolution is intended to illustrate the process of the movement in the same way as the vibrations of a tuning fork are best observed when printed on the drum of a kymograph. In order to understand the vertical oscillations of the hip line itself, one must imagine the midpoint of this line to be in the direction of the axes of the wave lines of graphs 1 and 9, to the side of the vertical line on which the movement is

Table 17. Co-ordinates of the centre of the right hip relative to the midpoint of the hip line

No.	Experiment I			Experiment II		
	x	y	z	x	y	z
1	− 0,45	+ 8,47	− 1,03	− 1,14	+ 8,39	− 0,77
2	0	+ 8,46	− 0,89	− 0,15	+ 8,42	− 0,62
3	+ 0,42	+ 8,45	− 0,71	+ 0,30	+ 8,43	− 0,60
4	+ 0,78	+ 8,45	− 0,59	+ 0,30	+ 8,44	− 0,65
5	+ 0,98	+ 8,43	− 0,53	+ 0,40	+ 8,46	− 0,67
6	+ 1,05	+ 8,44	− 0,59	+ 0,41	+ 8,47	− 0,59
7	+ 0,99	+ 8,42	− 0,77	+ 0,51	+ 8,46	− 0,54
8	+ 0,92	+ 8,39	− 0,99	+ 0,62	+ 8,46	− 0,66
9	+ 0,98	+ 8,38	− 1,07	+ 0,56	+ 8,43	− 0,90
10	+ 1,04	+ 8,38	− 1,01	+ 0,54	+ 8,41	− 1,07
11	+ 1,15	+ 8,37	− 0,95	+ 0,81	+ 8,40	− 1,03
12	+ 1,03	+ 8,41	− 0,73	+ 0,96	+ 8,39	− 1,01
13	+ 0,77	+ 8,44	− 0,58	+ 0,90	+ 8,40	− 0,83
14	+ 0,57	+ 8,45	− 0,47	+ 0,76	+ 8,44	− 0,60
15	+ 0,23	+ 8,47	− 0,62	+ 0,48	+ 8,47	− 0,58
16	− 0,23	+ 8,47	− 0,83	+ 0,25	+ 8,46	− 0,60
17	− 0,51	+ 8,43	− 0,92	− 0,01	+ 8,47	− 0,73
18	− 0,68	+ 8,42	− 0,93	− 0,33	+ 8,46	− 0,80
19	− 0,71	+ 8,43	− 0,86	− 0,61	+ 8,46	− 0,81
20	− 0,77	+ 8,43	− 0,75	− 0,83	+ 8,42	− 0,83
21	− 0,93	+ 8,42	− 0,58	− 0,95	+ 8,42	− 0,70
22	− 1,11	+ 8,41	− 0,47	− 1,10	+ 8,41	− 0,53
23	− 1,16	+ 8,40	− 0,49	− 1,23	+ 8,40	− 0,43
24	− 1,24	+ 8,38	− 0,59	− 1,24	+ 8,40	− 0,46
25	− 1,08	+ 8,40	− 0,74	− 1,37	+ 8,38	− 0,57
26	− 0,90	+ 8,42	− 0,84	− 1,11	+ 8,39	− 0,75
27	− 0,68	+ 8,44	− 0,96	− 0,88	+ 8,39	− 0,85
28	− 0,35	+ 8,44	− 0,96	− 0,65	+ 8,38	− 0,65
29	− 0,05	+ 8,44	− 0,77	− 0,43	+ 8,35	− 0,51
30	+ 0,23	+ 8,48	− 0,59	− 0,08	+ 8,30	− 0,48
31	+ 0,33	+ 8,48	− 0,54	+ 0,42	+ 8,28	− 0,48

The relative co-ordinates of the centre of the left hip joint differ from those of the right only by the plus/minus sign and so do not appear

I. Versuch
(ohne Belastung)

II. Versuch
(ohne Belastung)

III. Versuch
(mit Belastung)

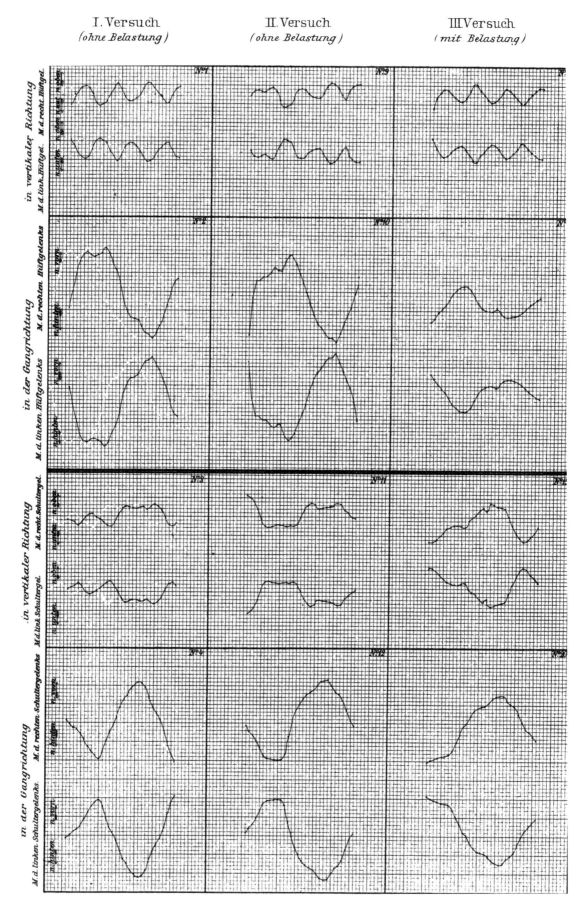

Fig. 28 a–d. Movement of **a** centres of shoulders relative to midpoint of shoulder line; **b** centres of hips

I. Versuch
(ohne Belastung)

II. Versuch
(ohne Belastung)

III. Versuch
(mit Belastung)

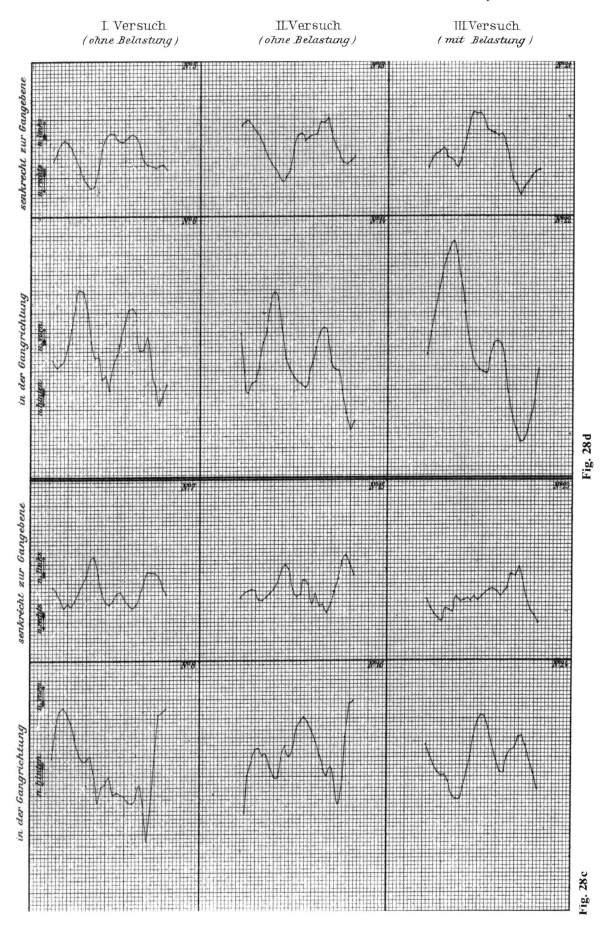

Fig. 28d

Fig. 28c

projected, for the centre of the right hip at a distance of about 8.5 cm to the left of the vertical and for the centre of the left hip at the same distance to the right of the vertical. If the midpoint of the hip line is thought of as connected with the points moving vertically up and down, one obtains a clear concept of the vertical oscillation of the right or left half of the hip line as seen from behind. The extremity of the hip line does not actually move on a straight line but on an arc with a radius of 8.5 cm and with the midpoint of the hip line as the centre. However, this cannot disturb the picture because of the smallness of the displacements. One could also obtain a view of the oscillation of the hip line by imagining the midpoint of this line as being 8.5 cm behind the plane of the graph. Depending on whether the upper or lower curve is used, one obtains the view of the rotation of the hip line from the right or from the left. The curves in Fig. 28 represent the actual sizes of the vertical up and down movements of the centres of the hips. The vertical oscillations of the hip line were obviously very small in our experimental subject. This is a further demonstration that the accuracy at which we aimed was actually achieved; otherwise, these small curves would not show such a constant course and the curves of the two experiments would not coincide to such a surprising extent. If the ordinate of one phase were changed by only 0.5 mm, the regular course of the curve would be considerably disturbed. In the interpolation of a wide gap, as at the beginning and end of experiment II, unavoidable errors immediately appear obvious in graph 9.

The results that appear from the graphs are remarkable. The curves are very similar to sine curves, which suggests that the hip line oscillates about its midpoint in the vertical plane, i.e. about a horizontal axis, and that these oscillations correspond to the simple oscillations of an elastic rod fixed at one extremity like a tuning fork. The period of such an oscillation is not equal to the duration or to a multiple of the duration of a step, but there are always three oscillations of the hip line during two steps. A step extends over 13 phases of movement, a double step over 26 phases. During 26 phases, for example from 4 to 30, three wave-lengths appear in the curves. We then look for the phases to which the maxima and minima, i.e. the moments at which the oscillation inverts its direction, belong. A maximum is found at I, 5, 14, $22\frac{1}{2}$ and II, 7, 15, $23\frac{1}{2}$ and a minimum at I, 9, $17\frac{1}{2}$, $27\frac{1}{2}$ and II, 10, 19, 27. If the discrepancies caused by inaccurate interpolation in experiment II did not exist, thus if instead of II, 7 and II, 27 we had II, 6 and II, $28\frac{1}{2}$, there would be complete correspondence of the results for both experiments.

At which phases do the vertical wave lines of the hip centres reach their maxima and minima? At which phase does each foot strike the ground? At the vertical maximum of a hip trajectory, which falls in the swinging period of the homolateral leg, the hip line is raised the highest on this side. It then goes down with further movement and is most inclined shortly, about 0.04 s, before heel strike. At heel strike, the hip line on the ipsilateral side is already on its way up and again reaches its maximum about 0.1 s after heel strike. It falls again and shows its greatest downward tilt to the right at the moment when the trajectory of the hip reaches its second vertical maximum. When the hip line is low on the right, it is of course raised on the left. In the trajectory of the left hip, the vertical maximum occurring in the period of swing corresponds to the maximum of the vertical wave line of the right hip, which occurs during the right single-support period. There appears to be a complete coincidence between right and left since, for this maximum of the left vertical wave line, the left hip is raised to its highest point. In fact, with further movement, until the next left maximum, the behaviour of the hip line as seen from the left corresponds exactly to what has been described earlier for the right side. To summarize, the hip line carries out three half oscillations about a horizontal axis during one single step in such a way that at each side the maximum elevation always coincides with the maximum of the vertical wave line of the hip of the swinging leg. Conversely, at the moment of the maximum of the trajectory of the hip, when the leg rests on the ground, the hip line of this side is at its lowest.

Giraud Teulon and Richer [20] disagree with one another with regard to the inclination of the hip line. The former asserts that the hip line is raised on the side of the swinging leg. The latter claims that the hip line falls towards the swinging leg. Our results enable us to conclude that both are partially right and wrong. Richer is correct for the first and last thirds of the swinging period; Giraud Reulon is correct for the middle third. Both are wrong in implicitly assuming a similar behaviour of the hip line in inclination during the whole period of swing.

The vertical oscillations of the hip line explain the discrepancies in the trajectories of the two hip centres.

Apart from the systematic one-sided inclination of the hip line caused either by the attitude of our experimental subject or by an inaccuracy in the fixation of the Geissler tubes, the beginning of the first segment of curve I, 5; II, 6 lies higher on the curve of the right hip than on that of the left (Figs. 26, 27). This results from the fact that in the position I, 5; II, 6 the hip line is elevated to the right. However, the curve is lower on the right than on the left until I, 9; II, 10 because the sideways movement of the right hip centre is increased by the simultaneous lowering of the hip line, whereas that of the left hip is decreased. The greater lowering of the whole first segment of the curve on the right must also be attributed to this circumstance. From I, 10; II, 11 and further on, the curve rises more quickly on the right than on the left. After two further phase intervals it has again reached the same height as on the left, and after two more intervals it is much higher on the right. Finally, at I, 17; II, 18 the right segment is again behind the left. This results from the fact that in the last stage of the elevation of the first curve segment, the hip line rises on the left and, therefore, causes a much more considerable increase in the segment of the left hip. The second segment of the curve does not fall so low to the right as it does to the left although its initial point is lower to the right. This is due to the simultaneous elevation of the hip line to the right. After having reached its lowest point at I, 23; II, 24, it climbs slowly at first, a little more steeply to the left than to the right, because now the hip line rises to the left. In the second half of the rising segment, the velocity predominates again on the right in such a way that, on the whole, the segment of the curve is higher on the right than on the left. This brings us again to the starting point of the first segment of the curve. Thus, the discrepancies in the curves for the two hips in Figs. 26 and 27 are explained by the vertical oscillations of the hip line.

In addition to the vertical plane, the hip line carries out rotations in the horizontal plane, i.e. rotations about a vertical axis passing through its midpoint. The magnitude and direction of this rotation are determined by the x-co-ordinates given in Table 17. To illustrate these rotations, the magnitudes of the x-co-ordinates must be drawn as ordinates separated from each other by 1 mm on the axis of the abscissa. This is shown in graph 2 of Fig. 28 for experiment I and in graph 10 for experiment II. The upper curve represents the centre of the right hip, the lower that of the left hip. Except for small variations at the beginning and end, the curves for the two experiments once again coincide as completely as one can expect from two experiments carried out independently of each other and from the small magnitude of the ordinates considered. The reason for the differences at the beginning and end has been explained above.

These diagrams present essentially the shape of a simple wave line (sine curve). They deviate somewhat from this shape just before each maximum and minimum, and after each maximum and minimum they are somewhat steeper than a sine curve. These small oscillations can be observed in both experiments and cannot, therefore, be attributed to inaccuracies in the measurements. They must be due to certain mechanical factors. The wave-length of these wave lines is exactly three times that in graphs 1 and 9 discussed above. The rotations of the hip line about the vertical axis thus more or less represent simple oscillations, the duration of which is equal to that of a double step.

---

[20] Richer P: La station et la marche chez l'homme sain et chez les malades myopathiques. Revue scientifique, 4th series, vol 2, no 4, p 104

A maximum of the right wave line is found at I, 11; II, 12 and a minimum at I, 24; II, 25. These correspond to the moments shortly before right or left heel strike.

In order to illustrate the oscillations of the hip line in the horizontal plane, one must imagine each wave line as projected on a vertical line at the side. The midpoint of the hip line is assumed to be fixed and in the direction of the axis of the wave line at the side of this vertical, 8.5 cm to the left of the vertical for the centre of the right hip and 8.5 cm to the right of the vertical for the centre of the left hip. If one imagines the midpoint of the hip line situated on the side and connected with the point moving up and down on the vertical, one gets a clear picture of the horizontal oscillations of the right and left halves of the hip line as seen from above. For the view from the right or left, one must assume the fixed midpoint of the hip line to be 8.5 cm behind the plane of the graph and use the upper curve in the first instance, the lower in the second.

From the magnitude of the ordinates in the diagram, it can be seen that the horizontal oscillations of the hip line are greater in range than the vertical. They are approximately five times greater. Whereas the proportion between the durations of the horizontal and vertical oscillations is the same for each individual, 1 : 3, the proportion between the amplitudes of the oscillations is different for each individual.

The following can be derived from the position of the maxima and minima in graphs 2 and 10 of the horizontal oscillation. The hip line always rotates in such a way that its extremity which goes forwards is that on the side of the swinging leg. Shortly after heel strike, the direction of rotation is inverted and the centre of the opposite hip moves forward. Slight deviations from the pure shape of the sine curve in the graphs indicate that the horizontal rotation of the hip line does not exactly represent a simple oscillation. These deviations result from the movement of the knee of the swinging leg. At present, discussion of the probable interdependence of the two phenomena would be too much of a digression.

The horizontal oscillations of the hip line cause practically no deviations in the projections of the hip trajectories in Fig. 26 since the centres of the hips move almost perpendicular to these projected curves during the oscillations. However, at certain places, they stretch the trajectory of a hip centre in the direction of gait, at other places they contract it. In each hip trajectory, therefore, the segment corresponding to the swinging of the homolateral leg corresponds to a stretched wave line. The wave-length of this segment is greater than that of the segment corresponding to the support of the leg by twice the total deviation of a hip joint resulting from the horizontal oscillation.

The oscillations of the hip line about both a horizontal and a vertical axis take place simultaneously and combine into a complicated type of oscillation. The process can be imagined by assuming that the axes are fixed to the pelvis and both change their direction with the rotation. It can also be assumed that the axes maintain both their vertical direction and the horizontal direction of gait for all the positions of the pelvis. They then change their relative position in the pelvis with the rotation. Since the amplitude of the two oscillations is small, the two ways of presenting the process are not very different. It is much easier to combine oscillations whose tridimensional directions are fixed in space. We shall, therefore, use the latter approach.

In the present case the problem consists of combining two oscillations about two axes perpendicular to each other. The oscillation about the vertical axis presents a period three times longer and an amplitude five times greater than that about the horizontal axis. It is assumed that the intersection of the two axes (midpoint of the hip line) is fixed. We obtain a complete insight into the resultant oscillation of the hip line if we know the trajectory of any point on this line. For clarity, the point should be chosen as far from the centre of oscillation as possible so that the trajectories are not too short. Each hip centre carries out horizontal excursions of about 2.5 cm (Fig. 28, graphs 2 and 10) and vertical excursions of only 0.5 cm (graphs 1 and 9). If we choose a point on the hip line, on the right, four times further from the midpoint of this line than is the centre of the right hip, the excursions of this point in the two

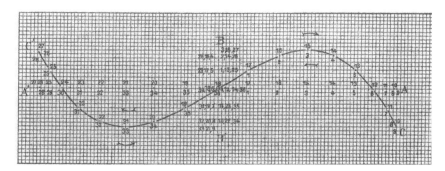

Fig. 29. Trajectory of a point on the prolongation of the hip line about a vertical ($A'OA$) and a horizontal (BOB') axis

perpendicular directions are four times greater than those of the hip centre, i.e. 10 cm and 2 cm. The trajectory of this point for the oscillation about the vertical axis is $A'OA$ (Fig. 29), and its trajectory corresponding to the oscillation about the horizontal axis BOB'. Whilst the point oscillates once between A and A' and back, it will have oscillated three times between B and B' and back.

It is initially supposed that we are dealing with single oscillations (sine oscillations) in the two cases. If the duration of the horizontal oscillation is divided into 36 equal parts, at the end of each of these the point would pass through the places marked by continuous numbering on each of its two trajectories. It is assumed that the point on the two trajectories passes simultaneously through the midposition. The time is reckoned from such a moment. If the point undergoes the two oscillation movements at the same time, it must follow the curved trajectory $C'OC$ drawn in Fig. 29 in such a way that it goes via $OCOC'O$ on this trajectory, during one horizontal oscillation counted from O. This can easily be seen in the figure. While the point oscillating vertically goes from O to B, it goes back only half the segment OA in the horizontal oscillation. Its curved trajectory rises to the midpoint of the upper side of the rectangle formed by OA and OB. Since the movement in the direction of OA slows down while the point comes back from B to O in the vertical direction, the next downward segment of the curve to the horizontal OA is shorter than that going up from O. This curve drops further to point C because of the further vertical movement downward from O to B'. The movement on OA in the second half of the trajectory becomes so slow that to complete this distance the same time is necessary as that taken to move vertically from B to B'. The point will only be halfway back from A to O while it will have gone in the other direction over the whole segment $B'B$. This can be deduced from the numbers in Fig. 29. Thus, the point will return on the same curved path on which it came to C. The further trajectory of the point needs no explanation since it can be easily followed in Fig. 29.

This resultant trajectory does not correspond exactly to the actual conditions. It is correct for the movement of a point on the hip line which passes simultaneously through the middle position during the two oscillations. But the horizontal oscillation deviates somewhat from the shape of a sine curve. The segment OA is traversed more slowly in the direction $\overline{OA}$ than in the opposite direction $\overline{AO}$. Correspondingly, the other segment OA' is traversed in the direction $\overline{OA'}$ with a lower velocity than in the direction $\overline{A'O}$. This is shown in graphs 2 and 10 of Fig. 28. While the point has moved vertically from O to B and then back from B to B' through O, in its horizontal oscillation from O it has not yet arrived at A. Therefore, point C to which the curve segment goes down must occur to the left of the position indicated in Fig. 29, roughly at the place assumed in Fig. 30. While the point comes back vertically from B' it must go horizontally over the missing segment to A in such a way that the trajectory describes an arc concave upward before joining the arc convex upward going back after O

Since the same conditions repeat themselves for reasons of symmetry in the further movement, the trajectory described by a point of the hip line presents the shape drawn in Fig. 30.

For clarity, we considered the trajectory of a point on the hip line which was four times further away from the midpoint of this line than was the centre of the right hip. The trajectory of the centre of the right hip, corresponding to the simultaneous oscillation about a vertical and a horizontal axis, must be exactly the same but its magnitude must be four times smaller. The truth of this is evident when considering the movement of the centre of the right hip relative to the midpoint of the hip line projected on the plane of gait. The hip line is on average perpendicular to this plane and deviates only slightly from this direction as a consequence of its small oscillations. Therefore, the projection of the trajectory of the centre of the hip on the plane of gait is almost the same in shape and magnitude as the trajectory itself. This projection is drawn using the x- and z-co-ordinates of Table 17 and is shown life-size for experiment I in Fig. 31. Figure 30 appears in reduced form in the upper half of Fig. 31 for the purposes of comparison. The numbers marking the different points of the lower curve in Fig. 31 signify the phases of walking whereas the numbers in Fig. 29 correspond to the arbitrary resolution of the duration of one horizontal oscillation into 36 equal parts. Except for minor differences, the lower curve of Fig. 31 corresponds in shape to the curve in Fig. 30 and is four times smaller. The above discussion thus has resolved mechanically the movement of the hip joint relative to the midpoint of the hip line and the resultant oscillation of the hip line, which have been given their simplest expression.

The movement of the pelvis during walking is not completely determined by the rotations of the hip line. It is possible and even probable that besides the movements which express themselves in rotations of the hip line, the pelvis also carries out rotations about the hip line. However, these cannot be directly deduced from our measurements since we unfortunately neglected to mark and use for measurement a point on the pelvis, outside the hip line, in addition to the two points on the latter.

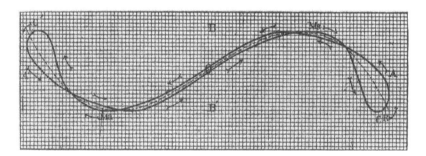

Fig. 30. Trajectory of a point on the prolongation of the hip line

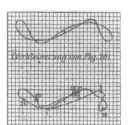

Fig. 31. Projection of the trajectory of the centre of the hip on the plane of gait (*lower curve*). The *upper curve* is a reduction of Fig. 30 and is included for comparison

Rotations of the Shoulder Line

The co-ordinates of the movement of a shoulder centre relative to the midpoint of the shoulder line are arrived at by subtracting the tridimensional co-ordinates of the former from those of the latter. These co-ordinates are given in Tables 14 and 15. The differences between these values are given for the centre of the right shoulder in Table 18. This table corresponds exactly to Table 17 for the hips. The relative co-ordinates of the centre of the left shoulder differ only by their plus/minus sign and are, therefore, not given.

It is useful to distinguish the rotations of the shoulder line about a horizontal axis parallel to the direction of gait from those about a vertical axis, as we did for the hip line. In the former case the centres of the shoulders move up and down, in the latter case they move backwards and forwards. Despite the longer distance between the centres of the shoulders and the midpoint of the shoulder line, the excursions in the two directions are not greater

Table 18. Co-ordinates of the centre of the right shoulder relative to the midpoint of the shoulder line

No.	Experiment I			Experiment II		
	x	y	z	x	y	z
1	+ 0,56	+ 17,37	− 0,08	− 0,08	+ 17,48	+ 0,56
2	+ 0,48	+ 17,37	− 0,12	− 0,22	+ 17,43	+ 0,51
3	+ 0,37	+ 17,37	− 0,15	− 0,32	+ 17,38	+ 0,40
4	+ 0,33	+ 17,43	− 0,23	− 0,49	+ 17,37	+ 0,20
5	+ 0,21	+ 17,47	− 0,14	− 0,66	+ 17,40	− 0,05
6	+ 0,05	+ 17,43	− 0,02	− 0,83	+ 17,38	− 0,25
7	− 0,12	+ 17,33	+ 0,06	− 0,93	+ 17,39	− 0,30
8	− 0,27	+ 17,25	+ 0,05	− 0,96	+ 17,30	− 0,25
9	− 0,38	+ 17,13	− 0,02	− 0,97	+ 17,19	− 0,21
10	− 0,47	+ 17,04	− 0,10	− 0,98	+ 17,10	− 0,21
11	− 0,28	+ 16,99	− 0,17	− 0,97	+ 17,05	− 0,24
12	+ 0,02	+ 16,98	− 0,25	− 0,82	+ 17,02	− 0,25
13	+ 0,26	+ 16,97	− 0,28	− 0,35	+ 17,10	− 0,22
14	+ 0,42	+ 16,99	− 0,17	+ 0,02	+ 17,13	− 0,25
15	+ 0,64	+ 16,98	+ 0,02	+ 0,32	+ 17,13	− 0,21
16	+ 0,91	+ 17,00	+ 0,18	+ 0,49	+ 17,14	− 0,02
17	+ 1,12	+ 17,06	+ 0,28	+ 0,65	+ 17,21	+ 0,18
18	+ 1,23	+ 17,04	+ 0,25	+ 0,76	+ 17,37	+ 0,30
19	+ 1,41	+ 16,87	+ 0,21	+ 0,83	+ 17,37	+ 0,39
20	+ 1,52	+ 16,81	+ 0,21	+ 0,89	+ 17,06	+ 0,32
21	+ 1,59	+ 16,76	+ 0,23	+ 1,00	+ 16,95	+ 0,20
22	+ 1,56	+ 16,69	+ 0,24	+ 1,11	+ 16,99	+ 0,22
23	+ 1,45	+ 16,75	+ 0,19	+ 1,15	+ 16,95	+ 0,24
24	+ 1,15	+ 16,85	+ 0,28	+ 1,15	+ 16,93	+ 0,26
25	+ 0,92	+ 16,89	+ 0,31	+ 1,00	+ 17,03	+ 0,29
26	+ 0,74	+ 16,93	+ 0,26	+ 0,82	+ 17,00	+ 0,34
27	+ 0,58	+ 17,18	+ 0,14	+ 0,78	+ 17,11	+ 0,32
28	+ 0,29	+ 17,24	− 0,05	+ 0,62	+ 17,31	+ 0,17
29	− 0,05	+ 17,30	− 0,21	+ 0,36	+ 17,32	+ 0,10
30	− 0,35	+ 17,39	− 0,25	+ 0,08	+ 17,33	− 0,01
31	− 0,56	+ 17,40	− 0,18	− 0,08	+ 17,39	− 0,03

than the corresponding ones for the centre of the hip. Therefore, only the changes of the z- and x-co-ordinates will give a clear picture of the rotations of the shoulder line about the horizontal and vertical axes.

The modifications of both co-ordinates are graphically represented exactly as were those of the hip line. Graphs 3 and 11 of Fig. 28 show the up and down movements of the centres of the shoulders and graphs 4 and 12 show the forward and backward movements in the first two experiments. The upper curves relate to the right side of the body, the lower curves correspond to the left side.

The curves charting the elevations and lowerings of the shoulders (graphs 3 and 11) do not show such a simple shape as those for the hips. If slight irregularities are neglected, they present the shape of a wave line in which alternately a down wave and an up wave are restricted in their vertical amplitude. This is shown in Fig. 32 which reproduces schematically the upper curve of graph 3 in Fig. 28. From the original shape marked by a dotted line, it appears that the oscillation period is exactly the same as for the vertical oscillations of the hip line. Therefore, three complete oscillations occur during two steps. As appears from a comparison with graph 1 in Fig. 28, the maxima in Fig. 32 are to be found at the same places at which the minima occur in the corresponding curve for the hip line and vice versa. The shoulder line thus oscillates about the horizontal axis in the opposite direction to the corresponding hip line.

If we consider the positions of the maxima and minima in the vertical projection of the trajectories of the shoulder centres (cf. Figs. 22, 23), the vertical oscillations of the shoulder line are seen to have the following behaviour during walking. At the vertical maximum of a shoulder trajectory, which occurs when the homolateral leg is swinging, the shoulder line is at its lowest on this side. The trajectory then rises a little but changes the direction of its oscillation before the heel of the swinging leg reaches the ground. It again attains its lowest position approximately 0.11 s after heel strike. It then rises more steeply at an angle about three times greater than before, and it reaches its highest position at the moment when the trajectory of the shoulder attains its next maximum. This occurs in the homolateral stance period. At the same time, the shoulder line is of course at its lowest on the contralateral side. Simultaneously, however, the trajectory of the contralateral shoulder is at its highest, which occurs in the swinging period of the contralateral leg. The oscillation process of the shoulder line thus described then repeats itself symmetrically.

The shoulder line carries out three half oscillations about the horizontal axis parallel to the direction of gait during a single step. These oscillations do not present the same amplitude, as occurs for the hip line. They show alternately smaller and larger amplitudes, as seen in Fig. 32. The left leg is swinging from phases 12 to 23 in experiment I. The shoulder line is lowered towards the supporting leg only at the beginning of the swinging period of the opposite leg. However, during most of this period it is inclined downwards on the side of the swinging leg.

The vertical oscillations of the hip line explain the differences in the shape of the trajectories of the two hip centres seen in Figs. 26 and 27. Similarly, the vertical oscillations of the shoulder line explain the differences in the trajectories of the shoulders. As a consequence of the lowering of the shoulder line to the right during the vertical maximum of the shoulder trajectory accompanying the swinging of the right leg, the beginning of the first segment of the curve I, 5; II, 6 lies lower to the right than to the left. The shoulder line tilts down to the right at right heel strike. Therefore, the first segment of the curve lies lower to the right than to the left (I, 11; II, 12). It then rises more on the right than on the left since soon after right

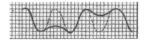

Fig. 32. Upward and downward movements of the centre of the shoulder during walking

heel strike the shoulder line is considerably raised to the right and remains at about the level it has reached for some time. Consequently, the point of intersection of the two segments of the curve (I, 17; II, 18) in the projection in Figs. 26 and 27 is higher to the right than to the left. Right and left sides thus exchange their roles, as was the case for the hips, and it cannot be otherwise in symmetrical gait.

The rotations of the shoulder line about the vertical axis are represented in graphs 4 and 12 of Fig. 28. They have almost exactly the shape of simple sine oscillations. Their period is three times and their amplitude ranges approximately $3\frac{1}{2}$ times that of the oscillations about the horizontal axis. Their maxima lie almost at the same places as the minima of the horizontal oscillations of the hip line, and vice versa. More precisely, the minimum and maximum in the curve of the shoulder line are found two phases or about 0.08 s earlier than the maximum and minimum of the hip line, since the latter ones are delayed by about this time due to their deviation from the single type of oscillation. Consequently, the shoulder line begins to oscillate forward shortly before the homolateral heel strike, and the direction of the oscillation is reversed only shortly before the opposite heel strike so that the shoulder line then oscillates forward on the other side.

To gain a clear concept of the way the shoulder line oscillates about the horizontal and vertical axes from the curves one should consider that the distance between the midpoint of this line and the two shoulder centres is approximately 17.5 cm.

Since the two oscillations occur simultaneously, the result is a rather complicated process of oscillation. In order to understand this, it is again useful to study the oscillations of a point further on the shoulder line, for instance four times further away than the distance between the midpoint of this line and the centre of the shoulder. The horizontal excursion of the shoulder is about 2.1 cm, and its vertical excursion about 0.6 cm. The point to be studied will thus oscillate on its horizontal trajectory between two points A and A' (Fig. 33) 8.4 cm apart, and simultaneously in a vertical direction between two points B and B' 2.4 cm apart. Both trajectories intersect at their midpoint O. The first horizontal oscillation constitutes a sine curve. If we again divide the oscillation time into 36 equal parts, at the end of each of the time intervals the point is at the places numbered from O. As the point goes through $OAOA'O$, because of the more complex vertical oscillation, it also goes through $OBDBOB'D'B'O$ whereby D and D' divide the segment BB' into three equal parts. As shown in graphs 3 and 4 in Fig. 28, the point passes through O almost simultaneously in its forward and upward oscillations (phase I, 15). Therefore, the corresponding numbering on the vertical line of Fig. 33 also starts from O. The combination of the two types of oscillation results in the curved trajectory $OCOC'O$. It can be seen in graphs 11 and 12 of Fig. 28 that the point has not yet exactly reached the midposition in its vertical movement when it passes through O on its horizontal trajectory. Hence the parts of the curved trajectory are somewhat separated from one another at O, as illustrated in Fig. 34. The trajectory of the centre of the right shoulder presents the same shape though four times smaller.

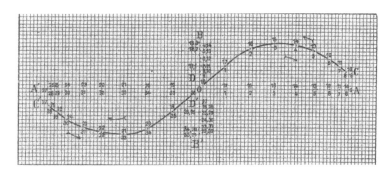

Fig. 33. Oscillations of a point on the prolongation of the shoulder line about a vertical ($A'OA$) and a horizontal (BOB') axis

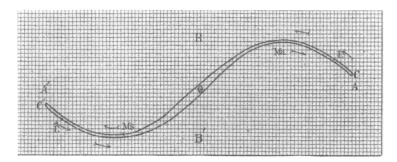

Fig. 34. Trajectory of a point on the prolongation of the shoulder line

Let us compare this trajectory with that of a point on the hip line in Fig. 30. There are essentially no great differences in shape. The extremities C and C' of the hip trajectory are lower and higher than those of the shoulder trajectory. The two opposite parts of the continuous curve separate somewhat at their extremities in the hip trajectory, whereas they separate in the middle in the shoulder trajectory. If one observes how these oscillations follow each other during the movement, one recognizes an exactly opposite behaviour of the two curves. To make this clearer, the places where the points on the hip line and shoulder line lie at right and left heel strikes are marked R and L in Figs. 30 and 34. The places which correspond to a vertical maximum of the absolute (not relative to the midpoint of the shoulder line) shoulder trajectory have also been marked . M_s signifies the maximum during homolateral leg swinging, M_a the maximum during homolateral stance. The position of these four points in the two figures shows that the two trajectories run in an exactly opposite manner. In order to obtain a clear picture of the complicated oscillation of each of the two lines, one has to imagine a plumb line perpendicular to the plane of the drawing in O. The midpoint of the hip line is assumed to be fixed on this plumb line at a distance of $4 \times 8.5\,\text{cm} = 34\,\text{cm}$ in Fig. 30, and the midpoint of the shoulder line fixed at a distance of $4 \times 17.5\,\text{cm} = 70\,\text{cm}$ in Fig. 34. The hip line moves then in such a way that its right half always intersects the curve in Fig. 30, and the shoulder line in such a way that its right half always intersects the curve in Fig. 34. It appears that the shoulder line moves backwards and upwards when the hip line moves forwards and downwards and vice versa. By describing the movement of the hip line in terms of forward, backward, upward and downward, we can obtain the simultaneous rotation of the shoulder line if we replace each of these four expressions by its opposite.

The amplitude of the oscillations is very different for the two lines although the sizes of the curves in Figs. 30 and 34 differ very little. This results from the fact that the midpoint of the shoulder line is about twice as far away from the plane of the drawing as is the midpoint of the hip line. The two curves only differ to a small extent in their horizontal and vertical amplitudes. The amplitude of the oscillation of the shoulder line about the vertical axis is approximately two-fifths and the amplitude of its oscillation about the horizontal axis is about three-fifths of the amplitude of the corresponding oscillation of the hip line. This last result has only an individual value. For instance, one can modify at will the proportion between the amplitudes of the oscillations about the vertical axis within certain limits.

From the simultaneous oscillations of the hip and shoulder lines, it follows that during walking the trunk rotates about the line connecting the midpoints of the hip and shoulder lines. The magnitude of the rotation corresponds approximately to the sum of the amplitudes of the two oscillations about the vertical axis. This sum would give an exact measure of the rotation of the trunk only if the trunk line remained vertical during all phases of the movement. This is not the case, however. The trunk line carries out many oscillations during walking. These will be studied in greater detail below.

Rotations of the Trunk Line

The trunk line will be considered as rotating about the midpoint of the hip line, which will be regarded as a fixed centre of rotation. The rotations of the trunk line thus are known if the movement of the midpoint of the shoulder line relative to the midpoint of the hip line has been determined. The co-ordinates of the midpoint of the shoulder line and those of the midpoint of the hip line are given in Tables 14 and 15 for the first two experiments. The differences between these values are given in Table 19. They are the co-ordinates of the movement of the midpoint of the shoulder line relative to the midpoint of the hip line.

The trunk line deviates only little from the vertical in its movements. It is, therefore, useful to differentiate its rotations about two horizontal axes, one in the direction of gait, the other perpendicular to the direction of gait. The shoulder line will move to the right and to the left in the former instance, and forward and backward in the latter instance. The amplitudes of

Table 19. Co-ordinates of the midpoint of the shoulder line relative to the midpoint of the hip line

No.	Experiment I			Experiment II		
	x	y	z	x	y	z
1	− 2,72	+ 0,59	+ 45,61	− 1,04	− 0,31	+ 46,14
2	− 2,91	+ 0,36	+ 46,13	− 2,07	− 0,43	+ 45,76
3	− 2,84	+ 0,15	+ 46,71	− 2,59	− 0,43	+ 45,53
4	− 2,71	+ 0,05	+ 46,72	− 2,32	− 0,31	+ 46,29
5	− 2,34	+ 0,15	+ 46,89	− 2,27	− 0,14	+ 46,92
6	− 1,91	+ 0,27	+ 47,10	− 1,89	− 0,03	+ 47,22
7	− 1,36	+ 0,50	+ 47,45	− 1,26	+ 0,21	+ 47,27
8	− 0,91	+ 0,77	+ 47,60	− 0,97	+ 0,43	+ 47,53
9	− 0,96	+ 1,02	+ 47,18	− 0,30	+ 0,63	+ 47,95
10	− 1,35	+ 1,18	+ 46,72	+ 0,03	+ 0,78	+ 47,80
11	− 2,24	+ 1,28	+ 46,23	− 0,14	+ 0,96	+ 47,16
12	− 2,67	+ 1,22	+ 45,90	− 0,90	+ 1,10	+ 46,87
13	− 2,66	+ 0,77	+ 46,10	− 1,67	+ 0,98	+ 46,47
14	− 3,24	+ 0,29	+ 46,12	− 1,91	+ 0,78	+ 46,60
15	− 3,14	− 0,02	+ 46,35	− 2,13	+ 0,15	+ 47,11
16	− 3,52	− 0,13	+ 46,70	− 2,29	− 0,08	+ 47,35
17	− 2,81	− 0,12	+ 46,97	− 2,36	− 0,12	+ 47,03
18	− 2,58	+ 0,04	+ 47,04	− 2,43	+ 0,04	+ 47,17
19	− 2,18	+ 0,09	+ 47,11	− 2,20	− 0,05	+ 47,17
20	− 1,76	+ 0,04	+ 47,23	− 1,78	− 0,39	+ 47,24
21	− 1,42	− 0,06	+ 47,16	− 1,40	− 0,21	+ 47,51
22	− 1,38	− 0,10	+ 46,70	− 1,04	− 0,43	+ 47,53
23	− 1,68	− 0,03	+ 46,13	− 0,89	− 0,41	+ 47,10
24	− 2,48	+ 0,22	+ 45,73	− 1,00	− 0,53	+ 46,71
25	− 2,46	+ 0,58	+ 45,69	− 1,76	− 0,18	+ 46,24
26	− 2,14	+ 0,68	+ 45,65	− 1,96	+ 0,15	+ 46,09
27	− 2,96	+ 0,69	+ 45,76	− 2,00	+ 0,33	+ 46,02
28	− 3,41	+ 0,73	+ 46,37	− 2,50	+ 0,60	+ 45,79
29	− 3,90	+ 0,69	+ 46,67	− 3,09	+ 0,66	+ 46,16
30	− 3,66	+ 0,67	+ 46,89	− 3,52	+ 0,57	+ 46,76
31	− 3,36	+ 0,81	+ 47,04	− 3,30	+ 0,48	+ 47,13

these two types of oscillation are even smaller than those of the oscillations of the shoulder and hip lines. The resolution of the rotations of the trunk line into two components will not, therefore, give essentially different results than if one assumes that both rotation axes through the midpoint of the hip line are fixed to the trunk, one in a sagittal, the other in a coronal direction.

The movements of the midpoint of the shoulder line which correspond to the rotation about the axis parallel to the direction of gait are represented by the modifications of the y-co-ordinate, the others by the modifications of the x-co-ordinate, as seen in Table 19. These modifications of the co-ordinates are represented in graphs 5 and 6 of Fig. 28 for experiment I and in graphs 13 and 14 for experiment II, in the manner described above.

These diagrams present some irregularities (e.g. graph 6) which have no essential mechanical significance since they do not coincide in the two experiments. They appear only to a much more limited extent in the curves of the rotation of the hip line. This may be due to the fact that the trunk line is five to six times longer than half the hip line. Consequently, small irregularities in the oscillation must appear five to six times greater in the curves of the trunk line than in those of the hip line. Apart from these irregularities, the graphs show that the trunk line carries out oscillations about the axis parallel to the direction of gait which are very similar in shape to the oscillations of the shoulder line about the axis parallel to the direction of gait. During two steps the trunk line carries out three oscillations of very different amplitudes, as does the shoulder line: large and very small amplitudes alternate. Shortly after right heel strike, the trunk line tilts to the right at a maximum. After left heel strike it tilts to the left at the maximum. The movement of the trunk line is not completely symmetrical. Its behaviour is different before right and left heel strike. After heel strike the trunk line oscillates towards the median plane and beyond. During the swinging of the other leg it comes back a little but soon returns to its initial movement so that it once again reaches its greatest inclination towards the other side after the contralateral heel strike. The total sideways deviation of the midpoint of the shoulder line resulting from this oscillation was only about 1.5 cm in the two experiments.

With regard to the oscillations about the axis perpendicular to the direction of gait, i.e. the forward and backward oscillations, the trunk line tilts forwards at a maximum always shortly before heel strike. Its greatest backward inclination always takes place a little before the moment when the trajectories of the hips and shoulders reach a vertical maximum. This occurs before the middle of the stance or swing period. In the last half of the leg swing, the trunk line and thus the whole trunk always move forward. However, the trunk line begins to move backwards again shortly before heel strike, as stated above. The duration of one oscillation about the axis perpendicular to the direction of gait equals the duration of one step. The total excursion of the midpoint of the shoulder line relative to the midpoint of the hip line in the direction of gait is 2.5 cm.

To obtain a clear picture of the resultant oscillation it was again expedient to combine the oscillations of a point of the trunk line further away from the midpoint of the hip line than was the midpoint of the shoulder line. Again, we chose a point that was four times as far from the midpoint of the hip line. This point thus carries out excursions four times as great than does the midpoint of the shoulder line in the two directions of oscillation. The total excursion BB' (Fig. 35) of the oscillation of this point perpendicular to the plane of gait (resulting from the rotation about the axis parallel to gait) is $4 \times 1.5 \, \text{cm} = 6 \, \text{cm}$, and the amplitude AA' of its oscillation in the direction of gait $4 \times 2.5 \, \text{cm} = 10 \, \text{cm}$. However, in the former type of oscillation, the point does not complete the whole distance $\overline{BB'}$ every time. After having traversed $\overline{BB'}$ once it comes back only 1 cm, turns again towards B and then, having reached B a second time, it starts to move again from B to B', etc. If the duration of two steps is divided into 36 equal parts the numbering of the segment BB' in Fig. 35 corresponds approximately to the way the point moves perpendicular to the plane of gait. Of course, only an outline of the oscillation in this direction is provided here since graphs 5 and 13 in Fig. 28

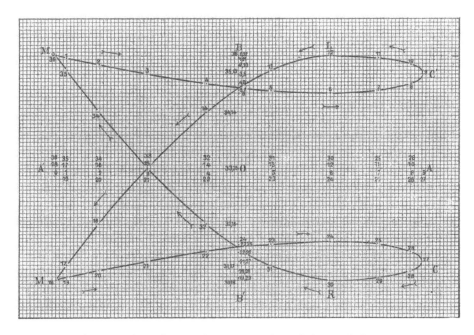

Fig. 35. Trajectory of a point on the prolongation of the trunk line

do not show complete symmetry. At the same time, the point oscillates twice in the direction AA' and back in a fairly regular manner. This corresponds to the numbering along AA' in Fig. 35. As can be seen from the curves, the moment when the point has come to the line at B and starts to swing back on the smaller segment is also the moment when the point has arrived at A' and is ready to swing back to A. Thus, the numbering begins at these two places.

The curved trajectory of the point of the trunk line (Fig. 35) is the result of the above. The points R, L and M signify the places where the point of the trunk line lies on its trajectory relative to the midpoint of the hip line at right (R) or left heel (L) strike and when the hips and shoulders pass through a vertical maximum (M) on their absolute trajectories. The trajectory of the midpoint of the shoulder line relative to the midpoint of the hip line presents the same shape as the curve in Fig. 35 but is four times smaller.

In order to obtain a clear concept of the rotations of the trunk line about the midpoint of the hip line from Fig. 35, one must imagine the midpoint of the hip line on a line through O, perpendicular to the plane of the figure, 4×48 cm $= 192$ cm behind this plane. The trunk line extended beyond the midpoint of the shoulder line would move along the curve of Fig. 35.

The deviations in the trajectories of the two hip and shoulder centres could be explained by the rotations of the hip and shoulder lines about their midpoints. Similarly, the differences in shape of the trajectories of the midpoints of the shoulder and hip lines (cf. Fig. 27) are due to the oscillations of the trunk line about the midpoint of the hip line. At right heel strike the trunk line is tilted to the right at its most. Since this moment almost corresponds in both curves to the minimum of the first segment of the curve, in the projection of the shoulder curves this minimum occurs further to the right than in the projection of the hip curves. The reverse is true for the left side. For the vertical maximum of the trajectories following right heel strike, the trunk line is already tilted to the left, as appears in Fig. 35. Therefore, the initial point of the first segment in the curves of Figs. 26 and 27 and the intersection of the two segments are closer to the middle in the projection of the trajectory of the midpoint of the shoulder line than in that of the midpoint of the hip line. This occurs on both sides. The shoulder curves, therefore, come a little closer together above than the hip curves. For most of the distance from R to M the trunk line tilts to the right. During the following segment to L it tilts to the left. Consequently, the upward part of the first segment of the shoulder curve and the downward part of the second are separated from each other as appears in Figs. 26 and 27. A further consequence is that the two points R and L are closer to each other in the shoulder curve than in the hip curve, etc.

Rotations of the Head

During walking the head does not remain completely immobile on the trunk. It moves slightly sideways, forwards and backwards. The movement results partly from rotation in the joint between the skull and atlas, partly from rotation in the joint between the atlas and axis and partly from flexion of the cervical spine. Strictly speaking, the movement cannot be considered as a simple rotation about a centre fixed in the trunk. However, in this case it would not be worthwhile determining the varying positions of the centre of rotation and the instant axis of rotation. Indeed, the oscillations of the head in relation to the trunk during walking are even much less important than, for instance, the pendulum movements of the arms. Both movements are not absolutely necessary for the human body to move forwards. They can be suppressed without essentially disturbing the gait. If they take place freely the movements of the head relative to the trunk, as well as those of the arms, complement the movements of the legs and trunk in a particular way. In certain phases of the movement the head will tilt forwards, in other phases it will tilt more to the right or left, etc. To establish when this happens and to answer other questions concerning the direction and change of this movement, we can simply assume that the head rotates about a fixed point located in the upper aspect of the trunk and consider the displacement of this centre of rotation. From our experiment we can only obtain an approximate view of the movement of the head relative to the trunk. For a more accurate analysis of this movement, measuring the co-ordinates of the vertex would not be sufficient.

To facilitate the analysis, the centre of rotation of the head will be assumed to be at the midpoint of the shoulder line. The rotations about the sagittal and coronal axes through this point can then be measured by the movements of the line connecting the midpoint of the shoulder line and the vertex, which will be called the „head line". This of course disregards the rotations of the head about this line itself.

The rotations of the head line about the midpoint of the shoulder line can be represented by the movement of the vertex in relation to the midpoint of the shoulder line. The co-ordinates of this movement are the differences between the co-ordinates of the vertex and those of the midpoint of the shoulder line given in Tables 14 and 15. However, this does not indicate the rotations of the head relative to the trunk alone. If the head were fixed to the trunk one would record in this way a movement of the vertex relative to the midpoint of the shoulder line. Consequently, one would be able to determine the rotation of the head line. This rotation would be that of the trunk line and one could consider the head line as a simple prolongation of the trunk line. The curves representing the movement of the vertex relative to the midpoint of the shoulder line would then present exactly the same shape as those of the trunk line in graphs 5, 6, 13 and 14 of Fig. 28. A difference, however, would exist because the ordinates of the graphs would appear smaller in the same proportion as the distance between the vertex and the midpoint of the shoulder line is shorter than the distance between the latter and the midpoint of the trunk line, i.e. 0.7.

To determine the rotation of the head line in relation to the trunk, one has to deduct the rotation of the trunk line from that of the head line, as mentioned above. Considering the small amplitude of the rotations, one obtains, with sufficient accuracy, the co-ordinates of the movement of the vertex relative to the whole trunk (and not only to the midpoint of the shoulder line) if one subtracts the differences between the co-ordinates of the midpoint of the shoulder line and those of the midpoint of the trunk line, reduced by a factor of 0.7, from the differences between the co-ordinates of the vertex and those of the midpoint of the shoulder line. The differences between the co-ordinates of the midpoints of the shoulder and trunk lines are given in Table 19. This calculation was carried out for the two relevant x- and y-co-ordinates. The results are listed in Table 20 for experiments I and II.

The curves resulting from Table 20 illustrate the rotations of the head line and are given in graphs 7 and 8 of Fig. 28 for experiment I, and in graphs 15 and 16 for experiment II, in the same manner as the graphs for the trunk line. These curves do not give an accurate picture of the movements of the head for the reasons cited above. However, they enable us to recognize the way the head oscillates. In the coronal and sagittal directions, the vertex moves in relation to the midpoint of the shoulder line in a way almost exactly opposite to the movement of the midpoint of the shoulder line in relation to the midpoint of the hip line. The head carries out rotations about the sagittal and coronal axes which are always in the opposite direction to the concomitant rotations of the trunk line. Consequently, one must

Table 20. Co-ordinates of the vertex relative to the trunk

No.	Experiment I		Experiment II	
	x	y	x	y
1	+ 3,98	− 0,37	− 0,43	+ 0,38
2	+ 4,83	− 0,26	+ 0,57	+ 0,25
3	+ 5,19	− 0,07	+ 1,53	+ 0,20
4	+ 5,28	+ 0,12	+ 1,86	+ 0,18
5	+ 5,09	+ 0,06	+ 2,21	+ 0,20
6	+ 4,83	+ 0,11	+ 2,23	+ 0,43
7	+ 4,45	− 0,07	+ 2,03	+ 0,44
8	+ 4,05	− 0,28	+ 2,02	+ 0,36
9	+ 3,90	− 0,48	+ 1,71	+ 0,26
10	+ 3,92	− 0,70	+ 1,49	+ 0,14
11	+ 4,01	− 1,09	+ 1,49	− 0,05
12	+ 3,46	− 1,15	+ 1,97	− 0,37
13	+ 2,80	− 0,73	+ 2,29	− 0,47
14	+ 3,21	− 0,24	+ 2,02	− 0,38
15	+ 3,27	0	+ 2,26	+ 0,16
16	+ 3,48	+ 0,04	+ 2,72	+ 0,34
17	+ 3,02	− 0,04	+ 2,91	+ 0,29
18	+ 3,07	− 0,21	+ 3,04	− 0,08
19	+ 3,01	− 0,27	+ 2,90	+ 0,01
20	+ 2,96	− 0,10	+ 2,66	+ 0,51
21	+ 2,82	+ 0,04	+ 2,40	+ 0,28
22	+ 2,79	+ 0,14	+ 2,03	+ 0,56
23	+ 2,93	+ 0,07	+ 1,69	+ 0,56
24	+ 3,25	− 0,16	+ 1,52	+ 0,75
25	+ 2,54	− 0,50	+ 1,76	+ 0,49
26	+ 1,81	− 0,79	+ 1,28	+ 0,24
27	+ 2,91	− 0,77	+ 0,81	− 0,12
28	+ 4,07	− 0,77	+ 1,47	− 0,59
29	+ 5,09	− 0,65	+ 2,53	− 0,72
30	+ 5,18	− 0,40	+ 3,39	− 0,51
31	+ 5,25	− 0,21	+ 3,47	− 0,20

imagine Fig. 35 rotated through 180° about an axis perpendicular to the plane of the diagram in order to have an overall view of the rotation of the head line resulting from the two oscillations.

Gait of the Loaded Man

In experiment III we provided the subject with the field equipment of a German infantryman to determine how the gait is influenced by a load. We chose military equipment mainly because it represents a load which most healthy strong men must carry for long distances at some time in their lives. The tridimensional co-ordinates for this experiment are given in Table 16 (pp. 78–79). We shall above all stress the differences which appear between the gaits of the loaded and unloaded man concerning the trajectories of the vertex and centres of the shoulders and hips on one hand and the rotations of the hip, shoulder, trunk and head lines on the other.

The projections of the trajectories of the seven points of the head and trunk are drawn life-size in Fig. 36 on a plane perpendicular to the direction of gait, as they were in Fig. 26 for experiments I and II. Thus, one can immediately recognize the characteristic differences in these trajectories in the loaded and unloaded subject. The curves in the horizontal direction are basically more stretched out though their height is not significantly altered. The widening of the curves indicates that the loaded subject carries out greater sideways oscillations of the whole upper body than does the unloaded subject.

The curves of Fig. 36 are much less regular than those of Fig. 26. The curve of the vertex comes back almost to its initial point whereas the end of the other curves is to the left of the starting point. This particularly applies to the hip curves. However, from this one cannot conclude that there is an irregularity of gait. It results from the fact that the direction of the axis of the trajectory of the vertex was used to rotate the system of co-ordinates to make the plane XZ parallel to the plane of gait. As shown in the curves of Fig. 36, the axis of the wave trajectory of the head point has not exactly the same direction as the axes of the trajectories of the shoulder and hip curves. Again, one must conclude that at the end of the measured period of walking, the trunk and the head were tilted rather more to the right than at the beginning. If this could have been predicted, it would have been better to make the direction of the X-axis coincide with that of the axis of the wave trajectory of the midpoint of the hip line, by appropriate rotation of the system of co-ordinates.

If the irregularities thus created as well as the widening of the curves in Fig. 36 are ignored, it appears that loading has almost no effect on the shape of the curves, the position of the maxima and minima, etc. We must, however, consider that in experiment III right heel strike occurs in phase 9 and left heel strike in phase 22, i.e. one phase earlier than in the first experiment.

An additional difference between the first two experiments and experiment III does not appear in the curves of Figs. 26 and 36 and concerns the length of the step. This is given, for instance, by the difference between the x-co-ordinates of the right ankle at right heel strike and those of the left ankle at left heel strike. It is also evident when the difference between the x-co-ordinates of any joint is halved, corresponding to two similar moments separated by the duration of a double step. In experiment I, the length of the step is 78 cm; in experiment II it is 77 cm. The average length of the step when the unloaded subject walks is thus 77.5 cm. In experiment III the length is only 72 cm. Consequently, the wave-shaped trajectories of the vertex, shoulder points and hip points in experiment III are both wider and shorter than in the first two experiments. Therefore, these trajectories for the loaded subject must also present a sharper curvature at every point.

To determine the differences in the rotations of the hip, shoulder, trunk and head lines between experiment III and the first two experiments, the curves representing the oscillations

III. Versuch *(mit Belastung.)*

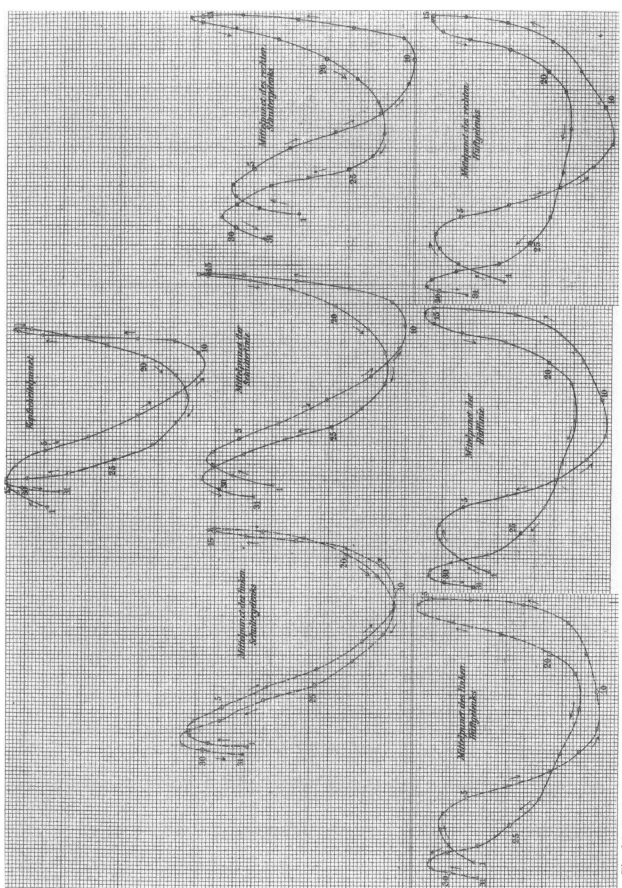

Fig. 36. Trajectories of different points during walking with load

of an extremity of each of these four lines are drawn exactly as before. The ordinates of these diagrams are calculated from the tridimensional co-ordinates of Table 16 as they were for the first two experiments. The results of this calculation are given in Table 21. The curves drawn from this table are graphs 17–24 of Fig. 28 adjacent to the corresponding graphs of the first two experiments. The following can be inferred from these results.

The oscillations of the hip line about the axis parallel to the direction of gait occur in the loaded subject exactly as in the unloaded subject (graph 17). However, the oscillations of the hip line about the vertical axis (graph 18) are of smaller amplitude in the gait of the loaded man than in that of the unloaded. The type of oscillation remains the same. Here also the hip line is generally rotated forwards on each side shortly after the homolateral heel strike. This occurs in phase 9 for the right leg, in phase 22 for the left in experiment III. The shortening of the oscillation amplitude was affected by the way the subject was loaded: the waist belt to

Table 21. Co-ordinates of an end point relative to the midpoint or another end point on the hip, shoulder, trunk and head lines in experiment III

No.	Hip line x	y	z	Shoulder line x	y	z	Trunk line x	y	z	Head line x	y
1	− 0,81	+ 8,45	+ 0,39	− 0,55	+ 16,83	− 0,66	+ 3,43	+ 2,22	+ 47,24	+ 0,82	− 2,11
2	− 0,70	+ 8,45	+ 0,65	− 0,49	+ 16,86	− 0,65	+ 3,99	+ 2,01	+ 47,33	+ 0,56	− 1,89
3	− 0,56	+ 8,45	+ 0,81	− 0,46	+ 16,90	− 0,59	+ 4,54	+ 1,89	+ 47,51	+ 0,31	− 1,72
4	− 0,41	+ 8,44	+ 0,90	− 0,41	+ 16,87	− 0,49	+ 4,98	+ 1,83	+ 47,45	+ 0,11	− 1,61
5	− 0,33	+ 8,45	+ 0,87	− 0,38	+ 16,89	− 0,40	+ 5,51	+ 1,74	+ 47,47	+ 0,03	− 1,52
6	− 0,19	+ 8,47	+ 0,77	− 0,36	+ 16,90	− 0,29	+ 5,89	+ 1,99	+ 47,60	+ 0,05	− 1,87
7	− 0,05	+ 8,48	+ 0,66	− 0,33	+ 16,85	− 0,23	+ 6,13	+ 2,01	+ 47,59	− 0,11	− 1,81
8	+ 0,07	+ 8,48	+ 0,61	− 0,22	+ 16,75	− 0,21	+ 6,34	+ 2,10	+ 47,25	− 0,50	− 1,80
9	+ 0,14	+ 8,48	+ 0,59	− 0,01	+ 16,76	− 0,21	+ 6,08	+ 2,23	+ 47,01	− 0,58	− 2,11
10	+ 0,16	+ 8,47	+ 0,68	+ 0,18	+ 16,79	− 0,23	+ 5,43	+ 1,89	+ 46,76	− 0,58	− 2,02
11	+ 0,16	+ 8,45	+ 0,86	+ 0,41	+ 16,75	− 0,28	+ 4,48	+ 1,53	+ 46,63	− 0,24	− 2,06
12	+ 0,07	+ 8,44	+ 1,01	+ 0,50	+ 16,66	− 0,08	+ 3,96	+ 1,13	+ 47,12	+ 0,17	− 2,09
13	− 0,11	+ 8,44	+ 1,01	+ 0,65	+ 16,65	0	+ 3,45	+ 0,84	+ 47,28	+ 0,52	− 2,03
14	− 0,29	+ 8,45	+ 0,85	+ 0,81	+ 16,66	+ 0,05	+ 3,17	+ 0,85	+ 47,42	+ 1,03	− 2,11
15	− 0,41	+ 8,45	+ 0,75	+ 0,91	+ 16,57	+ 0,18	+ 3,02	+ 0,86	+ 47,41	+ 1,42	− 2,07
16	− 0,48	+ 8,46	+ 0,61	+ 0,95	+ 16,54	+ 0,25	+ 2,96	+ 0,99	+ 47,46	+ 1,58	− 2,15
17	− 0,50	+ 8,47	+ 0,55	+ 0,98	+ 16,47	+ 0,17	+ 2,94	+ 1,22	+ 47,46	+ 1,55	− 2,24
18	− 0,50	+ 8,46	+ 0,58	+ 1,04	+ 16,42	+ 0,37	+ 3,30	+ 1,32	+ 47,12	+ 1,24	− 2,32
19	− 0,44	+ 8,46	+ 0,67	+ 1,15	+ 16,36	+ 0,33	+ 3,66	+ 1,33	+ 47,18	+ 0,80	− 2,37
20	− 0,53	+ 8,44	+ 0,81	+ 1,21	+ 16,33	+ 0,29	+ 3,76	+ 1,41	+ 47,15	+ 0,41	− 2,29
21	− 0,64	+ 8,43	+ 0,92	+ 1,25	+ 16,36	+ 0,28	+ 3,69	+ 1,38	+ 46,94	+ 0,15	− 2,18
22	− 0,67	+ 8,43	+ 0,89	+ 1,24	+ 16,43	+ 0,23	+ 3,21	+ 1,67	+ 46,58	+ 0,16	− 2,29
23	− 0,66	+ 8,44	+ 0,78	+ 1,10	+ 16,42	+ 0,02	+ 2,41	+ 2,01	+ 46,27	+ 0,54	− 2,39
24	− 0,63	+ 8,45	+ 0,65	+ 1,02	+ 16,47	− 0,24	+ 1,78	+ 2,54	+ 46,37	+ 0,70	− 2,72
25	− 0,58	+ 8,46	+ 0,54	+ 0,94	+ 16,62	− 0,42	+ 1,49	+ 2,70	+ 46,53	+ 0,75	− 2,70
26	− 0,53	+ 8,47	+ 0,53	+ 0,75	+ 16,68	− 0,56	+ 1,16	+ 2,94	+ 46,88	+ 1,02	− 2,92
27	− 0,44	+ 8,46	+ 0,67	+ 0,54	+ 16,85	− 0,64	+ 1,21	+ 2,77	+ 46,99	+ 1,07	− 2,59
28	− 0,34	+ 8,43	+ 0,78	+ 0,37	+ 16,72	− 0,61	+ 1,45	+ 2,61	+ 47,25	+ 0,80	− 2,14
29	− 0,29	+ 8,45	+ 0,86	+ 0,29	+ 16,70	− 0,49	+ 1,82	+ 2,48	+ 47,49	+ 0,41	− 1,82
30	− 0,20	+ 8,45	+ 0,94	+ 0,17	+ 16,69	− 0,41	+ 2,36	+ 2,38	+ 47,45	+ 0,08	− 1,67
31	− 0,16	+ 8,45	+ 0,94	+ 0,05	+ 16,59	− 0,26	+ 3,03	+ 2,30	+ 47,32	− 0,31	− 1,48

which three full cartridge pouches were fixed must hinder the rotations of the pelvis about the vertical axis. Whether any type of loading generally restricts the width of the oscillation in the horizontal plane cannot be deduced from our experiments.

The oscillations of the shoulder line must be suppressed on the side of the shoulder supporting the rifle, and indeed graphs 19 and 20 of Fig. 28 indicate a considerable one-sidedness. Except for this, the shape of the oscillation is the same as in the first two experiments. For instance, graph 19 can be thought of as deriving from a regular sine curve, as was evident in Fig. 32 for the corresponding graphs 3 and 11. Figure 37 shows this clearly. The one-sidedness of the oscillation about the axis parallel to the direction of gait consists of an obvious tendency of the hip line to rise abnormally to the right or to drop to the left. This is due to the weight of the rifle on the left shoulder. Only at left heel strike (phase 22) is the hip line very quickly lowered to the right and just as quickly lifted to the left. This elevation on the left side, however, does not last long. After the left leg has passed over the position in which the left hip joint lies on the plumb line through the left ankle joint, the elevation on the right begins again and there is lowering on the left. Except for the one-sided progressive lowering and the quick elevation on the left side at left heel strike, the shoulder line carries out the alternate small and large oscillations in experiment III as described for the first two experiments.

The oscillations about the vertical axis (phase 20) display a corresponding one-sided behaviour like the oscillations about the axis in the direction of gait. The latter tend to drop the shoulder line to the left; the former have a less pronounced tendency to push forward the right side of the shoulder line and thus rotate the left side backwards. This is because the rifle is held in the left hand and its centre of gravity lies between the hand and the shoulder, exerting a backward compression on the shoulder.

The oscillations of the trunk line are little altered by the load, as is apparent from graphs 21 and 22 of Fig. 28. The amplitude of the rotations about the axis parallel to the direction of gait is increased. Conversely, the rotation about the axis perpendicular to the plane of gait, the alternate tilt of the trunk backwards and forwards, takes place in a somewhat one-sided manner: the upper extremity of the trunk line does not move as far forward before left heel strike as it does before right heel strike.

The rotations of the head line in relation to the trunk are almost the opposite of those of the trunk line. This also applies to the gait of the loaded man. This can be seen in graphs 23 and 24 of Fig. 28.

The results of experiment III cannot be simply applied to all types of gait involving loading. These results only concern loading with the military equipment. Since this load is somewhat one-sided because of the rifle carried on the shoulder, asymmetries arise in the movements of the trunk and particularly the shoulders. These would not appear if the load were distributed symmetrically, for instance if the subject carried a rifle simultaneously on each shoulder.

Another notable difference appeared in the oscillations of the hip line about the vertical axis in experiment III compared with the gait of the unloaded subject. This result also does not necessarily apply to all types of loading since the restriction of the forward and backward oscillation of the hip line could be due to the position of the straps supporting the cartridge pouches.

Of the details of walking analysed so far, the only deviations of general significance concern the extent of the trajectories of the head, shoulders and hips in the direction of gait

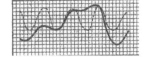

Fig. 37. Upward and downward movements of the centre of the right shoulder during walking

and in the direction perpendicular to the plane of gait. The main differences between the gait of the loaded and unloaded subject appear when the velocities and accelerations with which the different joint centres travel along their trajectories are considered. This will be dealt with in Chapter 2.

Conclusions

The new method described above consists of examining a movement in terms of a tridimensional system of co-ordinates using two-sided chronophotography. The examples considered sufficiently demonstrate that this method represents a useful tool for analysing the movements of human and animal bodies. In analysing the movement of bodies, it can perform a similar function to the microscope in examining the forms and construction of the bodies. As every improvement in the microscope must entail progress in the knowledge of the structure of different organisms, any improvement in our method of recording movement inevitably widens our knowledge of the laws of movement, as carried out by organisms.

The method could doubtless be further improved. We have by no means reached the limits of accuracy which can be attained in transposing the process of movement into a tridimensional system of co-ordinates.

The accuracy could for instance be increased by placing the two cameras in such a way that their optical axes would be perpendicular to one another instead of intersecting at 60°. Since it is not feasible to place an optical axis in the direction of gait, it would perhaps be better to choose 45° as the angle formed by the direction of gait and the optical axes of the four cameras.

If the table of co-ordinates could be dispensed with, this would further improve the method. When we began the experiments we did not possess the new co-ordinate measurer (p. 39). We, therefore, had to photograph directly the network of co-ordinates as in a previous work, in order to enable us to read directly the bidimensional co-ordinates of the points on the photographic plates. With the co-ordinate measurer the network of co-ordinates becomes superfluous. It is only necessary to project onto the plates a particular point in space and to use a determined direction and a determined length. When using the Geissler tubes, this projection could be carried out shortly before or after the experiments, simultaneously on all photographic plates. Choosing appropriate distances between the cameras and the origin of the co-ordinates would also make it possible to simplify the equations, etc.

All the above is of importance in the method. Future studies will have to take these and similar considerations into account when photographing the movements of man or animals.

In the first part of this work we have described the photographic recording of human gait and transposed the movement into a tridimensional system of co-ordinates. With some examples, we have demonstrated that the resulting tables of co-ordinates can provide the basis for the solution of all problems dealing in any way with the laws of movement of human gait. A detailed analysis of the laws of movement will constitute the second part of the research on "The Human Gait".

CHAPTER 2

The Movement of the Total Centre of Gravity and the External Forces

Introduction

Our studies on the gait of unloaded and loaded humans began several years ago. We recorded the process of this movement in a tridimensional rectangular system of co-ordinates (Chap. 1). The accuracy achieved was such that the trajectories followed by the different points of the human body could be thoroughly represented by the co-ordinates. These co-ordinates can also provide the starting point to answer many questions associated with gait. For example, the trajectories of the centres of the joints, the vertex and the tips of the feet were presented in Chap. 1. The rotations and deformations of the trunk were deduced from the rotations of the hip, shoulder and trunk lines, i.e. the lines connecting the two hip centres, the two shoulder centres and the midpoints of the hip and shoulder lines.

The great accuracy of the method revealed even slight unevennesses in the behaviour of the two sides of the body. For example, it appeared that the experimental subject lowered his whole body a few millimetres more before right heel strike than before left heel strike. This asymmetrical behaviour of the two legs is not surprising when one considers that, in the structure of the human body, perfect symmetry is attained extremely rarely, if at all. Though people appear symmetrical there are significant discrepancies in the length of the bones. When we measured the length of the segments of the limbs in order to determine the centre of gravity, we found differences of up to 1 cm in the length of the thighs of the same individual[1]. For this research we had chosen cadavers that were particularly well built and had a normal appearance. Correspondingly, differences also appeared in the position of the centres of gravity of two symmetrical limbs. Such discrepancies in the size and distribution of the mass in the lower extremities of course entail differences in the behaviour of the legs during walking. It can be predicted that in most people a precise examination of the gait would reveal such asymmetries as we found in our experimental subject. This man had been a soldier and had a very symmetrical body structure (see the figures in Braune and Fischer op. cit. which show different positions of the subject used for the studies on gait) and apparently a very regular gait. Therefore, it can be assumed that the asymmetries found in him are not greater than in many other individuals who seem to walk normally. In subsequent direct observations of the gait of this man, these asymmetries in the movement of the two sides of the body were not apparent, either to the author or to others. Of course, for our experiments we wanted an individual whose behaviour was normal. However, this demand could only be met if a great deal of time had been spent taking photographs and

---

[1] Braune W, Fischer O (1889) Über den Schwerpunkt des menschlichen Körpers mit Rücksicht auf die Ausrüstung des deutschen Infanteristen. Abhandlungen der mathematisch-physischen Klasse der Königlich Sächsischen Gesellschaft der Wissenschaften, vol 15, nr 7. English translation (1984) On the centre of gravity of the human body. Springer, Berlin Heidelberg New York Tokyo

carrying out the measurements and calculations on a large number of individuals and then using those who presented the greatest symmetry for further investigation. Unless this huge task is performed, one must be satisfied with deducing the most probable norm from the results of the measurements of the individual under consideration, taking into account the concordant results of the two sides of the body.

The few examples shown in Chap. 1 demonstrate that the differences in behaviour of the two sides of the body concern mainly the amplitude rather than the type of movement. There is virtually complete concordance between the two sides, as is further confirmed in this, the second part of our research. It appears that, in deducing normal gait, one only commits errors concerning the tridimensional amplitude of the different parts of the movement. The movements involved in the gaits of different individuals will also generally differ as fas as this characteristic is concerned. However, the succession and type of the different movements of the limbs are almost the same for all people. This results from the fact that the same kinds of forces are acting to promote the movement. It seems probable that the same groups of muscles come into action whatever the individual differences in gait. Therefore, the results obtained from researches on the mechanics of walking and the locomotor apparatus in one individual will qualitatively possess a general value. The data from Chap. 1 thus can be used for further analysis of walking.

The forward movement of the human body in walking, running, jumping or any other kind of locomotion is caused by the action of forces that are partly internal, partly external. The internal forces are produced in the body itself. They comprise above all muscular forces, elastic tensile and compressive stresses in tendons, ligaments and articular cartilage. The external forces are gravity, ground reaction force, friction with the ground and air resistance. For a flying bird, there is neither ground reaction force nor friction with the ground and, therefore, air resistance is most important in forward movement whereas the influence of air resistance in human walking is of secondary importance. It impedes forward movement in an immobile atmosphere or in the case of head winds. It helps forward movement when the wind blows in the direction of gait with a velocity greater than the average velocity of gait.

If external forces played no role whatsoever, a man could change position by contracting his muscles but would be unable to move his body forwards. If he were immobile, floating freely in space, muscular effort would be unable to move his centre of gravity by 1 mm in space. If some external cause were to set the centre of gravity into movement, the man would continue moving in space after the cause had ceased to act. His centre of gravity would follow a straight course with constant velocity. Any movements of the limbs due to muscular contractions would be unable to alter this state.

If there were no friction, as would be the case in walking on a perfectly smooth surface, voluntary movement forwards would also be impossible. A man then would only be able to bring the centre of gravity of his body further from or closer to the ground within certain limits by voluntary muscular contractions and thereby modify any movement imposed on the body by external forces.

In any circumstances, the movement of the centre of gravity follows the following basic rules. The centre of gravity of a body or system of bodies on which internal and external forces act always moves as if all the masses were united in this centre of gravity and as if the external forces were applied directly to this point. Internal forces have no effect on the movement of the centre of gravity because they cancel each other out. For example, if a man stands motionless on a perfectly flat, horizontal, frictionless surface, which is only possible if his line of gravity intersects the support area [2], he exerts with his feet normal compression on the ground. The magnitude of this compression is equal to the weight of his body. Generally this compression is distributed over different parts of the feet. The ground surface, if it is

---

[2]   With regard to the terms "line of gravity" and "support area" see Braune and Fischer, op. cit.

rigid and unyielding, will exert a vertical compression upwards on the feet. According to the principle of the equality of action and reaction, the ground reaction force will have exactly the same magnitude as the weight of the body. It must thus be considered an external force. In this case the centre of gravity of the human body is subjected to the action of two forces, gravity and the ground reaction force. These two forces are equal and opposite. They thus cannot move the centre of gravity.

One can change the compression on the ground by means of muscular action. In many cases one can increase this compression by extension of the knees and hips, decrease it by flexion of these joints and even cancel it completely by quick flexion. The compression need not always remain vertical or perpendicular to the ground surface. Let us assume that the muscular contractions are such that the modified compression acts vertically on the ground. The ground reaction force will change as the compression is increased or decreased. Consequently, the centre of gravity of the body will no longer be subjected to two equal and opposite forces which cancel each other and it will be compelled to move. Since the resultant force acting on the centre of gravity is still vertical, the centre of gravity can only move vertically, upwards or downwards. If the compression on the ground were completely removed, the centre of gravity would start the same downward movement as in free fall.

If the compression on the ground is no longer perpendicular but oblique, it can be resolved into two components, one vertical and one horizontal. Only the vertical component produces compression on the ground and causes an equal and opposite ground reaction force. The horizontal component displaces the feet on the ground as long as there is no friction to oppose this movement. Whatever its cause, this movement of the feet cannot displace the centre of gravity forwards, backwards or sideways if the ground is perfectly flat because no external forces are provoked the direction of which would deviate from the vertical. It may well happen that the line of gravity moves out of the support area. Then the circumstances change: the centre of gravity falls until other parts of the body provide a support. When the line of gravity moves out of the support area, the compression on the ground disappears so that the centre of gravity must fall. To consider this interesting mechanical process further would be too great a digression.

If the ground surface is flat but not horizontal, the compression on the ground due to gravity is less than the body weight since only the component of the vertical gravity force perpendicular to the ground surface exerts compression. The component parallel to the ground surface will move the whole body on the oblique plane, in this case without any friction. The movement of the centre of gravity takes place in the direction of the second component. The action of muscles again cannot change anything in this movement of the centre of gravity as long as the ground surface is frictionless. In this case also, one can change the distance between the centre of gravity and the ground surface at will, within limits depending on the dimensions of the body. One can now increase or decrease the compression perpendicular to the ground and thus make the ground reaction force different from the component of the gravity force perpendicular to the ground. Depending on the predominance of one or the other, the centre of gravity will move further from or nearer to the ground surface. The component of movement parallel to the ground surface is not changed.

These few examples show that without friction voluntary displacement of the human body is impossible. They also show how friction is involved in the movement. As mentioned above, it is possible to change voluntarily the magnitude or direction of the compression on the ground surface by means of muscular contraction. If the body were built in such a way that, whatever the movements of the limbs, only compression perpendicular to the ground could be exerted, friction on the ground would be useless for locomotion. Friction intervenes when compression is oblique to the ground and thus possesses a component parallel to the latter. If the ground is sufficiently rough, this component is counterbalanced by friction. The effect of friction is thus exactly the same as that of the ground reaction force in the direction perpendicular to the ground. It acts as an external force equal and opposite to the component of the compression which is parallel to the ground surface. This force will be designated below as "force of friction". It hinders sliding of the foot and gives the centre of gravity propulsion in the direction of the force of friction. Taking advantage of friction, appropriate contraction of the muscles can move the centre of gravity of the body parallel to the ground surface or, if the centre of gravity is already moving, muscle contraction can accelerate or decelerate it or change its direction, or do both.

When a man walks in an enclosed area with no noticeable movement of air, the resistance of the air opposes the movement forwards, the more so the greater his velocity. During walking the significant velocities have the direction of gait. Therefore, the action of air resistance can be regarded, at a first approximation, as that of a force parallel to the ground and opposite to the direction of gait.

The external forces thus contribute, each in its own way, to moving the centre of gravity of the human body. If we assume that the ground is a horizontal surface, any acceleration or deceleration of the movements of the centre of gravity in a vertical direction must be attributed to the combined action of gravity and normal ground reaction force, in a horizontal direction to the combined action of friction and air resistance. We can modify at will the ground reaction force and the force of friction within certain limits by means of the muscles, and thus we can impose voluntarily on the body the direction and velocity of gait.

The exact tridimensional movement of the centre of gravity during walking can be established if its position, velocity and acceleration have been determined for any moment of at least one double step. The magnitude of the normal ground reaction force, the force of friction and the air resistance can then be calculated for the whole process of movement. Knowing the ground reaction force and the force of friction thus provides the starting point for evaluating the activity of the main groups of muscles involved in gait. To this end, one must acquire a thorough knowledge of the movements of the joints occurring during a double step, using the tridimensional co-ordinates of the joint centres.

This chapter deals with the tridimensional movement of the centre of gravity and the external forces involved in this process. In the first section the different methods used to determine the trajectory of the centre of gravity will be discussed. In the second section the co-ordinates of this trajectory will be calculated, and in the third the velocities and accelerations of the movement of the centre of gravity deduced. In the fourth and final section the external forces that induce the movement of the centre of gravity will be calculated based on the above data.

Methods of Determining the Trajectory of the Centre of Gravity

From the measurements given in Chap. 1, as many points of the tridimensional trajectory of the centre of gravity of the human body during walking can be obtained as there are phases of movement whose co-ordinates have been established. These points are so close to each other that the whole trajectory of the centre of gravity can be determined perfectly clearly.

There are two methods of locating the centre of gravity for a precise phase of movement: one can start either from the centres of gravity of the different segments of the body or from their principal points. In either case there are then three possibilities:

1. the total centre of gravity can be found by geometrical construction;
2. the co-ordinates of the centre of gravity can be calculated from the co-ordinates of the joint centres;
3. a mechanism can be devised to indicate automatically the position of the centre of gravity of the human body in any attitude.

Determination of the Trajectory of the Centre of Gravity from the Centres
of Gravity of the Different Body Segments

When analysing human gait we considered the body to be composed of 12 different segments. Therefore, we have to determine the centres of gravity of these 12 segments and then combine them to obtain the centre of gravity of the whole body, otherwise referred to as the total centre of gravity of the body.

The positions of the centres of gravity of the feet are obtained directly from photographs. Those of the other centres of gravity must be deduced for each phase from the position of the joint centres. As shown in the previous study on the position of the centre of gravity in the human body in men of the same build as our experimental subject, the centres of gravity of the thigh, lower leg and upper arm lie virtually on the long axis of the limb segment, i.e. on the line connecting the centres of the two adjacent joints. The position of the centre of gravity divides this line into certain proportions which are similar for all individuals of the same build.

The distance between the centre of gravity and the proximal joint and the distance between the centre of gravity and the distal joint are in a ratio of $0.44 : 0.56$ in the thigh, $0.42 : 0.58$ in the lower leg and $0.47 : 0.53$ in the upper arm (see p. 45 of Braune and Fischer op. cit.). The centre of gravity of the lower arm + hand system also lies on the long axis of the forearm when the hand is semi-pronated. The ratio of the distance of this centre of gravity from the centre of the elbow to its distance from the wrist is $0.66 : 0.34$. This was deduced as an average from the masses for cadavers III and IV indicated on pp. 42 and 43 of Braune and Fischer op. cit. Measurements on cadaver IV, the one most similar to our experimental subject, indicated that the centre of gravity of the trunk lies almost on the line connecting the midpoints of the hip and shoulder lines (trunk line), as described in Chap. 1, when the subject is erect. It lies at 25.8 cm (p. 48 op. cit.) from the hip line whereas the distance between the midpoint of the latter and the midpoint of the shoulder line is 49 cm (p. 21 op. cit.). The ratio of the distance of the centre of gravity of the trunk from the midpoint of the hip line to the distance of the centre of gravity of the trunk from the midpoint of the shoulder line is $0.53 : 0.47$. The distance of the centre of gravity of the trunk from the occipito-atlantal joint is 40.2 cm (p. 48 op. cit.), and the latter lies approximately in the prolongation of the trunk line. Therefore, the centre of gravity of the trunk divides the distance between the midpoint of the hip line and the centre of the occipito-atlantal joint in a ratio of $0.39 : 0.61$. This last measurement can be used to determine, at least approximately, the position of the centre of the occipito-atlantal joint in the different phases of movement. This was not possible from the photographs of gait. To this end, one has only to find a point on the prolongation of the trunk line whose distance from the midpoint of the shoulder line and the distance between the midpoints of the shoulder and hip lines are in a ratio of $17 : 49$ or $0.35 : 1$.

There are two ways of finding the centre of gravity of the head, both of which are inaccurate. However, they can give a good approximation of the position of the centre of gravity of the head since they both use the vertex which is not far from the centre of gravity. It can be assumed that this centre of gravity lies on the line connecting the vertex and the centre of the occipito-atlantal joint. Measurements on the photographs in our study on the centre of gravity indicate that the centre of gravity of the head divides this line in the ratio of $0.26 : 0.74$, being closer to the occipito-atlantal joint than to the vertex. One can also determine the position of the centre of gravity on the line connecting the midpoint of the shoulder line and the vertex. Measurements on the above-mentioned photographs in connection with the photographs of gait in Chap. 1 show that the ratio of the distance of the centre of gravity of the head from the midpoint of the shoulder line to that of the centre of gravity from the vertex is $0.70 : 0.30$ in experiments I and II and $0.68 : 0.32$ in experiment III. The difference results from the altered position of the shoulders due to the load in experiment III. Both calculations of the centre of gravity of the head are less accurate than those of the other centres of gravity. However, they indicate with sufficient precision the shape of the trajectory of the centre of gravity of the head. Any errors are the same for all phases. They may at most cause a slight displacement but not a deformation of this tridimensional trajectory. It would even be sufficiently accurate to identify the shape of the trajectory of the centre of gravity with that of the vertex. The latter trajectory would then be displaced downwards and backwards by a distance equal to that between the centre of gravity of the head and the vertex. Both ways of determination improve accuracy since they take some account of the movements of the head relative to the trunk. In the present case, determining the centre of gravity of the head only contributes to establishing the centre of gravity of the whole body. Even more considerable errors here would have negligible influence on the trajectory of the total centre of gravity because of the relatively small mass of the head.

The above data and the lines on which the centres of gravity of the body segments have to be found provide the means for determining the co-ordinates of the 12 partial centres of gravity and entering these centres of gravity into the projections of the different phases of movement in Figs. 22 and 23 of Chap. 1. Even if the long axes of the limbs appear generally shorter in the projection, they are still divided by the centre of gravity in the same proportions.

Having determined the trajectories of the 12 partial centres of gravity, one can combine these for each phase of movement to obtain the total centre of gravity. This is arrived at by a simple calculation (see p. 124). This construction demands a step by step procedure. Two centres of gravity are combined to obtain one common centre of gravity. This is then combined with a third centre of gravity, etc. The order here is irrelevant. However, the centres of gravity of certain systems, for instance, the lower leg + foot system, the lower extremities, the upper extremities, the trunk + head system, etc. are of particular importance in analysing the whole process of movement. It is, therefore, expedient to proceed as follows. On each side the centres of gravity of the upper arm and forearm + hand system are combined to obtain the centre of gravity of the arm, then the centres of gravity of the trunk and head are combined to obtain the centre of gravity of the trunk + head system. The common centre of gravity of the two arms is then calculated and combined with that of the trunk + head system to obtain the centre of gravity of the upper part of the body, supported by the two legs. For each side of the body the centres of gravity of the lower leg and foot are combined to obtain that of the lower leg + foot system. This is combined with the centre of gravity of the thigh to obtain that of the whole leg. Finally, the common centre of gravity of the two legs is calculated and combined with that of the trunk + head + both arms system to obtain the centre of gravity of the whole body.

In order to carry out these combinations of the centres of gravity it is necessary to know the proportions of the different masses. These are given on p. 37 of our study on the centre of gravity for cadaver IV which corresponds to our experimental subject. These figures will be used often in the present work and are, therefore, cited again here though in a different form. If the mass of the whole body is designated as 1, the masses of the different body segments are in the following proportions:

Head	0.07055	Trunk + head	0.49755
Trunk	0.4270	Lower leg + foot	0.07055
Thigh	0.1158	Whole leg	0.18635
Lower leg	0.052675	Whole arm	0.064875
Foot	0.017875	Both legs	0.3727
Upper arm	0.03365	Both arms	0.12975
Forearm + hand	0.031225	Trunk + head + both arms	0.6273

The common centre of gravity of two masses lies on the line connecting their individual centres of gravity and divides this line in inverse proportion to the masses. One thus finds the centre of gravity of an arm by dividing the line connecting the centre of gravity of the upper arm and that of the forearm + hand system in the proportion 0.031225 : 0.03365 or, more easily, 0.481 : 0.519. The centre of gravity of the trunk + head system divides the line connecting the centre of gravity of the trunk and that of the head in the proportion 0.142 : 0.858. The centre of gravity of the lower leg + foot system divides the line connecting the centre of gravity of the lower leg and that of the foot in the proportion 0.253 : 0.747. The centre of gravity of a leg divides the line connecting the centre of gravity of the thigh and that of the lower leg + foot system in the proportion 0.379 : 0.621. The common centre of gravity of the two arms lies, of course, in the middle of the line connecting the centres of gravity of the two arms. The same is true for the legs. To determine the centre of gravity of the trunk + head + arms system it is necessary to divide the line connecting the centre of gravity of the trunk + head system and the common centre of gravity of the two arms in the ratio of 0.207 : 0.793. Finally, the centre of gravity of the whole body lies on the line connecting the centre of gravity of the trunk + head + arms system and the common centre of gravity of the two legs and divides this line in the proportion 0.373 : 0.627.

We thus have all the means at hand to establish the position of the centre of gravity of the whole body and of the centres of gravity of individual important body segments for each phase of the movement, either by construction from the projections of gait as given in Figs. 22 and 23 in Chap. 1 for experiments I and II, or by calculation from the co-ordinates of the joint centres, or by means of an appropriate mechanical contrivance.

With regard to the construction, it must be noted that the proportions given above are also true for the projections of the lines connecting the partial centres of gravity. If two different projections of this construction have been carried out, one obtains two projections of the trajectories of the different centres of gravity. These are sufficient to establish the tridimensional trajectories which usually are double curves. If the two projections are that on the plane of gait and that on the horizontal ground, as in Figs. 22 and 23 of Chap. 1, the tridimensional co-ordinates of the total centre of gravity and of the partial centres of gravity can be read directly. The accuracy attained in this manner is the greater the larger the scale of the projections. It is, therefore, recommended to draw a life-size or at least a half life-size construction.

If a calculation is used to arrive at the co-ordinates of the total centre of gravity, one must first determine the co-ordinates of the partial centres of gravity. To this end, for the thigh, lower leg and upper arm, the differences between the co-ordinates of the distal and proximal joint centres (Tables 14–16, Chap. 1) must be calculated, multiplied respectively by 0.44, 0.42 and 0.47 (see p. 121) and the products added to the co-ordinates of the proximal joint centre for each of the three body segments. To obtain the co-ordinates of the centre of gravity of the forearm + hand system, the differences of the co-ordinates of the wrist centre and those of the elbow centre are multiplied by 0.66 and the products are added to the co-ordinates of the elbow centre. The co-ordinates of the centre of gravity of the trunk are found by multiplying the differences of the co-ordinates of the midpoint of the shoulder line and those of the midpoint of the hip line by 0.53 and adding the products to the co-ordinates of the midpoint of the hip line. Finally, the co-ordinates of the centre of gravity of the head are found from the co-ordinates of the vertex and midpoint of the shoulder line, given in Tables 14–16 of Chap. 1. Their differences are multiplied by 0.7 in experiments I and II, by 0.68 in experiment III. The products are added to the co-ordinates of the midpoint of the shoulder line. The co-ordinates of the centre of gravity of the foot are given in these tables.

From these co-ordinates of the partial centres of gravity, there are two ways to arrive at the co-ordinates of the total centre of gravity of the human body. The simplest way consists of multiplying each of the co-ordinates of the 12 partial centres of gravity by the proportional figure corresponding to the body segment considered and adding algebraically the 12 products. The 12 body segments are numbered; x_i, y_i, z_i signify the three co-ordinates of the centre of gravity. The figure of the mass proportion of the body segment numbered i is designated as μ_i. Then the three co-ordinates x_0, y_0, and z_0 of the total centre of gravity can be calculated as follows:

$$x_0 = \mu_1 x_1 + \mu_2 x_2 + \mu_3 x_3 \cdots + \mu_{12} x_{12} = \sum_1^{12} {}^i \mu_i x_i,$$

$$y_0 = \mu_1 y_1 + \mu_2 y_2 + \mu_3 y_3 \cdots + \mu_{12} y_{12} = \sum_1^{12} {}^i \mu_i y_i,$$

$$z_0 = \mu_1 z_1 + \mu_2 z_2 + \mu_3 z_3 \cdots + \mu_{12} z_{12} = \sum_1^{12} {}^i \mu_i z_i.$$

These equations are simpler and easier for the calculation than their counterparts in the work on the centre of gravity (Braune and Fischer, op. cit. p. 33) because they do not use the masses themselves but rather their proportional figures. Since the total mass of the body is 1, the numerator disappears from the previous equations.

The calculation can also closely follow the construction of the total centre of gravity from the partial ones. This method is more awkward but offers the advantage of providing both the co-ordinates of the centres of gravity of important systems, such as of the whole extremities or the trunk + head + arms system, and the co-ordinates of the total centre of gravity. For this calculation, the proportional figures indicated above are used in each case to determine the position of the common centre of gravity on the line connecting two partial centres of gravity.

For example, to obtain the co-ordinates of the centre of gravity of the arm, one has to calculate the differences between the co-ordinates of the centre of gravity of the forearm + hand system and those of the centre of gravity of the upper arm. These differences are then multiplied by 0.481 and added to the co-ordinates of the centre of gravity of the upper arm. The co-ordinates of the centre of gravity of the trunk + head system are obtained by multiplying the differences between the co-ordinates of the centre of gravity of the head and those of the centre of gravity of the trunk by 0.142 and adding the products to the co-ordinates of the centre of gravity of the trunk. Multiplying by 0.253 the differences between the co-ordinates of the centre of gravity of the foot and those of the centre of gravity of the lower leg, and adding the products to the co-ordinates of the centre of gravity of the lower leg gives the co-ordinates of the centre of gravity of the lower leg + foot system. If we subtract from these the co-ordinates of the centre of gravity of the thigh, multiply the differences by 0.379 and add the products to the co-ordinates of the centre of gravity of the thigh, we obtain the co-ordinates of the centre of gravity of the whole leg. The co-ordinates of the common centre of gravity of the two arms are simply the arithmetic averages of the corresponding co-ordinates of the centres of gravity of the two arms. Similarly, the co-ordinates of the common centre of gravity of the two legs are the arithmetic averages of the corresponding co-ordinates of the centres of gravity of the two legs. If the co-ordinates of the centre of gravity of the trunk + head system are subtracted from the corresponding co-ordinates of the common centre of gravity of the two arms, the differences multiplied by 0.207 and the products added to the co-ordinates of the centre of gravity of the trunk + head system, the co-ordinates of the centre of gravity of the trunk + head + two arms system are obtained. If these co-ordinates are subtracted from the corresponding co-ordinates of the common centre of gravity of the two legs, the differences multiplied by 0.373 and the products added to the co-ordinates of the trunk + head + two arms system, the co-ordinates of the total centre of gravity of the human body are finally obtained.

Based on the partial centres of gravity, it is possible to build an articulated mechanical device which automatically indicates the position of the total centre of gravity for each position of the human body. Such a device correlates very closely with the construction of the total centre of gravity from the partial ones. The construction is carried out mechanically

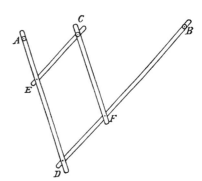

Fig. 1. Articulated device used to determine the centre of gravity of two body segments

and relies on the division of different segments into certain proportions. The whole device combines different partial devices, each of which divides a segment that is variable in length within certain limits, according to particular proportions whatever the length of the segment. Figure 1 represents such an articulated device. The four rods AD, DB, EC and FC are hinged at D, E, F and C. The axes of the hinges are parallel to each other. The position of the joint axes and the dimensions of the different rods are such that:

$$AE : ED = DF : FB, \quad \text{further,} \quad EC = DF \quad \text{and} \quad FC = DE$$

The quadrangle $ECFD$ thus represents a parallelogram. The three points A, C and B always lie in line and for each position of the joints we have:

$$AC : CB = AE : ED = DF : FB$$

The lengths of the two rods AD and DB can to a certain extent be chosen arbitrarily. Their sum must be at least, and their difference at most, equal to the shortest distance which can separate the two points A and B. In the special case where these two points approach each other and coincide, AD and DB must be equal. Otherwise, it is not compulsory that they be equal. If the device is attached to a firm surface at point A such that it can articulate about an axis parallel to the axes of the other hinges, it represents a classic pantograph. If B is moved on a curve, C draws a similar but reduced curve in the proportion $AC : AB$. If points A and B of the device are articulated at the centres of gravity of two bodies, point C represents the common centre of gravity of the two bodies for any mutual position of these two bodies, as long as segments AE and ED on one hand, and DF and FB on the other, are inversely proportional to the masses of the two bodies. Generally, points A and B of the device must be articulated with the centres of gravity of the two bodies by ball and socket joints. Only when the bodies themselves articulate with each other by hinge joints can the connections A and B be simple hinges. The axes of the latter, the four axes of the device and the axis of the hinge between the two bodies must be parallel.

Of the joints in the human body, few can be considered as hinges in a strict mechanical sense. Most of them present different degrees of freedom or allow oscillations of their axes, at least within certain limits. However, if we look at the projection of the movement of the human body on any plane, only that part of the movement of the joint appears which consists of a rotation of the projections of the two articulated body segments about an axis perpendicular to the plane of projection. Generally, the dimensions of the projections of the body segments vary with this rotation, and the variation is the greater the more the actual movement of the joint deviates from a pure rotation about the axis perpendicular to the plane of projection. If the axis of the actual movement of the joint is close to the axis perpendicular to the plane of projection, the projections of the body segments will not significantly vary in magnitude during the movement. This occurs in one projection only, during walking. Here, the rotations in all the joints take place about axes in approximately the same direction, perpendicular to the plane of gait. The long axes of all the body segments thus remain almost parallel to the plane of gait during their movement. They appear very close to their true magnitudes in their projection on the plane of gait and their lengths undergo only relatively small changes during the movement.

A model of the projection of the human body on the median plane is built from sheet-metal or cardboard with its main joints thus articulated. By changing the position of the joints almost all phases of the movements of the human body during walking can be accurately represented in their projection on the plane of gait. To this end, the 12 segments into which the human body has been divided only need to be connected by hinges whose axes are parallel to each other and perpendicular to the plane of gait. A mechanical device that automatically shows the projection of the total centre of gravity on the plane of gait can be easily fixed to this model. In this particular case, the different parts of this device only need to be connected with each other and with the projections of the partial centres of gravity by hinges all the axes of which are parallel to the axes of the joints of the model.

Figure 2 represents such a model of the projection of the human body with the device for determining the centre of gravity. The author made this model out of cardboard at a scale of $1:2$ [3]. It can be made in metal by E. Zimmermann, precision mechanic of Leipzig. The device for determining the centre of gravity comprises different elements of the type shown in Fig. 1. In Fig. 2 the 12 segments of the body are numbered and the partial centres of gravity are designated by the letter S with the number of the corresponding segment as subset figure. The trunk is numbered 1, the three segments of the right lower limb proceeding in a distal direction are 2, 4 and 6, those of the left lower limb 3, 5 and 7, the two segments of the right upper limb 8 and 10, those of the left upper limb 9 and 11, the head 12. For example, S_1 signifies the centre of gravity of the trunk, S_7 the centre of gravity of the left foot. The joints between the different body segments are designated by the letter G with the two figures representing the adjacent segments subset. For example, $G_{1,2}$ signifies the right hip, $G_{9,11}$ the left elbow, etc. The common centre of gravity of different body segments is indicated by multiple subset figures designating all the body segments participating in the common centre of gravity. Where possible, these are given in abbreviated form, such as 2–7 for the common centre of gravity of the two legs or 8–11 for that of the two arms. For example, $S_{2,4,6}$ signifies the centre of gravity of the right leg, $S_{5,7}$ that of the left lower leg + foot, $S_{1,12}$ that of the trunk + head, $S_{1,8-12}$ that of the part of the human body supported by the legs. The centre of gravity of the whole body is designated as S_0.

Figure 2 shows the different segments of the body and the lengths of the various strips of cardboard at a scale of $1:10$. This enables anyone to construct the whole mechanical device. Except for the total centre of gravity S_0, all the centres of gravity with one digit relate to one segment and, therefore, maintain the same position in this segment of the model. On the other hand, all the centres of gravity with multiple digits correspond to a system and generally change their position in relation to the different body segments when proceeding from one phase to another, as does the total centre of gravity S_0. The sizes of the different partial devices are of course given exactly by the proportions listed above. For example, the device between S_5 and S_7 combines the centre of gravity of the left lower leg with that of the foot into the centre of gravity $S_{5,7}$ of the left lower leg + foot system. The different strips of cardboard or better of sheet-metal are chosen such that $S_5 S_{5,7} : S_{5,7} S_7 = 0.253 : 0.747$ (see p. 123). The centre of gravity $S_{5,7}$ is combined with S_3 by the partial device which links it with the centre of gravity S_3 of the left thigh to give the centre of gravity $S_{3,5,7}$ of the whole left leg. The lengths of the strips of cardboard must here be such that $S_3 S_{3,5,7} : S_{3,5,7} S_{5,7} = 0.379 : 0.621$. The device which combines the centre of gravity of the left leg $S_{3,5,7}$ with

---

[3]   See no. 262 of the main catalogue of mathematical and mathematical-physical models, devices and instruments, published by the Deutsche Mathematiker-Vereinigung in 1892

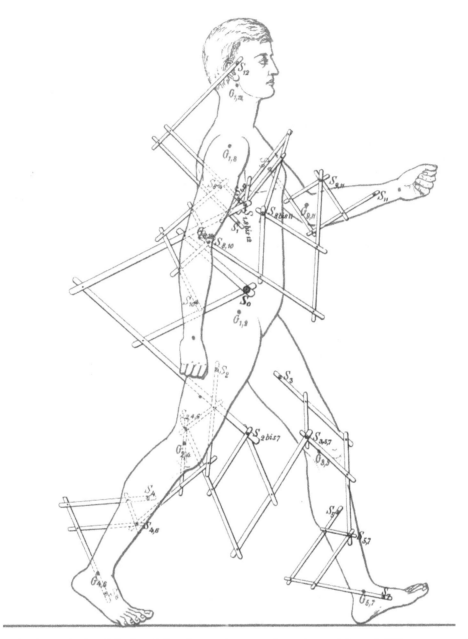

Fig. 2. Articulated device giving the total centre of gravity and that of different parts of the human body automatically

that of the right leg $S_{2,4,6}$ to give the common centre of gravity of the two legs S_{2-7} must be built in such a way that this centre of gravity always lies exactly centrally between the legs. Since the centres of gravity of the two legs can be superimposed in the projection, the two longer strips must be of the same length; the two shorter strips must, therefore, be half the length of the longer strips. The device between S_1 and S_{12} combines the centre of gravity of the trunk with that of the head to give the centre of gravity of the head + trunk system. The lengths of its cardboard strips must be chosen such that the following always applies: $S_1 S_{1,12} : S_{1,12} S_{12} = 0.142 : 0.858$. The device articulated at the centre of gravity S_{2-7} of the two legs and at the centre of gravity $S_{1,8-12}$ of the trunk + head + two arms system gives the centre of gravity S_0 of the whole body. The sizes of its parts must be in the following proportion, which remains true for any position of the human body:

$$S_{1,8-12}S_0 : S_0S_{2-7} = 0.373 : 0.626 \text{ (see p. 123)}$$

Further explanation would be superfluous since Fig. 2 is self-explanatory. The device automatically gives the centre of gravity of the whole body based on the partial centres of gravity, taking into account exactly the proportions listed above, in the projection on the plane of gait. It works with great accuracy when precisely built. The only inaccuracy affecting its results originates in the fact that the projections of the long axes of the limbs on the plane of gait change somewhat in length during the course of a step. This source of error can be eliminated by making the position of the joints variable to a certain extent and by ensuring that the partial centre of gravity of each body segment always divides the long axis of the latter in the proportion given above (p. 121).

Given the projection of the movement on the plane of gait, it is possible to use the device thus described to draw the trajectory of the total centre of gravity and those of the partial centres of gravity in this projection. To this end, the device is placed on the different phases of the movement. Or, by fixing small pencils to the articulations of the device, the drawing can even be performed automatically. If the gait is projected on millimetre paper, two co-ordinates of the centres of gravity, the x- and z-co-ordinates, can be read directly. The device does not give the y-co-ordinate, of course, since it works only in one projection. To obtain the y-co-ordinate one would have to use a second device giving the projection on the horizontal ground or on a plane perpendicular to the direction of gait. In these two projections, however, the long axes of the limbs appear at times very much shortened and undergo considerable variations in length during walking. These variations cannot be ignored as they can be in the projection on the plane of gait. Therefore, a device to obtain the centre of gravity as described above would become very complicated. If we want to determine the tridimensional position of the centre of gravity rather than one projection of the latter, we must use a tridimensional model of the human body, such as a skeleton with articulated main joints, and adapt to this model a device built according to the principles outlined above. This device could be constructed as shown in Fig. 2. The extremities of each element of the device must be connected to the bones at the centres of gravity and to the adjacent elements through ball and socket joints. Some of the bones, for instance some ribs or part of the skull, will have to be discarded so as not to hinder the movement of the different elements of the device. Otherwise, the construction of such a tridimensional device for determining the centre of gravity should not encounter technical difficulties. The author has hitherto only lacked the means to construct it. Such a device would work as easily as the one actually built out of cardboard strips. If a body segment is moved, for instance the right forearm + hand system, all the centres of gravity will not change place, but only $S_{8,10}$, S_{8-11}, S_{8-12} and S_0. The displacements of these four centres of gravity are the smaller the greater the corresponding system of masses. The centre of gravity S_0 of the whole body will thus always carry out the smallest movement. The force with which this movement is carried out is the greater the smaller the displacement, according to the principle of virtual velocities. Therefore, only a very small effort is required to move the whole device by a rotation of the forearm at the elbow. The total centre of gravity will always react to this rotation with great force.

Determining the Trajectory of the Centre of Gravity from the Principal Points of the Different Segments of the Human Body

In addition to the centres of gravity, other fixed points in the different elements play an important role in the mechanics of the human body, as in that of any system in which the component parts are connected to each other by joints. The position of each of the centres of gravity depends only on the distribution of mass in the individual body segment. It has no relation whatever to the masses of the other elements of the human body. In each body segment, however, there is a second fixed point which depends not only on the construction of this segment but also on the connection of this segment with all the others. This point is thus more closely related to the total centre of gravity of the human body than the partial centre of gravity itself. I shall call this point the principal point of the body segment. It can be imagined as the centre of gravity of a system of articulated masses, obtained by concentrating in the centre of each adjacent joint (of the body segment) the masses of all the parts of the body which are connected directly or indirectly with the body segment in question through this joint[4]. For example, the principal point of the right thigh is obtained by assuming the masses of the right lower leg and foot to be concentrated in the centre of the right knee and the masses of all the other parts of the body – trunk, left leg, both arms and head – to be concentrated in the centre of the right hip. The centre of gravity of the thigh loaded in this manner is then determined. The system of masses thus constituted by the thigh has the weight of the whole body. I shall refer to this as the reduced thigh system. Similarly, the principal point of the trunk represents the centre of gravity of the reduced trunk system. This system is obtained by loading the trunk in the centre of each shoulder with the mass of the ipsilateral arm, in the centre of each hip with the mass of the ipsilateral leg, and in the centre of the occipito-atlantal joint with the mass of the head.

Whereas, in the trunk, the principal point lies in the vicinity of the centre of gravity, that of the thigh must be relatively distant from the centre of gravity. It must lie fairly close to the hip joint because of the huge concentration of mass in the centre of this joint. The same is true for all the segments of the extremities. Throughout, the principal point lies closer to the proximal joint than does the centre of gravity of the limb under consideration.

The principal points are thus regarded as the centres of gravity of the 12 reduced systems of the human body. Their precise position is easily determined for each body segment. The position of the common centre of gravity of several masses depends on the proportions of the latter and not on their absolute magnitudes. Therefore, to determine the positions of the principal points, the figures on p. 122 can be used. They give the proportions of the masses of the body segments in relation to the total mass of the human body.

The principal point of a thigh thus represents the common centre of gravity of three masses concentrated in the centre of the hip, the centre of the knee and the centre of gravity of the thigh, the proportional figures of which are 0.81365, 0.07055 and 0.1158 respectively. These three mass points are successively combined in the manner detailed above, taking into account that the centre of gravity of the thigh lies on the line connecting the centres of the hip and knee and divides this line in the ratio of 0.44 : 0.56. The principal point of the thigh which also lies on this connecting line is found to divide it in the ratio of 0.122 : 0.878. If l is the distance between the centres of the hip and knee, the principal point of the thigh lies on the long axis of the latter at a distance of $0.122 \cdot l$ from the centre of the hip and $0.878 \cdot l$ from the centre of the knee.

The principal point of a lower leg is the common centre of gravity of three masses concentrated at the centre of the knee, the centre of the ankle and the centre of gravity of the lower leg, the proportional figures of which are 0.92945, 0.017875 and 0.052675 respectively. Since the centre of gravity of the lower leg lies on the line connecting the centres of the knee and ankle and divides the line in the ratio of 0.42 : 0.58, the principal point of the lower leg also lies on this line which it divides in the ratio of 0.04 : 0.96. If l is the length of the lower leg between the centres of the knee and ankle, the principal point of the lower leg lies at a distance of $0.04 \cdot l$ from the centre of the knee and $0.96 \cdot l$ from the centre of the ankle.

The principal point of a foot appears as the common centre of gravity of only two masses, one concentrated at the centre of the ankle and the other at the centre of gravity of the foot.

---

[4]   Fischer O (1894) Die Arbeit der Muskeln und die lebendige Kraft des menschlichen Körpers. Abhandlungen der mathematisch-physischen Klasse der Königlich Sächsischen Gesellschaft der Wissenschaften, vol 20, nr 1

Their proportional figures are 0.982125 and 0.017875. The principal point of the foot thus divides the line connecting these two points in the ratio of 0.018 : 0.982. If r designates the distance between the centre of the ankle and the centre of gravity of the foot, the principal point of the foot is at a distance of $0.018 \cdot r$ from the former and $0.982 \cdot r$ from the latter. It thus lies very close to the centre of the ankle.

The principal point of an upper arm is the common centre of gravity of three masses concentrated one at the centre of the shoulder, one at the centre of the elbow and one at the centre of gravity of the upper arm. The proportional figures of these masses are 0.935125, 0.031225 and 0.03365 respectively. The centre of gravity of the upper arm lies on the line connecting the centres of the shoulder and elbow and divides this line in the ratio of 0.47 : 0.53. The principal point of the upper arm also lies on this line and divides it in the ratio of 0.047 : 0.953. For a length l of the upper arm measured between the centre of the shoulder and the centre of the elbow, the principal point of the upper arm lies at a distance of $0.047 \cdot l$ from the former and $0.935 \cdot l$ from the latter.

The principal point of the forearm + hand system on each side is the common centre of gravity of two masses one of which must be imagined as concentrated at the centre of the elbow and the other at the centre of gravity of the forearm + hand system. The proportional figures of these masses are 0.968775 and 0.031225 respectively. With the hand semi-pronated, the centre of gravity of the forearm + hand system lies on the line connecting the centre of the elbow and the centre of the wrist and divides this line in the ratio of 0.66 : 0.34. The principal point of the forearm + hand system also lies on this line which it divides in the ratio of 0.021 : 0.979. If l designates the distance between the centre of the elbow and the centre of the wrist, the principal point of the forearm + hand system lies at a distance of $0.021 \cdot l$ from the centre of the elbow and $0.979 \cdot l$ from the centre of the wrist, i.e. very close to the former. Determining the principal point of the forearm alone and the hand alone is of little use in the present analysis of gait since the hand did not move with respect to the forearm and the two were regarded as a single rigid body segment. In analysing other types of movements of the human body, for example those of a pianist, not only the individual principal points of the forearm and hand would need to be considered but also those of the phalanges.

The principal point of the trunk represents the common centre of gravity of no less than six masses which must be thought of as concentrated at the centres of the two hips, the two shoulders, the occipito-atlantal joint and at the centre of gravity of the trunk. The mass concentrated at the centre of each hip has a proportional figure of 0.18635, the mass at the centre of each shoulder a proportional figure of 0.064875, the mass at the centre of the occipito-atlantal joint a proportional figure of 0.07055, and the mass at the centre of gravity of the trunk a proportional figure of 0.427. To facilitate the determination of the principal point the two masses concentrated at the centres of the hips can be replaced by one lying at the midpoint of the hip line with a proportional figure of 0.3727, and the two masses concentrated at the centres of the shoulders by one lying at the midpoint of the shoulder line with a proportional figure of 0.12975. Only four masses then remain to be combined into one common centre of gravity. These four masses lie on one straight line, the trunk line, and, therefore, the principal point of the trunk will also be found on this line. The ratio of the distance between the centre of the occipito-atlantal joint and the midpoint of the hip line to the distance between the midpoint of the hip line and the midpoint of the shoulder line is 1.35 : 1. The centre of gravity of the trunk divides the line connecting the midpoint of the shoulder line and the midpoint of the hip line in the ratio of 0.53 : 0.47. Calculation shows that the principal point of the trunk divides this same connecting line in the ratio of 0.451 : 0.549. The principal point of the trunk divides the line connecting the midpoint of the hip line and the centre of the occipito-atlantal joint in the ratio of 0.333 : 0.667. If h represents the distance between the midpoint of the shoulder line and the midpoint of the hip line, and l the distance between the latter and the centre of the occipito-atlantal joint, the principal point of the trunk lies at a distance of 0.451 h or 0.333 l from the midpoint of the hip line, a distance of 0.549 h from the midpoint of the shoulder line and a distance of 0.667 · l from the centre of the occipito-atlantal joint. Both 0.541 · h and 0.333 · l give the distance between the principal point of the trunk and the midpoint of the hip line. They must be equal of course, since 1.35 $l = h$ and 0.451 = 1.35 · 0.333. If b is half the distance between the centres of the hips and a half the distance between the centres of the shoulders when the body stands erect with the hip and shoulder lines nearly perpendicular to the trunk line, the principal point of the trunk is at a distance $\sqrt{0.203 \cdot h^2 + b^2}$ or $\sqrt{0.111 \cdot l^2 + b^2}$ from the centre of each hip and at a distance $\sqrt{0.301 \cdot l^2 + a^2}$ from the centre of each shoulder. These distances of course become inaccurate as soon as one of the two lines becomes inclined to the trunk line. They can thus be considered only as average values.

The movement of the head relative to the trunk during walking cannot be considered, strictly speaking, as a simple rotation about a point fixed in the trunk since the cervical spine flexes a little during this movement. This has been stressed in Chap. 1. However, a fixed centre of rotation will be assumed to enable us to take into consideration, at least approximately, the movements fairly complicated of the head relative to the midpart of the trunk. We can then come closer to the true movement and even take into account up to a certain extent the flexion of the cervical spine if we displace the fixed centre of rotation from the centre of the occipito-atlantal joint distally to the region of the distal end of the cervical spine. The movements of the head have been studied relative to the midpoint of the shoulder line for this reason in Chap. 1. There the rotations of the line connecting the midpoint of the shoulder line and the vertex, which was called "head line", relative to the trunk were analysed. This assumption concerning the position of the centre of rotation of the head changes the position of the principal point of the trunk only a little since it means that the head is articulated with the trunk at the midpoint of the shoulder line and otherwise represents a rigid mass. To determine the principal point of the trunk the mass of the head with its proportional figure of 0.07055 must then be imagined as concentrated in the midpoint of the shoulder line. This displaces the principal point of the trunk on the trunk line distally. The principal point of the trunk then divides the line connecting the midpoint of the hip line and the midpoint of the shoulder line in the ratio of 0.427 : 0.573. The distance of the principal point from the midpoint of the hip line is thus 0.427 · h and that from the midpoint of the shoulder line 0.573 h. The difference between these data and the previous ones is small: 0.024, less than 2.5 % of the distance between the midpoint of the hip line and the midpoint of the shoulder line. The actual movements of the head in relation to the trunk can be determined only by assuming an instant centre of rotation moving between the lower end of the cervical spine and the occipito-atlantal joint. The small difference resulting from the extreme positions of the centre of rotation shows that neither of the two assumptions entails an error which may jeopardize the accuracy. In the second hypothesis, the principal point of the trunk would lie at a distance of $\sqrt{0.182 h^2 + b^2}$ from the centre of each shoulder and a distance of $\sqrt{0.328 h^2 + a^2}$ from the centre of each hip.

Two different positions for the principal point of the head are also obtained depending on whether the centre of rotation is assumed to be in the occipito-atlantal joint or on the shoulder line. In the former case, the principal point of the head represents the common centre of gravity of two masses, one concentrated at the centre of the occipito-atlantal joint, the other at the centre of gravity of the head; their proportional figures are 0.92945 and 0.07055 respectively. The principal point, like the centre of gravity of the head, lies approximately on the line connecting the centre of the occipito-atlantal joint and the vertex. The centre of gravity of the head divides this line in the ratio of 0.26 : 0.74 and, as results from calculation, the principal point of the head divides this same line in the ratio of 0.018 : 0.982. If the length of this line is l, the distance of the principal point from the centre of the occipito-atlantal joint is 0.018 · l, and from the vertex 0.982 · l.

In the latter case the centre of rotation coincides with the midpoint of the shoulder line. The principal point then lies elsewhere. Its position depends mainly on the position of the joint connections of the head with the other segments of the body. The principal point then represents the common centre of gravity of two masses the proportional figures of which are 0.92945 and 0.07055. The first is concentrated in the midpoint of the shoulder line. The centre of gravity of the head lies on the line connecting the midpoint of the shoulder line and the vertex, dividing this line in the ratio of $0.7 : 0.3$ in experiments I and II, in the ratio of $0.68 : 0.32$ in experiment III. Therefore, the principal point of the head divides this line in the ratio of $0.049 : 0.951$ in experiments I and II, of $0.048 : 0.952$ in experiment III, and it lies on the head line. If h is the distance between the midpoint of the shoulder line and the vertex, the principal point of the head is at a distance of $0.049 \cdot h$ from the midpoint of the shoulder line and $0.951 \cdot h$ from the vertex in experiment I and II, at a distance of $0.048 \cdot h$ from the midpoint of the shoulder line and $0.952 \cdot h$ from the vertex in experiment III. In this second hypothesis it thus lies very close to the shoulder line.

The principal points of the 12 body segments thus were determined for the individual whose mass proportions for the different body segments are given on p. 122. The data concerning the distances from the principal points to the adjacent joint centres are given quite generally so that the distances can be easily calculated in a particular case after the dimensions of the individual under investigation have been directly measured.

The distances between the centres of the joints of a body segment and the principal point of the latter are called "principal distances". The thigh, the lower leg and the arm thus possess two principal distances, the forearm + hand system, the foot and the head each possesses one and the trunk has five.

A construction based on the principal points and principal distances enables us to position the total centre of gravity of the human body. Starting from the principal point of any one of the 12 body segments, the principal distances of the other 11 segments are then put together in any order, ensuring that their direction in the body is towards this principal point. The final point of the irregular polygon thus constructed will always coincide with the total centre of gravity of the human body. If this construction is carried out not in three dimensions but in the projection of the body on any plane, we combine of course the projections of the different principal distances. We then obtain the projection of the total centre of gravity on this plane. I have given the detailed proof of this law in a previous study[5].

If, for example, we start from the principal point of the trunk, the tridimensional polygon is drawn by combining the lines connecting the centre of each hip with the principal point of the ipsilateral thigh, the centre of each knee with the principal point of the ipsilateral lower leg, the centre of each ankle with the principal point of the ipsilateral foot, the centre of each shoulder with the principal point of the ipsilateral upper arm, the centre of each elbow with the principal point of the ipsilateral forearm + hand system, and the centre of the occipito-atlantal joint with the principal point of the head. Instead of the last distance it is possible to take the longer distance between the midpoint of the shoulder line and the principal point of the head, which is in accordance with the assumption that the centre of rotation of the head is the midpoint of the shoulder line. In this case, we have to start from the most distal principal point of the trunk, in accordance with our hypothesis. In both cases of course the same final result is obtained. The centre of gravity of the head divides the line connecting the centre of the occipito-atlantal joint and the vertex in the ratio of $0.26 : 0.74$, and the line connecting the midpoint of the shoulder line and the vertex in the ratio of $0.7 : 0.3$. The former connecting line is in the ratio of $0.68 : 1$, and the latter in the ratio of $1.68 : 1$ to the line connecting the

---

5   Fischer O (1894) Die Arbeit der Muskeln und die lebendige Kraft des menschlichen Körpers. Abhandlungen der mathematisch-physischen Klasse der Königlich Sächsischen Gesellschaft der Wissenschaften, vol 20, nr 1, p 72

midpoint of the shoulder line and the centre of the occipito-atlantal joint. The line connecting the midpoint of the shoulder line and the centre of the occipito-atlantal joint is in the ratio of 0.35:1 to the distance between the midpoints of the hip and shoulder lines. Therefore, the line connecting the centre of the occipito-atlantal joint and the vertex is in the ratio of 0.24:1 to the line connecting the midpoints of the hip and shoulder lines. The line connecting the midpoint of the shoulder line and the vertex is in the ratio of 0.59:1 to the line connecting the midpoints of the hip and shoulder lines. It thus appears that the two different principal distances of the head, $0.018 \cdot l$ and $0.049 \cdot h$ (see p. 131 and 132), are in the ratios of 0.004:1 and 0.028:1 to the distance between the midpoints of the hip and shoulder lines. The principal distance is thus smaller by 2.4% of the distance between the midpoints of the hip and shoulder lines when the centre of rotation of the head is assumed to be in the occipito-atlantal joint than when it is assumed to be on the shoulder line. In the former case, however, the principal point of the trunk is higher on the trunk line than in the latter case by exactly this distance (see p. 131). For a normal erect position, the location of the total centre of gravity of the body will be exactly the same in the two cases. For other positions of the body it will be almost in the same place.

It is also possible to start from any other principal point in order to determine the total centre of gravity. For example, it is expedient to start from the principal point of the foot resting on the ground if the movements of the centre of gravity during the single-support period of gait are to be analysed. Other principal distances then have to be considered: for the lower leg of the supporting limb, the line connecting the centre of the ankle and the principal point of the lower leg; for the thigh of the same leg, the line connecting the centre of the knee and the principal point of the thigh. The principal distance of the supporting foot is now omitted, but a principal distance of the trunk is included. The latter is the line connecting the centre of the hip of the supporting leg and the principal point of the trunk. The principal distances of the other eight body segments remain the same as when starting from the principal point of the trunk.

As an example of this method of determining the total centre of gravity, the construction for phase 11 of experiment I was carried out to life-size in Fig. 3 in two projections, the plane of gait and the plane perpendicular to the direction of gait, starting from the principal point of the trunk. If we start from the principal point of the trunk, the principal distances to be used are relatively short. Therefore, the polygon occupies a relatively small space even if the life-size principal distances are used. This is a particular advantage in starting from the principal point of the trunk. The order in which the principal distances were used is shown in Fig. 3. To demonstrate that this order does not affect the final result, the construction was also carried out in dotted lines using the principal distances in a different order. To give an idea of the position of the principal point of the trunk H_1 and the total centre of gravity S_0 of the body, the outlines of the two femoral heads have also been drawn in the projection on the plane of gait.

Construction of the total centre of gravity using the principal points presents great advantages over the method involving the partial centres of gravity (p. 121). First, it is much quicker. If the position of the principal points in the different body segments has been determined, establishing the total centre of gravity only requires geometric addition, the combination of 11 straight lines given by the principal points. Using the partial centres of gravity, all the lines connecting the different centres of gravity have to be divided according to certain proportions. Determining the principal points does not require more work than establishing the partial centres of gravity. Both require division of the long axes of the different body segments according to given proportions. More important than this practical advantage when using the principal points to find the total centre of gravity is that the principal distances provide a direct measurement of the influence which the movement of the different body segments and systems of body segments exerts on the position of the total centre of gravity. The position of the total centre of gravity in position 11 in experiment I has

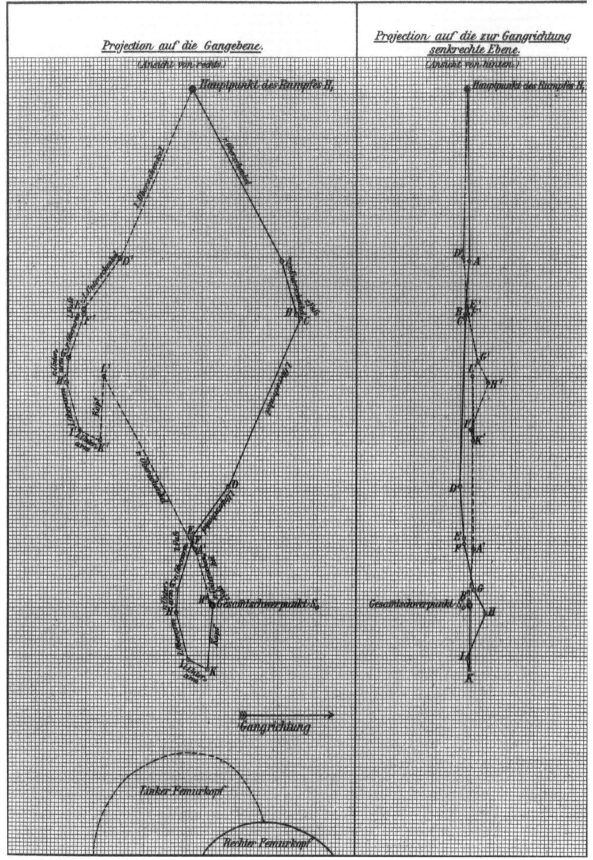

Fig. 3. Determination of the total centre of gravity of the human body using the life-size principal distance (phase 11 of experiment I)

been determined by using the principal points and principal distances (Fig. 3). If, for example, the head is moved, only the last $\overline{KS_0}$ of the 11 principal distances will change its direction, i.e. it will turn about the point K. For each rotation of the head about the midpoint of the shoulder line (cf. p. 131), the total centre of gravity S_0 of the human body thus describes a circle the radius of which is the principal distance of the head. The centre of this circle can be derived by subtracting the principal distance of the head from S_0. If the whole right leg is rotated about an axis through the hip whilst the position of the knee and ankle in phase 11 of experiment I is maintained, the total centre of gravity S_0 will also describe a circle as long as all the other body segments do not move. The radius of this circle is the combination of the three principal distances of the right thigh, lower leg and foot which maintain their orientation relative to each other: this radius is equal to their geometric sum. This can be seen immediately if the other construction in dotted lines in Fig. 3 is considered. The last three lines of the polygon are the principal distances of the segments of the leg. If the positions of the knee and ankle do not change, these lines will maintain their angulations to one another. With rotation of the whole leg, the centre of gravity S_0 will describe a circle about L' the radius of which is the distance $L'S_0$. The centre L' of this circle is obtained by subtracting successively from the total centre of gravity S_0 the principal distances of the three segments of the right leg. Similarly, for any movement of the forearm + hand system, the total centre of gravity describes a circle the radius of which is the principal distance of this system. For any movement of the arm with the elbow in a fixed position, the total centre of gravity describes a circle the radius of which equals the geometric sum of the proximal principal distances of the two segments of the arm.

As indicated above, the principal points depend closely on the division of the body into segments. If the whole leg is considered as a rigid mass in order to analyse a particular movement, because the movement does not entail any change in the position of the knee and ankle, it would be pointless to distinguish separate principal points for the three segments of the leg. In this case, only one principal point for the whole leg need be assumed; its position depends on the principal points of the three segments. The principal point of the leg is obtained by combining the three proximal principal distances of the thigh, lower leg and foot, starting from the centre of the hip and taking into account the actual directions of these principal distances. This results simply from the fact that for any movement of the leg when rigid, the total centre of gravity describes a circle the radius of which is the geometric sum of the proximal principal distances of the three segments of the leg. For the position of the leg in phase 11 of experiment I, the line $\overline{L'S_0}$ in Fig. 3 represents the magnitude and direction of the principal distance of the whole right leg. Therefore, the principal point of the whole leg must be found in the same direction and at the same distance from the centre of the hip.

The principal point of the whole leg can be regarded as the centre of gravity of two masses, one concentrated at the centre of gravity of the leg and the other at the centre of the hip joint. The former is equal to the mass of the whole leg and the latter to the mass of the human body minus the mass of the leg. It is thus very easy to deduce the centre of gravity of the whole leg from its principal point. It is only necessary to lengthen the line connecting the centre of the hip and the principal point of the whole leg in the same ratio as that of the total mass of the body to the mass of the whole leg. The same result is achieved by increasing by the same proportion the proximal principal distances of the thigh, lower leg and foot and adding these values to one another geometrically. These increased principal distances provide fixed points in the three segments of the leg. These points must be considered principal points if the leg is imagined as completely separated from the rest of the body and treated as a free body system. This results from the fact that these points play the same role in the construction of the centre of gravity of the whole leg as do the actual principal points in the construction of the total centre of gravity of the human body. These point can be designated as "partial principal points" of the whole leg. Correspondingly, we obtain as partial principal points of the lower leg + foot system, points whose distances from the centre of the knee and from the centre of the ankle are to the proximal principal distances of the lower leg and foot in the same ratio as that of the mass of the whole body to the mass of the lower leg + foot system. The partial principal points of the arm would lie on the long axes of the arm and forearm at distances from the shoulder and elbow which are in the same proportion to the proximal principal distances of the two segments of the arm as the mass of the whole body is to the mass of the arm. In certain circumstances, it is recommended that the partial principal points be used instead of the actual principal points, without regarding the arm as separate from the rest of the body. This is the case, for example, when the trunk and shoulder girdle are taken to be fixed and the arm rotates as an independent system about a fixed point in space, the centre of the shoulder joint fixed to the trunk. Then the masses of all other body segments and even the total centre of gravity of the body can be disregarded and only the behaviour of the centre of gravity of the whole arm considered. To this end, only the partial principal points are needed. Thus, in previous studies[6] dealing with the action of muscles on the arm, the trunk and shoulder girdle were assumed to be fixed by external forces and only the partial principal points and principal distances of the two segments of the arm were used. The mass of the whole arm is proportionally 0.064875 of the total mass of the human body (see p. 122). Therefore, the proximal partial principal distances of the arm are about 15.5 times as long as the actual principal distances of the arm articulated with the body.

From the principal points it is possible to obtain not only the total centre of gravity of the body but also, in a very simple manner, the centres of gravity of the different systems. However, it must not be ignored that the considerable magnifications of the principal distances that are necessary somewhat influence the accuracy of the results. Still, with regard to the determination of the centres of gravity of the different systems, the construction of the total centre of gravity by means of the principal points is as good as using the partial centres of gravity of the body segments.

Constructing the total centre of gravity from the principal points is much more simple than from the different centres of gravity. Calculation using the principal points and principal distances is also much simpler. At first, of course, the co-ordinates of the 12 principal points have to be determined. This calculation corresponds to that of the co-ordinates of the different centres of gravity (see p. 123). However, it is necessary to use the proportional figures for the position of the principal points, given above, instead of those for the partial centres of gravity. Subsequently, only subtractions and additions are required in the calculation. By subtraction one obtains the projections of the principal distances on the three axes of co-ordinates. Addition of these differences and the co-ordinates of the principal points gives the co-ordinates of the total centre of gravity. Starting from the principal point of the trunk, the projections of the 11 principal distances on the three axes of co-ordinates are obtained. From the co-ordinates of each of the 11 other principal points, the corresponding co-ordinates of the centre of the proximal joint adjacent to the body segment under consideration are deduced. The differences of the x-, y- and z-co-ordinates shall be generally designated as h_i', h_i'' and h_i''' in which i indicates the number of the body segment under consideration. ξ_1, η_1, ζ_1 are the three co-ordinates of the principal point of the trunk. The following equations thus can be used to calculate the co-ordinates x_0, y_0 and z_0 of the total centre of gravity of the human body:

$$x_0 = \xi_1 + h_2' + h_3' + h_4' \cdots + h_{12}' = \xi_1 + \sum_{2}^{12}{}^i h_i',$$

[6] cf. Fischer O (1895, 1897) Beiträge zu einer Muskeldynamik I und II, und Beiträge zur Muskelstatik I. Abhandlungen der Königl. Sächsischen Gesellschaft der Wissenschaften, vol 22, nr 2; vol 23, nr 6; and vol 23, nr 4

$$y_0 = \eta_1 + h_2'' + h_3'' + h_4'' \cdots + h_{12}'' = \eta_1 + \sum_{2}^{12} {}^i h_i'',$$

$$z_0 = \zeta_1 + h_2''' + h_3''' + h_4''' \cdots + h_{12}''' = \zeta_1 + \sum_{2}^{12} {}^i h_i'''.$$

This results from the fact that the projection of a polygon on a straight line is equal to the algebraic sum of the projections of the different sides of this polygon on the straight line.

Once again, an articulated model of the human body can be built which carries out automatically the construction of the total centre of gravity using the principal distances. It only requires a few rods or strips of metal or cardboard of the same lengths as the principal distances considered. These rods or strips articulate with each other. One end of the chain of rods thus built articulates with the principal point of the trunk. A special device must ensure that the long axes of the different rods always remain parallel to their corresponding principal distances inside the body, for any attitude of the latter. This can be achieved mechanically only by means of a large number of joints. For a system of five articulated bodies the number of auxiliary joints is such that even if the friction of the joints of the device is slight, this device will work with much more difficulty than that based on the partial centres of gravity [7]. An automatic device to carry out the construction of the centre of gravity from the principal distances would thus be too complicated if the human body is resolved into 12 segments. However, by keeping to the articulated chain of 11 elements formed by the principal distances, a very simple device is obtained, both theoretically interesting and practically useful, for finding the total centre of gravity. But the total centre of gravity does not appear automatically when the model of the body is brought into a particular attitude. In every case the different parts of the chain have to be orientated parallel to their corresponding principal distances. If the model of the human body to which the chain is articulated only represents the projection of the body on the plane of gait, the different parts of the chain can be articulated by simple hinge joints. In a tridimensional model of the human body, however, the articulations of the different rods of the chain must be ball and socket joints.

Trajectory of the Total Centre of Gravity of the Human Body

The different ways of determining the trajectory of the centre of gravity described in the previous section show that an articulated device is more advantageous than calculation because it provides a clear illustration of the processes involved. However, calculation generally ensures greater accuracy. Errors in calculation can be avoided by checking each step whereas slight inaccuracies in the construction of the mechanical device are unavoidable. Analysis of the velocities and accelerations of the total centre of gravity are possible only if its trajectory is exactly known. Therefore, in the present study, for the benefit of further research, the laborious and time-consuming process of calculation will be followed. There are two possibilities: either the co-ordinates of the partial centres of gravity and then the co-ordinates of the total centre of gravity are calculated from the co-ordinates of the joint centres, determined in Chap. 1, by one of the two methods described in the previous section; or the partial centres of gravity are ignored, the co-ordinates of the

---

[7]   cf. Fischer O (1892) Ein zweiter Mechanismus zur Bestimmung der Lage des Schwerpunktes eines Systems von in beliebiger Weise durch Gelenke mit einander verbundenen Körpern. Nr. 264a des Nachtrags-Katalogs mathematischer und mathematisch-physikalischer Modelle, Apparate und Instrumente. Deutsche Mathematiker-Vereinigung

principal points of the 12 body segments are established and the co-ordinales of the total centre of gravity calculated using the equations cited above.

Though the latter method leads more quickly to the objective than the former, it will not be used in the present study for purely economic reasons. The analysis of the movement of the different segments of the human body during walking is not dealt with in this part of the work but will constitute the subject of later chapters. It requires knowledge of the trajectories of the partial centres of gravity. If the co-ordinates of the principal points were used to calculate those of the total centre of gravity, it would be necessary to calculate, at a later stage, the co-ordinates of the partial centres of gravity from those of the joint centres or principal points. Instead, geometric constructions will be used to calculate the co-ordinates of the centres of gravity of the different body segments, then the co-ordinates of the centres of gravity of the most important systems of body segments, and finally the co-ordinates of the total centre of gravity. Thus the final result will be obtained in a roundabout way, but this involves less work than if the co-ordinates of the total centre of gravity were calculated first using the quicker method, and the co-ordinates of the partial centres of gravity were calculated later separately. I performed the calculation exactly as indicated on p. 123–124, using the co-ordinates of the joint centres shown in Tables 14–16 of Chap. 1. Each step of the calculation was carefully checked so as virtually to exclude any error. Since at this time I had no further assistance and my double profession left me little time to carry out such voluminous work, the calculation of all the co-ordinates of the partial centres of gravity for the three experiments took me several years. That is why this second part of the analysis of human gait (Chap. 2) was published so long after the first. Tables 1 and 2 give the co-ordinates for experiments I and II.

In order to give an overall view of the behaviour of the different centres of gravity in the gait of the unloaded man, Fig. 4, based on the x- and z-co-ordinates, shows in red the projection of the trajectories of all the partial centres of gravity and in blue the projection of the trajectories of the centres of gravity of some systems and of the total centre of gravity on the plane of gait at a scale of 1 : 10 for experiments I and II. Since the projection of the phases of movement on the horizontal ground is not sufficiently clear (see Figs. 22, 23 of Chap. 1), they have not been represented for this preliminary overall view.

Figure 4 shows a remarkable result with regard to the co-ordination of the upper and lower extremities. Whereas all the centres of gravity of the limb segments carry out considerable vertical movements, the trajectory of the common centre of gravity of the two legs, which is between the two trajectories of the centres of gravity of the legs, remains approximately at the same level above the horizontal ground during walking. Moreover, the centres of gravity of the trunk, head, trunk + head + arms system and whole body describe trajectories of the same shape as those of the hip centres, shoulder centres and vertex.

The projection of the trajectory of the centre of gravity of the whole body on the vertical plane perpendicular to the direction of gait (plane $Y-Z$), for a uniform gait, must be a closed curve. The total centre of gravity describes this curve once during a double step. This life-size projection, however, shows that the extremities of the curve do not coincide. Its end lies to the left of its beginning by 0.25 cm in experiment I, 0.37 cm in experiment II. Correspondingly, in the projection of the trajectory of the centre of gravity on the horizontal ground (plane $X-Y$), these points which are separated by the length a double step do not lie at the same distance from the plane of gait. The average forward direction of the total centre of gravity thus does not coincide exactly with the X-axis of the system of co-ordinates. In both experiments, it deviates from the latter by a small angle, counter-clockwise as seen from above. In both cases, this angle is so small that it can be ignored in the further analysis of the process of movement; it is $0°5'3''$ in experiment I and $0°8'17''$ in experiment II. The X-axis of the system of co-ordinates can thus continue to be considered as the direction of gait since, after one or several steps, another angle, perhaps even in the opposite direction, may appear. It is impossible in walking for a person to maintain a fixed direction to within minutes of an angle during the whole process of movement.

Table 1. Co-ordinates of the centres of gravity for experiment I

No.	Centre of gravity of upper arm						Centre of gravity of forearm+hand						No.
	Right			Left			Right			Left			
	x	y	z	x	y	z	x	y	z	x	y	z	
1	39,92	+19,05	117,75	36,34	−23,11	117,93	55,09	+25,31	88,79	43,13	−32,47	87,24	1
2	44,79	+18,61	119,07	43,15	−23,16	119,64	55,54	+25,89	88,10	53,58	−30,97	89,88	2
3	49,08	+18,38	120,38	49,83	−23,21	121,31	55,36	+26,81	88,42	63,67	−29,47	93,18	3
4	53,19	+18,43	121,26	56,21	−23,41	122,32	55,24	+27,83	89,15	72,88	−28,11	96,35	4
5	57,54	+18,62	121,79	62,69	−23,52	122,56	56,11	+28,69	89,88	81,51	−26,82	99,09	5
6	62,21	+18,77	121,77	69,14	−23,55	122,13	58,15	+29,20	90,26	89,37	−25,78	101,06	6
7	67,37	+18,97	121,09	75,59	−23,36	121,19	61,24	+30,03	90,12	96,63	−25,00	101,98	7
8	73,38	+19,21	119,74	82,39	−23,04	119,79	65,71	+30,87	89,44	103,86	−24,40	101,74	8
9	79,56	+19,38	118,34	88,75	−22,54	118,50	71,30	+31,46	88,37	110,21	−23,95	100,35	9
10	85,88	+19,80	117,33	94,76	−21,86	117,71	78,14	+31,87	87,25	115,75	−23,81	98,27	10
11	93,16	+20,35	116,71	100,97	−21,02	117,29	87,13	+31,97	86,71	121,00	−23,89	95,40	11
12	100,79	+20,90	116,58	106,86	−20,28	117,43	97,40	+31,95	85,12	125,01	−24,30	92,45	12
13	108,65	+21,33	116,92	112,82	−19,68	117,93	109,00	+31,86	85,19	127,94	−25,38	89,96	13
14	115,73	+21,58	117,96	117,85	−19,30	118,68	120,77	+31,32	86,41	129,71	−26,08	88,81	14
15	123,04	+21,81	119,78	122,55	−19,14	119,96	133,10	+30,54	89,50	130,72	−27,35	89,09	15
16	129,15	+22,10	121,38	126,25	−19,24	121,00	142,99	+29,83	92,99	131,41	−28,53	89,78	16
17	136,26	+22,53	122,48	130,72	−19,54	121,73	153,96	+29,02	97,09	132,69	−29,79	90,41	17
18	142,70	+22,90	122,72	135,08	−19,70	121,83	163,22	+28,08	100,34	134,77	−30,44	90,61	18
19	149,67	+23,02	122,38	140,13	−19,83	121,36	172,63	+27,32	102,89	137,92	−31,41	90,45	19
20	156,45	+22,96	121,55	145,59	−20,01	120,41	180,16	+26,59	104,08	142,07	−32,12	89,87	20
21	163,01	+22,59	120,45	151,50	−20,24	119,16	187,30	+26,10	103,97	147,23	−32,86	88,89	21
22	169,23	+21,94	119,24	157,89	−20,57	118,02	193,57	+25,83	102,70	153,60	−33,31	87,84	22
23	174,99	+21,40	118,28	164,53	−21,01	117,21	198,73	+25,69	100,28	161,10	−33,50	86,79	23
24	180,53	+20,74	117,85	171,77	−21,61	116,78	202,96	+25,32	97,08	169,99	−33,50	86,06	24
25	185,80	+20,16	117,81	179,13	−22,06	116,86	205,85	+25,61	93,64	179,57	−33,57	85,78	25
26	191,56	+19,53	118,07	187,08	−22,46	117,38	207,92	+26,05	90,73	190,81	−32,84	86,32	26
27	196,57	+19,10	118,59	194,01	−22,88	118,42	208,90	+26,54	88,55	201,42	−31,97	87,78	27
28	201,21	+18,61	119,61	201,04	−22,99	120,14	209,13	+27,04	88,17	211,99	−30,94	90,57	28
29	204,89	+18,39	120,54	207,32	−23,12	121,67	208,93	+27,56	88,54	221,25	−29,86	93,64	29
30	208,67	+18,37	121,28	213,74	−23,31	122,60	208,99	+27,92	89,24	230,32	−28,77	96,58	30
31	212,64	+18,81	121,56	219,83	−23,48	122,75	210,03	+28,78	89,92	238,32	−27,63	99,02	31

No.	Centre of gravity of whole arm						Centre of gravity of trunk			Centre of gravity of head			No.
	Right			Left									
	x	y	z	x	y	z	x	y	z	x	y	z	
1	47,22	+22,06	103,82	39,61	−27,61	103,17	38,01	−1,72	108,55	38,19	−1,41	153,48	1
2	49,96	+22,11	104,17	48,17	−26,92	105,33	43,90	−1,75	109,92	44,48	−1,59	154,87	2
3	52,10	+22,43	105,01	56,49	−26,22	107,78	49,72	−1,76	111,22	50,63	−1,66	156,14	3
4	54,18	+22,95	105,82	64,23	−25,67	109,83	55,42	−1,75	112,00	56,52	−1,62	156,78	4
5	56,85	+23,46	106,44	71,74	−25,11	111,27	61,15	−1,71	112,09	62,47	−1,52	156,78	5
6	60,26	+23,79	106,61	78,87	−24,62	112,00	66,82	−1,66	111,52	68,36	−1,32	156,18	6
7	64,42	+24,29	106,19	85,71	−24,15	111,95	72,53	−1,54	110,37	74,34	−1,11	155,05	7
8	69,69	+24,82	105,17	92,72	−23,69	111,11	78,78	−1,37	108,84	80,74	−0,83	153,67	8
9	75,59	+25,19	104,92	99,07	−23,22	109,77	85,08	−1,13	107,73	86,89	−0,49	152,53	9
10	82,16	+25,61	102,86	104,86	−22,80	108,36	91,40	−0,69	107,14	92,85	−0,05	151,86	10
11	90,26	+25,94	102,28	110,60	−22,40	106,76	98,50	0	106,97	99,16	+0,47	151,70	11
12	99,16	+26,22	101,45	115,59	−22,21	105,41	105,28	+0,59	107,33	105,14	+0,95	152,14	12
13	108,82	+26,39	101,66	120,09	−22,42	104,48	111,74	+1,16	107,84	111,15	+1,39	152,98	13
14	118,15	+26,26	102,78	123,55	−22,56	104,31	117,61	+1,48	108,88	116,75	+1,59	154,20	14
15	127,88	+26,01	105,22	126,48	−23,09	105,11	123,49	+1,68	110,47	122,76	+1,66	155,71	15
16	135,81	+25,82	107,72	128,73	−23,71	105,98	128,76	+1,74	111,53	127,82	+1,64	156,74	16
17	144,79	+25,65	110,75	131,67	−24,47	106,67	134,41	+1,71	112,13	133,83	+1,57	157,24	17
18	152,57	+25,39	111,96	134,93	−24,87	106,81	139,72	+1,61	112,01	139,39	+1,50	156,95	18
19	160,71	+25,09	113,01	139,07	−25,40	106,49	145,42	+1,54	111,25	145,44	+1,43	156,11	19
20	167,85	+24,71	113,15	143,90	−25,83	105,72	151,16	+1,37	110,09	151,54	+1,34	154,92	20
21	174,69	+24,28	112,52	149,45	−26,31	104,60	157,08	+1,18	108,87	157,69	+1,15	153,71	21
22	180,94	+23,81	111,77	155,83	−26,70	103,50	163,29	+0,79	108,00	163,91	+0,79	152,72	22
23	186,41	+23,46	109,62	162,88	−27,02	102,58	169,67	+0,28	107,50	170,11	+0,31	152,08	23
24	191,32	+22,94	107,86	170,91	−27,33	102,00	176,53	−0,39	107,47	176,42	−0,30	151,96	24
25	195,44	+22,78	106,18	179,34	−27,60	101,91	183,00	−1,06	107,76	182,41	−0,85	152,33	25
26	199,43	+22,67	104,92	188,87	−27,45	102,44	189,67	−1,41	108,32	188,88	−1,31	153,08	26
27	202,50	+22,68	104,14	197,57	−27,25	103,68	195,71	−1,66	109,17	194,91	−1,54	154,10	27
28	205,02	+22,66	104,49	206,31	−26,81	105,92	201,80	−1,75	110,35	201,28	−1,59	155,37	28
29	206,83	+22,80	105,15	214,02	−26,36	108,19	207,18	−1,73	111,32	207,00	−1,53	156,33	29
30	208,82	+22,96	105,87	221,71	−25,94	110,08	212,59	−1,64	112,02	212,70	−1,28	156,92	30
31	211,38	+23,61	106,34	228,71	−25,48	111,34	217,87	−1,63	111,98	218,32	−1,00	156,59	31

Table 1 (continued)

No.	Centre of gravity of trunk + head			Centre of gravity of arms			Centre of gravity of trunk + head + arms			No.
	x	y	z	x	y	z	x	y	z	
1	38,04	− 1,68	114,93	44,42	− 2,78	103,50	39,36	− 1,91	112,56	1
2	43,98	− 1,73	116,30	49,07	− 2,41	104,75	45,03	− 1,87	113,91	2
3	49,85	− 1,75	117,60	54,30	− 1,95	106,40	50,77	− 1,79	115,28	3
4	55,58	− 1,73	118,36	59,21	− 1,36	107,83	56,33	− 1,65	116,18	4
5	61,34	− 1,68	118,44	64,30	− 0,83	108,86	61,95	− 1,50	116,46	5
6	67,04	− 1,61	117,86	69,57	− 0,42	109,31	67,56	− 1,36	116,09	6
7	72,79	− 1,48	116,71	75,07	+ 0,07	109,07	73,26	− 1,16	115,13	7
8	79,06	− 1,29	115,21	81,21	+ 0,57	108,14	79,51	− 0,90	113,75	8
9	85,34	− 1,04	114,09	87,33	+ 0,99	107,35	85,75	− 0,62	112,69	9
10	91,61	− 0,60	113,49	93,51	+ 1,41	105,61	92,00	− 0,18	111,86	10
11	98,59	+ 0,07	113,32	100,43	+ 1,77	104,52	98,97	+ 0,42	111,50	11
12	105,26	+ 0,64	113,69	107,38	+ 2,01	103,43	105,70	+ 0,92	111,57	12
13	111,66	+ 1,19	114,25	114,46	+ 1,99	103,07	112,24	+ 1,36	111,94	13
14	117,49	+ 1,50	115,32	120,85	+ 1,85	103,55	118,19	+ 1,57	112,88	14
15	123,39	+ 1,68	116,89	127,18	+ 1,46	105,17	124,17	+ 1,63	114,46	15
16	128,63	+ 1,73	117,95	132,27	+ 1,06	106,85	129,38	+ 1,59	115,65	16
17	134,33	+ 1,69	118,54	138,23	+ 0,59	108,71	135,14	+ 1,46	116,51	17
18	139,67	+ 1,59	118,39	143,75	+ 0,26	109,39	140,51	+ 1,31	116,53	18
19	145,42	+ 1,52	117,62	149,89	− 0,16	109,75	146,35	+ 1,17	115,99	19
20	151,21	+ 1,37	116,46	155,88	− 0,56	109,44	152,18	+ 0,97	115,01	20
21	157,17	+ 1,18	115,24	162,07	− 1,02	108,56	158,18	+ 0,72	113,86	21
22	163,38	+ 0,79	114,35	168,39	− 1,45	107,64	164,42	+ 0,33	112,96	22
23	169,73	+ 0,28	113,83	174,65	− 1,78	106,10	170,75	− 0,15	112,23	23
24	176,51	− 0,38	113,79	181,12	− 2,20	104,93	177,46	− 0,76	111,96	24
25	182,92	− 1,03	114,09	187,39	− 2,41	104,05	183,85	− 1,32	112,01	25
26	189,56	− 1,40	114,68	194,15	− 2,39	103,68	190,51	− 1,60	112,40	26
27	195,60	− 1,64	115,55	200,04	− 2,29	103,91	196,52	− 1,77	113,14	27
28	201,73	− 1,73	116,74	205,67	− 2,08	105,21	202,55	− 1,80	114,35	28
29	207,15	− 1,70	117,71	210,43	− 1,78	106,67	207,83	− 1,72	115,42	29
30	212,61	− 1,59	118,40	215,27	− 1,49	107,98	213,16	− 1,57	116,24	30
31	217,93	− 1,54	118,31	220,05	− 0,94	108,84	218,37	− 1,42	116,35	31

No.	Centre of gravity of thigh						Centre of gravity of lower leg						No.
	Right			Left			Right			Left			
	x	y	z	x	y	z	x	y	z	x	y	z	
1	38,90	+ 8,39	65,71	44,73	− 10,45	69,60	25,64	+ 9,50	34,77	51,58	− 8,37	33,20	1
2	47,59	+ 8,20	67,22	48,51	− 10,40	70,26	37,99	+ 9,06	35,52	52,47	− 8,36	33,43	2
3	55,81	+ 8,08	68,85	52,06	− 10,15	70,83	50,49	+ 8,65	35,92	53,18	− 8,08	33,63	3
4	63,37	+ 7,87	70,14	55,53	− 9,89	71,29	62,64	+ 8,17	35,86	53,88	− 7,78	33,80	4
5	70,31	+ 7,57	70,64	59,23	− 9,62	71,29	74,51	+ 7,62	35,48	54,86	− 7,40	33,89	5
6	76,62	+ 7,34	70,19	63,19	− 9,48	70,92	85,72	+ 7,27	34,88	56,17	− 7,15	33,88	6
7	82,17	+ 6,99	68,68	67,51	− 9,35	70,13	95,56	+ 6,91	34,12	57,90	− 6,93	33,74	7
8	88,11	+ 6,55	66,76	72,54	− 9,30	69,22	104,68	+ 6,56	33,52	60,26	− 6,90	33,71	8
9	94,84	+ 6,59	65,89	78,10	− 9,39	68,67	112,23	+ 6,42	33,00	63,35	− 7,28	33,97	9
10	102,15	+ 7,13	65,99	84,28	− 9,50	68,31	118,59	+ 6,53	32,43	67,44	− 7,91	34,43	10
11	109,35	+ 7,79	65,92	91,83	− 9,54	68,16	124,13	+ 6,94	31,89	73,39	− 8,77	35,24	11
12	116,21	+ 8,92	66,81	100,21	− 9,77	67,92	128,73	+ 7,97	32,28	81,65	− 9,72	36,00	12
13	121,87	+ 9,56	67,26	109,06	− 9,47	67,63	131,90	+ 8,13	32,38	92,11	− 10,01	36,87	13
14	125,95	+ 10,02	67,80	117,71	− 9,08	68,36	132,91	+ 8,29	32,27	103,51	− 9,92	37,64	14
15	129,36	+ 9,98	68,46	126,09	− 8,67	70,01	133,38	+ 7,92	32,23	115,95	− 9,46	38,14	15
16	132,45	+ 9,70	68,79	133,46	− 8,45	71,34	133,75	+ 7,42	32,20	126,78	− 9,05	38,16	16
17	135,65	+ 9,39	68,91	140,97	− 8,24	72,33	134,31	+ 7,01	32,12	139,50	− 8,16	37,75	17
18	138,95	+ 9,08	68,77	147,60	− 8,08	72,53	134,97	+ 6,70	32,04	150,88	− 8,17	37,06	18
19	142,80	+ 8,93	68,26	153,88	− 7,89	71,88	136,07	+ 6,55	31,91	162,16	− 7,87	36,11	19
20	147,03	+ 8,66	67,46	159,46	− 7,65	70,50	137,73	+ 6,27	31,70	172,17	− 7,68	35,23	20
21	151,45	+ 8,55	66,84	164,92	− 7,34	68,85	139,69	+ 6,22	31,81	180,49	− 7,50	34,60	21
22	156,70	+ 8,43	66,57	171,49	− 7,55	68,12	142,62	+ 6,42	32,24	187,65	− 7,53	34,06	22
23	162,88	+ 8,31	66,32	179,01	− 8,32	68,40	146,76	+ 6,77	32,72	194,06	− 7,95	33,54	23
24	170,10	+ 8,17	66,18	185,40	− 9,09	68,27	152,47	+ 7,21	33,38	198,69	− 8,53	33,06	24
25	178,20	+ 8,17	65,87	191,61	− 9,99	68,83	160,51	+ 7,86	33,98	202,88	− 9,07	33,25	25
26	187,49	+ 8,08	65,58	197,72	− 10,56	69,53	171,67	+ 8,12	34,45	206,58	− 9,46	33,62	26
27	196,56	+ 7,71	66,12	202,28	− 10,71	70,10	183,45	+ 7,78	34,86	208,03	− 9,45	33,72	27
28	205,22	+ 7,51	67,38	206,39	− 10,75	70,66	195,68	+ 7,39	35,43	209,14	− 9,41	33,86	28
29	212,89	+ 7,40	68,81	210,14	− 10,50	71,04	207,04	+ 7,05	35,71	210,10	− 9,07	33,95	29
30	220,20	+ 7,47	70,00	213,57	− 10,35	71,33	218,97	+ 6,95	35,59	210,98	− 8,91	34,07	30
31	226,45	+ 7,42	70,35	217,13	− 10,35	71,20	229,83	+ 7,01	35,14	212,03	− 8,85	34,05	31

Table 1 (continued)

No.	Centre of gravity of foot Right x	y	z	Left x	y	z	Centre of gravity of lower leg + foot Right x	y	z	Left x	y	z	No.
1	4,86	+7,39	16,85	56,87	−6,23	6,07	20,38	+8,97	30,24	52,92	−7,83	26,34	1
2	19,27	+7,18	15,87	56,94	−6,23	6,10	33,25	+8,58	30,55	53,60	−7,82	26,52	2
3	35,58	+6,99	13,67	56,93	−6,20	6,11	46,72	+8,23	30,29	54,13	−7,60	26,67	3
4	52,80	+6,78	11,17	56,98	−6,19	6,14	60,15	+7,82	29,61	54,66	−7,38	26,80	4
5	70,97	+6,48	9,38	56,99	−6,16	6,16	73,61	+7,31	28,88	55,40	−7,09	26,87	5
6	88,88	+6,32	8,82	56,99	−6,13	6,18	86,52	+7,03	28,29	56,38	−6,89	26,87	6
7	105,53	+6,70	10,03	57,00	−6,10	6,20	98,08	+6,86	28,03	57,67	−6,72	26,77	7
8	119,84	+7,66	12,64	57,05	−6,09	6,34	108,52	+6,84	28,24	59,45	−6,70	26,79	8
9	127,92	+7,73	12,71	57,54	−6,35	6,81	116,20	+6,75	27,87	61,88	−7,04	27,10	9
10	131,33	+6,82	10,00	58,50	−6,78	7,60	121,81	+6,60	26,76	65,18	−7,62	27,64	10
11	134,81	+6,30	7,76	60,42	−7,13	9,19	126,83	+6,78	25,79	70,11	−8,36	28,65	11
12	135,68	+6,04	6,77	63,94	−7,15	12,50	130,49	+7,48	25,83	77,17	−9,07	30,05	12
13	135,91	+6,12	6,68	72,08	−7,68	16,41	132,91	+7,62	25,88	87,04	−9,42	31,69	13
14	135,92	+6,09	6,71	83,42	−7,63	17,61	133,67	+7,73	25,80	98,43	−9,34	32,57	14
15	135,93	+6,04	6,74	98,76	−7,15	16,33	134,03	+7,44	25,78	111,60	−8,88	32,62	15
16	135,93	+6,01	6,77	113,08	−6,68	14,03	134,30	+7,06	25,77	123,31	−8,45	32,06	16
17	135,94	+5,96	6,80	131,24	−6,55	11,35	134,72	+6,74	25,71	137,41	−7,75	31,07	17
18	135,95	+5,90	6,83	148,50	−6,72	9,36	135,22	+6,50	25,66	150,28	−7,80	30,05	18
19	135,97	+5,86	6,87	166,68	−6,91	8,54	136,04	+6,38	25,87	163,30	−7,63	29,13	19
20	136,00	+5,80	6,94	183,98	−7,49	9,79	137,29	+6,15	25,44	175,16	−7,63	28,79	20
21	136,28	+5,82	7,17	197,72	−8,69	12,43	138,83	+6,12	25,58	184,85	−7,80	28,99	21
22	136,83	+5,97	7,74	205,44	−8,97	12,30	141,16	+6,31	26,04	192,15	−7,89	28,55	22
23	137,76	+6,15	8,66	208,35	−8,36	9,43	144,48	+6,61	26,63	197,68	−8,05	27,44	23
24	139,59	+5,98	10,54	211,40	−8,14	7,34	149,21	+6,90	27,60	201,91	−8,43	26,55	24
25	143,27	+5,87	13,23	212,62	−7,60	6,39	156,15	+7,36	28,73	205,34	−8,70	26,45	25
26	151,42	+5,71	16,45	212,64	−7,76	6,24	166,55	+7,51	29,90	208,11	−9,03	26,69	26
27	162,92	+5,15	16,80	212,66	−7,77	6,30	178,26	+7,11	30,29	209,20	−9,02	26,78	27
28	177,34	+4,66	15,53	212,68	−7,79	6,37	191,04	+6,70	30,40	210,04	−9,00	26,91	28
29	192,31	+4,53	13,35	212,70	−7,80	6,43	203,31	+6,41	30,05	210,76	−8,75	26,99	29
30	209,06	+4,66	10,88	212,72	−7,81	6,50	216,46	+6,37	29,34	211,42	−8,63	27,09	30
31	226,00	+5,14	9,00	212,75	−7,83	6,58	228,86	+6,54	28,53	212,21	−8,59	27,10	31

No.	Centre of gravity of whole leg Right x	y	z	Left x	y	z	Centre of gravity of legs x	y	z	Total centre of gravity of body x	y	z	No.
1	31,88	+8,61	52,27	47,83	−9,46	53,20	39,86	−0,43	52,74	39,55	−1,36	90,25	1
2	42,16	+8,34	53,32	50,44	−9,42	53,72	46,30	−0,54	53,52	45,50	−1,37	91,38	2
3	52,36	+8,14	54,24	52,84	−9,18	54,09	52,60	−0,52	54,17	51,45	−1,32	92,49	3
4	62,15	+7,85	54,78	55,20	−8,94	54,43	58,68	−0,55	54,63	57,21	−1,24	93,21	4
5	71,56	+7,47	54,81	57,78	−8,66	54,45	64,67	−0,60	54,63	62,96	−1,16	93,40	5
6	80,37	+7,22	54,31	60,61	−8,50	54,23	70,49	−0,64	54,27	68,65	−1,09	93,03	6
7	88,20	+6,94	53,27	63,78	−8,35	53,70	75,99	−0,71	53,49	74,28	−0,99	92,14	7
8	95,85	+6,66	52,16	67,58	−8,31	53,14	81,72	−0,83	52,65	80,33	−0,87	90,96	8
9	102,94	+6,65	51,48	71,95	−8,50	52,91	87,45	−0,93	52,20	86,38	−0,74	90,13	9
10	109,60	+6,93	51,12	77,04	−8,79	52,90	93,32	−0,93	52,01	92,49	−0,46	89,54	10
11	115,22	+7,41	50,71	83,60	−9,09	53,19	99,41	−0,84	51,95	99,13	−0,05	89,29	11
12	121,62	+8,37	51,28	91,48	−9,50	53,57	106,55	−0,57	52,43	106,02	+0,36	89,51	12
13	126,05	+8,82	51,58	101,09	−9,45	54,01	113,57	−0,32	52,80	112,74	+0,73	89,88	13
14	128,88	+9,15	51,88	110,40	−9,18	54,80	119,64	−0,02	53,34	118,73	+0,98	90,67	14
15	131,13	+9,02	52,28	120,60	−8,75	55,84	125,87	+0,14	54,06	124,80	+1,07	91,93	15
16	133,15	+8,70	52,49	129,61	−8,45	56,45	131,38	+0,13	54,47	130,13	+1,05	92,83	16
17	135,30	+8,39	52,55	139,62	−8,05	56,69	137,46	+0,17	54,62	136,01	+0,98	93,43	17
18	137,54	+8,10	52,43	148,52	−7,97	56,43	143,03	+0,07	54,43	141,45	+0,85	93,37	18
19	140,24	+7,96	52,19	157,45	−7,80	55,68	148,85	+0,08	53,94	147,28	+0,76	92,85	19
20	143,34	+7,71	51,53	165,41	−7,64	54,69	154,38	+0,04	53,11	153,00	+0,62	91,92	20
21	146,67	+7,63	51,20	172,47	−7,51	53,74	159,57	+0,06	52,47	159,07	+0,47	90,96	21
22	150,81	+7,63	51,21	179,32	−7,68	53,12	165,07	−0,03	52,17	164,66	+0,20	90,29	22
23	155,91	+7,67	51,28	186,09	−8,22	52,88	171,00	−0,28	52,08	170,84	−0,20	89,72	23
24	162,18	+7,69	51,56	191,66	−8,84	52,46	176,92	−0,58	52,01	177,26	−0,69	89,60	24
25	169,84	+7,86	51,79	196,81	−9,50	52,77	183,33	−0,82	52,28	183,66	−1,13	89,73	25
26	179,55	+7,86	52,06	201,66	−9,98	53,29	190,61	−1,06	52,68	190,55	−1,40	90,12	26
27	189,62	+7,48	52,54	204,90	−10,07	53,68	197,26	−1,30	53,11	196,80	−1,59	90,75	27
28	199,85	+7,20	53,36	208,77	−10,09	54,08	204,31	−1,45	53,72	203,21	−1,67	91,74	28
29	209,26	+7,02	54,12	210,37	−9,84	54,35	209,82	−1,41	54,24	208,57	−1,60	92,60	29
30	218,78	+7,05	54,59	212,76	−9,70	54,56	215,77	−1,33	54,58	214,13	−1,48	93,24	30
31	227,36	+7,09	54,50	215,27	−9,68	54,49	221,32	−1,30	54,50	219,47	−1,38	93,28	31

Table 2. Co-ordinates of the centres of gravity for experiment II

| No. | Centre of gravity of upper arm | | | | | | Centre of gravity of forearm + hand | | | | | | No. |
| | Right | | | Left | | | Right | | | Left | | | |
	x	y	z	x	y	z	x	y	z	x	y	z	
1	46,26	+20,37	118,14	42,69	−22,77	116,50	65,78	+26,64	93,82	44,53	−33,65	85,23	1
2	51,34	+20,18	118,52	49,59	−22,69	117,05	68,24	+26,63	91,41	54,45	−32,25	85,70	2
3	56,16	+19,46	119,45	56,48	−22,48	118,43	69,64	+26,37	89,87	64,81	−30,33	87,50	3
4	60,88	+19,05	120,55	63,31	−22,38	120,28	70,33	+26,53	89,33	75,51	−28,34	90,54	4
5	64,95	+18,87	121,32	69,69	−22,43	121,80	70,79	+26,99	89,39	85,18	−26,74	94,18	5
6	69,27	+18,71	121,68	76,35	−22,64	122,73	71,63	+27,47	89,48	94,57	−25,03	97,29	6
7	74,04	+18,79	121,61	83,21	−22,80	122,85	73,32	+27,76	89,43	103,60	−23,64	100,32	7
8	78,62	+18,81	121,14	89,23	−22,78	122,32	75,75	+28,31	89,11	111,01	−22,63	102,22	8
9	84,54	+18,94	120,14	96,28	−22,56	121,26	79,64	+28,80	88,47	119,04	−21,94	103,38	9
10	90,50	+19,09	118,90	102,67	−22,17	120,00	84,42	+29,41	87,60	125,88	−21,51	103,12	10
11	97,26	+19,41	117,75	109,08	−21,60	118,87	90,47	+30,07	86,68	132,34	−21,48	101,83	11
12	104,08	+19,85	117,21	114,92	−20,90	118,24	98,19	+30,54	86,10	137,70	−21,62	99,69	12
13	111,69	+20,45	117,03	120,57	−20,30	117,98	107,36	+30,99	85,64	142,31	−22,49	96,90	13
14	118,99	+20,94	117,35	125,80	−19,84	118,40	117,26	+31,08	85,66	145,62	−23,01	94,21	14
15	126,47	+21,34	118,22	131,20	−19,60	119,18	128,65	+31,28	86,38	148,11	−24,22	91,80	15
16	133,43	+21,57	119,76	136,05	−19,43	120,23	139,91	+30,90	88,47	149,53	−25,73	90,79	16
17	140,16	+21,90	121,30	140,47	−19,55	121,08	150,76	+30,33	91,29	150,34	−27,23	90,51	17
18	146,86	+22,39	122,27	144,86	−19,79	121,56	161,41	+29,63	94,23	151,27	−28,80	90,50	18
19	153,87	+22,66	122,56	149,49	−20,17	121,54	171,87	+28,75	97,05	152,65	−30,57	90,28	19
20	160,93	+22,54	122,24	154,43	−20,49	121,22	181,59	+27,69	99,33	155,04	−31,72	90,15	20
21	167,48	+22,44	121,62	159,41	−20,57	120,58	189,74	+26,88	100,65	158,32	−32,87	89,74	21
22	173,94	+22,13	120,77	164,84	−20,88	119,59	197,11	+26,17	100,91	162,61	−33,63	89,08	22
23	180,92	+21,56	119,73	171,58	−21,13	118,44	204,28	+25,82	99,94	168,84	−34,23	88,18	23
24	187,00	+21,03	118,78	178,21	−21,45	117,70	209,75	+25,36	98,08	176,02	−34,32	87,32	24
25	192,78	+20,48	118,25	185,30	−21,88	117,22	214,11	+25,53	95,88	184,67	−33,72	86,52	25
26	198,13	+19,99	118,11	192,53	−22,17	117,19	217,15	+25,79	93,00	194,08	−33,53	86,18	26
27	203,62	+19,55	118,36	199,88	−22,41	117,64	219,40	+25,91	90,49	204,70	−32,64	86,64	27
28	208,52	+19,26	118,96	206,70	−22,58	118,68	220,65	+26,63	88,90	215,19	−31,38	88,15	28
29	212,93	+18,86	120,15	213,33	−22,75	120,32	221,40	+26,65	88,93	225,20	−30,29	90,92	29
30	216,83	+18,47	121,20	219,75	−23,02	121,88	221,87	+27,12	89,39	234,75	−28,79	94,16	30
31	221,10	+18,42	122,03	226,45	−23,35	122,87	222,70	+27,65	89,87	244,17	−27,49	97,17	31

| No. | Centre of gravity of whole arm | | | | | | Centre of gravity of trunk | | | Centre of gravity of head | | | No. |
| | Right | | | Left | | | | | | | | | |
	x	y	z	x	y	z	x	y	z	x	y	z	
1	55,65	+23,39	106,44	43,58	−28,00	101,46	44,05	−0,92	107,63	42,73	−0,96	152,43	1
2	59,47	+23,28	105,48	51,93	−27,29	101,97	50,30	−1,11	108,45	48,71	−1,35	153,33	2
3	62,64	+22,78	105,22	60,49	−26,26	103,55	56,18	−1,22	109,79	54,76	−1,49	154,65	3
4	65,43	+22,65	105,53	69,18	−25,25	105,98	61,97	−1,29	110,98	60,81	−1,47	156,04	4
5	67,76	+22,78	105,96	77,14	−24,50	108,51	67,45	−1,35	111,78	66,81	−1,35	156,95	5
6	70,41	+22,92	106,19	85,11	−23,79	110,49	73,07	−1,46	112,10	72,82	−1,18	157,20	6
7	73,69	+23,10	106,13	93,02	−23,20	112,00	78,91	−1,44	111,82	79,04	−0,93	156,77	7
8	77,24	+23,38	105,73	99,71	−22,71	112,65	84,09	−1,42	110,95	84,57	−0,76	155,89	8
9	82,18	+23,68	104,91	107,23	−22,26	112,66	90,27	−1,29	109,49	91,18	−0,50	154,54	9
10	87,58	+24,05	103,84	113,83	−21,85	111,88	96,31	−1,21	108,23	97,38	−0,27	153,30	10
11	93,99	+24,54	102,81	120,27	−21,54	110,67	103,00	−0,79	107,50	103,90	+0,09	152,46	11
12	101,25	+24,99	102,25	125,88	−21,25	109,32	109,68	−0,27	107,18	110,20	+0,53	152,11	12
13	109,61	+25,52	101,93	131,03	−21,35	107,84	116,70	+0,37	107,38	116,70	+0,98	152,30	13
14	118,16	+25,82	102,11	135,33	−21,36	106,76	123,00	+0,84	107,86	122,58	+1,33	152,96	14
15	127,52	+26,12	102,90	139,33	−21,82	106,01	129,11	+1,29	108,92	128,65	+1,55	154,08	15
16	136,55	+26,06	104,71	142,53	−22,46	106,07	134,84	+1,46	110,25	134,54	+1,62	155,43	16
17	145,26	+25,95	106,87	145,22	−23,24	106,38	140,49	+1,53	111,33	140,26	+1,62	156,49	17
18	153,86	+25,87	108,78	147,94	−24,12	106,62	146,14	+1,63	111,93	145,94	+1,61	156,98	18
19	162,53	+25,59	110,29	151,01	−25,17	106,50	151,92	+1,54	111,91	151,84	+1,50	156,81	19
20	170,87	+25,02	111,22	154,72	−25,89	106,28	157,71	+1,31	111,32	157,86	+1,30	156,10	20
21	178,19	+24,58	111,53	158,89	−26,49	105,75	163,20	+1,30	110,33	163,53	+1,29	155,09	21
22	185,08	+24,07	111,22	163,77	−27,01	104,91	168,86	+1,11	109,23	169,28	+1,09	154,00	22
23	192,16	+23,61	110,21	170,26	−27,43	103,88	175,60	+0,77	108,32	175,93	+0,77	153,04	23
24	197,94	+23,11	108,82	177,16	−27,64	103,09	182,07	+0,31	107,88	182,17	+0,33	152,50	24
25	203,04	+22,91	107,49	185,00	−27,58	102,45	188,98	−0,31	107,77	188,52	−0,14	152,31	25
26	207,28	+22,78	106,03	193,28	−27,63	102,27	195,48	−0,90	107,97	194,50	−0,61	152,60	26
27	211,21	+22,61	104,95	202,20	−27,33	102,73	201,92	−1,25	108,50	200,57	−1,01	153,31	27
28	214,35	+22,80	104,50	210,78	−26,81	104,00	207,82	−1,38	109,52	206,45	−1,22	154,40	28
29	217,00	+22,71	105,13	219,04	−26,38	106,18	213,59	−1,53	110,80	212,40	−1,40	155,69	29
30	219,25	+22,63	105,90	226,97	−25,80	108,55	219,16	−1,70	111,79	218,16	−1,51	156,70	30
31	221,87	+22,86	106,56	234,97	−25,34	110,51	224,83	−1,76	112,35	224,09	−1,43	157,22	31

Table 2 (continued)

No.	Centre of gravity of trunk + head			Centre of gravity of arms			Centre of gravity of trunk + head + arms			No.
	x	y	z	x	y	z	x	y	z	
1	43,86	− 0,93	113,99	49,62	− 2,31	103,95	45,05	− 1,22	111,91	1
2	50,07	− 1,14	114,82	55,70	− 2,01	103,73	51,24	− 1,32	112,52	2
3	55,98	− 1,26	116,16	61,57	− 1,74	104,39	57,14	− 1,36	113,72	3
4	61,84	− 1,32	117,38	67,31	− 1,30	105,76	62,97	− 1,32	114,97	4
5	67,36	− 1,35	118,19	72,45	− 0,86	107,24	68,41	− 1,25	115,92	5
6	73,03	− 1,42	118,50	77,76	− 0,44	108,34	74,01	− 1,22	116,40	6
7	78,93	− 1,37	118,20	83,36	− 0,05	109,07	79,85	− 1,10	116,31	7
8	84,16	− 1,33	117,33	88,48	+ 0,34	109,19	85,05	− 0,98	115,65	8
9	90,40	− 1,18	115,89	94,71	+ 0,71	108,79	91,29	− 0,79	114,42	9
10	96,46	− 1,00	114,63	100,71	+ 1,10	107,86	97,34	− 0,57	113,23	10
11	103,13	− 0,67	113,88	107,13	+ 1,50	106,74	103,96	− 0,22	112,40	11
12	109,75	− 0,16	113,56	113,57	+ 1,87	105,79	110,54	+ 0,26	111,95	12
13	116,70	+ 0,46	113,76	120,32	+ 2,09	104,89	117,45	+ 0,80	111,92	13
14	122,94	+ 0,91	114,26	126,75	+ 2,23	104,44	123,73	+ 1,18	112,23	14
15	129,04	+ 1,33	115,33	133,43	+ 2,15	104,46	129,95	+ 1,50	113,08	15
16	134,80	+ 1,48	116,67	139,54	+ 1,80	105,39	135,78	+ 1,55	114,34	16
17	140,46	+ 1,54	117,74	145,24	+ 1,36	106,63	141,45	+ 1,50	115,44	17
18	146,11	+ 1,63	118,33	150,90	+ 0,88	107,70	147,10	+ 1,47	116,13	18
19	151,91	+ 1,53	118,29	156,77	+ 0,21	108,40	152,92	+ 1,26	116,24	19
20	157,73	+ 1,31	117,68	162,80	− 0,44	108,75	158,78	+ 0,95	115,83	20
21	163,25	+ 1,30	116,69	168,54	− 0,96	108,64	164,35	+ 0,83	115,02	21
22	168,92	+ 1,11	115,59	174,43	− 1,47	108,07	170,06	+ 0,58	114,03	22
23	175,65	+ 0,77	114,67	181,21	− 1,91	107,05	176,80	+ 0,22	113,09	23
24	182,08	+ 0,31	114,22	187,55	− 2,27	105,96	183,21	− 0,22	112,51	24
25	188,91	− 0,29	114,09	194,02	− 2,34	104,97	189,97	− 0,71	112,20	25
26	195,34	− 0,86	114,31	200,28	− 2,43	104,15	196,36	− 1,18	112,21	26
27	201,73	− 1,22	114,86	206,71	− 2,36	103,84	202,76	− 1,46	112,58	27
28	207,63	− 1,36	115,89	212,57	− 2,01	104,25	208,65	− 1,49	113,48	28
29	213,42	− 1,51	117,17	218,01	− 1,84	105,66	214,37	− 1,58	114,79	29
30	219,02	− 1,67	118,17	223,11	− 1,59	107,23	219,87	− 1,65	115,91	30
31	224,72	− 1,71	118,72	228,42	− 1,24	108,54	225,49	− 1,61	116,61	31

No.	Centre of gravity of thigh						Centre of gravity of lower leg						No.
	Right			Left			Right			Left			
	x	y	z	x	y	z	x	y	z	x	y	z	
1	39,73	+ 9,37	65,22	52,72	− 9,23	68,69	23,12	+ 9,54	33,71	63,64	− 7,73	33,11	1
2	49,37	+ 9,34	65,68	57,45	− 9,45	69,27	34,15	+ 9,61	34,42	65,66	− 7,88	33,22	2
3	58,26	+ 9,09	67,03	61,58	− 9,64	70,19	46,18	+ 9,29	35,12	66,93	− 8,02	33,39	3
4	66,38	+ 8,80	68,39	65,42	− 9,60	70,70	58,67	+ 8,84	35,73	67,83	− 7,84	33,47	4
5	73,75	+ 8,57	69,56	68,93	− 9,56	71,04	70,36	+ 8,55	36,00	68,52	− 7,68	33,58	5
6	80,75	+ 8,26	70,35	72,48	− 9,51	71,14	82,29	+ 8,22	35,76	69,24	− 7,48	33,68	6
7	87,37	+ 7,93	70,35	76,34	− 9,47	70,92	94,07	+ 7,88	35,03	70,35	− 7,33	33,68	7
8	92,87	+ 7,62	69,32	80,01	− 9,41	70,35	103,80	+ 7,65	34,21	71,59	− 7,15	33,60	8
9	98,54	+ 7,09	67,27	84,65	− 9,28	69,41	113,61	+ 7,23	33,30	73,50	− 7,01	33,52	9
10	104,60	+ 6,83	65,87	89,63	− 9,24	68,74	121,49	+ 6,78	32,79	76,09	− 7,10	33,66	10
11	112,52	+ 7,13	65,99	95,91	− 9,38	68,33	129,12	+ 6,53	32,56	80,13	− 7,67	34,00	11
12	119,90	+ 7,51	65,93	102,84	− 9,49	68,05	135,13	+ 6,35	32,09	85,33	− 8,44	34,64	12
13	126,67	+ 8,51	66,31	111,37	− 9,68	67,81	139,66	+ 7,13	31,87	93,38	− 9,45	35,47	13
14	132,82	+ 9,08	67,03	119,91	− 9,53	67,40	143,47	+ 7,44	32,22	103,14	− 9,85	36,29	14
15	137,27	+ 9,58	67,42	129,02	− 9,02	67,92	145,06	+ 7,46	32,13	115,12	− 9,79	37,25	15
16	141,03	+ 9,60	68,05	137,18	− 8,77	69,36	145,93	+ 7,26	32,11	126,96	− 9,53	38,00	16
17	144,52	+ 9,48	68,61	144,93	− 8,63	70,99	146,54	+ 7,00	32,02	138,80	− 9,21	38,20	17
18	147,90	+ 9,20	68,77	152,51	− 8,51	72,05	147,05	+ 6,58	31,95	150,87	− 8,91	37,92	18
19	151,42	+ 9,01	68,70	159,70	− 8,28	72,40	147,75	+ 6,32	31,91	163,00	− 8,50	37,16	19
20	155,11	+ 8,85	67,53	166,24	− 7,95	71,87	148,72	+ 6,15	31,75	174,57	− 8,05	36,08	20
21	158,94	+ 8,64	67,53	171,58	− 7,60	70,50	150,04	+ 5,92	31,61	184,03	− 7,59	35,09	21
22	163,20	+ 8,54	66,93	176,69	− 7,18	68,94	151,95	+ 5,82	31,68	191,96	− 7,13	34,52	22
23	168,98	+ 8,51	66,58	183,73	− 7,21	68,26	155,26	+ 6,12	32,07	199,81	− 6,66	34,14	23
24	175,32	+ 8,51	66,31	191,16	− 7,80	68,38	159,63	+ 6,54	32,57	206,17	− 6,78	33,62	24
25	182,72	+ 8,33	66,07	197,63	− 8,48	68,35	165,62	+ 6,96	33,18	210,82	− 7,14	33,18	25
26	191,16	+ 8,46	65,68	204,16	− 9,39	68,97	174,23	+ 7,88	33,81	215,40	− 7,73	33,50	26
27	200,49	+ 8,29	65,46	209,91	− 9,86	69,52	185,49	+ 8,24	34,38	218,72	− 8,00	33,67	27
28	209,47	+ 7,97	66,56	214,40	−10,07	70,11	197,58	+ 8,17	35,09	220,30	− 8,06	33,77	28
29	217,80	+ 7,83	68,18	218,48	−10,03	70,69	209,40	+ 8,12	35,93	221,37	− 7,88	33,85	29
30	225,56	+ 7,69	69,58	222,16	− 9,90	71,10	220,93	+ 8,06	36,40	222,23	− 7,62	33,94	30
31	233,06	+ 7,60	70,59	225,59	− 9,80	71,35	233,01	+ 7,95	36,28	223,05	− 7,50	34,03	31

Table 2 (continued)

No.	Centre of gravity of foot						Centre of gravity of lower leg + foot						No.
	Right			Left			Right			Left			
	x	y	z	x	y	z	x	y	z	x	y	z	
1	4,45	+6,98	14,22	71,10	−6,02	6,01	18,40	+8,89	28,78	65,53	−7,30	26,25	1
2	13,88	+6,77	16,63	71,09	−6,02	6,02	29,02	+8,89	29,92	67,02	−7,41	26,34	2
3	26,56	+6,22	16,45	71,08	−6,00	6,03	41,22	+8,51	30,40	67,98	−7,51	26,47	3
4	42,00	+6,04	14,81	71,06	−5,98	6,03	54,45	+8,13	30,44	68,65	−7,37	26,53	4
5	57,84	+6,04	12,53	71,03	−5,95	6,05	67,19	+7,91	30,06	69,16	−7,24	26,61	5
6	75,38	+6,30	10,15	71,04	−5,95	6,07	80,54	+7,73	29,28	69,71	−7,09	26,69	6
7	94,03	+6,57	8,65	71,05	−5,91	6,10	94,06	+7,55	28,36	70,53	−6,97	26,70	7
8	110,29	+6,84	8,76	71,06	−5,90	6,12	105,44	+7,45	27,77	71,46	−6,83	26,65	8
9	127,15	+7,69	10,80	71,09	−5,90	6,18	117,04	+7,35	27,61	72,89	−6,73	26,60	9
10	137,70	+8,16	12,49	71,56	−6,09	6,46	125,59	+7,13	27,65	74,94	−6,84	26,78	10
11	142,90	+7,06	10,43	72,21	−6,34	7,10	132,61	+6,66	26,96	78,13	−7,33	27,19	11
12	146,27	+5,69	8,41	73,73	−6,75	8,31	137,95	+6,18	26,10	82,40	−8,01	27,98	12
13	147,94	+5,40	6,88	76,82	−6,85	11,25	141,75	+6,69	25,55	89,19	−8,79	29,34	13
14	148,36	+5,31	6,50	83,59	−7,48	15,10	144,71	+6,90	25,71	98,19	−9,25	30,93	14
15	148,37	+5,27	6,52	94,73	−7,57	17,29	145,90	+6,91	25,65	109,96	−9,23	32,20	15
16	148,39	+5,23	6,54	108,94	−7,19	16,60	146,55	+6,75	25,64	122,40	−8,94	32,59	16
17	148,40	+5,19	6,56	124,65	−6,71	14,37	147,01	+6,54	25,58	135,22	−8,58	32,17	17
18	148,41	+5,15	6,61	141,72	−6,57	11,65	147,39	+6,22	25,54	148,56	−8,32	31,27	18
19	148,43	+5,14	6,63	160,20	−6,75	9,33	147,92	+6,02	25,51	162,29	−8,06	30,12	19
20	148,46	+5,11	6,66	178,86	−6,97	8,35	148,65	+5,89	25,40	175,66	−7,78	29,06	20
21	148,51	+5,05	6,71	195,31	−7,38	9,43	149,65	+5,70	25,31	186,88	−7,54	28,60	21
22	148,81	+5,11	6,93	208,80	−7,85	11,97	151,16	+5,64	25,42	196,22	−7,31	28,81	22
23	149,37	+5,12	7,56	217,38	−7,65	12,25	153,77	+5,87	25,87	204,26	−6,91	28,60	23
24	150,27	+5,32	8,55	220,39	−6,53	9,45	157,26	+6,23	26,49	209,77	−6,72	27,50	24
25	152,28	+5,17	10,54	223,56	−5,58	7,66	162,24	+6,51	27,45	214,04	−6,75	26,72	25
26	156,39	+5,30	13,41	224,65	−5,50	6,81	169,72	+7,23	28,65	217,74	−7,17	26,75	26
27	164,98	+5,55	16,50	224,69	−5,65	6,67	180,30	+7,56	29,86	220,23	−7,41	26,84	27
28	176,93	+5,64	16,99	224,70	−5,66	6,68	192,36	+7,53	30,51	221,41	−7,45	26,92	28
29	191,01	+5,71	15,91	224,71	−5,67	6,67	204,75	+7,51	30,86	222,22	−7,32	26,97	29
30	206,33	+6,05	13,81	224,72	−5,68	6,66	217,24	+7,55	30,68	222,86	−7,13	27,04	30
31	223,72	+6,38	11,14	224,73	−5,68	6,66	230,66	+7,55	29,92	223,48	−7,04	27,11	31

No.	Centre of gravity of whole leg						Centre of gravity of legs			Total centre of gravity of body			No.
	Right			Left									
	x	y	z	x	y	z	x	y	z	x	y	z	
1	31,65	+9,19	51,41	57,57	−8,50	52,61	44,61	+0,35	52,01	44,89	−0,63	89,57	1
2	41,66	+9,17	52,13	61,08	−8,68	53,00	51,37	+0,25	52,57	51,29	−0,73	90,16	2
3	51,80	+8,87	53,15	64,01	−8,83	53,62	57,91	+0,02	53,39	57,43	−0,85	91,22	3
4	61,86	+8,55	54,01	66,64	−8,75	53,96	64,25	−0,10	53,98	63,45	−0,86	92,22	4
5	71,26	+8,32	54,59	69,02	−8,68	54,20	70,14	−0,18	54,40	69,06	−0,85	92,97	5
6	80,67	+8,06	54,78	71,43	−8,59	54,29	76,05	−0,27	54,54	74,77	−0,87	93,33	6
7	89,91	+7,79	54,43	74,14	−8,52	54,16	82,03	−0,37	54,30	80,66	−0,83	93,18	7
8	97,63	+7,56	53,57	76,77	−8,43	53,79	87,20	−0,44	53,68	85,85	−0,78	92,54	8
9	105,55	+7,19	52,24	80,19	−8,31	53,19	92,87	−0,56	52,72	91,88	−0,70	91,41	9
10	112,56	+6,94	51,38	84,06	−8,33	52,84	98,31	−0,70	52,11	97,70	−0,62	90,43	10
11	120,13	+6,95	51,20	89,17	−8,60	52,74	104,65	−0,83	51,97	104,22	−0,45	89,86	11
12	126,74	+7,01	50,83	95,09	−8,93	52,86	110,92	−0,96	51,85	110,68	−0,20	89,53	12
13	132,39	+7,82	50,86	102,96	−9,34	53,23	117,68	−0,76	52,05	117,54	+0,22	89,59	13
14	137,33	+8,25	51,37	111,68	−9,42	53,58	124,51	−0,59	52,48	124,02	+0,52	89,94	14
15	140,54	+8,57	51,59	121,80	−9,10	54,38	131,17	−0,27	52,98	130,41	+0,84	90,66	15
16	143,12	+8,52	51,98	131,58	−8,83	55,42	137,35	−0,16	53,70	136,37	+0,91	91,72	16
17	145,46	+8,37	52,30	141,25	−8,61	56,28	143,36	−0,12	54,29	142,16	+0,90	92,63	17
18	147,71	+8,07	52,39	151,01	−8,44	56,59	149,36	−0,19	54,49	147,94	+0,85	93,14	18
19	150,09	+7,88	52,33	160,68	−8,20	56,38	155,39	−0,16	54,36	153,84	+0,73	93,16	19
20	152,66	+7,73	51,99	169,81	−7,89	55,65	161,24	−0,08	53,82	159,70	+0,57	92,70	20
21	155,42	+7,53	51,53	177,38	−7,58	54,62	166,40	−0,03	53,08	165,11	+0,51	91,92	21
22	158,64	+7,44	51,20	184,09	−7,23	53,73	171,37	+0,11	52,47	170,55	+0,40	91,07	22
23	163,22	+7,51	51,15	191,51	−7,10	53,23	177,37	+0,21	52,19	177,01	+0,22	90,37	23
24	168,48	+7,65	51,22	198,21	−7,39	52,89	183,35	+0,13	52,06	183,26	−0,09	89,96	24
25	174,96	+7,64	51,43	203,85	−7,82	52,57	189,41	−0,09	52,00	189,76	−0,48	89,75	25
26	183,03	+7,99	51,65	209,31	−8,55	52,97	196,17	−0,28	52,31	196,29	−0,84	89,87	26
27	192,84	+8,01	51,97	213,82	−8,93	53,34	203,33	−0,46	52,66	202,97	−1,09	90,23	27
28	202,99	+7,80	52,90	217,06	−9,08	53,74	210,03	−0,64	53,32	209,16	−1,17	91,04	28
29	212,85	+7,71	54,04	219,90	−9,00	54,12	216,38	−0,65	54,08	215,12	−1,23	92,15	29
30	222,41	+7,67	54,84	222,43	−8,85	54,40	222,42	−0,59	54,62	220,82	−1,25	93,05	30
31	232,15	+7,58	55,18	224,79	−8,75	54,58	228,47	−0,59	54,88	226,60	−1,23	93,58	31

For further analysis of the trajectory of the total centre of gravity, however, the system of co-ordinates must be orientated in such a way that, in the projection on the plane perpendicular to the direction of gait the ends of the trajectory of this centre of gravity meet each other. This is achieved by rotating the initial system of co-ordinates counter-clockwise about the vertical Z-axis by the above-mentioned angle. This can change x- and y-co-ordinates somewhat whereas the z-co-ordinates remain unaltered. The angle of rotation is designated as ε and the new co-ordinates as x', y' and z'. In Fig. 20 of Chap. 1, the new axes were rotated clockwise in relation to the previous ones. Consequently, ε has the opposite sign. The following equations thus pertain:

$$x' = x \cos \varepsilon - y \sin \varepsilon$$

$$y' = y \cos \varepsilon + x \sin \varepsilon$$

$$z' = z$$

$$\sin(0°5'31'') = 0.0016 \text{ and } \cos(0°5'31'') = 1 \text{ (to the fifth decimal place)}$$
$$\sin(0°8'17'') = 0.00241 \text{ and } \cos(0°8'17'') = 1 \text{ (to the fifth decimal place)}$$

The following equations can be used to calculate the new co-ordinates:

Experiment I

$$x' = x - 0.0016 \cdot y$$

$$y' = y + 0.0016 \cdot x$$

$$z' = z$$

Experiment II

$$x' = x - 0.00241 \cdot y$$

$$y' = y + 0.00241 \cdot x$$

$$z' = z$$

In these equations $0.0016 \cdot y$ and $0.00241 \cdot y$ would have a value greater than 0.005 cm if y were equal to or greater than 3.13 cm in one case, or 2.07 cm in the other. In the whole trajectory of the total centre of gravity the y-co-ordinates are smaller than that value, and the new co-ordinates can only be given exactly to the second decimal place. Therefore, the x-co-ordinates of the total centre of gravity retain the same values in the two experiments. After rotation of the system of co-ordinates through this small angle, only the y-co-ordinates are altered somewhat to the second decimal place. They take the values shown in Table 3.

Using these y-co-ordinates, the trajectory of the total centre of gravity can be projected on the plane perpendicular to the direction of gait, and a closed curve is obtained (see Figs. 6, 7, upper middle drawing). In the further analysis of the movement of the total centre of gravity, these co-ordinates will thus be used instead of the y-co-ordinates of Tables 1 and 2, without the prime sign being added to the y.

The calculation of the co-ordinates of the total centre of gravity in experiment III was more complicated than in the first two experiments. A way had to be found of establishing the co-ordinates of the centre of gravity of the left foot and centre of the left wrist which were not obtained directly (see Table 16, Chap. 1). In addition, the fact that the walking individual in experiment III carried military equipment entailed a modification of the calculation.

The co-ordinates of the centre of gravity of the left foot were obtained as follows. If from Tables 14–16 of Chap. 1 the differences between the co-ordinates of the centre of gravity of the foot and those of the centre of the ankle are calculated for the two extremities in experiments I and II and for the right extremity in experiment III, a high degree of conformity in the behaviour of these differences can be seen. This is most apparent if diagrams are drawn representing these differences related to time. Since the phases of movement follow each other at exactly the same time intervals, lengths numbered by the numbers of the successive phases can be chosen as abscissae. The ordinates will be the life-size differences between the co-ordinates of the centre of gravity of the foot and those of the

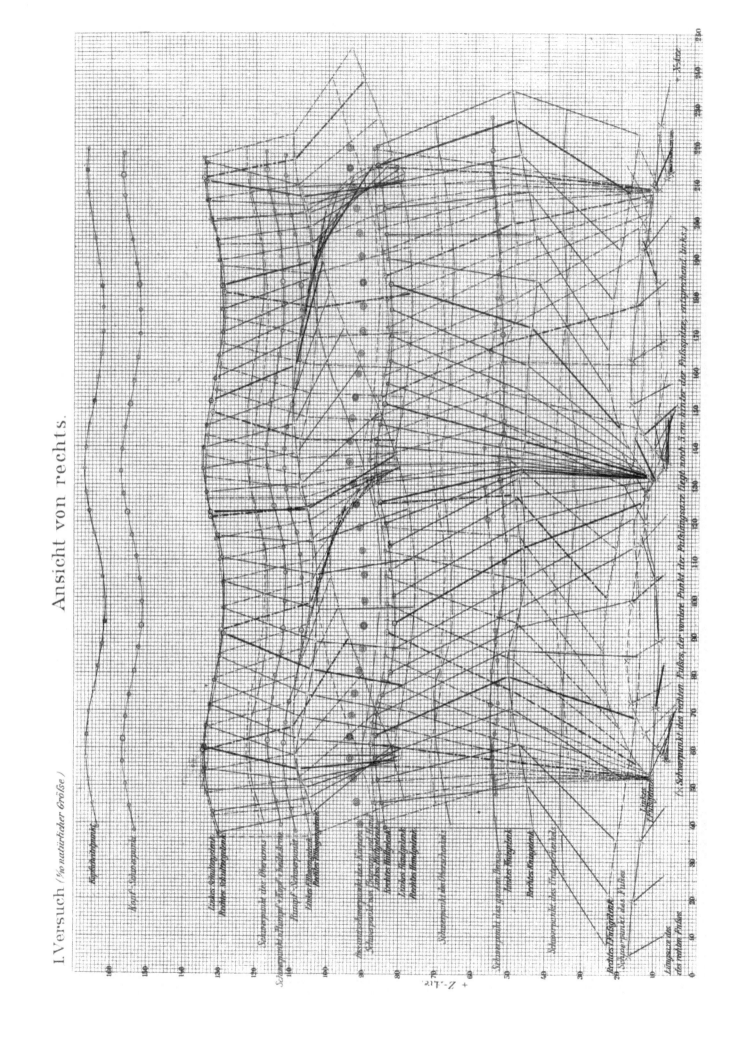

Ansicht von rechts.

I. Versuch (1/10 natürlicher Größe)

II Versuch (1/10 naturlicher Grosse)

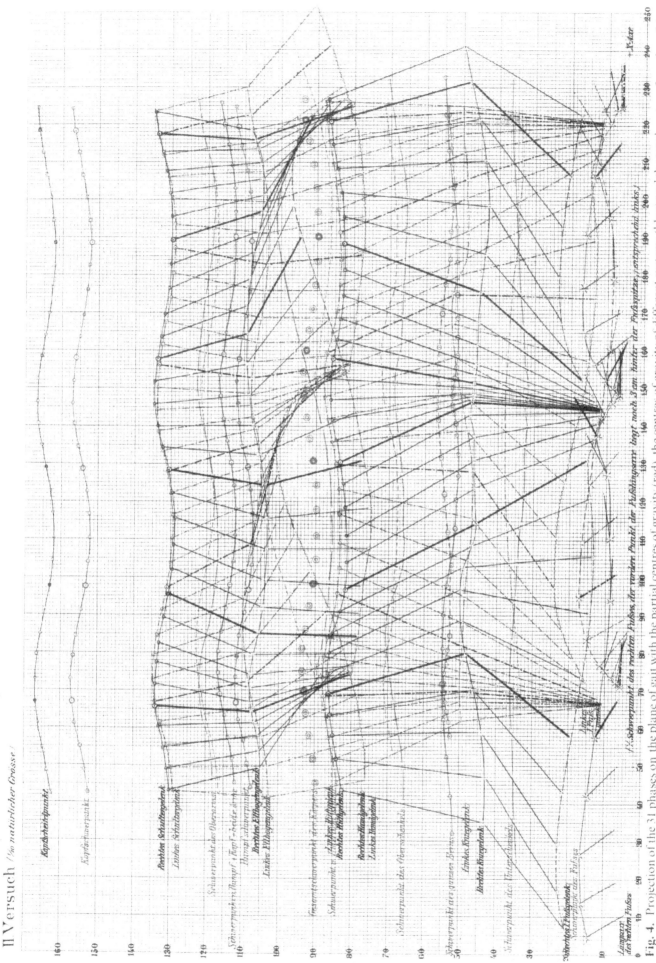

Fig. 4. Projection of the 31 phases on the plane of gait with the partial centres of gravity of different systems (blue) and the centres of gravity (red), the centres of gravity of the total centre of gravity of the body (⊙)

Table 3. Corrected y-co-ordinates of the total centre of gravity of the body

No.	Experiment I	Experiment II
1	− 1,30	− 0,52
2	− 1,30	− 0,61
3	− 1,24	− 0,71
4	− 1,15	− 0,71
5	− 1,06	− 0,68
6	− 0,98	− 0,69
7	− 0,87	− 0,64
8	− 0,74	− 0,57
9	− 0,60	− 0,48
10	− 0,31	− 0,38
11	+ 0,11	− 0,20
12	+ 0,53	+ 0,07
13	+ 0,91	+ 0,50
14	+ 1,17	+ 0,82
15	+ 1,27	+ 1,15
16	+ 1,26	+ 1,24
17	+ 1,20	+ 1,24
18	+ 1,08	+ 1,21
19	+ 1,00	+ 1,10
20	+ 0,86	+ 0,95
21	+ 0,72	+ 0,91
22	+ 0,46	+ 0,81
23	+ 0,07	+ 0,65
24	− 0,41	+ 0,35
25	− 0,84	− 0,02
26	− 1,10	− 0,37
27	− 1,28	− 0,60
28	− 1,35	− 0,67
29	− 1,27	− 0,71
30	− 1,14	− 0,72
31	− 1,03	− 0,68

centre of the ankle. For example, Fig. 5 represents the five diagrams for the x-co-ordinates. The same axis of the abscissa was always used for the two diagrams for the right and left sides of the body in the same experiment. Of course, these two diagrams are separated from each other, on the axis of the abscissa, by the time interval of one step. The diagrams of the same side of the body in different experiments, which are drawn one above another in Fig. 5, must also be displaced somewhat in relation to each other since the phases of movement designated by the same numbers do not correspond exactly in the three experiments to the same moments in the process of movement. Moreover, some discrepancies exist in the length of the diagrams corresponding to the different experiments along the axis of the abscissa, since the time of one step was not absolutely the same in all experiments. Except for these discrepancies in position and length and slight discrepancies in height, the five diagrams are almost identical. It is thus reasonable to attribute the same shape to the missing diagram of the differences in the x-co-ordinates of the lower extremity in experiment III. With regard to the duration of a step, experiment III coincides almost exactly with experiment I, as can be

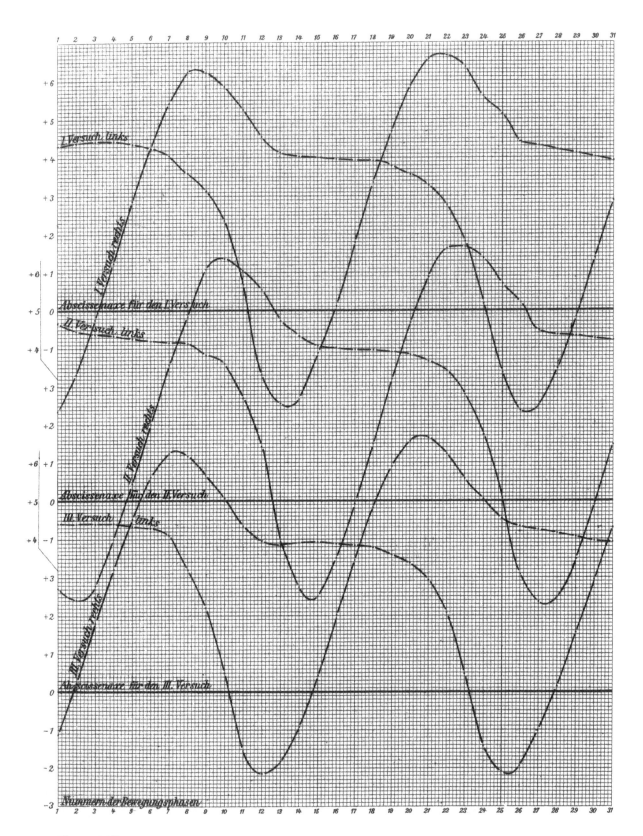

Fig. 5. Difference between the x-co-ordinates of the centre of gravity of the foot and those of the centre of the ankle

seen in the diagrams of the right side. Further, it can be observed in these diagrams that the negative ordinate values are somewhat lower in experiment III than in experiment I. Considering this and the fact that phase III, 1 corresponds almost exactly to phase I, $2\frac{1}{3}$, the diagrams of the x-co-ordinates of the left extremity in experiment III can be deduced directly from the co-ordinates in experiment I. The same applies of course to the diagrams of the two other co-ordinates. Of the three diagrams thus obtained, the diagram of the x-co-ordinates is represented in Fig. 5. The differences between the co-ordinates of the centre of gravity of the left foot and those of the centre of the left ankle in experiment III can be read directly from the diagrams. Adding these differences to the co-ordinates of the centre of the ankle gives the co-ordinates of the centre of gravity of the left foot which are listed in Table 6.

The co-ordinates of the centre of the left wrist in experiment III were directly measured only for seven phases of movement (phase 3 and phases 8–13; see Table 16, Chap. 1). The differences between the co-ordinates of the centre of the wrist and those of the centre of the elbow were calculated for these seven phases. With these data the gaps can be filled in the three diagrams representing these differences in relation to time, in the manner explained above. In this case, comparison with the corresponding diagrams of the right side of the body in experiment III and of both sides in the other two experiments cannot be used because in experiment III the left forearm was used to carry the rifle and virtually maintained the same direction in space. This circumstance, however, makes it possible to find the missing co-ordinates, although only approximately. Because of the behaviour of the left arm, the differences in the co-ordinates to be considered remain almost constant during the movement. Thus, it would not be too inaccurate to add to these differences the average values of the seven groups of three co-ordinates. Somewhat greater accuracy can be obtained by considering the small discrepancies between the seven known differences in each of the three co-ordinates and the average values. The rhythm of the movement during walking must be expressed in small oscillations, as in all the diagrams of this type; there is thus little choice if one wants to fill the gaps in the diagrams. On these diagrams (not published) the missing differences in the co-ordinates can be read. Adding these differences to the corresponding co-ordinates of the centre of the left elbow gives the co-ordinates of the centre of the left wrist (Table 4). These co-ordinates were added to Table 16, Chap. 1.

Of course it is not claimed that these co-ordinates have the same accuracy as those of the joint centres measured directly, or even those of the centre of gravity of the left foot obtained in an indirect manner. However, the co-ordinates are given to the tenth of a millimetre for uniformity of the further calculations. Errors in the co-ordinates of Table 4 would exert only an insignificant influence on the determination of the total centre of gravity of the human body. The position of the total centre of gravity depends on the direction of the longitudinal axis of the forearm, not on the exact position of the wrist. The direction of the longitudinal axis of the forearm is influenced only minimally by small deviations in the position of the centre of the wrist. The forearm in turn is only of slight importance in determining the position of the total centre of gravity. The extent of its contribution is shown by the very small proximal principal distance of the forearm + hand system, as explained above (pp. 128–137).

Determining the missing co-ordinates of the centre of gravity of the left foot and of the centre of the left wrist makes it possible to calculate the co-ordinates of the centre of gravity of the whole body for experiment III just as for the other two experiments, without considering the equipment carried. The co-ordinates of the tip of the left foot, still missing, are not used here. The co-ordinates of the centres of gravity of the body segments and of systems of body segments were calculated and are given with those of the total centre of gravity in Table 6.

Finally, to determine the co-ordinates of the total centre of gravity of the human body with loading, it is necessary to find the centre of gravity of the equipment alone. The various pieces of equipment are all more or less fixed to the trunk. Therefore, their common centre of gravity will maintain approximately the same position in relation to the trunk during gait. If

Table 4. Co-ordinates of the centre of the left wrist (complement to Table 16, in Chap. 1, p. 78)

No.	Experiment III		
	x	y	z
1	64,11	− 21,67	100,43
2	69,27	− 21,21	101,39
3	74,32	− 20,96	101,71
4	79,34	− 20,67	101,39
5	84,40	− 20,22	100,35
6	89,71	− 19,72	98,90
7	95,31	− 19,26	97,36
8	101,43	− 18,83	96,24
9	107,46	− 18,19	95,58
10	113,57	− 17,66	95,67
11	119,60	− 17,16	96,51
12	125,26	− 16,76	97,53
13	130,39	− 16,79	98,79
14	135,09	− 16,76	99,88
15	139,69	− 16,78	100,37
16	144,29	− 16,81	100,24
17	149,21	− 16,84	99,50
18	154,37	− 17,02	98,37
19	159,75	− 17,27	97,18
20	165,36	− 17,60	96,09
21	171,19	− 18,02	95,58
22	177,29	− 18,60	95,48
23	183,78	− 19,23	96,14
24	190,23	− 19,97	97,44
25	196,51	− 20,65	98,75
26	202,44	− 20,79	99,93
27	207,97	− 20,78	100,82
28	213,12	− 20,90	101,39
29	218,35	− 21,23	101,35
30	223,45	− 21,32	100,60
31	228,52	− 21,41	99,31

this position in relation to the trunk is known for one attitude of the human body, it is possible to deduce the whole trajectory of the centre of gravity of the military equipment from the movements of the trunk. The position of the centre of gravity of the military equipment in relation to the trunk can be determined using the data from the study on the centre of gravity previously mentioned. In this work the tridimensional co-ordinates of the total centre of gravity were calculated for the experimental subject standing to attention, bearing full military equipment complete with rifle. The co-ordinates and weights of the different pieces of equipment are given in Table 5. From these data the three co-ordinates of the common centre of gravity of all the equipment can be derived:

$$x = -5.77, \quad y = -2.11, \quad z = 112.33$$

The three co-ordinates of the centre of gravity of the trunk were:

$$x = +1.2, \quad y = +0.4, \quad z = 112$$

Table 5. Co-ordinates and weight of equipment

Equipment	Co-ordinates			weight (kg)
	x	y	z	
Knapsack with back cartridge pouch (with 40 live cartridges)	$-15,5$	$+0,4$	$114,5$	$12,25$
Front cartridge pouch { right (with 30 live cartridges) { left	$+16$ $+16$	$+12,5$ $-12,5$	$95,5$ $95,5$	$1,59$ $1,59$
Bayonet and spade	$-7,5$	-18	75	$1,57$
Full haversack with water bottle	$-7,5$	$+18$	75	$1,57$
Rifle	$+6$	$-11,5$	143	$4,7$

It appears that the centre of gravity of the equipment lies close to that of the trunk. Despite the different inclinations of the trunk during gait, detailed in Chap. 1, the centre of gravity of the whole equipment will thus roughly describe a trajectory which is congruent with that of the centre of gravity of the trunk and which appears to be only a short distance posterior to the left of and above the latter. The posterior difference is 6.97 cm, the difference to the left 2.51 cm, and the upward difference 0.33 cm. Thus, at a good approximation, the co-ordinates of the centre of gravity of the equipment are obtained for the 31 phases of experiment III by adding -6.97 to the x-co-ordinates, -2.51 to the y-co-ordinates and $+0.33$ to the z-co-ordinates of the centre of gravity of the trunk given in Table 6. The co-ordinates of the common centre of gravity of the equipment were thus calculated and are given in the penultimate column in Table 6.

Knowing these co-ordinates and those of the total centre of gravity of the unloaded human body, it is easy to find those of the total centre of gravity of the loaded body. The latter co-ordinates are designated as X_0, Y_0, Z_0, those of the total centre of gravity without loading as x_0, y_0, z_0, and those of the centre of gravity of the complete equipment as x_{13}, y_{13}, z_{13}. Thus:

$$X_0 = x_0 + \mu(x_{13} - x_0)$$

$$Y_0 = y_0 + \mu(y_{13} - y_0)$$

$$Z_0 = z_0 + \mu(z_{13} - z_0)$$

μ is the ratio of the mass of the whole equipment to the sum of the masses of the body and equipment. The equipment weighed 23.27 kg, and the experimental subject 58.7 kg. This relation is thus:

$$\mu = \frac{23.27}{81.97} = 0.284$$

Using these equations, the co-ordinates of the total centre of gravity of the subject and equipment are calculated, and they are given in Table 6 with those of the centres of gravity of the body segments and systems of body segments for experiment III.

A projection of the trajectories of the partial centres of gravity was not carried out for experiment III in the same manner as for the first two experiments since the projection of the whole movement could not be taken from the data in Chap. 1. The advantage of such a projection would not be worth the effort required to produce it. The differences between the trajectories of the centre of gravity in the walking of the loaded and unloaded individual are too small to be evident in a projection at a scale of 1 : 10. Roughly the same pictures would be obtained as in Fig. 4. Characteristic differences from normal walking caused by the heavy load only appear if the movements of mechanically important points of the human body in the various directions are measured life-size. The small rotation of the system of co-ordinates about the Z-axis, which would make the average forward direction of the total centre of gravity coincide exactly with the direction of the X-axis was also ignored. Indeed, it appeared that determining the average direction of the centre of gravity was much more unreliable in experiment III than in the other experiments because the heavy load made the subject deviate considerably from the indicated direction during a double step.

The tridimensional position of a point is determined clearly by its three rectangular co-ordinates. Similarly, the movement of a point in space can be established in every detail after this movement has been projected on three tridimensional directions perpendicular to each other. The various tables of co-ordinates give the projections of the movement of all the joint centres and centres of gravity on the direction of gait, on the direction perpendicular to that of gait, and on the vertical.

There is an essential difference between the movement of the individual points in the direction of gait and in the other two directions perpendicular to the direction of gait. In the direction perpendicular to the plane of gait (that of the y-co-ordinates), which will be referred to as the transverse direction, and in the vertical direction (that of the z-co-ordinates), all the points carry out periodical movements during walking. Each of these periods is as a rule equal to the duration of a double step, in some cases equal to the duration of a single step. For example, every point oscillates sideways on both sides of a vertical plane parallel to and corresponding to the average position of the plane of gait: the point leaves the plane on one side, returns, passes through the plane and leaves it on the other side before returning and passing through the plane, and the cycle starts again. Points lying in the median plane of the human body such as the midpoints of the hip and shoulder lines, the vertex, the centre of gravity and principal point of the trunk, the centre of gravity of the whole body, carry out this movement symmetrically relative to the plane of gait. Similarly, all the points oscillate vertically in relation to a horizontal plane corresponding to their middle position. Generally, all these oscillations are not as simple as, for example, the oscillations of a pendulum. However, what they have in common with the latter is that they repeat themselves exactly after a double step if the individual walks regularly. This repetition even occurs in the vertical direction after a single step for all the points lying in the median plane of the body, as for example the total centre of gravity of the body.

If, however, the movement of the different points of the body in the direction of gait is considered, there is no possibility of such an oscillation taking place relative to a plane fixed in space perpendicular to the direction of gait. All the points continue moving forwards. As soon as they have gone through a plane perpendicular to the direction of gait which shall be termed the "coronal plane of space", they continue moving away from this plane without returning. After a double step the points pass through another coronal plane of space parallel to the first and distant from the first by the length of a double step. However, it shall be assumed that the plane perpendicular to the direction of gait is not fixed but moves forwards, parallel to itself, at a constant velocity equal to the average velocity of gait. Any point of the body will then always return to the same place relative to this coronal plane after a double step. If this imaginary moving plane is correctly chosen, the point of the body will oscillate equally forwards and backwards relative to the moving plane exactly as it moves in other directions relative to fixed planes. The coronal plane moving uniformly forwards at the velocity of gait thus also corresponds to the average position of the total centre of gravity.

The average velocity of gait at which the coronal plane moves forwards can be easily calculated if the length and duration of a step or double step are known. Approximate values of these magnitudes are given on p. 86 for the first two experiments. They result from the projections of gait at a scale of 1 : 10. Somewhat greater accuracy can be attained using life-size projections. However, this process involves a small source of error since the magnitudes have to be measured from the displacement of a single part of the body, the foot, and the displacement of the body as a whole is not taken into account. The concept of the average velocity of gait implies that the movements of the different body segments are ignored and

Table 6. Co-ordinates of the centres of gravity for experiment III

No.	Centre of gravity of upper arm Right			Left			Centre of gravity of forearm Right			Left			No.
	x	y	z	x	y	z	x	y	z	x	y	z	
1	36,23	+18,23	120,30	35,91	−22,60	121,98	47,54	+24,60	87,42	54,56	−23,57	103,10	1
2	40,77	+18,69	121,35	41,36	−22,44	123,04	48,47	+26,36	87,68	59,78	−23,23	104,10	2
3	45,22	+19,30	122,09	46,79	−22,37	123,55	49,74	+28,07	88,10	64,94	−23,10	104,50	3
4	49,65	+19,88	122,39	52,14	−22,09	123,46	51,45	+29,77	88,52	70,04	−22,85	104,27	4
5	54,30	+20,42	122,13	57,51	−21,70	122,70	53,89	+31,32	88,55	75,15	−22,39	103,31	5
6	59,37	+20,99	121,34	62,97	−21,15	121,50	57,37	+32,60	88,11	80,47	−21,87	101,94	6
7	64,88	+21,51	120,23	68,55	−20,62	120,15	61,96	+33,77	87,29	86,05	−21,36	100,45	7
8	71,03	+22,08	119,06	74,39	−19,92	118,98	67,88	+34,80	86,26	92,08	−20,84	99,32	8
9	77,61	+22,77	118,20	80,15	−19,14	118,18	75,24	+35,44	85,37	98,00	−20,14	98,60	9
10	84,46	+23,37	117,93	86,11	−18,44	118,02	83,75	+35,69	84,90	104,09	−19,50	98,59	10
11	91,42	+23,92	118,23	91,87	−17,73	118,40	93,42	+35,61	85,22	110,09	−18,90	99,26	11
12	98,13	+24,21	119,33	97,34	−17,21	119,25	103,66	+35,05	86,40	115,69	−18,43	100,20	12
13	104,74	+24,41	120,81	102,44	−16,95	120,51	114,32	+34,12	88,56	120,80	−18,33	101,45	13
14	111,10	+24,54	122,44	107,25	−16,78	121,82	124,65	+32,82	91,51	125,53	−18,24	102,60	14
15	117,54	+24,57	123,76	112,15	−16,65	122,62	134,76	+31,14	94,74	130,20	−18,20	103,19	15
16	123,46	+24,46	124,41	117,06	−16,61	122,77	143,74	+29,35	97,72	134,89	−18,21	103,14	16
17	129,50	+24,34	124,29	122,21	−16,59	122,44	152,18	+27,56	100,12	139,88	−18,22	102,48	17
18	135,45	+24,06	123,71	127,45	−16,71	121,33	159,78	+25,92	101,90	145,09	−18,40	101,42	18
19	141,45	+23,62	122,79	132,87	−16,93	120,27	166,72	+24,57	102,73	150,51	−18,66	100,27	19
20	147,25	+23,06	121,67	138,43	−17,34	119,14	172,79	+23,55	102,42	156,13	−19,02	99,18	20
21	152,98	+22,50	120,72	144,24	−17,90	118,37	178,30	+23,00	100,84	161,94	−19,48	98,60	21
22	158,44	+21,74	119,98	150,13	−18,63	117,96	182,85	+22,64	98,57	167,98	−20,11	98,40	22
23	163,41	+21,03	119,49	156,18	−19,44	118,25	186,40	+22,82	95,71	174,40	−20,82	98,93	23
24	168,28	+20,39	119,43	162,27	−20,25	119,14	188,76	+23,59	92,76	180,80	−21,64	100,13	24
25	172,85	+19,87	119,68	168,09	−21,03	120,22	190,01	+24,47	90,27	186,98	−22,41	101,41	25
26	177,10	+19,35	120,45	173,74	−21,43	121,58	190,49	+25,36	88,85	192,87	−22,64	102,64	26
27	181,02	+19,06	121,31	179,31	−21,84	122,74	190,46	+26,44	88,55	198,41	−22,73	103,63	27
28	184,75	+18,83	122,11	184,66	−22,15	123,56	190,72	+27,38	88,77	203,64	−22,94	104,29	28
29	188,79	+18,73	122,63	190,19	−22,56	123,74	191,56	+28,24	89,04	208,97	−23,35	104,34	29
30	193,12	+18,64	122,53	195,61	−22,81	123,20	193,36	+28,65	89,08	214,14	−23,51	103,64	30
31	197,77	+18,52	121,98	201,00	−22,87	122,09	196,10	+29,45	88,76	219,26	−23,57	102,40	31

No.	Centre of gravity of whole arm Right			Left			Centre of gravity of trunk			Centre of gravity of head			No.
	x	y	z	x	y	z	x	y	z	x	y	z	
1	41,67	+21,29	104,48	44,88	−23,07	112,90	33,65	−2,68	111,26	37,48	−2,02	154,51	1
2	44,47	+22,38	105,15	50,22	−22,82	113,93	38,99	−2,34	112,26	43,15	−1,73	155,37	2
3	47,39	+23,52	105,74	55,52	−22,72	114,39	44,27	−2,03	112,74	48,77	−1,43	155,76	3
4	50,52	+24,64	106,10	60,75	−22,46	114,23	49,55	−1,67	112,73	54,34	−1,03	155,57	4
5	54,10	+25,66	105,98	65,99	−22,03	113,37	54,83	−1,28	112,08	60,07	−0,66	154,84	5
6	58,41	+26,57	105,36	71,39	−21,50	112,09	60,24	−0,86	110,96	65,85	−0,25	153,72	6
7	63,48	+27,41	104,39	76,97	−20,98	110,67	65,76	−0,41	109,69	71,48	+0,26	152,45	7
8	69,51	+28,20	103,28	82,90	·20,36	109,52	71,60	+0,14	108,69	77,26	+0,91	151,38	8
9	76,47	+28,86	102,41	88,74	−19,62	108,76	77,73	+0,81	108,06	83,09	+1,47	150,69	9
10	84,12	+29,30	102,04	94,76	−18,95	108,67	84,11	+1,57	108,03	88,85	+1,98	150,71	10
11	92,38	+29,54	102,35	100,63	−18,29	109,19	90,39	+2,34	108,45	94,47	+2,39	151,25	11
12	100,79	+29,42	103,49	106,17	−17,80	110,09	96,17	+2,87	109,33	99,93	+2,52	152,25	12
13	109,35	+29,08	105,30	111,27	−17,61	111,34	101,69	+3,19	110,62	105,31	+2,60	153,60	13
14	117,62	+28,52	107,56	116,04	−17,48	112,58	106,83	+3,32	111,96	110,53	+2,67	154,83	14
15	125,82	+27,73	109,80	120,83	−17,40	113,27	111,96	+3,36	112,91	115,78	+2,76	155,59	15
16	133,21	+26,81	111,57	125,64	−17,38	113,33	117,02	+3,29	113,13	120,89	+2,77	155,73	16
17	140,41	+25,89	112,66	130,71	−17,37	112,84	122,20	+3,14	112,89	126,03	+2,76	155,33	17
18	147,15	+24,95	113,22	135,93	−17,52	111,75	127,26	+2,98	111,96	131,22	+2,65	154,40	18
19	153,60	+24,08	113,14	141,35	−17,76	110,65	132,56	+2,74	110,83	136,56	+2,39	153,29	19
20	159,53	+23,30	112,41	146,94	−18,15	109,54	138,08	+2,27	109,65	141,92	+2,05	152,22	20
21	165,16	+22,74	111,16	152,75	−18,66	108,86	143,99	+1,73	108,95	147,58	+1,56	151,44	21
22	170,18	+22,17	109,68	158,72	−19,34	108,55	150,05	+0,94	108,63	153,20	+0,96	151,06	22
23	174,47	+21,89	108,05	164,94	−20,10	108,96	156,15	0	108,83	158,80	+0,27	151,29	23
24	178,13	+21,93	106,60	171,18	−20,92	110,00	162,25	−0,87	109,34	164,48	−0,32	151,90	24
25	181,10	+22,08	105,53	177,18	−21,69	111,17	167,97	−1,44	110,05	169,89	−0,72	152,81	25
26	183,54	+22,24	105,25	182,94	−22,01	112,47	173,48	−1,86	111,12	175,27	−1,06	154,05	26
27	185,56	+22,61	105,55	188,50	−22,27	113,55	178,80	−2,14	112,13	180,68	−1,28	155,11	27
28	187,62	+22,94	106,07	193,79	−22,53	114,29	183,86	−2,43	112,94	185,78	−1,41	155,80	28
29	190,12	+23,30	106,47	199,22	−22,94	114,41	189,09	−2,67	113,02	191,09	−1,55	155,91	29
30	193,24	+23,45	106,44	204,52	−23,15	113,79	194,26	−2,86	112,59	196,55	−1,74	155,41	30
31	196,97	+23,78	106,00	209,78	−23,21	112,62	199,39	−3,02	111,74	202,04	−1,85	154,45	31

Table 6 (continued)

No.	Centre of gravity of trunk + head			Centre of gravity of arms			Centre of gravity of trunk + head + arms			No.
	x	y	z	x	y	z	x	y	z	
1	34,19	− 2,59	117,40	43,28	− 0,89	108,69	36,07	− 2,24	115,60	1
2	39,58	− 2,25	118,38	47,35	− 0,22	109,54	41,19	− 1,83	116,55	2
3	44,91	− 1,94	118,85	51,46	+ 0,40	110,07	46,27	− 1,46	117,03	3
4	50,23	− 1,58	118,81	55,64	+ 1,09	110,17	51,35	− 1,03	117,02	4
5	55,57	− 1,19	118,15	60,05	+ 1,82	109,68	56,50	− 0,57	116,40	5
6	61,04	− 0,77	117,03	64,90	+ 2,54	108,73	61,84	− 0,08	115,31	6
7	66,57	− 0,31	115,76	70,23	+ 3,22	107,53	67,33	+ 0,42	114,06	7
8	72,40	+ 0,25	114,75	76,21	+ 3,92	106,40	73,19	+ 1,01	113,02	8
9	78,49	+ 0,90	114,11	82,61	+ 4,62	105,59	79,34	+ 1,67	112,35	9
10	84,78	+ 1,63	114,09	89,44	+ 5,18	105,36	85,74	+ 2,36	112,28	10
11	90,97	+ 2,35	114,53	96,51	+ 5,63	105,77	92,12	+ 3,03	112,72	11
12	96,70	+ 2,82	115,42	103,48	+ 5,81	106,79	98,10	+ 3,44	113,63	12
13	102,20	+ 3,11	116,72	110,31	+ 5,74	108,32	103,88	+ 3,65	114,98	13
14	107,36	+ 3,23	118,05	116,83	+ 5,52	110,07	109,32	+ 3,70	116,40	14
15	112,50	+ 3,27	118,97	123,33	+ 5,17	111,54	114,74	+ 3,66	117,43	15
16	117,57	+ 3,22	119,18	129,43	+ 4,72	112,45	120,03	+ 3,53	117,79	16
17	122,74	+ 3,09	118,92	135,56	+ 4,26	112,75	125,39	+ 3,33	117,64	17
18	127,82	+ 2,93	117,99	141,54	+ 3,72	112,49	130,66	+ 3,09	116,85	18
19	133,13	+ 2,69	116,86	147,48	+ 3,16	111,90	136,10	+ 2,79	115,83	19
20	138,63	+ 2,24	115,69	153,24	+ 2,58	110,98	141,65	+ 2,31	114,72	20
21	144,50	+ 1,71	114,98	158,96	+ 2,04	110,01	147,49	+ 1,78	113,95	21
22	150,50	+ 0,94	114,66	164,45	+ 1,42	109,12	153,39	+ 1,04	113,51	22
23	156,53	+ 0,04	114,86	169,71	+ 0,90	108,51	159,26	+ 0,22	113,55	23
24	162,57	− 0,79	115,38	174,66	+ 0,51	108,30	165,07	− 0,52	113,91	24
25	168,24	− 1,34	116,12	179,14	+ 0,20	108,35	170,48	− 1,02	114,51	25
26	173,73	− 1,75	117,22	183,24	+ 0,12	108,86	175,70	− 1,36	115,49	26
27	179,07	− 2,02	118,23	187,03	+ 0,17	109,55	180,72	− 1,57	116,43	27
28	184,13	− 2,29	119,03	190,71	+ 0,21	110,18	185,49	− 1,77	117,20	28
29	189,37	− 2,51	119,11	194,67	+ 0,18	110,44	190,47	− 1,95	117,32	29
30	194,59	− 2,70	118,67	198,88	+ 0,15	110,12	195,48	− 2,11	116,90	30
31	199,77	− 2,85	117,80	203,38	+ 0,29	109,31	200,52	− 2,20	116,04	31

No.	Centre of gravity of thigh						Centre of gravity of lower leg						No.
	Right			Left			Right			Left			
	x	y	z	x	y	z	x	y	z	x	y	z	
1	31,50	+ 5,81	69,85	35,76	− 11,62	68,75	21,07	+ 6,48	37,06	38,76	− 9,48	31,69	1
2	38,56	+ 6,28	71,32	39,03	− 11,20	69,17	32,63	+ 6,93	37,19	39,43	− 9,19	31,74	2
3	45,30	+ 6,83	72,29	42,30	− 10,77	69,30	44,26	+ 7,62	36,95	40,15	− 8,84	31,75	3
4	51,56	+ 7,20	72,70	45,69	− 10,37	69,23	55,41	+ 8,28	36,42	40,98	− 8,49	31,71	4
5	57,07	+ 7,49	72,15	49,18	− 9,98	68,81	65,72	+ 8,94	35,69	41,93	− 8,22	31,63	5
6	62,27	+ 7,65	70,73	52,88	− 9,67	68,15	74,96	+ 9,63	34,92	43,06	− 7,89	31,57	6
7	67,44	+ 7,74	69,18	57,09	− 9,35	67,37	82,65	+ 10,04	34,45	44,70	− 7,67	31,40	7
8	73,91	+ 8,30	68,45	62,07	− 9,10	66,74	89,66	+ 10,60	33,81	47,69	− 7,84	31,38	8
9	81,06	+ 9,34	68,21	67,92	− 8,92	66,42	95,88	+ 11,48	33,16	51,67	− 8,22	32,14	9
10	86,86	+ 10,20	68,13	74,91	− 8,54	66,16	99,83	+ 12,02	32,54	57,21	− 8,44	32,78	10
11	93,60	+ 11,23	68,92	83,50	− 8,34	65,90	104,24	+ 12,65	32,91	65,97	− 8,90	33,55	11
12	99,18	+ 11,87	69,65	92,10	− 7,64	65,78	107,36	+ 12,90	33,29	76,46	− 8,49	34,57	12
13	102,97	+ 12,07	70,42	100,56	− 7,02	66,91	108,23	+ 12,78	33,41	87,99	− 7,55	35,58	13
14	106,06	+ 12,07	71,21	108,07	− 6,94	68,59	108,54	+ 12,07	33,65	99,22	− 6,96	36,45	14
15	109,34	+ 11,94	71,89	115,29	− 6,93	70,07	109,11	+ 12,43	33,97	110,62	− 6,61	36,70	15
16	112,92	+ 11,74	72,01	121,86	− 6,99	70,94	110,14	+ 12,18	34,10	121,61	− 6,59	36,41	16
17	116,59	+ 11,51	71,89	127,98	− 7,21	71,13	111,18	+ 12,02	34,18	132,34	− 6,94	35,75	17
18	120,05	+ 11,27	71,53	133,24	− 7,33	70,51	112,16	+ 11,83	34,24	142,02	− 7,42	34,81	18
19	124,16	+ 10,99	70,83	138,08	− 7,32	69,01	113,85	+ 11,59	34,16	150,76	− 7,92	33,78	19
20	128,65	+ 10,59	70,26	142,94	− 7,28	67,47	116,01	+ 11,47	34,36	157,68	− 8,41	32,96	20
21	133,87	+ 10,27	70,08	149,06	− 7,75	66,71	119,03	+ 11,51	34,89	163,93	− 8,68	32,17	21
22	140,06	+ 9,71	69,93	155,71	− 8,66	66,75	123,32	+ 11,55	35,51	169,50	− 9,18	31,72	22
23	147,05	+ 8,91	69,91	161,74	− 9,98	67,13	129,22	+ 11,16	36,18	173,68	− 10,05	31,29	23
24	155,03	+ 8,22	69,66	167,97	− 11,02	67,72	137,42	+ 10,83	37,02	177,82	− 10,76	31,84	24
25	163,20	+ 7,57	69,41	173,03	− 11,66	68,31	147,37	+ 10,07	37,60	180,33	− 11,03	32,04	25
26	171,30	+ 6,86	70,09	177,17	− 11,96	68,87	158,48	+ 8,89	38,15	181,65	− 10,98	32,15	26
27	178,56	+ 6,34	71,30	180,56	− 12,02	69,37	169,30	+ 7,56	38,51	182,31	− 10,92	32,25	27
28	185,25	+ 5,94	72,21	183,71	− 12,01	69,68	179,88	+ 6,51	38,23	182,93	− 10,78	32,38	28
29	191,75	+ 5,60	72,68	187,13	− 12,03	69,68	190,80	+ 5,58	37,49	183,79	− 10,63	32,46	29
30	197,72	+ 5,31	72,62	190,52	− 12,04	69,38	201,52	+ 4,86	36,48	184,77	− 10,52	32,44	30
31	203,16	+ 4,92	71,90	194,07	− 12,09	68,85	211,78	+ 4,27	35,39	185,99	− 10,46	32,36	31

Table 6 (continued)

No.	Centre of gravity of foot Right x	y	z	Left x	y	z	Centre of gravity of lower leg + foot Right x	y	z	Left x	y	z	No.
1	4,70	+ 4,89	14,92	41,83	− 8,57	5,60	16,93	+ 6,08	31,46	39,54	− 9,25	25,09	1
2	20,51	+ 5,41	12,75	41,81	− 8,46	5,59	29,56	+ 6,55	31,01	40,03	− 9,01	25,12	2
3	37,60	+ 6,15	10,34	41,78	− 8,32	5,56	42,58	+ 7,25	30,22	40,56	− 8,71	25,12	3
4	55,06	+ 7,38	8,56	41,74	− 8,16	5,52	55,32	+ 8,05	29,37	41,17	− 8,41	25,08	4
5	72,54	+ 8,96	8,14	41,62	− 8,03	5,49	67,45	+ 8,95	28,72	41,85	− 8,17	25,02	5
6	88,73	+10,97	9,55	41,54	− 7,76	5,39	·78,44	+ 9,97	28,50	42,68	− 7,86	24,95	6
7	101,14	+12,73	12,03	41,31	− 7,40	5,13	87,33	+10,72	28,78	43,84	− 7,60	24,75	7
8	107,76	+13,44	11,29	41,76	− 7,57	5,19	94,24	+11,32	28,11	46,19	− 7,77	24,75	8
9	110,76	+13,03	8,72	42,77	− 7,79	6,27	99,64	+11,87	26,98	49,42	− 8,11	25,59	9
10	113,19	+12,82	6,44	44,07	− 7,35	8,06	103,21	+12,22	25,94	53,89	− 8,16	26,53	10
11	113,75	+12,65	5,97	47,86	− 7,23	11,64	106,65	+12,65	26,09	61,39	− 8,48	28,01	11
12	113,75	+12,58	5,98	56,03	− 7,27	15,66	108,98	+12,82	26,38	71,29	− 8,18	29,79	12
13	113,75	+12,51	6,00	67,75	− 6,19	16,69	109,63	+12,71	26,48	82,87	− 7,21	30,80	13
14	113,75	+12,45	6,03	81,46	− 4,74	15,93	109,86	+12,61	26,66	94,73	− 6,40	31,26	14
15	113,77	+12,42	6,06	96,64	− 3,87	13,72	110,29	+12,43	26,91	107,08	− 5,92	30,89	15
16	113,77	+12,41	6,05	112,86	− 4,23	11,30	111,06	+12,24	27,00	119,40	− 5,99	30,06	16
17	113,77	+12,37	6,11	129,76	− 5,19	9,57	111,84	+12,11	27,08	131,69	− 6,50	29,13	17
18	113,78	+12,35	6,14	146,23	− 6,30	8,94	112,57	+11,96	27,13	143,09	− 7,14	28,26	18
19	113,79	+12,29	6,16	161,84	− 8,11	10,16	113,83	+11,77	27,08	153,56	− 7,97	27,80	19
20	114,07	+12,34	6,49	173,47	−10,70	12,05	115,52	+11,69	27,31	161,67	− 8,99	27,67	20
21	114,65	+12,50	7,09	179,85	−10,95	11,44	117,92	+11,76	27,86	167,96	− 9,25	26,93	21
22	115,69	+12,58	8,07	182,99	−10,70	9,26	121,39	+11,81	28,57	172,91	− 9,56	26,04	22
23	117,64	+12,05	9,94	184,79	−10,11	7,34	126,29	+11,39	29,54	176,49	−10,07	25,23	23
24	121,62	+11,44	13,23	185,84	− 9,90	6,35	133,42	+10,98	31,00	179,85	−10,54	25,39	24
25	129,39	+10,25	16,68	185,63	−10,13	6,19	142,82	+10,12	32,31	181,67	−10,80	25,50	25
26	140,65	+ 8,37	17,29	185,53	−10,07	6,29	153,97	+ 8,76	32,87	182,63	−10,75	25,61	26
27	154,02	+ 6,40	16,16	185,50	−10,06	6,41	165,43	+ 7,27	32,86	183,12	−10,70	25,71	27
28	168,19	+ 4,83	13,78	185,49	−10,04	6,51	176,92	+ 6,08	32,04	183,58	−10,59	25,83	28
29	184,18	+ 3,45	10,99	185,49	−10,07	6,64	189,13	+ 5,04	30,79	184,22	−10,49	25,93	29
30	201,05	+ 2,48	8,66	185,46	−10,06	6,70	201,40	+ 4,26	29,44	184,94	−10,40	25,93	30
31	218,22	+ 2,12	7,77	185,53	−10,01	6,76	213,41	+ 3,73	28,40	185,87	−10,35	25,88	31

No.	Centre of gravity of whole leg Right x	y	z	Left x	y	z	Centre of gravity of legs x	y	z	No.
1	25,98	+ 5,91	55,30	37,19	−10,72	52,20	31,59	− 2,41	53,75	1
2	35,15	+ 6,38	56,04	39,41	−10,37	52,48	37,28	− 2,00	54,26	2
3	44,27	+ 6,99	56,35	41,64	− 9,99	52,56	42,96	− 1,50	54,46	3
4	52,99	+ 7,52	56,28	43,98	− 9,63	52,50	48,49	− 1,06	54,39	4
5	61,00	+ 8,04	55,69	46,40	− 9,29	52,21	53,70	− 0,63	53,95	5
6	68,40	+ 8,53	54,72	49,01	− 8,98	51,78	58,71	− 0,23	53,25	6
7	74,98	+ 8,87	53,87	52,07	− 8,69	51,22	63,53	+ 0,09	52,55	7
8	81,62	+ 9,44	53,16	56,05	− 8,60	50,83	68,84	+ 0,42	52,00	8
9	88,10	+10,30	52,58	60,91	− 8,61	50,95	74,51	+ 0,85	51,77	9
10	93,06	+10,97	52,14	66,94	− 8,40	51,14	80,00	+ 1,29	51,64	10
11	98,55	+11,77	52,69	75,12	− 8,39	51,54	86,84	+ 1,69	52,12	11
12	102,89	+12,23	53,25	84,21	− 7,84	52,14	93,55	+ 2,20	52,70	12
13	105,49	+12,31	53,77	93,86	− 7,09	53,22	99,68	+ 2,61	53,50	13
14	107,50	+12,27	54,33	103,01	− 6,74	54,44	105,26	+ 2,77	54,39	14
15	109,70	+12,13	54,84	112,18	− 6,55	55,22	110,94	+ 2,79	55,03	15
16	112,22	+11,93	54,95	120,93	− 6,61	55,45	116,58	+ 2,66	55,20	16
17	114,79	+11,74	54,91	129,39	− 6,94	55,21	122,09	+ 2,40	55,06	17
18	117,22	+11,53	54,70	136,97	− 7,26	54,50	127,10	+ 2,14	54,60	18
19	120,24	+11,29	54,25	143,95	− 7,57	53,39	132,10	+ 1,86	53,82	19
20	123,67	+11,01	53,98	150,04	− 7,93	52,39	136,86	+ 1,54	53,19	20
21	127,82	+10,83	54,08	156,22	− 8,32	51,63	142,02	+ 1,26	52,86	21
22	132,98	+10,51	54,25	162,23	− 9,00	51,32	147,61	+ 0,76	52,79	22
23	139,18	+ 9,85	54,61	167,33	−10,01	51,25	153,26	− 0,08	52,93	23
24	146,84	+ 9,27	55,01	172,47	−10,84	51,68	159,66	− 0,79	53,35	24
25	155,48	+ 8,54	55,35	176,30	−11,33	52,09	165,89	− 1,40	53,72	25
26	164,73	+ 7,58	55,98	179,24	−11,50	52,47	171,99	− 1,96	54,23	26
27	173,58	+ 6,69	56,73	181,53	−11,52	52,82	177,56	− 2,42	54,78	27
28	182,09	+ 5,99	56,99	183,66	−11,47	53,06	182,88	− 2,74	55,03	28
29	190,76	+ 5,39	56,80	186,03	−11,45	53,10	188,40	− 3,03	54,95	29
30	199,11	+ 4,91	56,25	188,41	−11,42	52,91	193,76	− 3,26	54,58	30
31	207,04	+ 4,47	55,41	190,96	−11,43	52,56	199,00	− 3,48	53,99	31

Table 6 (continued)

No.	Total centre of gravity of body without equipment			Total centre of gravity of equipment			Total centre of gravity of body with equipment			No.
	x	y	z	x	y	z	x	y	z	
1	34,40	−2,30	92,53	26,68	−5,19	111,59	32,21	−3,12	97,94	1
2	39,73	−1,89	93,32	32,02	−4,85	112,59	37,54	−2,73	98,79	2
3	45,04	−1,47	93,69	37,30	−4,54	113,07	42,84	−2,34	99,19	3
4	50,28	−1,04	93,66	42,58	−4,18	113,06	48,09	−1,93	99,17	4
5	55,46	−0,59	93,11	47,86	−3,79	112,41	53,30	−1,50	98,59	5
6	60,67	−0,14	92,16	53,27	−3,37	111,29	58,57	−1,06	97,59	6
7	65,91	+0,30	91,12	58,79	−2,92	110,02	63,89	−0,61	96,49	7
8	71,57	+0,79	90,26	64,63	−2,37	109,02	69,60	−0,11	95,59	8
9	77,54	+1,36	89,75	70,76	−1,70	108,39	75,61	+0,49	95,04	9
10	83,60	+1,96	89,66	77,14	−0,94	108,36	81,77	+1,14	94,97	10
11	90,15	+2,53	90,12	83,42	−0,17	108,78	88,24	+1,76	95,42	11
12	96,40	+2,98	90,90	89,20	+0,36	109,66	94,36	+2,24	96,23	12
13	102,31	+3,26	92,05	94,72	+0,68	110,95	100,15	+2,53	97,42	13
14	107,81	+3,35	93,27	99,86	+0,81	112,29	105,55	+2,63	98,67	14
15	113,32	+3,34	94,15	104,99	+0,85	113,24	110,95	+2,63	99,57	15
16	118,74	+3,21	94,44	110,05	+0,78	113,46	116,27	+2,52	99,84	16
17	124,16	+2,98	94,30	115,23	+0,63	113,22	121,62	+2,31	99,67	17
18	129,33	+2,74	93,63	120,29	+0,47	112,29	126,76	+2,10	98,93	18
19	134,61	+2,44	92,70	125,59	+0,23	111,16	132,05	+1,81	97,94	19
20	139,86	+2,02	91,77	131,11	−0,24	109,98	137,37	+1,38	96,94	20
21	145,45	+1,59	91,16	137,02	−0,78	109,28	143,06	+0,92	96,31	21
22	151,23	+0,94	90,86	143,08	−1,57	108,96	148,92	+0,23	96,00	22
23	157,02	+0,11	90,94	149,18	−2,51	109,16	154,79	−0,63	96,11	23
24	163,05	−0,62	91,32	155,28	−3,38	109,67	160,84	−1,40	96,53	24
25	168,77	−1,16	91,84	161,00	−3,95	110,38	166,56	−1,95	97,11	25
26	174,32	−1,58	92,64	166,51	−4,37	111,45	172,10	−2,37	97,98	26
27	179,54	−1,89	93,43	171,83	−4,65	112,46	177,35	−2,67	98,83	27
28	184,52	−2,13	94,01	176,89	−4,94	113,27	182,35	−2,93	99,48	28
29	189,70	−2,35	94,06	182,12	−5,18	113,35	187,55	−3,15	99,54	29
30	194,84	−2,54	93,65	187,29	−5,37	112,92	192,70	−3,34	99,12	30
31	199,95	−2,68	92,90	192,42	−5,53	112,07	197,81	−3,49	98,34	31

only the forward movement of the whole mass of the body is considered, which is expressed by the trajectory of the total centre of gravity. Since further study demands that the magnitude of the average velocity be known as accurately as possible, the movement of the total centre of gravity has to be determined. The periodicity of this movement can be recognized in its trajectory as well as in the diagrams of its co-ordinates in relation to time. This periodicity provides a relatively reliable means of determining the length and duration of a double step. The results are given in Table 7. The average velocity of a double step in this table is obtained by dividing the length of a double step by its duration.

The coronal plane relative to which the total centre of gravity carries out periodical oscillations in the direction of gait, has to move forward at this velocity. To determine the position of the centre of gravity relative to this plane at each of the 31 phases, it is necessary to know how far this plane moves forward in the time between two successive phases. In 1 s, 26.09 phases occur in the three experiments (pp. 15–16). Dividing the average velocity by

Table 7. Length and duration of a double step

	Experiment I	Experiment II	Experiment III
Length of a double step	155,75 cm	153,6 cm	143,8 cm
Duration of a double step	0,99 s	0,97 s	0,9895 s
Average velocity of gait	157,32 cm s^{-1}	158,35 cm s^{-1}	145,33 cm s^{-1}

26.09 gives the length in centimetres which the plane travels between two successive phases. In experiment I this length is 6.03 cm, in experiment II 6.07 cm, and in experiment III 5.57 cm.

In order to find the position of this plane in relation to the fixed tridimensional system of co-ordinates for each of the 31 phases, its position has to be determined for at least one phase. The moving plane always remains parallel to the plane YZ of the system of co-ordinates. Therefore, its position in space can be established as soon as its distance from the plane YZ is found, i.e. the distance at which the moving plane intersects the X-axis in the system of co-ordinates. This distance gives the x-co-ordinates for all the points of the moving plane. For example, if in experiment I the total centre of gravity passed through the moving plane in the first phase of movement, the position of this plane in the first phase would be given by $x = 39.55$ cm (Table 1). In the second phase, the moving plane would lie at a distance of 39.55 cm + 6.03 cm = 45.58 cm from the plane YZ of the system of co-ordinates; in the third phase it would be at a distance of 39.55 cm + 2 . 6.03 = 51.61 cm, etc. As results from the values of its x-co-ordinates given in Table 1, the total centre of gravity would lie in the second phase at a distance of 45.5 cm − 45.58 cm = −0.08 cm, in the third phase at a distance of 51.45 cm − 61.16 cm = −0.61 cm from the moving plane. The negative sign indicates that the total centre of gravity lies behind the plane moving forwards. It should thus be possible to determine the position of the total centre of gravity relative to the moving plane for every phase.

However, in the first phase, the total centre of gravity does not exactly lie in the coronal plane about which it oscillates equally forwards and backwards during walking. If it did so it would be pure coincidence. As a rule, the plane will be in front of or behind the centre of gravity in the first phase. To find exactly where it does lie in the first phase, the position of the total centre of gravity is determined in relation to any coronal plane moving forwards at the average velocity of gait, for each phase. To this end, the plane in which the total centre of gravity lies in the first phase can be chosen. Generally, the total centre of gravity will oscillate in front of and behind this plane but it will go further on one side than on the other. It is easy then to find the distance which separates this plane chosen arbitrarily from that corresponding to the mid-position of the total centre of gravity, in one direction or the other. This enables the position of the mid-plane to be established in each of the 31 phases. The distances between the moving mid-coronal plane and the plane YZ were found in this manner for the 31 phases and are given in Table 8 for the three experiments. The same distances would also have been obtained if the initial system of co-ordinates had not been rotated through an angle of some minutes in experiments I and II since the x-co-ordinates were unaffected by this rotation.

The distance of the total centre of gravity from the mid-plane moving forwards perpendicular to the direction of gait is obtained for each phase by subtracting the distance between the moving plane and the plane YZ, given in Table 8, from the x-co-ordinates of the total centre of gravity. The results are given in Table 9.

These distances can be regarded as a particular type of co-ordinates. If the rectangular system of co-ordinates is imagined not as fixed in space but as moving uniformly forwards at the average velocity of gait, the distances in Table 9 represent the co-ordinates of the centre of gravity parallel to the direction of gait in the moving system of co-ordinates. This presupposes that the plane of co-ordinates perpendicular to the direction of gait corresponds to the mid-coronal plane, as described above. Despite displacement of the whole system of co-ordinates, the two other planes of co-ordinates – the plane of gait and that of the horizontal ground – remain unchanged. The y-co-ordinates (and the y'-co-ordinates given in Table 3 for the first two experiments) and z-co-ordinates have the same values in the moving system of co-ordinates as in the fixed system. The y-co-ordinates give the distances of the points from the plane of gait, and the z-co-ordinates the distances from the horizontal floor. Only the x-co-ordinates change and, for example, for the total centre of gravity take the

Table 8. Distances of the mid-coronal plane, moving forwards at the average velocity of gait, from the plane YZ (cm)

No.	Experiment I	Experiment II	Experiment III
1	39,25	44,25	31,88
2	45,28	50,32	37,45
3	51,31	56,39	43,02
4	57,34	62,46	48,59
5	63,37	68,53	54,16
6	69,40	74,60	59,73
7	75,43	80,67	65,30
8	81,46	86,74	70,87
9	87,49	92,81	76,44
10	93,52	98,88	82,01
11	99,55	104,95	87,58
12	105,58	111,02	93,15
13	111,61	117,09	98,72
14	117,64	123,16	104,29
15	123,67	129,23	109,86
16	129,70	135,30	115,43
17	135,73	141,37	121,00
18	141,76	147,44	126,57
19	147,79	153,51	132,14
20	153,82	159,58	137,71
21	159,85	165,65	143,28
22	165,88	171,72	148,85
23	171,91	177,79	154,42
24	177,94	183,86	159,99
25	183,97	189,93	165,56
26	190,00	196,00	171,13
27	196,03	202,07	176,70
28	202,06	208,14	182,27
29	208,09	214,21	187,84
30	214,12	220,28	193,41
31	220,15	226,35	198,98

Table 9. Distances of the total centre of gravity of the body from the mid-coronal plane moving forwards at the average velocity of gait (cm)

No.	Experiment I	Experiment II	Experiment III [a]
1	+ 0,30	+ 0,64	+ 0,33
2	+ 0,22	+ 0,97	+ 0,09
3	+ 0,14	+ 1,04	− 0,18
4	− 0,13	+ 0,99	− 0,50
5	− 0,41	+ 0,53	− 0,86
6	− 0,75	+ 0,17	− 1,16
7	− 1,15	− 0,01	− 1,41
8	− 1,13	− 0,89	− 1,27
9	− 1,11	− 0,93	− 0,83
10	− 1,03	− 1,18	− 0,24
11	− 0,42	− 0,73	+ 0,66
12	+ 0,44	− 0,34	+ 1,21
13	+ 1,13	+ 0,45	+ 1,43
14	+ 1,09	+ 0,86	+ 1,26
15	+ 1,13	+ 1,18	+ 1,09
16	+ 0,43	+ 1,07	+ 0,84
17	+ 0,28	+ 0,79	+ 0,62
18	− 0,31	+ 0,50	+ 0,19
19	− 0,51	+ 0,33	− 0,09
20	− 0,82	+ 0,12	− 0,34
21	− 0,78	− 0,54	− 0,22
22	− 1,22	− 1,17	+ 0,07
23	− 1,07	− 0,78	+ 0,37
24	− 0,68	− 0,60	+ 0,85
25	− 0,31	− 0,17	+ 1,00
26	+ 0,55	+ 0,29	+ 0,97
27	+ 0,77	+ 0,90	+ 0,65
28	+ 1,15	+ 1,02	+ 0,08
29	+ 0,48	+ 0,91	− 0,29
30	+ 0,01	+ 0,54	− 0,71
31	− 0,68	+ 0,25	− 1,17

a) Overall centre of gravity of body and equipment

values given in Table 9. To distinguish these from the co-ordinates used hitherto, they shall be referred to as ξ-co-ordinates.

All the other points of the human body the co-ordinates of which are known can be easily introduced into this new mobile system of ξyz-co-ordinates. To determine the new ξ-co-ordinates, it is only necessary to subtract the distances given in Table 8 from the previous x-co-ordinates, as was done for the total centre of gravity.

Introducing the movement of the total centre of gravity into the mobile system of ξyz-co-ordinates has an important mechanical significance. After the human body has been set in motion in the direction of gait at the average velocity of gait, according to the law of inertia, this movement will maintain its direction and velocity if it is not subjected to external forces, including gravity. The human body would thus be at rest in relation to the new system of co-ordinates which carries out exactly the same movement. The action of one or more external forces which do not counterbalance each other will entail a movement of the body relative to the new system of co-ordinates. The aim of the present study is to determine the external forces acting on the human body during gait when the body has attained its average velocity. Thus, further considerations can be based on the new system of co-ordinates. This offers the significant advantage of dealing with relatively small excursions of the body in all directions and separating off that part of the movement that is independent of external forces once the subject is walking forward.

In uniform gait the total centre of gravity of the human body must describe a closed tridimensional curve relative to the ξyz-co-ordinates during a double step. This curve is symmetrical relative to the plane of gait if the gait is perfectly regular. Conversely, there is no plane of symmetry either horizontal or perpendicular to the direction of gait. Of course, a horizontal plane can be found which divides the tridimensional curve into two equal parts of equal height. Similarly, there is a plane perpendicular to the direction of gait in front of which and behind which the curve extends equally. These two planes form with the plane of gait the middle planes described above for the periodical displacement of the total centre of gravity in the three different directions. They intersect each other at a point whereby the tridimensional curve is equidistant on either side in each of the three directions. This point gives the mid-position of the total centre of gravity in the mobile system of ξyz-co-ordinates. It will be designated as the "core point of the trajectory of the total centre of gravity". The total centre of gravity carries out periodical oscillations about the core point simultaneously in the three different directions, thereby forming its closed tridimensional curve or trajectory.

The core point is of course a fixed point in the system of ξyz-co-ordinates. To represent the displacements of the total centre of gravity in the system of co-ordinates moving forwards at the average velocity of gait, it is expedient to make the core point the initial point of this system, maintaining the directions of the axes of co-ordinates. This will be discussed below in some detail. In the mobile system of ξyz-co-ordinates previously described, the core point lies on the z-axis in experiments I and III. It lies at 0.25 cm to the right of the plane ξZ in experiment II, as will be seen below. The plane ξZ in experiment II thus does not coincide exactly with the plane of gait: it is parallel to the latter, 0.25 cm to its left. This small discrepancy – like the deviation of the x-axis of the system of co-ordinates from the direction of gait – is partly due to the fact that, before the co-ordinates of the total centre of gravity were determined, the plane and direction of gait were found from the displacement of certain joint centres. But above all, the discrepancy results from the small rotation of the system of co-ordinates about the Z-axis itself, as mentioned above, rather than about a line parallel to the Z-axis in the plane XZ, as would have been necessary in experiment II. Before this slight rotation there was, however, a deviation of the plane XZ from the plane of gait to the opposite side. Finally, the core point is at a height of 94.45 cm above the ground in experiments I and II, as will be demonstrated. In experiment III it lies about 6 cm higher because of the equipment carried; its height above the ground is 97.55 cm. Even without the equipment, the total centre of gravity of the human body in experiment III lies somewhat higher, about 0.5 cm, than in the other two experiments.

Though the core point of the trajectory of the centre of gravity represents a fixed point in the mobile system of ξyz-co-ordinates, this system moves regularly in the immobile system of xyz-co-ordinates at the average velocity of gait and in the direction of gait. Like the total centre of gravity, the core point does not of course remain in a constant position in the human body during gait but it moves backwards and forwards in the abdomen.

The shape of the double-curved trajectory of the total centre of gravity in the mobile system can be represented by its projections on the three planes of co-ordinates. The projections are drawn for the three experiments in Figs. 6–8. To facilitate comparison of the results of the three experiments, the mobile system of co-ordinates has been used in all cases. In this system, the core point K of the trajectory of the centre of gravity represents the origin.

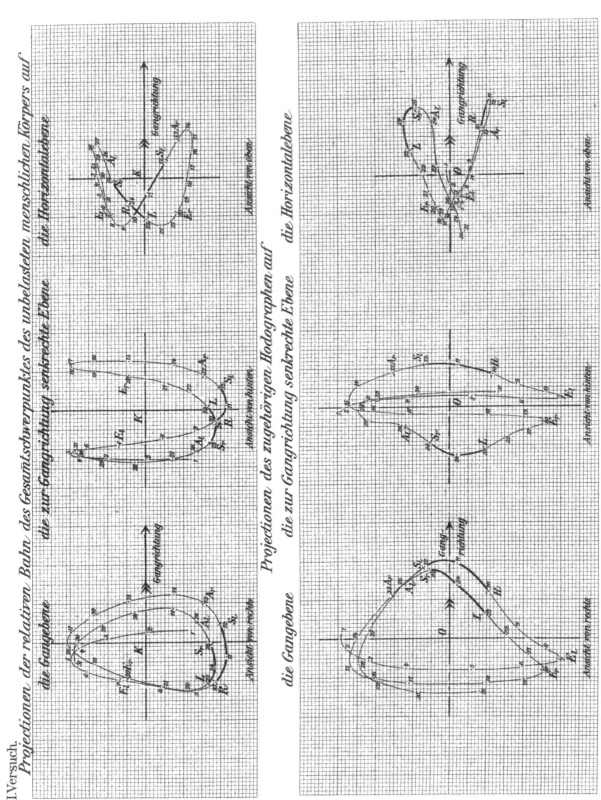

Fig. 6. Projections of the relative trajectory of the total centre of gravity of the unloaded human body (above) and of the corresponding hodographs (below) on the different planes in experiment I

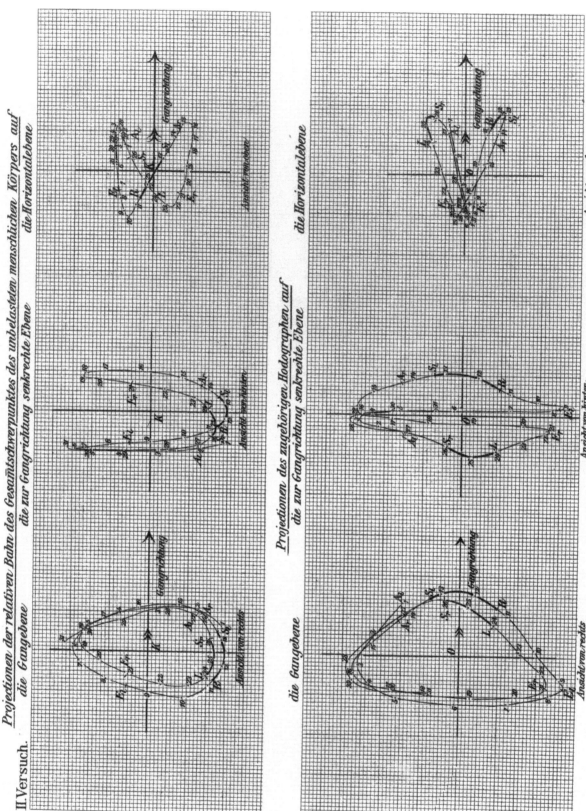

Fig. 7. Projections of the relative trajectory of the total centre of gravity of the unloaded human body (above) and of the corresponding hodographs (below) on the different planes in experiment II

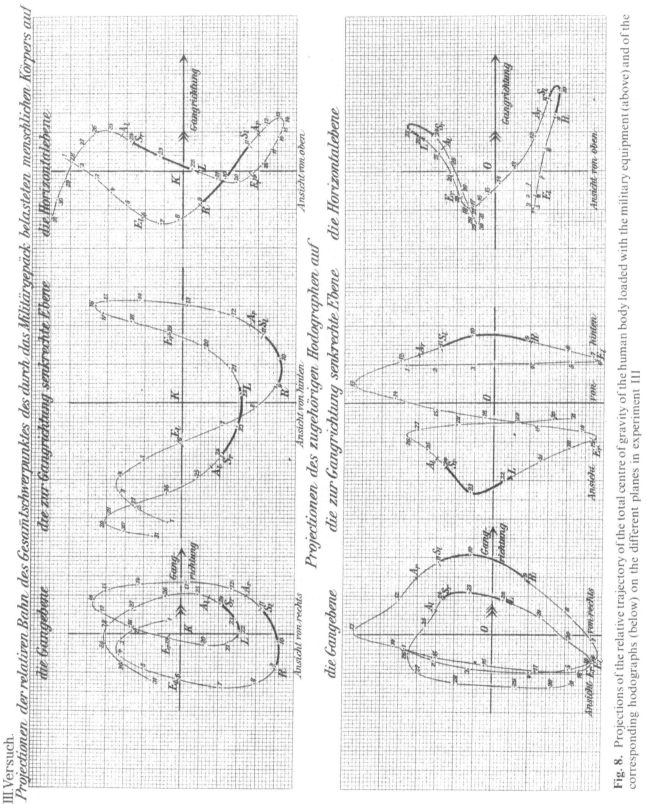

Fig. 8. Projections of the relative trajectory of the total centre of gravity of the human body loaded with the military equipment (above) and of the corresponding hodographs (below) on the different planes in experiment III

Otherwise this system is exactly the same as the system of ξyz-co-ordinates. The first diagram of each series represents the life-size projection on the plane of gait, as seen from the right-hand side, the second the projection on the plane perpendicular to the direction of gait, as seen from behind, and the third the projection on the horizontal plane, as seen from above. The diagrams of experiments I and II dealing with the gait of the unloaded individual concur well with each other. The small irregularities in the projection on the plane of gait and on the horizontal plane are due to the fact that, whereas the period of the oscillations of the body in the plane of gait lasts exactly the time of the whole process of movement, the width of these oscillations is more influenced by external circumstances than is the case with the oscillations in the two other directions perpendicular to that of gait. As predicted, the trajectory of the total centre of gravity, like those of the joint centres in Chap. 1, shows that the two lower extremities of our experimental subject display a particular asymmetrical behaviour.

Figures 6 and 7 show how the projections of the trajectory of the total centre of gravity of the unloaded subject walking regularly and symmetrically were arrived at. Fig. 9 represents these projections. From a comparison of the middle diagram of Fig. 9 with Fig. 27 of Chap. 1 (p. 93), the trajectory of the total centre of gravity, as seen from behind, appears to be very similar to that of the vertex whereas it looks very different from that of the midpoint of the hip line and it is even more unlike that of the midpoint of the shoulder line. This should be noted because the displacement of the total centre of gravity is often identified with that of the midpoint of the hip line or that of a point in the sacrum. Another consequence of this difference is that the total centre of gravity undergoes considerable changes in position inside the body during walking. If the projection of the trajectory of the total centre of gravity on the plane perpendicular to the direction of gait is very similar to that of the vertex, the former is nonetheless much less elongated than the latter; it is more than 1 cm lower and almost 1 cm narrower. Careful analysis discloses that the total centre of gravity carries out smaller displacements in any direction in the system moving forward at the average velocity of gait than any other point of the human body.

The second and third projections of Fig. 9 are described only once during each double step. The first projection is described twice during the same time.

It is not difficult by means of the three projections to obtain a clear image of the double-curved trajectory of the total centre of gravity. However, an immediate picture of this tridimensional pathway can be obtained directly, using the stereoscopic double-image of Fig. 10. If two pictures are directly combined with the naked eye or using a stereoscope, the

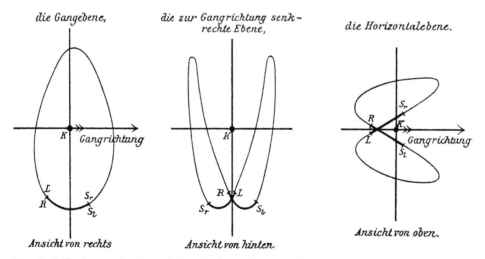

Fig. 9. Life-size projection of the relative trajectory of the centre of gravity of the subject without loading

curve appears to be tridimensional, as seen from behind. It passes through the plane of gait twice at the same point, briefly before reaching its lowest point on either side. Between two passages across the plane of gait, the total centre of gravity describes an oval figure which is tall and a little oblique forwards and laterally, as represented by the first diagram in Fig. 9.

The three projections of the trajectory of the total centre of gravity in experiment III (Fig. 8) show that the excursions of the centre of gravity are greater, in the gait of the loaded than in the unloaded subject. Except for some irregularities and asymmetries, the projections of Fig. 8 present roughly the same shape as their counterparts in Figs. 6 and 7 though the greater width and height in Fig. 8 are striking. Whereas the second and third diagrams of Figs. 6 and 7 represent closed curves as a result of the previously mentioned slight rotation of the system of co-ordinates about the z-axis, the diagrams of Fig. 8 are not completely closed. Closure could not have been achieved in this case by appropriate rotation of the initial system of co-ordinates because the experimental subject, as a consequence of his heavy military equipment, did not exactly maintain the correct direction during the double step under consideration, as mentioned above. Finally, the asymmetrical behaviour of the two legs is more obvious in the projections of experiment III than in the other experiments. Disregarding all these irregularities, the three projections of the trajectory of the total centre of gravity for a perfectly regular gait of the subject carrying military equipment are obtained (Fig. 11). Here also, the first projection is described twice, the two other projections only

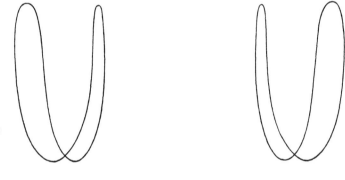

Fig. 10. Stereoscopic picture of the relative trajectory of the centre of gravity of the subject without loading

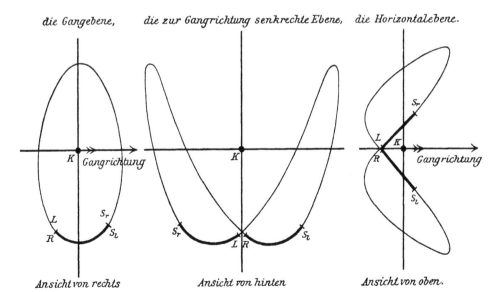

Fig. 11. Life-size projection of the relative trajectory of the centre of gravity of the subject bearing military equipment

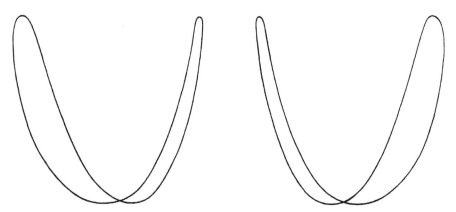

Fig. 12. Stereoscopic picture of the trajectory of the centre of gravity of the subject bearing military equipment

once during one double step. Except for an increase in size, the tridimensional trajectory of experiment III corresponds to that of the first two experiments, as appears by comparing Fig. 11 with Fig. 9. A stereoscopic double picture can immediately illustrate the shape of the double-curved trajectory of the centre of gravity, as shown in Fig. 12.

The relationship of the displacement of the centre of gravity on its double-curved trajectory to the whole process of movement has now to be determined. It is necessary to know where the centre of gravity is on its trajectory at either heel strike, what segment of trajectory it traverses during the double-support period of gait, during the swinging of a leg, etc. To this end, it is important to determine as precisely as possible the moment of heel strike, that of toe-off, etc. The phases closest to these particular moments are given on p. 86 for experiments I and II. The times have been determined to 0.02 s since two successive phases are separated by a time interval of 0.038 s.. The accuracy of this determination could be much better; it could be improved tenfold so that the error would be less than 0.002 s.

To determine as exactly as possible the moments R and L of right and left heel strike, the procedure is as follows. A life-size projection of the centre of the ankle on the plane of gait is drawn in a number of phases from shortly before to shortly after heel strike, for each lower extremity. The segment of the trajectory of the centre of the ankle before heel strike can then be distinguished from that immediately after heel strike. The latter is almost an arc the centre of which coincides with the back of the heel which first strikes the ground. The two segments of the curve do not follow each other smoothly but form an angle. Their intersection corresponds to the moment of heel strike. It is easy to estimate this moment to within one-tenth of a phase if the time interval between two phases is divided into equal parts separating additional imaginary phases. In this way, it was found that the two moments R and L occur at the phase numbers given in Table 10 for the three experiments.

The data on p. 86 were thus fairly accurate.

In order to obtain more precisely the moments S_r and S_l of right and left toe-off, the point on the long axis of the foot designated as its tip is used. The moment of toe-off is better marked on the trajectory of this point than on that of the ankle. The latter is still involved in its forward movement during push-off whereas the tip of the foot undergoes only slight displacement. Therefore, the segment of the trajectory of the tip of the foot before toe-off can again be clearly distinguished from that after the leg has begun to swing. Before toe-off the segment of the trajectory of the tip of the foot falls down and forwards, and the points of the trajectory corresponding to the successive phases are very close to each other. After toe-off it climbs up and forwards and the points of the trajectory are some distance from each other. The two segments of the trajectory form an angle the apex of which indicates toe-off.

Table 10. Phase numbers at right and left heel strike

	Experiment I	Experiment II	Experiment III
Phase number of moment R (right heel strike)	10,0	11,33	8,9
Phase number of moment L (left heel strike)	23,0	24,0	21,75

Table 11. Phase numbers at right and left toe-off

	Experiment I	Experiment II	Experiment III
Phase number of moment S_r (right toe-off)	25,0	26,0	23,9
Phase number of moment S_l (left toe-off)	12,1	13,33	–

Table 12. Phase numbers at beginning and end of foot-flat

	Experiment I	Experiment II	Experiment III
A_r	13,0	14,33	11,5
E_r	20,0	21,0	19,0
A_l	25,9	27,0	24,5
E_l	7,0	8,5	6,0

Except for the left lower extremity in experiment III in which the trajectory of the tip of the foot was not available, toe-off was determined in this manner for both legs with the utmost accuracy. This analysis also disclosed a small error in Table 14 of Chap. 1. The z-co-ordinate of the tip of the left foot is undoubtedly somewhat too large for phase 12. The segment of the trajectory before the apex of the angle appears too small, and the segment after the apex too flat, compared with the corresponding curve of experiment II and those of the right extremities. Full concurrence appears when 3.78 cm is replaced by 3.48 cm. Because of the otherwise striking agreement between the two experiments, this correction is not arbitrary but obviously necessary. Of course it is difficult now to establish how this inaccuracy occurred. It is just as possible that in transcribing the table a "4" was written as a "7", as that the inaccuracy was caused by an actual error in the calculation. After correction of this z-co-ordinate, the moments S_r and S_l coincide with the phase numbers in Table 11.

In experiment III, the phase difference between left heel strike and right toe-off is 2.1. This difference in the first two experiments coincides almost exactly with the phase differences between R and S_l. Therefore, without knowing the trajectory of the tip of the left foot, 11.0 can be given as a good approximation of the phase number for S_l.

After heel strike, the foot rotates about the heel until all the sole rests on the ground. The foot remains some time in this foot-flat position, and then, before toe-off, it rotates again about a point of the forefoot which moves progressively forwards (push-off). Besides heel strike and toe-off, it is important to know the moments which delineate foot-flat. These moments can be determined by life-size projections of the trajectories of the ankle and tip of the foot on the plane of gait.

The beginning of this time interval is designated as A_r or A_l, and the end as E_r or E_l according to the right or left extremity. The phase numbers in Table 12 are obtained.

From these data, the important moments during gait can be marked on the trajectory of the total centre of gravity. This has been done in Figs. 6–8.

In this part of the study not only the trajectory of the total centre of gravity in walking must be considered, but also how the movement proceeds in time with regard to velocity and acceleration. Therefore, the time that elapses between one determined moment and the different phases must be established. In the three experiments, time can be counted from the first phase, and 26.09 vibrations of the contact-breaking tuning fork occur during 1 s. Therefore, up to the second phase, $1/26.09$ s has elapsed. More generally, up to the nth phase a time of $(n-1)/26.09$ s has elapsed (see p. 16). Consequently, the successive phases correspond to the times given in Table 13.

From Table 13 it is easy to deduce the time of any portion of phase by interpolation. We thus obtain the time data in Table 14 for the important moments of gait. This table gives the duration of important portions of the movement of the lower extremities. Between heel strike of the right foot and the next heel strike of the left foot 0.498 s elapsed in experiment I, 0.486 s in experiment II and 0.493 s in experiment III. It appears that the right swing period begins 0.494 s, 0.485 s and 0.495 s respectively, later than the left. If the gait were absolutely symmetrical, the two times would be the same in each experiment. They would coincide with the duration of a single step which is respectively 0.495 s, 0.485 s and 0.4948 s for the three experiments. Though there is no complete agreement the differences between these three times in the same experiment are very small, even with the legs behaving asymmetrically, as was the case in our subject. They attain a maximum of 0.004 s in the first experiment, and in the other experiments they are even smaller. If, without knowing the trajectory of the total centre of gravity, the arithmetic average of the two first times were considered as the duration of a single step, the maximum error would thus be 0.001 s. The error in determining the duration of a step using the trajectory of the centre of gravity could in fact have been equal to that. The two ways of determining the duration of a single step are completely independent of each other. The results coincide absolutely and act as mutual checks. The concurrence is proof of the accuracy of our experiments. Thus it is now known for certain that the other times to be deduced from Table 14 are also accurate to 0.001 s, and presenting the data to the third decimal place is fully justified.

By subtracting R from S_r the duration of the right stance period is obtained. The duration of left stance can also be found by adding the duration of a double step to the time given for S_l and then subtracting the time given for L. The duration of the rotation of the foot about the heel from heel-strike to foot-flat is the interval between R and A_r or L and A_l. The interval between A_r and E_r gives the duration of the right foot-flat. This time can also be determined for the left foot if the duration of a double step is added to the time given for E_l. The intervals E_rS_r and E_lS_l indicate the duration of push-off. The interval between S_l and L measures the duration of swing of the left leg whereas that of the right leg is obtained by adding the duration of a double step to the time given for R and subtracting the time given for S_r from the sum. Finally, the intervals RS_l and LS_r give the duration of the double-support periods. All the intervals thus calculated are given in Table 15 as well as the duration of a single step and that of a double step. The data in this table do not depend on the choice of the initial time, unlike those of Table 14; the values are thus absolute.

Velocities and Accelerations of the Total Centre of Gravity

In order to calculate the velocities and accelerations with which a point moves, it is necessary to know the relationship between the tridimensional trajectory of the point and time. If the trajectory of the point is introduced into a system of tridimensional rectangular co-ordinates, the three co-ordinates of the mobile point have to be correlated with time. In many cases, particularly when the circumstances of the movement are known exactly, it is possible to find equations which express the mathematical relationship between both the three co-ordinates and the time elapsed from a given moment. Velocities and accelerations are then

Table 13. Correspondence between phases and time

Phase number	Time (s)	Phase number	Time (s)	Phase number	Time (s)	Phase number	Time (s)
1	0,000	9	0,307	17	0,613	25	0,920
2	0,038	10	0,345	18	0,652	26	0,958
3	0,077	11	0,383	19	0,690	27	0,997
4	0,115	12	0,422	20	0,728	28	1,035
5	0,153	13	0,460	21	0,767	29	1,073
6	0,192	14	0,498	22	0,805	30	1,112
7	0,230	15	0,537	23	0,843	31	1,150
8	0,268	16	0,575	24	0,882		

Table 14. Correspondence between important moments and time

Moment	Experiment I		Experiment II		Experiment III	
	Phase number	Time (s)	Phase number	Time (s)	Phase number	Time (s)
E_l	7,0	0,230	8,5	0,287	6,0	0,192
R	10,0	0,345	11,33	0,396	8,9	0,303
S_l	12,1	0,426	13,33	0,473	11,0	0,383
A_r	13,0	0,460	14,33	0,511	11,5	0,402
E_r	20,0	0,728	21,0	0,767	19,0	0,690
L	23,0	0,843	24,0	0,882	21,75	0,796
S_r	25,0	0,920	26,0	0,958	23,9	0,878
A_l	25,9	0,955	27,0	0,997	24,5	0,901

Table 15. Duration of the components of a double step

Duration of	Experiment I		Experiment II		Experiment III	
	Time (s)	Arithmetic average	Time (s)	Arithmetic average	Time (s)	Arithmetic average
A double step	0,990		0,970		0,9895	
A step	0,495		0,485		0,4948	
Right stance	0,575	0,574	0,562	0,5615	0,575	0,5758
Left stance	0,573		0,561		0,5765	
Right heel support	0,115	0,1135	0,115	0,115	0,099	0,102
Left heel support	0,112		0,115		0,105	
Right foot-flat	0,268	0,2665	0,256	0,258	0,288	0,2843
Left foot-flat	0,265		0,260		0,2805	
Right push-off	0,192	0,194	0,191	0,1885	0,188	0,1985
Left push-off	0,196		0,186		0,191	
Swinging of right leg	0,415	0,416	0,408	0,4085	0,4145	0,4138
Swinging of left leg	0,417		0,409		0,413	
Double-support period RS_l	0,081	0,079	0,077	0,0765	0,080	0,081
Double-support period LS_r	0,077		0,076		0,082	

arrived at mathematically by repeated differentiations of determined equations. However, the law of movement of the total centre of gravity of the human body during walking is much too complicated to be expressed in relatively simple equations. The same is generally true for all movements of organisms. All mathematical equations only approximately relate to the actual movement. Hence, the velocities, and even more the accelerations, arrived at by differentiation of these equations would be so inaccurate that they would be unable to represent the regular process of the movement and the forces involved in a reliable manner.

Consequently, in physiological mechanics, another way of deducing the velocities and accelerations must be chosen which ensures an accuracy at least equal to that of the empirical determination of the trajectory. This implies geometry. Diagrams representing the time dependence of the co-ordinates of the total centre of gravity, as described previously, form the starting point. These diagrams are termed "displacement curves". If the number of direct determinations of the co-ordinates is sufficient, these diagrams can be drawn very accurately, and they express the law of movement much more reliably than any approximating equations.

The displacement curves of the total centre of gravity should not be confused with its trajectory. To draw these curves, the co-ordinates of the total centre of gravity were entered as ordinates, and the time from the first phase of the movement as abscissae. To achieve the desired accuracy it is appropriate to represent 1 s by a length of 10 cm; 1 cm on the abscissa thus corresponds to 0.1 s, and 1 mm to 0.01 s. The ordinate corresponding to each millimetre on the abscissa cannot be given. This would be possible only if the phases illustrated by the photography followed each other at an interval of 0.01 s. In 1 s only 26.09 phases are photographed, not 100. Therefore, only the ordinates corresponding to the abscissae given in Table 13 are known. The figures in this table express the abscissae in decimetres. The 31 points which we can thus obtain for each displacement curve are sufficient to draw the whole displacement curve. In so doing, it should be borne in mind that this displacement is normally regular and smooth: irregularities could only occur if, while walking, the human body was subjected to heavy blows. Consequently, gross errors in the determination of the co-ordinates would immediately appear in the curves. Moreover, in many cases, the displacement curves can be drawn with greater accuracy than could be achieved by finding the various points empirically. Therefore, it may occasionally be possible by smoothing the displacement curves to correct the co-ordinates to some extent.

This is confirmed by the upper diagrams in Figs. 13–15. These show the displacement curves corresponding to the three co-ordinates of the total centre of gravity, drawn as described above, for the three experiments. The displacement curves of the z-co-ordinates (pp. 173, 176, 179) represent the vertical movement of the total centre of gravity; they coincide exactly with the points found empirically. It is not possible to improve the smoothness of the curves by displacing one of the experimental points by even 1 mm. The displacement curves of the y-co-ordinate (pp. 172, 175, 178) represent the movement sideways of the total centre of gravity. It was necessary only in a few places to bypass the experimental points so as to avoid small bumps in the curve which certainly do not correspond to actual irregularities. However, the displacement curves corresponding to the movement of the total centre of gravity in the direction of gait, relative to the mobile system of co-ordinates (pp. 171, 174, 177), deviated significantly in several places from the experimental points. The curves connecting these points would obviously present far too many irregularities to express accurately the law of the movement in the direction of gait. If the irregularities were part of the movement, they would repeat themselves periodically in the same diagram and appear correspondingly in the first two experiments carried out in the same circumstances. This is not the case. If, independently for each experiment, the curve is smoothed, a periodical repetition appears in each of the two curves as well as a surprising concurrence. These curves can thus be considered as a good approximation of the expression of the law of the movement under consideration.

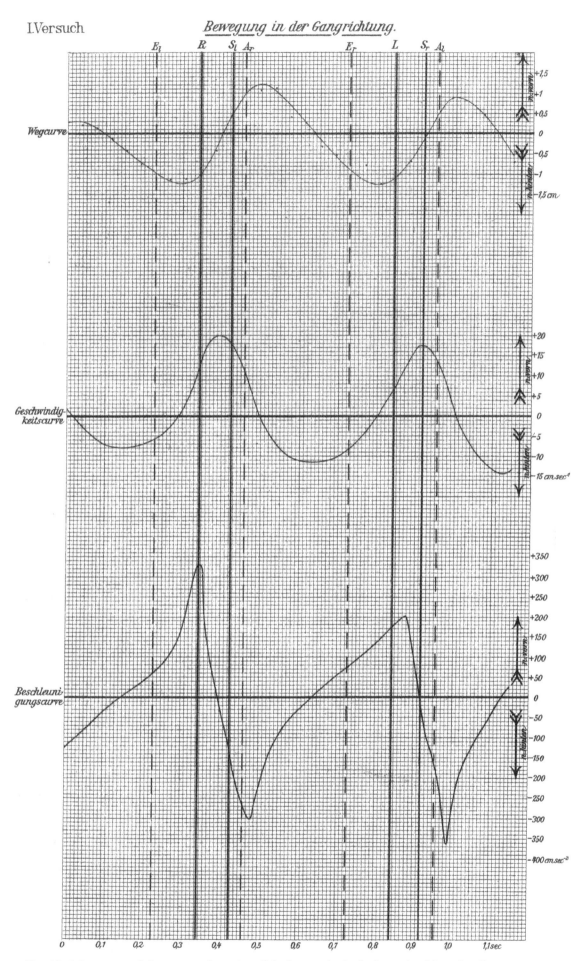

Fig. 13. Movement of the centre of gravity of the human body during gait without loading

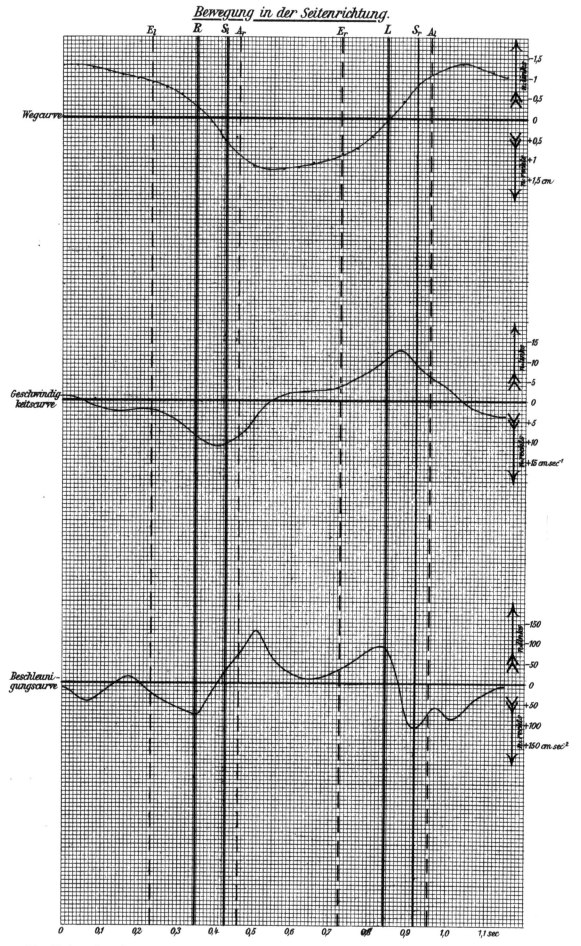

Fig. 13 (continued)

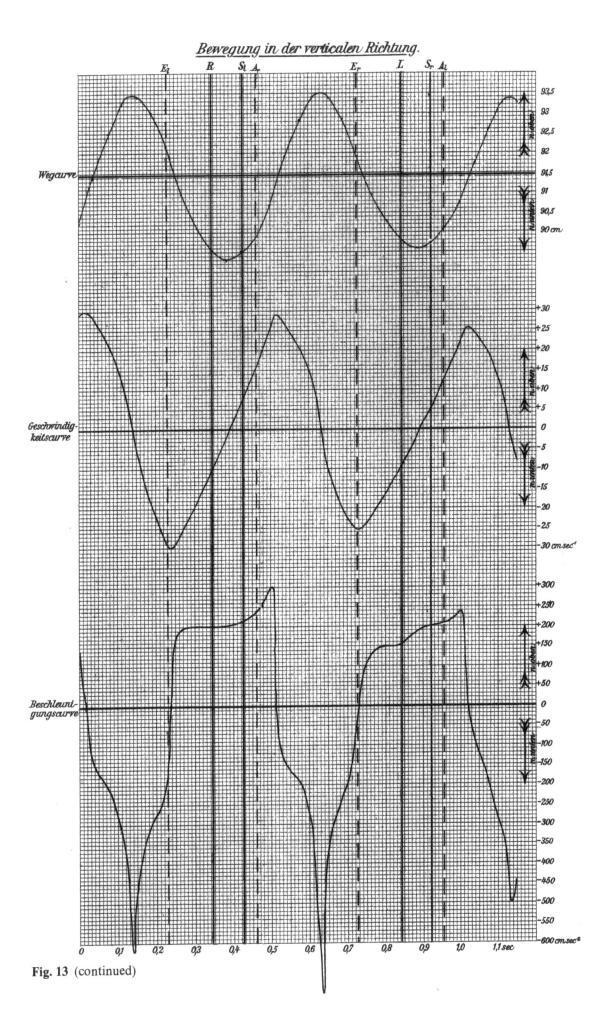

Fig. 13 (continued)

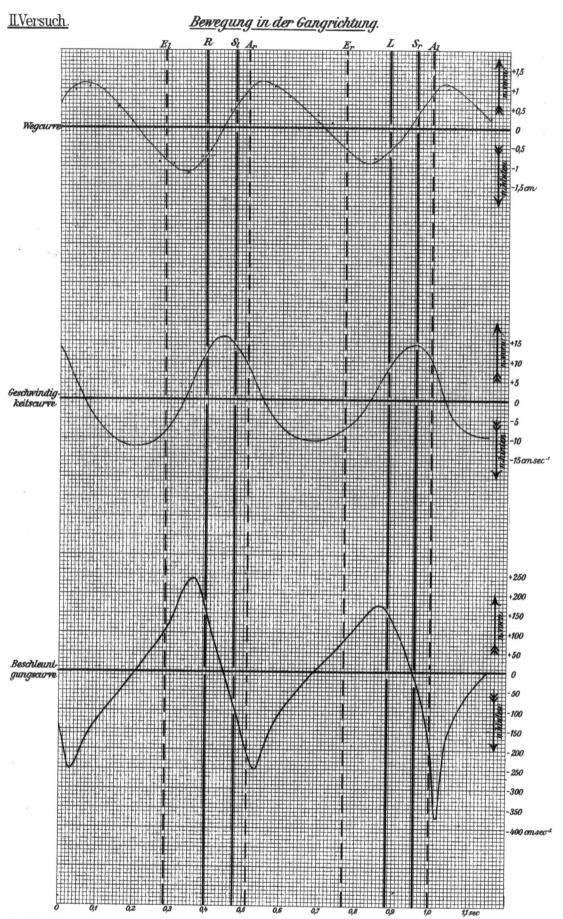

Fig. 14. Movement of the centre of gravity of the human body during gait without loading

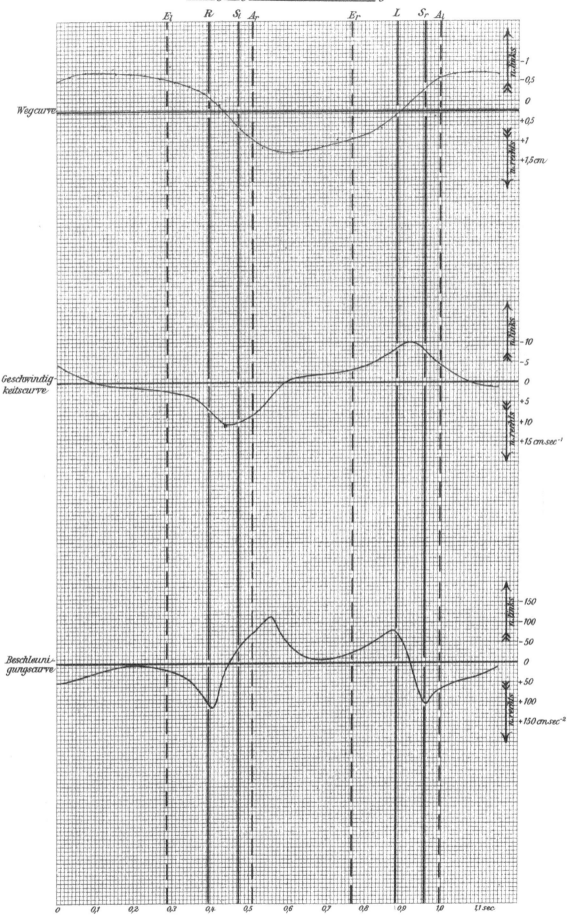

Fig. 14 (continued)

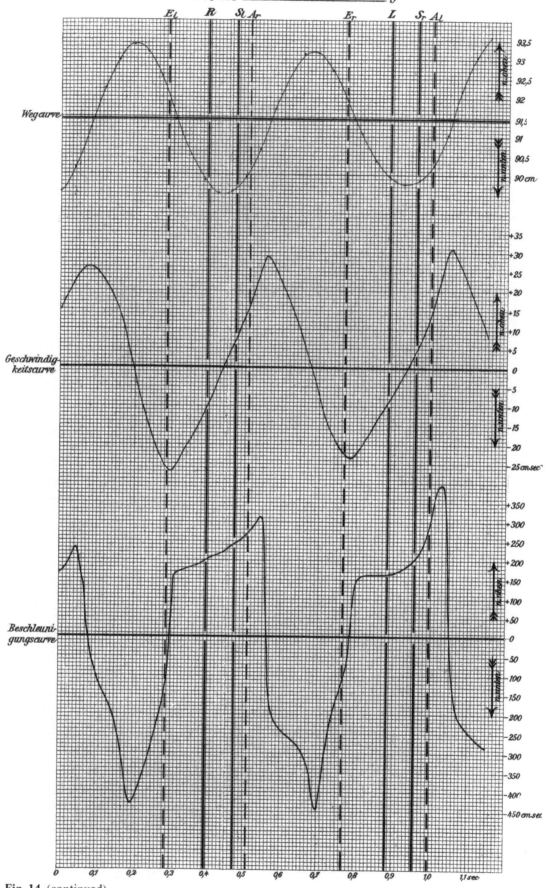

Fig. 14 (continued)

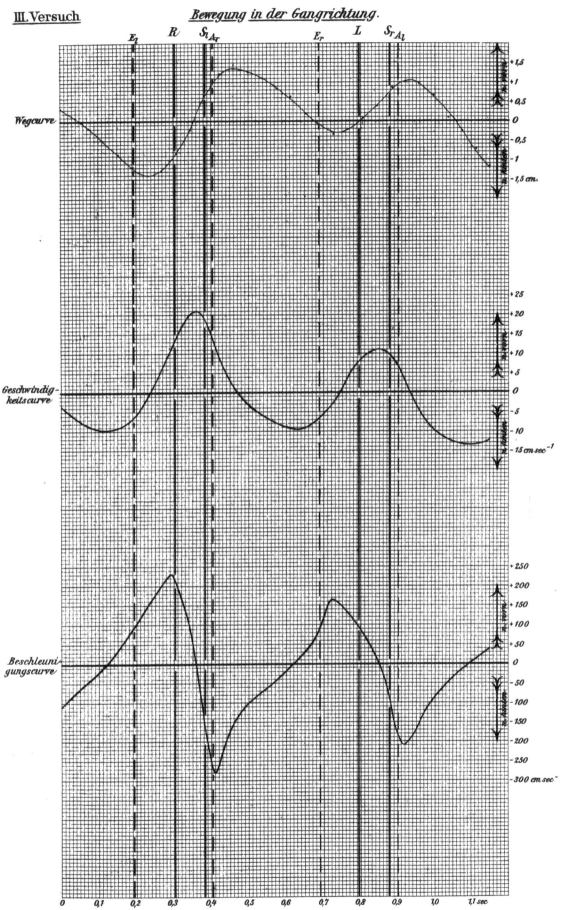

Fig. 15. Movement of the centre of gravity of the human body during gait loaded with military equipment

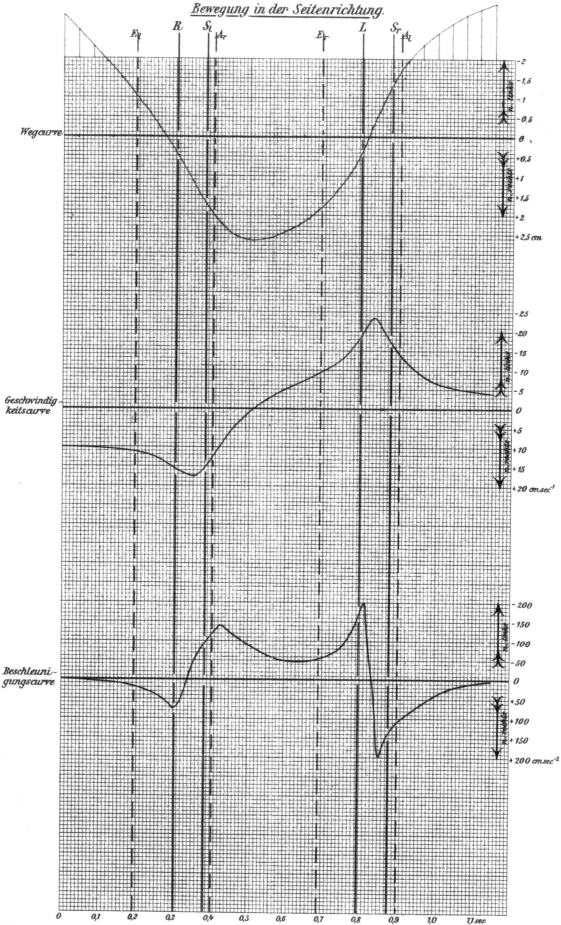

Fig. 15 (continued)

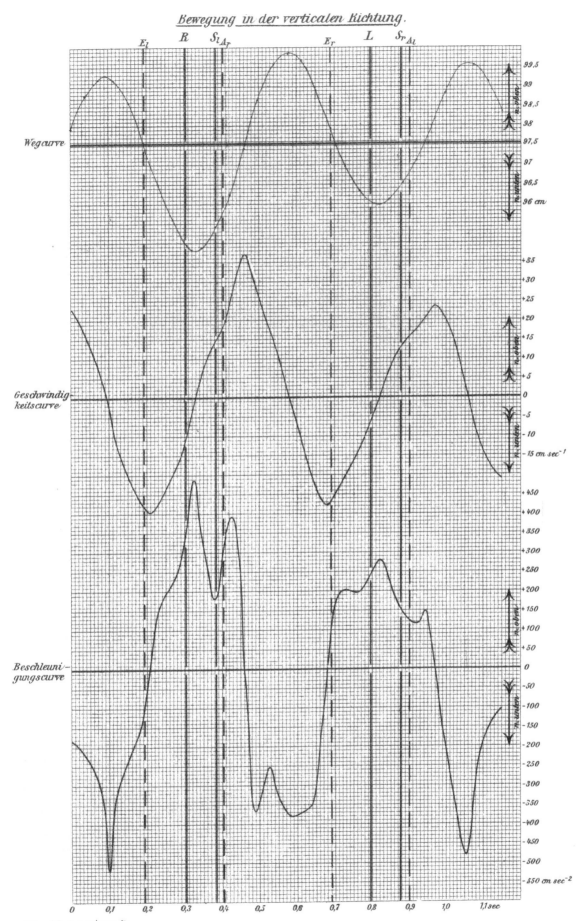

Fig. 15 (continued)

The displacement curves give a clear picture of the movement of the total centre of gravity in the direction of the co-ordinate under consideration. A point can be imagined passing along the displacement curve in such a way that its projection on the horizontal axis of the abscissa moves forwards at a constant velocity of 1 dm/s. Its projection on the vertical axis of the ordinate moves at the same velocity and acceleration as the total centre of gravity in the direction of the co-ordinate. A displacement curve thus expresses the whole law of the movement in the direction under consideration. Consequently, all the peculiarities of the movement appear in this curve, and it must be possible to deduce the velocities and accelerations of the moving point from the shape of the curve.

The velocities express themselves in the continuous change of direction of the displacement curve. At any time, the velocity is proportional to the trigonometric tangent of the angle formed by the abscissa and the tangent to the displacement curve at the point considered. In Figs. 13–15, 1 dm represents 1 s, the trigonometric tangent of this angle, therefore, directly gives the velocity if the unity of velocity is 1 dm/s. (The mobile point in case of uniform movement would travel a distance of 1 dm in 1 s.) However, in the present study, the lengths are expressed in centimetres; it is, therefore, appropriate to use 1 cm/s as the unit of velocity. In mechanics, it is usual to write $cm\,s^{-1}$. All the trigonometric tangents of the different points of the nine displacement curves must then be multiplied by a factor of 10.

In practice, the velocity in $cm\,s^{-1}$ can be found for any time in the following way. The tangent to the displacement curve is drawn at the point corresponding to the time under consideration. This tangent is made the hypotenuse of a right-angle triangle; the sides of the right angle are horizontal and vertical, and the horizontal side is 1 dm long. The length in centimetres of the vertical side directly gives the velocity in $cm\,s^{-1}$. This way of measuring the velocity is facilitated by drawing the displacement curve on a millimetre grid. After drawing the tangent, which must be performed carefully, the measurement consists of counting the number of centimetres. From any point on the tangent, 10 cm is measured horizontally on the millimetre paper and then the vertical distance is measured in centimetres between the end of this horizontal line and the tangent. The displacement curve must of course be drawn very accurately on sufficiently large millimetre paper.

The velocities were measured in this manner for a sufficient number of points on the nine displacement curves. The results of these measurements can in turn be illustrated by diagrams drawn in very much the same way as the displacement curves. The abscissa gives the time elapsed since the first phase of the movement, 1 mm corresponding to 0.01 s. The lengths in the ordinate are proportional to the velocities. If the unit of velocity was represented by 1 cm, the diagram would be much too large vertically and, therefore, inappropriate for an overall view. It is more convenient to assign a length of 1 mm to the unit of velocity $cm\,s^{-1}$ This corresponds better to the accuracy which can be achieved in determining the velocity in the present case. Diagrams of this type represent the velocity plotted against time and are called "velocity curves".

Figures 13–15 give the velocity curves corresponding to the movements in the three directions for each of the experiments. They are beneath the displacement curves. Close relationships appear in each case between the two curves. The velocity curve intersects the axis of the abscissa every time the displacement curve reaches a maximum or minimum. At each maximum or minimum, the tangent to the displacement curve is indeed parallel to the axis of the abscissa with which it forms an angle of 0°. The tangent of such an angle is 0. The velocity is zero when the direction of the movement is reversed. Such a reversal in the direction of the movement appears in the displacement curve as a maximum or minimum. Each velocity curve itself reaches a maximum or minimum when the corresponding displacement curve changes from convex to concave or vice versa. Such a point of the displacement curve is called a "turning point". On the displacement curve, shortly before this turning point, the tangent to the displacement curve alters its direction in relation to the axis of the abscissa in such a way that it seems to rotate in a particular manner. At the turning point the tangent reverses the direction of its rotation. If the angles formed by the tangents with the axis of the abscissa progressively increased prior to, they progressively decrease after the turning point, and vice versa. Since the trigonometric tangents increase and decrease with their angles, the velocity will reach a maximum or minimum at each turning point of the displacement curve.

The shape of a displacement curve thus immediately indicates when the velocity of the total centre of gravity goes through zero in the direction under consideration and, consequently, when the movement changes direction. It also shows when the velocity is at its greatest in one or the other direction. Since these moments are particularly important for the movement of the total centre of gravity, when drawing the velocity curve it is necessary to direct special attention to these points on the displacement curve and the corresponding points on the abscissa; at the turning points the direction of the tangent must be measured as accurately as possible. This provides some essential points on the velocity curve which roughly characterize its course. The velocity between these moments is then precisely measured. How many of these measurements should be carried out to determine the whole course of the velocity curve? This depends on the shape of the displacement curve. If the latter is as regular as the movement of the total centre of gravity in the vertical direction, fewer measurements are necessary than for the movement in the other two principal directions. In the present case, the movement sideways required the greatest number of measurements. Generally, it is important that attention is not restricted to the phases initially determined in the experiment. A main advantage of graphic representation lies in the fact that it shows the velocity at any moment. It is irrelevant whether this moment had been recorded by photography earlier in the experiment or not.

In order to determine the acceleration of the total centre of gravity in the three directions for any time, one proceeds from the velocity curves. A few characteristic values of the acceleration can be directly derived from the displacement curves though it is generally quicker to use the velocity curves. This results from the fact that the accelerations are related to the velocities in the same way as the latter are related to the co-ordinates. The velocities provide a measure of the increase in the co-ordinates, and the accelerations can be regarded as measuring the increase in the velocities. If a point travels along the velocity curve in such a way that its projection on the horizontal axis of the abscissa moves forward at a constant velocity of $1\,\mathrm{dm\,s^{-1}}$, its projection on the axis of the ordinate will move at a velocity which at any time indicates the acceleration of the total centre of gravity in the direction of the co-ordinate under consideration. Therefore, the acceleration must be expressed by the continuously changing direction of the velocity curve. The acceleration is proportional to the trigonometric tangent of the angle formed by the axis of the abscissa and the tangent to the velocity curve at the point considered. Since 1 s is represented by 1 dm in the displacement curves of Figs. 13–15 and by 1 mm in the velocity curves, the trigonometric tangent will show $\frac{1}{100}$ the acceleration if the unit is the acceleration of a movement uniformly accelerated the velocity of which is increased by $1\,\mathrm{cm\,s^{-1}}$. This unit of acceleration is $\mathrm{cm\,s^{-2}}$. Therefore, all the trigonometric tangents corresponding to the individual points of the nine velocity curves have to be multiplied by a factor of 100 to express the acceleration in $\mathrm{cm\,s^{-2}}$

The accelerations are established from a sufficient number of points on the nine velocity curves, in the same manner as explained above in determining velocities. From these measurements, it is possible to draw graphs plotting acceleration against time. In the abscissa, 1 dm corresponds to 1 s. The unit of acceleration $\mathrm{cm\,s^{-2}}$ is represented by a length of 0.1 mm so as to avoid the curves becoming intractable. This again corresponds to the accuracy which can be reached in deducing the accelerations. By progressive deduction of new graphs resulting from the determination of the directions of tangents in previous curves, unavoidable sources of error accumulate. These graphs in which acceleration is plotted against time are called "acceleration curves".

The acceleration curves of the movements in the three directions are given in Figs. 13–15 for the three experiments, beneath the corresponding displacement and velocity curves. The acceleration curves derive from the velocity curves, just as the latter derive from the displacement curves. Therefore, the same close relationships exist between acceleration and velocity curves as between velocity and displacement curves. An acceleration curve always intersects the axis of the abscissa when the velocity curve reaches a maximum or minimum. These points on the velocity curves correspond to turning points on the displacement curves. The latter indicate when the acceleration of the movement of the total centre of gravity is zero. Moreover, the acceleration is at its greatest or smallest when the corresponding velocity curve travels through a turning point. These moments are not particularly obvious on the displacement curves.

The displacement, velocity and acceleration curves illustrate all aspects of the movement of the total centre of gravity for each of the three components of the movement. To visualize how this movement of the total centre of gravity relates to that of the lower extremities, the moments of heel strike R, L and of toe-off S_r, S_l have to be marked on the curves. It is also useful to indicate the beginning A_r, A_l and end of foot-flat or beginning of push-off E_r, E_l. This was done in Figs. 13–15 for the three experiments. The vertical lines which correspond to these moments, according to Table 14, are thicker than the lines of the millimetre grid, and interrupted lines indicate the less important moments A and E. The designation of the moments corresponding to each of these vertical lines is written above these lines.

Besides the movement of the total centre of gravity with its velocities and accelerations, these vertical lines provide a clear picture of the process of walking in its significant stages. In the three experiments, the first phase happened to occur during the left single-support period whilst the right leg was swinging forward. In experiments I and III, left foot-flat occurred shortly before. A relatively long time elapses, 0.230 s in experiment I, 0.192 in experiment III, before left push-off (E_l). The left foot-flat period lasts 0.265 s in experiment I, and 0.2805 s in experiment III, according to Table 15. In experiment II, the left foot is rotating about its heel during the first phase. The double step lasts 0.970 s in this experiment, and the next left foot-flat A_l occurs 0.997 s after the first phase, according to Table 14. Consequently, if the gait is perfectly uniform, the first moment A_l must have occurs at 0.027 s. Actually this time may be a little shorter. The exact moment A_l cannot be determined here by the above-mentioned method because it lies too close to the first phase. Therefore, it was not marked by a vertical line in the diagrams of experiment II in Fig. 14. Left push-off occurs between the two subsequent moments E_l and S_l, and tooe-off occurs at S_l. Moment R corresponds to right heel strike. The foot initially rotates about its heel until foot-flat at moment A_r. Shortly before that, the left leg begins to swing. Between A_r and E_r, right foot-flat occurs. At E_r, push-off causes the foot to rotate about a point that moves gradually forwards in the forefoot. At S_r, the right foot starts to swing. Right foot-flat thus takes place between R and S_r. At the moment L prior to S_r the left heel strikes the ground, at A_l the foot is flat on the ground, etc. Once the whole process of the movement is understood by means of the vertical lines, with further study of the diagram it becomes easy to correlate each phase of the movement of the total centre of gravity with the different times of gait.

Let us first consider the diagrams of experiments I and II carried out under the same circumstances in Figs. 13 and 14. All the corresponding curves appear similar at first glance. Although the two experiments were carried out independently of each other, and although the co-ordinates of the centre of gravity were calculated and the velocities and accelerations in experiment II deduced without considering the results of experiment I, Fig. 14 very closely resembles Fig. 13. It must be remembered that at the beginning the two experiments do not coincide exactly: the starting time of experiment I is about 0.05 s later than that of experiment II. All the numbers on the abscissa in experiment I will thus be about 5 mm smaller than those of experiment II. The vertical lines, however, are independent of this incidental difference in choice of the initial time. Closer observation discloses a slight difference in the acceleration curves. The last peak of the acceleration curve of the vertical displacement of experiment I is somewhat different in shape from that of experiment II. The minima are lower in the former than in the latter. However, the maxima as well as the minima of the two curves lie at the same places. The same is true for the acceleration curves of the displacement sideways, even if small differences in shape appear at the two extremities. In the acceleration curves of the displacement in the direction of gait, small discrepancies also appear in the timing of the peaks. In experiment I, the first peak occurs at right heel strike, the second during the double-support period. In experiment II, the two peaks occur before heel strike. It is certainly not incorrect to place the peak at heel strike. These small discrepancies are not surprising in the displacement in the direction of gait since here the displacement curves were not reliably constructed from the points directly measured as they were for the other two components of the movement. Moreover, as a rule, less accuracy is to be expected in the acceleration curves than in the velocity and displacement curves. If these slight discrepancies are ignored, it is possible to deduce from the curves of Figs. 13 and 14 the following laws of the movement of the total centre of gravity in the three principal directions in the gait of the unloaded man.

Movement of the Total Centre of Gravity in the Direction of Gait

During a single step, the total centre of gravity of the human body oscillates backwards and forwards in the direction of gait, in relation to a plane perpendicular to the direction of gait moving uniformly forwards at the average velocity of gait. The greatest anterior and posterior distance from this plane is on average 12 mm in our subject. The centre of gravity is at its furthest behind the moving plane shortly before heel strike on one side and during push-off on the other. Then its velocity relative to the moving plane is zero and is thus equal to the velocity of the moving plane. At the same time it experiences forward acceleration. Subsequently, the centre of gravity increases its relative velocity and moves closer to the moving plane. It reaches the moving plane shortly before toe-off of the supporting leg, after heel strike of the opposite side. The centre of gravity gains velocity and passes through the plane, moving anteriorly from behind, at its greatest velocity. This increase in velocity is not uniform, as shown by the acceleration curve. The acceleration initially increases considerably until heel strike and then rapidly decreases during the double-support period so that it is zero at or shortly before toe-off of the opposite leg. Simultaneously, the total centre of gravity reaches the moving plane from behind. The behaviour of the two legs is not identical, as was to be expected from the asymmetry mentioned above. The right leg begins swinging almost exactly at the same time as the centre of gravity passes through the moving plane. The left foot leaves the ground only after the centre of gravity has passed to the front of the moving plane and the velocity decreases. The acceleration is then directed posteriorly, i.e. it has become a deceleration. The absolute velocity, however, remains greater than the average velocity of gait. The centre of gravity, therefore, moves further forwards from the moving plane. It reaches its greatest forward distance briefly after the beginning of foot-flat. It moves then at the average velocity of gait. The deceleration initially increases considerably and is at its greatest shortly before the centre of gravity is furthest from the moving plane. The centre of gravity then approaches the moving plane from the front, and its velocity is negative relative to this plane; absolutely, its velocity is smaller than the average velocity of gait. By this time the deceleration has decreased. At the beginning of the last third of foot-flat, the centre of gravity again passes behind the moving plane. Almost simultaneously with its second passage through the moving plane, it reaches its greatest backward relative velocity and its smallest absolute velocity so that deceleration changes to acceleration. While the centre of gravity moves backwards from the moving plane with decreasing relative velocity, its acceleration increases. Consequently, its relative velocity soon reaches zero and its absolute velocity is again equal to the average velocity of gait. Shortly before heel strike, the centre of gravity reverses its displacement relative to the moving plane and approaches the latter from behind. The whole sequence then begins again. The duration of this oscillation of the total centre of gravity relative to the coronal plane moving forwards at the average velocity of gait coincides with the duration of a single step.

Movement of the Total Centre of Gravity Sideways

The total centre of gravity carries out the following oscillation perpendicular to the plane of gait, i.e., relative to the vertical plane in the direction of gait. During the single-support period, it lies always on the side of the plane of gait towards the supporting leg. Its passage from one side to the other is somewhat different for the two legs in our subject. The passage from right to left coincides exactly with left heel strike. The passage from left to right does not occur until after right heel strike, at a time which corresponds approximately to the middle of the double-support period. Its velocity reaches a maximum, in both instances, shortly after the passage through the plane of gait, and at the same time the acceleration becomes zero. The centre of gravity is at the furthest from the plane of gait on either side at the end of the first third of foot-flat. Since the movement then changes direction, the velocity is of course zero. From the middle of the double-support period, this velocity progressively decreases but not uniformly; the deceleration is increasing at the beginning and decreasing later. The maximum of deceleration occurs at different times for the two sides. On the right, it occurs shortly before the centre of gravity is at its furthest from the plane of gait. When the centre of gravity lies to the left of the plane of gait, the maximum of deceleration coincides with the beginning of the right leg swing in the two experiments. Whilst the centre of gravity approaches the plane of gait on either side, its velocity progressively increases. The velocity remains almost constant for a short while, and the acceleration is close to zero at the beginning of the last third of foot-flat. The velocity then increases again with increasing acceleration until the centre of gravity passes through the plane of gait and a little while after this, but then with decreasing acceleration. Finally, the velocity reaches a maximum in the second half of the double-support period and the acceleration gives way to deceleration. The duration of oscillation of the total centre of gravity relative to the plane of gait is equal to that of one double step, unlike the displacement in the direction of gait which was equal to one step.

Vertical Movement of the Total Centre of Gravity

The total centre of gravity of the human body also carries out very regular vertical oscillations relative to a horizontal plane 91.45 cm above the ground in experiments I and II. It goes down through this plane after the only supporting foot has started to lift from the ground by rotating about its forefoot. Shortly before this, its velocity from above downwards was maximum and its acceleration was zero. It then moves downwards with decreasing velocity until it has reached its lowest point approximately in the middle of the double-support period. Then its velocity is zero. This velocity hass decreased relatively quickly from the start. The deceleration of the downwards displacement (upwards acceleration) has reached a considerable value when the centre of gravity passes through the horizontal plane. Afterwards, the centre of gravity reverses its movement and rises to the horizontal plane with increasing velocity until it reaches this plane at the end of the first quarter of foot-flat of the opposite leg. Shortly before, the velocity, and before this the upwards acceleration have reached a maximum. The acceleration then falls to zero almost instantly as required by the maximum of velocity. It turns over quickly into negative acceleration, i.e. downwards acceleration, in such a way that the velocity is again somewhat reduced when the centre of gravity passes upwards through the horizontal plane. The centre of gravity then rises above the horizontal plane with further decreasing velocity until it reaches its highest point at the end of the second third of foot-flat and its velocity is again zero. Hereby, the downwards acceleration (deceleration of the upwards movement) has increased at first quickly, then more slowly and finally more quickly again. It has reached its maximum, significantly smaller in experiment II than in experiment I, when the centre of gravity is at its highest. Then the centre of gravity reverses its movement. During its subsequent downwards displacement, the downwards acceleration decreases again, so quickly that it has reached zero before the centre of gravity moving downwards has reached the horizontal plane. At the same time, the velocity of this downwards movement has increased quickly and soon reaches its greatest value in this direction. Shortly afterwards, the centre of gravity goes downwards through the horizontal plane and the whole sequence starts again. The duration of the vertical oscillation of the centre of gravity is equal to that of a single step, like that of the oscillation in the direction of gait. Comparing this tridimensional movement of the centre of gravity with that of the centre of gravity of the man carrying military equipment (Fig. 15), essential discrepancies in the extension of the different components of the movement and small displacements sideways of the maxima and minima can be observed in the three types of curves. Moreover, the vertical accelerations behave in a somewhat unstable manner. This results from the unusually heavy load: the muscles did not reach their correct tension immediately. Otherwise, the curves of experiment III present the same characteristics as those of the first two experiments.

An overall view of the regular process of the movement of the total centre of gravity has now been obtained. In order to use this movement as the starting point for further research, it is important to determine the exact values of the velocities and accelerations. These can be read in Figs. 13–15 for any moment. It is useful to determine these values for the 31 phases since the co-ordinates of the joint centres and of the partial centres of gravity, and thus the attitude of the body, are known exactly for these phases. Table 13 (p. 169) is used for this purpose. This table gives the times (abscissa. on the graphs) corresponding to the different phases. The co-ordinates of the total centre of gravity relative to the mobile system of co-ordinates (which are known) are read again from the displacement curves for the 31 phases. In this way, more accurate values are obtained which are used to deduce the velocities and accelerations. The initial point of the mobile system of co-ordinates will no longer be constituted by a point on the horizontal ground plane, but by the core point of the trajectory of the total centre of gravity. The core point is the intersection of the three middle planes perpendicular to each other, about which the total centre of gravity oscillates in the three directions within the system of co-ordinates moving forward at the average velocity of gait. The three middle planes thus represent the planes of the co-ordinates. The initial point of the co-ordinates coincides with the middle point of the oscillations of the total centre of gravity in the three principal directions.

The co-ordinates entered into this system are the vertical distances between the three displacement curves of each experiment and the middle horizontal line. In the displacement curves related to the forward displacement in the direction of gait, this middle horizontal line coincides with the axis of the abscissa initially used. The same is true in experiments I and III for the middle horizontal line related to the transverse displacement. Only in experiment II does this middle line lie 2.5 mm beneath the initial axis of the abscissa. Otherwise, the greatest excursion of the total centre of gravity to the right would be greater than that to the left. For the displacement curves of the vertical movement, the distances between the centre of gravity and the horizontal ground plane were initially used as ordinates. Therefore, the axis of the abscissa of the diagram could not be drawn in the figures. The middle horizontal line is 91.45 cm above the ground in experiments I and II and, as a result of the heavy load, it is 97.55 cm above the ground in experiment III. This can be seen in Figs. 13–15. The initial point of the mobile system of co-ordinates thus lies in the plane XZ of the fixed system of co-ordinates 91.45 cm above the X-axis in experiment I. In experiment II, it lies 0.25 cm to the right of the plane XZ of the fixed system of co-ordinates, also 91.45 cm above the X-axis. In experiment III, it again lies in the plane XZ of the fixed system of co-ordinates, but 97.55 cm above the X-axis, Moreover, in phase 1, the initial point of the mobile system of co-ordinates lies at a distance of 39.25 cm from the plane YZ in experiment I. This distance is 44.25 cm in experiment II, 31.88 cm in experiment III (Table 8). The distance increases by increments of 6.03 cm between two successive phases in experiment I, 6.07 cm in experiment II, and 5.57 cm in experiment III (cf. pp. 157–158). ξ, η and ζ represent the co-ordinates of the mobile system and will be called "*relative co-ordinates*". From these it is possible to deduce, for each phase, the x, y', z-co-ordinates of the fixed system as they appear after the slight rotation about the Z-axis mentioned on p. 145.

If, finally, for experiments I and II, the small rotation is carried out about the Z-axis in the opposite direction, we obtain the x, y, z-co-ordinates of the total centre of gravity in the fixed system of co-ordinates used in Chap. 1. The partial and the total centres of gravity are expressed in these co-ordinates in the beginning of the present part of the work as well. These "*absolute co-ordinates*" of the total centre of gravity, deduced backwards from the relative co-ordinates ξ, η, ζ, are more accurate than those given in Tables 1, 2 and 6. In all further studies they will be used instead of the previous ones. They are thus given in Table 16 besides the relative co-ordinates read on the displacement curves.

Table 16 also gives the "*relative velocities*" and the accelerations of the total centre of gravity in the mobile system of co-ordinates. The "*absolute velocities*" in the fixed system of co-ordinates are exactly the same as the relative ones as far as the transverse and vertical displacements are concerned. In the direction of gait, however, the absolute velocities also given in Table 16 are greater than the relative ones by the constant average velocity of gait. Since a difference of a constant velocity does not change the values of the accelerations, all the accelerations given in Table 16 remain the same as in the fixed system.

The moments when the distance from the middle plane or the velocity or acceleration of any component of the movement reaches a maximum in either direction are also of particular interest. They can be read on the curves in Figs. 13–15. It can be seen that all the maxima and minima of the displacement curves correspond to zero on the velocity curves; all the maxima and minima of the velocity curves correspond to zero on the acceleration curves. For each of these moments, the diagrams directly indicate the time elapsed since the first phase. To establish in which of the 31 phases such a moment occurs, the time has to be transposed into phases. This is facilitated by Table 17 which gives all the times from 0 to 1.15 s in 0.01 s increments and the corresponding phases. The decimal places after the numbers of the phases indicate where the time considered belongs in the time interval. Thus, the interval between two successive phases is divided into 99 equal parts. In a way Table 17 represents the reverse of Table 13 (p. 169); it can be used to locate accurately an event which occurs at a given time from the first phase in the process of movement illustrated by the 31 phases in Figs. 22 and 23 of Chap. 1.

In Table 18, the moments of the maxima of the displacements from the middle plane and the moments of the maxima of the velocities and accelerations can be found for the various directions. Each time is accompanied by the corresponding phase number, according to Table 17. The phase numbers are always given to one decimal place, which is approximately in accordance with the accuracy attained in all the experiments. Thus, nine subphases occur in the interval between two successive phases, dividing the interval into ten equal parts. The decimal point indicates the number of the subphase.

Tables 16 and 18 give all the details of the movement of the total centre of gravity of the human body. These tables thus constitute the initial point for all further studies based on the movement of the total centre of gravity.

These tables can be used to find the projections of the relative trajectory of the centre of gravity on the three planes. For the projection on the plane of gait the ξ- and ζ-co-ordinates

Table 16. Relative and absolute co-ordinates, velocities and accelerations of the total centre of gravity of the human body

Experiment I

| No. | Co-ordinates of total centre of gravity (cm) | | | | | | Velocity of total centre of gravity (cm s^{-1}) | | | | Acceleration of total centre of gravity (cm s^2) | | | No. |
| | Relative co-ordinates | | | Absolute co-ordinates | | | Direction of gait | | Transverse direction | Vertical direction | Direction of gait | Transverse direction | Vertical direction | |
	ξ	η	ζ	x	y	z	Relative	Absolute						
1	+0,30	−1,29	−1,21	39,55	−1,35	90,24	+ 2,0	159,3	− 1,0	+29,1	−125	+ 13	+140	1
2	+0,27	−1,29	−0,05	45,55	−1,36	91,40	− 2,5	154,8	− 0,2	+28,2	− 92	+ 33	−142	2
3	+0,10	−1,26	+1,03	51,41	−1,34	92,48	− 5,7	151,6	+ 1,5	+22,0	− 59	+ 44	−208	3
4	−0,15	−1,20	+1,77	57,19	−1,29	93,22	− 7,7	149,6	+ 2,6	+11,7	− 23	+ 14	−340	4
5	−0,46	−1,10	+1,98	62,91	−1,20	93,43	− 8,0	149,3	+ 2,7	− 5,9	+ 7	− 10	−452	5
6	−0,73	−1,00	+1,58	68,67	−1,11	93,03	− 7,3	150,0	+ 2,1	−19,7	+ 32	− 8	−287	6
7	−0,97	−0,90	+0,69	74,46	−1,02	92,14	− 5,6	151,7	+ 2,4	−29,7	+ 59	+ 26	−170	7
8	−1,16	−0,77	−0,46	80,30	−0,90	90,99	− 2,9	154,4	+ 3,8	−26,0	+101	+ 48	+197	8
9	−1,23	−0,57	−1,31	86,26	−0,71	90,14	+ 2,9	160,2	+ 5,9	−18,1	+189	+ 67	+2'00	9
10	−1,02	−0,31	−1,88	92,50	−0,46	89,57	+12,0	169,3	+ 8,5	−10,6	+328	+ 78	+201	10
11	−0,39	+0,08	−2,15	99,16	−0,08	89,30	+19,5	176,8	+10,9	− 2,0	+ 61	+ 33	+205	11
12	+0,40	+0,54	−1,99	105,98	+0,37	89,46	+18,4	175,7	+11,0	+ 6,8	−118	− 23	+214	12
13	+0,99	+0,89	−1,57	112,60	+0,71	89,88	+11,2	168,5	+ 9,0	+15,4	−268	− 67	+237	13
14	+1,22	+1,16	−0,76	118,86	+0,97	90,69	0	157,3	+ 5,2	+25,0	−230	−120	+294	14
15	+1,03	+1,27	+0,43	124,70	+1,07	91,88	− 7,0	150,3	+ 0,6	+26,6	−123	− 84	−140	15
16	+0,70	+1,26	+1,37	130,40	+1,05	92,82	−10,2	147,1	− 1,1	+20,6	− 60	− 38	−190	16
17	+0,28	+1,18	+1,98	136,01	+0,96	93,43	−11,5	145,8	− 1,9	+11,0	− 22	− 17	−390	17
18	−0,17	+1,11	+1,94	141,59	+0,88	93,39	−11,4	145,9	− 2,3	−10,0	+ 12	− 10	−355	18
19	−0,56	+1,02	+1,35	147,23	+0,78	92,80	−10,4	146,9	− 2,8	−20,2	+ 44	− 18	−223	19
20	−0,90	+0,87	+0,47	152,92	+0,63	91,92	− 8,2	149,1	− 3,9	−25,3	+ 74	− 35	− 25	20
21	−1,14	+0,69	−0,49	158,71	+0,44	90,96	− 4,2	153,1	− 5,7	−21,3	+105	− 57	+137	21
22	−1,23	+0,43	−1,16	164,65	+0,17	90,29	+ 0,9	158,2	− 7,9	−15,5	+136	− 83	+150	22
23	−1,07	+0,05	−1,65	170,84	−0,22	89,80	+ 6,4	163,7	−11,0	− 9,5	+174	− 90	+157	23
24	−0,68	−0,39	−1,86	177,26	−0,67	89,59	+13,8	171,1	−12,5	− 1,9	+202	+ 12	+186	24
25	−0,07	−0,84	−1,73	183,90	−1,13	89,72	+17,4	174,7	− 9,3	+ 5,3	− 23	+110	+201	25
26	+0,53	−1,09	−1,34	190,53	−1,39	90,11	+13,4	170,7	− 4,9	+13,0	−177	+ 67	+211	26
27	+0,88	−1,28	−0,71	196,91	−1,60	90,74	+ 2,0	159,3	− 3,3	+21,9	−305	+ 74	+238	27
28	+0,79	−1,35	+0,29	202,85	−1,67	91,74	− 6,5	150,8	0	+24,4	−160	+ 76	− 72	28
29	+0,47	−1,28	+1,15	208,56	−1,61	92,60	−10,9	146,4	+ 2,2	+19,5	− 90	+ 44	−190	29
30	−0,01	−1,15	+1,79	214,11	−1,49	93,24	−13,9	143,4	+ 3,3	+ 8,9	− 28	+ 20	−335	30
31	−0,52	−1,03	+1,84	219,63	−1,38	93,29	−13,5	143,8	+ 3,7	− 8,2	+ 23	+ 6	−444	31

are required, for the projection on the plane perpendicular to the direction of gait the η- and ζ-co-ordinates, for the projection on the horizontal plane the ξ- and η-co-ordinates. The upper parts of Figs. 6–8 (pp. 161–163) were drawn in this manner. In these projections as well as in the corresponding Figs. 9 and 11 (pp. 164–165), heel strike and toe-off are marked by R, L, S_r, S_l. The arc which the centre of gravity describes during the double-support period is indicated by a thicker line. The beginning and end of foot-flat are indicated by A and E.

Table 16 (continued)

Experiment II

No.	Relative co-ordinates ξ	η	ζ	Absolute co-ordinates x	y	z	Direction of gait Relative	Absolute	Transverse direction	Vertical direction	Direction of gait	Transverse direction	Vertical direction	No.
1	+0,67	−0,74	−1,90	44,92	−0,60	89,55	+13,2	171,6	−4,4	+14,7	−136	+50	+166	1
2	+1,06	−0,89	−1,26	51,38	−0,76	90,19	+5,0	163,4	−2,7	+21,5	−243	+41	+220	2
3	+1,11	−0,96	−0,20	57,50	−0,85	91,25	−2,5	155,9	−0,9	+26,1	−163	+30	+13	3
4	+0,94	−0,97	+0,72	63,40	−0,87	92,17	−7,2	151,2	+0,3	+23,1	−109	+20	−136	4
5	+0,63	−0,96	+1,54	69,16	−0,88	92,99	−10,4	148,0	+1,1	+16,2	−66	+12	−218	5
6	+0,17	−0,94	+1,93	74,77	−0,87	93,38	−12,0	146,4	+1,3	+2,2	−17	+7	−430	6
7	−0,29	−0,90	+1,71	80,38	−0,84	93,16	−11,7	146,7	+1,6	−12,5	+32	+6	−345	7
8	−0,67	−0,83	+1,07	86,07	−0,79	92,52	−9,6	148,8	+1,9	−23,2	+81	+11	−210	8
9	−1,00	−0,75	0	91,81	−0,72	91,45	−5,7	152,7	+2,5	−25,9	+146	+21	+160	9
10	−1,13	−0,63	−0,88	97,75	−0,62	90,57	+1,8	160,2	+3,5	−19,4	+226	+41	+177	10
11	−0,83	−0,45	−1,56	104,12	−0,48	89,89	+10,0	168,4	+5,5	−12,3	+218	+81	+194	11
12	−0,33	−0,15	−1,92	110,69	−0,17	89,53	+15,3	173,7	+9,5	−4,1	+79	+78	+216	12
13	+0,33	+0,24	−1,90	117,42	+0,21	89,55	+15,4	173,8	+10,3	+4,3	−45	−20	+235	13
14	+0,87	+0,57	−1,52	124,03	+0,52	89,93	+11,1	169,5	+8,7	+12,8	−167	−62	+260	14
15	+1,18	+0,85	−0,79	130,41	+0,79	90,66	+3,0	161,4	+5,6	+22,9	−245	−95	+306	15
16	+1,11	+1,00	+0,25	136,41	+0,92	91,70	−4,5	153,9	+1,5	+27,2	−151	−90	−190	16
17	+0,84	+1,01	+1,14	142,21	+0,92	92,59	−8,2	150,2	−0,8	+18,0	−86	−39	−250	17
18	+0,52	+0,95	+1,64	147,96	+0,84	93,09	−9,8	148,6	−1,7	+8,8	−38	−12	−283	18
19	+0,14	+0,87	+1,72	153,65	+0,75	93,17	−10,4	148,0	−2,0	−2,9	0	−7	−420	19
20	−0,22	+0,81	+1,27	159,36	+0,68	92,72	−9,6	148,8	−2,4	−15,7	+39	−13	−250	20
21	−0,56	+0,69	+0,50	164,09	+0,54	91,95	−7,6	150,8	−3,2	−22,2	+80	−26	−105	21
22	−0,81	+0,55	−0,35	170,91	+0,39	91,10	−3,9	154,5	−4,4	−21,1	+122	−41	+155	22
23	−0,83	+0,35	−1,04	176,96	+0,17	90,41	+1,3	159,7	−6,1	−15,5	+163	−62	+162	23
24	−0,60	+0,08	−1,50	183,26	−0,11	89,95	+7,8	166,2	−8,7	−8,8	+157	−74	+163	24
25	−0,19	−0,25	−1,70	189,74	−0,46	89,75	+12,3	170,7	−10,0	−2,3	+85	−2	+175	25
26	+0,36	−0,59	−1,59	196,36	−0,81	89,86	+14,1	172,5	−7,6	+4,0	−20	+104	+203	26
27	+0,92	−0,83	−1,22	202,99	−1,07	90,23	+10,3	168,7	−4,3	+14,3	−185	+67	+295	27
28	+1,11	−0,94	−0,44	209,25	−1,19	91,01	−1,1	157,3	−2,0	+28,2	−205	+49	+385	28
29	+0,79	−0,97	+0,70	215,00	−1,22	92,15	−6,3	152,1	−0,4	+27,5	−95	+40	−200	29
30	+0,55	−0,97	+1,57	220,83	−1,25	93,02	−8,9	149,5	+1,0	+17,8	−39	+27	−256	30
31	+0,22	−0,93	+2,12	226,57	−1,23	93,57	−9,5	148,9	+1,3	+7,9	−2	+13	−284	31

Tables 16 and 18 can also be used to illustrate the velocities and accelerations of the total centre of gravity in a particularly appropriate manner. Let us imagine a point O in space supported by segments of lines parallel to the successive velocities of the total centre of gravity, the lengths of which indicate the magnitudes of these velocities. These segments lie on the surface of a cone and their extremities form a curve on this surface. This curve is called a "hodograph" according to Hamilton. In combination with the point O, it gives a clear picture of the relative velocities of the total centre of gravity and a precise information on the accelerations of the centre of gravity in the different phases of its tridimensional trajectory.

Table 16 (continued)

Experiment III
(with equipment)

No.	Co-ordinates of total centre of gravity (cm)						Velocity of total centre of gravity (cm s^{-1})				Acceleration of total centre of gravity (cm s^{-2})			No.
	Relative co-ordinates			Absolute co-ordinates			Direction of gait		Transverse direction	Vertical direction	Direction of gait	Transverse direction	Vertical direction	
	ξ	η	ζ	x	y	z	Relative	Absolute						
1	+0,32	-3,12	+0,38	32,20	-3,12	97,93	- 3,5	141,8	+ 9,9	+23,5	-110	0	-185	1
2	+0,10	-2,73	+1,23	37,55	-2,73	98,78	- 6,8	138,5	+10,0	+15,7	- 73	+ 1	-222	2
3	-0,19	-2,34	+1,65	42,83	-2,34	99,20	- 8,9	136,4	+10,1	+ 6,2	- 39	+ 3	-308	3
4	-0,55	-1,92	+1,64	48,04	-1,92	99,19	- 9,7	135,6	+10,3	-10,3	- 3	+ 5	-350	4
5	-0,89	-1,51	+1,02	53,27	-1,51	98,57	- 8,8	136,5	+10,7	-21,0	+ 41	+ 10	-230	5
6	-1,19	-1,06	+0,04	58,54	-1,06	97,59	- 6,5	138,8	+11,2	-28,0	+ 93	+ 18	-110	6
7	-1,39	-0,59	-1,06	63,91	-0,59	96,49	- 1,1	144,2	+12,1	-27,0	+152	+ 31	+150	7
8	-1,28	-0,12	-1,97	69,59	-0,12	95,58	+ 5,9	151,2	+13,7	-20,0	+205	+ 49	+220	8
9	-0,84	+0,49	-2,51	75,60	+0,49	95,04	+14,6	159,9	+16,2	- 9,1	+200	+ 70	+370	9
10	-0,21	+1,13	-2,58	81,80	+1,13	94,97	+20,6	165,9	+17,6	+ 5,9	+ 70	- 20	+350	10
11	+0,66	+1,75	-2,13	88,24	+1,75	95,42	+18,8	164,1	+14,7	+14,5	-155	- 95	+180	11
12	+1,17	+2,22	-1,32	94,32	+2,22	96,23	+ 9,0	154,3	+11,0	+25,0	-250	-140	+390	12
13	+1,34	+2,52	-0,13	100,06	+2,52	97,42	+ 1,0	146,3	+ 5,1	+37,0	-156	-115	0	13
14	+1,27	+2,64	+1,14	105,56	+2,64	98,69	- 3,0	142,3	+ 1,2	+26,5	-102	- 90	-350	14
15	+1,10	+2,63	+2,02	110,96	+2,63	99,57	- 5,9	139,4	- 1,7	+15,1	- 71	- 67	-290	15
16	+0,88	+2,52	+2,32	116,31	+2,52	99,87	- 7,9	137,4	- 3,8	+ 2,1	- 41	- 53	-375	16
17	+0,54	+2,33	+2,12	121,54	+2,33	99,67	- 9,1	136,2	- 5,8	-11,9	- 9	- 49	-370	17
18	+0,19	+2,10	+1,38	126,76	+2,10	98,93	- 9,1	136,2	- 7,3	-24,0	+ 31	- 49	-280	18
19	-0,10	+1,80	+0,35	132,04	+1,80	97,90	- 6,9	138,4	- 9,3	-27,0	+ 93	- 54	+100	19
20	-0,30	+1,42	-0,59	137,41	+1,42	96,96	- 1,1	144,2	-11,5	-20,5	+165	- 70	+202	20
21	-0,21	+0,90	-1,24	143,07	+0,90	96,31	+ 4,8	150,1	-14,9	-13,0	+128	-107	+200	21
22	+0,04	+0,23	-1,54	148,89	+0,23	96,01	+ 9,0	154,3	-20,0	- 3,0	+ 77	-193	+262	22
23	+0,38	-0,63	-1,45	154,80	-0,63	96,10	+10,9	156,2	-22,9	+ 6,1	+ 16	+150	+240	23
24	+0,77	-1,40	-1,03	160,76	-1,40	96,52	+ 9,1	154,4	-16,0	+13,2	-104	+131	+147	24
25	+1,01	-1,95	-0,42	166,57	-1,95	97,13	+ 2,4	147,7	-12,0	+17,8	-210	+ 95	+116	25
26	+0,97	-2,37	+0,42	172,10	-2,37	97,97	- 4,9	140,4	- 8,9	+22,4	-155	+ 68	+ 81	26
27	+0,64	-2,67	+1,29	177,34	-2,67	98,84	- 9,7	135,6	- 6,8	+20,2	- 90	+ 46	-195	27
28	+0,21	-2,93	+1,94	182,48	-2,93	99,49	-12,3	133,0	- 5,4	+ 9,9	- 49	+ 27	-400	28
29	-0,27	-3,15	+2,01	187,57	-3,15	99,56	-13,3	132,0	- 4,9	- 6,4	- 19	+ 16	-360	29
30	-0,75	-3,34	+1,55	192,66	-3,34	99,10	-13,4	131,9	- 4,2	-16,2	+ 11	+ 9	-160	30
31	-1,21	-3,49	+0,79	197,77	-3,49	98,34	-12,3	133,0	- 4,0	-21,2	+ 39	+ 5	-106	31

While the total centre of gravity describes its relative trajectory, the end point of the velocity segments simultaneously describes the hodograph at a particular velocity which usually changes continuously. The velocity of this end point at each place on the hodograph indicates the direction and magnitude of the acceleration of the centre of gravity at the corresponding point of its trajectory. If the trajectory of the centre of gravity were plane, the hodograph would be in the form of a plane curve. Since the trajectory of the centre of gravity is double-curved, the hodograph appears as a tridimensional curve. Here, the hodograph, like the trajectory of the centre of gravity, can only be represented by its projections on

Table 17. Correspondence between time and phases

Time (s)	Phase number	Time (s)	Phase number	Time (s)	Phase number	Time (s)	Phase number	Time (s)	Phase number
0	1,00	0,25	7,52	0,50	14,05	0,75	20,57	1,00	27,09
0,01	1,26	0,26	7,78	0,51	14,31	0,76	20,83	1,01	27,35
0,02	1,52	0,27	8,04	0,52	14,57	0,77	21,09	1,02	27,61
0,03	1,78	0,28	8,31	0,53	14,83	0,78	21,35	1,03	27,87
0,04	2,04	0,29	8,57	0,54	15,09	0,79	21,61	1,04	28,13
0,05	2,30	0,30	8,83	0,55	15,35	0,80	21,87	1,05	28,39
0,06	2,57	0,31	9,09	0,56	15,61	0,81	22,13	1,06	28,66
0,07	2,83	0,32	9,35	0,57	15,87	0,82	22,39	1,07	28,92
0,08	3,09	0,33	9,61	0,58	16,13	0,83	22,65	1,08	29,18
0,09	3,35	0,34	9,87	0,59	16,39	0,84	22,92	1,09	29,44
0,10	3,61	0,35	10,13	0,60	16,65	0,85	23,18	1,10	29,70
0,11	3,87	0,36	10,39	0,61	16,91	0,86	23,44	1,11	29,96
0,12	4,13	0,37	10,65	0,62	17,18	0,87	23,70	1,12	30,22
0,13	4,39	0,38	10,91	0,63	17,44	0,88	23,96	1,13	30,48
0,14	4,65	0,39	11,18	0,64	17,70	0,89	24,22	1,14	30,74
0,15	4,91	0,40	11,44	0,65	17,96	0,90	24,48	1,15	31,00
0,16	5,17	0,41	11,70	0,66	18,22	0,91	24,74		
0,17	5,44	0,42	11,96	0,67	18,48	0,92	25,00		
0,18	5,70	0,43	12,22	0,68	18,74	0,93	25,26		
0,19	5,96	0,44	12,48	0,69	19,00	0,94	25,52		
0,20	6,22	0,45	12,74	0,70	19,26	0,95	25,79		
0,21	6,48	0,46	13,00	0,71	19,52	0,96	26,05		
0,22	6,74	0,47	13,26	0,72	19,78	0,97	26,31		
0,23	7,00	0,48	13,52	0,73	20,05	0,98	26,57		
0,24	7,26	0,49	13,78	0,74	20,31	0,99	26,83		
0,25	7,52	0,50	14,05	0,75	20,57	1,00	27,09		

Time (s)	Phase number
0,001	0,03
0,002	0,05
0,003	0,08
0,004	0,10
0,005	0,13
0,006	0,16
0,007	0,18
0,008	0,21
0,009	0,23
0,010	0,26

planes. Of course, the projection of the hodograph on any plane represents the hodograph of the projection of the movement of the centre of gravity on the same plane. The projections of a hodograph on the three principal planes, combined with the corresponding projections of the relative trajectory of the centre of gravity, provide an even clearer picture of the movement of the centre of gravity than the displacement, velocity and acceleration curves of Figs. 13–15. The velocity at which the hodograph travels in one projection also indicates the direction and magnitude of the acceleration of the centre of gravity for every point in this projection.

In Figs. 6–8 (pp. 161–163), the corresponding projection of the hodograph has been drawn under each projection of the relative trajectory of the total centre of gravity and provided with the indications R, L, S_r, S_l, A_r, A_l, E_r and E_l. If the gait were absolutely regular, the hodograph of the relative movement of the centre of gravity would be a tridimensional closed curve which would be described exactly once during a double step. In this ideal case, in the projections of the hodograph on the three planes, the extremities would meet each other exactly and not approximately as in Figs. 6–8. In particular, the projection on the plane of gait would almost have the shape of a right-angle triangle with smoothed angles and a vertical hypothenuse, described twice during one double step.

The projections of the hodograph show that the line connecting any point on the hodograph with the fixed point O is parallel to the tangent to the projection of the trajectory

Table 18. Moments of the maxima of displacements from the middle plane, velocities and accelerations for the three directions

Experiment I

Time (s)	Phase number	Distance from plane moving forward at average velocity of gait	Time (s)	Phase number	Relative velocity (cm s^{-1})	Time (s)	Phase number	Acceleration (cm s^{-2})
				Movement in direction of gait				
				Anterior maxima				
0,015	1,4	+ 0,33	0,396	11,3	+ 20,0	0,347	10,1	+ 330
0,498	14,0	+ 1,23	0,916	24,9	+ 17,4	0,880	24,0	+ 202
1,003	27,2	+ 0,91						
				Posterior maxima				
0,291	8,6	− 1,23	0,144	4,8	− 18,0	0,478	13,5	− 303
0,800	21,9	− 1,23	0,638	17,7	− 21,6	0,990	26,8	− 365
			1,130	30,5	− 24,3			
				Movement in transverse direction				
				Right maxima				
0,547	15,3	+ 1,27	0,402	11,5	+ 11,3	0,345	10,0	+ 78
						0,918	25,0	+ 110
				Left maxima				
0,045	2,2	− 1,31	0,878	23,9	− 12,6	0,509	14,3	− 129
1,036	28,0	− 1,35				0,835	22,8	− 92
				Movement in vertical direction				
				Upper maxima				
0,141	4,7	+ 2,01	0,019	1,5	+ 30,1	0,505	14,2	+ 301
0,634	17,5	+ 2,07	0,513	14,4	+ 29,1	0,997	27,0	+ 239
1,130	30,5	+ 1,91	1,020	27,6	+ 25,3			
				Lower maxima (minima)				
0,391	11,2	− 2,15	0,240	7,3	− 29,9	0,140	4,7	− 620
0,891	24,3	− 1,87	0,731	20,1	− 25,3	0,635	17,6	− 727
						1,137	30,7	− 500

at this point. For example, the segment OR in the projection of the hodograph on the plane of gait is oblique forwards and downwards. Correspondingly, the projection of the relative trajectory of the centre of gravity on the plane of gait also runs forwards and downwards at the time R. In the projection on the plane perpendicular to the direction of gait, the same

Table 18   (continued)

Experiment II

Time (s)	Phase number	Distance from plane moving forward at average velocity of gait	Time (s)	Phase number	Relative velocity (cm s⁻¹)	Time (s)	Phase number	Acceleration (cm s⁻²)

Movement in direction of gait

Anterior maxima

0,063	2,7	+ 1,15	0,495	13,9	+ 15,2	0,363	10,5	+ 241
0,550	15,4	+ 1,19	0,950	25,8	+ 14,3	0,862	23,5	+ 173
1,030	27,9	+ 1,13						

Posterior maxima

0,338	9,8	− 1,13	0,205	6,4	− 12,0	0,030	1,8	− 250
0,834	22,8	− 0,88	0,690	19,0	− 10,4	0,532	14,9	− 250
			1,150	31,0	− 9,4	1,015	27,5	− 372

Movement in transverse direction

Right maxima

0,593	16,5	+ 1,02	0,447	12,7	+ 10,4	0,407	11,6	+ 113
						0,959	26,0	+ 105

Left maxima

0,105	3,7	− 0,98	0,921	25,0	− 10,1	0,561	15,6	− 114
1,080	29,2	− 0,99				0,873	23,8	− 78

Movement in vertical direction

Upper maxima

0,197	6,1	+ 1,93	0,080	3,1	+ 26,1	0,043	2,1	+ 232
0,680	18,7	+ 1,73	0,563	15,7	+ 28,7	0,545	15,2	+ 312
			1,050	28,4	+ 31,0	1,030	27,9	+ 393

Lower maxima (minima)

0,441	12,5	− 1,95	0,297	8,8	− 27,0	0,193	6,0	− 430
0,934	25,4	− 1,71	0,784	21,5	− 23,3	0,698	19,2	− 442

segment of the hodograph runs to the right and downwards. Consequently, the tangent to the trajectory at point R also runs to the right and downwards. The projections of the hodograph show more clearly than those of the trajectory that the centre of gravity moves relatively forwards, to the right and downwards at right heel strike. Similarly, it appears that the centre of gravity moves forwards, to the left and downwards at left heel strike. The centre

Table 18 (continued)

Experiment III

Time (s)	Phase number	Distance from plane moving forward at average velocity of gait	Time (s)	Phase number	Relative velocity (cm s⁻¹)	Time (s)	Phase number	Acceleration (cm s⁻²)

Movement in direction of gait
Anterior maxima

Time (s)	Phase number	Distance	Time (s)	Phase number	Relative velocity	Time (s)	Phase number	Acceleration
0,467	13,2	+ 1,34	0,359	10,4	+ 21,0	0,290	8,6	+ 229
0,932	25,3	+ 1,03	0,849	23,2	+ 11,1	0,723	19,9	+ 166

Posterior maxima

Time (s)	Phase number	Distance	Time (s)	Phase number	Relative velocity	Time (s)	Phase number	Acceleration
0,237	7,2	− 1,40	0,118	4,1	− 9,7	0,410	11,7	− 280
0,735	20,2	− 0,30	0,622	17,2	− 9,4	0,916	24,9	− 211
			1,097	29,6	− 13,5			

Movement in transverse direction
Right maxima

Time (s)	Phase number	Distance	Time (s)	Phase number	Relative velocity	Time (s)	Phase number	Acceleration
0,515	14,4	+ 2,66	0,338	9,8	+ 17,7	0,303	8,9	+ 72
						0,852	23,2	+ 199

Left maxima

Time (s)	Phase number	Distance	Time (s)	Phase number	Relative velocity	Time (s)	Phase number	Acceleration
—	—	—	0,833	22,7	− 23,5	0,424	12,1	− 139
						0,809	22,1	− 297

Movement in vertical direction
Upper maxima

Time (s)	Phase number	Distance	Time (s)	Phase number	Relative velocity	Time (s)	Phase number	Acceleration
0,094	3,5	+ 1,74	0,461	13,0	+ 37,0	0,322	9,4	+ 486
0,582	16,2	+ 2,33	0,970	26,3	+ 23,4	0,823	22,5	+ 282
1,058	28,6	+ 2,04						

Lower maxima (minima)

Time (s)	Phase number	Distance	Time (s)	Phase number	Relative velocity	Time (s)	Phase number	Acceleration
0,329	9,6	− 2,63	0,207	6,4	− 29,4	0,100	3,6	− 520
0,819	22,4	− 1,58	0,679	18,7	− 32,8	0,587	16,3	− 380
						1,055	28,5	− 480

of gravity moves forwards horizontally to one side or the other approximately in the middle of the double-support period of gait. At S_l the centre of gravity moves forwards, to the right and upwards, and at S_r forwards, to the left and upwards. This appears in the first two projections of the hodograph and is confirmed by the third projection, on the horizontal plane. The loading of the subject does not alter these facts, as shown in Fig. 8.

The projections of the hodograph demonstrate that the velocity of the movement of the centre of gravity is relatively small at heel strike in the projection on the plane of gait. It is relatively large at phases 7 and 20 in experiment I, $8\frac{1}{2}$ and 21 in experiment II, 6 and 19 in experiment III. These correspond exactly to the moments E, when the only supporting foot begins its push-off, according to Table 14. From then on, in the projection on the plane of gait, the velocity progressively decreases until heel strike of the opposite foot and increases again until foot-flat. The velocity in this projection then decreases again until it reaches a relative minimum almost at the end of the second third of foot-flat. It then increases until the end of this period. This example sufficiently shows how, by a thorough study of Figs. 6–8, one can obtain an overall view of the behaviour of the velocities of the centre of gravity in the projections of the movement of the latter on the three planes.

Even more interesting than the velocities of the projected movement are the absolute values of the resultant velocity of the total centre of gravity at any point on its tridimensional double-curved trajectory during its relative displacement. These values can easily be deduced from Table 16 for the 31 phases. The velocities in the three directions, as given in Table 16, represent the three components of the resultant velocity that are perpendicular to each other. The rule of the parallelogram of the velocities, generalized to the three dimensions, is used to obtain the resultant velocity. For each phase, the squares of the three components are added. The square root of the sum is the resultant velocity. This resultant velocity can be considered as the diagonal of a rectangular parallelepiped the three sides of a corner of which are the velocities in the direction of gait, the transverse direction and the vertical direction. This calculation carried out for the 31 phases of the three experiments gives the data of Table 19.

From the data on this table, a figure can be drawn which illustrates the velocity of the total centre of gravity during its whole relative course (Fig. 16a). Here again the most important moments of the movement are marked by thick vertical lines to allow easy comparison between the figures belonging to the different experiments. Time is less important than the correlation between velocity and the 31 phases. The numbers of the phases are, therefore, marked on the axis of the abscissa above and below the vertical lines separated by a distance of $\frac{1}{2}$ cm. The values of the resultant velocity are entered as ordinates. A distance of 1 mm corresponds to a velocity of $1\,\mathrm{cm\,s^{-1}}$.

These figures combined with the projections of the trajectory and the hodograph (Figs. 6–8) clearly illustrate the velocities of the total centre of gravity. Whereas Figs. 6–8 give a precise information on the direction of the velocities and on the values of their projections on the three principal planes, Fig. 16a presents the magnitudes of the resultant velocity. These diagrams correspond fairly well in the first two experiments. They show that the absolute values of the velocities increase during a period equal to the duration of a single step. From E_r on, Fig. 16a presents the same shape as from E_l on. With regard to the direction of the velocity, the process repeats itself after a double step. During each period, the figure shows two very pronounced maxima and minima. One maximum coincides exactly with the moment E, the other approximately with the end of the first third of foot-flat. One minimum corresponds fairly well with heel strike, the other with the end of the second third of foot-flat. The resultant velocity is almost the same as its projection on the plane of gait. A less pronounced maximum occurs close to the end of the double-support period. Finally a slightly pronounced minimum can be recognized either at the very beginning of or shortly before foot-flat. Figures 6–8 show that the velocity of the centre of gravity is relatively small when the latter is close to the highest or lowest points of its trajectory and moves almost horizontally. The velocity is relatively great when the centre of gravity moves nearly vertically upwards or downwards.

The gait of the loaded man presents almost the same characteristics. The range of the absolute values of the velocity of the total centre of gravity is somewhat wider.

Figure 16 is very clear and provides all the details concerning the behaviour of the velocity much better than any explanation. The curves will not, therefore, be further analysed here though it is recommended that they be studied.

Table 19. Velocity of the movement of the total centre of gravity relative to a plane moving forward at the average velocity of gait (cm s^{-1}; absolute magnitudes)

No.	Experiment I	Experiment II	Experiment III
1	29,2	20,2	25,7
2	28,3	22,2	19,8
3	22,8	26,2	14,8
4	14,2	24,2	17,5
5	10,3	19,3	25,1
6	21,1	12,3	30,8
7	30,3	17,2	29,6
8	26,4	25,2	24,9
9	19,2	26,6	23,6
10	18,1	19,8	27,7
11	22,4	16,8	27,9
12	22,5	18,5	28,8
13	21,1	19,0	37,0
14	25,5	19,0	26,7
15	27,5	23,8	16,3
16	23,0	27,6	9,0
17	16,0	19,8	16,1
18	15,3	13,3	26,7
19	22,9	11,0	29,4
20	26,9	18,6	23,5
21	22,4	23,7	20,3
22	17,4	21,9	22,1
23	15,9	16,7	26,1
24	18,7	14,6	22,7
25	20,4	16,0	21,6
26	19,3	16,5	24,6
27	22,2	18,1	23,4
28	25,4	28,3	16,7
29	22,4	28,2	15,5
30	16,8	19,9	21,4
31	16,2	12,4	24,8

Table 20. Acceleration of the movement of the total centre of gravity (cm s^{-2}; absolute magnitudes)

No.	Experiment I	Experiment II	Experiment III
1	188	220	215
2	172	330	234
3	221	166	310
4	341	175	350
5	452	228	234
6	289	430	145
7	182	346	216
8	227	225	305
9	283	218	426
10	393	290	358
11	216	303	256
12	245	243	484
13	364	240	194
14	392	315	376
15	204	403	306
16	203	259	381
17	391	267	373
18	355	286	286
19	228	420	147
20	86	253	270
21	182	135	260
22	219	201	334
23	251	238	283
24	275	238	223
25	230	195	258
26	283	229	188
27	394	355	220
28	191	439	404
29	215	225	361
30	337	260	161
31	445	284	109

The projections of the hodographs in Figs. 6–8 provide information both about the velocities and about the accelerations. From the direction of the projections of the hodographs, it is possible to recognize immediately the direction of the acceleration in the projection of the movement of the centre of gravity on the plane under consideration.

Unlike the direction of velocity, the direction of acceleration is generally different from that of movement. It is the same only if the movement is linear. When a mobile point describes a tridimensional curved trajectory, the acceleration always deviates from the tangent towards the concave side of the trajectory. This can be seen in Figs. 6–8. At point R, for instance, in the projection on the plane of gait, the tangent to the trajectory points forwards and downwards and the trajectory itself presents an upward forward concavity. In the projection of the corresponding hodograph, which indicates the direction of the

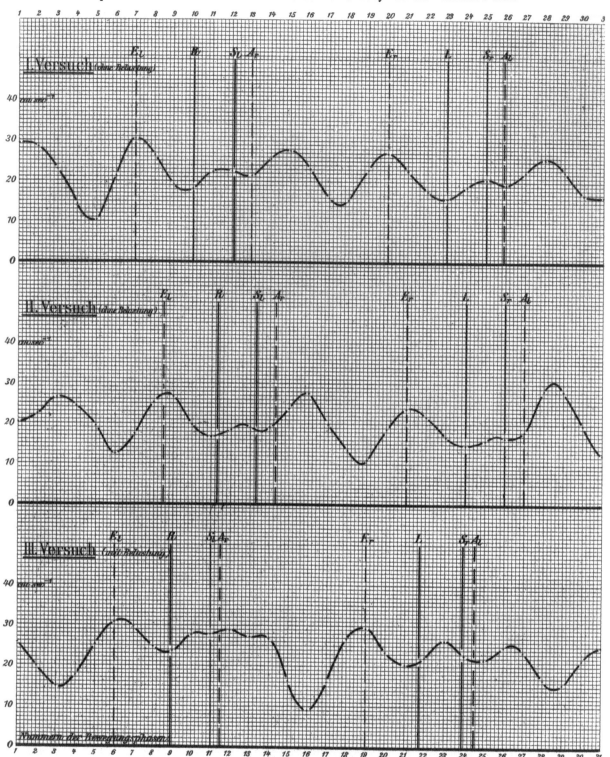

Fig. 16a   **Fig. 16a, b.** Velocity (**a**) and acceleration (**b**) of the movement of the centre of gravity relative moving forward at the average velocity of gait; absolute values to a plane

acceleration, the tangent at point R points forwards and upwards. The tridimensional trajectory of the centre of gravity, at point R, runs forwards downwards and to the right and presents an upward concavity. Consequently, the tangent to the tridimensional hodograph, at point R, shows an upward deviation from the tangent to the trajectory. This deviation is

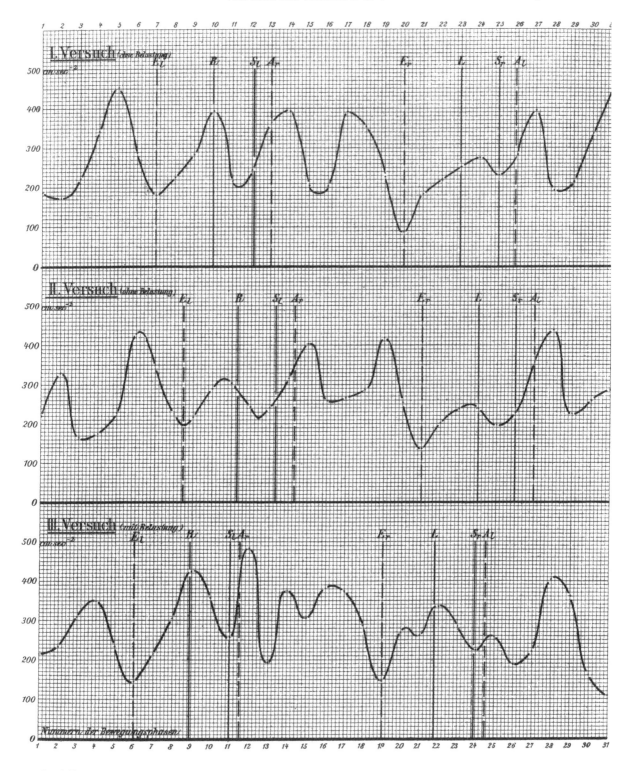

Fig. 16b

generally not such that the direction of the acceleration is perpendicular to the tangent of the trajectory; both usually intersect at an acute or obtuse angle.

The acceleration can always be imagined as resolved into two components, one coinciding with the tangent to the trajectory, the other perpendicular to this tangent. The component perpendicular to the tangent passes through the centre of curvature of the part of the trajectory under consideration. The former component is called "tangential acceleration", the latter "normal acceleration". Each of the two components plays a particular role in the movement.

Tangential acceleration only affects the velocity of the movement of the centre of gravity on its trajectory. Normal acceleration only determines the curvature of the trajectory. If normal acceleration is great, the curvature of the part of the trajectory under consideration will be relatively pronounced; if it is small, the trajectory will be much less curved. If the trajectory is flat, the normal acceleration is zero. Of course the curvature of the trajectory also depends on the magnitude of the velocity. With a high velocity normal acceleration will cause a smaller curvature of the trajectory than with a low velocity. If the velocity is constant, the tangential acceleration is zero and the resultant acceleration is perpendicular to the tangent to the trajectory. If the movement is accelerated on its trajectory, the tangential acceleration has a finite magnitude and the same direction as the movement. If at the same time there is normal acceleration and the trajectory is thus curved, the resultant acceleration forms an acute angle with the tangent to the side of the movement. If the movement is decelerated, the tangential acceleration has an opposite direction to the movement, and the resultant acceleration forms an obtuse angle with the tangent as long as the normal acceleration is not zero. Knowing the direction of the resultant acceleration allows the quantitative proportion between the tangential and normal acceleration to be determined. The direction of the resultant acceleration of the centre of gravity can easily be recognized on the projections of the hodographs. Thorough study of the hodograph projected on the three planes thus gives the quantitative proportion between the two components of the acceleration.

At present we shall not go further into this subject though another use of the hodograph will be mentioned. The magnitude of the resultant acceleration, as well as that of the resultant velocity, can be deduced from their rectangular components given in Table 16. To this end, the squares of the accelerations in the direction of gait, in the transverse direction and in the vertical direction, are added and the square root of their sum is determined. This was done for the 31 phases in the three experiments. The results of this calculation are given in Table 20.

These values are illustrated in Fig. 16b in the same manner as the absolute values of the resultant velocity. A length of $1/10$ mm corresponds to an acceleration of $1\,\mathrm{cm\,s^{-2}}$, as in Figs. 13–15. As shown by the graphs, the resultant acceleration reaches a maximum almost exactly at all the places where the resultant velocity is relatively small, and vice versa.

The velocities given in Table 19 relate to the relative movement of the total centre of gravity. In the absolute movement, the average velocity of gait is an additional component in the direction of gait. The values of the resultant acceleration are true for the relative as well as for the absolute movement of the centre of gravity.

On the External Forces

The external forces which act on the human body in walking are proportional to the accelerations of the total centre of gravity, as explained in the introduction. If the forces are measured by weight, they are expressed in terms of terrestrial mass. An external force would then be equal to the weight of the human body if it gave the total centre of gravity an acceleration of $981.11\,\mathrm{cm\,s^{-2}}$. This is the acceleration due to gravity in Leipzig. If the acceleration of the centre of gravity is greater or smaller than the acceleration due to gravity, the external force which provoked this acceleration is greater or smaller than the body weight. The external force is designated by K, the acceleration which this force causes by γ, and the body weight by G. Thus:

$$\frac{K}{G} = \frac{\gamma}{981.111}$$

$$K = \frac{G}{981.11} \cdot \gamma$$

The subject weighed 58.7 kg, and his military equipment 23.27 kg. If the external forces are expressed in kilograms, G is equal to 58.7 in experiments I and II and 81.97 in experiment III. In the present instance, the external forces are determined from the accelerations γ of the total centre of gravity:

$K = 0.05983 \cdot \gamma$ for experiments I and II

$K = 0.08355 \cdot \gamma$ for experiment III

These equations are valid for the resultant of all the external forces as well as for the sums of their components in the three principal directions. In the former case, γ is replaced by the values of the resultant acceleration given in Table 20, in the latter case by the components of the resultant acceleration in the three directions given in Table 16.

The magnitude of the resultant of the external forces should not be directly determined since its direction is not yet known. The external forces are distributed in a particular manner in the three principal directions, as explained in the introduction. Therefore, it seems more appropriate to calculate the sums of the components of the external forces in the direction of gait, in the transverse direction and in the vertical direction. These sums also give the direction of the resultant of the external forces. This calculation based on data in Table 16 was carried out for the 31 phases of the three experiments, and the results are presented in Table 21.

Components of the force of friction and air resistance must be considered in the direction of gait. Friction acts in the direction of gait at several phases and in the opposite direction at other phases. Air resistance is always directed posteriorly. It acts in such a way that the sums of the components in the direction of gait have a smaller absolute value than friction when they are positive and a greater absolute value when they are negative. Since no accurate studies have been carried out to determine the magnitude of air resistance, it is not possible to establish in the sum of the components which part belongs to friction acting in the direction of gait. However, air resistance can be considered very small compared with the other external forces in the present instance in which the subject walked in an enclosed area. At a good approximation, the sums of the components thus calculated for the direction of gait can, therefore, be considered as the components of the force of friction acting in this direction.

Friction is the only external force to act in the transverse direction. Air resistance can be ignored since the velocities in the transverse direction are small. The data for the transverse direction in Table 21 can thus be regarded as being even more accurate in representing the transverse components of friction.

In the vertical direction, of course, the effect of air resistance can also be ignored. However, two other forces come into effect: normal ground reaction force and gravity. The latter is constant and known, and thus it is easy to calculate the normal ground reaction force from the sum of the vertical components of the external forces for each of the 31 phases. The force of gravity acts downward and must be considered negative. It is, therefore, only necessary to add the body weight to the vertical components of Table 21 to obtain the values of the normal ground reaction force. It must be taken into account that in experiment III the weight of the loaded body is 81.97 kg. The results of the calculation are given in Table 22 for the 31 phases of the three experiments.

Normal ground reaction force is equal and opposite to the compression exerted on the ground by the feet. Table 22 thus also gives the values of this compression which is directed downward and is, therefore, negative.

Table 21. Components of the external forces (kg)

| No. | Experiment I | | | Experiment II | | | Experiment III | | | No. |
	Direction of gait[a]	Transverse direction[b]	Vertical direction[c]	Direction of gait	Transverse direction	Vertical direction	Direction of gait	Transverse direction	Vertical direction	
1	− 7,48	+ 0,78	+ 8,38	− 8,14	+ 2,99	+ 9,93	− 9,19	0	− 15,46	1
2	− 5,50	+ 1,97	− 8,50	−14,54	+ 2,45	+ 13,16	− 6,10	+ 0,08	− 18,55	2
3	− 3,53	+ 2,63	− 12,44	− 9,75	+ 1,79	+ 0,78	− 3,26	+ 0,25	− 25,73	3
4	− 1,38	+ 0,84	− 20,34	− 6,52	+ 1,20	− 8,14	− 0,25	+ 0,42	− 29,24	4
5	+ 0,42	− 0,60	− 27,04	− 3,95	+ 0,72	− 13,04	+ 3,43	+ 0,84	− 19,22	5
6	+ 1,91	− 0,48	− 17,17	− 1,02	+ 0,42	− 25,73	+ 7,77	+ 1,50	− 9,19	6
7	+ 3,53	+ 1,56	− 10,17	+ 1,91	+ 0,36	− 20,64	+12,70	+ 2,59	+ 12,53	7
8	+ 6,04	+ 2,87	+ 11,79	+ 4,85	+ 0,66	− 12,56	+17,13	+ 4,09	+ 18,38	8
9	+11,32	+ 4,01	+ 11,97	+ 8,74	+ 1,26	+ 9,57	+16,71	+ 5,85	+ 30,91	9
10	+19,62	+ 4,67	+ 12,03	+13,52	+ 2,45	+ 10,59	+ 5,85	− 1,67	+ 29,24	10
11	+ 3,65	+ 1,97	+ 12,27	+13,04	+ 4,85	+ 11,61	−12,95	− 7,94	+ 15,04	11
12	− 7,06	− 1,38	+ 12,80	+ 4,73	+ 4,67	+ 12,92	−20,88	−11,70	+ 32,58	12
13	−16,03	− 4,01	+ 14,18	− 2,69	− 1,20	+ 14,06	−13,03	− 9,61	0	13
14	−13,76	− 7,18	+ 17,59	− 9,99	− 3,71	+ 15,56	− 8,52	− 7,52	− 29,24	14
15	− 7,36	− 5,03	− 8,38	−14,66	− 5,68	+ 18,31	− 5,93	− 5,60	− 24,23	15
16	− 3,59	− 2,27	− 11,37	− 9,03	− 5,38	− 11,37	− 3,43	− 4,43	− 31,33	16
17	− 1,32	− 1,02	− 23,33	− 5,15	− 2,33	− 14,96	− 0,75	− 4,09	− 30,91	17
18	+ 0,72	− 0,60	− 21,24	− 2,27	− 0,72	− 16,93	+ 2,59	− 4,09	− 23,39	18
19	+ 2,63	− 1,08	− 13,34	0	− 0,42	− 25,13	+ 7,77	− 4,51	+ 8,36	19
20	+ 4,43	− 2,09	− 1,50	+ 2,33	− 0,78	− 14,96	+13,79	− 5,85	+ 16,88	20
21	+ 6,28	− 3,41	+ 8,20	+ 4,79	− 1,56	− 6,28	+10,69	− 8,94	+ 16,71	21
22	+ 8,14	− 4,97	+ 8,97	+ 7,30	− 2,45	+ 9,27	+ 6,43	−16,13	+ 21,89	22
23	+10,41	− 5,38	+ 9,39	+ 9,75	− 3,71	+ 9,69	+ 1,34	+12,53	+ 20,05	23
24	+12,09	+ 0,72	+ 11,13	+ 9,39	− 4,43	+ 9,75	− 8,69	+10,95	+ 12,28	24
25	− 1,38	+ 6,58	+ 12,03	+ 5,09	− 0,12	+ 10,47	−17,55	+ 7,94	+ 9,69	25
26	−10,59	+ 4,01	+ 12,62	− 1,20	+ 6,22	+ 12,15	−12,95	+ 5,68	+ 6,77	26
27	−18,25	+ 4,43	+ 14,24	−11,07	+ 4,01	+ 17,65	− 7,52	+ 3,84	− 16,29	27
28	− 9,57	+ 4,55	− 4,31	−12,27	+ 2,93	+ 23,03	− 4,09	+ 2,26	− 33,42	28
29	− 5,38	+ 2,63	− 11,37	− 5,68	+ 2,39	− 11,97	− 1,59	+ 1,34	− 30,08	29
30	− 1,68	+ 1,20	− 20,04	− 2,33	+ 1,62	− 15,32	+ 0,92	+ 0,75	− 13,37	30
31	+ 1,38	+ 0,36	− 26,56	− 0,12	+ 0,78	− 16,99	+ 3,26	+ 0,42	− 8,86	31

[a] Positive forwards; [b] Positive to the right; [c] Positive upwards

Finally, it is of interest to know the greatest and smallest values of the normal compression on the ground. These are easily calculated from Table 18, which indicates the maxima and minima of the vertical acceleration of the total centre of gravity. The maximum and minimum values of normal compression are given in Table 23.

The variations of the compression on the ground in walking is illustrated in Fig. 17. This figure is based on the data of Tables 22 and 23 and constructed in the manner of Fig. 16. A force of 1 kg is represented by a length of 1 mm. The body weight is indicated by a horizontal line.

As can be seen from Fig. 17, in walking without a load, at time *E* when the only foot on the ground begins its push-off, the compression is almost equal to body weight. The compression then undergoes a rapid and substantial increase, subsequently remaining

No.	Experiment I	Experiment II	Experiment III
1	67,08	68,63	66,51
2	50,20	71,86	63,42
3	46,26	59,48	56,24
4	38,36	50,56	52,73
5	31,66	45,66	62,75
6	41,53	32,97	72,78
7	48,53	38,06	94,50
8	70,49	46,14	100,35
9	70,67	68,27	112,88
10	70,73	69,29	111,21
11	70,97	70,31	97,01
12	71,50	71,62	114,55
13	72,88	72,76	81,97
14	76,29	74,26	52,73
15	50,32	77,01	57,74
16	47,33	47,33	50,64
17	35,37	43,74	51,06
18	37,46	41,77	58,58
19	45,36	33,57	90,33
20	57,20	43,74	98,85
21	66,90	52,42	98,68
22	67,67	67,97	103,86
23	68,09	68,39	102,02
24	69,83	68,45	94,25
25	70,73	69,17	91,66
26	71,32	70,85	88,74
27	72,94	76,35	65,68
28	54,39	81,73	48,55
29	47,33	46,73	51,89
30	38,66	43,38	68,60
31	32,14	41,71	73,11

Table 22. Normal ground reaction force ($= -$ compression force exerted by the feet perpendicular to the ground)

Table 23. Maxima and minima of the compression force perpendicular to the ground (kg)

	Experiment I		Experiment II		Experiment III	
	No.	Compression force	No.	Compression force	No.	Compression force
Maxima	14,2	76,71	2,1	72,58	9,4	122,58
	27,9	72,94	15,2	77,37	22,5	105,53
			27,9	82,21		
Minima	4,7	21,61	6,0	32,97	3,6	38,52
	17,6	15,20	19,2	32,26	16,3	50,22
	30,7	28,78			28,5	41,87

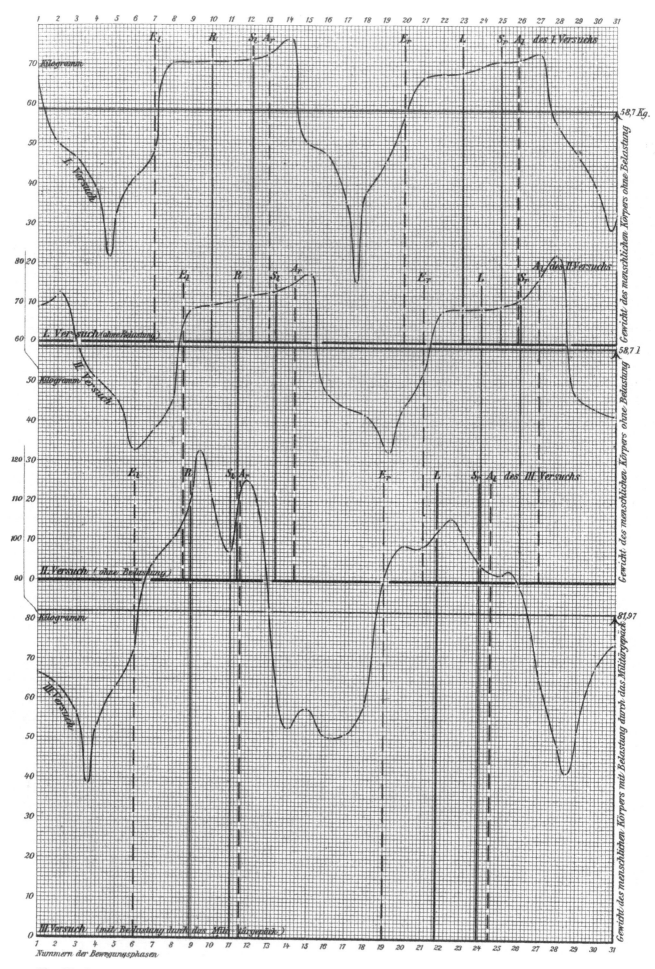

Fig. 17. Compression force perpendicular to the ground (kg)

almost constant for a long time, until foot-flat of the opposite side. It then increases further and falls rapidly. In the second half of foot-flat it is less than half the body weight in experiment I and virtually half the body weight in experiment II. Precisely at this moment, the centre of gravity is at its highest, as seen in Figs. 6–8. The compression remains low for only a short time and then increases quickly. At the end of foot-flat, the compression is again equal to body weight.

In the gait of the loaded subject the process is similar. However, here the compression also oscillates during the double-support period.

The data of Tables 22 and 23 give the ground reaction force and also the normal compression exerted by the feet on the ground. If air resistance is ignored, the components in the direction of gait and in the transverse direction, given in Table 21, can be regarded as the components of the compression of the feet on the ground in these two directions. The corresponding components of friction are equal but opposite to these components of compression. The plus-minus signs, therefore, have to be reversed. After this has been done, the data in Table 21 give the components of compression exerted by the feet on the ground in the direction of gait and in the transverse direction, and the data in Table 22 the vertical components of this compression. The compression results from muscular activity combined with gravity. It is, therefore, imperative that this compression be determined for an understanding of muscular activity in walking.

Conclusions

Chapter 2, the second part of this study of human gait, has dealt with the movement of the centre of gravity of the human body, an important problem in the mechanics of walking. The tridimensional double-curved trajectory which the centre of gravity describes during a double step has been determined as well as the velocities and accelerations of the centre of gravity in its trajectory. The movement of the centre of gravity was resolved into three rectangular components, one in the direction of gait, one in a direction perpendicular to that of gait (transverse) and one vertical.

The first component is of particular significance. The centre of gravity, like any other point of the body, always moves forward in the direction of gait. In the transverse and vertical directions, however, it carries out periodical movements the period of which is equal to the duration of a double step in the former, and to that of a single step in the latter. During one step the centre of gravity thus describes a movement equal to the length of the step. However, this forward movement is not uniform: the velocity oscillates in magnitude periodically. The average velocity of gait is the velocity with which the length of a step would be travelled in a uniform movement in the duration of a step. The actual velocity of the total centre of gravity of the human body in the direction of gait is sometimes greater, sometimes smaller than this average velocity of gait. Let us imagine a space moving forwards regularly in the direction of gait at the average velocity of gait. If the movement of the total centre of gravity is entered into this space, the centre of gravity will carry out a periodical movement in the moving space in the direction of gait. The components of the movement in the other two directions are the same whether the space is mobile or not. The total centre of gravity thus describes a closed, tridimensional double-curved trajectory relative to the space moving forwards at the average velocity of gait. It travels over this trajectory once during a double step (Figs. 9–12). If the movement of the total centre of gravity of the human body during walking could be observed directly, an observer in a carriage moving uniformly forwards at the side of the walking man, and at the average velocity of gait, would see the centre of gravity describing its closed tridimensional trajectory.

The double curved trajectory of the relative movement of the centre of gravity easily gives a clear picture of the absolute movement of the latter in immobile space. The total centre of gravity must be imagined travelling over the closed tridimensional curve with its variable velocity while this tridimensional curve itself moves forwards uniformly at the average

The transverse and vertical components of the velocity with which the centre of gravity describes the closed trajectory in its relative movement are the same as for the absolute movement in immobile space. The component of the velocity in the direction of gait is at any time greater by the average velocity of gait in the immobile than in the mobile space. The components of acceleration are all exactly the same for the relative movement on the closed tridimensional trajectory in the mobile system as for the absolute movement of the centre of gravity in immobile space. Rather than use the absolute movement of the centre of gravity, it is more convenient to employ the relative movement for further research because the excursions are smaller in the three dimensions. This is what has been done in the present work.

Precise and thorough knowledge of the movement of the centre of gravity of a body is of the utmost importance in analysing movement. It enables one to deduce the direction and magnitude of the external forces which act during walking. The centre of gravity of a body or a system of bodies on which any internal and external forces act, always moves as if all the masses were united in it and as if all the external forces were directly applied to it. The internal forces exert no influence on the movement of the centre of gravity because they always act in pairs, counterbalancing each other.

The external forces acting during walking are gravity, normal ground reaction force, friction with the ground and air resistance. These forces are involved, each in a particular manner, in the forward movement of the total centre of gravity of the human body. The movement in the direction of gait is effected by friction in this direction, which acts as a counterforce opposing the horizontal displacement of the foot, and air resistance which can be considered as a force opposing the movement of the whole body. As long as it hinders gliding of the foot, friction equals the horizontal component of the force with which the foot compresses the ground. Air resistance depends on the velocity of the forward movement. The transverse movement of the total centre of gravity is induced exclusively by the friction in this direction. Since there are not great transverse velocities, air resistance which essentially depends on velocity plays a very small role. The vertical movement of the total centre of gravity is the resultant of gravity and normal ground reaction force.

It is possible to alter the normal ground reaction force and friction at will, within certain limits, by means of the muscles. Both depend on the compression exerted on the ground by the feet. It is this that permits walking and being able to impose direction and velocity on the body in the first place.

Through the accelerations which the total centre of gravity undergoes in the three directions at each moment, the magnitude of the normal ground reaction force and friction can be calculated if air resistance is ignored. The exact effect of air resistance is not known, but in an enclosed area it cannot be significant compared with the other external forces. Air resistance has been ignored, therefore, in the present study.

Finally, knowledge of the ground reaction force and friction can provide the basis for an evaluation of the activity of the main muscular groups that are involved in walking. Here, thorough understanding of the articular movements taking place during a double step must be acquired using the tridimensional co-ordinates of the joint centres. This will constitute the subject of the next part of the analysis of human gait in Chap. 3.

The present chapter is divided into four sections. The first discusses the different methods by which the trajectory of the centre of gravity can be determined. There are two possibilities: either one starts from the partial centres of gravity of the body segments, or one determines the total centre of gravity using the principal points of the different body segments. Both ways are dealt with at length. The second section comprises a calculation of the co-ordinates of the total centre of gravity for the 31 phases of the three experiments. In the third section, the velocities and accelerations of the total centre of gravity are deduced. A calculation of the components of the external forces which act on the human body during walking is presented in the fourth section.

CHAPTER 3

Considerations on the Further Goals of the Research and Overall View of the Movements of the Lower Extremities

Considerations on the Further Process and Final Goals of the Investigation

In Chap. 2 the movement of the total centre of gravity of the human body during walking was analysed. This provided a clear insight into the movement of the body as a whole. The external forces which act on the whole human body in the different phases of gait were also established. Some of these forces are independent of the will, e. g. gravity. The magnitude of others can be modified at will, within certain limits, e. g. the ground reaction force and the frictional resistance of the ground. Air resistance can be classed in either of these groups: though no influence can be exerted on air currents, it is possible to change the velocity of the forward movement of the body, on which the magnitude of air resistance mainly depends.

Ground reaction force and frictional resistance of the ground can be modified at will because they depend on the compression that is exerted on the ground in each phase of gait. The ground reaction force is perpendicular to the ground surface, and frictional resistance, or simply friction, is equal and opposite to the component of compression that is parallel to the ground surface. This compression depends on the state of movement of the body which in turn is a function of the changes in muscular tension.

The tension of the muscles involved need not necessarily be actively induced by innervation since each muscle always displays some degree of purely elastic tension when stretched beyond its natural length. This tension is greater the closer the muscle comes to its maximum length. Whether the muscular tension is due to innervation or to passive elongation or shortening of the muscle is irrelevant for the state of movement of the body, other circumstances being equal. Any change in the tension of a muscle, or in a joint ligament, acts as paired forces, equal and opposite, applied at different segments of the body. These forces are internal with respect to the body and as such are unable to influence directly the movement of the total centre of gravity. During walking they can, however, influence this movement by altering the compression exerted on the ground by the feet.

The change in compression results from the fact that the two equal and opposite muscular forces represent external forces for the body segments situated between the points of muscular insertion and generally move these body segments. In particular these forces tend to rotate the two body segments to which they are applied by acting on them as a couple (in Poinsot's sense). One force is applied directly to the body segment whereas the other, equal and opposite to the first, is exerted at the centre of the joint that directly or indirectly connects this body segment with that bearing the other muscular insertion. The force which is not applied to the body segment itself exerts compression on the centre of the connecting joint that is as great as if the force were directly applied at the centre of this joint.

The points of application of the two muscular forces coincide with the points of insertion or the centres of the insertion areas only when the pulling action of the muscle is direct. This is not the case when the muscle, as mostly occurs, is deflected from its direct course by an osseous prominence, a ligament or a tendinous sheath. Those points on the obstacles between which the muscle is unhindered in its course from one body segment to the other must then be regarded as the points of application of the muscular force. The length of this section of muscle is irrelevant, likewise whether or not all the muscular fibres insert on this part of the muscle. Even if there are no muscular fibres but only part of the terminal tendon in this area, the points of application and the directions of the two muscular forces depend only on the part of the tendon which extends unhindered between the two body segments. This is the case for most of the muscles of the wrist and joints of the fingers and toes. Thus, the engine, so to speak, lies at some distance from the place where the work is carried out and the force has to be transmitted there, as often occurs in machinery.

With respect to a polyarticular muscle, the two muscular forces generally exert a torque on the body segments on which the muscle is not directly inserted but which lie between the two insertion areas. Each of the two equal and opposite muscular forces acts on the centre of the joint of the body segment in question, adjacent to the point of insertion, as if the force were applied directly to this joint centre with the same magnitude and direction. If the muscle is deflected several times between its insertions, as occurs with the long flexors and extensors of the fingers, it usually spans the various joints in different directions. The whole muscle then acts as a chain of muscles linked to each other so that muscular contraction occurs in different directions but each section of muscle possesses the same tension.

As a result of their active or passive tension, the muscles generally tend to rotate all the body segments lying between their insertions by acting as couples.

It can also be demonstrated that the external forces, gravity and reaction forces such as ground reaction force and friction, when they tend to rotate a body segment, act on the latter as couples. This will be explained further with particular examples.

The couples resulting from the internal and external forces generally rotate the body segments on which they act. There is no rotation when all the couples acting on a body segment balance each other. The whole body is then immobile as long as it is in contact with the ground or some other fixed object, and not oscillating freely in the air for a short time as in jumping. If the limbs cannot turn, the body can only move as a whole, all of its parts maintaining the same acceleration and velocity. However, this is impossible as soon as one point of the body, for instance part of the foot, contacts the ground and, consequently, cannot take part in the movement.

The body is assumed to be immobile initially. If a certain muscle, stimulated by a nerve, increases its tension, the moments of all the couples with which this muscle acts on the body segments between its insertions increase. In other words, new couples are added to the previously existing ones. They correspond to the change in tension of the muscle. The equilibrium between the couples acting on the body segments is disrupted and there is movement.

The type and course of this movement, however, depend not only on the moments of the new couples but also on the magnitude and distribution of the masses inside the body segments, that is on the position of their centres of gravity and on their moments of inertia. Above all, they depend on the position of every body segment within the body and in particular on the type of joint connection with the other body segments. Finally, the movement will also be modified 1. by the change in position of other muscles relative to the joints which results from the movement of the joints, 2. by the change in tension as the muscles become passively stretched or shortened and 3. by the change in the couples generated by the external forces, which generally occurs in addition to the tridimensional change in position of the different body segments.

Whatever happens, the movement will not usually be restricted to the body segments which lie between the insertions of the stimulated muscle the couples of which undergo primary alteration. As a rule, other body segments will also be involved in the movement. This is a general rule and it is irrelevant whether or not the contraction of the muscle entails a considerable change in the couples of the other muscles and external forces. Not only parallel displacement of the adjacent body segments occurs, as one would expect, but generally also rotation. A body segment, by its movement, exerts a compression on the other body segment with which it it articulated, so that the articulated extremity takes part in the movement. This compression, however, acts on the adjacent body segment as an external force which usually tends to rotate the body.

Consequently, a muscle acts not only on the body segments and joints which it spans, but by its contraction also causes movement of other segments which appear to be completely beyond its sphere of action. I analysed and demonstrated this with particular examples in an earlier study[1]

If the tone of several muscles is simultaneously changed voluntarily, or if a certain number of muscles are innervated so that the couples resulting from the internal (passive and active) and external forces and acting on each body segment do not exactly balance each other, the different body segments rotate precisely as if only one couple were acting on each of them. This couple is equal to the resultant of all the couples applied to this body segment.

Thus it becomes easy to understand how it is possible to change the compression of the feet on the ground and hence the ground reaction force and frictional resistance. It is only necessary to move the feet in a given direction by contracting the appropriate muscles. As long as the acceleration of the feet does not go beyond a certain limit, this movement is hindered either by the ground reaction force and frictional resistance or by gravity and frictional resistance, depending on whether the acceleration imparted to the feet is directed towards the ground or away from it. The effect of muscular action at first consists of a change in the compression exerted on the ground, which in turn provokes a corresponding modification of the ground reaction force and friction and, consequently, a change in the acceleration of the total centre of gravity.

The compression exerted on the ground by the feet at each moment during walking represents the final visible effect of muscular effort in locomotion. It obviously depends on the constantly varying tension of the muscles during the movement. Conversely, knowing the compression alone does not, of course, provide any information as to the activity of the different muscles during walking. The same compression can be exerted on the ground in an infinite number of ways by contracting different individual muscles or combining the activity of various groups of muscles. This problem can be approached only if the movements which the muscles enforce on the different body segments during walking are precisely known, that is to say, a thorough understanding of the whole process of gait is required. It is not sufficient to establish the successive positions of the body and the movement of the total centre of gravity. The angular velocities and angular accelerations of the different body segments at the successive phases, the movements of the partial centres of gravity, etc. must also be determined.

Well determined relationships exist between all the values characteristic of the law of movement in walking, or any other movement of the human body, and the internal and external forces acting on the body. They are expressed in the differential equations of the movement. These differential equations should, in principle, enable the activity of the muscles to be deduced from the process of movement. However, the rotations of the different body segments depend only on the couples which result from the combination of all the couples applied to each body segment. Therefore, the differential equations of the movement can only give the moments of the resultant couples for each body segment. The resultant couples must then be resolved into their components to determine the tension of the different muscles and above all the involvement of the different muscles in walking. To this end, it is, of course, important to know exactly with which torques the external forces and each muscle act on the various body segments in the different successive positions of the body during gait.

---

[1]   Fischer O (1895) Beiträge zu einer Muskeldynamik: I. Ueber die Wirkungsweise eingelenkiger Muskeln. Abhandlungen der Königl. Sächsischen Gesellschaft der Wissenschaften, mathematisch-physische Klasse, vol 22, no 2

This indicates roughly how the study must proceed. Owing to the complex mechanical circumstances presented by the human body with its many parts, this approach appears extremely difficult. However, I am convinced that it is the only way which can lead to the goal. I do not believe that one can obtain absolutely reliable results concerning the activity of the muscles during walking and other movements of the human body without taking into account the equations of the movement. I do not deny the value of the numerous studies on the action of the different muscles by applying local faradism to the muscles, stimulations of determined aspects of convolutions of the brain or observations of abnormal movements in pathological conditions, etc. These investigations and many theoretical considerations have often thrown light on the function of muscles in particular circumstances. However, one must always remember that in most cases the results concern one special instance and cannot be applied as such to another case. Moreover, as mentioned above, the contraction of a muscle generally not only moves the body segment in which the muscle is inserted but has consequences far beyond this narrow range. The result of the contraction, even if carried out when the body is at rest, essentially depends on the position of the body. This position indeed determines the mechanical reaction of the body segments to the acting forces. For example, one cannot determine the action of the gastrocnemius muscle by faradizing the muscle when the leg is hanging from the trunk. In muscular contraction when the foot is pressing on the ground, the mechanical circumstances are completely different and the action of the muscle also differs. It is of particular importance here whether or not the other foot is also pressing on the ground. Strictly speaking, the action of the muscle changes for each new position of the body. The different instances that would have to be considered in a complete study on the function of a muscle are so numerous that it would be impossible to examine them all empirically. However, this would be a sine qua non if one were to extend the results of such experimental investigations to complex movements of the human body that involve continuously varying positions of the joints, as in walking.

Clinical observation, the analysis of impaired movements due to palsy or atrophy of individual muscles and groups of muscles cannot throw sufficient light on the activity of the muscles in normal circumstances. Such observations can give a general idea of the movements which a muscle induces but they do not provide information for instance on the tone which a muscle must display to influence the movement in a particular manner.

Even if it were possible to determine the action of the different muscles for each position of the body, this would not yet provide a sufficient basis for establishing the muscular activity in walking and many other movements of the human body. The effect of contraction of a muscle also depends on the actual movement of the various body segments. There is a difference in muscular contraction according to whether it occurs during a movement or in a position of rest. Every movement of a body segment usually also provokes movement of the other segments. In certain circumstances, the same movement that a muscle can induce in a joint previously at rest may result from external forces acting on a body segment at some distance from the joint. It would thus be a mistake to conclude that the movement of the joint is due to a particular muscle the contraction of which is known to produce this movement.

It is necessary, therefore, to know the whole process of the movement, i.e. how the different segments of the human body influence each other in their movements, how the action of a muscle is changed by the position and movement of the whole body, by the simultaneous activity of other muscles or by the activity of external forces, in order to establish reliably the role of individual muscles or main groups of muscles in the performance of complex movements of the whole body. This knowledge can be obtained only by means of the differential equations of the movements (movement equations). These are the complete expression of the relationships between the movement of the whole body, the successive positions of the different limbs, the velocities and accelerations of their movement, the mechanical behaviour of the body segments characterized by the position of their centres of gravity, their masses, moments of inertia and connecting joints on the one hand and all the external and internal forces acting on the body on the other. The movement equations cannot of course replace experimental research into the function of the muscles using anatomical specimens and living subjects. In fact they are based partly on the results of these experiments.

Because of the extremely complicated mechanical circumstances presented by the human body which consists of many articulated segments, the movement equations for walking are long and complex. The number of such equations is great even if the body is only divided into its large segments and if the segments are assumed to behave as rigid masses during the movement, which is not exactly true. In a previous study [2], I demonstrated a method for putting the movement equations into a relatively simple form. I deduced these equations, assuming plane movements of the human body formed by 12 articulated segments. The lower extremities were each assumed to consist of three rigid segments, the upper extremities of two. The head was considered to be articulated with the trunk by one joint. Such a resolution is sufficient at a first approximation in the analysis of the movements of locomotion.

The movement equations provide valuable information for further research. They throw light on the reciprocal influence of the limbs in movement. Moreover, each movement equation can easily be applied both to movements in one plane and more generally to any kind of movement.

The particular case of a movement in one plane will first be considered. It is assumed that all the limbs move parallel to one plane, the vertical plane of gait. This is not exactly what occurs in walking since the body also oscillates sideways, perpendicular to the plane of gait, always displacing the total centre of gravity towards the side of the supporting foot. This transverse displacement, however, is much smaller than the movement parallel to the plane of gait.

During the movement in one plane, the position of the body is at all times clearly determined by the rectangular co-ordinates of any point of the body in the plane of gait and by as many positional angles as the number of segments into which the body has been resolved. In the present case, there would be 2 co-ordinates and 12 positional angles. The total centre of gravity can be conveniently chosen as the point the co-ordinates of which are to be established. The position of each limb can be determined by the angle that the long axis of this limb forms with the vertical. In addition to the 2 co-ordinates and 12 positional angles, the velocities and accelerations must also be considered in order to describe completely the state of movement of the body. The two velocities with which the total centre of gravity moves in the direction of the two rectangular axes of co-ordinates and the 12 angular velocities with which the long axes of the 12 body segments change their direction in space have to be established. Similarly, the components of the acceleration of the total centre of gravity in the directions of the two axes of co-ordinates must be known as well as the 12 angular accelerations of the rotation of the 12 body segments. The movement equations then represent the relationship which exist between the 2 co-ordinates, the 12 positional angles, the 14 velocities and the 14 accelerations on one hand and the external and internal forces on the other.

The number of movement equations is generally also 14. Twelve of these comprise all the positional angles, angular velocities, angular accelerations and couples of forces acting on the body segments. Each of the two remaining movement equations comprises one of the two components of the acceleration of the total centre of gravity and the sum of the components of the external forces acting in the same direction as the component of acceleration. When a point of the body maintains its position in space for some time, as when the forward foot rotates about a point of its heel, the movement of the total centre of gravity can be deduced

---

[2]   Fischer O (1894) Abhandlungen der Königl. Sächsischen Gesellschaft der Wissenschaften, mathematisch-physische Klasse, vol 20, no 1

from the rotation of the 12 body segments. The two differential equations of the movement of the total centre of gravity are essentially a consequence of the 12 other movement equations and can thus be excluded as independent equations. When a whole body segment remains immobile for some time, e.g. the supporting foot during foot-flat, the number of equations is further reduced by one. Only 11 equations are then required to represent the state of movement of the body relative to the forces.

The two equations of the movement of the total centre of gravity are simple and need no further discussion. They are the mathematical expression of the fact that the total centre of gravity of a body or of a system of bodies always moves as if the mass of the whole system were concentrated at this centre and as if all the external forces were applied to it. The law is also generally true for any tridimensional movement of the total centre of gravity and was used in Chap. 2 to deduce the magnitude of the ground reaction force and frictional resistance from the accelerations of the total centre of gravity.

The other 12 equations are less obvious at first sight. Each of them deals with the movement of all the 12 segments of the body. However, in each of these equations one body segment is given particular consideration: the moment of inertia of this one segment and only the couples acting on it appear in the equation. Precise examination of the different expressions appearing in the equations shows that each of the 12 equations can be regarded essentially as the movement equation of a single body segment. The velocities and accelerations of all the other segments are taken into account only as far as they affect the compression or tension on this segment and thus modify its movement.

Each body segment is affected by the movement of the others due to the movement of the centres of gravity of all the systems of segments with which the first one is articulated. One body segment acts on another because by its movements it influences the movement of the centre of gravity of the articulated system of body segments to which it belongs. For example, the movement of the left lower leg affects the trunk by influencing the movement of the centre of gravity of the whole left leg. The movement of the head can exert an effect on the rotation of the right thigh by influencing the movement of the centre of gravity of the trunk + head + arms + left leg system.

The centre of gravity of such a system of segments is, of course, subject to the same laws as the centre of gravity of the whole body. It moves as if all the external forces acting on the system were applied to it and as if the masses of all the segments of the system were concentrated in it. The external forces initially comprise gravity, air resistance and, if one or both feet resting on the ground belong to the system, ground reaction force and frictional resistance. However, other forces are involved which can be considered as external forces for the system. Every muscle that has only one insertion in the system will act on it as an external force as a result of its tension at the insertion. Moreover, compression (or tension) is generally exerted at the centre of each joint that links the system with other parts of the body, and this compression also acts as an external force. Its magnitude and direction at any time must be such that together with all the other external forces, it imparts to the centre of gravity of the system the acceleration which it displays at this time.

According to the law of equality of action and reaction, the system in turn exerts an equal and opposite compression on the adjacent body segment at each connecting joint. This compression exerted by the system on the adjacent body segment, as a result of the movement of its centre of gravity and of the external forces applied to it, represents the influence exerted by the system on the movement of the body segment under consideration.

The same compression in magnitude and direction is obtained in another way. If the system is thought of as separate from the rest of the body, there usually remains another system which when added to the first reconstitutes the whole body. This latter system can be reduced to one body segment. Compression must be exerted on the complementary system by the first system through the centre of the connecting joint. The magnitude of this compression is such that, combined with all the external forces acting on the complementary system, it gives the centre of gravity of this system its acceleration. It will be demonstrated below that, taking this fact into account, one arrives at the same compression in the joint as with the law of the equality of action and reaction.

Let us consider all the compression effects acting on the joints of a body segment and add them to the forces applied directly to this segment. The segment subjected to all these forces moves as if it were no longer connected with the rest of the body. For a plane movement the movement equation of this body segment is one of the 12 differential equations mentioned above. The equations of each body segment for any movement of the human body can thus be reduced to the simple equations of the movement of a rigid body.

The right thigh will be considered as an example. The influence of air resistance and of the very slight friction in the joints can be ignored. The thigh thus moves as a free body. The following forces act on it:

1. A force applied at its centre of gravity equal to its weight and directed vertically downwards.
2. At each muscular insertion, a force equal to the tension of the muscle and acting towards the other insertion. The direction of the tension and position of the insertion at the thigh depend on that part of the muscle extending freely and directly to the adjacent body segment (cf. pp. 205–206).
3. A compressive force acting at the centre of the hip joint, equal and opposite to the compression exerted by the thigh on the trunk + head + arms + left leg system. This force in magnitude and direction is such that, combined with the other external forces acting on this system, it imparts to the partial centre of gravity of the system its acceleration. It must be considered whether the left foot is on the ground or not. If the left foot is on the ground, the ground reaction force and frictional resistance produced by the compression of the left foot on the ground must be included among the external forces. The same results are obtained if the compressive force acting on the thigh at the centre of the hip has such a magnitude and direction that, combined with all the external forces applied at the right leg, it imparts to the centre of gravity of the latter its acceleration. Here, it is necessary to know whether the right foot rests on the ground or not. If the right foot is on the ground, the ground reaction force and frictional resistance must be counted among the external forces acting on the leg.
4. A compressive force acting on the thigh at the centre of the knee. Its magnitude and direction are equal and opposite to the compression exerted by the thigh on the right lower leg + right foot system. This compression is such that, combined with all the external forces acting on the right lower leg and foot, it imparts to the centre of gravity of this system its acceleration. The magnitude and direction of the compression acting on the right thigh at the knee can also be considered such that this compression, combined with all the external forces acting on the whole body minus those acting on the right lower leg and foot, imparts to the partial centre of gravity of the system (body − right lower leg − right foot) its acceleration.

Subjected to these four types of external forces, the thigh will move as a rigid free body. Its centre of gravity will move as if the whole mass of the thigh were concentrated in it and the external forces were applied to it. Simultaneously, the thigh will rotate about its centre of gravity as if the latter were fixed and the torques of the external forces were acting about axes passing through it.

The compression in the hip and knee depends on the absolute movement of the centres of gravity of the adjacent systems of body segments. This movement is the combination of two movements, one relative to the centre of the connecting joint, the other coinciding with the movement of this joint centre. Each of the two movements causes a particular type of compression in the joint.

We consider first the movement of the centre of gravity of the system which is the same as the movement of the centre of the joint, for example the movement component of the centre of gravity of the trunk + head + arms + left leg system which is equal to the movement of the centre of the right hip. This component acts on the proximal extremity of the right thigh as if the whole mass of this system were concentrated at the centre of the hip and all the external forces acting on the system were applied to this centre. Then the centre of gravity of the system would coincide with the centre of the hip and move exactly like the latter. Similarly, the movement component of the common centre of gravity of the right lower leg and foot, which coincide with the movement of the centre of the knee must act on the distal extremity of the right thigh as if the masses of the right lower leg and foot were concentrated at the centre of the knee and all the external forces acting on these two body segments were applied to the centre of the knee. If the masses of the adjacent systems are thought of as concentrated at the centres of the hip and knee joints, the thigh is loaded by these masses. Then, only the compressive forces remain which are induced by the relative movement of each centre of gravity of the two adjacent systems relative to the centre of the connecting joint under consideration. In an earlier paper I termed this system of masses the "reduced thigh system" [3]. Its centre of gravity is the point on the long axis of the thigh, designated as "principal point of the thigh" (see p. 129). This system of masses moves as a rigid body under the action of the compressions on the two joints, resulting from the relative movement of the two partial centres of gravity, and under the action of all the other external forces exerted partly at the centres of the joints and partly at other points of the thigh. The mass of this body is equal to the mass of the whole human body and its moments of inertia are essentially different from those of the thigh.

Introduction of the reduced thigh system and the other corresponding reduced systems results in much simpler movement equations, because now the movements of the other body segments only appear in these equations as far as they influence the relative movements of the partial centres of gravity in relation to the centres of the connecting joints.

These are, in principle, the methods which enable the movement equations to be obtained in an elementary manner, without going through the general differential equations of Lagrange. It only remains to express the compression exerted on each articular extremity of the body segment under consideration by the acceleration of the centre of gravity of the adjacent system and the external forces acting on the latter. This compression is the key to deducing the movement equations.

If the acceleration of the centre of gravity of a system of mass m is γ, the resultant of all the external forces (including all the reaction forces), displaced parallel to itself toward the centre of gravity, must be of the magnitude $m\gamma$ and have the same direction as γ. This force designated as the effective force by d'Alembert must, therefore, represent the geometric sum of all the external forces. The compression exerted on the system at its connecting joint, which is part of the external forces, must represent the geometric difference between the effective force $m\gamma$ and the geometric sum of all the other external forces. Therefore, action and reaction being equal, the compression exerted by the system on the adjacent body segment is equal to the geometric difference between the geometric sum of all the other external forces acting on the system and the effective force. In other words, the effective force is the geometric sum of all the other external forces and a force $-m\gamma$ (inertia force) equal in magnitude but opposite in direction to the effective force. The compression exerted on the articular extremity of any body segment by a system as a result of the movement of the centre of gravity of the latter and the external forces exerted on the system is obtained by displacing all the external forces of the system (except the joint compression) and a force $-m\gamma$ equal

[3] Fischer O (1894) Die Arbeit der Muskeln und die lebendige Kraft des menschlichen Körpers. Abhandlungen der Königl. Sächsischen Gesellschaft der Wissenschaften, mathematisch-physische Klasse, vol 20, no 1, p 61

and opposite to the effective force, parallel to themselves toward the centre of the connecting joint and by combining these forces into their resultant. This is in accordance with the fact that any force which is not applied directly to a body segment acts on the latter as if it were applied to the centre of the joint of the body segment that is closest to its actual point of application within the human body. Another external force must then be imagined which is applied to the centre of gravity of the system, equal and opposite to the effective force. This immediately indicates the total compression on the articular extremity of a body segment.

This rule can be applied to the right thigh. As a result of its connection with the other body segments, the centre of the proximal extremity of the femur in the hip joint undergoes a compression as great as if all the external forces applied to the trunk + head + arms + left leg system, including the ground reaction force and frictional resistance exerted on the left foot, were applied to this point (with their magnitude and direction) and as if an additional force were also acting. The magnitude of this additional force would be the product of the acceleration of the centre of gravity and the mass of this system hanging from the hip, and its direction would be opposite to that of the acceleration of the centre of gravity. Moreover, a compression is exerted on the other extremity of the right thigh, at the centre of the knee. This is as great as if all the external forces applied to the right lower leg and foot, including any ground reaction force and frictional resistance, were applied to the centre of the knee with the same magnitude and direction, and as if this point were also subject to an additional force. The direction of this additional force is opposite to that of the acceleration of the common centre of gravity of the lower leg and foot and its magnitude is the product of the acceleration of this centre of gravity and the sum of the masses of the lower leg and foot.

The compression exerted on the proximal extremity of the right thigh can also be determined in another way. Combined with the external forces acting on the right leg, this compression imparts to the centre of gravity of the right leg its acceleration. This acceleration is designated as γ' and the mass of the whole leg as m'. The compression is the geometric difference between the effective force $m'\gamma'$ and the geometric sum of the external forces acting on the right leg, including any ground reaction force and frictional resistance.

It can be easily demonstrated that the compression on the thigh established by the latter method does not differ from that obtained by the former. $\Sigma(K)$ signifies the geometric sum of the external forces acting on the trunk + head + arms + left leg system, excluding the compression on the hip. The geometric sum of the external forces acting on the right leg, excluding the compression on the hip, is designated as $\Sigma(K')$. The compression exerted on the thigh at the centre of the hip joint is then either $[\Sigma(K)] - (m\gamma)$ or $(m'\gamma') - [\Sigma(K')]$. The brackets indicate geometric addition or subtraction. The geometric sum of the effective forces of the two complementary systems is equal to the effective force of the whole body. If the acceleration of the centre of gravity of the whole body is γ_0 and the mass of the whole human body m_0, we have:

$$(m\gamma) + (m'\gamma') = m_0\gamma_0$$

The effective force $m_0\gamma_0$ is also equal to the resultant of all the external forces displaced to the total centre of gravity, in other words, their geometric sum. This is equal to the geometric sum of $\Sigma(K)$ and $\Sigma(K')$. When, among the external forces of one of the two systems, there are some which do not represent external forces for the whole body, such as the tension of the muscles which bridge the two systems, these forces exert no influence on the value of the geometric sum of $\Sigma(K)$ and $\Sigma(K')$ since each of these forces acting on one system is balanced by the equal and opposite force acting on the other. Therefore, the effective force $m_0\gamma_0$ can be replaced by the geometric sum $[\Sigma(K)] + [\Sigma(K')]$. The previous equation can thus be written:

$$(m\gamma) + (m'\gamma') = [\Sigma(K)] + [\Sigma(K')]$$

Consequently, $(m'\gamma') - [\Sigma(K')] = [\Sigma(K)] - (m\gamma)$. This equation is generally valid for the compressive forces on any joint. Circumstances dictate which of the two geometric differences will be used in a given case. Generally, of course, preference will be given to the form for which data have already been determined. For example, if the acceleration of the centre of gravity of the right leg is known, but not that of the other system, the form $(m'\gamma') - [\Sigma(K')]$ will be chosen.

It can be easily demonstrated that the muscles and external forces tend to rotate the different bodies by acting as couples in the Poinsot manner.

The muscles which act directly on a body segment either have one insertion in this segment or they are inserted in two body segments adjacent to the first. For muscles of the former type, the force that is equal and opposite to the muscular force applied directly to the body segment is found among the components of the compression acting on the centre of one joint of the body segment. For muscles of the latter type, the two equal and opposite forces forming the couple act as compressive components at different joints limiting the body segment. All the couples resulting from the muscles acting directly on a body segment can be combined into one resultant couple the moment of which will be designated as the resultant torque of the muscles.

Gravity acts with several components on each body segment. First, it pulls the partial centre of gravity vertically downwards with a force equal to the weight of the body segment. Then a component equal to the weight of the adjacent system and directed vertically downwards takes part in the compression exerted on the centre of each of the joints at the ends of the body segment. All these forces of gravity can be combined into one resultant which is equal to the weight of the human body. The point of application of this resultant is the principal point of the body segment under consideration. When the body is not free in space, this weight is always balanced by an equal and opposite force, and the two form a couple. The second force of the couple is provided by the ground reaction force. If the body is supported by one foot, the ground reaction force can be resolved into two components: one represents the vertical component of the effective force $m_0\gamma_0$; the other is equal to the body weight but is directed vertically upwards. The latter component must always be one of the components of compression acting at the centre of the joint which within the body is closest to the supporting foot. If, for instance, only the right foot is on the ground, the compressive component equal and opposite to the weight of the body acts at the centre of the right knee; if the left foot only is on the ground, it acts on the right thigh at the centre of the hip joint. If both feet are on the ground, none of the components of the ground reaction force is generally equal and opposite to the weight of the body; only a component of the ground reaction force to the body weight normally acts on each foot. In any case, the sum of the two vertical components directed upwards is equal and opposite to the weight of the human body. Each of these two components acts as compression on another joint of the body segment. Since the two compressive forces have the same direction, they can be combined into one resultant which is equal to the body weight, directed vertically upwards and applied to a particular point in the body segment under consideration. This resultant forms a couple with the weight acting on the principal point of the body segment. Since the component of the ground reaction force is also induced by gravity, the moment of the couple with which this component and gravity act on a body segment in a given case will be termed the torque of gravity.

The resultant torque of the muscles of a body segment and that of gravity give a clear idea of the influence which muscles and gravity exert on the movement of an individual body segment. Except for air resistance, it only remains to consider the forces due to the accelerations of the partial and total centres of gravity. These also form couples which tend to rotate the body segment. For example, for the right thigh, the force $-m\gamma$ is one of the components of compression acting on the centre of the right hip. m is the mass of the trunk + head + arms + left leg system and γ is the acceleration of the corresponding centre of gravity. The negative sign signifies that the force is opposite to the acceleration γ. The force $-m''\gamma''$ acts on the centre of the knee. m'' signifies the mass of the right lower leg + foot system and γ'' the acceleration of its centre of gravity. These two components of compression are always present even when the body is free in space. If one foot contacts the ground, the additional compressive force $m_0\gamma_0$ acts on the knee or hip depending on whether the right or left foot is on the ground. This force results from the reaction and frictional resistance of the ground. m_0 again signifies the mass of the whole body, and γ_0 the acceleration of the total centre of gravity. If both feet are on the ground, the compressive force $m_0\gamma_0$ is generally divided into two components, one acting on the knee, the other on the hip. This force results partly from the ground reaction force, partly from friction with the ground.

If only the right foot rests on the ground, the two components $m_0\gamma_0$ and $-m''\gamma''$ act on the thigh at the centre of the right knee. The resultant of these two forces is the effective force of the trunk + head + arms + left leg + right thigh system. This system complements the right lower leg + foot system to reconstitute the whole body. This effective force can be resolved into two components $m\gamma$ and $m_2\gamma_2$ in which m and γ are as above, m_2 is the mass of the right thigh and γ_2 the acceleration of its centre of gravity. The component of compression $m\gamma$ applied to the knee is equal and opposite to $-m\gamma$ acting on the hip, and the two form a couple. Only the component of compression $m_2\gamma_2$ acting at the centre of the right knee does not appear to be part of a couple. The couples cannot influence the movement of the centre of gravity of a body because one of the two forces will always give the centre of gravity an opposite acceleration to the other. Therefore, the centre of gravity of the right thigh must always move as if only the force $m_2\gamma_2$ were acting on it and as if the mass m_2 of the thigh were concentrated at this centre of gravity. This is in accordance with the fact that γ_2 represents its acceleration. Besides accelerating the centre of gravity, the compressive force $m_2\gamma_2$ acting at the centre of the knee joint will also tend to rotate the thigh if its direction does not pass exactly through the centre of gravity S_2 of the thigh. This tendency of the force to rotate the thigh can finally be considered as the moment of a couple. It is only necessary to imagine the effective force $m_2\gamma_2$ and the equal and opposite force $-m_2\gamma_2$ as applied to the centre of gravity S_2. These two forces together cannot change the movement of the thigh since they balance each other. The force $-m_2\gamma_2$ acting at the centre of the knee and the equal and opposite component of compression $+m_2\gamma_2$ can be regarded as a couple. Then, only the effective force $m_2\gamma_2$ remains. It gives the thigh a translation movement of acceleration γ_2, i.e. a displacement without rotation, whereas the two couples $(+m\gamma, -m\gamma)$ and $(+m_2\gamma_2, -m_2\gamma_2)$ tend only to rotate the body segment in opposite directions about an axis passing through the centre of gravity S_2. These two couples can be combined into one the moment of which will be termed the resultant torque of the effective forces.

If the left foot only contacts the ground, again the component of compression $-m''\gamma''$ acts at the centre of the knee, but now the component $m_0\gamma_0$ due to the ground reaction force and frictional resistance must be added to the component $-m\gamma$ exerted on the right hip. The resultant of these two forces is the effective force for the whole right leg, designated above (p. 213) as $m'\gamma'$. This can be resolved into the two components $m''\gamma''$ and $m_2\gamma_2$. The component of compression $m''\gamma''$ forms a couple with the equal and opposite component of compression $-m''\gamma''$ acting on the centre of the knee. If the effective force $m_2\gamma_2$ and the equal and opposite force $-m_2\gamma_2$ are thought of as acting on the centre of gravity S_2 of the thigh, $-m_2\gamma_2$ can be imagined to form a couple with the component of compression $m_2\gamma_2$ acting at the centre of the hip. Again, only the effective force exerted at S_2 remains, which tends to impart a translation movement of acceleration γ_2 to the thigh. The two couples $(+m''\gamma'', -m''\gamma'')$ and $(+m_2\gamma_2, -m_2\gamma_2)$ tend only to rotate the thigh about an axis passing through its centre of gravity. Combining their moments gives the resultant torque of the effective forces. The moment of the second couple $(+m_2\gamma_2, -m_2\gamma_2)$ is not identical with that designated in the same way in the previous instance where one of the forces of the couple acted at the centre of gravity S_2 and the other at the centre of the knee. Now, one is applied to the centre of gravity S_2 and the other to the centre of the hip.

Finally if both feet contact the ground, forces are added to the two components of compression $-m\gamma$ and $-m''\gamma''$ acting at the centres of the two joints. These two additional forces are such that, displaced to one point, they give the resultant force $m_0\gamma_0$. In order to illustrate the couples in this instance, the six forces $+m\gamma$, $-m\gamma$, $+m''\gamma''$, $-m''\gamma''$, $+m_0\gamma_0$ and $-m_0\gamma_0$ in opposite pairs can be imagined applied to the centre of gravity S_2 in such a way that they balance each other. Of these forces, the first forms a couple with the component of compression $-m\gamma$ acting at the centre of the hip, and the third a couple with the component $-m''\gamma''$ acting at the centre of the knee. If the sixth is resolved into two components equal and opposite to the components of compression due to the ground reaction force and frictional resistance and acting at the hip and knee, two more couples are obtained. The four couples can be combined into one resultant the moment of which represents the resultant torque of the effective forces. The three remaining forces acting at the centre of gravity S_2, $-m\gamma$, $-m''\gamma''$ and $+m_0\gamma_0$, together give the acceleration force $m_2\gamma_2$ which tends to impart a movement of translation of acceleration γ_2 to the thigh.

Having determined the couples which influence the rotation of a body segment and which result from all the external forces acting on this body segment, it is now possible to write the equations of the movement of the body segment in the simplest form. In a plane movement, all the body segments move in a plane parallel to the plane of gait and the planes in which all the couples act are also parallel to the plane of gait. The equation of the movement of each body segment can be expressed in the following simple form:

The product of the moment of inertia of the body segment in relation to the axis perpendicular to the plane of gait and passing through its centre of gravity, and of the angular acceleration of the body segment about this axis is equal to the sum of the resultant torque of the muscles, the resultant torque of gravity and the resultant torque of the effective forces.

This equation is valid generally for each segment of the human body, even if this segment is, like the trunk, linked to other segments by more than two joints. In such instances, the considerations applied to the particular example of the thigh and which led to the torque of the effective forces, have only to be generalized. The equation is also valid when the movement is not plane. In this more common occurrence, it can be applied for each of the projections of the movement on one of the three planes of co-ordinates.

This clearly describes the way which has to be followed for further analysis of the gait, based on the precise recording of the movement as presented in Chap. 1.

Starting from the co-ordinates of the joint centres, the successive tridimensional positions of each body segment, the angular velocities with which the body segments modify their direction and the angular accelerations of these rotations at each phase must be found. The moments of inertia of the different body segments of the subject can be deduced from previous studies [4]. The successive values of one side of the movement equation of each body segment, i.e. the products of the moments of inertia and angular accelerations, can thus be given. From the successive positions of the different body segments, the successive values of the "torques of gravity" are found. Then the velocities and accelerations of the movement of each of these centres of gravity are deduced from the co-ordinates of the different centres of gravity in Chap. 2. The values of the resultant torque of the effective forces for each body segment can then be calculated. The values of all the expressions involved in the movement equation are obtained with the exception of the "resultant torque of the muscles". Therefore,

---

[4]   cf. Braune W, Fischer O (1892) Bestimmung der Trägheitsmomente des menschlichen Körpers und seiner Glieder. Abhandlungen der Königl. Sächsischen Gesellschaft der Wissenschaften, mathematisch-physische Klasse, vol 18, no 8

the movement equations can be used to calculate the value of the resultant torque of the muscles for each body segment and every moment of the movement. The necessary basis for further research into the activity of the muscles in human gait is thus provided.

Further studies will deal with the movement of the lower extremities in the way thus indicated. The remainder of this chapter will provide an overall view of the successive positions of the legs and the different body segments. Later (Chap. 5), the angular accelerations of the segments of the lower extremities as well as the accelerations of their centres of gravity will be deduced. The movement equations will also be used to determine the values of the resultant torques of the muscles acting on the different body segments.

General Behaviour of the Legs

In the course of a double step, each leg displays a twofold behaviour. Over a period longer than a step, the leg stands on the ground, exerting compression. During a second period much shorter than a step since the first period is longer than a step, the leg leaves the ground and is suspended from the pelvis, swinging forwards about the centre of the hip. It is not yet clear whether this oscillation of the leg occurs in the manner of a pendulum subject only to gravity or whether the muscles play a significant role. Precise determination of the laws of the movement carried out by the different segments of the leg during this oscillation is alone able to provide the basis for a satisfactory solution to this problem by means of the movement equations.

During both the stance and the swinging periods the leg continuously changes its shape. This is apparent from Figs. 22 and 23 of Chap. 1. It appears even more clearly in Fig. 1 b of this chapter, since the contours of the bones of the different body segments are shown and the cross-hatching of the bones during the swinging period makes it easier to distinguish the two periods. For clarity, the lower leg is represented only by the tibia, and the patella has been omitted. The presentation of Fig. 1 b offers a further advantage over Fig. 22 and 23 of Chap. 1 in that it does not display the chance phase division produced by photographically recording the experiments, where the intervals do not relate precisely to the duration of the step. Rather, it shows phases obtained by dividing the duration of a double step exactly into 20 equal parts. Everything belonging to the right extremity is shown black and everything belonging to the left extremity appears red. Phase 1 corresponds to right toe-off. Phase 21 is a repetition of phase 1 from which it is separated by the length of a double step. Symmetrical behaviour of the two legs is assumed. Therefore, phase 11 corresponds to left toe-off. Right heel strike occurs between phases 9 and 10, somewhat closer to phase 9 than to phase 10. Left heel strike occurs between phases 19 and 20.

Representation of the movement of the whole body, rather than just the movement of the lower extremities, in 20 phases was adopted in a previous model of gait [5]. The 20 phases were obtained as follows.

The position of the moving body at any time can be accurately determined using the displacement curves of Chap. 2. The displacement curves give the co-ordinates of any point of the human body with respect to time. The figures in the abscissa are proportional to the time elapsed whereas the values of the co-ordinates under consideration are entered as ordinates. In the displacement curve for each of the three co-ordinates for all the joint centres of the lower extremities, 1 dm represents 1 s, as in the displacement curves of the total centre of gravity. The co-ordinates can be measured on each displacement curve for moments separated by an interval of 1/1000 s. In experiment I a douple step lasted 0.990 s and was thus

[5]   Fischer O (1895) Beschreibung eines neuen Modells zur Veranschaulichung der Bewegungen beim Gange des Menschen. Archiv für Anatomie und Physiologie, Anatomische Abteilung, p 257

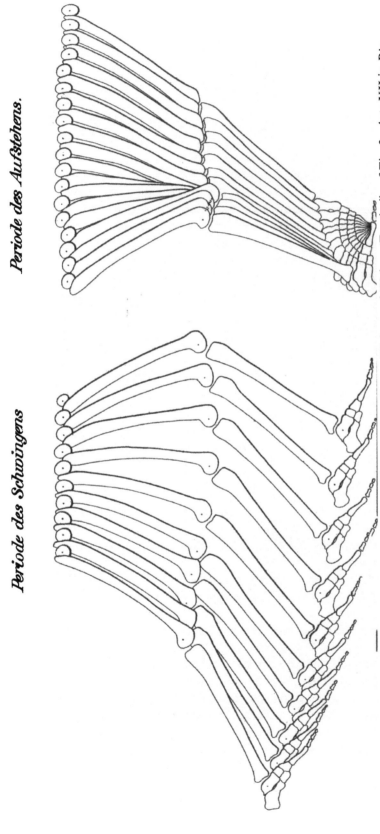

Periode des Aufstehens.

Periode des Schwingens

Fig. 1. a Successive positions of the right leg during one double step, according to Weber and Weber. (Facsimile of Fig. 3, plate XII in *Die Mechanik der menschlichen Gehwerkzeuge*). **b** Successive positions of both legs during one double step. (The right leg is shown in black and the left in red. The phases of movement making up the swing period for each leg are distinguished from the others by cross-hatching. The interval between two consecutive phases is precisely one-tenth of the duration of a single step)

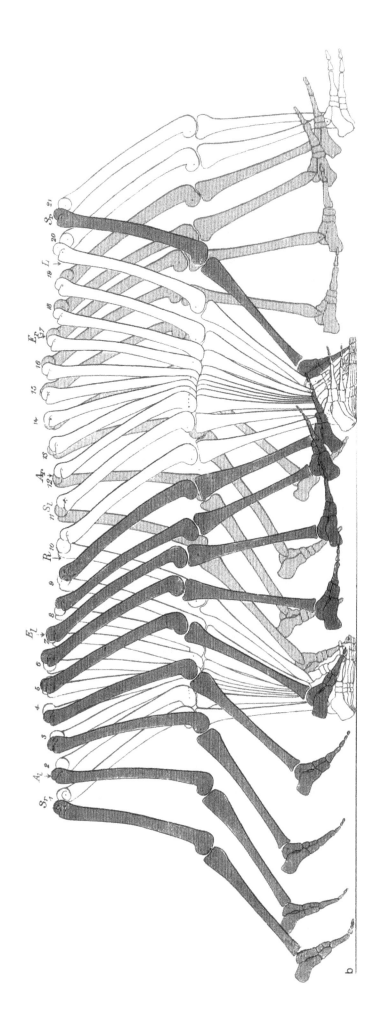

represented by 99 mm in the abscissa. If this length is divided in the abscissa into 20 equal parts, 4.95 mm will correspond to the time interval between two successive phases of the 20 phases making up the double step. Phases 1 and 21 correspond to right toe-off. In the photographs of experiment I, this moment was phase 25. We must thus proceed on the abscissa of the previous displacement curves by increments of 4.95 mm and measure the corresponding ordinate values to find the values of the new co-ordinates.

If phase 1 on the photograph is taken to be point 0 in the abscissa, phase 25 lies at a distance of 92 mm (cf. p. 169). Then the abscissae of the displacement curves given in Table 1 correspond to the 21 phases which divide a double step into 20 equal parts, where numbers 1 and 21 coincide with right toe-off.

The points on the axis of the abscissa belonging to the first two phases would lie to the left of point 0. They have, therefore, been displaced to the right by a distance of 99 mm corresponding to a double step. Due to the periodical repetition of the whole process of the movement, the last values of all y- and z-co-ordinates become the same as those of the corresponding phases 1 and 2 after 0.99 s have elapsed, although in reality they belong to phase 21 and subsequent phases. The x-co-ordinate, however, is greater by the length of a double step, 155.75 cm, in phase 21 than in phase 1. The same is true for all following phases. Hence, it is possible to deduce the values of x for the first two phases from the values in phases 21 and 22. Moreover, there are two possible ways of deducing phases 3, 4 and 5. This provides a means of checking the accuracy of the results for the first two phases.

Table 1 makes it possible to determine how the phases indicated in Fig. 1 correspond to particular positions of the whole leg, or between which phases these positions are to be found. Table 14 of Chap. 2 (p. 169) is also required here. Right heel strike (R) occurs between phases 9 and 10, as mentioned above. Table 14 of the previous chapter indicates that the corresponding abscissa is 34.5 mm. Therefore, right heel strike occurs at phase 9.4, i.e. four-tenths of the interval between 9 and 10. According to this Table 14, right foot-flat (A_r) occurs when the abscissa of the displacement curve is 46 mm. Therefore, in Fig. 1 this occurs between phases 11 and 12, at phase 11.7. Finally, right push-off starts, the foot rotates about a more anterior point of the sole and gradually leaves the ground (E_r) when the abscissa is 72.8 mm. This corresponds to phase 17.1. The corresponding process for the left foot occurs exactly ten phases earlier or later because of the symmetry of the movement.

Table 1. New phases and time

Phase no.[a]	Values on abscissa (mm)[b]	Phase no.[a]	Values on abscissa (mm)[b]
1	92,00	11	42,50
2	96,95	12	47,45
3	2,90 or 101,90	13	52,40
4	7,85 or 106,85	14	57,35
5	12,80 or 111,80	15	62,30
6	17,75	16	67,25
7	22,70	17	72,20
8	27,65	18	77,15
9	32,60	19	82,10
10	37,55	20	87,05
		21	92,00

[a] 1 double step = 20 phases;
[b] 1 s = 100 mm on the abscissa

The various characteristic moments are indicated in Fig. 1 b in the same manner as before (e.g. Table 14 of Chap. 2), i.e. by the point which the femoral head of the extremity under consideration would occupy. S_r indicates right toe-off, S_l left toe-off, R right heel strike, L left heel strike, A_r right foot-flat, A_l left foot-flat, E_r the beginning of right push-off, and E_l the beginning of left push-off.

The following features become immediately apparent in the projection of the process of the movement of the legs on the plane of gait in Fig. 1 b. At the beginning of the swinging period, one leg is a long way behind the other so that the long axis of the thigh, as well as that of the lower leg, is posterior and inclined to the vertical. However, the angle formed by the lower leg and the vertical is greater than that formed by the thigh and the vertical since the knee is somewhat flexed. The foot still contacts the ground through its extreme tip. The thigh rotates forwards, i.e. counter-clockwise as seen from the right. Shortly after the contralateral foot-flat, the thigh passes through the vertical and keeps rotating in the same direction until shortly before push-off of the contralateral side. Then, the thigh moves forwards parallel to itself in the projection on the plane of gait, for a short while. It then rotates a little backwards but rotates considerably forwards before heel strike. At this moment, its long axis forms its greatest anterior angle with the vertical.

The lower leg behaves differently. At the beginning of the swinging period, it initially rotates backwards, i.e. clockwise as seen from the right. This backward rotation continues until the long axis of the thigh passes through the vertical. Then, the lower leg rotates forwards. This movement is carried out with a velocity significantly greater than the simultaneous forward rotation of the thigh. The lower leg continues rotating forwards during the short backward rotation of the thigh. Almost when the thigh begins its final forward rotation, the lower leg reverses its rotation and rotates backwards with a greater velocity than that of the forward rotation of the thigh. Consequently, at heel strike, the long axis of the lower leg no longer forms its greatest anterior angle with the vertical. Similarly, at the beginning of the swinging period it has not reached its greatest posterior angle with the vertical.

As a consequence of the different behaviour of the thigh and lower leg during the swinging period, the knee does not maintain a constant position, but flexes and extends. At the beginning, it flexes since the thigh rotates forwards, and the lower leg backwards. Shortly after the long axis of the thigh is vertical, the knee attains its greatest flexion in walking. Then, as a consequence of the greater velocity of the forward rotation of the lower leg, the knee extends relatively energetically. This extension lasts until the lower leg reverses its rotation shortly before heel strike. Finally, the knee again flexes significantly in such a way that, at heel strike, it is slightly flexed. This process appears clearly in Fig. 2 which represents the movement of the right leg relative to the centre of the hip during the swinging period. The phases are those of Fig. 1 b. The last position of the leg is shown dotted. It corresponds to phase 10 of Fig. 1 b and thus belongs to the stance period.

During the swinging period, the foot moves generally like the lower leg in the projection on the plane of gait, as can be seen in Fig. 1 b. It initially rotates backwards somewhat, i.e. clockwise, and then rotates counter-clockwise with greater velocity. Finally, it rotates clockwise again shortly before heel strike. The movement of the thigh and lower leg in the projection on the plane of gait is not very different from their tridimensional movement because the lateral rotations of these two body segments are relatively small. This projection of the movement of the foot, however, does not give a complete picture of its behaviour. An analysis of the movement of the foot in at least one other projection is thus necessary to determine its tridimensional movement with sufficient accuracy.

If the movement of the foot in the projection on the plane of gait coincided with that of the lower leg, no change in the position of the ankle would occur, at least in this projection. However, the ankle does move even in the swinging period. At the beginning of this period, the foot is plantar-flexed as a consequence of push-off. During the swinging period

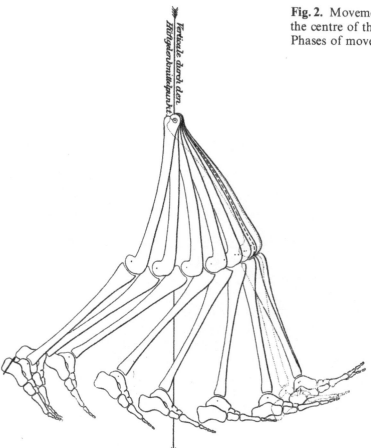

Fig. 2. Movement of the right leg relative to the centre of the hip joint during swing. Phases of movement as in Fig. 1

dorsiflexion occurs and the foot returns to its midposition in relation to the lower leg since this position requires minimum muscle action. This process will be explained below. However, it does not appear as clearly in Fig. 1b as the processes hitherto described.

Until shortly before the end of the single-support period, the thigh carries out roughly the opposite rotations to those of the swinging period. At heel strike, it briefly rotates clockwise. This is not apparent in Fig. 1b because heel strike occurs between phases 9 and 10 and is not drawn. It thus cannot be compared with the next phase (10). After this clockwise rotation the thigh carries out a small counter-clockwise rotation which is not very different from a displacement of the thigh parallel to itself. Then, the thigh rotates clockwise again with relatively greater velocity. This rotation continues until exactly contralateral heel strike. During the following double-support period the thigh finally rotates counter-clockwise. Thus, each thigh forms the greatest posterior angle with the vertical exactly when the ipsilateral foot is on the ground and the opposite heel strikes the ground. The thigh forms the greatest anterior angle with the vertical at ipsilateral heel strike.

Throughout the whole stance period, the lower leg rotates in the same clockwise direction. Initially, this rotation is relatively quick so that the lower leg which was a long way in front of the vertical at heel strike becomes vertical at foot-flat. The rotation is relatively slow during foot-flat and becomes fast again during push-off.

The different behaviour of the thigh and lower leg gives rise to movements of the knee. At the beginning of the stance period, the knee completes the flexion started during the swinging period, until foot-flat. This results from the fast clockwise rotation of the lower leg hitherto. The thigh initially rotates in the same direction but at a lower velocity and then rotates in the opposite direction. During foot-flat, the knee extends. In this period, the two body segments rotate in the same direction though the thigh rotates faster than the lower leg. Finally, during push-off the knee flexes considerably since the thigh initially rotates in the same direction as the lower leg, but more slowly, and during the double-support period it rotates in the opposite direction.

The movement of the foot during the stance period does not appear clearly in Fig. 1b because the different phases are superimposed on each other and the scale of the drawing is too small. However, the behaviour of the foot during the stance period is of decisive importance for the movement of the whole leg during this period whereas the foot plays only a passive role during the swinging period. Therefore, the movement of the right foot during the stance period is represented in a special way in Figs. 3–5 depicting the three experiments. These three figures represent the life-size projection of the movement of the ankle, centre of gravity and tip of the foot on the plane of gait. The previous phases used in the photographs are used here rather than the new ones presented in Fig. 1b. Only the two phases which delineate the stance period and correspond to the moments R and S_r have been added where they did not coincide with one of the photograph phases. This was at least partly the case in Figs. 4 and 5 whereas in Fig. 3 corresponding to experiment I, the original phases 10 and 25 coincided with moments R and S_r. The three points of the foot are connected by interrupted lines in the derived delineating phases whereas the connections are uninterrupted in all the other phases. Finally, in Figs. 3–5 the trajectories of the three points of the foot are indicated by interrupted curves. The bones were not drawn as in Fig. 1b, because otherwise the clarity would have been impaired. It is not difficult to represent the position of the foot at each phase, starting from the three given points. The centre of gravity of the foot lies below the third cuneiform close to its anterior aspect [6] and the point designated as the tip of the foot (cf. p. 67) lay 3 cm behind the actual tip in the experimental subject.

In Figs. 3–5 it is now easy to distinguish the three parts of the stance period. It is best to consider the trajectory of the centre of gravity of the foot. The foot initially rotates about a nearly fixed point of its heel, then remains almost immobile during foot-flat and finally leaves the ground. While the foot rotates about the heel, its centre of gravity describes an arc that is concave downwards and backwards. After foot-flat, the centre of gravity remains virtually at the same place until push-off. In this last part of the stance period, the centre of gravity of the foot describes a curve that is concave downwards and forwards. This curve is not a true arc because the point about which the foot rotates does not remain in the same place but moves progressively forwards on the sole. The curve can be regarded approximately as part of a cycloid described by a point of a circle rolling over a straight line. Whereas the centre of gravity of the foot travels along its trajectory with decreasing velocity during its rotation about the heel, its movement on the cycloid trajectory becomes increasingly faster during push-off.

The centre of the ankle describes roughly the same trajectories as the centre of gravity of the foot. Only the arc and the part of the cycloid are more extended than those of the trajectory of the centre of gravity because the ankle is always further away from the instant point of rotation. During foot-flat the centre of the ankle does not exactly maintain its position but moves a few millimetres forwards and upwards. During the same period the centre of gravity of the foot does not remain immobile though its excursion appears even smaller. This is partly due to the fact that the foot does not behave as a rigid mass during stance but is deformed somewhat in its tarsal joints. The small excursion of the centre of

---

[6] Braune W, Fischer O (1889) Ueber den Schwerpunkt des menschlichen Körpers mit Rücksicht auf die Ausrüstung des deutschen Infanteristen. Abhandlungen der mathematisch-physischen Klasse der Königlich Sächsischen Gesellschaft der Wissenschaften, vol 15, nr 7. English translation (1985) On the centre of gravity of the human body. Springer, Berlin Heidelberg New York Tokyo

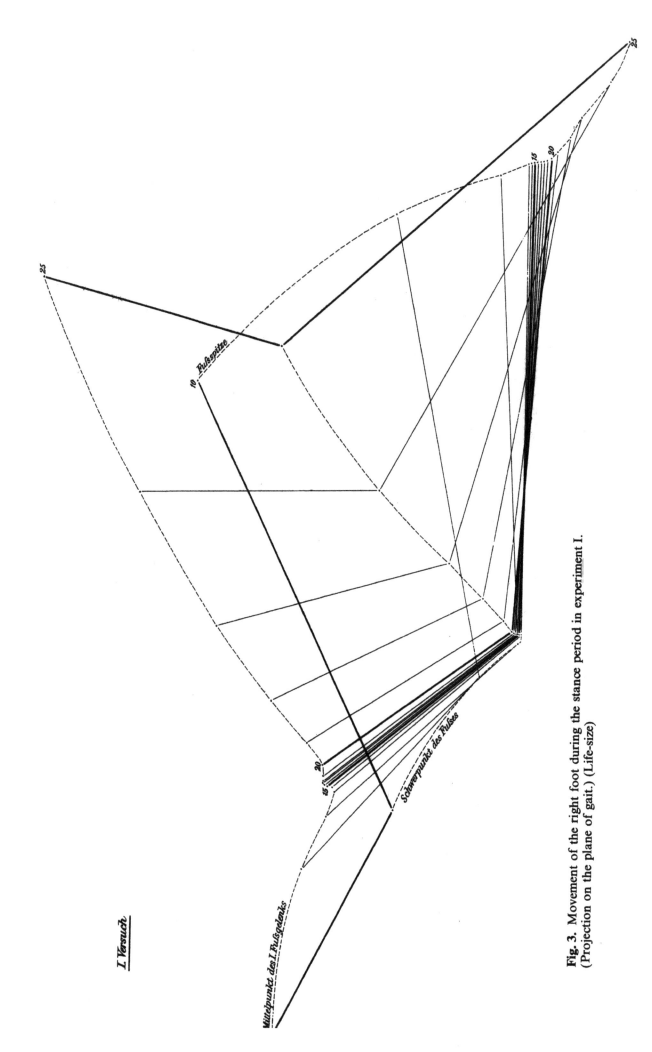

I Versuch

Fußspitze

Schwerpunkt des Fußes

Mittelpunkt des I. Fußgelenks

Fig. 3. Movement of the right foot during the stance period in experiment I. (Projection on the plane of gait.) (Life-size)

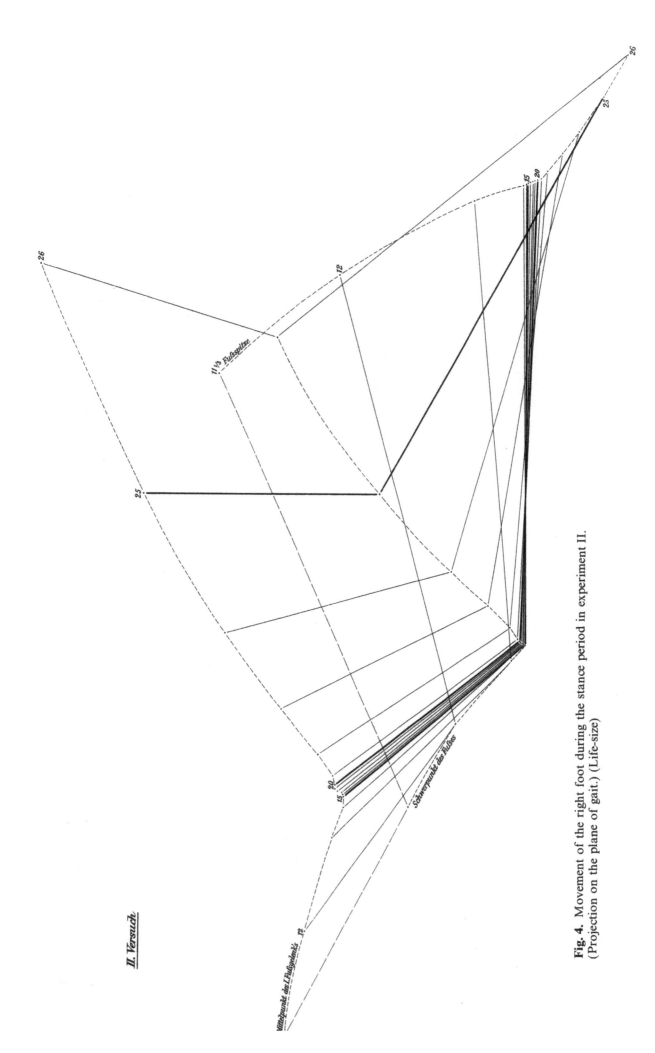

II. Versuch

Mittelpunkt des I. Fußgelenks
Fußspitze
Schwerpunkt des Fußes

Fig. 4. Movement of the right foot during the stance period in experiment II.
(Projection on the plane of gait.) (Life-size)

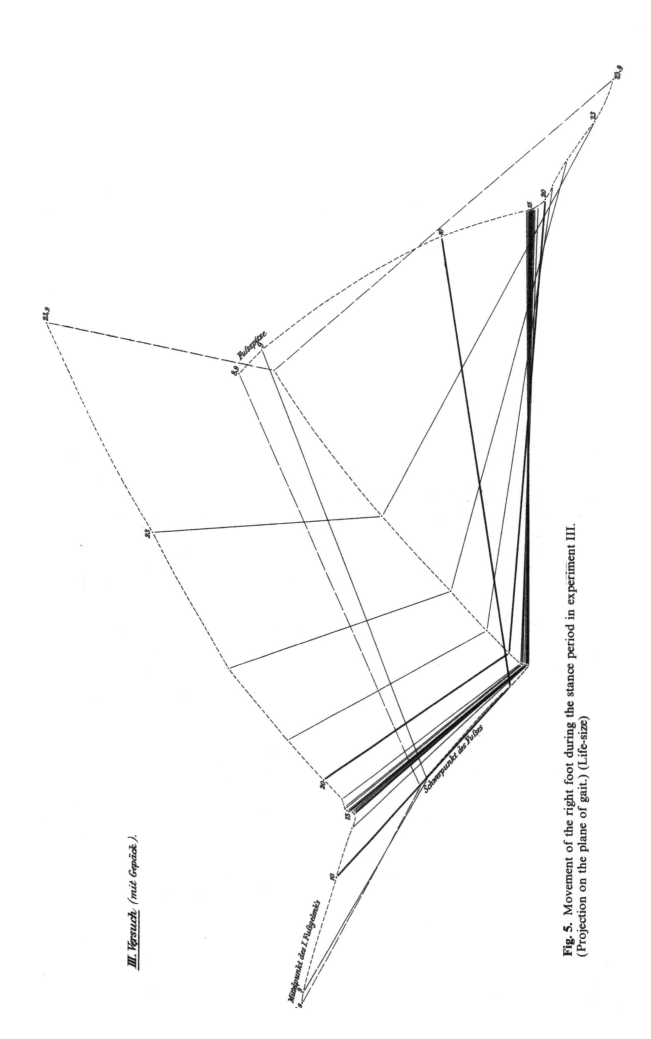

III. Versuch *(mit Gepäck).*

Fußspitze

Schwerpunkt des Fußes

Mittelpunkt des I. Fußgelenks

Fig. 5. Movement of the right foot during the stance period in experiment III. (Projection on the plane of gait.) (Life-size)

gravity of the foot also results from progressive forward displacement of the point of maximum compression on the sole, which causes successively a small lowering of the anterior aspect and a minor lifting of the posterior aspect of the foot.

The tip of the foot also describes an arc during the rotation of the foot about the heel. This arc, however, is even more extended than that described by the centre of the ankle as a consequence of the greater distance from the centre of rotation. During foot-flat the tip of the foot continues moving further down the arc for a distance almost twice that travelled upwards by the centre of gravity of the foot. During this period the whole foot rotates about a point on the line connecting its centre of gravity and the designated tip. This point lies at the union of the posterior third and the anterior two-thirds of the connecting line. Finally, during push-off the tip of the foot describes again part of a cycloid curve. This curve is much shorter than the cycloid curves of the other two points of the foot as a consequence of the vicinity of the tip of the foot to the forward moving centre of rotation. Moreover, the curve runs, not forward and upward as do the other two curves, but forward and downward in a less regular way.

If the movement of the foot as a whole is considered during the stance period, it can be seen that the foot rotates from beginning to end in the same direction, clockwise as seen from the right. This rotation is relatively fast to start with, and then the velocity progressively decreases until it becomes hardly noticeable during foot-flat. During push-off, however, the velocity of the rotation increases again very rapidly.

A comparison of Figs. 3, 4 and 5 shows complete concordance between experiments I and II if it is taken into account that photography showed phases of movement in experiment II that were not identical with those in experiment I. The results are only slightly different in experiment III, when the subject was carrying the heavy military equipment. A small difference appears in the extension of the three curves described during the rotation of the foot about its heel. These curves are somewhat shorter than in the first two experiments.

During the stance period, the ankle moves in different ways. Initially, it undergoes plantar flexion in the projection on the plane of gait since the foot rotates more quickly than does the lower leg in the same direction. This plantar flexion, however, is of short duration as will be shown. Approximately in the middle of the double-support period, it is replaced by dorsi-flexion. This dorsiflexion continues into the first part of the push-off period. Finally, it is in turn replaced by a very energetic plantar flexion which persists until the beginning of toe-off.

The trajectories of the different joint centres can be deduced from the movements of the segments of the leg. The trajectory of the centre of the ankle during ipsilateral stance, explained in detail above, may be used as a starting point. From this, knowing the movement of the lower leg, the trajectory of the centre of the knee is easily determined. These data combined with knowledge of the trajectory of the thigh give the trajectory of the centre of the hip during stance. From the rotations of the hip line described on p. 94, the trajectory of the centre of the opposite hip can be obtained. As long as the two legs behave absolutely symmetrically, the trajectory of the centre of the hip during a double step can be determined, since the contralateral leg swings during the single-support period. Before and after this swinging there is a period of double support because, at the beginning of the stance period on one leg, the contralateral leg briefly remains on the ground, and shortly before the end of the stance period the contralateral foot has contacted the ground again.

Knowing the trajectory of the centre of the hip during the swinging period of the leg under consideration, the trajectories of the knee, ankle, centre of gravity of the foot and tip of the foot can be deduced. The trajectory of the knee during this period results from the displacement of the centre of the hip and the simultaneous rotation of the thigh. The trajectory of the ankle results from the trajectory of the centre of the knee and the simultaneous rotation of the lower leg. The trajectory of the centre of gravity of the foot and that of the tip of the foot result from the displacement of the centre of the ankle and the simultaneous tridimensional rotation of the foot.

The trajectory described by the centre of the ankle during the stance period comprises three distinct parts, an arc forwards and downwards, an arc of only a few millimetres upwards and a cycloid forwards and upwards. The three parts correspond to the three parts of the stance period: rotation about the heel, foot-flat and push-off. If the lower leg maintained its direction during the stance period, i.e. if it remained parallel to itself during its movement, then each point of it, and particularly the centre of the knee, would follow a trajectory exactly parallel to the trajectory of the centre of the ankle. The discrepancy between the two trajectories thus results exclusively from the rotation of the lower leg.

If the rotation were exactly parallel to the plane of gait, the centre of the knee would be forced into a circular trajectory. Whilst the foot rotates about its heel, this circular trajectory would run forwards and upwards since at heel strike the lower leg is considerably inclined anterior to the vertical and returns to the vertical only at the end of the rotation about the heel. However, the simultaneous trajectory of the centre of the ankle runs forwards and downwards. Therefore, the centre of the knee moves almost horizontally during this time. This is apparent in Fig. 1 b and even more so in Figs. 22 and 23 of Chap. 1. Only at the end of this period does the trajectory rise somewhat because then the movement of the centre of the ankle on its circular trajectory is very slow and the lower leg continues rotating with noticeable velocity.

During foot-flat the displacement of the ankle is so small compared with the rotation of the lower leg that it can be ignored. During this period, therefore, the knee describes an arc initially in a horizontal direction, as a consequence of the vertical position of the lower leg, and then in a forward and downward direction. The shape of this trajectory is altered to a certain extent because the rotation of the lower leg is not exactly parallel to the plane of gait. This trajectory is not as long as that segment of the trajectory described by the centre of the knee during the rotation about the ankle although the duration of this rotation is shorter. This results from the fact that the lower leg rotates relatively slowly during foot-flat.

During the last part of the stance period (push-off), the centre of the ankle describes a forward and upward cycloid trajectory. Simultaneously, the centre of the knee moves relative to the ankle as a result of the rotation of the lower leg on a circular trajectory. This trajectory has a marked forward and downward inclination since initially the lower leg is considerably inclined to the vertical. The result is the same as in the first part of the stance period. Under the influence of the two movements, the centre of the knee roughly describes a horizontal trajectory. This trajectory does not move under the horizontal until the second half of this period because then the downward movement of the knee relative to the ankle predominates over the upward movement of the ankle on its cycloid trajectory. The extent of the trajectory in this third part of the stance period is about the same as that of the first two parts combined.

During walking, the centre of the hip follows a double-curved trajectory which appears as an undulating line in the projection on the plane of gait as well as in the projection on the horizontal ground. In the horizontal projection, the wave-length coincides with the length of a double step. During a double step, in the projection on the plane of gait each hip centre describes two waves which differ in length and shape. These waves coincide only in the trajectory of the midpoint of the hip line. The wave-length of this trajectory in the projection on the plane of gait is equal to the length of a step, i.e. half the wave-length in the horizontal projection. The inequality of the two parts of the wave line of a hip centre in the projection on the plane of gait results from the rotations of the hip line. In the stance period of a leg, the hip centre travels somewhat more than a wave-length. At heel strike the ipsilateral hip centre is at the lowest part of a wave. At about the middle of foot-flat, it reaches the highest point of the next wave, and at heel strike of the contralateral leg it is again at the lowest point. In the double-support period the ipsilateral hip centre rises again on its wave line. This is evident in Fig. 1 b and in Figs. 22 and 23 of Chap. 1. It remains now to deduce the wave trajectory of the hip centre in the projection on the plane of gait, first during the stance period, from the movement of the centre of the knee and the simultaneous rotation of the thigh.

At heel strike, the thigh is at its furthest anterior from the vertical, and the lower leg is not far either from its greatest anterior inclination to the vertical. The centre of the hip is then at its lowest. Whilst the foot rotates about the heel, the centre of the knee moves initially horizontally and then upwards somewhat. In this period, the thigh rotates only slightly and moves forwards almost parallel to itself. Therefore, the trajectory of the hip centre cannot differ much from that of the knee centre. The first segment of the rising part of the wave line is thus about the same up to the ascending nodal point. During foot-flat the knee describes a small arc. The thigh rotates clockwise from a considerable posterior inclination of its upper extremity to an anterior inclination. If the knee remained fixed in space, the centre of the hip would describe an arc. Because of the simultaneous forward movement of the knee, this arc is extended and an up wave follows a down wave. During the following push-off period the thigh rotates further, initially in the same direction but with decreasing velocity, until this rotation ceases upon heel strike of the contralateral leg. The thigh then rotates in the opposite direction. At the same time, the knee moves forward almost horizontally. Therefore, the centre of the hip moves down at a progressively decreasing gradient until finally the movement becomes horizontal, like that of the knee, at heel strike of the contralateral leg. The lowest point of the next down wave is then reached. In the part of push-off that coincides with the double-support period, the thigh rotates in the opposite direction. Since the upper extremity of the thigh is considerably inclined forwards, this rotation must induce an elevation of the hip which cannot be cancelled by the small descent in the simultaneous trajectory of the knee. This is actually what occurs. The segment of the down wave follows the lowest point and does not extend to the next ascending nodal point of the wave line.

The trajectory of the centre of the contralateral hip can easily be deduced from that of the ipsilateral hip during the stance period and from the rotations of the hip line (connecting the two hip centres). This was discussed in Chap. 1. Here it is only necessary to note that the differences in the shapes of the trajectories described simultaneously by the two hips are not very important because the excursions of each hip in the directions of the three axes of co-ordinates (direction of gait, transverse direction and vertical direction) are much greater than the movement of each hip in relation to the other. It is the movement of the hip line parallel to the horizontal plane, i.e. its rotation about the vertical axis, which exerts the greatest influence on the differences between the two hip trajectories. At heel strike there is a considerable anterior inclination of the hip line on this side so that the hip centre of the other side remains relatively far behind. Of course, the reverse occurs upon heel strike of the contralateral leg. Consequently, the segment of the hip trajectory between heel strike of one leg and that of the other leg appears more extended on the side of the swinging leg. This can be seen in Fig. 1 b. Between R and L the wave trajectory of the left hip (in red) is longer than that of the right.

The swinging period does not take up the whole interval between two successive lowest points on the wave line. The initial part of this segment belongs to the period of double support.

From the trajectory of the hip during the swinging period it is possible to deduce that of the knee during the same period if the simultaneous rotation of the thigh is known. Initially, the thigh is inclined to the vertical with its lower extremity behind. Seen from the right, it rotates counter-clockwise. Therefore, relative to the hip, the knee moves at first on a forward and downward circular trajectory. Simultaneously, the hip rises somewhat on its wave trajectory. The centre of the knee initially continues moving forwards and slightly downwards for a short while during push-off. The thigh soon passes through the vertical and continues rotating counter-clockwise during the first two-thirds of the swinging period whereas during the first half of this period the hip simultaneously moves forwards and upwards. Therefore, the knee quickly reaches the lowest point of its trajectory and then moves upwards. It keeps moving upwards for a time after the hip has passed through the highest point of its wave line because the rotation of the thigh in the same direction raises the knee relative to the hip more than the latter is simultaneously lowered. Finally, the two movements compensate each other in such a way that the centre of the knee reaches its highest position. The next segment of the trajectory of the centre of the knee is not very different from the corresponding segment of the hip trajectory since the thigh only rotates a little in one and then in the other direction so that it moves forwards almost parallel to itself. The short backward rotation of the thigh described above results in the centre of the knee reaching its lowest point a little before heel strike whereas the centre of the hip is at its lowest exactly at the moment of heel strike.

On the whole, during the swinging period the centre of the knee describes a trajectory the shape of which is a distorted undulating line in the direction of gait.

The trajectory of the centre of the ankle during the swinging period behaves in almost exactly the reverse manner to that of the ipsilateral hip. When the latter presents a maximum the former is near a minimum, and vice versa. This opposite behaviour of the two trajectories results exclusively from the rotations of the thigh and lower leg.

At first, the hip trajectory is at a minimum and the knee moves forward almost horizontally, only slightly downwards. The trajectory of the ankle rises and briefly prolongs the cycloid trajectory of the end of the stance period as a result of the backward rotation of the lower leg which continues for a short while. The trajectory of the ankle, however, soon reaches its highest point. The lower leg then forms the maximum posterior angle with the vertical. During the following forward rotation of the lower leg, the centre of the ankle describes a continuously downward trajectory despite the simultaneous rising of the knee. This is because the lowering due to the rotation of the lower leg relative to the knee is greater than the simultaneous elevation of the knee joint. The ankle reaches its lowest point approximately when the lower leg is vertical. This almost coincides with the highest position of the knee and occurs only shortly before the last maximum of the hip trajectory. The trajectory of the knee falls again but the lower leg rotates further forward at a relatively greater velocity. Consequently, the trajectory of the ankle rises, though only slightly, as long as the lower leg continues to rotate in the same direction. Shortly before heel strike the lower leg rotates in the opposite direction in a clockwise manner, as seen from the right. This final rotation induces a downward movement of the centre of the ankle since the knee rises only slightly. This results in another maximum in the trajectory of the centre of the ankle shortly before heel strike whereas the trajectory of the hip presents a minimum exactly at the moment of heel strike.

We thus have an overall view of the movement of the different segments of the legs, their changes in position and the characteristic trajectories of the joint centres. This will help in understanding further details of the behaviour of the lower extremities in walking. However, this preliminary overall view does not yet provide a complete picture of the movement of the legs since analysis has only been made from the side, i.e. in the projection of the movement on the plane of gait.

Before going further in this study, it will be of interest to check the validity of the description of the movement of the lower extremities made by Weber and Weber in their work on the mechanics of walking[7]. Their study had to be based partly on only few experimental data, partly on theoretical considerations since chronography was not available to these researchers. This critical analysis is necessary since the laws of the movement given by Weber and Weber are the foundation of their theory of human gait.

---

[7]   Weber W (1894) Werke, vol 6, Berlin

Critical Analysis of the Leg Movements According to Weber and Weber

Fig. 1a is an exact copy of Fig. 3 from Plate XII in *Mechanics* of the *Human Locomotor Apparatus* by Weber and Weber. The only difference from the original is that the two parts of the picture respresenting the positions of the leg during stance and swing have been interchanged to facilitate comparison with the phases of the movement represented in Fig. 1b. The numbering of the successive phases by Weber and Weber has been omitted.

It is immediately apparent that the representation of Weber and Weber barely corresponds to the actual conditions. This is not really surprising, however, since the two authors had little in the way of empirical data concerning the behaviour of the legs during walking. Except for some measurements of step length, the length of the leg at different degrees of flexion, the length of the extended back leg at different walking velocities, the lifting of the foot, the height of the femoral heads above the ground and the magnitude of their vertical oscillations, the duration of a step, the duration of the swinging and stance periods of a leg, etc., they had to rely throughout only on a theoretical approach. Interpretations of the structure of certain tissues based on their properties could hardly be expected to correspond exactly to the facts before the microscope enabled histologists to see the elements of these tissues directly. Neither could it be expected that the view on the successive positions of the legs expressed by Weber and Weber based on their few direct measurements, would coincide with the facts exactly, before instantaneous photography enabled researchers to record the successive phases of movement directly. Weber and Weber could only express a hypothesis which was in agreement with the results of their measurements. It is thus surprising that despite the limited means at their disposal for analysing the complicated process of walking they discovered so many facts correctly. Although instantaneous photography failed to confirm much of Weber and Weber's representation of walking, their *Mechanics of the Human Locomotor Apparatus* remains a classic work as the very first attempt at a precise study in the field of movement physiology.

Comparison of the series of pictures in Fig. 1 reveals that Weber and Weber had misconceived the behaviour of the lower leg. According to them, the latter rotates relatively slightly and always remains inclined posterior to the vertical, even at heel strike. In fact, at heel strike the lower leg extends forwards. Weber and Weber misconstrued the positions of the lower leg because of another assumption with regard to the action of the forces on the body. They believed they had to make this assumption for theoretical reasons.

Weber and Weber considered it absolutely necessary that when a leg begins to swing the other leg is vertical (§121). By a vertical position of the leg, they meant that the vertical through the centre of the hip passed through the ground support area of the foot (see remark to §8). They held the fulfilment of this prerequisite to be so important for the regularity of gait that they made it one of the three principles of their theory of walking. They called it the "principle of the initial position" (§125). The principle of the initial position is not fulfilled even approximately in walking, except for the very first step. Fig. 1b, Figs. 22 and 23 of Chap. 1 and any of the numerous series of instantaneous pictures of a walking man taken by Marey, Muybridge or Anschütz show that at toe-off the contralateral leg is not vertical but extends far forward. The vertical through the centre of the hip of the front leg passes a long way behind the foot. It reaches the ground at a point further behind the tip of the foot of the front leg than it is in front of the tip of the foot of the leg that is starting its swing. Under these circumstances, of course, the Webers' principle of the initial position cannot be true, not even approximately. One of the three pillars of the Webers' theory of walking thus collapses.

Heel strike occurs shortly before swinging of the contralateral leg begins. Therefore, according to the principle of the initial position, the leg could not be far from the vertical at heel strike. Moreover, having measured the height of the femoral head above the ground during walking, Weber and Weber concluded that the vertical position of the leg had to be linked with considerable flexion of the knee. In this way, they arrived at their misconception about the position of the lower leg at heel strike and about its attitude during the swinging period.

This error in the position of the lower leg at the beginning of stance seems to have led the Webers to a false assumption regarding the simultaneous behaviour of the foot. According to their concept, when the foot contacts the ground it does so with the entire sole whereas in fact the heel strikes the ground while the tip of the foot is considerably raised. This can be easily observed in normal gait. If the lower leg were inclined with its upper extremity forwards, as assumed by the Webers for the beginning of stance, such an initial position of the foot would be made impossible by the limited dorsiflexion of the ankle. Therefore, this concept impelled Weber and Weber to regard contact with the ground as being performed by the sole.

Another source of error resulted from the views of the Webers on the vertical oscillations, which they observed directly and measured. They assumed (§24) that the trunk and femoral head move downwards with each step immediately before a leg becomes vertical and move upwards as soon as the leg is vertical. This appears in the stance picture of the Webers in Fig. 1 a. Moreover, they assumed that the femoral head always moves in the same horizontal plane. This does not agree at all with the facts. The hip joint moves steadily upwards from heel strike, when the leg is still a long way anterior to the vertical, and has reached its highest point when the leg is vertical. It then moves downwards until shortly before the leg starts swinging. During the swinging period the hip joint at first moves upwards again until about mid-swing when the centre of gravity of the foot passes through the vertical through the hip. After this it moves downwards again until heel strike. The contralateral hip and the trunk move similarly despite the vertical oscillations of the hip line which are small compared with the upwards and downwards movements of the hips.

According to Weber and Weber, the femoral heads must move forwards along almost a straight line (§92) and, shortly before the stance leg becomes vertical, they may at most move downwards a little (§121). Upon becoming vertical the leg would restore the horizontal forwards movement of the upper body to its previous level. From this concept, the Webers deduced a second basic principle of gait (§92): "The striking force of the stance leg which supports and propels the human body during walking is vertical and must be almost equal to the weight of the body, whatever the position of the leg; consequently the body cannot go up or down to any considerable extent". They called this condition "the principle of the measure of the effort" (§125). This vertical component of the striking force of the leg, which expresses itself in the normal compression on the ground, was deduced from the accelerations of the total centre of gravity in Chap. 2 of the present work. The results of this research are given for the three experiments in Tables 22 and 23 of Chap. 2 (p. 201). The body weighed 58.7 kg in experiments I and II. As appears in Table 23, however, the normal compression on the ground varied between approximately 15 and 82 kg. In experiment III the body weight increased to about 82 kg as a result of the military equipment. In this instance the normal compression on the ground varied roughly between 39 and 123 kg. Therefore, normal compression, in other words the vertical component of the striking force of the leg, during a double step falls below half the body weight and increases to 150% of this weight. It thus appears that the Webers' principle of the measure of the effort is also erroneous.

Finally, the third of the principles which provided the basis for Weber and Weber's theory of walking which they called the "principle of the direction of striking" (§125), is also disputable. According to this principle, the line of action of the striking force always passes through the total centre of gravity of the body and through the foot of the stance leg. In other words, the line of action of the compression exerted on the ground always passes through the total centre of gravity of the body. The components of this compression in the direction of gait and in the transverse direction are given in Table 21 and its vertical component in Table 22 of Chap. 2 for each of the 31 phases of the movement. It is only necessary to reverse the plus-minus signs given in these tables, as was explained on p. 203. Careful analysis shows it is by no means true that the resultant of these three components always passes through the total centre of gravity of the body.

Consequently, the three pillars on which Weber and Weber built their theory of walking have crumbled. The Webers' theory is no longer in agreement with the facts as discovered using instantaneous photography. The theory corresponds to an ideal case which does not occur in man because it assumes uniform horizontal movement of the total centre of gravity instead of the actual movement of the latter, involving both acceleration and deceleration on a double-curved trajectory. However, under the assumption of such ideal circumstances, their study represents an exact approach to a natural process. Thus, it can still be considered exemplary for any other theory of walking based on subsequently established data.

So as to impart a better understanding of the new data concerning the change in position of the legs during walking, some other erroneous details in the Webers' representation of gait will be considered.

Weber and Weber assert (§8) that the stance leg is at its shortest when the centre of the femoral head is on the vertical passing through the heel. This is only approximately true since, during stance, the knee is maximally flexed at foot-flat, as explained above. In this position, the vertical through the centre of the hip still passes behind the supporting foot (Fig. 1 b). The knee then starts to extend whereas Weber and Weber assumed (§14) that the extension occurs only when the trunk is so far forward that the supporting part of the foot is behind the vertical through the femoral head. Moreover, Weber and Weber asserted (§14) that the knee continues extending until the end of the stance period. This contradicts the finding described above that the knee steadily flexes throughout push-off so that at toe-off the knee attains its greatest flexion during stance.

Weber and Weber assumed (§17) that the knee is flexed during the longest part of the swinging period and does not extend until the end of this period, near heel strike. "The leg is then lengthened by extending the knee until heel strike". In reality, the knee is briefly flexed at the beginning of the swinging period. It extends during most of this period and finally flexes again at the end of the swinging period until heel strike.

Positional Angles of the Segments of the Lower Extremities

The previous analysis gave an overall view of the change in position of each leg during a double step. Further research requires the successive positions of the legs to be established by means of more accurate data. This will enable the torques of gravity and the torques of the muscles to be calculated excluding as yet unknown physiological factors, and the angular velocities and accelerations of the rotations of the different body segments to be deduced.

The position of each body segment at any phase of the movement is determined by the rectangular co-ordinates of all the adjacent joint centres given in Chap. 1. The relationships between the movements of the body segments and the forces acting on them almost exclusively concern the rotations of the body segments and constitute the basis for further investigation. The analysis would be unnecessarily long and complicated if only the rectangular co-ordinates were used to find the angular velocities and accelerations. It appears more convenient to introduce angles which clearly indicate the direction of the long axis of each of the segments of the lower extremities. The direction of the long axes of all the segments of a leg generally determines completely the position of the whole leg, since as a rule the rotation of any segment about its long axis is given by the direction of the long axes of the adjacent segments. The positional angles, as they will be designated, must of course be calculated from the rectangular co-ordinates since the latter clearly indicate the direction of the long axis of a body segment. Introducing the positional angles offers a great advantage: in order to determine the direction of the long axis of a segment of the leg, only two angles independent of each other are necessary whereas six rectangular co-ordinates are generally required, three for each joint centre.

Different pairs of angles can be used. We can take the angles formed by the projection of the long axis on two different planes and an axis of co-ordinates in each of these planes. For instance, the long axis can be projected on the plane of gait and on the plane perpendicular to the direction of gait and, in each case, the angle formed by this projection and the vertical is measured. Two independent angles can also be obtained in another way: the angle formed by the long axis and the vertical, and the angle formed by the vertical plane passing through the long axis and the vertical plane of gait are determined. Both methods present their particular advantages and thus both will be used.

To deduce the angles, the long axis of the body segment and the centre of the proximal joint are displaced to the origin of a rectangular system of co-ordinates, the positive axes X, Y and Z of which point forward in the direction of gait, transversely to the right and vertically upward. This system of co-ordinates need not necessarily be the same as that into which the whole process of walking has been transposed, but the systems must be parallel to each other. The co-ordinates of the centre P of the distal joint in this system of co-ordinates (Fig. 6) are $x_d - x_p, y_d - y_p$, and $z_d - z_p$, if x_p, y_p, z_p are the co-ordinates of the proximal joint centre and x_d, y_d, z_p those of the distal joint centre in the initial system. The values of these co-ordinates are given for the centres of all the joints in the three experiments in Tables 14–16 of Chap. 1.

The projection of the point P on the plane of co-ordinates parallel to the plane of gait (plane XZ) is designated as P', the projection of P on the plane of co-ordinates perpendicular to the direction of gait (plane YZ) as P'', and the projection of P on the Z-axis as V. OP', OP'' and OV thus represent the projections of the long axis OP on the plane of gait, the plane perpendicular to the direction of gait and the vertical axis of co-ordinates. The angles formed by the projections OP' and OP'' with the vertical pointing downward from O are respectively φ and ψ. The vertical plane comprising OP (plane of the triangle OPV) forms the angle α with the plane of gait. The long axis OP itself forms the angle ε with the vertical pointing downward from O. Two of the four angles clearly give the tridimensional orientation of the long axis. Thus, φ and ψ, for example, or α and ε can be considered as the positional angles determining the direction of the long axis.

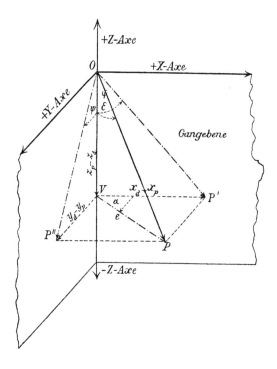

Fig. 6. Positional angles of segment OP

To determine the pair φ, ψ, the following equations, apparent from Fig. 6, are used:

$$tg\,\varphi = \frac{x_d - x_p}{z_p - z_d} \quad \text{and} \quad tg\,\psi = \frac{y_d - y_p}{z_p - z_d}$$

In the denominator of these equations, the vertical co-ordinates of the centre of the distal joint are subtracted from those of the centre of the proximal joint since φ and ψ are the angles formed by the projection of the long axis pointing downwards on the two planes of co-ordinates and the negative Z-axis. φ can be positive or negative depending on whether the long axis lies anterior or posterior to the vertical pointing downwards in the projection on the plane of gait. ψ is positive or negative depending on whether the long axis is to the right or to the left of the vertical pointing downwards in the projection on the plane perpendicular to the direction of gait.

To calculate α the following equation is used:

$$tg\,\alpha = \frac{y_d - y_p}{x_d - x_p}$$

α is positive if the vertical plane comprising the long axis appears as rotated clockwise in relation to the anterior part of the plane of gait when seen from above, up to an angle of 180°. Beyond 180°, α is taken as negative.

The ε-angle belongs to the rightangle triangle OPV. If e signifies the distance PV between point P and the vertical axis Z, then:

$$tg\,\varepsilon = \frac{e}{z_p - z_d}$$

From the triangle $PP'V$ it appears that:

$$e = \frac{x_d - x_p}{\cos \alpha} = \frac{y_d - y_p}{\sin \alpha}$$

If one of these values is introduced into the equation for $tg\,\varepsilon$ and the equation for $tg\,\varphi$ and $tg\,\psi$ are taken into account, the following equation for calculating ε from the angles φ and α or ψ and α is obtained:

$$tg\,\varepsilon = \frac{tg\,\varphi}{\cos \alpha} = \frac{tg\,\psi}{\sin \alpha}$$

ε needs no sign.

Using these equations, the four angles φ, ψ, α and ε were calculated for the long axes of the thigh and of the lower leg as well as for the line connecting the centre of the ankle with the centre of gravity of the foot. The calculations were carried out for both lower extremities, for all the 31 phases of movement, and for the three experiments. The results are given in Tables 2–4. The angles corresponding to a body segment present as an index the number used previously for this body segment. The right thigh, lower leg and foot are designated respectively as 2, 4 and 6, their left counterparts as 3, 5 and 7.

Figure 7–9 represent the changes of the positional angles φ and ψ with time in the three experiments. They illustrate the continuous tridimensional variations in the direction of the different long axes of the leg and provide the possibility to determine the direction of the long axes even for those phases of the movement which have not been photographically recorded in the experiments. The successive points at 5-mm intervals on the axis of the abscissa correspond to the 31 phases. The vertical distance upwards or downwards from each of these

Table 2. Positional angles of the segments of the lower extremities in experiment I

No.	Thigh								No.
	Right				Left				
	φ_2	ψ_2	α_2	ε_2	φ_3	ψ_3	α_3	ε_3	
1	− 0°20′	+6°18′	+ 92°58′	6°20′	+17°0′	+0°11′	+ 0°34′	17°0′	1
2	+ 7°4′	+5°32′	+ 38°0′	8°56′	+10°47′	−0°2′	− 0°10′	10°47′	2
3	+13°49′	+4°57′	+ 19°25′	14°37′	+ 4°25′	+0°28′	+ 5°58′	4°27′	3
4	+19°9′	+4°9′	+ 11°47′	19°32′	− 1°52′	+1°10′	+148°5′	2°11′	4
5	+23°22′	+3°18′	+ 7°37′	23°33′	− 7°30′	+2°2′	+164°54′	7°46′	5
6	+26°7′	+2°39′	+ 5°23′	26°13′	−12°26′	+2°35′	+168°24′	12°41′	6
7	+26°43′	+1°24′	+ 2°46′	26°44′	−16°41′	+3°8′	+169°39′	16°56′	7
8	+26°35′	−0°12′	− 0°25′	26°35′	−20°37′	+3°15′	+171°26′	20°50′	8
9	+27°41′	−0°26′	− 0°49′	27°41′	−23°17′	+2°27′	+174°19′	23°24′	9
10	+30°19′	+0°15′	+ 0°26′	30°19′	−24°17′	+0°44′	+178°22′	24°17′	10
11	+28°38′	+0°21′	+ 0°39′	28°38′	−23°44′	−1°51′	−175°48′	23°48′	11
12	+28°44′	+2°7′	+ 3°52′	28°48′	−19°4′	−4°43′	−166°33′	19°34′	12
13	+27°3′	+1°21′	+ 2°38′	27°5′	−11°26′	−6°10′	−151°52′	12°55′	13
14	+20°30′	+0°51′	+ 2°18′	20°31′	− 3°36′	−6°44′	−118°6′	7°37′	14
15	+13°19′	−0°36′	− 2°32′	13°20′	+ 4°4′	−6°31′	− 58°3′	7°40′	15
16	+ 6°48′	−1°55′	− 15°44′	7°4′	+ 9°7′	−6°17′	− 34°28′	11°1′	16
17	+ 0°50′	−2°41′	− 72°36′	2°49′	+16°5′	−5°40′	− 19°0′	16°57′	17
18	− 4°48′	−3°4′	−147°26′	5°42′	+20°38′	−4°37′	− 12°6′	21°4′	18
19	−10°9′	−3°17′	−162°12′	10°39′	+23°25′	−3°33′	− 8°9′	23°38′	19
20	−14°18′	−3°48′	−165°25′	14°45′	+23°20′	−2°8′	− 4°57′	23°25′	20
21	−18°19′	−3°45′	−168°48′	18°39′	+21°34′	−0°27′	− 1°9′	21°34′	21
22	−20°57′	−2°53′	−172°29′	21°7′	+22°13′	+0°6′	+ 0°14′	22°13′	22
23	−21°53′	−1°23′	−176°34′	21°55′	+25°37′	−0°47′	− 1°37′	25°38′	23
24	−21°20′	+1°3′	+177°18′	21°21′	+22°23′	−0°44′	− 1°46′	22°23′	24
25	−16°32′	+3°51′	+167°13′	16°56′	+21°58′	−0°51′	− 2°6′	21°59′	25
26	− 7°45′	+4°37′	+149°20′	9°0′	+21°19′	−1°24′	− 3°36′	21°21′	26
27	− 0°9′	+4°10′	+ 91°56′	4°11′	+15°20′	−0°54′	− 3°15′	15°22′	27
28	+ 6°44′	+3°59′	+ 30°32′	7°49′	+ 9°0′	−0°40′	− 4°9′	9°1′	28
29	+12°15′	+3°35′	+ 16°5′	12°44′	+ 2°56′	+0°9′	+ 3°1′	2°56′	29
30	+17°47′	+3°25′	+ 10°33′	18°4′	− 2°34′	+0°27′	+170°9′	2°36′	30
31	+21°49′	+3°32′	+ 8°47′	22°3′	− 7°37′	+0°39′	+175°11′	7°38′	31

No.	Lower leg								No.
	Right				Left				
	φ_4	ψ_4	α_4	ε_4	φ_5	ψ_5	α_5	ε_5	
1	− 57°6′	− 9°11′	−174°2′	57°15′	+ 2°30′	+7°4′	+ 70°38′	7°29′	1
2	− 52°8′	− 7°36′	−174°4′	52°17′	+ 0°12′	+7°10′	+ 88°22′	7°11′	2
3	− 42°56′	− 6°27′	−173°4′	43°9′	− 1°43′	+6°36′	+104°31′	6°49′	3
4	− 31°13′	− 5°15′	−171°23′	31°30′	− 3°22′	+5°50′	+119°54′	6°43′	4
5	− 17°31′	− 4°25′	−166°15′	18°0′	− 5°36′	+5°8′	+137°25′	7°34′	5
6	− 2°50′	− 3°45′	−126°57′	4°42′	− 8°35′	+4°51′	+150°40′	9°50′	6
7	+ 12°48′	− 2°15′	− 9°50′	12°59′	−12°25′	+4°39′	+159°45′	13°13′	7
8	+ 26°22′	+ 0°21′	+ 0°42′	26°22′	−17°10′	+4°38′	+165°19′	17°43′	8
9	+ 28°9′	− 0°5′	− 0°8′	28°9′	−22°42′	+4°47′	+168°41′	23°6′	9
10	+ 19°39′	− 2°48′	− 7°48′	19°49′	−29°8′	+5°12′	+170°44′	29°27′	10
11	+ 15°32′	− 3°55′	− 13°50′	15°58′	−36°14′	+5°56′	+171°56′	36°30′	11
12	+ 6°39′	− 6°26′	− 44°5′	9°13′	−44°31′	+8°16′	+171°36′	44°50′	12
13	− 0°21′	− 7°9′	− 92°48′	7°9′	−52°3′	+9°47′	+172°20′	52°18′	13
14	− 2°48′	− 7°46′	−109°44′	8°14′	−53°8′	+9°44′	+172°40′	53°21′	14
15	− 4°4′	− 7°5′	−119°46′	8°9′	−46°57′	+8°26′	+172°7′	47°14′	15
16	− 5°2′	− 6°1′	−129°52′	7°49′	−38°42′	+7°40′	+170°28′	39°5′	16
17	− 6°29′	− 5°13′	−141°14′	8°17′	−26°45′	+6°34′	+167°8′	27°21′	17
18	− 8°16′	− 4°40′	−150°39′	9°28′	−14°42′	+5°26′	+160°5′	15°35′	18
19	− 11°0′	− 4°27′	−158°9′	11°49′	− 0°26′	+4°15′	+ 95°54′	4°17′	19
20	− 14°57′	− 3°52′	−165°50′	15°24′	+15°15′	+2°32′	+ 9°12′	15°27′	20
21	− 18°57′	− 3°54′	−168°45′	19°17′	+28°12′	+0°1′	+ 0°2′	28°12′	21
22	− 24°18′	− 4°4′	−171°2′	24°33′	+29°35′	−0°3′	− 0°5′	29°35′	22
23	− 31°9′	− 4°36′	−172°25′	31°23′	+20°30′	+2°21′	+ 6°15′	20°37′	23
24	− 38°35′	− 6°29′	−171°53′	38°52′	+18°27′	+3°0′	+ 8°56′	18°40′	24
25	− 47°30′	− 9°40′	−171°8′	47°51′	+11°50′	+4°21′	+ 19°56′	12°34′	25
26	− 56°1′	− 11°41′	−172°4′	56°17′	+ 4°13′	+5°32′	+ 52°46′	6°56′	26
27	− 56°44′	− 10°27′	−173°6′	56°56′	+ 0°50′	+5°30′	+ 81°28′	5°33′	27
28	− 51°14′	− 9°43′	−172°10′	51°29′	− 1°42′	+5°30′	+107°11′	5°45′	28
29	− 42°36′	− 8°30′	−170°46′	42°59′	− 3°50′	+4°46′	+128°46′	6°7′	29
30	− 31°33′	− 7°36′	−167°43′	32°8′	− 5°45′	+4°28′	+142°12′	7°15′	30
31	− 18°19′	− 6°32′	−160°54′	19°18′	− 8°5′	+4°27′	+151°19′	9°11′	31

Table 2 (continued)

No.		Right				Left			No.
	φ_6	ψ_6	α_6	ε_6	φ_7	ψ_7	α_7	ε_7	
1	−23°11′	−1°57′	−175°28′	23°14′	+42°57′	−7°59′	−8°35′	43°16′	1
2	−14°46′	−0°59′	−176°15′	14°48′	+42°45′	−8°30′	−9°11′	43°7′	2
3	−2°12′	+1°3′	+154°22′	2°27′	+42°25′	−8°41′	−9°29′	42°48′	3
4	+11°4′	+2°36′	+13°4′	11°21′	+41°53′	−8°24′	−9°21′	42°16′	4
5	+26°13′	+4°14′	+8°33′	26°28′	+40°48′	−9°1′	−10°25′	41°16′	5
6	+39°48′	+4°53′	+5°51′	39°57′	+39°51′	−10°3′	−11°59′	40°28′	6
7	+53°6′	+8°7′	+6°6′	53°15′	+38°13′	−10°55′	−13°46′	39°2′	7
8	+65°23′	+19°9′	+9°3′	65°39′	+33°49′	−10°15′	−15°7′	34°45′	8
9	+67°31′	+27°22′	+12°5′	67°58′	+28°17′	−8°26′	−15°25′	29°10′	9
10	+61°58′	+21°44′	+11°59′	62°29′	+19°47′	−6°11′	−16°45′	20°35′	10
11	+49°2′	+8°45′	+7°37′	49°17′	+5°20′	−2°19′	−23°30′	5°48′	11
12	+42°18′	+4°21′	+4°46′	42°24′	−13°16′	+1°35′	+173°17′	13°21′	12
13	+40°19′	+7°1′	+8°15′	40°36′	−19°54′	−0°20′	−179°4′	19°54′	13
14	+39°15′	+7°4′	+8°38′	39°34′	−19°6′	∓0°0′	∓180°0′	19°6′	14
15	+38°37′	+7°36′	+9°29′	39°0′	−9°17′	+0°41′	+175°53′	9°18′	15
16	+38°4′	+8°18′	+10°34′	38°32′	+1°0′	+0°25′	+22°37′	1°5′	16
17	+37°37′	+8°48′	+11°22′	38°10′	+15°28′	−2°23′	−8°35′	15°38′	17
18	+37°45′	+9°33′	+12°16′	38°24′	+29°21′	−5°59′	−10°33′	29°47′	18
19	+37°6′	+9°51′	+12°56′	37°49′	+43°24′	−8°16′	−8°44′	43°44′	19
20	+35°29′	+9°48′	+13°37′	36°16′	+58°3′	−11°53′	−7°28′	58°17′	20
21	+32°13′	+10°10′	+15°53′	33°14′	+70°47′	−27°39′	−10°21′	71°4′	21
22	+26°4′	+9°8′	+18°11′	27°15′	+72°10′	−33°34′	−12°3′	72°32′	22
23	+16°11′	+7°32′	+24°30′	17°42′	+65°36′	−23°58′	−11°24′	66°2′	23
24	+1°36′	+5°34′	+74°3′	5°47′	+51°51′	−9°26′	−7°26′	52°6′	24
25	−14°23′	+3°53′	+165°9′	14°52′	+46°19′	−2°20′	−2°14′	46°20′	25
26	−23°18′	+0°22′	+179°8′	23°18′	+41°52′	−5°27′	−6°5′	42°2′	26
27	−21°55′	−4°7′	−169°50′	22°14′	+41°13′	−5°34′	−6°21′	41°24′	27
28	−13°31′	−3°45′	−164°45′	13°59′	+40°40′	−6°24′	−7°27′	40°55′	28
29	−1°29′	−1°19′	−138°35′	1°59′	+39°46′	−7°2′	−8°27′	40°4′	29
30	+12°12′	+1°27′	+6°40′	12°16′	+38°54′	−7°34′	−9°21′	39°16′	30
31	+25°56′	+4°22′	+8°55′	26°13′	+38°2′	−8°18′	−10°35′	38°30′	31

points gives the opening of the φ- or ψ-angle, formed by the long axis with the vertical, at the phase corresponding to this point; 1 mm equals 1°. The same axis of the abscissa is used for the values of φ for the three segments of the same leg. This makes comparison easier. The same is true for the values of the ψ-angle for the three segments of each extremity. The left half of each of Figs. 7–9 shows the diagrams of the positional angles of the right leg, the right half those of the left leg. The diagrams of the φ-angles are above, those of the ψ-angles below. Moreover, at the top two additional diagrams related to the joint angles are drawn, using the same axis of the abscissa. Therefore, different colours have been used to distinguish the different segments of the leg. The diagrams related to the thigh are black, those to the lower leg red, and those to the foot blue. Two colours have been used for the diagrams related to the joint angles. These colours correspond to those of the segments adjacent to the joint. Finally, heel strike (R, L), toe-off (S_r, S_l), foot-flat (A_r, A_l) and the beginning of push-off (E_r, E_l) are marked by vertical lines as in the figures of Chap. 2.

The curves of the first two experiments carried out in the same circumstances coincide almost perfectly. Significant differences appear in the diagrams of experiment III in which the subject was carrying the regulation equipment.

The upper curves of Figs. 7–9 confirm more reliably all the data deduced from Fig. 1 b for the projection of the movement on the plane of gait. They complement these data by adding precise angles. At the same time they take into account the circumstances of walking with a load.

At first glance, it appears that each thigh is at its furthest anterior to the vertical at ipsilateral heel strike and furthest posterior to the vertical at contralateral heel strike. All the upper diagrams for the right thigh reach their maximum at R and their minimum at L. The

Table 3. Positional angles of the segments of the lower extremities in experiment II

No.	Thigh								No.
	Right				Left				
	φ_2	ψ_2	α_2	ε_2	φ_3	ψ_3	α_3	ε_3	
1	− 12° 14'	+ 5° 46'	+ 155° 0'	13° 27'	+ 24° 36'	− 0° 20'	− 0° 43'	24° 36'	1
2	− 6° 0'	+ 5° 45'	+ 136° 14'	8° 17'	+ 20° 47'	− 0° 35'	− 1° 32'	20° 47'	2
3	+ 1° 19'	+ 5° 13'	+ 75° 53'	5° 22'	+ 15° 5'	− 0° 46'	− 2° 51'	15° 6'	3
4	+ 9° 23'	+ 4° 54'	+ 27° 24'	10° 33'	+ 8° 44'	− 0° 6'	− 0° 36'	8° 44'	4
5	+ 15° 45'	+ 4° 46'	+ 16° 28'	16° 23'	+ 2° 21'	+ 0° 37'	+ 14° 34'	2° 26'	5
6	+ 21° 15'	+ 4° 21'	+ 11° 4'	21° 37'	− 4° 4'	+ 1° 23'	+ 161° 11'	4° 18'	6
7	+ 24° 42'	+ 3° 40'	+ 7° 55'	24° 55'	− 9° 28'	+ 1° 53'	+ 168° 53'	9° 39'	7
8	+ 25° 51'	+ 2° 55'	+ 6° 1'	25° 59'	− 13° 51'	+ 2° 27'	+ 170° 11'	14° 3'	8
9	+ 25° 29'	+ 1° 1'	+ 2° 8'	25° 30'	− 18° 30'	+ 2° 50'	+ 171° 36'	18° 41'	9
10	+ 25° 40'	− 0° 10'	− 0° 22'	25° 40'	− 21° 33'	+ 2° 35'	+ 173° 29'	21° 40'	10
11	+ 29° 9'	+ 0° 7'	+ 0° 12'	29° 9'	− 22° 41'	+ 1° 10'	+ 177° 13'	22° 43'	11
12	+ 29° 41'	− 0° 6'	+ 0° 10'	29° 41'	− 22° 34'	− 0° 58'	− 177° 40'	22° 35'	12
13	+ 27° 40'	+ 0° 55'	+ 1° 45'	27° 41'	− 18° 39'	− 4° 3'	− 168° 8'	19° 1'	13
14	+ 27° 25'	+ 0° 46'	+ 1° 28'	27° 25'	− 11° 33'	− 5° 17'	− 155° 37'	12° 39'	14
15	+ 22° 19'	− 0° 22'	− 0° 53'	22° 19'	− 2° 32'	− 6° 4'	− 112° 37'	6° 34'	15
16	+ 16° 1'	− 1° 15'	− 4° 21'	16° 3'	+ 4° 48'	− 6° 15'	− 52° 30'	7° 52'	16
17	+ 9° 17'	− 1° 56'	− 11° 40'	9° 29'	+ 11° 10'	− 6° 12'	− 28° 48'	12° 42'	17
18	+ 2° 38'	− 2° 52'	− 47° 25'	3° 53'	+ 16° 53'	− 6° 1'	− 19° 8'	17° 49'	18
19	− 3° 28'	− 3° 21'	− 135° 58'	4° 49'	+ 21° 24'	− 5° 12'	− 13° 5'	21° 55'	19
20	− 8° 55'	− 3° 37'	− 158° 2'	9° 36'	+ 23° 56'	− 3° 55'	− 8° 46'	24° 11'	20
21	− 13° 27'	− 4° 2'	− 163° 35'	14° 0'	+ 23° 35'	− 2° 12'	− 5° 2'	23° 40'	21
22	− 17° 8'	− 4° 11'	− 166° 38'	17° 35'	+ 21° 36'	− 0° 24'	− 1° 1'	21° 37'	22
23	− 19° 42'	− 3° 6'	− 171° 25'	19° 54'	+ 22° 31'	+ 0° 45'	+ 1° 48'	22° 31'	23
24	− 20° 16'	− 1° 40'	− 175° 30'	20° 20'	+ 25° 47'	+ 0° 2'	+ 0° 4'	25° 47'	24
25	− 19° 17'	+ 0° 34'	+ 178° 24'	19° 18'	+ 22° 21'	+ 0° 22'	+ 0° 55'	22° 21'	25
26	− 13° 56'	+ 3° 31'	+ 166° 5'	14° 20'	+ 23° 6'	− 0° 5'	− 0° 12'	23° 6'	26
27	− 5° 11'	+ 4° 15'	+ 140° 39'	6° 41'	+ 21° 26'	− 0° 13'	− 0° 32'	21° 26'	27
28	+ 3° 5'	+ 4° 5'	+ 52° 58'	5° 7'	+ 16° 14'	∓ 0° 0'	∓ 0° 0'	16° 14'	28
29	+ 9° 38'	+ 4° 24'	+ 24° 24'	10° 34'	+ 9° 56'	+ 0° 40'	+ 3° 50'	9° 57'	29
30	+ 15° 13'	+ 4° 41'	+ 16° 47'	15° 52'	+ 3° 42'	+ 1° 25'	+ 20° 54'	3° 58'	30
31	+ 20° 24'	+ 4° 40'	+ 12° 22'	20° 51'	− 1° 57'	+ 1° 42'	+ 139° 4'	2° 35'	31

No.	Lower leg								No.
	Right				Left				
	φ_4	ψ_4	α_4	ε_4	φ_5	ψ_5	α_5	ε_5	
1	− 50° 57'	− 11° 57'	− 170° 16'	51° 22'	+ 7° 9'	+ 5° 41'	+ 38° 25'	9° 6'	1
2	− 56° 33'	− 13° 30'	− 170° 59'	56° 53'	+ 2° 29'	+ 6° 13'	+ 68° 18'	6° 41'	2
3	− 54° 34'	− 11° 52'	− 171° 30'	54° 52'	− 0° 34'	+ 6° 36'	+ 94° 50'	6° 36'	3
4	− 47° 17'	− 10° 2'	− 170° 43'	47° 39'	− 2° 46'	+ 6° 13'	+ 113° 52'	6° 48'	4
5	− 37° 17'	− 8° 17'	− 169° 11'	37° 47'	− 4° 25'	+ 5° 47'	+ 127° 20'	7° 16'	5
6	− 24° 39'	− 6° 31'	− 166° 2'	25° 18'	− 6° 2'	+ 5° 18'	+ 138° 45'	8° 0'	6
7	− 9° 49'	− 5° 4'	− 152° 53'	11° 0'	− 8° 42'	+ 5° 6'	+ 149° 46'	10° 2'	7
8	+ 4° 33'	− 3° 46'	− 39° 38'	5° 54'	− 11° 42'	+ 4° 53'	+ 157° 36'	12° 37'	8
9	+ 21° 27'	− 0° 56'	− 2° 22'	21° 28'	− 15° 42'	+ 4° 35'	+ 164° 5'	16° 18'	9
10	+ 29° 10'	+ 0° 3'	+ 0° 6'	29° 10'	− 20° 33'	+ 4° 37'	+ 167° 51'	20° 59'	10
11	+ 22° 14'	− 2° 41'	− 6° 32'	22° 21'	− 27° 10'	+ 5° 3'	+ 170° 14'	27° 30'	11
12	+ 15° 54'	− 4° 31'	− 15° 31'	16° 28'	− 34° 5'	+ 5° 40'	+ 171° 39'	34° 22'	12
13	+ 10° 1'	− 6° 38'	− 33° 23'	11° 56'	− 42° 25'	+ 7° 39'	+ 171° 38'	42° 43'	13
14	+ 1° 32'	− 7° 12'	− 78° 2'	7° 22'	− 50° 32'	+ 8° 52'	+ 172° 41'	50° 46'	14
15	− 2° 4'	− 7° 34'	− 105° 14'	7° 50'	− 53° 41'	+ 8° 51'	+ 173° 28'	53° 51'	15
16	− 4° 18'	− 7° 11'	− 120° 53'	8° 21'	− 48° 44'	+ 8° 16'	+ 172° 44'	48° 58'	16
17	− 5° 52'	− 6° 43'	− 131° 7'	8° 53'	− 39° 44'	+ 7° 40'	+ 170° 47'	40° 6'	17
18	− 7° 13'	− 5° 54'	− 140° 49'	9° 17'	− 28° 29'	+ 6° 50'	+ 167° 32'	29° 4'	18
19	− 9° 3'	− 5° 24'	− 149° 16'	10° 30'	− 15° 25'	+ 5° 40'	+ 160° 13'	16° 20'	19
20	− 11° 30'	− 5° 9'	− 156° 4'	12° 32'	− 0° 58'	+ 4° 18'	+ 102° 39'	4° 25'	20
21	− 14° 36'	− 4° 45'	− 162° 17'	15° 17'	+ 13° 52'	+ 2° 45'	+ 10° 59'	14° 7'	21
22	− 18° 32'	− 4° 44'	− 166° 6'	19° 3'	+ 27° 0'	+ 0° 46'	+ 1° 30'	27° 0'	22
23	− 24° 37'	− 5° 14'	− 168° 41'	25° 3'	+ 28° 46'	+ 1° 8'	+ 2° 4'	28° 47'	23
24	− 31° 46'	− 6° 4'	− 170° 16'	32° 9'	+ 20° 12'	+ 3° 45'	+ 10° 7'	20° 29'	24
25	− 39° 35'	− 7° 40'	− 170° 45'	39° 57'	+ 18° 13'	+ 4° 29'	+ 13° 24'	18° 42'	25
26	− 48° 49'	− 10° 46'	− 170° 33'	49° 13'	+ 10° 23'	+ 6° 3'	+ 30° 0'	11° 57'	26
27	− 56° 56'	− 11° 40'	− 172° 21'	57° 10'	+ 3° 51'	+ 6° 48'	+ 60° 31'	7° 48'	27
28	− 57° 1'	− 9° 38'	− 173° 43'	57° 11'	+ 0° 10'	+ 7° 3'	+ 88° 41'	7° 3'	28
29	− 51° 17'	− 8° 24'	− 173° 15'	51° 29'	− 2° 29'	+ 6° 42'	+ 110° 16'	7° 8'	29
30	− 42° 9'	− 6° 54'	− 172° 23'	42° 24'	− 4° 29'	+ 6° 9'	+ 126° 7'	7° 35'	30
31	− 29° 49'	− 5° 37'	− 170° 15'	30° 11'	− 6° 22'	+ 5° 53'	+ 137° 16'	8° 38'	31

Table 3 (continued)

No.	Foot								No.
	Right				Left				
	φ_6	ψ_6	α_6	ε_6	φ_7	ψ_7	α_7	ε_7	
1	− 20° 11′	+ 2° 19′	+ 173° 45′	20° 18′	+ 44° 16′	− 6° 5′	− 6° 15′	44° 27′	1
2	− 22° 51′	− 0° 17′	− 179° 20′	22° 51′	+ 43° 15′	− 7° 6′	− 7° 32′	43° 30′	2
3	− 19° 23′	− 4° 17′	− 168° 0′	19° 47′	+ 42° 30′	− 7° 3′	− 7° 41′	42° 45′	3
4	− 8° 40′	− 2° 8′	− 166° 14′	8° 55′	+ 41° 56′	− 7° 7′	− 7° 54′	42° 12′	4
5	+ 3° 54′	− 0° 16′	− 3° 54′	3° 55′	+ 41° 14′	− 6° 41′	− 7° 37′	41° 29′	5
6	+ 17° 54′	+ 2° 46′	+ 8° 32′	18° 6′	+ 40° 36′	− 6° 45′	− 7° 52′	40° 52′	6
7	+ 32° 59′	+ 5° 36′	+ 8° 36′	33° 17′	+ 40° 6′	− 6° 55′	− 8° 12′	40° 23′	7
8	+ 45° 54′	+ 6° 49′	+ 6° 36′	46° 6′	+ 39° 37′	− 7° 39′	− 9° 13′	39° 59′	8
9	+ 59° 40′	+ 12° 10′	+ 7° 12′	59° 52′	+ 36° 38′	− 7° 25′	− 9° 55′	37° 3′	9
10	+ 67° 20′	+ 27° 5′	+ 12° 3′	67° 47′	+ 33° 10′	− 7° 39′	− 11° 37′	33° 42′	10
11	+ 62° 14′	+ 24° 4′	+ 13° 14′	62° 52′	+ 23° 30′	− 4° 36′	− 10° 30′	23° 51′	11
12	+ 53° 58′	+ 12° 27′	+ 9° 8′	54° 19′	+ 11° 14′	− 1° 46′	− 8° 52′	11° 22′	12
13	+ 44° 13′	+ 7° 10′	+ 7° 22′	44° 28′	− 8° 23′	+ 2° 31′	+ 163° 21′	8° 45′	13
14	+ 41° 14′	+ 5° 47′	+ 6° 35′	41° 25′	− 17° 56′	+ 1° 14′	+ 176° 11′	17° 59′	14
15	+ 39° 32′	+ 6° 30′	+ 7° 51′	39° 48′	− 20° 10′	+ 1° 26′	+ 176° 7′	20° 12′	15
16	+ 39° 7′	+ 6° 36′	+ 8° 5′	39° 24′	− 12° 16′	+ 1° 55′	+ 171° 17′	12° 24′	16
17	+ 38° 51′	+ 7° 3′	+ 8° 44′	39° 10′	− 0° 45′	+ 1° 49′	+ 112° 15′	1° 58′	17
18	+ 38° 11′	+ 7° 37′	+ 9° 39′	38° 35′	+ 12° 28′	− 0° 5′	− 0° 23′	12° 28′	18
19	+ 37° 37′	+ 8° 13′	+ 10° 36′	38° 6′	+ 27° 46′	− 3° 51′	− 7° 18′	27° 57′	19
20	+ 37° 13′	+ 8° 42′	+ 11° 24′	37° 46′	+ 42° 32′	− 7° 3′	− 7° 41′	42° 48′	20
21	+ 35° 50′	+ 8° 48′	+ 12° 6′	36° 27′	+ 57° 30′	− 12° 40′	− 8° 9′	57° 46′	21
22	+ 33° 2′	+ 9° 52′	+ 14° 59′	33° 56′	+ 69° 26′	− 21° 46′	− 8° 31′	69° 39′	22
23	+ 26° 24′	+ 7° 26′	+ 14° 43′	27° 10′	+ 72° 56′	− 33° 57′	− 11° 40′	73° 16′	23
24	+ 15° 32′	+ 6° 19′	+ 21° 41′	16° 40′	+ 65° 47′	− 21° 50′	− 10° 13′	66° 7′	24
25	+ 0° 27′	+ 3° 28′	+ 82° 42′	3° 30′	+ 53° 57′	− 1° 39′	− 1° 12′	53° 57′	25
26	− 16° 27′	+ 0° 37′	+ 177° 54′	16° 27′	+ 48° 27′	− 1° 22′	− 1° 13′	48° 27′	26
27	− 22° 59′	− 2° 39′	− 173° 45′	23° 6′	+ 43° 56′	− 3° 57′	− 4° 6′	44° 0′	27
28	− 21° 45′	− 4° 47′	− 168° 10′	22° 11′	+ 43° 5′	− 4° 41′	− 5° 0′	43° 12′	28
29	− 13° 14′	− 3° 36′	− 165° 0′	13° 41′	+ 42° 35′	− 5° 14′	− 5° 42′	42° 43′	29
30	− 0° 52′	− 0° 37′	− 145° 0′	1° 4′	+ 41° 50′	− 5° 46′	− 6° 26′	42° 0′	30
31	+ 12° 45′	+ 2° 25′	+ 10° 33′	12° 58′	+ 41° 4′	− 6° 2′	− 6° 55′	41° 17′	31

opposite is true for the diagrams of the left thigh. Only in experiment III is a slight discrepancy observed: the left thigh (upper black diagram in Fig. 9) reaches its maximum shortly after L and its minimum shortly after R. Moreover, the first maximum is preceded by a slightly greater maximum shortly before E_r. The latter maximum is only about 2° higher than the former. This discrepancy in the behaviour of the left thigh in experiment III may result from the fact that the body is more loaded on the left side than on the right owing to the rifle carried on the left shoulder. The angle between the extreme positions of the left thigh also appears somewhat abnormal in experiment III. This angle rounded to the nearest half degree is respectively 52°30′, 51° and 50° for the right leg in experiments I–III; for the left leg, it is 50°, 49° and 44°30′. The greatest deviation of the right thigh anterior to the vertical at R is 30°30′, 30°30′ and 22°30′, and posterior to the vertical at L 22°, 20°30′ and 27°30′. The greatest deviation of the left thigh anterior to the vertical at L is 25°30′, 26° and 22°30′ and posterior to the vertical at R 24°30′, 23° and 22°. Carrying the equipment thus results in a reduction of the deviation of both thighs anterior to the vertical at ipsilateral heel strike. The deviation posterior to the vertical is greater for the right thigh and somewhat smaller for the left. Here also, asymmetry appears in the behaviour of the two legs in experiment III.

The greatest deviation of the long axis of the lower leg (red curve) anterior or posterior to the vertical does not coincide with the greatest inclination of the thigh. The greatest inclination of the lower leg anterior to the vertical occurs earlier than that of the thigh. Its greatest inclination posterior to the vertical occurs considerably later than that of the thigh. The greatest inclination of the lower leg anterior to the vertical occurs at the end of the first third of contralateral push-off, the greatest inclination posterior to the vertical occurs after contralateral foot-flat. Both occur outside the period of ipsilateral stance; the former occurs approximately as long before as the latter occurs after this period.

Table 4. Positional angles of the segments of the lower extremities in experiment III

No.	Thigh Right φ_2	ψ_2	α_2	ε_2	Thigh Left φ_3	ψ_3	α_3	ε_3	No.
1	+ 1°38'	+4°10'	+ 68°35'	4°29'	+10°21'	+2°17'	+ 12°20'	10°36'	1
2	+ 8°11'	+4°17'	+ 27°30'	9°13'	+ 4°49'	+2°12'	+ 24°27'	5°17'	2
3	+13°59'	+5°0'	+ 19°22'	14°47'	− 0°22'	+2°19'	+ 98°53'	2°20'	3
4	+17°46'	+5°5'	+ 15°31'	18°24'	− 5°21'	+2°20'	+156°25'	5°50'	4
5	+19°20'	+4°32'	+ 12°45'	19°47'	−10°3'	+2°12'	+167°43'	10°17'	5
6	+18°42'	+3°57'	+ 11°31'	19°3'	−14°43'	+2°24'	+170°57'	14°54'	6
7	+17°21'	+2°40'	+ 8°29'	17°31'	−18°23'	+2°6'	+173°42'	18°30'	7
8	+19°30'	+2°51'	+ 8°1'	19°40'	−20°31'	+1°15'	+176°44'	20°33'	8
9	+22°26'	+4°32'	+ 10°53'	22°48'	−21°49'	−0°15'	−179°23'	21°49'	9
10	+19°6'	+4°12'	+ 11°59'	19°30'	−20°32'	−2°15'	−174°2'	20°38'	10
11	+19°3'	+4°34'	+ 13°2'	19°31'	−14°22'	−4°44'	−162°3'	15°4'	11
12	+17°47'	+4°14'	+ 12°58'	18°13'	− 6°10'	−4°47'	−142°14'	7°47'	12
13	+11°17'	+3°10'	+ 15°30'	11°42'	+ 1°57'	−4°17'	− 65°27'	4°42'	13
14	+ 4°10'	+2°36'	+ 31°55'	4°55'	+ 8°37'	−4°27'	− 27°8'	9°40'	14
15	− 2°5'	+2°2'	+135°38'	2°54'	+14°55'	−4°38'	− 16°54'	15°34'	15
16	− 7°2'	+1°46'	+166°0'	7°15'	+19°52'	−4°30'	− 12°18'	20°17'	16
17	−12°13'	+1°55'	+171°12'	12°21'	+23°6'	−4°22'	− 10°10'	23°25'	17
18	−17°12'	+1°54'	+173°53'	17°17'	+24°29'	−4°8'	− 9°1'	24°45'	18
19	−21°3'	+1°48'	+175°19'	21°7'	+23°37'	−3°10'	− 7°13'	23°47'	19
20	−24°26'	+2°22'	+174°48'	24°31'	+21°7'	−1°13'	− 3°9'	21°9'	20
21	−26°45'	+3°13'	+173°39'	26°53'	+21°16'	−1°4'	− 2°45'	21°18'	21
22	−27°5'	+4°44'	+170°49'	27°23'	+22°20'	−0°59'	− 2°25'	22°21'	22
23	−25°15'	+5°48'	+167°52'	25°45'	+20°48'	−1°39'	− 4°20'	20°51'	23
24	−19°44'	+7°12'	+160°36'	20°50'	+20°14'	−1°28'	− 3°58'	20°17'	24
25	−11°37'	+6°51'	+149°42'	13°24'	+17°43'	−1°9'	− 3°35'	17°45'	25
26	− 3°33'	+6°12'	+119°48'	7°8'	+12°37'	−0°15'	− 1°8'	12°37'	26
27	+ 2°54'	+5°8'	+ 60°36'	5°53'	+ 6°32'	+0°10'	+ 1°25'	6°32'	27
28	+ 8°42'	+4°37'	+ 27°53'	9°49'	+ 0°56'	+0°43'	+ 37°26'	1°10'	28
29	+13°43'	+4°1'	+ 16°4'	14°15'	− 4°14'	+1°20'	+162°27'	4°26'	29
30	+17°19'	+3°33'	+ 11°14'	17°38'	− 8°54'	+1°45'	+168°57'	9°4'	30
31	+19°27'	+2°35'	+ 7°17'	19°35'	−12°53'	+2°2'	+171°11'	13°2'	31

No.	Lower leg Right φ_4	ψ_4	α_4	ε_4	Lower leg Left φ_5	ψ_5	α_5	ε_5	No.
1	−43°56'	−4°28'	−175°22'	44°2'	− 3°36'	+4°44'	+127°18'	5°56'	1
2	−34°18'	−4°2'	−174°6'	34°27'	− 5°26'	+4°19'	+141°35'	6°56'	2
3	−22°25'	−3°51'	−170°45'	22°40'	− 7°27'	+3°51'	+152°46'	8°22'	3
4	− 9°4'	−2°31'	−164°33'	9°24'	− 9°45'	+3°38'	+159°47'	10°23'	4
5	+ 5°45'	−0°26'	− 4°19'	5°46'	−12°25'	+3°25'	+164°48'	12°51'	5
6	+20°32'	+2°11'	+ 5°49'	20°38'	−15°29'	+3°19'	+168°12'	15°48'	6
7	+31°35'	+5°20'	+ 8°38'	31°53'	−19°50'	+3°28'	+170°28'	20°5'	7
8	+30°44'	+5°7'	+ 8°33'	31°1'	−24°21'	+3°13'	+172°54'	24°31'	8
9	+23°33'	+2°7'	+ 4°52'	23°38'	−30°16'	+3°17'	+174°23'	30°24'	9
10	+21°13'	+1°15'	+ 3°12'	21°15'	−38°20'	+4°14'	+174°39'	38°27'	10
11	+13°9'	−0°40'	− 2°49'	13°10'	−48°8'	+6°30'	+174°10'	48°17'	11
12	+ 6°9'	−1°35'	− 14°25'	6°21'	−56°10'	+6°34'	+175°35'	56°15'	12
13	+ 4°1'	−1°30'	− 20°32'	4°17'	−56°22'	+7°22'	+175°5'	56°27'	13
14	+ 3°16'	−1°13'	− 20°23'	3°30'	−50°37'	+9°37'	+172°5'	50°53'	14
15	+ 1°50'	−0°53'	− 25°29'	2°2'	−41°29'	+9°57'	+168°47'	42°2'	15
16	− 0°37'	−0°42'	−131°16'	0°56'	−29°48'	+8°35'	+165°14'	30°39'	16
17	− 3°3'	−0°39'	−168°4'	3°7'	−16°14'	+6°59'	+157°11'	17°32'	17
18	− 5°19'	−0°25'	−175°34'	5°20'	− 1°38'	+5°4'	+107°51'	5°19'	18
19	− 8°58'	−0°5'	−179°27'	8°58'	+14°8'	+2°2'	+ 8°2'	14°16'	19
20	−13°5'	+0°17'	+178°49'	13°6'	+26°5'	−2°54'	− 5°55'	26°12'	20
21	−18°3'	+0°38'	+178°2'	18°4'	+26°17'	−2°17'	− 4°37'	26°21'	21
22	−24°28'	+1°1'	+177°47'	24°29'	+20°14'	−0°37'	− 1°41'	20°14'	22
23	−31°10'	+1°9'	+178°6'	31°11'	+15°23'	+2°4'	+ 7°27'	15°31'	23
24	−39°35'	+0°23'	+179°32'	39°35'	+ 8°12'	+3°17'	+ 21°40'	8°49'	24
25	−46°53'	−0°9'	−179°51'	46°53'	+ 2°10'	+3°57'	+ 61°22'	4°30'	25
26	−47°8'	−1°31'	−178°36'	47°8'	− 1°13'	+4°2'	+106°44'	4°13'	26
27	−41°25'	−3°21'	−176°12'	41°29'	− 2°49'	+3°55'	+125°49'	4°49'	27
28	−33°5'	−4°53'	−172°32'	33°19'	− 4°17'	+3°37'	+139°54'	5°36'	28
29	−21°50'	−5°37'	−166°13'	22°25'	− 6°24'	+3°19'	+152°39'	7°12'	29
30	− 8°42'	−6°0'	−145°33'	10°31'	− 8°48'	+3°12'	+160°6'	9°21'	30
31	+ 5°23'	−5°22'	− 44°55'	7°35'	−11°53'	+3°20'	+164°32'	12°19'	31

Table 4 (continued)

No.	Right				Left				No.
	φ_6	ψ_6	α_6	ε_6	φ_7	ψ_7	α_7	ε_7	
1	− 10° 8'	− 3° 10'	− 162° 47'	10° 36'	+ 41° 48'	− 9° 41'	− 10° 49'	42° 19'	1
2	+ 2° 10'	− 2° 10'	− 45° 0'	3° 4'	+ 41° 38'	− 9° 58'	− 11° 11'	42° 11'	2
3	+ 16° 38'	− 0° 51'	− 2° 51'	16° 39'	+ 41° 21'	− 10° 19'	− 11° 41'	41° 57'	3
4	+ 30° 24'	+ 0° 56'	+ 1° 36'	30° 24'	+ 41° 7'	− 11° 20'	− 12° 56'	41° 51'	4
5	+ 44° 0'	+ 2° 19'	+ 2° 23'	44° 2'	+ 40° 19'	− 11° 52'	− 13° 55'	41° 10'	5
6	+ 57° 11'	+ 8° 0'	+ 5° 11'	57° 18'	+ 39° 28'	− 11° 43'	− 14° 6'	40° 20'	6
7	+ 68° 13'	+ 18° 26'	+ 7° 35'	68° 23'	+ 36° 45'	− 10° 13'	− 13° 35'	37° 32'	7
8	+ 68° 8'	+ 22° 57'	+ 9° 39'	68° 25'	+ 28° 27'	− 8° 15'	− 14° 59'	29° 17'	8
9	+ 59° 42'	+ 13° 3'	+ 7° 43'	59° 55'	+ 18° 10'	− 5° 30'	− 16° 21'	18° 52'	9
10	+ 46° 59'	+ 4° 7'	+ 3° 56'	47° 4'	+ 3° 50'	− 1° 27'	− 20° 48'	4° 6'	10
11	+ 42° 10'	+ 3° 7'	+ 3° 26'	42° 13'	− 12° 18'	− 0° 10'	− 179° 15'	12° 18'	11
12	+ 39° 59'	+ 3° 46'	+ 4° 29'	40° 4'	− 17° 53'	− 1° 38'	− 174° 57'	17° 57'	12
13	+ 39° 14'	+ 4° 1'	+ 4° 55'	39° 20'	− 15° 16'	− 1° 54'	− 173° 4'	15° 23'	13
14	+ 39° 38'	+ 3° 17'	+ 3° 58'	39° 42'	− 7° 50'	− 1° 2'	− 172° 34'	7° 54'	14
15	+ 39° 21'	+ 4° 33'	+ 5° 32'	39° 29'	+ 3° 35'	− 0° 58'	− 14° 41'	3° 42'	15
16	+ 38° 43'	+ 3° 40'	+ 4° 34'	38° 48'	+ 17° 39'	− 4° 21'	− 13° 26'	18° 6'	16
17	+ 38° 25'	+ 7° 12'	+ 9° 3'	38° 46'	+ 31° 46'	− 8° 1'	− 12° 49'	32° 25'	17
18	+ 38° 21'	+ 8° 12'	+ 10° 19'	38° 48'	+ 46° 54'	− 9° 41'	− 9° 5'	47° 16'	18
19	+ 36° 27'	+ 8° 33'	+ 11° 30'	37° 1'	+ 62° 35'	− 16° 44'	− 8° 52'	62° 52'	19
20	+ 33° 52'	+ 8° 37'	+ 12° 43'	34° 32'	+ 72° 39'	− 32° 59'	− 11° 28'	72° 58'	20
21	+ 28° 14'	+ 7° 46'	+ 14° 16'	28° 59'	+ 73° 43'	− 38° 5'	− 12° 54'	74° 6'	21
22	+ 18° 53'	+ 6° 6'	+ 17° 21'	19° 43'	+ 66° 25'	− 25° 48'	− 11° 55'	66° 53'	22
23	+ 4° 15'	+ 4° 25'	+ 46° 12'	6° 7'	+ 52° 59'	− 10° 31'	− 7° 58'	53° 15'	23
24	− 12° 31'	+ 4° 23'	+ 160° 58'	13° 13'	+ 46° 46'	− 3° 59'	− 3° 45'	46° 49'	24
25	− 19° 5'	+ 2° 5'	+ 173° 59'	19° 11'	+ 43° 9'	− 6° 38'	− 7° 5'	43° 23'	25
26	− 17° 46'	− 1° 13'	− 176° 12'	17° 48'	+ 42° 0'	− 6° 45'	− 7° 30'	42° 14'	26
27	− 9° 21'	− 1° 57'	− 168° 22'	9° 33'	+ 41° 23'	− 6° 53'	− 7° 48'	41° 39'	27
28	+ 1° 16'	− 1° 11'	− 42° 53'	1° 44'	+ 40° 43'	− 7° 0'	− 8° 7'	41° 0'	28
29	+ 14° 23'	− 1° 8'	− 4° 24'	14° 26'	+ 40° 2'	− 7° 48'	− 9° 15'	40° 25'	29
30	+ 27° 35'	− 0° 31'	− 0° 59'	27° 35'	+ 39° 8'	− 8° 22'	− 10° 14'	39° 35'	30
31	+ 41° 44'	∓ 0° 0'	∓ 0° 0'	41° 44'	+ 38° 41'	− 8° 52'	− 11° 2'	39° 13'	31

The line connecting the centre of the ankle and the centre of gravity of the foot behaves similarly.

The greatest forward inclination of the right lower leg is respectively 30°30′, 29°, 34°, that of the left lower leg 30°, 30°, 27°30′ for experiments I–III. The greatest backward inclination of the right lower leg is 57°30′, 59°, 48°30′, that of the left lower leg 54°, 54°, 57°30′. Here also the two legs behave asymmetrically in experiment III. The angle formed by the two extreme positions of the long axis of the right lower leg is 88°, 88°, 82°32′ and by the two extreme positions of the long axis of the left lower leg 84°, 84°, 85°. Thus, once again the first two experiments are in concordance.

The angle formed by the line connecting the centre of the ankle and the centre of gravity of the right foot, and the vertical in the extreme position is respectively 67°30′, 67°30′, 69°30′ when the foot is forward. For the left foot, this angle is 72°30′, 73°, 74°. It is 24°, 23°30′, 19°30′ for the right foot and 21°, 21°, 18° for the left when the foot is inclined posteriorly. The angle between the extreme positions of the foot in the projection on the plane of gait is 91°30′, 91°, 89° for the right foot and 93°30′, 94, 92° for the left.

The last angles do not differ much from the corresponding angles of the lower leg. In Figs. 7–9 the upper diagrams for the foot appear very similar to the corresponding diagrams for the lower leg, except during foot-flat. The diagrams for the thigh are quite different from the others. The discrepancies between the diagrams of the different segments of the same leg result partly from the fact that the leg does not really move strictly in one plane parallel to that of gait, but these discrepancies are largely the consequence of the simultaneous movement in the knee and ankle.

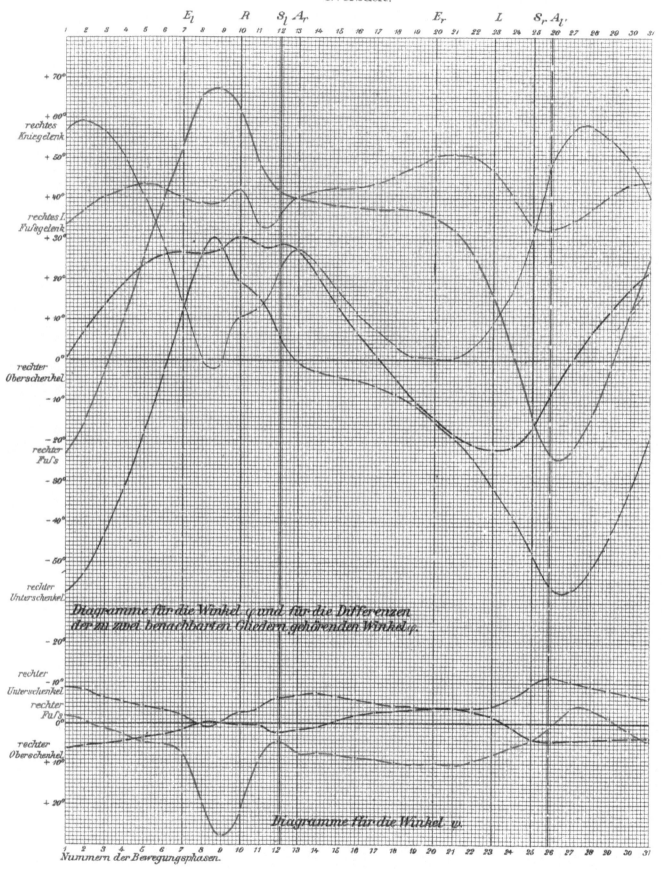

I.Versuch.

E_l R S_l A_r E_r L S_r $A_{l'}$

rechtes
Kniegelenk

rechtes I.
Fußgelenk

rechter
Oberschenkel

rechter
Fuß

rechter
Unterschenkel

Diagramme für die Winkel φ und für die Differenzen
der zu zwei benachbarten Gliedern gehörenden Winkel φ.

rechter
Unterschenkel
rechter
Fuß

rechter
Oberschenkel

Diagramme für die Winkel ψ.

Nummern der Bewegungsphasen.

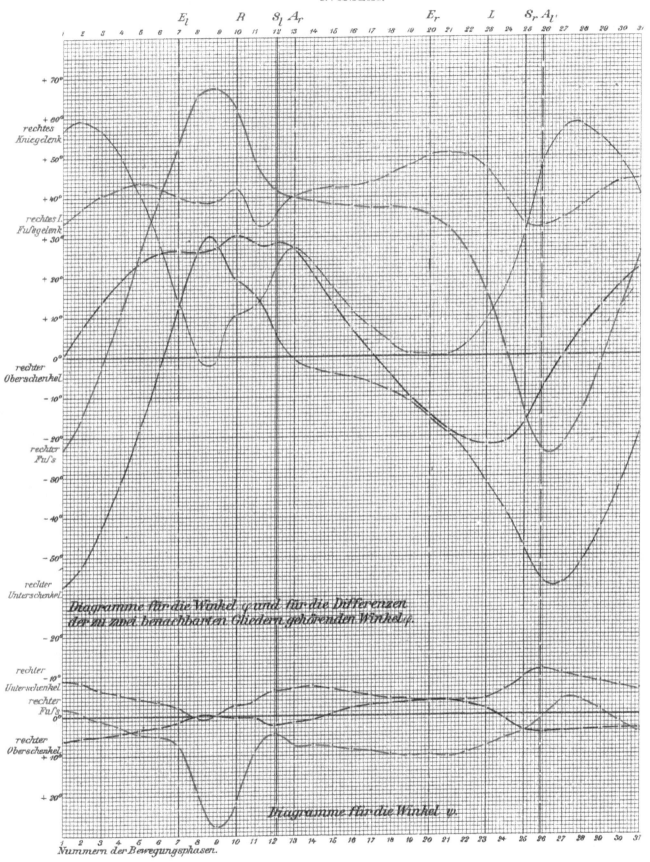

I.Versuch.

Diagramme für die Winkel φ und für die Differenzen der zu zwei benachbarten Gliedern gehörenden Winkel φ.

Diagramme für die Winkel ψ.

Nummern der Bewegungsphasen.

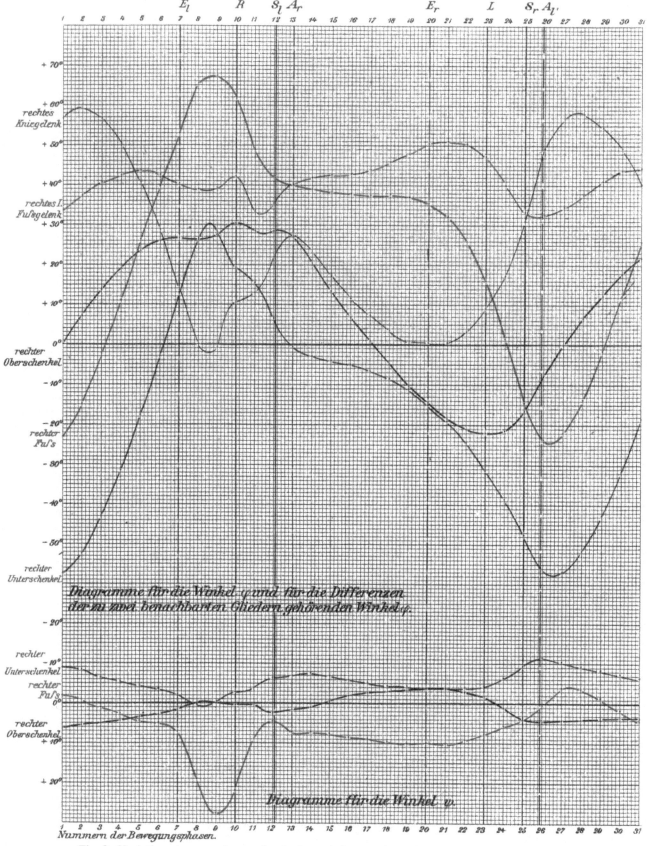

Fig. 8. Variations in φ- and ψ-angles during gait in experiment II

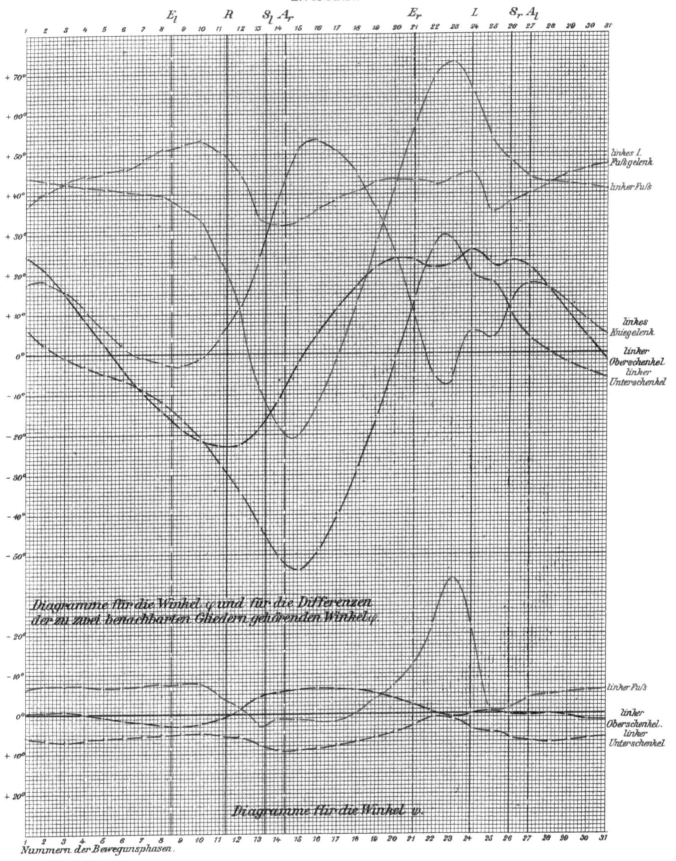

Fig. 8 (continued)

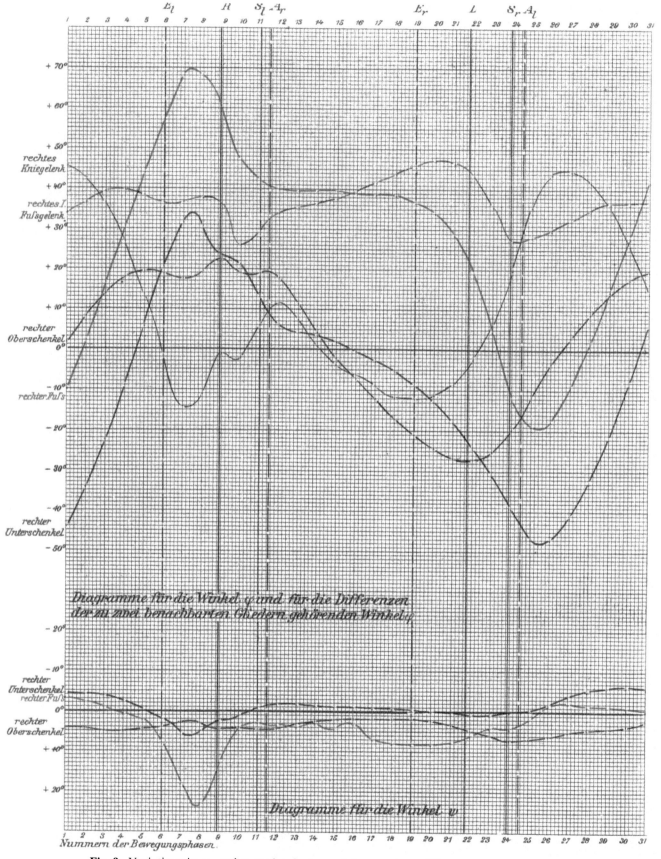

Fig. 9. Variations in φ- and ψ-angles during gait in experiment III

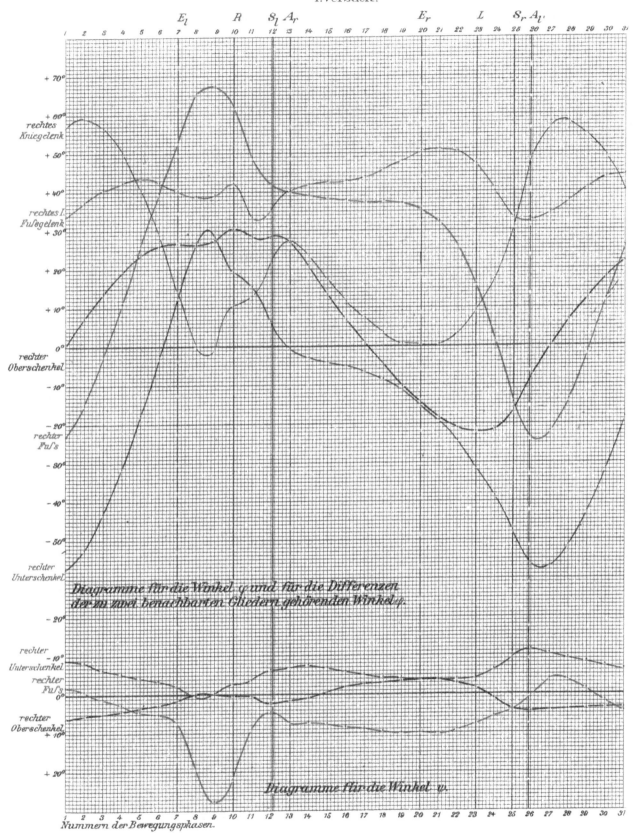

I. Versuch.

How much the plane movement is disturbed by the sideways movement of the different body segments can be seen in the lower diagrams in Figs. 7–9. These diagrams give the ψ-angle formed by the projection of each long axis on the plane perpendicular to the direction of gait and the vertical, during the movement. The diagrams of the first two experiments show what follows.

The long axis of the thigh is parallel to the plane of gait at ipsilateral heel strike as well as shortly before and after. The ψ-angle is then almost zero. "Lateral inclination" signifies an inclination of the long axis of a segment of the leg in relation to the plane of gait, with the distal end of this axis further away laterally from the plane of gait than the proximal end; "medial inclination" is the opposite. Thus, shortly after heel strike, the thigh is inclined at first a little laterally and then medially. This medial inclination persists almost throughout the whole of stance. Only at the end, shortly before toe-off, does the thigh tilt laterally, and it remains thus inclined during swing until shortly before heel strike when it becomes parallel to the plane of gait.

The left thigh behaves almost in a similar fashion in experiment III whereas the right thigh is laterally inclined during the whole movement. In the latter case, however, the variations in magnitude of the ψ-angle at least correspond to the sideways movements of the right thigh in experiments I and II.

The long axis of the lower leg (red diagrams) behaves in a manner exactly opposite to that of the thigh, relative to the plane of gait. In the first two experiments, both are medially inclined almost the whole time. However, the variations in the ψ-angle demonstrate movements of the projection of the long axis of the lower leg on the plane perpendicular to the direction of gait which are almost exactly the reverse of those of the long axis of the thigh. Only in experiment III is the behaviour of the right lower leg different from that in the first two experiments as well as from that of the left lower leg. In experiment III the movements of the left lower leg do not correspond to those of the right.

The foot line (blue diagrams) is inclined laterally in all instances almost throughout the movement. The maximum of this lateral inclination is considerable and occurs shortly before ipsilateral heel strike. Thereafter, the lateral inclination quickly diminishes until shortly before foot-flat. It then increases again but remains moderate until the end of stance. During foot-flat it remains fairly constant and progressively diminishes during push-off. At the beginning of the swinging period the foot line quickly comes close to a position parallel to the plane of gait, takes a medial inclination for some time, becomes again parallel to the plane of gait and then turns laterally to reach the maximum mentioned above shortly before heel strike. In gait with a load, the foot behaves approximately as in experiments I and II.

The deviations in the diagrams of the segments of a leg are mostly due to the movements of the knee and ankle. If the movement were plane and all the segments moved exactly parallel to the plane of gait, the positions of the knee and ankle could easily be deduced at any time from the upper diagrams in Figs. 7–9. When two segments are linked by a joint, the angle formed by the long axis of the distal segment and the prolongation of the long axis of the proximal segment is called the joint angle. The two long axes can also be displaced and their proximal extremities superimposed; the angle thus formed is the joint angle. The ordinates of the upper diagrams are proportional to the angle formed by the projection of a long axis on the plane of gait and the vertical. Therefore, if the movement is plane, the extent of the joint angle at any time corresponding to a point on the abscissa is given by the vertical distance between the diagrams of the two segments at this level of the abscissa. This distance gives the difference in inclination of the projections of the two long axes on the vertical and thus represents the measure of the angle formed by the two long axes, as long as the movement is plane. In the upper diagrams, the vertical distance between the black and the red curves thus gives the knee angle, the vertical distance between the red and the blue curves gives the angle of the ankle joint at any time. From the changes in these distances, the movements of the joints under consideration can be determined.

The black diagram lies above the red, except in a few places. Therefore, in the projection on the plane of gait, the long axis of the thigh most often forms a greater anterior angle and a smaller posterior angle with the vertical than the long axis of the lower leg. This means that the knee is mostly flexed. Only shortly before heel strike and at the beginning of push-off does the red curve reach the black or even extend somewhat beyond it. The two long axes then form a straight line or even an obtuse angle open in front. The latter case may well occur since the condyles lie some distance behind the diaphysis of the femur. In extreme extension of the knee, when the lower leg appears to be in the prolongation of the thigh, the axis of the knee and thus the centre of the knee lie behind the straight line connecting the centres of the hip and ankle. In and near extension of the knee, the line connecting the centres of the hip and knee, designated as "long axis of the thigh" thus forms an angle open in front with the line connecting the centres of the knee and ankle, also called the long axis of the lower leg. It is questionable whether this angle can become as large as it appears for the right leg in experiment III (Fig. 9). The Geissler tube fixed to the thigh, and which should have been located in the prolongation of the axis of the knee, may have moved a little too far backwards shortly before the experiment. The forward inclination of the long axis of the thigh may then have appeared somewhat too small and that of the long axis of the lower leg somewhat too large. This would explain why, compared with all the other diagrams of the same type, the upper black diagram of the left of Fig. 9 seems to be excessively displaced downwards, and the corresponding red diagram upwards. Whatever the cause, such an unintentional backward displacement of the centre of the knee does not influence the shape of the diagram. It only changes its position, as can be seen.

Table 5. Knee and ankle joint angles in experiment I

No.	Knee						Ankle						No.
	Right			Left			Right			Left			
	$\alpha_4-\alpha_2$	$\varphi_2-\varphi_4$	$\beta_{2,4}$	$\alpha_5-\alpha_3$	$\varphi_3-\varphi_5$	$\beta_{3,5}$	$\alpha_6-\alpha_4$	$\varphi_6-\varphi_4$	$\beta_{4,6}$	$\alpha_7-\alpha_5$	$\varphi_7-\varphi_5$	$\beta_{5,7}$	
1	+ 93° 0'	+56° 46'	+57° 48'	+ 70° 4'	+14° 30'	+16° 1'	− 1° 26'	33° 55'	34° 2'	− 79° 13'	40° 27'	42° 23'	1
2	+147° 56'	+59° 12'	+59° 58'	+ 88° 32'	+10° 35'	+12° 47'	− 2° 11'	37° 22'	37° 30'	− 97° 33'	42° 33'	44° 31'	2
3	+167° 31'	+56° 45'	+57° 29'	+ 98° 33'	+ 6° 8'	+ 8° 40'	− 32° 34'	40° 44'	41° 6'	−114° 0'	44° 8'	45° 55'	3
4	+176° 50'	+50° 22'	+51° 1'	− 28° 11'	+ 1° 30'	+ 4° 54'	−175° 33'	42° 17'	42° 49'	−129° 15'	45° 15'	46° 45'	4
5	−173° 52'	+40° 53'	+41° 29'	− 27° 29'	− 1° 54'	− 3° 38'	+174° 48'	43° 44'	44° 25'	−147° 50'	46° 24'	47° 49'	5
6	−132° 20'	+28° 57'	+29° 34'	− 17° 44'	− 3° 51'	− 4° 27'	+132° 48'	42° 38'	43° 15'	−162° 39'	48° 26'	49° 55'	6
7	− 12° 36'	+13° 55'	+14° 20'	− 9° 54'	− 4° 16'	− 4° 31'	+ 15° 56'	40° 18'	40° 53'	−173° 31'	50° 38'	52° 11'	7
8	+ 1° 7'	+ 0° 13'	+ 0° 31'	− 6° 7'	− 3° 27'	− 3° 43'	+ 8° 21'	39° 1'	39° 40'	+179° 34'	50° 59'	52° 28'	8
9	+ 0° 41'	− 0° 28'	− 0° 38'	− 5° 38'	− 0° 35'	− 2° 15'	+ 12° 13'	39° 22'	40° 42'	+175° 54'	50° 59'	52° 14'	9
10	− 8° 14'	+10° 40'	+11° 2'	− 7° 38'	+ 4° 51'	+ 6° 12'	+ 19° 47'	42° 19'	44° 9'	+172° 31'	48° 55'	49° 55'	10
11	− 14° 29'	+13° 6'	+13° 43'	− 12° 16'	+12° 30'	+14° 3'	+ 21° 27'	33° 30'	34° 48'	+164° 34'	41° 34'	42° 7'	11
12	− 47° 57'	+22° 5'	+23° 33'	− 21° 51'	+25° 27'	+27° 27'	+ 48° 51'	35° 39'	36° 52'	+ 1° 41'	31° 15'	31° 30'	12
13	− 95° 26'	+27° 24'	+28° 35'	− 35° 48'	+40° 37'	+42° 19'	+101° 3'	40° 40'	42° 27'	+ 8° 36'	32° 9'	32° 43'	13
14	−112° 2'	+23° 18'	+24° 45'	− 69° 14'	+49° 32'	+51° 0'	+118° 22'	42° 3'	43° 59'	+ 7° 20'	34° 2'	34° 28'	14
15	−117° 14'	+17° 23'	+18° 31'	−129° 50'	+51° 1'	+52° 24'	+129° 15'	42° 41'	44° 32'	+ 3° 46'	37° 40'	37° 57'	15
16	−114° 8'	+11° 50'	+12° 29'	−155° 4'	+47° 49'	+49° 15'	+140° 26'	43° 6'	44° 48'	−147° 51'	39° 42'	40° 0'	16
17	− 68° 38'	+ 7° 19'	+ 7° 42'	−173° 52'	+42° 50'	+44° 14'	+152° 36'	44° 6'	45° 39'	−175° 43'	42° 13'	42° 57'	17
18	− 3° 13'	+ 3° 28'	+ 3° 48'	+172° 11'	+35° 20'	+36° 34'	+162° 55'	46° 1'	47° 31'	−170° 38'	44° 3'	45° 13'	18
19	+ 4° 3'	+ 0° 51'	+ 1° 26'	+104° 3'	+23° 51'	+25° 0'	+171° 5'	48° 6'	49° 31'	−104° 38'	43° 50'	44° 58'	19
20	− 0° 25'	+ 0° 39'	+ 0° 38'	+ 14° 9'	+ 8° 5'	+ 9° 12'	+179° 27'	50° 26'	51° 40'	− 16° 40'	42° 48'	43° 38'	20
21	+ 0° 3'	+ 0° 38'	+ 0° 43'	+ 1° 11'	− 6° 38'	− 6° 38'	−175° 22'	51° 10'	52° 28'	− 10° 23'	42° 35'	43° 29'	21
22	+ 1° 27'	+ 3° 21'	+ 3° 29'	− 0° 19'	− 7° 22'	− 7° 22'	−170° 47'	50° 22'	51° 37'	− 11° 58'	42° 35'	43° 48'	22
23	+ 4° 9'	+ 9° 16'	+ 9° 38'	+ 7° 52'	+ 5° 7'	+ 5° 53'	−163° 5'	47° 20'	48° 34'	− 17° 39'	45° 6'	46° 37'	23
24	+ 10° 49'	+17° 15'	+18° 17'	+ 10° 42'	+ 3° 56'	+ 5° 16'	−114° 4'	40° 11'	41° 31'	− 16° 22'	33° 24'	34° 29'	24
25	+ 21° 39'	+30° 58'	+32° 34'	+ 22° 2'	+10° 8'	+11° 18'	− 23° 43'	33° 7'	34° 38'	− 22° 10'	34° 29'	34° 57'	25
26	+ 38° 36'	+48° 16'	+49° 28'	+ 56° 22'	+17° 6'	+18° 24'	− 8° 48'	32° 43'	33° 23'	− 58° 51'	37° 39'	38° 49'	26
27	+ 94° 58'	+56° 35'	+57° 24'	+ 84° 43'	+14° 30'	+15° 50'	+ 3° 16'	34° 49'	34° 45'	− 87° 49'	40° 23'	41° 30'	27
28	+157° 18'	+57° 58'	+58° 45'	+111° 20'	+10° 42'	+12° 19'	+ 7° 25'	37° 43'	37° 39'	−114° 38'	42° 22'	43° 34'	28
29	+173° 9'	+54° 51'	+55° 38'	+125° 45'	+ 6° 46'	+ 8° 11'	+ 32° 11'	41° 7'	41° 19'	−137° 13'	43° 36'	44° 43'	29
30	−178° 16'	+49° 20'	+50° 12'	− 27° 57'	+ 3° 11'	+ 5° 6'	+174° 23'	43° 45'	44° 21'	−151° 33'	44° 39'	45° 45'	30
31	−169° 41'	+40° 8'	+41° 11'	− 23° 52'	+ 0° 28'	+ 3° 47'	+169° 49'	44° 15'	45° 20'	−161° 54'	46° 7'	47° 18'	31

To obtain precise values for the joint angles in the case of a plane movement, it is necessary only to subtract the corresponding φ angles of the adjacent long axes from each other. For the knee, the φ-angle of the lower leg is subtracted from the φ-angle of the thigh, since the black curve mostly lies above the red. But for the ankle, the φ-angle of the lower leg is subtracted from the φ-angle of the foot since the red curve lies below the blue throughout the movement. These differences are given in Tables 5–7.

The two diagrams of the joint angles have been added to the upper diagrams of the long axes in Figs. 7–9. They have each been drawn in two colours corresponding to those of the segments adjacent to the joint. The black and red diagram relates to the knee, the blue and red to the ankle.

The diagrams concerning the knee confirm the conclusion drawn from Fig. 1 b. The knee always displays its greatest extension shortly before heel strike and then quickly flexes until heel strike. Subsequently it flexes somewhat more slowly or even not at all for a short time. It then continues flexing again at the initial velocity until foot-flat.

During foot-flat, the knee extends steadily until it almost reaches its greatest extension at the end of the period. During push-off, it flexes again until a long way into the swinging period. It then extends until its maximum extension shortly before heel strike.

During a double step, the ankle moves much less than the knee, as can be seen in the blue and red diagrams. At heel strike, it is in neutral position. Up to this moment it has slightly dorsiflexed for a short time. It then begins to plantar flex but soon reverts to dorsiflexion so that at foot-flat it is again in neutral position. This is fairly precise for the right foot in

Table 6. Knee and ankle joint angles in experiment II

No.	Knee						Ankle						No.
	Right			Left			Right			Left			
	$\alpha_4-\alpha_2$	$\varphi_2-\varphi_4$	$\beta_{2,4}$	$\alpha_5-\alpha_3$	$\varphi_3-\varphi_5$	$\beta_{3,5}$	$\alpha_6-\alpha_4$	$\varphi_6-\varphi_4$	$\beta_{4,6}$	$\alpha_7-\alpha_5$	$\varphi_7-\varphi_5$	$\beta_{5,7}$	
1	+34°44'	+38°43'	+40°50'	+39°8'	+17°27'	+18°24'	−15°59'	30°46'	32°13'	−44°40'	37°7'	38°24'	1
2	+52°47'	+50°33'	+52°9'	+69°50'	+18°18'	+19°28'	−8°21'	33°42'	34°23'	−75°50'	40°46'	42°16'	2
3	+112°37'	+55°53'	+57°5'	+97°41'	+15°39'	+17°14'	+3°30'	35°11'	35°8'	−102°31'	43°4'	44°33'	3
4	+161°53'	+56°40'	+57°45'	+114°28'	+11°30'	+13°6'	+4°29'	38°37'	38°46'	−121°46'	44°42'	46°5'	4
5	+174°21'	+53°2'	+54°6'	+112°46'	+6°46'	+8°30'	+165°17'	41°11'	41°35'	−134°57'	45°39'	46°51'	5
6	−177°6'	+45°54'	+46°54'	−22°26'	+1°58'	+4°21'	+174°34'	42°33'	43°21'	−146°37'	46°38'	47°43'	6
7	−160°48'	+34°31'	+35°29'	−19°7'	−0°46'	−3°16'	+161°29'	42°48'	43°50'	−157°58'	48°48'	49°48'	7
8	−45°39'	+21°18'	+22°14'	−12°35'	−2°9'	−3°14'	+46°14'	41°21'	42°11'	−166°49'	51°19'	52°20'	8
9	−4°30'	+4°2'	+4°25'	−7°31'	−2°48'	−3°17'	+9°34'	38°13'	38°48'	−174°0'	52°20'	53°17'	9
10	+0°28'	−3°30'	−3°30'	−5°38'	−1°0'	−2°9'	+11°57'	38°10'	39°30'	−179°28'	53°43'	54°41'	10
11	−6°44'	+6°55'	+7°23'	−6°59'	+4°29'	+5°37'	+19°46'	40°0'	42°15'	+179°16'	50°40'	51°21'	11
12	−15°21'	+13°47'	+14°25'	−10°41'	+11°31'	+12°48'	+24°39'	38°4'	39°46'	+179°29'	45°19'	45°44'	12
13	−35°8'	+17°39'	+19°6'	−20°14'	+23°46'	+25°34'	+40°45'	34°12'	36°6'	−8°17'	34°2'	34°5'	13
14	−79°30'	+25°53'	+26°59'	−31°42'	+38°59'	+40°24'	+84°37'	39°42'	34°35'	+3°30'	32°36'	32°50'	14
15	−104°21'	+24°23'	+25°22'	−73°55'	+51°9'	+52°18'	+113°5'	41°36'	43°22'	+2°39'	33°31'	33°41'	15
16	−116°32'	+20°19'	+21°7'	−134°46'	+53°32'	+54°43'	+128°58'	43°25'	45°3'	−1°27'	36°28'	36°34'	16
17	−119°27'	+15°9'	+15°51'	−160°25'	+50°54'	+52°13'	+139°51'	44°43'	46°15'	−58°32'	38°59'	39°6'	17
18	−93°24'	+9°51'	+10°16'	−173°20'	+45°22'	+46°48'	+150°28'	45°24'	46°51'	−167°55'	40°57'	41°20'	18
19	−13°18'	+5°35'	+5°55'	+173°18'	+36°49'	+38°11'	+159°52'	46°40'	48°4'	−167°31'	43°11'	44°2'	19
20	+1°58'	+2°35'	+2°56'	+111°25'	+24°54'	+26°6'	+167°28'	48°43'	50°4'	−110°20'	43°30'	44°29'	20
21	+1°18'	+1°9'	+1°21'	+16°1'	+9°43'	+10°47'	+174°23'	50°26'	51°41'	−19°8'	43°38'	44°35'	21
22	+0°32'	+1°24'	+1°28'	+2°31'	−5°24'	−5°29'	−178°55'	51°34'	52°59'	−10°1'	42°26'	43°12'	22
23	+2°44'	+4°55'	+5°15'	+0°16'	−6°15'	−6°16'	−176°36'	51°1'	52°12'	−13°44'	44°10'	45°33'	23
24	+5°14'	+11°30'	+12°2'	+10°3'	+5°35'	+6°35'	−168°3'	47°18'	48°34'	−20°20'	45°35'	47°13'	24
25	+10°51'	+20°18'	+21°15'	+12°29'	+4°8'	+5°41'	−106°33'	40°2'	41°4'	−14°36'	35°44'	36°4'	25
26	+23°22'	+34°53'	+36°24'	+30°12'	+12°43'	+14°3'	−11°33'	32°22'	33°13'	−31°13'	38°4'	38°37'	26
27	+47°0'	+51°45'	+52°45'	+61°3'	+17°35'	+18°52'	−1°24'	33°57'	34°5'	−64°37'	40°5'	41°9'	27
28	+133°19'	+60°6'	+60°46'	+88°41'	+16°4'	+17°31'	+5°33'	35°16'	35°9'	−93°41'	42°55'	44°6'	28
29	+162°21'	+60°55'	+61°37'	+106°26'	+12°25'	+13°46'	+8°15'	38°3'	37°59'	−115°58'	45°4'	46°12'	29
30	+170°50'	+57°22'	+58°7'	+105°13'	+8°11'	+9°26'	+27°23'	41°17'	41°27'	−132°33'	46°19'	47°24'	30
31	+177°23'	+50°13'	+51°1'	−1°48'	+4°25'	+6°3'	−179°12'	42°34'	43°9'	−144°11'	47°26'	48°30'	31

experiments I and II. The left foot, however, has not yet completely reached its neutral position at this point. During foot-flat, the ankle dorsiflexes further. It continues dorsiflexing through the first quarter of push-off and then resumes plantar flexion. This plantar flexion stops at toe-off. During the subsequent swinging period, the ankle mainly dorsiflexes although it plantar flexes briefly during the second half of the swing before finally resuming dorsiflexion. All these movements during the swinging period are relatively slow.

It should be noted that the differences in the φ-angles of adjacent body segments and the joint diagrams determine the position and movement of the joint only in their projection on the plane of gait. They do not describe the actual movements and positions of the joints since the movement of the lower extremities is not exactly plane.

The analysis could be completed by finding the corresponding differences in the ψ-angles and drawing the relevant diagrams. This would show the projection of the joint movement on the plane perpendicular to the direction of gait. From this projection, combined with that on the plane of gait, the actual tridimensional movement of the joints could be determined. This, however, was not carried out because the projection of the joint movement on the plane perpendicular to the direction of gait is much less important than that on the plane of gait and, in addition, it is not always easy to represent the joint movement from its projections. It is more appropriate to calculate, for each phase, the β-angle formed in space by the long axis of the distal segment with the prolongation of the long axis of the proximal segment, i.e. the actual value of the joint angle in space.

Table 7. Knee and ankle joint angles in experiment III

| No. | Knee | | | | | | Ankle | | | | | | No. |
| | Right | | | Left | | | Right | | | Left | | | |
	$\alpha_4 - \alpha_2$	$\varphi_2 - \varphi_4$	$\beta_{2,4}$	$\alpha_5 - \alpha_3$	$\varphi_3 - \varphi_5$	$\beta_{3,5}$	$\alpha_6 - \alpha_4$	$\varphi_6 - \varphi_4$	$\beta_{4,6}$	$\alpha_7 - \alpha_5$	$\varphi_7 - \varphi_5$	$\beta_{5,7}$	
1	+116° 3'	+45° 34'	+46° 8'	+114° 58'	+13° 57'	+14° 9'	+12° 35'	33° 48'	33° 45'	-138° 7'	45° 24'	46° 52'	1
2	+158° 24'	+42° 29'	+43° 8'	+117° 8'	+10° 15'	+10° 27'	+129° 6'	36° 28'	36° 27'	-152° 46'	47° 4'	48° 26'	2
3	+169° 53'	+36° 24'	+37° 18'	+53° 53'	+7° 5'	+7° 15'	+167° 54'	39° 3'	39° 6'	-164° 27'	48° 48'	51° 51'	3
4	+179° 56'	+26° 50'	+27° 48'	+3° 22'	+4° 24'	+4° 34'	+166° 9'	39° 28'	39° 35'	-172° 43'	50° 52'	52° 10'	4
5	-17° 4'	+13° 35'	+14° 22'	-2° 55'	+2° 22'	+2° 38'	+6° 42'	38° 15'	38° 19'	-178° 43'	52° 44'	54° 1'	5
6	-5° 42'	-1° 50'	-2° 29'	-2° 45'	+0° 46'	+1° 9'	-0° 38'	36° 39'	36° 40'	+177° 42'	54° 57'	56° 7'	6
7	+0° 9'	-14° 14'	-14° 22'	-3° 14'	+1° 27'	+1° 52'	-1° 3'	36° 38'	36° 30'	+175° 57'	56° 35'	57° 35'	7
8	+0° 32'	-11° 14'	-11° 21'	-3° 50'	+3° 50'	+4° 13'	+1° 6'	37° 24'	37° 24'	+172° 7'	52° 48'	53° 40'	8
9	-6° 1'	-1° 7'	-2° 30'	-6° 14'	+8° 27'	+9° 0'	+2° 51'	36° 9'	36° 19'	+169° 16'	48° 26'	49° 3'	9
10	-8° 47'	-2° 7'	-3° 30'	-11° 19'	+17° 48'	+18° 36'	+0° 44'	25° 46'	25° 49'	+164° 33'	42° 10'	42° 25'	10
11	-15° 51'	+5° 54'	+7° 42'	-23° 47'	+33° 46'	+34° 54'	+6° 15'	29° 1'	29° 9'	+6° 35'	35° 50'	36° 5'	11
12	-27° 23'	+11° 38'	+12° 54'	-42° 11'	+50° 0'	+50° 40'	+18° 54'	33° 50'	34° 7'	+9° 28'	38° 17'	38° 37'	12
13	-36° 2'	+7° 16'	+8° 36'	-119° 28'	+58° 19'	+58° 51'	+25° 27'	35° 13'	35° 30'	+11° 51'	41° 6'	41° 29'	13
14	-52° 18'	+0° 54'	+3° 55'	-160° 47'	+59° 14'	+60° 4'	+24° 21'	36° 22'	36° 32'	+15° 21'	42° 47'	43° 18'	14
15	-161° 7'	-3° 55'	-4° 52'	-174° 19'	+56° 24'	+57° 32'	+31° 1'	37° 31'	37° 45'	+176° 32'	45° 4'	45° 44'	15
16	+62° 44'	-6° 25'	-6° 53'	+177° 32'	+49° 40'	+50° 55'	+135° 50'	39° 20'	39° 28'	-178° 40'	47° 27'	48° 45'	16
17	+20° 44'	-9° 10'	-9° 30'	+167° 21'	+39° 20'	+40° 42'	+177° 7'	41° 28'	41° 53'	-170° 0'	48° 0'	49° 46'	17
18	+10° 33'	-11° 53'	-12° 5'	+116° 52'	+26° 7'	+27° 32'	-174° 7'	43° 40'	44° 6'	-116° 56'	48° 32'	49° 51'	18
19	+5° 14'	-12° 5'	-12° 13'	+15° 15'	+9° 29'	+10° 39'	-169° 3'	45° 25'	45° 51'	-16° 54'	48° 27'	49° 19'	19
20	+4° 1'	-11° 21'	-11° 29'	-2° 46'	-4° 58'	-5° 10'	-166° 6'	46° 57'	47° 20'	-5° 33'	46° 34'	46° 55'	20
21	+4° 23'	-8° 42'	-8° 58'	-1° 52'	-5° 1'	-5° 6'	-163° 46'	46° 17'	46° 35'	-8° 17'	47° 26'	48° 6'	21
22	+6° 58'	-2° 37'	-4° 12'	+0° 44'	+2° 6'	+2° 9'	-160° 26'	43° 21'	43° 32'	-10° 14'	46° 11'	47° 3'	22
23	+10° 14'	+5° 55'	+7° 17'	+11° 47'	+5° 25'	+6° 27'	-131° 54'	35° 25'	35° 32'	-15° 25'	37° 36'	38° 27'	23
24	+18° 56'	+19° 51'	+20° 50'	+25° 38'	+12° 2'	+12° 53'	-18° 34'	27° 4'	27° 20'	-25° 25'	38° 34'	39° 0'	24
25	+30° 27'	+35° 16'	+35° 50'	+64° 57'	+15° 33'	+16° 21'	-6° 10'	27° 48'	27° 52'	-68° 27'	40° 59'	41° 54'	25
26	+61° 36'	+43° 35'	+44° 5'	+107° 52'	+13° 50'	+14° 28'	+2° 24'	29° 22'	29° 21'	-114° 14'	43° 13'	44° 6'	26
27	+123° 12'	+44° 19'	+44° 56'	+124° 24'	+9° 21'	+10° 4'	+7° 50'	32° 4'	32° 3'	-133° 37'	44° 12'	45° 5'	27
28	+159° 35'	+41° 47'	+42° 38'	+102° 28'	+5° 13'	+5° 58'	+129° 39'	34° 21'	34° 27'	-148° 1'	45° 0'	45° 50'	28
29	+177° 43'	+35° 33'	+36° 40'	-9° 48'	+2° 10'	+2° 55'	+161° 49'	36° 13'	36° 24'	-161° 54'	46° 26'	47° 18'	29
30	-156° 47'	+26° 1'	+27° 36'	-8° 51'	-0° 6'	-1° 27'	+144° 34'	36° 17'	36° 37'	-170° 20'	47° 56'	48° 49'	30
31	-52° 12'	+14° 4'	+16° 3'	-6° 39'	-1° 0'	-1° 36'	+44° 55'	36° 21'	36° 41'	-175° 34'	50° 34'	51° 30'	31

The Angles of the Knee and Ankle Joints

The joint angle β obviously can only be calculated from the φ- and ψ-angles of the two body segments linked by the joint. The joint angle can be established if the directions of the long axes of the two adjacent segments are known and these directions are clearly defined by the two positional angles φ and ψ. However, the equation that is used in such a calculation is very inconvenient for practical purposes. The α- and ε-angles have already been calculated for each long axis and phase (Tables 2–4). To calculate β it is more convenient to establish an equation which makes use of these two other positional angles. This appears relatively simple and more convenient for practical calculation than using the equation with φ and ψ.

The two body segments linked by the joint are thought of as being displaced, each parallel to itself, to the origin O of a system of co-ordinates parallel to the system used hitherto (Fig. 10). From point V on the negative Z-axis, a horizontal plane thus parallel to the plane XY, intersects the proximal long axis at P and the distal long axis at D. The angle formed by the proximal long axis and the vertical is ε_p, that formed by the distal long axis and the vertical is ε_d. α_p and α_d are the angles formed by the vertical planes through the proximal and distal long axes and the anterior part of the plane of gait. β signifies the joint angle which is the angle formed by the two long axes originating in O.

In Fig. 10: $\measuredangle VOP = \varepsilon_p$ $\measuredangle VOD = \varepsilon_d$ $\measuredangle DVP = \alpha_d - \alpha_p$ $\measuredangle DOP = \beta$.

$\measuredangle DVP$ represents the angle formed by the planes in which the two triangles VOD and VOP are.

According to an established law of spherical trigonometry, the following relationship exists between β and the three angles ε_p, ε_d, $\alpha_d - \alpha_p$:

$$\cos \beta = \cos \varepsilon_d \cos \varepsilon_p + \sin \varepsilon_d \sin \varepsilon_p \cos (\alpha_d - \alpha_p)$$

Using this equation, the knee angles $\beta_{2,4}$ and $\beta_{3,5}$ and the angles of the ankle $\beta_{4,6}$ and $\beta_{5,7}$ were calculated and are given in Tables 5–7. These tables also give the values of the difference $\alpha_d - \alpha_p$.

Comparing the values of β with those of the difference between the two corresponding adjacent φ-angles does not show any great dissimilarities. This results from the fact that the lower extremities in walking move almost parallel to the plane of gait. If they moved exactly parallel to the plane of gait, the two angles would be the same since the difference between the

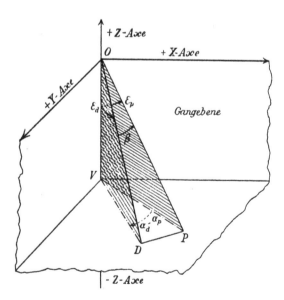

Fig. 10. Determination of the angle formed by two body segments

two adjacent φ-positional angles actually represents the joint angle in the projection on the plane of gait. If all the long axes kept moving parallel to the plane of gait, the projected movement would no longer be different from the actual movement. The joint diagrams of Figs. 7–9 discussed above thus can be used to represent at a good approximation the values of the actual joint angles. If a more thorough analysis is to be carried out based on the joint angles, the accurate values of the β-angles given in Tables 5–7 must be used rather than Figs. 7–9.

Summary

This chapter began by discussing further studies to be carried out on human gait. All studies basically pursue the same goal which consists of determining the involvement of the different muscles in the process of walking. To this end, thorough knowledge of the changing state of movement of the human body during a double step is required. Only then can there be hope of reaching the goal by means of the differential equations of the movement.

The movement equations for the human body are indeed very complicated. However, they permit a relatively simple interpretation which is very valuable for further research, as was explained in detail in the first section of this chapter. This clearly demonstrates the reciprocal influence of the different body segments in their movements. Above all, it also enables the equations themselves to be established without great difficulty in a given case. Subsequent sections of this chapter make a further contribution to our understanding of the laws of movement involved in human gait and deal with the movement of the lower extremities.

Transposing the whole process of movement into a tridimensional rectangular system of co-ordinates, as described in Chap. 1, enables the movements of the different body segments to be represented in detail without further experiments. Chapter 1 gave the trajectories of the different joint centres and other points of the human body that are important in walking. Chapter 2 described the trajectory of the total centre of gravity with the correlated velocities and accelerations. The present chapter has given an overall view of the behaviour of the lower extremities during a double step. The rotations carried out by the three segments of each leg in the projection of the whole process of movement on the plane of gait have been determined. A description has also been given of the simultaneous movements in the knees and ankles.

Comparing our results with the representation of the successive positions of the legs given by the brothers Wilhelm and Eduard Weber in their *Mechanics of the Human Locomotor Apparatus* shows that their conceptions have become obsolete as a result of the data established by instantaneous photography. In particular, the three principles stated by the two researchers which form the basis of their theory of walking are far from being acceptable. They considered it absolutely necessary in regular gait that the front leg be vertical at contralateral toe-off, when the opposite leg starts swinging (principle of the initial position). Actually, at this moment the vertical through the centre of the hip of the front leg intersects the ground a long way behind the supporting foot. This vertical is closer to the tip of the back foot than to that of the front foot. Weber and Weber also assumed that the striking force of the stance leg, supposed to be vertical, was always equal to the body weight (principle of the measure of the effort). It appears, however, that this vertical component of the striking force falls at times to less than half the body weight and at others increases by almost half the body weight. According to the third principle, the line of action of the compression exerted on the ground through the supporting foot always passes through the centre of gravity of the body (principle of the direction of the striking). The validity of this principle cannot be demonstrated either.

The three principles of Weber and Weber correspond to the ideal case of a linear horizontal movement of the total centre of gravity at constant velocity whereas the centre of gravity actually describes a double-curved trajectory, sometimes accelerating, sometimes decelerating. However, the Webers were unable to determine the exact movement of the centre of gravity with the means at their disposal. They lacked above all instantaneous photography for establishing the successive phases of the movement directly. Although the three basic principles of the Webers' theory of walking can no longer be accepted, the importance of "*Mechanics of the Human Locomotor Apparatus*" for science must not be underestimated. It will always remain a classic work, one that initiated exact studies in the physiology of movement.

Representing the process of movement in the projection on the plane of gait gives an overall view of the successive positions of the lower extremities and of the concomitant movements of the joints. However, it does not provide a sufficient basis for further analysis of the movements of the legs. To this end, and in particular to deduce the angular velocities and accelerations with which the different segments of the leg modify their tridimensional direction, the precise positional angles must be known which clearly determine the tridimensional position of each segment. Here, two angles are sufficient for each body segment because of the interdependence of the different segments. Therefore, the angles which the projection of the different segments of the leg on the plane of gait and on the plane perpendicular to the direction of gait form with the vertical have been calculated for all the phases of the three experiments. The joint angles of the knees and ankles have also been calculated. These are the angles formed at each phase by the long axis of the distal and the prolongation of the long axis of the proximal segment adjacent to the joint.

CHAPTER 4

On the Movement of the Foot and the Forces Acting on It

Introduction

R esearch of the interplay of the forces involved in walking requires a precise description of the movement of all the segments of the human body. The field of mechanics that is concerned with the analysis of movement from a purely geometric point of view, without dealing either with the masses and moments of inertia of the bodies in movement or with the forces exerted, is called "kinematics". The first goal of the analysis thus consists in describing the kinematics of human gait.

The kinematic problem is not yet completely solved when the successive positions of the human body during a double step have been determined by means of instantaneous photography, as was done in the earlier parts of the study. It is still necessary to find out the velocities and accelerations at which the different body segments move in space from one position to the next. Only after this problem has been solved can the analysis be extended to the causes of the movement and the forces in action during walking be considered. This is no longer merely kinematics, but kinetics of human gait.

Based on the kinematics of gait and knowledge of the masses, the moments of inertia and the external forces, in kinetics the movement equations can be used to disclose the forces which act in the human body during gait. These equations are complicated [1]. However, they can be simplified as explained in Chap. 3. To simplify the problems, as a first approximation, walking will be considered as a plane movement, disregarding the side-to-side oscillations. In a sense, a particular equation appears to be linked with each body segment. There are only two further movement equations relating to the plane movement of the total centre of gravity parallel to the plane of gait. If the simplifying assumption of a plane movement is rejected, complete characterization of the movement of the total centre of gravity requires three equations since it is a movement with three degrees of freedom. In this general instance, however, more than one equation corresponds to each body segment, though never more than three, because, as a consequence of the particular nature of joint connections in the human body, each body segment possesses a maximum of three degrees of freedom in relation to the adjacent segment.

The equations relating to the total centre of gravity are used to determine the external forces exerted on the human body during walking, at each instant of a double step, insofar as these forces are not known beforehand, as is gravity, for each attitude of the body. The main prerequisite for their application is that the acceleration of the total centre of gravity is known for each instant of the movement because the three equations set the product of the mass of the human body and the acceleration of its centre of gravity equal to the resultant of all the external forces at any given moment, if these forces are regarded as displaced parallel to themselves towards the centre of gravity. Since the tridimensional movement of the total centre of gravity possesses three degrees of freedom, its acceleration in any given direction is

---

[1] Abhandlungen der Königlich Sächsischen Gesellschaft der Wissenschaften, mathematisch-physische Klasse vol 20, no 1

given by three components. Correspondingly, the three equations relating to the total centre of gravity set the products of the total mass and the components of the acceleration of the centre of gravity in three directions perpendicular to each other (direction of gait, transverse direction and vertical direction) equal to the sums of the components of the external forces in these directions. If the movement of the centre of gravity is regarded as plane, basically only its projection on the plane of gait is considered; the components of the acceleration of the centre of gravity and those of the external forces in the transverse direction are disregarded. Besides gravity, the ground reaction force and the frictional resistance of the ground are the only external forces that need then be considered if the relatively small air resistance set up during walking in a closed space is disregarded. The three movement equations for the total centre of gravity are, therefore, sufficient for determination of the vertical ground reaction force and the components of the frictional resistance in the direction of gait and in the transverse direction from the three components of the acceleration of the total centre of gravity and the total mass or the body weight. These problems have been dealt with in Chap. 2.

Whereas the movement equations of the total centre of gravity constitute the means of determining the external forces, the other movement equations allow deduction of the internal forces acting in the human body during walking. In particular, they provide an insight into the distribution of the work of the muscles in locomotion. As explained in Chap. 3, the muscles act as force couples on the body segments which they span. They tend to rotate these segments. The external forces acting on the whole body also generally exert torques on the different body segments. Finally, each segment is subject to an additional turning action resulting from its connection with the other segments of the body. In effect, each movement of the other segments induces compression in one of the joints of the segment considered. In any movement, three types of torques are thus generally acting on each segment of the human body. They cause the body segment to rotate as if it were separate from the rest of the body and freely suspended from its centre of gravity whilst, of course, this centre of gravity itself generally moves forward simultaneously. The equations for the rotatory movement of a rigid body about a fixed point are thus applicable to the body segment. This is the significance of the other movement equations, as explained in Chap. 3: in the case of a plane movement one equation relates to each body segment whereas two or three equations are applicable to each when the movement is tridimensional.

Let us consider a plane movement of the body. According to the equation for one of the body segments, the product of the moment of inertia of this segment in relation to the axis through its centre of gravity perpendicular to the plane of gait, and the angular acceleration about this same axis is equal to the sum of the three types of torques resulting from the muscles, the external forces and the movement of the other body segments. The quantities in these equations can be determined, except for the resultant torque of the muscles, if the kinematics of human gait are known. The equations thus can be used to calculate the resultant torque of the muscles for each body segment. A muscle always tends to rotate several, or at least two, body segments. Consequently, the unknown tension of each muscle participates in the resultant muscular torques related to different body segments. For the kinematic analysis of gait the human body was subdivided into 12 segments which could be regarded as virtually rigid bodies during walking. Therefore, the kinetics of gait on a kinematic basis can only deal with the muscles connecting different segments among these 12. It cannot be extended to muscles that are contained entirely in a single segment, e.g., the forearm + hand system which is regarded as one rigid body. The assumption that each of the 12 segments behaves as a rigid mass presupposes that the muscles which have both their insertions on the same body segment do not take any active part in the movement involved in walking. This is the case for the muscles originating in the forearm (and inserted into the wrist or hand). If all the other muscles could be combined in 12 groups in such a way that one group acted in the same way on the different segments, it would be possible to deduce the resultant tension of each group from the 12 resultant torques of the muscles for each instant of the movement. In terms of algebra, this comes down to the solution of 12 simple equations with 12 unknown quantities. Thus, for determination of the muscular tensions no particular minimum principle such as that of the smallest effort, is necessary. The further analysis will show whether and how far such a grouping of the muscles acting in the same way is possible.

Interpretation of the equations of movement relating to a particular body segment clearly shows what kinematic data the kinetic analysis requires in each case. It indicates the characteristics of the movement of the segment which must be considered in the analysis. Therefore, before analysing the movement of a body segment, it is necessary to explain the meaning of the corresponding equations for the different periods of gait, and particularly the forces acting on this segment. The influence exerted on this segment by the movement of the other segments of the human body must be defined.

This chapter (Chap. 4) will deal with the movement of the foot. In the first section the forces exerted on the foot at the different periods of a double step will be identified, at first without determination of their magnitude. From these the movement equations for the movement of the foot can be deduced simply and clearly and a good view of the interaction of the different forces involved is possible with no need for the general Lagrangian differential equations. Consequently, the further analysis will be easily understood even by those who are not familiar with the methods of differential and integral calculation and who do not know what a differential equation is. The method of deducing the movement equations has been discussed in Chap. 3. The basic rules can, therefore, now be applied specifically to the foot. The other sections of this chapter deal with the kinematics of the foot.

On the Forces Acting on the Foot and Their Torques

The forces which move the foot in walking can be divided into four types depending on whether they result from muscular contraction, from gravity, from the ground reaction or from the articular connection of the foot with the other segments of the body. These four types of forces result in the three groups of torques mentioned above.

The muscles act by way of active contraction or elastic elongation. Each muscle spanning the ankle acts directly on the foot only when the force is pulling from the insertion toward the origin. An equal force pulling in the opposite direction, i.e. from the origin toward the insertion of the muscle, contributes to compression in the ankle joint.

Gravity finds its expression in a force applied at the centre of gravity of the foot, pulling vertically downwards. Its magnitude is equal to the weight of the foot. But the weight of the remainder of the body also exerts its effect on the movement of the foot, acting as one of the compressive forces exerted on the ankle. This force acts vertically downwards and is equal to the weight of the body minus the foot.

The reaction of the ground to the compression exerted on it can, of course, act directly as an external force on the foot only when the foot is on the floor. However, even when it has no contact with the floor, the ground reaction force exerted directly only on the opposite foot nonetheless has an indirect influence, being one of the components of the compression in the ankle during foot swing. As explained in Chap. 2, if the air resistance is disregarded, the ground reaction force can be resolved into two components. One is equal to the weight of the whole body but acts vertically upwards whereas the other is equal to the effective force of the total centre of gravity. For each instant of the movement, this effective force is given by the product of the body mass and the acceleration of the total centre of gravity. The ground reaction force is either exerted on one foot or is distributed in a particular way between the two feet when both of them are in contact with the floor. During the single-support period it can also be distributed over different points of the sole of the foot. This occurs, for instance, during foot-flat.

The compression in the ankle must also be considered as an external force acting on the foot if the movement of the foot is dealt with as if the foot was a free body separate from the remainder of the body. Several of the components of this compression have been mentioned. Only the component resulting from the movement of the centre of gravity of the system that remains when the foot is regarded as separate from the body must be added. This component is equal but opposite to the effective force of the centre of gravity of this system. If a force is imagined applied at this centre of gravity, which is equal and opposite to its effective force, and this force is added to all the other external forces exerted on the system, the compression acting on the ankle is obtained by displacing all the external forces exerted on the system parallel to themselves toward the centre of the ankle joint, as explained in Chap. 3. This is strictly true only as long as the movement in walking can be considered as plane, however. If the movement is tridimensional, strictly speaking all the external forces of the system cannot generally be displaced to a single point on the axis of the ankle; the different components of the compression are applied, as a rule, at different points on the articular axis. There is no reason to go into these rather complicated conditions in more detail as long as the assumption of a plane movement of walking can be retained. To determine the magnitude of the compressive force in the ankle without further investigation of the origin of the different components, the effective force of the centre of gravity of the foot can be used. The compression exerted on the foot at the ankle by the rest of the body must be precisely so great that, when combined with all the other external forces exerted on the foot, it imparts its acceleration to the centre of gravity of the foot. One corollary of this, however, is that one of the components of compression is equal to the effective force of the centre of gravity of the foot whereas the others are equal and opposite to the other external forces acting on the foot, displaced to the ankle. Of course, the compressive force obtained in this way is the same as that deduced from the external forces of the system mentioned above. The second way of deducing the compression in the ankle leads directly to the three groups of couples with which all the forces acting on the foot tend to rotate the foot about an axis through its centre of gravity. This will appear in the further investigation.

The forces which act on the foot at the different periods of a double step and the way they combine to give couples will now be presented with reference to the right foot. The results also apply to the left foot, however. If the gait is absolutely regular and symmetrical, the movements of the right foot are repeated by the left foot after a time interval of one step, and are a symmetrical reflection of them in relation to the plane of gait. If the movement is considered as plane and the side-to-side oscillations perpendicular to the plane of gait are disregarded, the movement of the right foot is exactly like that of the left foot which occurs one step later.

Once the forces thus exerted have been found, the movement equations concerning the rotation about the axis through the centre of gravity can easily be established for each case.

During the right swing period there is no direct ground resistance to the right foot. The circumstances are thus particularly simple. In the hypothesis of a plane movement parallel to the plane of gait, only the rotations of the foot about an axis through its centre of gravity and perpendicular to the plane of gait need be considered. Two types of muscles are involved: those that tend to rotate the foot clockwise (as seen from the right) and those that tend to rotate the foot counter-clockwise. The movement in the ankle is not taken into account since, to establish the movement equations, the foot has to be considered as a rigid body separated from the rest of the body. Rotation of the right foot clockwise at the ankle is not always accompanied by plantar flexion, nor is rotation counter-clockwise by any means always accompanied by dorsiflexion. The overall view of the movements of the lower extremity (Chap. 3, Figs. 7–9) shows rather, for instance, that shortly before right heel strike there is dorsiflexion in the right ankle while the foot itself is rotating clockwise. Furthermore, during foot-flat, the foot does not rotate in space but the ankle dorsiflexes. During the last half of the swing period, the ankle plantar flexes whereas the foot is rotating counter-clockwise in space.

Consequently, for the purposes of further discussion it does not matter how a muscle acts on the ankle, but in what direction and with what torque this muscle tends to rotate the foot in space about an axis perpendicular to the plane of gait.

All the muscles which pass behind the axis of the ankle in the projection of the foot on the plane of gait tend to rotate the right foot clockwise, as seen from the right, whereas those which pass in front of this axis rotate the foot counter-clockwise (p. 262). If the resultant pull passes exactly through the centre of the joint it does not cause any rotation. We cannot yet decide which muscles are acting at each instant of a double step and with what tension. We shall attempt to explain how the muscles tend to rotate the foot if they are involved in the movement at all. It is thus sufficient to select one representative muscle from each of the two groups. The results thus obtained can then be applied mutatis mutandis to each of the muscles of the group. The calf muscles will be used to represent the muscles which tend to rotate the foot clockwise about the axis through its centre of gravity. This group of muscles acts directly on the foot as a force applied at point A of the calcaneus, in the direction of the Achilles tendon (Figs. 1–10). The magnitude of this force which is the tension of the calf muscles throughout their physiological cross section, will be referred to from this point onward as M_w. The tibialis anticus will be used to represent the muscles which tend to rotate the foot counter-clockwise about the axis through its centre of gravity and perpendicular to the plane of gait. This muscle acts directly on the foot as a force applied not at the insertion of the muscle on the first cuneiform and first metatarsal but rather at point C (Figs. 1–10) where the tendon of the muscle leaves the inferior extensor retinaculum to pass over the superior extensor retinaculum. The direction of the force applied at C is given by that of the small fraction of the tendon spanning the two ligaments. It has no relation whatsoever with the course of the muscular fibres. The total tension of the tibialis anticus produced over the physiological cross section of the muscle will be called M_t. Both forces M_w and M_t are given in Figs. 1–10. However, the length of the vectors by which they are represented has nothing to do with their magnitude. It is probable that the posterior muscles act on the foot with greater force at some phases of gait, and the anterior muscles at others. It can also be assumed that generally only one of the two groups is actively contracting whilst the muscles of the other group are acting on the foot only through their elastic tension. In this instance the difference between the two forces must be relatively great. The further analysis consists in determining the tensions of the different muscles or groups of muscles so that their magnitudes obviously cannot yet be evaluated perfectly correctly. The reader must thus avoid drawing erroneous conclusions about the magnitudes of the muscular forces from the figures.

Besides the muscular forces M_w and M_t, the weight of the foot itself acts as an external force during the swing period. If the centre of gravity of the foot is S_6 and its weight G_6, a force of magnitude G_6 applied at S_6 pulls vertically downwards. In Fig. 1, this force is drawn a little larger than it should be in proportion to the weights G_0 of the whole body, and G_6' of the body minus the foot. If the true proportions were observed, G_6 would have to be too small or G_0 and G_6' too large. But at this point only the rotatory action of the different weights is of interest. The true magnitude of the forces and of the torques they set up is irrelevant at this stage although in this case both are known from the start.

Finally, in the swing period the third and last type of forces acting on the right foot, compression in the ankle joint, must be considered. The projection of the axis of the ankle for the movement of the right foot assumed to be plane, is designated as F and no longer as $G_{4,6}$, as it was in Chap. 2 for other reasons. This avoids confusion with the sign G which designates weight in this instance where the right foot is considered in isolation without reference to the rest of the body. The compressive force applied at F will be represented by its different components. These components consist in all the external forces which act on the body minus the right foot. These external forces are displaced to F. We must remember to count among these external forces the ground reaction force on the left foot and especially the

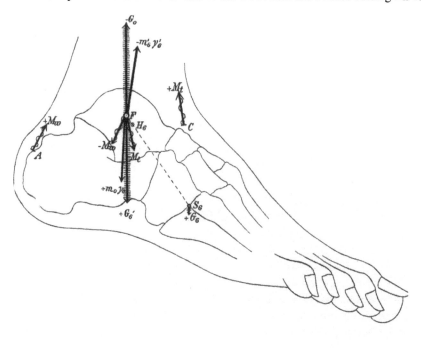

Fig. 1. Forces acting on the right foot during the swinging period

effective force applied at the centre of gravity of the body minus the right foot, after its sign has been reversed.

m'_6 is the mass of the body minus the right foot and γ'_6 the acceleration of its centre of gravity S'_6. The effective force of this system is thus $m'_6 \gamma'_6$. The component of compression at F resulting from the movement of S'_6 is thus $-m'_6 \gamma_6$. The negative sign indicates that this force is opposite to the effective force.

The weight of the partial system also acts at the centre of gravity S'_6 and pulls this centre of gravity vertically downwards. This weight is equal to that of the body minus the weight G_6 of the foot and will be designated as G'_6. The second component of the compression at F thus is the weight $+G'_6$ acting vertically downwards.

The ground reaction force acting on the left foot during the swinging period has two components. One is equal and opposite to the weight of the body G_0. It acts vertically upwards. The other is equal to the effective force of the total centre of gravity and has the same sign. Its magnitude is $m_0 \gamma_0$ if m_0 is the mass of the whole body and γ_0 its acceleration. The ground reaction force thus gives the third and fourth components $-G_0$ and $+m_0 \gamma_0$ of the compression at F. The negative sign in front of G_0 means that this force must be considered as acting vertically upwards.

Finally, each muscle acting on the foot gives an additional component of compression at F. The origin of each of these muscles is in the system which remains when the right foot is removed. Its insertion is in the right foot itself. Therefore, the muscular force applied at the origin of the muscle and pulling towards the insertion represents an external force which is always equal and opposite to the muscular force applied directly to the foot. Both have the same magnitude but their directions are opposite. The two representatives of the anterior and posterior groups of muscles thus each give a component of compression at point F. One is equal and opposite to the force M_w applied at A. The other is equal and opposite to the force M_t applied at C. These two components are thus designated by $-M_w$ and $-M_t$ if the muscular forces acting at A and C have positive signs.

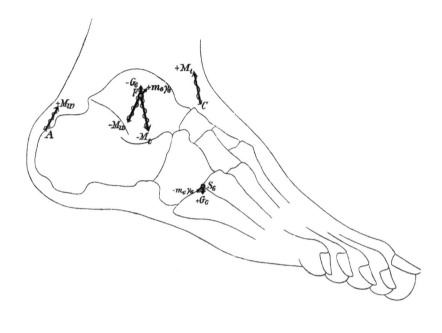

Fig. 2. Couples acting on the right foot during the swinging period

At each instant of the swinging period, the compression exerted by the system at the centre F of the right ankle thus results from six components $-M_w$, $-M_t$, $+G'_6$, $-G_0$, $+m_0\gamma_0$ and $-m'_6\gamma'_6$. This is represented in Fig. 1. The two weights $+G'_6$ and $-G_0$ can be combined into one resultant, as can the two effective forces $+m_0\gamma_0$ and $-m'_6\gamma'_6$. The weight $+G'_6$ acting vertically downwards is smaller than the force $-G_0$ acting vertically upwards by the weight G_6 of the right foot. Their resultant is a force acting vertically upwards and equal to the weight G_6 of the right foot. It is designated as $-G_6$. The resultant of the effective forces $+m_0\gamma_0$ and $-m'_6\gamma'_6$ is the effective force $m_6\gamma_6$ of the centre of gravity of the right foot. This results from the fact that the total centre of gravity S_0 of the whole body can be regarded as the centre of gravity of the two masses m_6 and m'_6 thought of as concentrated at the centres of gravity S_6 of the foot and S'_6 of the body – right foot system. Therefore, the effective force $m_0\gamma_0$ applied at S_0 represents the resultant of the two effective forces $m_6\gamma_6$ and $m'_6\gamma'_6$ applied at S_6 and S'_6. This constitutes another expression of the relationship mentioned above between the effective forces. If $m_0\gamma_0$ is the resultant of $m_6\gamma_6$ and $m'_6\gamma'_6$, then $m_6\gamma_6$ is also the resultant of $m_0\gamma_0$ and the force $-m'_6\gamma'_6$ which is equal and opposite to the effective force $m'_6\gamma'_6$.

The six components of the compression at F are thus reduced to four: $-M_w$, $-M_t$, $-G_6$ and $+m_6\gamma_6$. This is illustrated in Fig. 2.

The same result is obtained directly by determining the compression at F in the second way mentioned above. The compressive force at F must be of such a magnitude that, combined with the external forces acting on the right foot, it imparts its acceleration γ_6 to the latter. The effective force $m_6\gamma_6$ then appears as the resultant of this compressive force and all the external forces applied directly to the right foot. Consequently, the compressive force at F results from the effective force $+m_6\gamma_6$ and the three external forces exerted on the right foot taken with opposite signs: $-M_w$, $-M_t$ and $-G_6$.

Figure 2 shows the couples with which the different forces tend to rotate the foot. Each type of force is represented in a different way. All the muscular forces are characterized by a wavy line on the vectors and the forces of gravity by short lines across them whereas the effective forces are represented by plain vectors. The muscular force $+M_w$ acting at A and its counterpart $-M_w$ acting at F constitute a couple that tends to rotate the foot clockwise. The moment arm of this couple is equal to the distance between the resultant pull of the calf muscles and the joint centre F. Its moment is the product of this distance and the tension M_w. The forces $+M_t$ and $-M_t$ acting at C and F form a second couple the moment of which is calculated in a similar way. This couple rotates the foot counter-clockwise. The moments of these two couples thus have opposite signs. They will be referred to below as the torques exerted by these muscles on the right foot. All the muscles which pass behind the joint axis tend to rotate the foot in the same direction as the calf group. All the muscles which lie in front of the joint axis tend to rotate the foot in the same direction as the tibialis anticus. If the torques of all the muscles are determined in this way, their algebraic sum represents the "resultant torque of the muscles", as described in Chap. 3 (p. 214). This expression involves as many unknown tensions as there are muscles. The torque of the calf group itself is the sum of two torques since the gastrocnemius also spans the knee whereas the soleus originates in the lower leg. The two are not necessarily at the same tension.

During the swinging period gravity acts with a relatively small torque. The active couple results from the forces $+G_6$ and $-G_6$ applied at S_6 and F. Its moment is the product of the weight of the foot and the distance between the verticals through F and S_6. In the position of the foot represented in Fig. 2, the couple tends to rotate the foot clockwise. The rotation can also be just the opposite, for instance during the first third of the swinging period. Then the vertical through the centre of gravity of the foot passes behind the axis of the ankle. This appears in Figs. 22–23 of Chap. 1. At the beginning of the second third of this period, the torque D_s due to gravity disappears for a short time since the centre of gravity of the foot lies on the vertical through and beneath the centre of the ankle. Later in the swinging period the vertical through the centre of gravity of the foot lies in front of the ankle, and gravity then tends to rotate the foot clockwise, as shown in Fig. 2.

After all the muscular forces and the forces due to gravity have been combined into couples, only the component of compression $+m_6\gamma_6$ applied at F remains. This force is the only one that can provoke an acceleration of the foot since all the other forces acting at the centre of gravity of the foot cancel each other out. They are all grouped in pairs of equal and opposite forces, and the centre of gravity of the foot moves as if all the external forces were applied to it. The force $+m_6\gamma_6$ imparts the acceleration γ_6 to the centre of gravity. Moreover, it also tends to rotate the right foot as long as its line of action does not pass through its centre of gravity. The rotating effect of this force can easily be represented as distinct from its effect on the movement of the centre of gravity by bringing to the centre of gravity S_6 the two equal and opposite forces $+m_6\gamma_6$ and $-m_6\gamma_6$. They balance each other and cannot in any way change the state of movement of the foot. Only the force $+m_6\gamma_6$ applied at S_6 can accelerate the centre of gravity. The force $-m_6\gamma_6$ which is also applied at S_6 and which is only represented in Fig. 2 forms a couple with $+m_6\gamma_6$. Its moment can be considered as the "torque of the effective forces" mentioned in Chap. 3 (p. 215). This torque D_e is the product of the magnitude of one of the forces and the distance between the two forces $+m_6\gamma_6$ and $-m_6\gamma_6$.

During the remainder of the double step, the muscles act in the same way as during the swinging period. Contraction of a muscle lying behind the axis of the ankle tends to rotate the foot clockwise. Contraction of a muscle lying in front of the axis of the ankle sets up a couple which tends to rotate the foot counter-clockwise. The successive periods differ only in the muscles involved and in their tension. From this point of view each phase is generally different from another even in the same period. To indicate the change in the tension of the muscles, the vectors representing the muscles are of different lengths in the next figures. However, these lengths do not and cannot give reliable information about the actual muscular tension. The two couples $(+M_w, -M_w)$ and $(+M_t, -M_t)$ can be found in all the figures. The algebraic sum of their moments always gives the resultant torque D_m of the muscles.

The torques of gravity and of the effective forces combine at each period in a particular way. The characteristic differences in arriving at these torques during the other periods must thus be considered.

The swinging period of the right foot is followed by a short period of double support. Then the circumstances are essentially different. Since the right foot is in contact with the floor, it exerts compression on the latter and is subject to a ground reaction force. The ground reaction force is small at first on the right foot but increases rapidly. For the left foot the opposite is true. It had to take up the whole of the ground reaction force as long as it alone was in contact with the floor. From right heel strike on, part of the ground reaction force passes to the right foot. In general, the greater the part of the ground reaction force taken over by the right foot, the smaller is that acting on the left foot. The sum of the two reaction forces exerted simultaneously on the feet is of course always the resultant ground reaction force. This force results from the combination of the effective force $+m_0\gamma_0$ of the total centre of gravity and the force $-G_0$ due to gravity, equal to the weight of the body and acting vertically upwards.

The swinging period of the left leg follows the double-support period. During this swinging period, the right foot, the only one on the ground, withstands the whole of the ground reaction force. It is predictable that already in the double-support period the right foot is taking an increasing proportion of the ground reaction force, progressively relieving the left foot until left toe-off. The exact distribution of the ground reaction force over the two feet at each moment cannot be determined until the precise movements of all the body segments during walking are known. For the moment, the two parts of the ground reaction force have to be considered separately during the further analysis of the process of the movement. The two components of that part of the reaction force exerted on the right foot are designated as $+(m_0\gamma_0)^r$ and $-G_0^r$ and those exerted on the left foot as $+(m_0\gamma_0)^l$ and $-G_0^l$. The letters r and l are obviously not exponents but indices. The resultant of the two forces $(m_0\gamma_0)^r$ and $(m_0\gamma_0)^l$ is the effective force $m_0\gamma_0$ and that of the two forces G_0^r and G_0^l is the weight of the body G_0.

During the double-support period, the right foot in front is in contact with the floor only at one point or through a small area underneath the heel. This is where a part of the ground reaction force is exerted on the right foot. A central point B in this small area of contact can be considered as the point of application of the two reaction components $+(m_0\gamma_0)^r$ and $-G_0^r$, as illustrated in Fig. 3. At point F, we no longer find the two components of compression $+m_0\gamma_0$ and $-G_0$, as in Fig. 1, but the two components $+(m_0\gamma_0)^l$ and $-G_0^l$.

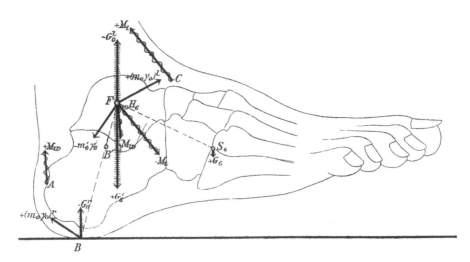

Fig. 3. Forces acting on the right foot during the double-support period

Otherwise the conditions are the same as for any phase of the right swing. We thus additionally find at the centre of the ankle F the compressive components $+G_6'$, $-m_6'\gamma_6'$, $-M_w$ and M_t. Moreover, the forces $+M_w$ and $-M_t$ act at the points A and C and the weight G_6 at the centre of gravity S_6, as previously. As mentioned above, the four muscular forces give the resultant torque D_m of the muscles.

In these new circumstances, the four forces due to gravity, as well as the effective forces now present in greater number, can be looked on as couples if the force $-m_6\gamma_6$ (see Fig. 4) is added at point S_6. This force is equal and opposite to the effective force of the centre of gravity of the foot. The purpose of this is to exclude the acceleration of the centre of gravity of the foot from the calculation and to bring out the rotating effect of the forces applied to the foot only. The force $-m_6\gamma_6$ will then cancel out the acceleration γ_6 of the centre of gravity of the foot without influencing the rotation of the foot about the axis through its centre of gravity and perpendicular to the plane of gait.

The two forces $-G_0'$ and $-G_6'$ applied at point F can be combined into one resultant which differs from the force $+G_0^r$ only by the weight G_6. On the one hand, G_6' is smaller than G_0 by G_6. On the other hand, subtraction of G_0^l from G_0 gives G_0^l. The forces due to gravity and applied at F can thus be replaced by two forces $-G_0^r$ and $-G_6$. This has been done in Fig. 4. It is then obvious that the forces $+G_0^r$ and $-G_0^r$ applied at F and B, and the forces $-G_6$ and $+G_6$ applied at F and S_6 form couples. The second of the two couples was already acting during the previous period. The double-support period differs from the swing period by the addition of the couple $(+G_0^r, -G_0^r)$. The moment of this couple is the product of the component G_0^r of the ground reaction force acting on the right foot and the distance of the vertical through point B from the centre F of the ankle joint. In the position of the foot shown in Fig. 4, both couples resulting from the gravity forces tend to rotate the right foot clockwise. Therefore, addition of their moments gives the torque of gravity D_s mentioned before.

The effective forces can also be similarly combined into two couples after the reversed effective force $-m_6\gamma_6$ has been added at S_6. The two forces $+(m_0\gamma_0)^l$ and $-m_6'\gamma_6'$ (Fig. 3) applied at F can be replaced by the two forces $-(m_0\gamma_0)^r$ and $+m_6\gamma_6$ (Fig. 4) since the latter combine to give the same resultant as the former. The two forces $(m_0\gamma_0)^l$ and $(m_0\gamma_0)^r$, as well as the two forces $m_6'\gamma_6'$ and $m_6\gamma_6$, combined into one resultant give the effective force $m_0\gamma_0$ of the total centre of gravity. Then the forces $-(m_0\gamma_0)^r$ and $+(m_0\gamma_0)^r$ applied at F and B form a couple, as do the forces $+m_6\gamma_6$ and $-m_6\gamma_6$ applied at F and S_6. These couples are equivalent to the four effective forces. The algebraic sum of their moments is equal to the

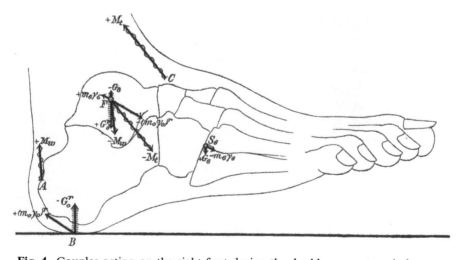

Fig. 4. Couples acting on the right foot during the double-support period

resultant torque D_e of the effective forces in the double-support period. Comparison with the resultant torque of the effective forces in the preceding swinging period shows that an additional couple $[-(m_0\gamma_0)^r, +(m_0\gamma_0)^r]$ is now present.

The same results can be obtained by deducing the components of the compression at F from the external forces applied directly at the right foot, knowing that all the external forces acting on the right foot combined with the compression exerted on the foot at F must give the centre of gravity S_6 its acceleration γ_6, after they have been displaced parallel to themselves to this centre of gravity. This leads to the components drawn in Fig. 4.

The right single-support period, during which the left leg swings from behind forwards, differs essentially from the last period in that all the ground reaction force now acts directly on the right foot. At the very beginning of the period, before foot-flat, the ground reaction must be represented by the gravity force $-G_0$ corresponding to the body weight acting vertically upwards, and by the effective force $+m_0\gamma_0$ of the total centre of gravity. These forces act at the instant contact point B underneath the heel. The two compressive components $-G_0^l$ and $+(m_0\gamma_0)^l$ at point F in Fig. 3 disappear. Besides the muscular forces only the two components $+G_6'$ and $-m_6'\gamma_6'$ remain at this point to exert compression on the foot in the ankle (see Figs. 5 and 7). If the force $+G_6'$ is replaced by the two components $+G_0$ and $-G_6$ and the reversed effective force $-m_6'\gamma_6'$ is replaced by the two components $-m_0\gamma_0$ and $+m_6\gamma_6$, again two couples are obtained for the forces due to gravity and applied at the foot and two for the effective forces, if the reversed effective force $-m_6\gamma_6$ has been added at the centre of gravity S_6 to allow selective representation of the rotatory influence of the forces. Figure 4 can be used to represent these couples if the upper indices r are omitted (see also Fig. 8).

The circumstances change at right foot-flat, when the compression of the foot is no longer restricted to one small area but is distributed over almost the whole of the plantar surface, since the foot does not behave as a rigid body at all. Under the effect of compression it becomes so deformed that most of the sole is in contact with the ground. However, the distribution of the compression and of the ground reaction force is not even. The heel and the heads of the first and fifth metatarsals will be exposed to greater compression from the floor than the middle aspect of the sole although the latter does not completely escape compression. This results from the arched architecture of the foot. Furthermore, it can be

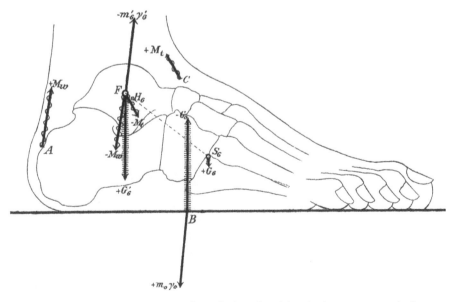

Fig. 5. Forces acting on the right foot during the right single-support period

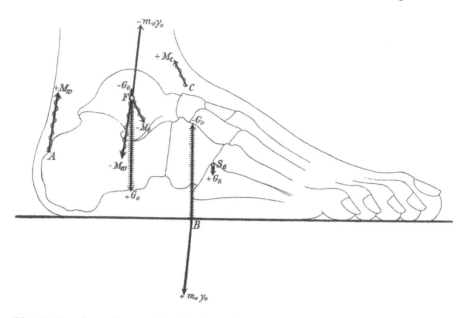

Fig. 6. Couples acting on the right foot during the right single-support period

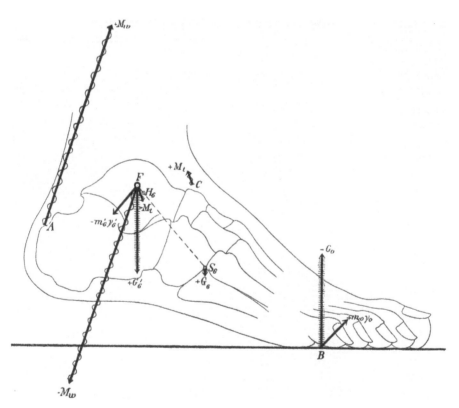

Fig. 7. Forces acting on the right foot at the beginning of push-off in the right single-support period

expected that, at the beginning of foot-flat the heel, and at the end of foot-flat the heads of the metatarsals will be exposed to the greatest compression. For the hypothesis of a plane movement during walking, a midpoint of the plantar surface can only be assumed at which the resultant of all the reaction forces distributed over the sole is applied. This point can be called the "compression point of the foot". During foot-flat it moves progressively forward.

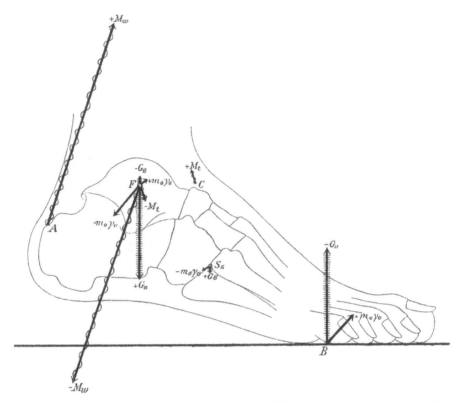

Fig. 8. Couples acting on the right foot at the beginning of push-off in the right single-support period

Its position cannot yet be determined precisely for each phase. It is represented in a middle position and marked B in Fig. 5. It can be considered as the point of application of the two components $-G_0$ and $+m_0\gamma_0$ of the ground reaction force. Right foot-flat is characterized mainly by the fact that the centre of gravity S_6 of the foot remains at nearly the same place and its acceleration γ_6 is 0. Therefore, the reversed effective force to be applied at S_6 disappears. Furthermore, the component $-m_6'\gamma_6'$ at F can be replaced by the reversed effective force $-m_0\gamma_0$ of the total centre of gravity, as in Fig. 6. If the force of gravity $+G_6'$ applied at F (Fig. 5) is replaced by the two components $+G_0$ and $-G_6$ (Fig. 6), again four forces due to gravity are obtained. They form two couples. There is only one couple formed by effective forces. It is formed by the forces $-m_0\gamma_0$ and $+m_0\gamma_0$ applied at F and B.

The moment of this last couple constitutes the torque D_e of the effective forces. The torque of gravity D_s is again the algebraic sum of the moments of the two couples $[+G_0, -G_0]$ and $[-G_6, +G_6]$. The resultant torque D_m of the muscles is obtained as before.

These three torques must present particularly simple relationships to each other during right foot-flat. Since the foot does not move, they must balance each other. This is only possible if their algebraic sum is 0.

Push-off follows foot-flat. Most of push-off falls in the single-support period but it continues for a short time after heel strike of the opposite foot. At the beginning of push-off the area of contact between the foot and the floor is very limited. The heel has left the ground and the foot rotates about an axis nearly perpendicular to the plane of gait. This axis initially lies behind the metatarsal head area but soon moves forwards to the head of the first metatarsal. The circumstances are now the same as shortly before foot-flat. However, the point B (Fig. 7) at which the ground reaction is exerted on the foot is no longer at the heel but at the metatarsal heads. Besides the muscular forces $-M_w$ and $-M_t$, the compressive components $+G_6'$ and $-m_6'\gamma_6'$ act at the centre of the ankle F, and the weight G_6 of the foot acts at the centre of gravity S_6. Moreover, the reversed effective force $-m_6\gamma_6$ must be

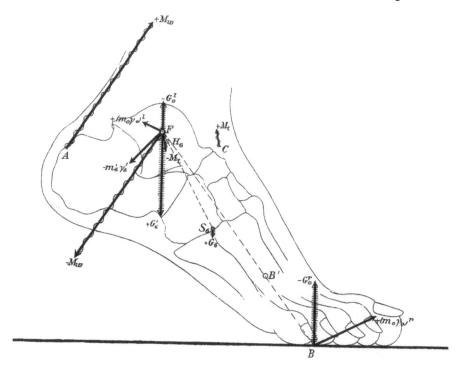

Fig. 9. Forces acting on the right foot during the double-support period between right single support and right swing

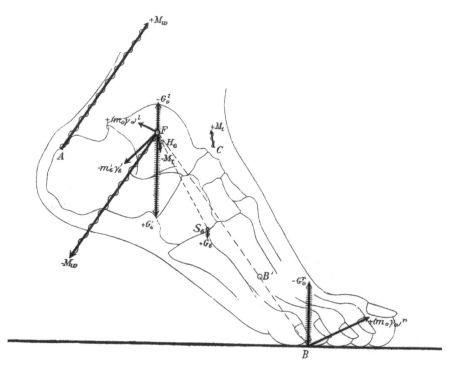

Fig. 10. Couples acting on the right foot during the double-support period between right single support and right swing

applied at S_6 to eliminate the acceleration γ_6 which the centre of gravity of the foot has resumed, and to represent the rotatory influence of the forces exerted on the foot. The force $+G_6'$ can be replaced by the two components $+G_0$ and $-G_6$ and the reversed effective force $-m_6'\gamma_6'$ by the two components $-m_0\gamma_0$ and $+m_6\gamma_6$. Again, the forces due to gravity and the effective forces each give two couples (Fig. 8). Their moments when combined give the torque of gravity D_s and the resultant torque D_e of the effective forces.

Between the right single-support period and the next right swinging period there is a double-support period. But now the left foot is in front and the right behind. In the previous double-support period an increasing proportion of the ground reaction force was taken over by the right foot, and the left foot was progressively relieved of this until the beginning of its swinging period. Of course, exactly the opposite occurs now. The expressions $-G_0^r$ and $+(m_0\gamma_0)^r$ are used for the components of the reaction force which act on the right foot, and $+G_0^l$ and $+(m_0\gamma_0)^l$ for those acting on the left foot. The forces drawn in Fig. 9 are thus obtained. Besides the compressive components of the previous period, the components $-G_0^l$ and $+(m_0\gamma_0)^l$ act at the centre F of the ankle. The forces due to gravity $+G_6'$ and $-G_0^l$ acting at F are replaced by the forces $+G_0^r$ and $-G_6$, and the effective forces $-m_6'\gamma_6'$ and $+(m_0\gamma_0)^l$ by $-(m_0\gamma_0)^r$ and $+(m_6\gamma_6)$ (Fig. 10). The forces $+G_0^r$ and $-G_0^r$ applied at F and B and the forces $-G_6$ and $+G_6$ applied at F and S_6 give the two couples that result in the torque D_s of gravity. The forces $-(m_0\gamma_0)^r$ and $+(m_0\gamma_0)^r$ applied at F and B and the forces $+m_6\gamma_6$ and $-m_6\gamma_6$ applied at F and S_6 give the two couples that combine to give the resultant torque D_e of the effective forces.

The Resultant Force Couples of the Muscles, of Gravity and of the Effective Forces

A review of Figs. 1–10 shows the results obtained for the different periods. As stated above, the muscles act on the foot in the same way throughout the movement. Of course, their tension can vary over a continuum extending from 0 to a maximum threshold. Each muscle evokes a force couple. One of the forces of the couple is applied at the insertion or at the point from which the muscle can pass freely into the lower leg or the thigh. The direction is given by the line joining the origin and the insertion. The other force of the couple which is equal and opposite to the first is applied at the centre F of the ankle. All the muscles of the posterior group, for which the calf group has been chosen as an example, tend to rotate the foot clockwise, as seen from the right, as long as they are tensed. All the muscles of the anterior group, for which the tibialis anticus was chosen as an example, tend to rotate the foot counter-clockwise. Whether the tension of a muscle results from nervous stimulation or from elastic elongation is irrelevant to the mechanical effect. The resultant torque D_m of the muscles, at each instant of the movement, is equal to the algebraic sum of the different couples. The moment of a couple is the product of the tension exerted by the muscle and the distance between the two forces of the couple. It is preceded by a plus or a minus sign, depending on whether the couple tends to rotate the foot clockwise or counter-clockwise. The force couples formed by all the muscles of both groups can also be combined into a single couple for each instant of the movement. The forces of this couple are equal to the algebraic sum of the tensions of the different muscles. The tension of each individual muscle is taken as positive or as negative depending on whether it tends to rotate the foot clockwise or counter-clockwise. This couple will be referred to below as the "resultant couple of the muscles". Its moment is D_m.

The couples through which gravity acts on the foot differ with the periods of gait. However, they obey a general principle. In Figs. 1, 3, 5, 7 and 9 a force $+G_6'$ is applied at the centre of the ankle F and a force $+G_6$ at the centre of gravity of the foot S_6. They pull vertically downwards. Addition of the weights G_6' (whole body minus the right foot) and G_6 (the right foot) gives the weight of the whole body G_0. $+G_0$ is thus the resultant of $+G_6$ and

$+G'_6$. It is applied at a point of the connecting line $\overline{FS}_6$, which divides this line in inverse proportion to the two forces. This point is designated by H_6 in Figs. 1, 3, 5, 7, 9 and represents the principal point of the right foot, as explained before. It remains in the same place in the foot as long as the centre of gravity S_6 of the right foot is assigned a fixed point in the foot. Since G_6 is much smaller than G'_6, the principal point H_6 lies very close to the centre of the ankle F, less than 1.5 mm away. In Figs. 1, 3, 5, 7, 9 the principal point has been drawn a little too far from F. Similarly, the weight $+G_6$ appears too large, if compared with the other forces due to gravity to be clearly distinguishable. Figures 1, 3, 5, 7, 9 indeed do not and cannot represent the actual magnitudes of the forces, but only the way they act. The two weights acting vertically downwards are opposed to the force $-G_0$ acting vertically upwards in Figs. 1, 5 and 7. This force $-G_0$ is applied at F during the swinging period (Fig. 1). During the right single-support period (Figs. 5 and 7) it is applied at a point B of the plantar surface, which moves progressively forward from the heel towards the metatarsal heads during this period. During the two short double-support periods (Figs. 3 and 9) two reaction components $-G_0^r$ and $-G_0^l$ acting vertically upwards are applied at B and F. They give the resultant $-G_0$ acting vertically upwards. This resultant is applied at a point B' on the connecting line $\overline{FB}$ and divides the latter in inverse proportion to the two forces. During double support, the point B' moves continuously on $\overline{FB}$ because the two components $-G_0^r$ and $-G_0^l$ of the ground reaction force change their magnitudes continuously. During the first double-support period, the right foot is in front of the left (Fig. 3). Then G_0^r increases and G_0^l decreases. Consequently, the point B' moves progressively from F to B in the heel. During the second double-support period the right foot is behind the left (Fig. 9). Then G_0^r decreases and G_0^l increases. Consequently, the point B' moves progressively from the point B at the metatarsal heads towards F. During both the first period, when B is at the heel, and the second, when B is at the metatarsal heads, the point of contact with the floor B moves continuously forward. Therefore, in both instances the line $\overline{FB}$ rotates somewhat counter-clockwise about the point F inside the foot. During the first period the point B' thus describes a curve that is convex backwards inside the foot (see Fig. 11) between F and the point M at which the heel compresses the floor at the end of the period. During the second period B' moves back inside the foot along a curve convex forwards and upwards from the point M' at the metatarsal heads which compresses the ground at the beginning of the period, to F. The exact shape and position of these two trajectories of the point B' cannot yet be given since the distribution of the ground reaction force over the two feet during the double-support periods cannot yet be determined. As soon as this is known, it will be possible to draw the trajectories of B'. The movement of this point B' over the two curves will provide a very simple way of

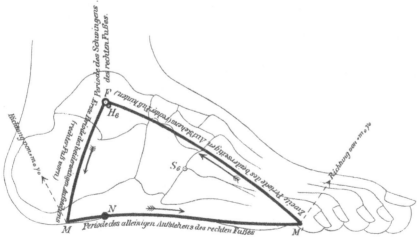

Fig. 11. Triangular trajectory described by the point of application of the ground reaction force $-G_0$

illustrating the distribution of the ground reaction force $-G_0$ over the two feet during the double-support periods.

In every case, all the forces due to gravity acting on the foot can be combined into one couple made up of forces that are equal and opposite, $+G_0$ and $-G_0$. This represents the "resultant couple of gravity". Its moment is D_s. The two forces of the resultant couple of gravity are equal to the weight G_0 of the whole body. One force $+G_0$ acts vertically downwards and is applied at the principal point H_6 of the foot throughout the movement. This principal point lies very close to the centre of the ankle F. The other force $-G_0$ acts vertically upwards and is applied at a point which continuously changes its position in the foot. During a double step it moves inside the foot over a closed triangular trajectory, as illustrated in Fig. 11. During the swinging period of the right foot the point of application of the second force of the couple coincides with the centre of the ankle F. During the following double-support period the right foot is in front of the left. The point of application of the force $-G_0$ moves from F over a curved trajectory inside the foot towards the point M of the heel which is in contact with the floor at the end of the period. During the right single-support period the point of application of the force $-G_0$ moves forwards from M, over the plantar surface, towards the point M' at the metatarsal heads where the right foot is in contact with the floor at left heel-strike. During the following double-support period the left foot is in front of the right. Then the point of application of the force $-G_0$ moves back from the point M' at the metatarsal heads to the centre of the ankle F over a curved trajectory inside the foot.

The point of application of the second force of the resultant couple of gravity does not move on its closed triangular trajectory at a constant velocity. Throughout the swinging period it coincides with F. Furthermore, the trajectory from the metatarsal heads to the centre of the ankle is significantly longer than that between the centre of the ankle and the heel whereas both are traversed in the same length of time. Finally, the right single-support period is much longer, approximately 7 times longer than either of the two double-support periods. The exact law of the movement of the point of application of $-G_0$ over its trajectory is not known so far and neither is the exact shape of the triangular trajectory itself. Both these points require further clarification. However, some conclusions on the rotatory action of gravity and its limitation can be drawn from the provisional trajectory drawn in Fig. 11.

The point of application of the force $-G_0$ must come to lie on the same vertical line as the point of application H_6 of the force $+G_0$ twice during a double step. When this occurs, the moment of the couple is 0 and gravity does not tend to rotate the foot. This occurs for the first time during the right single-support period when the point of application of $-G_0$ is at N on the sole of the foot (Fig. 11). The second time is shortly before the end of the second double-support period. The point of application of the force $-G_0$ then almost coincides with H_6 itself, but also approximately with the centre of the ankle F which is only 1.5 mm away. The points N and H_6 divide the closed trajectory into two parts each of which corresponds to a well-determined direction of rotation of the resultant couple of gravity. When the point of application lies in the posterior part of its trajectory, between H_6 and N, gravity tends to rotate the right foot clockwise. This occurs during the right swinging period, at the beginning of the first double-support period, at the very beginning of the right single-support period and at the very end of the second double-support period. When the point of application of $-G_0$ lies on the front part of its trajectory, gravity tends to rotate the foot counter-clockwise. This occurs during most of the right single-support period and almost throughout the second double-support period. The moment arm of the resultant couple of gravity twice reaches a maximum, once each time that the point of application of the force $-G_0$ passes through one of the lower corners of its triangular trajectory. When it passes through the posterior corner M of the trajectory, gravity tends to cause maximum clockwise rotation of the foot. When it passes through the front corner M' of its triangular trajectory, gravity tends to cause maximum counter-clockwise rotation of the foot.

Shortly before right toe-off, the torque D_s is 0. During the swinging period its magnitude is positive but very small. During the following double-support period, after right heel strike, the torque remains positive and increases progressively until it reaches its maximum at the end of this period. At the beginning of the right single-support period, D_s initially remains positive but steadily decreases so that it soon reaches 0 again. Later in this period it is negative, and increases progressively until it reaches a maximum (with negative sign) at left heel-strike. The latter maximum of the torque D_s for a counter-clockwise rotation is significantly greater in absolute terms than the earlier maximum for a clockwise rotation. This is seen in Fig. 11. After left heel-strike and during the following double-support period, D_s decreases quickly in absolute value until it reaches 0 again shortly before right toe-off.

All the forces due to gravity could be combined into one resultant couple the forces of which are equal to the total weight of the human body. Similarly, the effective forces can be combined in every instance into one couple made up of forces which are equal to the effective force $m_0\gamma_0$ exerted at the total centre of gravity. The moment of this couple is the resultant torque D_e of the effective forces. The couple itself will be referred to below as the "resultant couple of the effective forces". The reversed effective force $-m_6\gamma_6$ of the centre of gravity of the foot, acting on the rotation, when combined with the compression component $-m_6'\gamma_6'$ applied at the centre of the ankle F (Figs. 1, 3, 5, 7 and 9), gives a resultant equal to $m_0\gamma_0$ the direction of which is opposite to that of the effective force of the total centre of gravity. This resultant is called $-m_0\gamma_0$. It can always be assumed that the point of application of this resultant lies on the line connecting the centre of gravity of the foot to the centre of the ankle or on the prolongation of this line since the point of application of a force can be displaced at will along the line of action of the force. The point of application of this force, however, does not keep the same position in the foot throughout the movement, as does that of the forces of gravity at point H_6. The exact position of this point of application of the force $-m_0\gamma_0$ can also not yet be known because it depends on the magnitude and direction of the acceleration γ_6 of the centre of gravity of the right foot which this analysis is meant to determine. However, it seems that the point of application of the force $m_0\gamma_0$ is to be looked for on the connecting line FS_6 or on its prolongation, generally in the close vicinity of F. Indeed, while the acceleration γ_6 of the centre of gravity of the foot can be much greater than the simultaneous accelerations γ_6' of the centre of gravity of the body – right foot system, the mass m_6' of this system is approximately 55 times that of the right foot m_6 in a person with normal build. Consequently, as a rule the effective force $m_6'\gamma_6'$ is very much greater than the effective force $m_6\gamma_6$, and the point of application of the resultant must thus lie very close to the point of application F of the former force. During the right foot-flat period γ_6 equals 0. Then the point of application of the reversed effective force $-m_0\gamma_0$ coincides with F. In Figs. 1, 5, and 7, the effective force $+m_0\gamma_0$ is opposed to the resultant $-m_0\gamma_0$. The force $+m_0\gamma_0$ is applied at F during the swinging period (Fig. 1). It is applied at a point B of the sole during the right single-support period (Figs. 5 and 7). Particularly at the beginning and at the end of this period it acts at M and M' with approximately the direction given in Fig. 11. So far, the circumstances are similar to those for the forces due to gravity. The resultant of the two forces $+(m_0\gamma_0)^r$ and $+(m_0\gamma_0)^l$ in the double-support period is equal to the effective force $m_0\gamma_0$. The position of its point of application is quite unpredictable as long as the exact distribution of the ground reaction force over the two feet is not yet known since the two components $+(m_0\gamma_0)^r$ and $+(m_0\gamma_0)^l$ change continuously not only in magnitude, like their counterparts the forces of gravity G_0^r and G_0^l, but also in direction.

The Movement Equations for the Feet

The tendency to rotate exerted by all the forces on the right foot at any time finds its complete expression in the resultant torque D_m of the muscles, the torque D_s of gravity, and the torque

D_e of the effective forces. The three torques must result in an angular acceleration of the foot about the axis through its centre of gravity and perpendicular to the plane of gait, which previous studies have shown to be a principal axis of inertia of the foot. This angular acceleration is of such a magnitude that the product of this and the moment of inertia of the foot thought of as separate from the remainder of the body, in relation to the axis through the centre of gravity, mentioned above, is equal to the algebraic sum of the three torques. Each of the torques has either a positive or a negative sign depending on whether it tends to rotate the foot clockwise or counter-clockwise. The angular acceleration actually set up is correspondingly counted as positive or negative. This is the meaning of the movement equation for the right foot. The symbol φ_6'' is used for the angular acceleration mentioned above, and $m_6 \varkappa_6^2$ for the moment of inertia of the right foot in relation to the axis through its centre of gravity and perpendicular to the plane of gait. $\varkappa_6$ signifies the corresponding radius of inertia. Thus, the movement equation for the right foot can be written:

$$m_6 \varkappa_6^2 \cdot \varphi_6'' = D_m + D_s + D_e.$$

The equation is valid for the whole process of movement involved in walking. The successive periods differ from each other only in that the torques are reached in different ways.

It is now obvious that the movement equation provides a means of calculating the resultant torque of all the muscles acting on the foot actively or passively during the movement. All the other magnitudes which appear in the equation either are known or at least can be determined for each instant of the movement. The mass of the body m_0 and that of the foot m_6 have been determined in previous works, as have the radius of inertia $\varkappa_6$ of the foot relative to the axis through the centre of gravity[2]. The other magnitudes can be deduced by referring the whole movement to a tridimensional rectangular system of co-ordinates. This has been done in Chap. 1. In particular, the values of the effective force $m_0 \gamma_0$ have been calculated for the course of a double step and set out in Chap. 2. The values remaining to be determined are the linear acceleration of the centre of gravity of the foot and the angular acceleration of its rotation about the axis through its centre of gravity, for all the phases. This investigation, of course, must be carried out not only for the right foot which was used above as an example, but for both feet. Once this has been done, it is easy to use the successive positions of the foot to determine the moment arms and the moments of the couples for the swinging and stance periods of one leg, except for the yet unknown tensions of the muscles involved in the torque D_m. The two short double-support periods require a separate study dealing with the distribution of the ground reaction force over the two feet. This must be postponed, however, since this point can be clarified only after a kinematic analysis of gait dealing with all the segments of the body. The angular accelerations of the foot and those of the other segments of the limbs also will be determined later. At this moment, it remains only to deduce the accelerations of the centres of gravity and the consequent effective forces of the two feet. This will be the subject of the next two sections of this chapter.

Overall View of the Activity of the Muscles of the Foot During Walking

The resultant torque of the muscles cannot be calculated precisely until the acceleration of the centre of gravity of the foot and the angular acceleration of the foot about the axis through its centre of gravity and perpendicular to the plane of gait have been determined for a double step. In the present instance, however, the action of the muscles on the foot in the different periods of gait can at least be evaluated, even without this knowledge. The

---

[2] cf. Abhandlungen der mathematisch-physischen Klasse der Königlich Sächsischen Gesellschaft der Wissenschaften, vol 15, no 7 and vol 18, no 8

circumstances are particularly simple because the mass m_6 of the foot is very small compared with that of the rest of the body. Therefore, the torques due to the weight G_6 of the foot and to the effective force $m_6\gamma_6$ of its centre of gravity, and also the product $m_6\varkappa_6^2 \cdot \varphi_6''$, are very small in relation to all the other torques which, when combined, give D_m, D_s and D_e. If the mass m_6 of the foot is disregarded, and thus not only the weight G_6 but also the acceleration force $m_6\gamma_6$ and the product $m_6\varkappa_6^2 \cdot \varphi_6''$ are set at 0, this entails only a slight error. Then, however, we are no longer dealing with a movement equation but with the conditions of equilibrium between the remaining forces that act on the foot. If there is equilibrium, all the torques acting on the body must cancel each other out. In other words, their algebraic sum must be 0. We are very close to the truth if we use the equation:

$$D_m + D_s + D_e = 0$$

to calculate D_m. The couples $[+G_6, -G_6]$ and $[+m_6\gamma_6, -m_6\gamma_6]$ which are shown in Figs. 2, 4, 6, 8 and 10 are disregarded. To determine D_s and D_e, only the couples due to the weight G_0 and the effective force $m_0\gamma_0$ of the whole human body are considered. The main implication when the couple $[+G_6, -G_6]$ is disregarded is that, for the resultant couple of gravity, the force $+G_0$ is considered to be applied at the centre F of the right ankle rather than at the point H_6 which is very close by. Correspondingly, ignoring the couple $[+m_6\gamma_6, -m_6\gamma_6]$ means that the force $-m_0\gamma_0$ of the resultant couple of the effective forces is assumed to be applied at the centre F of the ankle instead of at a point very close to F on the connecting line FS_6.

The data below are thus obtained concerning the involvement of the muscular groups spanning the ankle in the different periods of gait. These data are a very close approximation.

During the swinging period the two torques D_s and D_e are 0 (Fig. 2). Therefore, the resultant torque D_m of the muscles must also nearly disappear throughout this period. This does not mean that the muscles attached to the right foot and spanning the ankle are not tensed during the swing period of the right foot. It actually means that the tensions of all the muscles that are active are such that the resultant torque of the muscles of the anterior group is equal, or at least nearly equal, to that of the muscles of the posterior group. The muscles of one group are antagonistic to those of the other group as far as rotation about the axis perpendicular to the plane of gait is concerned. Thus, in the swinging period the muscles of one group are balanced by their antagonists. No other conclusion can be drawn from the purely mechanical condition that D_m is nearly 0, even if the muscular equilibrium probably results from an absence of nervous stimulation of the muscles involved.

This in no way decides whether muscles are acting during the swinging period or whether the leg behaves as a pendulum moved by gravity, as was assumed by Weber and Weber. This approach, however, can lead to a decisive answer to this controversial question if we make a corresponding study of the forces exerted on the other two segments of the lower extremity and the torques which they set up and relate these to the angular accelerations with which these segments rotate about the axes through the centres of gravity and perpendicular to the plane of gait during walking. The movement equations obtained in this way for the lower leg and the thigh make it possible to calculate the resultant torques D_m of all the muscles acting on the lower leg and on the thigh. If D_m is approximately 0 also for the thigh and the lower leg throughout the swinging period, this shows that the swing of the leg results solely from gravity and from the effective forces due to the displacement of the body. If the two resultant torques of the muscles are not negligible compared with the torques of gravity and the torques of the effective forces, then the concept of Weber and Weber who considered the movement of the leg swing as a pendulum-type swing, was erroneous. This question will be answered later in our analysis of human gait.

In the right single-support period, the equation

$$D_m + D_s + D_e = 0$$

gives the precise relationship between the torques because throughout this period the angular acceleration φ_6'' is practically 0. It is, therefore, not necessary to disregard the mass m_6 of the foot to obtain an insight into the activity of the muscles in this instance. Exact determination of D_m requires that the precise position of the point at which the ground reaction force is applied be known for each moment of this period. But, even without that knowledge which we have not yet acquired, important data concerning the action of the muscles can be deduced.

At the beginning of the right single-support period, the ground reaction force is applied at a point M of the heel (Fig. 11) that lies behind the vertical through the centre F of the ankle. Therefore, the resultant couple of gravity $[+G_0, -G_0]$ tends to rotate the foot clockwise. The earlier analysis of the magnitude of the ground reaction force (Figs. 6–8, 13–15 of Chap. 2) has shown that the acceleration γ_0 and the effective force of the total centre of gravity are directed backwards and upwards at the beginning of this period. The component $+m_0\gamma_0$ of the ground reaction, therefore, has the same direction at point M (Fig. 11). The other force $-m_0\gamma_0$ of the resultant couple of the effective forces is applied at a point in the vicinity of F. Therefore, this couple also tends to rotate the foot clockwise. Consequently, according to the relationship between the three torques, the resultant couple of the muscles must necessarily tend to rotate the foot counter-clockwise, as do the anterior muscles. Thus, at the beginning of the right single-support period, either only muscles of the anterior group, as represented by the tibialis anticus, can contract or, at least, the torque with which the anterior muscles act on the foot must be significantly greater than the torque of the posterior muscles. The same is true during the preceding double-support period (Fig. 4) if it is confirmed that the force $+(m_0\gamma_0)^r$ applied at B is also directed backwards and upwards.

At the end of the right single-support period, when the foot is in contact with the floor at M' (Fig. 11), the conditions are the exact reverse. The resultant couple of gravity tends to rotate the foot counter-clockwise. Furthermore, the acceleration γ_0 and the effective force $+m_0\gamma_0$ of the total centre of gravity are directed forwards and upwards, as shown by the analysis in Chap. 2. The equal and opposite force $-m_0\gamma_0$ is again applied in the vicinity of the centre F of the ankle. Therefore, the resultant couple of the effective forces also tends to rotate the foot counter-clockwise. Consequently, at this instant, either only the posterior muscles represented by the calf group contract, or at least they act on the foot with a much greater torque than the muscles of the anterior group. The circumstances are similar in the following double-support period as long as it is justified to assume that the force $+(m_0\gamma_0)^r$ is directed forwards and upwards, as shown in Fig. 10.

Since at the end of the right single-support period the muscles of the posterior group, and at the beginning of the same period the muscles of the anterior group, are principally active, the muscles of these two groups must exactly balance each other at an instant during this period.

If we assume that the muscles of only one of the two groups are stimulated at any one time, we obtain the following picture of the activity of the muscles inserted into the foot and spanning the ankle. This picture is only approximate. During the forward swing of the leg, either none of the muscles mentioned contract or they contract only slightly. From heel strike onwards, the muscles of the anterior group, e.g. the tibialis anticus, immediately contract. Shortly after foot-flat, the muscles of the anterior group give way to those of the posterior group, especially the calf group. These remain contracted until toe-off.

This gives a rough general picture of the activity of the muscles acting on the foot but no clue as to the exact degree of tension of the different muscles in the successive phases of the movement. The further analysis will deal not only with the masses and the moments of inertia, but also, especially, with the angular accelerations of the foot. It will also extend to the other segments of the lower limbs and it will show how far this picture is correct.

Velocities and Accelerations of the Centre of Gravity of the Foot

The trajectories of the centres of gravity of the two feet have been determined above and entered into the system of co-ordinates. Their co-ordinates for the three experiments are given in Tables 1, 2 and 6 (pp. 141, 144, and 156) of Chap. 2 together with those of the trajectories of all the other centres of gravity. The velocities and accelerations of the centres of gravity of the two feet still need to be determined for each moment of the movement. The method of deducing the velocities and accelerations from the co-ordinates has been explained in Chap. 2 (p. 168 ff.).

First a displacement curve must be drawn for each of the three co-ordinates, plotting the co-ordinates against time. The distances along the abscissa of the different points of such a displacement curve are in direct proportion to the time that has elapsed since a given initial time (phase 1 in the three experiments), 1 s corresponding to 1 dm. Thus, 1 mm in the abscissa represents 1/100 s. The figures on the ordinate of the curve indicate the value of the co-ordinate at the time given on the abscissa. The upper curves in the second and third parts of Figs. 12–14 thus represent the displacement curves of the centres of gravity of the two feet for the y-co-ordinate (transverse direction) and for the z-co-ordinate (vertical direction). The curves related to the centre of gravity of the right foot are drawn in black and those related to that of the left foot in red.

As mentioned above, these displacement curves constitute an appropriate way of improving the accuracy somewhat over the initial determination of the co-ordinates. For physical reasons (Chap. 2, p. 170 ff.) they must not display any irregularities. Therefore, they must be drawn between the 31 points for which the co-ordinates are known in such a way that they run as close as possible to these co-ordinates and have a smooth overall course with no irregularities. Figures 12–14 show how close the curves are to the points that were previously determined. The values for the displacement curves on the ordinate allow correction of the co-ordinates of the centre of gravity of the foot at the 31 phases captured by instantaneous photography. They also give the co-ordinates at any time during the movement. The displacement curves thus provide an interpolation of the co-ordinates for the intervals between successive phases. The corrected y- and z-co-ordinates are given in Tables 1–6. Comparison with their counterparts in Tables 1, 2 and 6 of Chap. 2 shows that in general the corrections involve only fractions of a millimetre.

The displacement curves for the x-co-ordinate (direction of gait) cannot be drawn in the same way as those for the other two co-ordinates since the x-co-ordinate of any point of the human body changes by the length of two steps, approximately 1.5 m, in the course of a double step. Therefore, to find out the velocities and accelerations of the total centre of gravity, the process of movement has been referred to a system of co-ordinates that moves forward evenly, and without any sort of turning movements, at the average velocity of gait (Chap. 2). This means that even the co-ordinates of the centre of gravity in the direction of gait are not very large and that they also recur periodically. The y- and z-co-ordinates are not modified by this displacement of the system of co-ordinates. The relative velocities of the movement in the direction of gait deduced from the mobile system of x-co-ordinates differ from the corresponding absolute velocities by the constant average velocity of gait. Consequently, the accelerations are exactly the same whether calculated from the relative velocities or from the absolute velocities.

If the movement of the centre of gravity of the foot is referred to the mobile system of co-ordinates, relatively large values are obtained for the co-ordinates relating to the direction of gait. This is because the movement of the foot in walking is generally very different from the uniform movement of the system of co-ordinates. During foot-flat, the foot moves backwards in relation to the coronal plane moving regularly forward to exactly the same degree as this plane moves forward. When the foot swings forwards after toe-off, it reaches a final velocity more than three times that of the system of co-ordinates. The foot thus does not

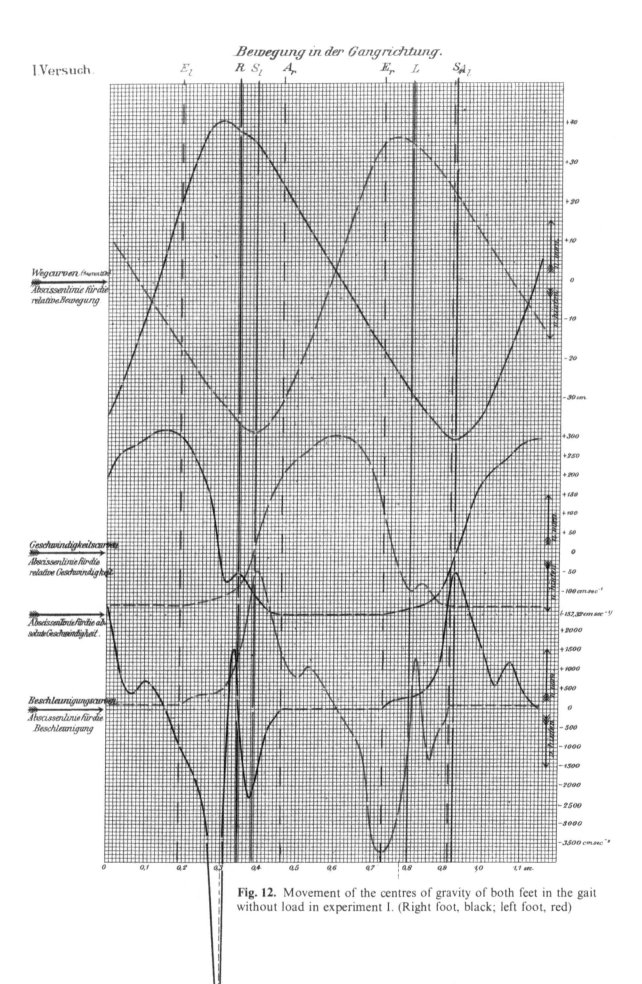

Fig. 12. Movement of the centres of gravity of both feet in the gait without load in experiment I. (Right foot, black; left foot, red)

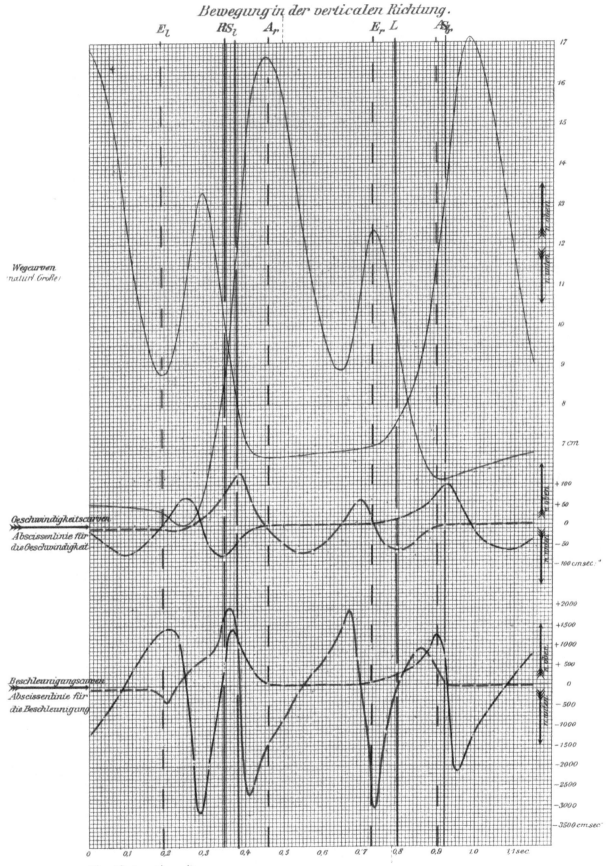

Fig. 12 (continued)

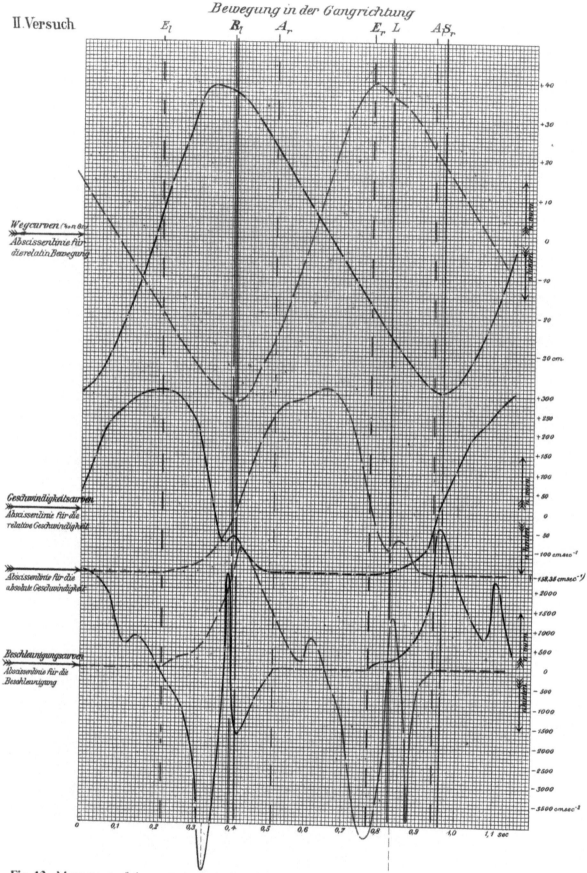

Fig. 13. Movement of the centres of gravity of both feet in the gait without load in experiment II. (Right foot, black; left foot, red)

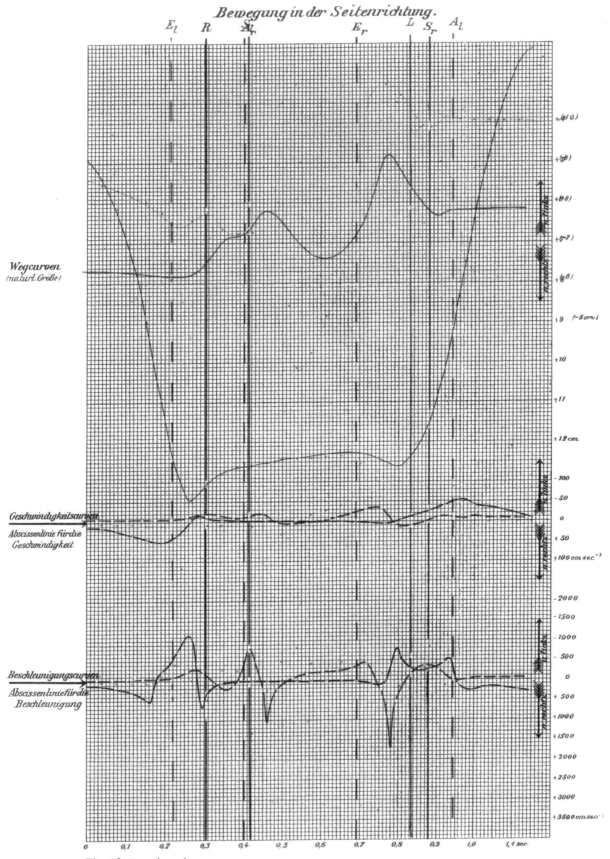

Fig. 13 (continued)

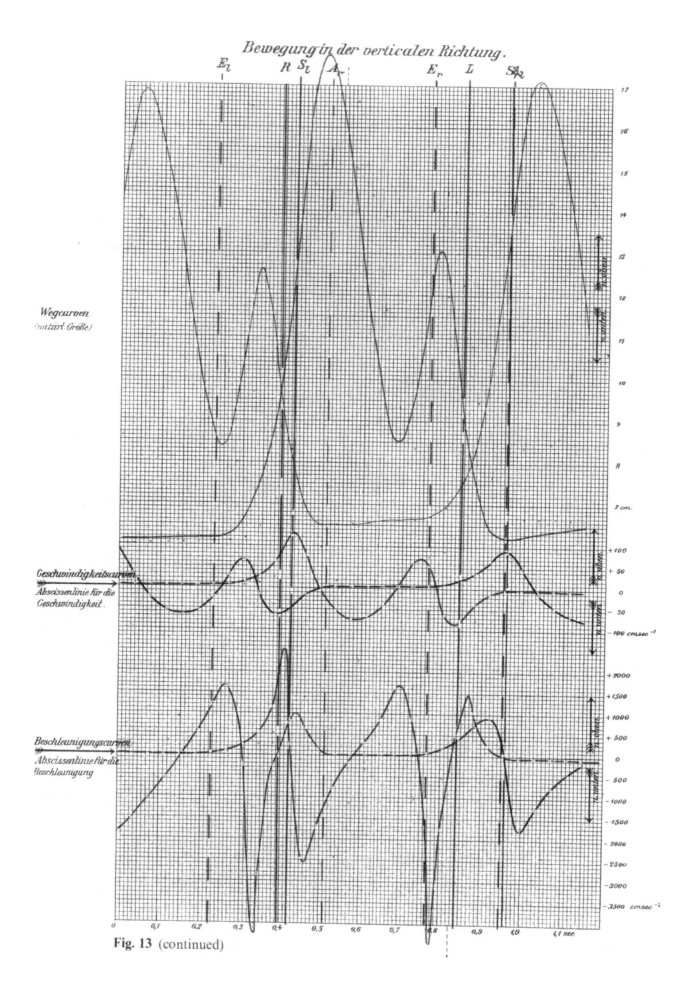

Fig. 13 (continued)

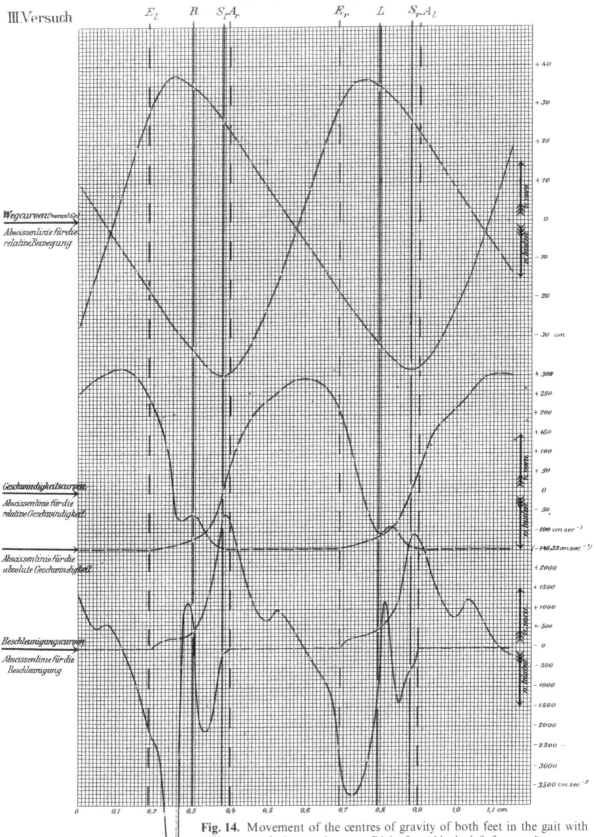

Fig. 14. Movement of the centres of gravity of both feet in the gait with military regulation equipment. (Right foot, black; left foot, red)

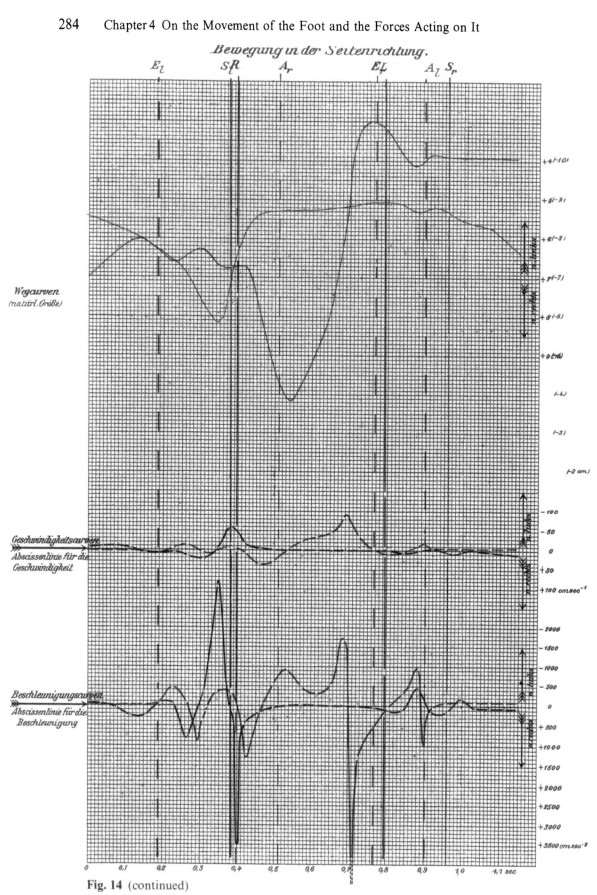

Fig. 14 (continued)

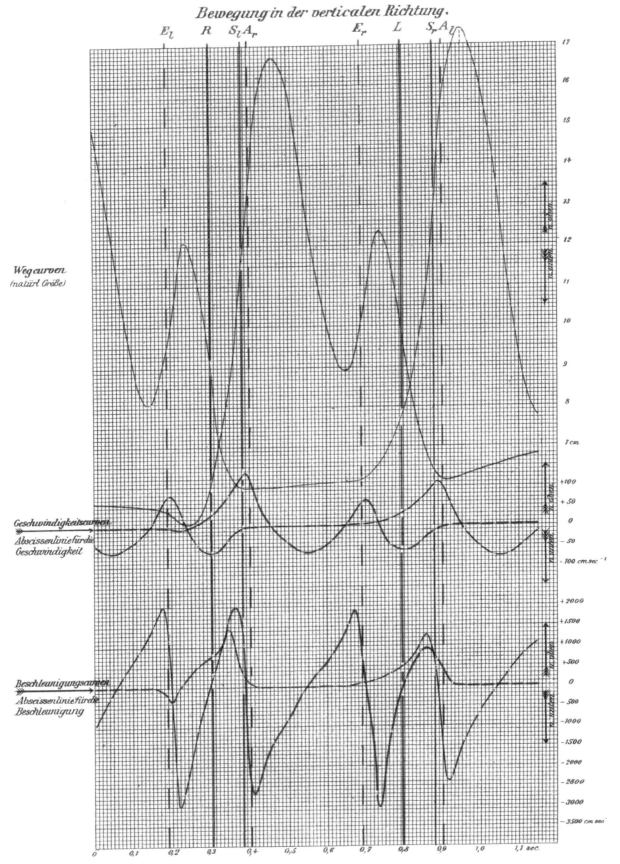

Fig. 14 (continued)

merely advance very quickly from behind towards the coronal plane of the system as it moves constantly forward, but also advances far beyond this plane. Consequently, the figures in the ordinate of the corresponding displacement curves are so large for both feet that they are out of proportion with the figures representing time along the abscissa. Therefore, to ensure the greatest possible accuracy, the displacement curves of the x-co-ordinate in the mobile system of co-ordinates were drawn in such a way that 1 s was represented by 1 m in the abscissa. This of course required very large sheets of millimetre paper. The displacement curves above in the first parts of Figs. 12–14 are 1/10 scale copies of the displacement curves actually used to determine the velocities and accelerations.

The co-ordinates in the mobile system of co-ordinates relating to the direction of gait are again called relative x-co-ordinates. They are found as described in Chap. 2 (p. 158). The

Table 1. Movement of the centre of gravity of the right foot in experiment I

No.	Co-ordinates (cm) x_6 relative	x_6 absolute	y_6	z_6	Velocities (cm s⁻¹) Direction of gait relative	Direction of gait absolute	Transverse direction	Vertical direction	Accelerations (cm s⁻²) Direction of gait	Transverse direction	Vertical direction	No.
1	−34,39	4,86	+7,42	16,85	+195	+352	−4	−13	+2670	−30	−1190	1
2	−26,01	19,27	+7,23	15,86	+261	+418	−6	−43	+890	−20	−760	2
3	−15,73	35,58	+6,98	13,67	+280	+437	−7	−68	+470	−10	−200	3
4	−4,54	52,80	+6,75	11,18	+306	+463	−7	−64	+620	−10	+460	4
5	+7,60	70,97	+6,49	9,35	+318	+475	−7	−32	+30	+30	+1030	5
6	+19,48	88,88	+6,31	8,82	+302	+459	0	+8	−900	+470	+1390	6
7	+30,10	105,53	+6,75	10,03	+251	+408	+25	+59	−1730	+230	+1130	7
8	+38,38	119,84	+7,66	12,63	+158	+315	+20	+53	−3380	−890	−1910	8
9	+40,43	127,92	+7,73	12,70	−61	+96	−19	−49	−3400	−640	−2000	9
10	+37,81	131,33	+6,80	9,99	−52	+105	−23	−75	+300	+230	+450	10
11	+35,26	134,81	+6,27	7,71	−100	+57	−11	−38	−3160	+400	+1120	11
12	+30,10	135,68	+6,05	6,79	−145	+12	−1	−8	−590	+110	+370	12
13	+24,30	135,91	+6,07	6,68	−157	0	+1	0	−20	+10	+30	13
14	+18,28	135,92	+6,10	6,70	−157	0	0	+2	0	−40	0	14
15	+12,26	135,93	+6,06	6,73	−157	0	−1	+2	0	−30	0	15
16	+6,23	135,93	+6,01	6,77	−157	0	−1	+2	0	−20	0	16
17	+0,21	135,94	+5,97	6,81	−157	0	−1	+2	0	−10	0	17
18	−5,81	135,95	+5,91	6,85	−157	0	−1	+2	0	+10	0	18
19	−11,82	135,97	+5,86	6,88	−157	0	−1	+2	0	+30	0	19
20	−17,82	136,00	+5,82	6,93	−157	0	0	+3	0	+50	+50	20
21	−23,57	136,28	+5,84	7,17	−150	+7	+2	+7	+240	+70	+140	21
22	−29,05	136,83	+5,98	7,74	−140	+17	+6	+17	+310	+70	+280	22
23	−34,15	137,76	+6,17	8,66	−125	+32	+1	+32	+480	−530	+470	23
24	−38,35	139,59	+6,00	10,53	−98	+59	−7	+58	+1150	+110	+930	24
25	−40,70	143,27	+5,87	13,23	−8	+149	−2	+98	+3450	0	+700	25
26	−38,58	151,42	+5,67	16,46	+108	+265	−8	+53	+2540	−330	−2110	26
27	−33,11	162,92	+5,17	16,89	+189	+346	−14	−17	+1490	−10	−1270	27
28	−24,72	177,34	+4,68	15,54	+224	+381	−9	−49	+620	+200	−650	28
29	−15,78	192,31	+4,52	13,35	+258	+415	0	−62	+1170	+250	−130	29
30	−5,06	209,06	+4,67	10,90	+289	+446	+9	−58	+350	+210	+360	30
31	+5,85	226,00	+5,14	9,01	+296	+453	+14	−36	+20	+70	+790	31

absolute x-co-ordinates for the three experiments are given in Tables 1, 2 and 6 of Chap. 2 (pp. 141, 144 and 156) and again in Tables 1–6 of the present chapter. The distances between the plane YZ in the mobile system and that in the fixed system are given in Table 8 of Chap. 2 (p. 159). These distances are subtracted from the absolute values for x. The relative x-co-ordinates obtained in this way for the centres of gravity of both feet are also given in Tables 1–6.

The displacement curves are used to derive the velocities of the centre of gravity of the foot, as has been previously described (Chap. 2, p. 180), by determining tangents to an adequate number of points on each displacement curve. The trigonometric tangent of the angle formed by the tangent to such a point and the abscissa is in direct proportion to the velocity of the centre of gravity of the foot in the direction of the co-ordinate, at the instant

Table 2. Movement of the centre of gravity of the left foot in experiment I

| No. | Co-ordinates (cm) | | | | Velocities (cm s^{-1}) | | | | Accelerations (cm s^{-2}) | | | No. |
| | x_7 | | y_7 | z_7 | Direction of gait | | Transverse direction | Vertical direction | Direction of gait | Transverse direction | Vertical direction | |
	relative	absolute			relative	absolute						
1	+17,62	56,87	−6,21	6,10	−157	0	−1	+1	0	+30	0	1
2	+11,66	56,94	−6,23	6,11	−157	0	0	+1	0	+20	0	2
3	+5,62	56,93	−6,21	6,13	−157	0	0	+1	0	+10	0	3
4	−0,36	56,98	−6,20	6,15	−157	0	+1	+1	0	+10	0	4
5	−6,38	56,99	−6,18	6,16	−157	0	+1	+2	0	0	+10	5
6	−12,41	56,99	−6,13	6,17	−157	0	+1	+3	0	−20	+10	6
7	−18,43	57,00	−6,10	6,20	−157	0	0	+4	0	−40	+30	7
8	−24,41	57,05	−6,10	6,35	−150	+7	−3	+6	+220	−180	+110	8
9	−29,95	57,54	−6,36	6,75	−139	+18	−11	+15	+350	−140	+260	9
10	−35,02	58,50	−6,80	7,60	−120	+37	−11	+31	+710	+140	+540	10
11	−39,13	60,42	−7,08	9,20	−78	+79	−3	+62	+1370	+170	+1140	11
12	−41,64	63,94	−7,21	12,40	−11	+146	−7	+126	+2440	−710	+790	12
13	−39,53	72,08	−7,69	16,40	+90	+247	−12	+81	+2870	+300	−2430	13
14	−34,22	83,42	−7,58	17,63	+187	+344	+8	−3	+2040	+270	−1690	14
15	−24,91	98,76	−7,16	16,33	+250	+407	+11	−52	+1240	−20	−780	15
16	−15,10	114,60	−6,78	14,03	+284	+441	+8	−71	+490	−130	−160	16
17	−4,49	131,24	−6,58	11,35	+295	+452	+3	−67	+250	−160	+340	17
18	+6,74	148,50	−6,63	9,25	+316	+473	−4	−40	+550	−190	+890	18
19	+18,89	166,68	−6,92	8,54	+313	+470	−13	+6	−590	−260	+1660	19
20	+30,16	183,98	−7,53	9,82	+257	+414	−25	+64	−2330	−370	+900	20
21	+37,87	197,72	−8,60	12,45	+128	+285	−26	+56	−4330	+310	−1650	21
22	+39,56	205,44	−9,05	12,30	−39	+118	+12	−59	−3610	+600	−1680	22
23	+36,44	208,35	−8,40	9,43	−91	+66	+17	−76	+1170	−90	+550	23
24	+33,46	211,40	−7,88	7,35	−81	+76	+9	−34	−1710	−300	+1090	24
25	+28,65	212,62	−7,63	6,40	−150	+7	−1	−8	−430	−200	+300	25
26	+22,64	212,64	−7,73	6,26	−157	0	−4	0	0	+40	+40	26
27	+16,63	212,66	−7,78	6,32	−157	0	−2	+2	0	+40	0	27
28	+10,62	212,68	−7,79	6,38	−157	0	−1	+2	0	+20	0	28
29	+4,61	212,70	−7,80	6,44	−157	0	0	+2	0	+10	0	29
30	−1,40	212,72	−7,81	6,50	−157	0	0	+2	0	+10	0	30
31	−7,40	212,75	−7,82	6,57	−157	0	0	+2	0	0	0	31

Table 3. Movement of the centre of gravity of the right foot in experiment II

No.	Co-ordinates (cm) x_6 relative	absolute	y_6	z_6	Velocities (cm s^{-1}) Direction of gait relative	absolute	Transverse direction	Vertical direction	Accelerations (cm s^{-2}) Direction of gait	Transverse direction	Vertical direction	No.
1	−39,80	4,45	+7,01	14,22	+46	+204	−8	+93	+2450	−60	−1890	1
2	−36,44	13,88	+6,66	16,60	+136	+294	−10	+29	+2290	−30	−1520	2
3	−29,83	26,56	+6,28	16,55	+223	+381	−9	−26	+1770	+60	−1060	3
4	−20,46	42,00	+6,05	14,78	+264	+422	−5	−57	+620	+240	−580	4
5	−10,69	57,84	+6,06	12,53	+291	+449	+5	−63	+750	+220	−10	5
6	+0,78	75,38	+6,32	10,20	+307	+465	+8	−55	+180	−60	+550	6
7	+13,36	94,03	+6,58	8,65	+304	+462	+5	−26	−430	−40	+1210	7
8	+23,55	110,29	+6,91	8,77	+278	+436	+16	+31	−1030	+750	+1590	8
9	+34,34	127,15	+7,70	10,80	+204	+362	+20	+65	−3480	−380	−250	9
10	+38,82	137,70	+8,15	12,55	+26	+184	−3	−13	−4020	−3100	−2850	10
11	+37,95	142,90	+6,90	10,51	−79	+79	−56	−62	+500	+80	−240	11
12	+35,25	146,27	+5,74	8,25	−90	+68	−15	−48	−1590	+420	+790	12
13	+30,85	147,94	+5,36	6,87	−134	+24	−5	−16	−810	+170	+680	13
14	+25,20	148,36	+5,28	6,51	−156	+2	−1	−1	−240	+60	+140	14
15	+19,14	148,37	+5,26	6,51	−158	0	0	+3	0	−10	0	15
16	+13,09	148,39	+5,24	6,53	−158	0	0	+3	0	−30	0	16
17	+7,03	148,40	+5,22	6,57	−158	0	−1	+3	0	−20	0	17
18	+0,97	148,41	+5,19	6,60	−158	0	−1	+3	0	−10	0	18
19	−5,08	148,43	+5,14	6,63	−158	0	−1	+4	0	0	+10	19
20	−11,12	148,46	+5,09	6,66	−158	0	−1	+4	0	+10	+20	20
21	−17,14	148,51	+5,06	6,71	−158	0	−1	+5	0	+60	+60	21
22	−22,91	148,81	+5,06	6,94	−151	+7	+1	+11	+210	+120	+180	22
23	−28,42	149,37	+5,13	7,53	−142	+16	+6	+19	+310	+110	+390	23
24	−33,61	150,27	+5,31	8,56	−126	+32	0	+37	+640	−450	+720	24
25	−37,65	152,28	+5,22	10,58	−89	+69	−2	+65	+1380	+160	+910	25
26	−39,61	156,39	+5,29	13,40	−1	+157	+6	+89	+3500	+140	+20	26
27	−37,09	164,98	+5,52	16,49	+107	+265	+4	+49	+2390	−160	−1780	27
28	−31,21	176,93	+5,60	17,08	+187	+345	+2	−11	+1370	+40	−1250	28
29	−23,20	191,01	+5,73	15,97	+226	+384	+5	−43	+800	+70	−770	29
30	−13,95	206,33	+6,01	13,81	+273	+431	+8	−67	+2030	+70	−420	30
31	−2,63	223,72	+6,38	11,14	+305	+463	+11	−78	+340	+50	−170	31

corresponding to the value on the abscissa for the point to which the tangent has been drawn. The data thus obtained make it possible to draw the velocity curve corresponding to each displacement curve. The time is plotted along the abscissa and the velocity in the direction considered along the ordinate. These velocity curves are shown for all three co-ordinates and all three experiments in Figs. 12–14 below the corresponding displacement curves. The unit of velocity of a body that moves forwards 1 cm in 1 s during uniform movement is represented by a length of 0.1 mm in the ordinate, and not by 1 mm as was the case for the movement of the total centre of gravity (Chap. 2, Figs. 13–15). If the velocities of the centre of gravity of the foot are to be compared with those of the total centre of gravity, the values along the ordinate of the velocity curves in Figs. 12–14 must thus be multiplied by a factor of 10.

Table 4. Movement of the centre of gravity of the left foot in experiment II

No.	Co-ordinates (cm) x_7 relative	x_7 absolute	y_7	z_7	Velocities (cm s⁻¹) Direction of gait relative	absolute	Transverse direction	Vertical direction	Accelerations (cm s⁻²) Direction of gait	Transverse direction	Vertical direction	No.
1	+26,85	71,10	−6,02	6,01	−158	0	0	+2	0	+10	0	1
2	+20,77	71,09	−6,01	6,02	−158	0	0	+2	0	+20	0	2
3	+14,69	71,08	−6,00	6,03	−158	0	0	+2	0	+20	0	3
4	+8,60	71,06	−5,98	6,04	−158	0	+1	+2	0	+30	0	4
5	+2,50	71,03	−5,96	6,05	−158	0	+1	+2	0	+20	0	5
6	−3,56	71,04	−5,94	6,07	−158	0	+1	+2	0	+10	0	6
7	−9,62	71,05	−5,92	6,09	−158	0	+1	+3	0	−10	+20	7
8	−15,68	71,06	−5,89	6,11	−158	0	0	+4	0	−30	+50	8
9	−21,72	71,09	−5,92	6,18	−155	+3	−2	+5	+230	−110	+120	9
10	−27,32	71,56	−6,07	6,44	−144	+14	−7	+11	+340	−160	+260	10
11	−32,74	72,21	−6,40	7,32	−129	+29	−10	+24	+460	−20	+520	11
12	−37,29	73,73	−6,74	8,30	−103	+55	−7	+50	+890	+270	+1080	12
13	−40,27	76,82	−6,87	11,25	−41	+117	−8	+108	+2630	−900	+1390	13
14	−39,57	83,59	−7,48	15,10	+75	+233	−16	+99	+3260	+330	−1500	14
15	−34,50	94,73	−7,62	17,29	+183	+341	+9	+18	+2120	+280	−2440	15
16	−26,36	108,94	−7,18	16,62	+236	+394	+13	−49	+920	+20	−1160	16
17	−16,62	124,65	−6,72	14,37	+266	+424	+9	−72	+770	−260	−330	17
18	−5,72	141,72	−6,58	11,65	+303	+461	0	−70	+1110	−180	+280	18
19	+6,69	160,20	−6,70	9,33	+335	+493	−5	−48	+340	−120	+830	19
20	+19,28	178,86	−6,98	8,33	+314	+472	−9	0	−1450	−70	+1600	20
21	+29,66	195,31	−7,39	9,45	+242	+400	−12	+74	−2270	−30	+1150	21
22	+37,08	208,80	−7,86	11,88	+138	+296	−9	+87	−3330	+380	−530	22
23	+39,59	217,38	−7,66	12,32	−60	+98	+27	−44	−4520	+560	−3000	23
24	+36,53	220,39	−6,54	9,46	−66	+92	+29	−77	+930	−280	+680	24
25	+33,63	223,56	−5,63	7,65	−104	+54	+12	−30	−2040	−640	+1020	25
26	+28,65	224,65	−5,52	6,80	−149	+9	−2	−6	−570	−120	+310	26
27	+22,62	224,69	−5,63	6,69	−158	0	−3	−2	0	+30	+10	27
28	+16,56	224,70	−5,66	6,68	−158	0	−1	−1	0	+40	0	28
29	+10,50	224,71	−5,67	6,68	−158	0	0	−1	0	+30	0	29
30	+4,44	224,72	−5,68	6,68	−158	0	0	−1	0	+10	0	30
31	−1,62	224,73	−5,69	6,67	−158	0	0	−1	0	+10	0	31

The values on the ordinate of a velocity curve give the velocity at any point in time during the movement. They have been measured for all the points corresponding to the 31 phases. The data thus obtained are given in Tables 1–6. For the movement in the direction of gait, the velocity curves give the velocities which the displacement of the centre of gravity of the foot displays relative to the mobile system of co-ordinates. The absolute velocities referred to the immobile system are obtained by adding the relative velocities and the average velocity of gait. This is 157.32 cm s⁻¹ in experiment I, 158.35 cm s⁻¹ in experiment II and 143.33 cm s⁻¹ in experiment III. These absolute velocities are also given in Tables 1–6.

In the same way as the velocity curves have been obtained from the displacement curves, the acceleration curves are derived from the velocity curves (Chap. 2, p. 181). These acceleration curves are shown below the velocity curves in Figs. 12–14 where 1 s is

Table 5. Movement of the centre of gravity of the right foot in experiment III (with loading)

No.	Co-ordinates (cm) x_6 relative	x_6 absolute	y_6	z_6	Velocities (cm s^{-1}) Direction of gait relative	Direction of gait absolute	Transverse direction	Vertical direction	Accelerations (cm s^{-2}) Direction of gait	Transverse direction	Vertical direction	No.
1	−27,18	4,70	+ 4,89	14,92	+252	+397	+12	− 42	+1380	+ 90	−1050	1
2	−16,94	20,51	+ 5,40	12,75	+287	+432	+16	− 64	+ 590	+130	− 280	2
3	− 5,42	37,60	+ 6,14	10,34	+307	+452	+23	− 56	+ 620	+190	+ 310	3
4	+ 6,47	55,06	+ 7,38	8,56	+319	+464	+32	− 31	+ 50	+290	+ 890	4
5	+18,38	72,54	+ 8,96	8,11	+300	+445	+44	+ 7	− 970	+480	+1510	5
6	+29,00	88,73	+10,99	9,60	+238	+383	+51	+ 76	−2240	− 240	+1150	6
7	+35,84	101,14	+12,73	12,03	+121	+266	+33	+ 22	−4190	− 730	−2730	7
8	+36,89	107,76	+13,45	11,20	− 68	+ 77	− 3	− 47	−2700	−1060	−1070	8
9	+34,32	110,76	+13,04	8,51	− 59	+ 86	−10	− 66	− 260	+320	+ 310	9
10	+31,18	113,19	+12,77	6,44	−116	+ 29	− 5	− 33	−2000	+ 70	+1440	10
11	+26,17	113,75	+12,65	5,99	−144	+ 1	− 3	− 2	− 160	+ 40	+ 290	11
12	+20,60	113,75	+12,58	5,98	−145	0	− 2	+ 2	0	+ 30	0	12
13	+15,03	113,75	+12,51	6,00	−145	0	− 2	+ 2	0	+ 20	0	13
14	+ 9,46	113,75	+12,47	6,02	−145	0	− 1	+ 2	0	+ 10	0	14
15	+ 3,91	113,77	+12,41	6,04	−145	0	− 1	+ 2	0	+ 10	0	15
16	− 1,66	113,77	+12,36	6,07	−145	0	− 1	+ 2	0	+ 10	0	16
17	− 7,23	113,77	+12,33	6,09	−145	0	− 1	+ 2	0	+ 10	0	17
18	−12,79	113,78	+12,28	6,11	−145	0	0	+ 3	0	+ 20	+ 20	18
19	−18,35	113,79	+12,29	6,17	−145	0	+ 1	+ 5	0	+ 30	+ 70	19
20	−23,64	114,07	+12,36	6,47	−137	+ 8	+ 2	+ 9	+ 260	+ 70	+ 150	20
21	−28,63	114,65	+12,49	7,08	−125	+ 20	+ 5	+ 20	+ 350	+ 30	+ 320	21
22	−33,16	115,69	+12,58	8,11	−108	+ 37	− 6	+ 37	+ 570	− 610	+ 560	22
23	−36,78	117,64	+12,15	10,02	− 77	+ 68	−17	+ 67	+1180	− 240	+1040	23
24	−38,37	121,62	+11,44	13,30	+ 5	+150	−28	+105	+2850	− 280	+ 450	24
25	−36,17	129,39	+10,15	16,68	+109	+254	−41	+ 56	+2380	− 430	−2320	25
26	−30,48	140,65	+ 8,34	17,33	+180	+325	−54	− 12	+1280	+110	−1370	26
27	−22,68	154,02	+ 6,40	16,09	+216	+361	−44	− 50	+ 780	+260	− 690	27
28	−14,08	168,19	+ 4,85	13,79	+251	+396	−36	− 75	+1140	+190	− 150	28
29	− 3,66	184,18	+ 3,45	10,90	+291	+436	−28	− 71	+ 670	+200	+ 330	29
30	+ 7,64	201,05	+ 2,48	8,67	+303	+448	−19	− 46	+ 40	+240	+ 810	30
31	+19,24	218,22	+ 2,12	7,77	+300	+445	− 8	− 12	− 190	+290	+1090	31

represented by 1 dm along the abscissa. But 1 mm in the ordinate corresponds to an acceleration of 100 cm s^{-2} whereas in the acceleration curves for the total centre of gravity, in Figs. 13–15 of Chap. 2, an acceleration of 100 cm s^{-2} was represented by a length of 10 mm. For comparison of the acceleration curves for the centres of gravity of both feet with those for the total centre of gravity the values along the ordinate of the former curves must, therefore, be multiplied by a factor of 10. Measurement of values in the ordinate corresponding to the 31 phases under discussion gives the accelerations shown in Tables 1–6.

In determining the velocities and the accelerations, the degree of accuracy possible must always be borne in mind. This is why the velocity data have been rounded to the nearest whole number and the acceleration data to the nearest ten in Tables 1–6 (pp. 286–291) which give an overall view of the movement of the centres of gravity of both feet in the three experiments.

Table 6. Movement of the centre of gravity of the left foot in experiment III (with loading)

No.	Co-ordinates (cm)				Velocities (cm s⁻¹)				Accelerations (cm s⁻²)			No.
	x_7		y_7	z_7	Direction of gait		Transverse direction	Vertical direction	Direction of gait	Transverse direction	Vertical direction	
	relative	absolute			relative	absolute						
1	+ 9,95	41,83	− 8,57	5,59	−145	0	+ 3	− 1	0	+ 20	0	1
2	+ 4,36	41,81	− 8,47	5,58	−145	0	+ 3	− 1	0	+ 30	0	2
3	− 1,24	41,78	− 8,32	5,57	−145	0	+ 5	− 1	0	+ 50	0	3
4	− 6,85	41,74	− 8,18	5,54	−145	0	+ 6	− 1	0	+ 60	0	4
5	− 12,54	41,62	− 7,98	5,48	−145	0	+ 8	− 1	0	+ 60	0	5
6	− 18,19	41,54	− 7,70	5,37	−145	0	+ 9	− 4	0	− 40	− 260	6
7	− 23,99	41,31	− 7,42	5,12	−139	+ 6	0	− 7	+ 220	− 410	+ 110	7
8	− 29,11	41,76	− 7,57	5,19	−130	+ 15	− 8	+ 7	+ 260	− 20	+ 510	8
9	− 33,67	42,77	− 7,75	6,26	−117	+ 28	+ 8	+ 34	+ 440	+ 480	+ 810	9
10	− 37,94	44,07	− 7,37	8,11	− 90	+ 55	+ 8	+ 76	+1 370	− 270	+1 620	10
11	− 39,72	47,86	− 7,23	11,65	+ 7	+152	− 2	+132	+3 380	− 260	+1 050	11
12	− 37,12	56,03	− 7,27	15,60	+121	+266	+11	+ 60	+2 590	+1 320	−2 440	12
13	− 30,97	67,75	− 6,23	16,71	+197	+342	+40	0	+1 280	+ 250	−1 370	13
14	− 22,83	81,46	− 4,76	15,89	+234	+379	+33	− 41	+ 720	− 490	− 840	14
15	− 13,22	96,64	− 3,89	13,74	+265	+410	+ 6	− 65	+ 930	− 830	− 140	15
16	− 2,57	112,86	− 4,23	11,28	+290	+435	−18	− 59	+ 400	− 420	+ 340	16
17	+ 8,76	129,76	− 5,10	9,55	+294	+439	−28	− 34	− 240	− 270	+ 800	17
18	+ 19,66	146,23	− 6,29	8,93	+273	+418	−41	+ 6	− 830	− 480	+1 530	18
19	+ 29,70	161,84	− 8,10	10,15	+217	+362	−79	+ 59	−2 500	−1 620	+ 700	19
20	+ 35,76	173,47	− 10,70	12,25	+ 86	+231	−32	+ 25	−3 740	+1 440	−2 700	20
21	+ 36,57	179,85	− 11,02	11,50	− 49	+ 96	+ 3	− 48	−2 830	+ 630	−1 130	21
22	+ 34,14	182,99	− 10,70	9,29	−103	+ 42	+16	− 60	+ 590	+ 160	+ 170	22
23	+ 30,37	184,79	− 10,11	7,32	− 99	+ 46	+12	− 40	−1 210	− 210	+ 830	23
24	+ 25,85	185,84	− 9,89	6,37	−140	+ 5	− 5	− 11	− 530	− 870	+ 690	24
25	+ 20,07	185,63	− 10,11	6,18	−145	0	0	+ 2	0	+ 120	+ 70	25
26	+ 14,40	185,53	− 10,09	6,29	−145	0	+ 1	+ 3	0	+ 10	0	26
27	+ 8,80	185,50	− 10,08	6,40	−145	0	+ 1	+ 3	0	− 10	0	27
28	+ 3,22	185,49	− 10,07	6,52	−145	0	+ 1	+ 3	0	0	0	28
29	− 2,35	185,49	− 10,06	6,62	−145	0	+ 1	+ 3	0	0	0	29
30	− 7,95	185,46	− 10,05	6,72	−145	0	+ 1	+ 3	0	0	0	30
31	− 13,45	185,53	− 10,03	6,78	−145	0	+ 1	+ 3	0	0	0	31

The phases captured by photography during the experiments, of course, do not generally correspond to maxima or minima of the co-ordinates, velocities or accelerations. However, these maxima and minima are very important for characterization of the movement. Therefore, it was necessary to make a specific analysis of the maxima of the different curves in Figs. 12–14 and determine the values corresponding to them on both abscissa and ordinate in each case. The results of this analysis are collected in Tables 7–12 (pp. 292–297) in a way similar to that used previously for the movement of the total centre of gravity (Chap. 2, Table 18, pp. 191–193). Next to the time (to 0.001 s) that has elapsed between the first of the 31 phases and the maximum or minimum considered, the number of the corresponding phase is given. This makes it easy to compare the data with those for the overall movement, which are presented in Figs. 22, 23 and 26 in Chap. 1. In this case the number of the phase is

Table 7. Maximum values for the movement of the centre of gravity of the right foot in experiment I

Time (s)	No. of phase of movement	Relative value for co-ordinate (cm)	Absolute value for co-ordinate (cm)	Time (s)	No. of phase of movement	Relative velocity (cm s⁻¹)	Absolute velocity (cm s⁻¹)	Time (s)	No. of phase of movement	Acceleration (cm s⁻²)

$$\text{Movement in the direction of gait}$$
$$\text{Anterior maxima}$$

Time (s)	No. of phase of movement	Relative value for co-ordinate (cm)	Absolute value for co-ordinate (cm)	Time (s)	No. of phase of movement	Relative velocity (cm s⁻¹)	Absolute velocity (cm s⁻¹)	Time (s)	No. of phase of movement	Acceleration (cm s⁻²)
0,297	8,8	+40,76	126,74	0,155	5,0	+318	+475	0,098	3,6	+ 760
				0,347	10,1	− 52	+105	0,336	9,8	+1530
								0,460—0,728	13,0—20,0	0
								0,923	25,1	+3490
								1,076	29,1	+1200

$$\text{Posterior maxima}$$

0,922	25,1	−40,72	143,55	0,320	9,4	− 75	+ 82	0,067	2,8	+ 430
				0,460—0,728	13,0—20,0	−157	0	0,298	8,8	−8100
								0,379	10,9	−2320
								0,460—0,728	13,0—20,0	0
								1,031	27,9	+ 610

$$\text{Transverse movement}$$
$$\text{Lateral (right) maxima}$$

0,290	8,6	—	+7,96	0,239	7,2	—	+ 25	0,205	6,4	+1450
0,501	14,1	—	+6,10	0,463	13,1	—	+ 1	0,373	10,7	+ 430
0,845	23,1	—	+6,17	0,813	22,2	—	+ 6	0,790	21,6	+ 100
				0,920	25,0	—	− 2	0,900	24,5	+ 370
								1,076	29,1	+ 260

$$\text{Medial (left) maxima}$$

0,192	6,0	—	+6,31	0,135	4,5	—	− 8	0,290	8,6	−1290
0,433	12,3	—	+6,04	0,333	9,7	—	− 25	0,505	14,2	− 50
0,734	20,2	—	+5,82	0,620	17,2	—	− 1	0,845	23,1	− 540
1,074	29,0	—	+4,52	0,877	23,9	—	− 7	0,955	25,9	− 340
				0,998	27,0	—	− 14			

$$\text{Vertical movement}$$
$$\text{Upper maxima}$$

0,289	8,5	—	13,26	0,250	7,5	—	+ 72	0,203	6,3	+1420
0,983	26,7	—	17,12	0,470—0,700	13,3—19,3	—	+ 2	0,370	10,7	+1370
				0,925	25,1	—	+100	0,902	24,5	+1300

$$\text{Lower maxima (minima)}$$

0,184	5,8	—	8,74	0,090	3,4	—	− 72	0,287	8,5	−3170
0,460	13,0	—	6,68	0,337	9,8	—	− 77	0,470—0,700	13,3—19,3	0
				0,470—0,700	13,3—19,3	—	+ 2	0,954	25,9	−2130
				1,083	29,3	—	− 63			

Table 8. Maximum values for the movement of the centre of gravity of the left foot in experiment I

Time (s)	No. of phase of movement	Relative value for co-ordinate (cm)	Absolute value for co-ordinate (cm)	Time (s)	No. of phase of movement	Relative velocity (cm s⁻¹)	Absolute velocity (cm s⁻¹)	Time (s)	No. of phase of movement	Acceleration (cm s⁻²)
					Movement in the direction of gait					
					Anterior maxima					
0,795	21,7	+39,77	204,08	0,667	18,4	+320	+477	0—0,230	0—7,0	0
				0,867	23,6	− 70	+ 87	0,442	12,5	+3080
								0,637	17,6	+ 780
								0,846	23,1	+1300
								0,955—1,15	25,9—31	0
					Posterior maxima					
0,426	12,1	−40,66	65,64	0—0,230	0—7,0	−157	0	0—0,230	0—7,0	0
				0,833	22,7	− 96	+ 61	0,606	16,8	+ 220
				0,955—1,15	25,9—31	−157	0	0,777	21,3	−4420
								0,891	24,3	−3990
								0,955—1,15	25,9—31	0
					Transverse movement					
					Lateral (left) maxima					
0,040	2,0	—	−6,23	0,327	9,5	—	− 12	0,284	8,4	− 270
0,475	13,4	—	−7,76	0,449	12,7	—	− 14	0,423	12,0	− 720
0,795	21,7	—	−9,12	0,748	20,5	—	− 29	0,727	20,0	− 380
				0,943	25,6	—	− 4	0,890	24,2	− 310
					Medial (right) maxima					
0,243	7,3	—	−6,09	0,150	4,9	—	+ 1	0,368	10,6	+ 270
0,630	17,4	—	−6,56	0,399	11,4	—	− 2	0,471	13,3	+1080
0,916	24,9	—	−7,62	0,533	14,9	—	+ 12	0,794	21,7	+1720
				0,833	22,7	—	+ 18	0,970	26,3	+ 60
					Vertical movement					
					Upper maxima					
0,495	13,9	—	17,64	0—0,150	0—4,9	—	+ 1	0,410	11,7	+2560
0,785	21,5	—	13,05	0,430	12,2	—	+129	0,701	19,3	+1720
				0,748	20,5	—	+ 73	0,871	23,7	+1500
				0,980—1,15	26,6—31	—	+ 2			
					Lower maxima (minima)					
0,686	18,9	—	8,52	0—0,150	0—4,9	—	+ 1	0—0,150	0—4,9	0
0,956	26,0	—	6,25	0,588	16,3	—	− 73	0,469	13,2	−2530
				0,832	22,7	—	− 80	0,783	21,4	−4400
				0,980—1,15	26,6—31	—	+ 2	0,980—1,15	26,6—31	0

Table 9. Maximum values for the movement of the centre of gravity of the right foot in experiment II

Time (s)	No. of phase of movement	Relative value for co-ordinate (cm)	Absolute value for co-ordinate (cm)	Time (s)	No. of phase of movement	Relative velocity (cm s⁻¹)	Absolute velocity (cm s⁻¹)	Time (s)	No. of phase of movement	Acceleration (cm s⁻²)

Movement in the direction of gait
Anterior maxima

Time (s)	No. of phase of movement	Relative value for co-ordinate (cm)	Absolute value for co-ordinate (cm)	Time (s)	No. of phase of movement	Relative velocity (cm s⁻¹)	Absolute velocity (cm s⁻¹)	Time (s)	No. of phase of movement	Acceleration (cm s⁻²)
0,351	10,2	+39,04	138,89	0,208	6,4	+308	+466	0,148	4,9	+ 780
				0,396	11,3	− 61	+ 97	0,388	11,1	+2410
								0,511—0,767	14,3—21,0	0
								0,962	26,1	+3570
								1,108	29,9	+2120

Posterior maxima

Time (s)	No. of phase of movement	Relative value for co-ordinate (cm)	Absolute value for co-ordinate (cm)	Time (s)	No. of phase of movement	Relative velocity (cm s⁻¹)	Absolute velocity (cm s⁻¹)	Time (s)	No. of phase of movement	Acceleration (cm s⁻²)
0,958	26,0	−39,61	156,39	0,382	11,0	− 79	+ 79	0,123	4,2	+ 580
				0,511—0,767	14,3—21,0	−158	0	0,325	9,5	−5260
								0,419	11,9	−1610
								0,511—0,767	14,3—21,0	0
								1,075	29,1	+ 800

Transverse movement
Lateral (right) maxima

Time (s)	No. of phase of movement	Relative value for co-ordinate (cm)	Absolute value for co-ordinate (cm)	Time (s)	No. of phase of movement	Relative velocity (cm s⁻¹)	Absolute velocity (cm s⁻¹)	Time (s)	No. of phase of movement	Acceleration (cm s⁻²)
0,343	10,0	—	+8,16	0,182	5,8	—	+ 8	0,133	4,5	+ 270
0,882	24,0	—	+5,31	0,293	8,7	—	+ 21	0,263	7,9	+ 820
				0,535	15,0	—	0	0,401	11,5	+3510
				0,854	23,3	—	+ 6	0,828	22,6	+ 170
				0,973	26,4	—	+ 6	0,937	25,4	+ 260
								1,082	29,2	+ 80

Medial (left) maxima

Time (s)	No. of phase of movement	Relative value for co-ordinate (cm)	Absolute value for co-ordinate (cm)	Time (s)	No. of phase of movement	Relative velocity (cm s⁻¹)	Absolute velocity (cm s⁻¹)	Time (s)	No. of phase of movement	Acceleration (cm s⁻²)
0,133	4,5	—	+6,03	0,051	2,3	—	− 10	0,212	6,5	− 110
0,797	21,8	—	+5,05	0,234	7,1	—	+ 5	0,349	10,1	−3180
0,930	25,3	—	+5,21	0,381	10,9	—	− 56	0,575	16,0	− 30
				0,700	19,3	—	− 1	0,878	23,9	− 490
				0,905	24,6	—	− 3	0,997	27,0	− 160
				1,023	27,7	—	+ 2			

Vertical movement
Upper maxima

Time (s)	No. of phase of movement	Relative value for co-ordinate (cm)	Absolute value for co-ordinate (cm)	Time (s)	No. of phase of movement	Relative velocity (cm s⁻¹)	Absolute velocity (cm s⁻¹)	Time (s)	No. of phase of movement	Acceleration (cm s⁻²)
0,057	2,5	—	16,86	0,304	8,9	—	+ 66	0,260	7,8	+1650
0,342	9,9	—	12,59	0,515—0,670	14,4—18,5	—	+ 3	0,438	12,4	+1020
1,026	27,8		17,13	0,959	26,0	—	+ 89	0,915	24,9	+ 920

Lower maxima (minima)

Time (s)	No. of phase of movement	Relative value for co-ordinate (cm)	Absolute value for co-ordinate (cm)	Time (s)	No. of phase of movement	Relative velocity (cm s⁻¹)	Absolute velocity (cm s⁻¹)	Time (s)	No. of phase of movement	Acceleration (cm s⁻²)
0,248	7,5	—	8,38	0,154	5,0	—	− 63	0,337	9,8	−4200
0,505	14,2	—	6,50	0,388	11,1	—	− 63	0,515—0,670	14,4—18,5	0
				0,515—0,670	14,4—18,5	—	+ 3	1,002	27,1	−1800

Table 10. Maximum values for the movement of the centre of gravity of the left foot in experiment II

Time (s)	No. of phase of movement	Relative value for co-ordinate (cm)	Absolute value for co-ordinate (cm)	Time (s)	No. of phase of movement	Relative velocity (cm s⁻¹)	Absolute velocity (cm s⁻¹)	Time (s)	No. of phase of movement	Acceleration (cm s⁻²)

Movement in the direction of gait
Anterior maxima

Time (s)	No. of phase	Rel. co-ord	Abs. co-ord	Time (s)	No. of phase	Rel. vel.	Abs. vel.	Time (s)	No. of phase	Acceleration
0,834	22,8	+39,84	144,85	0,699	19,2	+337	+495	0—0,287	0—8,5	0
				0,889	24,2	— 63	+ 95	0,493	13,9	+3 280
								0,652	18,0	+1 110
								0,874	23,8	+1 410
								0,997—1,15	27,0—31	0

Posterior maxima

Time (s)	No. of phase	Rel. co-ord	Abs. co-ord	Time (s)	No. of phase	Rel. vel.	Abs. vel.	Time (s)	No. of phase	Acceleration
0,474	13,4	—40,47	78,81	0—0,287	0—8,5	—158	0	0—0,287	0—8,5	0
				0,858	23,4	— 88	+ 70	0,599	16,6	+ 710
				0,997—1,15	27,0—31	—158	0	0,832	22,7	—7 320
								0,915	24,9	—2 540
								0,997—1,15	27,0—31	0

Transverse movement
Lateral (left) maxima

Time (s)	No. of phase	Rel. co-ord	Abs. co-ord	Time (s)	No. of phase	Rel. vel.	Abs. vel.	Time (s)	No. of phase	Acceleration
0,521	14,6	—	—7,68	0,386	11,1	—	— 10	0,343	10,0	— 170
0,822	22,4	—	—8,01	0,483	13,6	—	— 20	0,463	13,1	—1 070
				0,779	21,3	—	— 13	0,619	17,1	— 280
				0,979	26,5	—	— 4	0,915	24,9	— 650

Medial (right) maxima

Time (s)	No. of phase	Rel. co-ord	Abs. co-ord	Time (s)	No. of phase	Rel. vel.	Abs. vel.	Time (s)	No. of phase	Acceleration
0,272	8,1	—	—5,88	0,210	6,5	—	+ 1	0,100	3,6	+ 30
0,651	18,0	—	—6,58	0,441	12,5	—	— 3	0,423	12,0	+ 280
0,944	25,6	—	—5,51	0,578	16,1	—	+ 13	0,516	14,5	+1 100
				0,863	23,5	—	+ 31	0,829	22,6	+1 710
								1,015	27,5	+ 50

Vertical movement
Upper maxima

Time (s)	No. of phase	Rel. co-ord	Abs. co-ord	Time (s)	No. of phase	Rel. vel.	Abs. vel.	Time (s)	No. of phase	Acceleration
0,546	15,3	—	17,38	0—0,200	0—6,2	—	+ 2	0,447	12,7	+1 890
0,830	22,7	—	12,65	0,478	13,5	—	+121	0,739	20,3	+1 850
				0,794	21,7	—	+ 96	0,903	24,6	+1 430
				1,020—1,15	27,6—31	—	— 1			

Lower maxima (minima)

Time (s)	No. of phase	Rel. co-ord	Abs. co-ord	Time (s)	No. of phase	Rel. vel.	Abs. vel.	Time (s)	No. of phase	Acceleration
0,728	20,0	—	8,33	0—0,200	0—6,2	—	+ 2	0—0,200	0—6,2	0
				0,631	17,5	—	— 76	0,533	14,9	—2 480
				0,869	23,7	—	— 80	0,827	22,6	—4 750
				1,020—1,15	27,6—31	—	— 1	1,020—1,15	27,6—31	0

Table 11. Maximum values for the movement of the centre of gravity of the right foot in experiment III (with loading)

Time (s)	No. of phase of movement	Relative value for co-ordinate (cm)	Absolute value for co-ordinate (cm)	Time (s)	No. of phase of movement	Relative velocity (cm s⁻¹)	Absolute velocity (cm s⁻¹)	Time (s)	No. of phase of movement	Acceleration (cm s⁻²)
				Movement in the direction of gait						
				Anterior maxima						
0,255	7,7	+37,30	106,22	0,117	4,1	+319	+464	0,077	3,0	+ 620
				0,303	8,9	− 58	+ 87	0,290	8,6	+1 170
				1,114	30,1	+303	+448	0,402—0,690	11,5—19,0	0
								0,888	24,2	+2 900
								1,045	28,3	+1 230
				Posterior maxima						
0,879	23,9	−38,39	121,21	0,277	8,2	− 75	+ 70	0,053	2,4	+ 470
				0,402—0,690	11,5—19,0	−145	0	0,261	7,8	−6 800
								0,338	9,8	−2 090
								0,402—0,690	11,5—19,0	0
								0,989	26,8	+ 780
				Transverse movement						
				Lateral (right) maxima						
0,265	7,9	—	+13,46	0,180	5,7	—	+ 53	0,161	5,2	+ 510
0,797	21,8	—	+12,61	0,773	21,2	—	+ 5	0,297	8,8	+ 720
								0,742	20,4	+ 90
								0,852	23,2	− 230
								0,984	26,7	+ 270
				Medial (left) maxima						
0,660	18,2	—	+12,27	0,288	8,5	—	− 17	0,258	7,7	−1 160
				0,953	25,9	—	− 54	0,550	15,4	+ 10
								0,797	21,8	− 790
								0,935	25,4	− 550
								1,043	28,2	+ 180
				Vertical movement						
				Upper maxima						
0,238	7,2	—	12,06	0,200	6,2	—	+ 82	0,177	5,6	+1 990
0,950	25,8	—	17,39	0,420—0,620	12,0—17,2	—	+ 2	0,344	10,0	+1 450
				0,887	24,2	—	+107	0,857	23,4	+1 270
				Lower maxima (minima)						
0,146	4,8	—	8,09	0,055	2,4	—	− 67	0,221	6,8	−3 000
0,395	11,3	—	5,97	0,298	8,8	—	− 67	0,420—0,620	12,0—17,2	0
				0,420—0,620	12,0—17,2	—	+ 2	0,917	24,9	−2 350
				1,046	28,3	—	− 77			

Table 12. Maximum values for the movement of the centre of gravity of the left foot in experiment III (with loading)

Time (s)	No. of phase of movement	Relative value for co-ordinate (cm)	Absolute value for co-ordinate (cm)	Time (s)	No. of phase of movement	Relative velocity (cm s^{-1})	Absolute velocity (cm s^{-1})	Time (s)	No. of phase of movement	Acceleration (cm s^{-2})
					Movement in the direction of gait					
					Anterior maxima					
0,751	20,6	+36,77	177,82	0,601	16,7	+296	+441	0—0,192	0—6,0	0
				0,827	22,6	— 88	+ 57	0,386	11,1	+ 3400
								0,531	14,9	+ 970
								0,811	22,2	+ 1180
								0,901—1,15	24,5—31	0
					Posterior maxima					
0,381	10,9	—39,73	47,52	0—0,192	0—6,0	—145	0	0—0,192	0—6,0	0
				0,800	21,9	—104	+ 41	0,500	14,1	+ 720
				0,901—1,15	24,5—31	—145	0	0,727	20,0	— 3740
								0,851	23,2	— 1390
								0,901—1,15	24,5—31	0
					Transverse movement					
					Lateral (left) maxima					
0,298	8,8	—	— 7,77	0,270	8,0	—	— 8	0,224	6,8	— 420
0,413	11,8	—	— 7,29	0,397	11,4	—	— 4	0,365	10,5	— 340
0,762	20,9	—	—11,03	0,700	19,3	—	— 86	0,530	14,8	— 850
0,919	25,0	—	—10,11	0,892	24,3	—	— 12	0,687	18,9	— 1640
								0,879	23,9	— 890
								0,997	27,0	— 10
								1,030—1,15	27,9—31	0
					Medial (right) maxima					
0,230	7,0	—	— 7,42	0,186	5,9	—	+ 9	0,135	4,5	+ 60
0,374	10,8	—	— 7,22	0,323	9,4	—	+ 11	0,297	8,8	+ 950
0,544	15,2	—	— 3,87	0,470	13,3	—	+ 40	0,428	12,2	+ 1400
0,873	23,8	—	— 9,87	0,820	22,4	—	+ 17	0,616	17,1	— 260
				0,965	26,2	—	+ 1	0,712	19,6	ca +20000
								0,902	24,5	+ 1100
					Vertical movement					
					Upper maxima					
0,460	13,0	—	16,71	0—0,170	0—5,4	—	+ 1	0—0,170	0—5,4	0
0,738	20,3	—	12,34	0,388	11,1	—	+133	0,362	10,4	+ 1970
				0,702	19,3	—	+ 63	0,667	18,4	+ 1890
				0,940—1,15	25,5—31	—	+ 3	0,858	23,4	+ 940
					Lower maxima (minima)					
0,249	7,5	—	5,09	0—0,170	0—5,4	—	+ 1	0,202	6,3	— 330
0,648	17,9	—	8,92	0,223	6,8	—	— 9	0,409	11,7	— 2690
0,908	24,7	—	6,16	0,547	15,3	—	— 66	0,732	20,1	— 3020
				0,798	21,8	—	— 61	0,940—1,15	25,5—31	0
				0,940—1,15	25,5—31	—	+ 3			

also given to one decimal place. The time interval between two successive phases must be regarded as divided into ten equal parts, thus adding nine new subphases numbered 1–9. The first place of decimals indicates the subphase.

The kinematics of the centre of gravity of the foot appears clearly in Figs. 12–14. If the significance of the different curves is borne in mind, these figures immediately provide an overall view of the changes in position and the magnitudes of and changes in velocities and accelerations of the centres of gravity of both feet during a double step. It must be remembered that the displacement and velocity curves for the direction of gait refer to the relative movement within the system of co-ordinates moving uniformly forward at the average velocity of gait. The variations in absolute velocities are easily obtained from the relative velocities. To this end, the axis of the abscissa of the velocity curve must be moved downwards by a distance corresponding to the average velocity of gait, in the first part of each of the three graphs. The position of this axis of the abscissa corresponding to the absolute velocities is indicated in the graphs. This displacement is not necessary for the acceleration curves since absolute and relative accelerations are the same.

The curves related to the centre of gravity of the left foot are drawn in red, as are the vertical lines marking the moments characteristic for the left foot. As previously (Chap. 2) R means right heel strike, S_r right toe-off, A_r beginning of right foot-flat, and E_r end of right foot-flat. The corresponding symbols for the left foot are L, S_l, A_l, E_l. The times at which these phenomena occur are given in Table 14 in Chap. 2 (p. 169).

A complete analysis of the law governing the movement of the centre of gravity of the foot, which appears in Figs. 12–14, would lead us much too far. It is also unnecessary since the details of this law are obvious to anyone who makes a thorough study of these numerous curves. Therefore, only the most important aspects of the law governing the movement will be highlighted.

It can be seen from the three graphs that the relative velocity of the centre of gravity of each foot in the direction of gait is nil at toe-off, just before the leg swings forward. This means that the centre of gravity has already attained the average velocity of gait in relation to the ground as a consequence of push-off. During the swinging period the velocity in the direction of gait increases very quickly until it reaches its maximum at the beginning of the second half of this period. This maximum is about twice the average velocity of gait for the relative movement, which means three times this average velocity for the absolute movement. When it occurs, the centre of the ankle is already further forwards than the hip. The increase of velocity in the direction of gait up to the maximum is not of course uniform. The acceleration is at its greatest at toe-off after which it decreases very quickly, to increase a little after a while but it decreases again soon after and falls to 0. This instant, of course, coincides with the maximum of velocity in the direction of gait. Later in the swinging period, the acceleration is directed backwards, i.e. it becomes deceleration. Consequently, the velocity of the absolute movement of the right foot decreases to half the average velocity of gait, but that of our subject's left foot falls to less than half the average velocity of gait. This minimum velocity occurs, remarkably, before heel strike. Only for experiment III in which the subject was carrying the regulation military equipment, does the minimum velocity for the left foot coincide with left heel strike. Thereafter, the velocity in the direction of gait increases slightly again, reaching a second maximum for the right foot at right heel strike in all three experiments, but for the left foot not until after left heel strike. This maximum is much smaller than the one that occurs in the swinging period. Corresponding to the behaviour of the velocity, the acceleration directed backwards initially increases considerably and then decreases again to 0. It reaches 0, of course, when the velocity attains the minimum mentioned above. Thereafter, the acceleration again becomes positive, soon reaches a maximum, and then falls back to 0 when the velocity reaches its second, lower, maximum.

After heel strike the absolute velocity decreases to 0 while the foot is rotating about its heel. This continuous diminution of velocity corresponds to an acceleration directed backwards which at first increases in absolute magnitude, then very quickly reaches a maximum and falls back to 0. At foot-flat both velocity and acceleration are 0. This situation prevails throughout foot-flat. As soon as push-off starts an acceleration directed forwards develops which increases continuously until toe-off. Then it reaches a maximum. Consequently, the centre of gravity of the foot moves forward with increasing absolute velocity. At the beginning of the swinging period the absolute velocity has reached the same value as the average velocity of gait, as mentioned above.

So far only the projection of the movement of the centre of gravity of the foot on the direction of gait has been considered. In the projection on the vertical direction the movement appears essentially different. After toe-off the centre of gravity of the foot continues to move upwards, as it did during push-off. Since the velocity of this movement is continuously decreasing and soon reaches 0, the centre of gravity attains its highest point shortly after the beginning of the swinging period. It falls again, first with increasing and then with decreasing velocity, until it has reached the lowest point of its trajectory shortly after the instant of the greatest velocity in the direction of gait, as described above. Then it moves again from the floor, at first with increasing and then with decreasing velocity. It is far from reaching the height attained at the beginning of the swinging period, however, soon moving towards the floor again until heel strike. At heel strike it moves further downwards and finally reaches its lowest point at foot-flat, where it remains for a short time. The velocity of the movement of the centre of gravity of the foot downwards is nearly at its greatest at heel strike. Whilst the foot rotates about its heel, the downward velocity decreases until it has reached 0 at foot-flat. At the beginning of the swinging period the acceleration is 0. Then it becomes negative, corresponding to the change in velocity. It is thus directed downwards. It increases at first in absolute value in that direction and then decreases, passes through 0 and becomes positive so that it is directed upwards again. In the same way, it first increases and then decreases, becomes negative and very quickly reaches an absolute maximum. Finally, it decreases again so that it has reached 0 before heel strike and is positive at heel strike. At the beginning of the following period, the acceleration of the centre of gravity of the foot increases a little at first but soon decreases and, of course, falls to 0 at foot-flat. During push-off, the centre of gravity is accelerated increasingly upwards. This acceleration does not reach its maximum until shortly before toe-off, subsequently decreasing very quickly, however, so that at toe-off it has fallen to 0, as described above. Correspondingly, the velocity of the movement of the centre of gravity of the foot upwards increases continuously during push-off and reaches its maximum at the beginning of swing.

The transverse movement remains to be considered. The displacement curves for the centres of gravity of the right and left feet are further apart in the second part of Fig. 12 than in Fig. 13. This means that in experiment I the double step was a little wider than in experiment II. The excursions sideways in walking are subject to much larger variations than those in any other direction. In experiment III (Fig. 14) in which the subject had to carry the heavy regulation equipment, the displacement of the total centre of gravity results in much greater movements of the feet sideways. The legs are much wider apart than without the load. Otherwise, certain similarities can be seen in the transverse movement in the three experiments. For instance, during the swinging period the centre of gravity of the foot alternately comes closer to and moves from the median plane, to come closer again just before heel strike. Whilst the foot rotates about its heel, the centre of gravity keeps moving closer to the median plane until it comes to a stop at foot-flat. During push-off, it moves slightly outwards again. However, the two feet do not display precisely symmetrical behaviour at this stage. The sideways displacements are obviously the most susceptible to individual variations so that these give the gait its individual characteristics more than the movements in the other directions.

The velocity of the sideways movement of the centre of gravity of the foot is at a maximum directed inwards at heel strike. It is preceded by a maximum directed outwards at the end of the swinging period. Otherwise, during the swinging period the centre of gravity displays a maximum of velocity outwards at the beginning and a maximum inwards a little while later. Between these two successive maxima of velocity in opposite directions, of course, the centre of gravity passes through an instant with no sideways velocity at all.

The acceleration reaches a maximum outwards at toe-off or very shortly before. Moreover, during the swinging period one more maximum of acceleration outwards and two maxima inwards occur, or two outwards and three inwards during walking with a load. Between two successive maxima of acceleration in opposite directions there is, of course, always an instant with no sideways acceleration at all. Heel strike is followed by a maximum of acceleration outwards before the centre of gravity stops at the beginning of foot-flat. Finally, the curves in the second part of Figs. 12–14 show a constant maximum of acceleration inwards which, for the centre of gravity of the right foot, coincides with left heel strike. For the centre of gravity of the left foot it coincides with right heel strike only in the experiment with a load whereas in the other two experiments it falls in the double-support period preceding the left swing.

Effective Forces and Momenta of Both Feet

Multiplication of each of the components of the acceleration of the centre of gravity of the foot in the direction of gait, in the transverse direction and in the vertical direction, by the mass of the foot gives the components of the effective force of the foot. Multiplication of the components of the velocity of the centre of gravity of the foot by the mass of the foot gives the components of the quantity of movement, referred to in mechanics as momentum, of the foot. To determine the resultant torque of the muscles, only the effective forces and their components are used. But the solution of other problems of the mechanics of human gait to be considered later also requires that the momenta of the feet are known. Therefore, their components have been calculated and are given with the components of the effective forces of the two feet in Tables 14–19 which also give the products of the mass of one foot and the co-ordinates of its centre of gravity. These products are important for the relationship between the centre of gravity of each foot and the total centre of gravity of the body.

For the further analysis, it is useful to express masses and forces in the terrestrial system of mass rather than as absolutes since it is usual in anatomy and physiology to measure forces by weights. If the unit of force is the gram and the unit of acceleration is 1 cm s^{-2}, the mass of a weight of 981.11 g must be used as the unit of mass. Then a force of 981.11 g gives a mass of this weight the acceleration of gravity. This is an acceleration of 981.11 cm s^{-2} in Leipzig. A force of 1 g would thus give rise to 1 unit of acceleration. Consequently, in this system the mass of a segment of the body is obtained by dividing the weight of this segment, expressed in grams, by 981.11. If the kilogram were taken as the unit of force, the unit of mass would be 1000 times greater and the mass figure would be 1000 times smaller than with a unit of force of 1 g. This latter method has been used in Chap. 2 to determine the mass of the whole body. Hence, the external forces are expressed directly in kilograms in Chap. 2. If the kilogram were retained as the unit for the different parts of the body the mass figures would be unmanageably small; for this reason the gram is the unit of force used from now on. Division of the weight figures in the work on the centre of gravity of the human body[3] by 981.11 gives the mass figures shown in Table 13 for the different parts of the body of the experimental subject.

---

[3] Braune W, Fischer O (1985) Abhandlungen der mathematisch-physischen Klasse der Königlich Sächsischen Gesellschaften der Wissenschaften, vol 15, no 7. English translation: On the centre of gravity of the human body, Table 14, p 37 and Table 34, p 73. Springer, Berlin Heidelberg New York Tokyo

Trunk (m_1)	25,54
Right thigh (m_2)	6,93
Left thigh (m_3)	6,93
Right lower leg (m_4)	3,15
Left lower leg (m_5)	3,15
Right foot (m_6)	1,07
Left foot (m_7)	1,07
Right upper arm (m_8)	2,02
Left upper arm (m_9)	2,02
Right forearm + hand (m_{10})	1,86$_5$
Left forearm + hand (m_{11})	1,86$_5$
Head (m_{12})	4,22
Whole body (without load)	59,83
Regulation equipment [a]	23,72
Whole body + load	83,55

Table 13. Mass of the subject's body segments, taking 1 g as the unit of force and $1\,\mathrm{cm\,s^{-2}}$ as the unit of acceleration

[a] cf. Chap. 2, p. 199

For instance, multiplication of the mass m_6 or m_7 of a foot by the acceleration of its centre of gravity or by one of its components gives the effective force or the corresponding component of the effective force of the foot in grams.

Tables 14–19 (pp. 302–307) give the results of these multiplications and those of the multiplications of the mass by the velocities and by the co-ordinates of the centres of gravity of both feet for the three experiments.

Finally, the maximum values of the co-ordinates, velocities and accelerations of the centre of gravity of the foot have been multiplied by the mass m_6 or m_7 of the foot. The results are given in Tables 20–25 in the same way as the maximum values themselves were displayed above.

Summary

Chapter 3 of this study of the human gait provided an overall view of the movements of the lower extremities. The present Chapter (Chap. 4) deals in more detail with the movement of the foot.

First, the forces that act on the foot during gait were determined. The simultaneous movement of the other segments of the body sets up compression or tension in the ankle, which can be regarded as an external force applied at the centre of the ankle joint. Other forces are exerted on the foot such as those originating in the contraction or elastic elongation of the muscles, the attractive force of the earth and the ground reaction force resulting from the compression by the foot. The foot moves under the action of all these forces as if it were completely separate from the remainder of the body. In general these forces, therefore, exert two types of action on the foot. They influence its forward movement by giving its centre of gravity an acceleration. In addition, they impart to the foot an angular acceleration about an axis through its centre of gravity. During the almost plane movement achieved in walking, this axis is perpendicular to the plane of gait. The rotatory action can be isolated by imagining a force equal and opposite to its effective force applied at the centre

Table 14. Products of the mass of the right foot and the co-ordinates, velocities and accelerations of its centre of gravity in experiment I

No.	$m_6 x_6$ relative	$m_6 x_6$ absolute	$m_6 y_6$	$m_6 z_6$	$m_6 \dfrac{dx_6}{dt}$ relative	$m_6 \dfrac{dx_6}{dt}$ absolute	$m_6 \dfrac{dy_6}{dt}$	$m_6 \dfrac{dz_6}{dt}$	$m_6 \dfrac{d^2 x_6}{dt^2}$	$m_6 \dfrac{d^2 y_6}{dt^2}$	$m_6 \dfrac{d^2 z_6}{dt^2}$	No.
1	−36,80	5,20	+7,94	18,03	+209	+377	− 4	− 14	+2860	− 30	−1270	1
2	−27,83	20,62	+7,74	16,97	+279	+447	− 6	− 46	+ 950	− 20	− 810	2
3	−16,83	38,07	+7,47	14,63	+300	+468	− 7	− 73	+ 500	− 10	− 210	3
4	− 4,86	56,50	+7,22	11,96	+327	+495	− 7	− 68	+ 660	− 10	+ 490	4
5	+ 8,13	75,94	+6,94	10,00	+340	+508	− 7	− 34	+ 30	+ 30	+1100	5
6	+20,84	95,10	+6,75	9,44	+323	+491	0	+ 9	− 960	+500	+1490	6
7	+32,21	112,92	+7,22	10,73	+269	+437	+27	+ 63	−1850	+250	+1210	7
8	+41,07	128,23	+8,20	13,51	+169	+337	+21	+ 57	−3620	−950	−2040	8
9	+43,26	136,87	+8,27	13,59	− 65	+103	−20	− 52	−3640	−680	−2140	9
10	+40,46	140,52	+7,28	10,69	− 56	+112	−25	− 80	+ 320	+250	+ 480	10
11	+37,73	144,25	+6,71	8,25	−107	+ 61	−12	− 41	−3380	+540	+1200	11
12	+32,21	145,18	+6,47	7,27	−155	+ 13	− 1	− 9	− 630	+120	+ 400	12
13	+26,00	145,42	+6,49	7,15	−168	0	+ 1	0	− 20	+ 10	+ 30	13
14	+19,56	145,43	+6,53	7,17	−168	0	0	+ 2	0	− 40	0	14
15	+13,12	145,45	+6,48	7,20	−168	0	− 1	+ 2	0	− 30	0	15
16	+ 6,67	145,45	+6,43	7,24	−168	0	− 1	+ 2	0	− 20	0	16
17	+ 0,22	145,46	+6,39	7,29	−168	0	− 1	+ 2	0	− 10	0	17
18	− 6,22	145,47	+6,32	7,33	−168	0	− 1	+ 2	0	+ 10	0	18
19	−12,65	145,49	+6,27	7,36	−168	0	− 1	+ 2	0	+ 30	0	19
20	−19,07	145,52	+6,23	7,42	−168	0	0	+ 3	0	+ 50	+ 50	20
21	−25,22	145,82	+6,25	7,67	−161	+ 7	+ 2	+ 7	+ 260	+ 70	+ 150	21
22	−31,08	146,41	+6,40	8,28	−150	+ 18	+ 6	+ 18	+ 330	+ 70	+ 300	22
23	−36,54	147,40	+6,60	9,27	−134	+ 34	+ 1	+ 34	+ 510	−570	+ 500	23
24	−41,03	149,36	+6,42	11,27	−105	+ 63	− 7	+ 62	+1230	+120	+1000	24
25	−43,55	153,30	+6,28	14,16	− 9	+159	− 2	+105	+3690	0	+ 740	25
26	−41,28	162,02	+6,07	17,61	+116	+284	− 9	+ 57	+2720	−350	−2260	26
27	−35,43	174,32	+5,53	18,07	+202	+370	−15	− 18	+1590	− 10	−1360	27
28	−26,45	189,75	+5,01	16,63	+240	+408	−10	− 52	+ 660	+210	− 700	28
29	−16,88	205,77	+4,84	14,28	+276	+444	0	− 66	+1250	+270	− 140	29
30	− 5,41	223,69	+5,00	11,66	+309	+477	+10	− 62	+ 370	+220	+ 390	30
31	+ 6,26	241,82	+5,50	9,64	+317	+485	+15	− 39	+ 20	+ 70	+ 850	31

of gravity of the foot. This additional force cancels out the acceleration of the centre of gravity without changing the rotation of the foot about its centre of gravity. If this reversed effective force is added to the other forces applied to the foot, all the forces can be combined into three kinds of couples as defined by Poinsot. One results from the muscular forces, the second from the forces of gravity, and the third from the effective forces. The algebraic sum of the moments of these couples must be equal to the product of the moment of inertia and the angular acceleration of the foot, referred to the axis through the centre of gravity of the foot and perpendicular to the plane of gait, if the movement is plane. This is the simple meaning of the differential equation concerning the foot in the system of differential

Table 15. Products of the mass of the left foot and the co-ordinates, velocities and accelerations of its centre of gravity in experiment I

No.	$m_7 x_7$ relative	absolute	$m_7 y_7$	$m_7 z_7$	$m_7 \dfrac{dx_7}{dt}$ relative	absolute	$m_7 \dfrac{dy_7}{dt}$	$m_7 \dfrac{dz_7}{dt}$	$m_7 \dfrac{d^2 x_7}{dt^2}$	$m_7 \dfrac{d^2 y_7}{dt^2}$	$m_7 \dfrac{d^2 z_7}{dt^2}$	No.
1	+18,85	60,85	−6,64	6,53	−168	0	− 1	+ 1	0	+ 30	0	1
2	+12,48	60,93	−6,67	6,54	−168	0	0	+ 1	0	+ 20	0	2
3	+ 6,01	60,92	−6,64	6,56	−168	0	0	+ 1	0	+ 10	0	3
4	− 0,39	60,97	−6,63	6,58	−168	0	+ 1	+ 1	0	+ 10	0	4
5	− 6,83	60,98	−6,61	6,59	−168	0	+ 1	+ 2	0	0	+ 10	5
6	−13,28	60,98	−6,56	6,60	−168	0	+ 1	+ 3	0	− 20	+ 10	6
7	−19,72	60,99	−6,53	6,63	−168	0	0	+ 4	0	− 40	+ 30	7
8	−26,12	61,04	−6,53	6,79	−161	+ 7	− 3	+ 6	+ 240	−190	+ 120	8
9	−32,05	61,57	−6,81	7,22	−149	+ 19	−12	+ 16	+ 370	−150	+ 280	9
10	−37,47	62,60	−7,28	8,13	−128	+ 40	−12	+ 33	+ 760	+150	+ 580	10
11	−41,87	64,65	−7,58	9,84	− 83	+ 85	− 3	+ 66	+1470	+180	+1220	11
12	−44,55	68,42	−7,71	13,27	− 12	+156	− 7	+135	+2610	−760	+ 850	12
13	−42,30	77,13	−8,23	17,55	+ 96	+264	−13	+ 87	+3070	+320	−2600	13
14	−36,62	89,26	−8,11	18,86	+200	+368	+ 9	− 3	+2180	+290	−1810	14
15	−26,65	105,67	−7,66	17,47	+267	+435	+12	− 56	+1330	− 20	− 830	15
16	−16,16	122,62	−7,25	15,01	+304	+472	+ 9	− 76	+ 520	−140	− 170	16
17	− 4,80	140,43	−7,04	12,14	+316	+484	+ 3	− 72	+ 270	−170	+ 360	17
18	+ 7,21	158,90	−7,09	9,90	+338	+506	− 4	− 43	+ 590	−200	+ 950	18
19	+20,21	178,35	−7,40	9,14	+335	+503	−14	+ 6	− 630	−280	+1780	19
20	+32,17	196,86	−8,06	10,51	+275	+443	−27	+ 68	−2490	−400	+ 960	20
21	+40,52	211,56	−9,20	13,32	+137	+305	−28	+ 60	−4630	+330	−1770	21
22	+42,33	219,82	−9,68	13,16	− 42	+126	+13	− 63	−3860	+640	−1800	22
23	+38,99	222,93	−8,99	10,09	− 97	+ 71	+18	− 81	+1250	−100	+ 590	23
24	+35,80	226,20	−8,43	7,86	− 87	+ 81	+10	− 36	−1830	−320	+1170	24
25	+30,66	227,50	−8,16	6,85	−161	+ 7	− 1	− 9	− 460	−210	+ 320	25
26	+24,22	227,52	−8,27	6,70	−168	0	− 4	0	0	+ 40	+ 40	26
27	+17,79	227,55	−8,32	6,76	−168	0	− 2	+ 2	0	+ 40	0	27
28	+11,36	227,57	−8,34	6,83	−168	0	− 1	+ 2	0	+ 20	0	28
29	+ 4,93	227,59	−8,35	6,89	−168	0	0	+ 2	0	+ 10	0	29
30	− 1,50	227,61	−8,36	6,96	−168	0	0	+ 2	0	+ 10	0	30
31	− 7,92	227,64	−8,37	7,03	−168	0	0	+ 2	0	0	0	31

equations given in a previous contribution. These movement equations were derived from the general Lagrangian differential equations and orginally appeared in a rather complicated form. Such a simple interpretation of these equations is, therefore, very important for the further analysis. The couples of gravity and those of the effective forces, and also the angular acceleration of the foot about the axis through its centre of gravity thus can be deduced and quantified from the previous description of the whole process of movement. The moments of inertia of the foot have been measured previously. The resultant torque of the muscles acting on the foot can thus be calculated by using the movement equation. In the present instance, this problem can be solved approximately in even a simpler way because the

Table 16. Products of the mass of the right foot and the co-ordinates, velocities and accelerations of its centre of gravity in experiment II

No.	$m_6 x_6$ relative	$m_6 x_6$ absolute	$m_6 y_6$	$m_6 z_6$	$m_6 \dfrac{dx_6}{dt}$ relative	$m_6 \dfrac{dx_6}{dt}$ absolute	$m_6 \dfrac{dy_6}{dt}$	$m_6 \dfrac{dz_6}{dt}$	$m_6 \dfrac{d^2 x_6}{dt^2}$	$m_6 \dfrac{d^2 y_6}{dt^2}$	$m_6 \dfrac{d^2 z_6}{dt^2}$	No.
1	−42,59	4,76	+7,50	15,22	+ 49	+218	− 9	+100	+2620	− 60	−2020	1
2	−38,99	14,85	+7,13	17,76	+146	+315	−11	+ 31	+2450	− 30	−1630	2
3	−31,92	28,42	+6,72	17,71	+239	+408	−10	− 28	+1890	+ 60	−1130	3
4	−21,89	44,94	+6,47	15,81	+282	+452	− 5	− 61	+ 660	+260	− 620	4
5	−11,44	61,89	+6,48	13,41	+311	+480	+ 5	− 67	+ 800	+240	− 10	5
6	+ 0,83	80,66	+6,76	10,91	+328	+497	+ 9	− 59	+ 190	− 60	+ 590	6
7	+14,30	100,61	+7,04	9,26	+325	+494	+ 5	− 28	− 460	− 40	+1290	7
8	+25,20	118,01	+7,39	9,38	+297	+467	+17	+ 33	−1100	+800	+1700	8
9	+36,74	136,05	+8,24	11,56	+218	+387	+21	+ 70	−3720	−410	− 270	9
10	+41,54	147,34	+8,71	13,43	+ 28	+197	− 3	− 14	−4300	−3320	−3050	10
11	+40,61	152,90	+7,38	11,25	− 85	+ 85	−60	− 66	+ 530	+ 90	− 260	11
12	+37,72	156,51	+6,14	8,83	− 96	+ 73	−16	− 51	−1700	+450	+ 850	12
13	+33,01	158,30	+5,74	7,35	−144	+ 26	− 5	− 17	− 870	+180	+ 730	13
14	+26,96	158,75	+5,65	6,97	−167	+ 2	− 1	− 1	− 260	+ 60	+ 150	14
15	+20,48	158,76	+5,63	6,97	−169	0	0	+ 3	0	− 10	0	15
16	+14,01	158,78	+5,61	6,99	−169	0	0	+ 3	0	− 30	0	16
17	+ 7,52	158,79	+5,59	7,03	−169	0	− 1	+ 3	0	− 20	0	17
18	+ 1,04	158,80	+5,55	7,06	−169	0	− 1	+ 3	0	− 10	0	18
19	− 5,44	158,82	+5,50	7,09	−169	0	− 1	+ 4	0	0	+ 10	19
20	−11,90	158,85	+5,45	7,13	−169	0	− 1	+ 4	0	+ 10	+ 20	20
21	−18,34	158,91	+5,41	7,18	−169	0	− 1	+ 5	0	+ 60	+ 60	21
22	−24,51	159,23	+5,41	7,43	−162	+ 7	+ 1	+ 12	+ 220	+130	+ 190	22
23	−30,41	159,83	+5,49	8,06	−152	+ 17	+ 6	+ 20	+ 330	+120	+ 420	23
24	−35,96	160,79	+5,68	9,16	−135	+ 34	0	+ 40	+ 680	−480	+ 770	24
25	−40,29	162,94	+5,59	11,32	− 95	+ 74	− 2	+ 70	+1480	+170	+ 970	25
26	−42,38	167,34	+5,66	14,34	− 1	+168	+ 6	+ 95	+3750	+150	+ 20	26
27	−39,69	176,53	+5,91	17,64	+114	+284	+ 4	+ 52	+2560	−170	−1900	27
28	−33,39	189,32	+5,99	18,28	+200	+369	+ 2	− 12	+1470	+ 40	−1340	28
29	−24,82	204,38	+6,13	17,09	+242	+411	+ 5	− 46	+ 860	+ 70	− 820	29
30	−14,93	220,77	+6,43	14,78	+292	+461	+ 9	− 72	+2170	+ 70	− 450	30
31	− 2,81	239,38	+6,83	11,92	+326	+495	+12	− 83	+ 360	+ 50	− 180	31

product of the moment of inertia and the angular acceleration, as a consequence of the small mass of the foot, is so small compared with the torques due to most of the forces that it can be disregarded without involving any great error. It then becomes apparent, however, that all the forcesing on the foot must be almost in equilibrium.

In the present chapter the different couples acting on the foot are given in detail for each period of gait. These data provide a good, although approximate, instructive overall view of the activity of the muscles inserted into the foot and spanning the ankle. It appears that during the swinging period the muscles probably virtually do not contract, or only weakly.

Table 17. Products of the mass of the left foot and the co-ordinates, velocities and accelerations of its centre of gravity in experiment II

No.	$m_7 x_7$ relative	$m_7 x_7$ absolute	$m_7 y_7$	$m_7 z_7$	$m_7 \dfrac{dx_7}{dt}$ relative	$m_7 \dfrac{dx_7}{dt}$ absolute	$m_7 \dfrac{dy_7}{dt}$	$m_7 \dfrac{dz_7}{dt}$	$m_7 \dfrac{d^2 x_7}{dt^2}$	$m_7 \dfrac{d^2 y_7}{dt^2}$	$m_7 \dfrac{d^2 z_7}{dt^2}$	No.
1	+28,73	76,08	−6,44	6,43	−169	0	0	+ 2	0	+ 10	0	1
2	+22,22	76,07	−6,43	6,44	−169	0	0	+ 2	0	+ 20	0	2
3	+15,72	76,06	−6,42	6,45	−169	0	0	+ 2	0	+ 20	0	3
4	+ 9,20	76,03	−6,40	6,46	−169	0	+ 1	+ 2	0	+ 30	0	4
5	+ 2,68	76,00	−6,38	6,47	−169	0	+ 1	+ 2	0	+ 20	0	5
6	− 3,81	76,01	−6,36	6,49	−169	0	+ 1	+ 2	0	+ 10	0	6
7	− 10,29	76,02	−6,33	6,52	−169	0	+ 1	+ 3	0	− 10	+ 20	7
8	− 16,78	76,03	−6,30	6,54	−169	0	0	+ 4	0	− 30	+ 50	8
9	− 23,24	76,07	−6,33	6,61	− 166	+ 3	− 2	+ 5	+ 250	− 120	+ 130	9
10	− 29,23	76,57	−6,49	6,89	− 154	+ 15	− 7	+ 12	+ 360	− 170	+ 280	10
11	− 35,03	77,26	−6,85	7,83	− 138	+ 31	− 11	+ 26	+ 490	− 20	+ 560	11
12	− 39,90	78,89	−7,21	8,88	− 110	+ 59	− 7	+ 54	+ 950	+ 290	+ 1160	12
13	− 43,09	82,20	−7,35	12,04	− 44	+ 125	− 9	+ 116	+ 2810	− 960	+ 1490	13
14	− 42,34	89,44	−8,00	16,16	+ 80	+ 249	− 17	+ 106	+ 3490	+ 350	− 1610	14
15	− 36,92	101,36	−8,15	18,50	+ 196	+ 365	+ 10	+ 19	+ 2270	+ 300	− 2610	15
16	− 28,21	116,57	−7,68	17,78	+ 253	+ 422	+ 14	− 52	+ 980	+ 20	− 1240	16
17	− 17,78	133,38	−7,19	15,38	+ 285	+ 454	+ 10	− 77	+ 820	− 280	− 350	17
18	− 6,12	151,64	−7,04	12,47	+ 324	+ 493	0	− 75	+ 1190	− 190	+ 300	18
19	+ 7,16	171,41	−7,17	9,98	+ 358	+ 528	− 5	− 51	+ 360	− 130	+ 890	19
20	+ 20,63	191,38	−7,47	8,91	+ 336	+ 505	− 10	0	− 1550	− 70	+ 1710	20
21	+ 31,74	208,98	−7,91	10,11	+ 259	+ 428	− 13	+ 79	− 2430	− 30	+ 1230	21
22	+ 39,68	223,42	−8,41	12,71	+ 148	+ 317	− 10	+ 93	− 3560	+ 410	− 570	22
23	+ 42,36	232,60	−8,20	13,18	− 64	+ 105	+ 29	− 47	− 4840	+ 600	− 3210	23
24	+ 39,09	235,82	−7,00	10,12	− 71	+ 98	+ 31	− 82	+ 1000	− 300	+ 730	24
25	+ 35,98	239,21	−6,02	8,19	− 111	+ 58	+ 13	− 32	− 2180	− 680	+ 1090	25
26	+ 30,66	240,38	−5,91	7,28	− 159	+ 10	− 2	− 6	− 610	− 130	+ 330	26
27	+ 24,20	240,42	−6,02	7,16	− 169	0	− 3	− 2	0	+ 30	+ 10	27
28	+ 17,72	240,43	−6,06	7,15	− 169	0	− 1	− 1	0	+ 40	0	28
29	+ 11,24	240,44	−6,07	7,15	− 169	0	0	− 1	0	+ 30	0	29
30	+ 4,75	240,45	−6,08	7,15	− 169	0	0	− 1	0	+ 10	0	30
31	− 1,73	240,46	−6,09	7,14	− 169	0	0	− 1	0	+ 10	0	31

At heel strike, however, muscles spanning the front of the ankle joint, e.g. the tibialis anticus, contract immediately. A little while after foot-flat, the anterior muscles are superseded by the calf muscles which pass behind the ankle. The contraction of the calf muscles persists until toe-off, when a new swinging period begins.

The remaining sections deal with the velocities and accelerations of the centre of gravity of the foot as well as the momenta and the effective forces of the foot. The results of the study are collected and displayed in tables.

Table 18. Products of the mass of the right foot and the co-ordinates, velocities and accelerations of its centre of gravity in experiment III (with load)

No.	$m_6 x_6$ relative	$m_6 x_6$ absolute	$m_6 y_6$	$m_6 z_6$	$m_6 \frac{dx_6}{dt}$ relative	$m_6 \frac{dx_6}{dt}$ absolute	$m_6 \frac{dy_6}{dt}$	$m_6 \frac{dz_6}{dt}$	$m_6 \frac{d^2 x_6}{dt_6}$	$m_6 \frac{d^2 y_6}{dt^2}$	$m_6 \frac{d^2 z_6}{dt^2}$	No.
1	−29,08	5,03	+ 5,23	15,96	+270	+425	+13	− 45	+1480	+ 100	−1120	1
2	−18,13	21,95	+ 5,78	13,64	+307	+462	+17	− 68	+ 630	+ 140	− 300	2
3	− 5,80	40,23	+ 6,57	11,06	+328	+484	+25	− 60	+ 660	+ 200	+ 330	3
4	+ 6,92	58,91	+ 7,90	9,16	+341	+496	+34	− 33	+ 50	+ 310	+ 950	4
5	+19,67	77,62	+ 9,59	8,68	+321	+476	+47	+ 7	−1040	+ 510	+1620	5
6	+31,03	94,94	+11,76	10,27	+255	+410	+55	+ 81	−2400	− 260	+1230	6
7	+38,35	108,22	+13,62	12,87	+129	+285	+35	+ 24	−4480	− 780	−2920	7
8	+39,47	115,30	+14,39	11,98	− 73	+ 82	− 3	− 50	−2890	−1130	−1140	8
9	+36,72	118,51	+13,95	9,11	− 63	+ 92	−11	− 71	− 280	+ 340	+ 330	9
10	+33,36	121,11	+13,66	6,89	−124	+ 31	− 5	− 35	−2140	+ 70	+1540	10
11	+28,00	121,71	+13,54	6,41	−154	+ 1	− 3	− 2	− 170	+ 40	+ 310	11
12	+22,04	121,71	+13,46	6,40	−155	0	− 2	+ 2	0	+ 30	0	12
13	+16,08	121,71	+13,39	6,42	−155	0	− 2	+ 2	0	+ 20	0	13
14	+10,12	121,71	+13,34	6,44	−155	0	− 1	+ 2	0	+ 10	0	14
15	+ 4,18	121,73	+13,28	6,46	−155	0	− 1	+ 2	0	+ 10	0	15
16	− 1,78	121,73	+13,23	6,49	−155	0	− 1	+ 2	0	+ 10	0	16
17	− 7,74	121,73	+13,19	6,52	−155	0	− 1	+ 2	0	+ 10	0	17
18	−13,69	121,74	+13,14	6,54	−155	0	0	+ 3	0	+ 20	+ 20	18
19	−19,63	121,76	+13,15	6,60	−155	0	+ 1	+ 5	0	+ 30	+ 70	19
20	−25,29	122,05	+13,23	6,92	−147	+ 9	+ 2	+ 10	+ 280	+ 70	+ 160	20
21	−30,63	122,68	+13,36	7,58	−134	+ 21	+ 5	+ 21	+ 370	+ 30	+ 340	21
22	−35,48	123,79	+13,46	8,68	−116	+ 40	− 6	+ 40	+ 610	− 650	+ 600	22
23	−39,35	125,87	+13,00	10,72	− 82	+ 73	−18	+ 72	+1260	− 260	+1110	23
24	−41,06	130,13	+12,24	3,53	+ 5	+161	−30	+112	+3050	− 300	+ 480	24
25	−38,70	138,45	+10,86	17,85	+117	+272	−44	+ 60	+2550	− 460	−2480	25
26	−32,61	150,50	+ 8,92	18,54	+193	+348	−58	− 13	+1370	+ 120	−1470	26
27	−24,27	164,80	+ 6,85	17,22	+231	+386	−47	− 54	+ 830	+ 280	− 740	27
28	−15,07	179,96	+ 5,19	14,76	+269	+424	−39	− 80	+1220	+ 200	− 160	28
29	− 3,92	197,07	+ 3,69	11,66	+311	+467	−30	− 76	+ 720	+ 210	+ 350	29
30	+ 8,17	215,12	+ 2,65	9,28	+324	+479	−20	− 49	+ 40	+ 260	+ 870	30
31	+20,59	233,50	+ 2,27	8,31	+321	+476	− 9	− 13	− 200	+ 310	+1170	31

Table 19. Products of the mass of the left foot and the co-ordinates, velocities and accelerations of its centre of gravity in experiment III (with load)

No.	$m_7 x_7$ relative	$m_7 x_7$ absolute	$m_7 y_7$	$m_7 z_7$	$m_7 \dfrac{dx_7}{dt}$ relative	$m_7 \dfrac{dx_7}{dt}$ absolute	$m_7 \dfrac{dy_7}{dt}$	$m_7 \dfrac{dz_7}{dt}$	$m_7 \dfrac{d^2 x_7}{dt_7}$	$m_7 \dfrac{d^2 y_7}{dt^2}$	$m_7 \dfrac{d^2 z_7}{dt^2}$	No.
1	+10,65	44,76	−9,17	5,98	−155	0	+3	−1	0	+20	0	1
2	+4,67	44,74	−9,06	5,97	−155	0	+3	−1	0	+30	0	2
3	−1,33	44,70	−8,90	5,96	−155	0	+5	−1	0	+50	0	3
4	−7,33	44,66	−8,75	5,93	−155	0	+6	−1	0	+60	0	4
5	−13,42	44,53	−8,54	5,86	−155	0	+9	−1	0	+60	0	5
6	−19,46	44,45	−8,24	5,75	−155	0	+10	−4	0	−40	−280	6
7	−25,67	44,20	−7,94	5,48	−149	+6	0	−7	+240	−440	+120	7
8	−31,15	44,68	−8,10	5,55	−139	+16	−9	+7	+280	−20	+550	8
9	−36,03	45,76	−8,29	6,70	−125	+30	+9	+36	+470	+510	+870	9
10	−40,60	47,15	−7,89	8,68	−96	+59	+9	+81	+1470	−290	+1730	10
11	−42,50	51,21	−7,74	12,47	+7	+163	−2	+141	+3620	−280	+1120	11
12	−39,72	59,95	−7,78	16,69	+129	+285	+12	+64	+2770	+1410	−2610	12
13	−33,14	72,49	−6,67	17,88	+211	+366	+43	0	+1370	+270	−1470	13
14	−24,43	87,16	−5,09	17,00	+250	+406	+35	−44	+770	−520	−900	14
15	−14,15	103,40	−4,16	14,70	+284	+439	+6	−70	+1000	−890	−150	15
16	−2,75	120,76	−4,53	12,07	+310	+465	−19	−63	+430	−450	+360	16
17	+9,37	138,84	−5,46	10,22	+315	+470	−30	−36	−260	−·290	+860	17
18	+21,04	156,47	−6,73	9,56	+292	+447	−44	+6	−890	−510	+1640	18
19	+31,78	173,17	−8,67	10,86	+232	+387	−85	+63	−2680	−1730	+750	19
20	+38,26	185,61	−11,45	13,11	+92	+247	−34	+27	−4000	+1540	−2890	20
21	+39,13	192,44	−11,79	12,31	−52	+103	+3	−51	−3030	+670	−1210	21
22	+36,53	195,80	−11,45	9,94	−110	+45	+17	−64	+630	+170	+180	22
23	+32,50	197,73	−10,82	7,83	−106	+49	+13	−43	−1290	−220	+890	23
24	+27,66	198,85	−10,58	6,82	−150	+5	−5	−12	−570	−930	+740	24
25	+21,47	198,62	−10,82	6,61	−155	0	0	+2	0	+130	+70	25
26	+15,41	198,52	−10,80	6,73	−155	0	+1	+3	0	+10	0	26
27	+9,42	198,49	−10,79	6,85	−155	0	+1	+3	0	−10	0	27
28	+3,45	198,47	−10,77	6,98	−155	0	+1	+3	0	0	0	28
29	−2,51	198,47	−10,76	7,08	−155	0	+1	+3	0	0	0	29
30	−8,51	198,44	−10,75	7,19	−155	0	+1	+3	0	0	0	30
31	−14,39	198,52	−10,73	7,25	−155	0	+1	+3	0	0	0	31

Table 20. Maximum values for the products of the mass of the right foot and the co-ordinates, velocities and accelerations of its centre of gravity in experiment I

Time (s)	No. of phase of movement	Mass × value of co-ordinate		Time (s)	No. of phase of movement	Mass × velocity		Time (s)	No. of phase of movement	Mass × acceleration
		relative	absolute			relative	absolute			
Movement in the direction of gait										
Anterior maxima										
0,297	8,8	+43,61	135,61	0,155	5,0	+340	+508	0,098	3,6	+ 810
				0,347	10,1	− 56	+112	0,336	9,8	+1640
								0,460–0,728	13,0—20,0	0
								0,923	25,1	+3730
								1,076	29,1	+1280
Posterior maxima										
0,922	25,1	−43,57	153,60	0,320	9,4	− 80	+ 88	0,067	2,8	+ 460
				0,460–0,728	13,0—20,0	−168	0	0,298	8,8	−8670
								0,379	10,9	−2480
								0,460—0,728	13,0—20,0	0
								1,031	27,9	+ 650
Transverse movement										
Lateral (right) maxima										
0,290	8,6	—	+ 8,52	0,239	7,2	—	+ 27	0,205	6,4	+1550
0,501	14,1	—	+ 6,53	0,463	13,1	—	+ 1	0,373	10,7	+ 460
0,845	23,1	—	+ 6,60	0,813	22,2	—	+ 6	0,790	21,6	+ 110
				0,920	25,0	—	− 2	0,900	24,5	+ 400
								1,076	29,1	+ 280
Medial (left) maxima										
0,192	6,0	—	+ 6,75	0,135	4,5	—	− 9	0,290	8,6	−1380
0,433	12,3	—	+ 6,46	0,333	9,7	—	− 27	0,505	14,2	− 50
0,734	20,2	—	+ 6,23	0,620	17,2	—	− 1	0,845	23,1	− 580
1,074	29,0	—	+ 4,84	0,877	23,9	—	− 7	0,955	25,9	− 360
				0,998	27,0	—	− 15			
Vertical movement										
Upper maxima										
0,289	8,5	—	14,19	0,250	7,5	—	+ 77	0,203	6,3	+1520
0,983	26,7	—	18,32	0,470—0,700	13,3—19,3	—	+ 2	0,370	10,7	+1470
				0,925	25,1	—	+107	0,902	24,5	+1390
Lower maxima (minima)										
0,184	5,8	—	9,35	0,090	3,4	—	− 77	0,287	8,5	−3390
0,460	13,0	—	7,15	0,337	9,8	—	− 82	0,470—0,700	13,3—19,3	0
				0,470—0,700	13,3—19,3	—	+ 2	0,954	25,9	−2280
				1,083	29,3	—	− 67			

Table 21. Maximum values for the products of the mass of the left foot and the co-ordinates, velocities and accelerations of its centre of gravity in experiment I

Time (s)	No. of phase of movement	Mass × value of co-ordinate		Time (s)	No. of phase of movement	Mass × velocity		Time (s)	No. of phase of movement	Mass × acceleration
		relative	absolute			relative	absolute			

Movement in the direction of gait
Anterior maxima

Time (s)	No. of phase of movement	Mass × value of co-ordinate		Time (s)	No. of phase of movement	Mass × velocity		Time (s)	No. of phase of movement	Mass × acceleration
0,795	21,7	+42,55	218,37	0,667	18,4	+342	+510	0—0,230	0—7,0	0
				0,867	23,6	− 75	+ 93	0,442	12,5	+3300
								0,637	17,6	+ 830
								0,846	23,1	+1390
								0,955—1,15	25,9—31	0

Posterior maxima

Time (s)	No. of phase of movement	Mass × value of co-ordinate		Time (s)	No. of phase of movement	Mass × velocity		Time (s)	No. of phase of movement	Mass × acceleration
0,426	12,1	−43,51	70,23	0—0,230	0—7,0	−168	0	0—0,230	0—7,0	0
				0,833	22,7	−103	+ 65	0,606	16,8	+ 240
				0,955—1,15	25,9—31	−168	0	0,777	21,3	−4730
								0,891	24,3	−4270
								0,955—1,15	25,9—31	0

Transverse movement
Lateral (left) maxima

Time (s)	No. of phase of movement	Mass × value of co-ordinate		Time (s)	No. of phase of movement	Mass × velocity		Time (s)	No. of phase of movement	Mass × acceleration
0,040	2,0	—	− 6,67	0,327	9,5	—	− 13	0,284	8,4	− 290
0,475	13,4	—	− 8,30	0,449	12,7	—	− 15	0,423	12,0	− 770
0,795	21,7	—	− 9,76	0,748	20,5	—	− 31	0,727	20,0	− 410
				0,943	25,6	—	− 4	0,890	24,2	− 330

Medial (right) maxima

Time (s)	No. of phase of movement	Mass × value of co-ordinate		Time (s)	No. of phase of movement	Mass × velocity		Time (s)	No. of phase of movement	Mass × acceleration
0,243	7,3	—	− 6,52	0,150	4,9	—	+ 1	0,368	10,6	+ 290
0,630	17,4	—	− 7,02	0,399	11,4	—	− 2	0,471	13,3	+1160
0,916	24,9	—	− 8,15	0,533	14,9	—	+ 13	0,794	21,7	+1840
				0,833	22,7	—	+ 19	0,970	26,3	+ 60

Vertical movement
Upper maxima

Time (s)	No. of phase of movement	Mass × value of co-ordinate		Time (s)	No. of phase of movement	Mass × velocity		Time (s)	No. of phase of movement	Mass × acceleration
0,495	13,9	—	18,87	0—0,150	0—4,9	—	+ 1	0,410	11,7	+2740
0,785	21,5	—	13,96	0,430	12,2	—	+138	0,701	19,3	+1840
				0,748	20,5	—	+ 78	0,871	23,7	+1610
				0,980—1,15	26,6—31	—	+ 2			

Lower maxima (minima)

Time (s)	No. of phase of movement	Mass × value of co-ordinate		Time (s)	No. of phase of movement	Mass × velocity		Time (s)	No. of phase of movement	Mass × acceleration
0,686	18,9	—	9,12	0—0,150	0—4,9	—	+ 1	0—0,150	0—4,9	0
0,956	26,0	—	6,69	0,588	16,3	—	− 78	0,469	13,2	−2710
				0,832	22,7	—	− 86	0,783	21,4	−4710
				0,980—1,15	26,6—31	—	+ 2	0,980—1,15	26,6—31	0

310 Chapter 4 On the Movement of the Foot and the Forces Acting on It

Table 22. Maximum values for the products of the mass of the right foot and the co-ordinates, velocities and accelerations of its centre of gravity in experiment II

Movement in the direction of gait

Anterior maxima

Time (s)	No. of phase	Mass × co-ord. relative	absolute	Time (s)	No. of phase	Mass × vel. relative	absolute	Time (s)	No. of phase	Mass × acceleration
0,351	10,2	+41,77	148,61	0,208	6,4	+330	+499	0,148	4,9	+ 830
				0,396	11,3	− 65	+104	0,388	11,1	+2580
								0,511—0,767	14,3—21,0	0
								0,962	26,1	+3820
								1,108	29,9	+2270

Posterior maxima

Time (s)	No. of phase	Mass × co-ord. relative	absolute	Time (s)	No. of phase	Mass × vel. relative	absolute	Time (s)	No. of phase	Mass × acceleration
0,958	26,0	−42,38	167,34	0,382	11,0	− 85	+ 85	0,123	4,2	+ 620
				0,511—0,767	14,3—21,0	−169	0	0,325	9,5	−5630
								0,419	11,9	−1720
								0,511—0,767	14,3—21,0	0
								1,075	29,1	+ 860

Transverse movement

Lateral (right) maxima

Time (s)	No. of phase	Mass × co-ord. relative	absolute	Time (s)	No. of phase	Mass × vel. relative	absolute	Time (s)	No. of phase	Mass × acceleration
0,343	10,0	—	+ 8,73	0,182	5,8	—	+ 9	0,133	4,5	+ 290
0,882	24,0	—	+ 5,68	0,293	8,7	—	+ 22	0,263	7,9	+ 880
				0,535	15,0	—	0	0,401	11,5	+3760
				0,854	23,3	—	+ 6	0,828	22,6	+ 180
				0,973	26,4	—	+ 6	0,937	25,4	+ 280
								1,082	29,2	+ 90

Medial (left) maxima

Time (s)	No. of phase	Mass × co-ord. relative	absolute	Time (s)	No. of phase	Mass × vel. relative	absolute	Time (s)	No. of phase	Mass × acceleration
0,133	4,5	—	+ 6,45	0,051	2,3	—	− 11	0,212	6,5	− 120
0,797	21,8	—	+ 5,40	0,234	7,1	—	+ 5	0,349	10,1	−3400
0,930	25,3	—	+ 5,57	0,381	10,9	—	− 60	0,575	16,0	− 30
				0,700	19,3	—	− 1	0,878	23,9	− 520
				0,905	24,6	—	− 3	0,997	27,0	− 170
				1,023	27,7	—	+ 2			

Vertical movement

Upper maxima

Time (s)	No. of phase	Mass × co-ord. relative	absolute	Time (s)	No. of phase	Mass × vel. relative	absolute	Time (s)	No. of phase	Mass × acceleration
0,057	2,5	—	18,04	0,304	8,9	—	+ 71	0,260	7,8	+1770
0,342	9,9	—	13,47	0,515—0,670	14,4—18,5	—	+ 3	0,438	12,4	+1090
1,026	27,8	—	18,33	0,959	26,0	—	+ 95	0,915	24,9	+ 980

Lower maxima (minima)

Time (s)	No. of phase	Mass × co-ord. relative	absolute	Time (s)	No. of phase	Mass × vel. relative	absolute	Time (s)	No. of phase	Mass × acceleration
0,248	7,5	—	8,97	0,154	5,0	—	− 67	0,337	9,8	−4490
0,505	14,2	—	6,96	0,388	11,1	—	− 67	0,515—0,670	14,4—18,5	0
				0,515—0,670	14,4—18,5	—	+ 3	1,002	27,1	−1930

Table 23. Maximum values for the products of the mass of the left foot and the co-ordinates, velocities and accelerations of its centre of gravity in experiment II

Time (s)	No. of phase of movement	Mass × value of co-ordinate		Time (s)	No. of phase of movement	Mass × velocity		Time (s)	No. of phase of movement	Mass × acceleration
		relative	absolute			relative	absolute			
				Movement in the direction of gait						
				Anterior maxima						
0,834	22,8	+42,63	154,99	0,699	19,2	+361	+530	0—0,287	0—8,5	0
				0,889	24,2	— 67	+102	0,493	13,9	+3510
								0,652	18,0	+1190
								0,874	23,8	+1510
								0,997—1,15	27,0—31	0
				Posterior maxima						
0,474	13,4	—43,30	84,33	0—0,287	0—8,5	—169	0	0—0,287	0—8,5	0
				0,858	23,4	— 94	+ 75	0,599	16,6	+ 760
				0,997—1,15	27,0—31	—169	0	0,832	22,7	—7830
								0,915	24,9	—2720
								0,997—1,15	27,0—31	0
				Transverse movement						
				Lateral (left) maxima						
0,521	14,6	—	— 8,22	0,386	11,1	—	— 11	0,343	10,0	— 180
0,822	22,4	—	— 8,57	0,483	13,6	—	— 21	0,463	13,1	—1140
				0,779	21,3	—	— 14	0,619	17,1	— 300
				0,979	26,5	—	— 4	0,915	24,9	— 700
				Medial (right) maxima						
0,272	8,1	—	— 6,29	0,210	6,5	—	+ 1	0,100	3,6	+ 30
0,651	18,0	—	— 7,04	0,441	12,5	—	— 3	0,423	12,0	+ 300
0,944	25,6	—	— 5,90	0,578	16,1	—	+ 14	0,516	14,5	+1180
				0,863	23,5	—	+ 33	0,829	22,6	+1830
								1,015	27,5	+ 50
				Vertical movement						
				Upper maxima						
0,546	15,3	—	18,60	0—0,200	0—6,2	—	+ 2	0,447	12,7	+2020
0,830	22,7	—	13,54	0,478	13,5	—	+129	0,739	20,3	+1980
				0,794	21,7	—	+103	0,903	24,6	+1530
				1,020—1,15	27,6—31	—	— 1			
				Lower maxima (minima)						
0,728	20,0	—	8,91	0—0,200	0—6,2	—	+ 2	0—0,200	0—6,2	0
				0,631	17,5	—	— 81	0,533	14,9	—2650
				0,869	23,7	—	— 86	0,827	22,6	—5080
				1,020—1,15	27,6—31	—	— 1	1,020—1,15	27,6—31	0

Table 24. Maximum values for the products of the mass of the right foot and the co-ordinates, velocities and accelerations of its centre of gravity in experiment III (with load)

Time (s)	No. of phase of movement	Mass × value of co-ordinate		Time (s)	No. of phase of movement	Mass × velocity		Time (s)	No. of phase of movement	Mass × acceleration
		relative	absolute			relative	absolute			
	**Movement in the direction of gait**
	Anterior maxima
0,255	7,7	+39,91	113,66	0,117	4,1	+341	+496	0,077	3,0	+ 660
				0,303	8,9	− 62	+ 93	0,290	8,6	+1 250
				1,114	30,1	+324	+479	0,402—0,690	11,5—19,0	0
								0,888	24,2	+3 100
								1,045	28,3	+1 320
	Posterior maxima
0,879	23,9	−41,08	129,69	0,277	8,2	− 80	+ 75	0,053	2,4	+ 500
				0,402—0,690	11,5—19,0	−155	0	0,261	7,8	−7 280
								0,338	9,8	−2 240
								0,402—0,690	11,5—19,0	0
								0,989	26,8	+ 830
	**Transverse movement**
	Lateral (right) maxima
0,265	7,9	—	+14,40	0,180	5,7	—	+ 57	0,161	5,2	+ 550
0,797	21,8	—	+13,49	0,773	21,2	—	+ 5	0,297	8,8	+ 770
								0,742	20,4	+ 100
								0,852	23,2	− 250
								0,984	26,7	+ 290
	Medial (left) maxima
0,660	18,2	—	+13,13	0,288	8,5	—	− 18	0,258	7,7	−1 240
				0,953	25,9	—	− 58	0,550	15,4	+ 10
								0,797	21,8	− 850
								0,935	25,4	− 590
								1,043	28,2	+ 190
	**Vertical movement**
	Upper maxima
0,238	7,2	—	12,90	0,200	6,2	—	+ 88	0,177	5,6	+2 130
0,950	25,8	—	18,61	0,420—0,620	12,0—17,2	—	+ 2	0,344	10,0	+1 550
				0,887	24,2	—	+114	0,857	23,4	+1 360
	Lower maxima (minima)
0,146	4,8	—	8,66	0,055	2,4	—	− 72	0,221	6,8	−3 210
0,395	11,3	—	6,39	0,298	8,8	—	− 72	0, 20—0,620	12,0—17,2	0
				0,420—0,620	12,0—17,2	—	+ 2	0,917	24,9	−2 510
				1,046	28,3	—	− 82			

Table 25. Maximum values for the products of the mass of the left foot and the co-ordinates, velocities and accelerations of its centre of gravity in experiment III (with load)

Time (s)	No. of phase of movement	Mass × value of co-ordinate		Time (s)	No. of phase of movement	Mass × velocity		Time (s)	No. of phase of movement	Mass × acceleration
		relative	absolute			relative	absolute			
					**Movement in the direction of gait**					
					Anterior maxima					
0,751	20,6	+39,34	190,27	0,601	16,7	+317	+472	0—0,192	0—6,0	0
				0,827	22,6	−94	+61	0,386	11,1	+3640
								0,531	14,9	+1040
								0,811	22,2	+1260
								0,901—1,15	24,5—31	0
					Posterior maxima					
0,381	10,9	−42,51	50,85	0—0,192	0—6,0	−155	0	0—0,192	0—6,0	0
				0,800	21,9	−111	+44	0,500	14,1	+770
				0,901—1,15	24,5—31	−155	0	0,727	20,0	−4000
								0,851	23,2	−1490
								0,901—1,15	24,5—31	0
					Transverse movement					
					Lateral (left) maxima					
0,298	8,8	—	−8,31	0,270	8,0	—	−9	0,224	6,8	−450
0,413	11,8	—	−7,80	0,397	11,4	—	−4	0,365	10,5	−360
0,762	20,9	—	−11,80	0,700	19,3	—	−92	0,530	14,8	−910
0,919	25,0	—	−10,82	0,892	24,3	—	−13	0,687	18,9	−1750
								0,879	23,9	−950
								0,997	27,0	−10
								1,030—1,15	27,9—31	0
					Medial (right) maxima					
0,230	7,0	—	−7,94	0,186	5,9	—	+10	0,135	4,5	+60
0,374	10,8	—	−7,73	0,323	9,4	—	+12	0,297	8,8	+1020
0,544	15,2	—	−4,14	0,470	13,3	—	+43	0,428	12,2	+1500
0,873	23,8	—	−10,56	0,820	22,4	—	+18	0,616	17,1	−280
				0,965	26,2	—	+1	0,712	19,6	ca.+21400
								0,902	24,5	+1180
					Vertical movement					
					Upper maxima					
0,460	13,0	—	17,88	0—0,170	0—5,4	—	+1	0—0,170	0—5,4	0
0,738	20,3	—	13,20	0,388	11,1	—	+142	0,362	10,4	+2110
				0,702	19,3	—	+67	0,667	18,4	+2020
				0,940—1,15	25,5—31	—	+3	0,858	23,4	+1010
					Lower maxima (minima)					
0,249	7,5	—	5,45	0—0,170	0—5,4	—	+1	0,202	6,3	−350
0,648	17,9	—	9,54	0,223	6,8	—	−10	0,409	11,7	−2880
0,908	24,7	—	6,59	0,547	15,3	—	−71	0,732	20,1	−3230
				0,798	21,8	—	−65	0,940—1,15	25,5—31	0
				0,940—1,15	25,5—31	—	+3			

Kinematics of the Swing of the Leg

Introduction

Among the movements carried out by the different segments of the human body during walking, none has aroused such controversy as the swinging of the leg. Weber and Weber claimed that this movement was purely that of a pendulum and resulted solely from gravity. The controversy has continued ever since. There are opponents and supporters of the Webers'[1] theory. Both have tried to provide the empirical bases to sustain their point of view. But no unequivocal confirmation or refutation of the Webers' theory has yet been brought forwards. The question thus remains open.

The reason why a problem that is apparently so simple has not yet been solved lies principally in the fact that it has been approached with inadequate means. There is no direct way of demonstrating whether certain muscles take part in the swinging of the leg and what tension these muscles must develop. The aforementioned experiments and observations of Duchenne are certainly important and interesting. They suggest that muscles are probably involved in this movement. However, they provide no unquestionable proof of this and do not give any indication as to the degree of tension of the muscles. It may well be that a movement takes place under the effect of gravity only, which in other circumstances would result from muscular activity. The swinging of the leg during walking occurs in completely different circumstances from those existing when certain muscles of the leg at rest are subjected to faradic stimulation. In the former instance the attachement of the leg itself, i.e. the centre of the hip joint, moves. This exerts a decisive influence on the way the leg swings. Moreover, the phenomena that appear in pathological movements cannot be directly related to the normal individual.

To decide whether the swinging of the leg is purely a pendulum movement or not, recourse to a roundabout approach is necessary. This approach consists in acquiring such an accurate knowledge of the movement that it becomes possible to decide what forces must act on the different segments of the leg to bring about this movement. It will then be apparent whether gravity is sufficient or whether other forces are necessary. In the present case, these other forces could be set up only by the contraction of muscles. It is thus necessary to know the relationships between the forces acting and the changes of position, velocities and accelerations of the different segments of the leg brought about by these forces. These relationships find their expression in the movement equations.

It might be considered sufficient to obtain a series of instantaneous pictures of the swinging leg and to compare these pictures with the corresponding phases of the movement of a pendulum that is mechanically equivalent to the leg. This cannot be achieved by calculation, however, because the phases of an articulated pendulum cannot be exactly deduced. The movements of a three-segment pendulum such as the leg, present a very

---

[1]   The various opinions on the swing of the leg have been touched on in the introduction to Chapter 1, and are, therefore, not repeated at this point

complicated mechanical problem. It is true that movement equations can be set up for such a pendulum, but integration of these differential equations comes up against insurmountable obstacles. The successive positions of the swinging pendulum can be deduced from the differential equations only by way of integration.

The construction of a three-segment pendulum equivalent to the leg articulated at the knee and ankle might also be considered so that the swinging of the leg could be compared with that of the model. The articulations of such a pendulum would have to be built like the three main joints of the lower extremity, and the three segments of the pendulum would have to correspond exactly to their counterparts in the leg as far as length, position of the centre of gravity and moments of inertia are concerned. These requirements can be met. To study and demonstrate the activity of the muscles of the arm, some time ago I actually had an articulated model of this kind constructed; from a mechanical point of view it was precisely equivalent to the arm articulated at the elbow [2]. With a model of the lower extremity made in the same way it would be possible to demonstrate empirically whether the swinging of the leg during walking must be regarded purely as a pendulum swing or not. The pendulum swing of the model would have to be recorded chronophotographically from two sides and entered into a tridimensional system of co-ordinates, precisely as for the movements of the human gait. It would then be possible to see whether the swinging of the leg is different from that of the pendulum, and if so in what ways.

Practical application of this method, however, meets with considerable difficulties. If the pendulum had to swing about a fixed point the experiment would be easy to carry out. But in walking the centre of the hip joint ifself moves, and this movement is not even uniform and linear; the hip moves along a curved trajectory, at times accelerated, at times decelerated. This makes the swinging of the pendulum completely different from that about a fixed point. If the model were to move in a way that would make an answer to the question of the pendulum swing of the leg possible, its point of attachment would have to move in exactly the same way as the centre of the hip joint during the swinging period. This is very difficult to achieve exactly, and would first require a thorough study of the movement of the centre of the hip.

It thus appears that, for an answer to the question of the pendulum swing, it is not sufficient to have a large number of successive phases of the swinging leg recorded on chronophotographic exposures taken during walking. A very thorough knowledge of the kinematics of the swinging leg, based on data obtained empirically, is also necessary. Without a precise empirical basis such questions cannot be answered. Theoretical speculations alone are not reliable.

This is best illustrated by the assumptions that Weber and Weber were forced to make concerning details of the movements involved in walking once their ancillary equipment was inadequate to provide empirical documentation. Many of these assumptions have proved to be erroneous, as has been explained in Chap. 3.

The same is true for the data given by von Meyer. These data are not based on measurements recorded in a walking subject but are derived almost exclusively from theoretical considerations. Consequently, they are often in direct contradiction to the facts shown by a precise recording of the process of movement. For instance, according to von Meyer, before its swinging period the posterior leg must remain in contact with the floor until the centre of gravity is on the vertical through the heel of the anterior foot. He writes: "The supporting leg is in contact with the floor through its plantar surface before the movement of the centre of gravity reaches so far forward that it is beyond the contact area of the sole with the floor; the gravity line is still behind the anterior foot and then the posterior

---

[2]   Fischer O (1895) Contributions to muscular dynamics. Abhandlungen der Königlichen Sächsischen Gesellschaft der Wissenschaften, mathematisch-physische Klasse, vol 22, no 2

foot must still help to support the centre of gravity. The weight is thus now supported by *both* legs. To enable the posterior leg to leave the floor and to move forward, it is necessary for the weight to be completely transmitted to the anterior (supporting) leg. This is made possible only by such a movement of the ankle of the anterior leg that the line of gravity finally passes through the heel". ... "When the weight is transmitted to the supporting foot, the posterior leg can leave the floor and be made to swing forward"[3].

Actually the swinging leg leaves the floor long before the total centre of gravity reaches a position above the heel of the anterior foot. The vertical through the centre of gravity lay about 25 cm behind the axis of the ankle at the beginning of the swinging period in our subject. This can be seen in Chap. 2, Fig. 4. In experiment I, right toe-off occurred in the 25th phase, in experiment II in the 26th phase. Left toe-off occurred shortly after the 12th phase in experiment I, between the 13th and 14th phases but closer to the 13th in experiment II (Table 14, p. 169 of Chap. 2). The conclusions derived from von Meyer's theoretical considerations concerning the beginning of the swinging period are thus far from being correct.

Moreover, the conclusions reached by von Meyer do not seem to me to be consistent. On the one hand he assumes, correctly, that the centre of gravity of the body is not supported for a period of time when the gravity line has moved in front of the tip of the foot of the only leg in contact with the floor. On the other hand, I find in his work the exactly opposite opinion, that the swinging leg must strike the floor at the instant when the gravity line arrives in front of the tip of the opposite foot, so that the centre of gravity is supported by the two feet together. The supporting area, besides the contact surfaces of the feet, also includes the strip of floor between the two feet. At least that is how I understand the following excerpts from his *Die Statik und Mechanik des menschlischen Knochengerüstes*: "At the end of its principal arc (by principal arc, von Meyer means the arc described by the centre of gravity of the body in the vertical plane of gait whilst the stance leg supports it directly), the centre of gravity finds no more support and falls forwards along a circle the centre of which is the foot, until the opposite leg, which in the meantime has struck the floor, hinders further fall. This falling arc is the *"anterior complementary arc"* (p. 308). As I understand it, the centre of gravity would not be supported for a period of time during the fall. This is also the meaning of von Meyer's Fig. 40 (p. 309) which shows the swinging leg reaching the floor only at the end of the anterior complementary arc V. Moreover, on p. 305 von Meyer claims "that what characterizes human gait, as compared with the gait of quadrupeds, is that the centre of gravity is not necessarily always supported". This is in contradiction to the following considerations published by von Meyer: "The time of the forward swing (of the posterior leg) must, of course, be the same as the time in which the principal arc is carried out by the supporting leg. Only during this time the weight is carried by the supporting leg in such a way that the other leg can be relieved" (p. 311). On p. 313 he writes: "If gait consists of an arched movement forwards of the trunk, alternately on one leg and then on the other, and if in this movement the whole body is supported only by the stance leg, the line of gravity must then fall through the foot of this supporting leg".

If this was the opinion of von Meyer, it would again be in contradiction with the facts revealed by chronophotography. At heel strike, the vertical through the centre of gravity has long passed in front of the tip of the opposite foot. The total centre of gravity in fact remains without direct support for a considerable time. In our experiments, a vertical line through the centre of gravity was approximately 20 cm in front of the tip of the stance leg at heel strike of the contralateral extremity. This can be seen from Chap. 2 Fig. 4 and from the data in Table 14 (p. 169).

---

[3] Meyer H (1873) Die Statik und Mechanik des menschlichen Knochengerüstes. Leipzig, p 308

These examples show that many an "of course" or a "necessary" of von Meyer is not at all appropriate. This may explain why von Meyer's concepts in the usual text books of physiology have not met with the appreciation which they deserve according to Fuchs[4]. In fact, von Meyer did not find anything that added to the measurements of Weber and Weber. After Weber and Weber, Carlet was the first to add to the reliable data concerning the human gait. Carlet, using new methods, measured many phenomena of walking and made graphic recordings of the movement of a point of the human body. Von Meyer's opinion[5] that the results published by Carlet were mostly meaningless and disproportionate to the apparatus used, and that most of his material was not new or was self-evident, therefore, does not seem to me to be quite justified.

The present chapter (Chap. 5) of the study of the human gait will deal principally with the purely kinematic side of the problem of the swinging leg and will lay the basis for the treatment of the kinetic problem which consists in finding the forces acting during the swinging period. This latter problem will be discussed in Chap. 6. The present task consists in investigating thoroughly what forces might be involved in the swinging period, how they tend to rotate the different segments of the leg, and how they relate to the movements of the thigh, lower leg and foot. This research must precede determination of the purely kinematic data, to establish how far the kinematic analysis has to go and where it can be limited for the present purpose. As the only data needed are the directions, angular velocities and accelerations of the long axes and the velocities and accelerations of the centres of gravity of the three segments of the leg for the successive phases of the swinging period, these data will be determined in the second section if they have not already been calculated, as is the case for the foot, for instance.

Before this, however, the question of the typical gait will be discussed thoroughly again in the first section.

On the Typical Walker's Step

According to von Meyer[6] "there can be no such thing as a typical gait, and the only typical feature recognizable in the gait of different individuals is that they move forward with the help of their legs". This concept, like so many of his statements, can only be regarded as the result of theoretical speculation not based on facts. Of course people can use their legs in very different ways to move from one place to another. They can walk slowly and can accelerate their pace up to a certain limit. They can walk with long or short steps. They can put their feet on the ground heel first or with the whole plantar surface, as in the military parade step. It is also possible to walk only on tiptoe. People can stroll about indoors or walk about the city streets to spend the time. They can also affect to walk with a dainty, light or tripping step. No one will deny any of this. If the purpose of walking is to go as far as possible without unnecessary stops and without excessive fatigue, however, one spontaneously adopts a certain regular pace and tends to avoid any affectation and, consequently, any premature tiredness. It is certainly a common experience that a brisk pace on country roads can be kept up much longer than strolling in city streets, even if the influence of the hard ground is disregarded. It is also well known that the parade step, walking on tiptoe or taking short steps is so tiring that it would never occur to anyone to adopt one of these ways of walking for a long excursion on foot. During walks in the mountains ladies forget that walking with long

[4] Fuchs RF (1901) Der Gang des Menschen. Biologisches Zentralblatt vol XXI, p 719

[5] v Meyer H (1881–1882) Die Mechanik des menschlichen Ganges. Biologisches Zentralblatt vol I, p 435

[6] Biologisches Zentralblatt, vol I, p 402

steps is not looked on as pretty and refrain from wearing anything that would prevent them from stepping out freely. When they are on manoeuvres, soldiers must be able to walk for a long time. They would not think of walking as they do on the parade ground. During various military exercises I have had to carry out, some of them on manoeuvres, I was able to observe hundreds of soldiers in walking, after chronophotography had made me familiar with the successive positions of the body during brisk walking. I could never distinguish any significant difference between the gait of the soldiers and that of a hiker. On the contrary, individual differences in gait which were obvious in slow walking in the streets of a garrison subsided during walking on country roads. The sequence and type of movements seemed to be the same in everyone and to correspond to the successive phases of movement which we found in chronophotographic pictures of gait, as far as it was possible to tell with the naked eye.

Even the differences in body build generally seen in different nationalities seem to have no essential influence on the movements involved in walking forward. The chronophotographic pictures which Marey obtained before us do not show any noticeable differences from ours. For example, the series of 40 phases of movement in a person walking, published in the Comptes rendus of 1884 (vol. 98, p. 1218) and reproduced in Fig. 1, can be compared with the photographs in Chap. 1 of the present volume or with Fig. 11 in this chapter (p. 355) which is derived from these pictures. This comparison is not restricted to observations with the naked eye, and it shows essential similarity between the two series. The only discrepancies are quantitative. For instance, the vertical movements of the individual of Marey are relatively greater than those of our subject. The sequence and type of movement, however, are the same. The description of the movement of the leg in Chap. 3 of our work could be applied nearly word for word to the serial pictures of Marey as long as it is restricted to one extremity and does not take into account the simultaneous movement of the opposite extremity which does not appear in Marey's pictures.

As far as the swinging of the leg is concerned, Marey's pictures show that at the beginning of this period the leg has a pronounced backward rotation. Both the long axis of the thigh and that of the lower leg are at a backward inclination to the vertical. The angle formed by the long axis of the lower leg with the vertical, however, is greater than that formed by the long axis of the thigh with the vertical since the knee is not extended but slightly flexed. The foot is still in contact with the ground through its tip. The thigh at first rotates forwards and then moves forwards perceptibly parallel to itself in the projection on the plane of gait

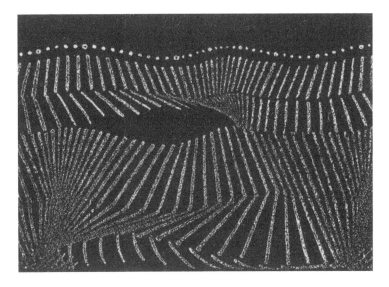

Fig. 1. Series of 40 phases of movement of a man walking, published by Marey in 1884

almost until heel strike. Shortly before heel strike it again rotates forwards briefly. This cannot be reliably observed on Marey's picture because of its small size and poor definition. This movement could not be directly deduced from our pictures either. It appeared from calculation of the angle formed by the long axis of the thigh and the vertical in the successive positions of the leg. For Marey, as for ourselves, the lower leg first rotates backward at the beginning of the swinging period. This backward rotation also lasts almost up to the instant when the long axis of the thigh passes through the vertical whereupon a forward rotation of the lower leg begins, its speed significantly exceeding that of the simultaneous forward rotation of the thigh. Only shortly before heel strike does the lower leg reverse the direction of its rotation. The different behaviour of the thigh and lower leg during the swinging period causes the knee to flex during the first part of the swing. The maximum flexion is not reached until the thigh, having passed the vertical, appears rotated somewhat forwards. After that, the knee extends rather vigorously until the lower leg reverses the direction of its rotation shortly before heel strike. Finally, the knee flexes again slightly so that the leg is not extended at heel strike.

The movement of the swinging leg of Marey's subject coincides with that of our subject. Any discrepancies shown by an accurate determination of the angles formed by the long axes of the thigh and lower leg with the vertical could only affect the extreme values of these angles. They would be only quantitative. Even without accurate measurement of the angles, it can be seen in Marey's serial pictures that the knee does not extend completely before heel strike as it does in our subject. In Marey's pictures the long axis of the lower leg reaches the prolongation of the long axis of the thigh whereas in our experiments it seems to rotate beyond this in such a way that the two long axes form an angle open forwards. On the other hand, both in Marey's pictures and in our own, the long axes of the thigh and lower leg form an angle open forward during the stance period. Such an attitude is not as impossible as has been assumed[7] by not taking into account the fact that in full extension the axis of the knee joint lies behind the line joining the centre of the hip and that of the ankle.

Comparison of the successive positions of the leg during the stance period or of the phases of movement of the upper extremity in the picture of Marey and in ours shows the same concordance in the type of movement. Consideration of other serial pictures published by Marey in the Comptes rendus of 1887 (vol. 105, p. 149) leads to the same conclusion.

The subjects studied by Marey have thus moved in essentially the same way as ours. This is another demonstration that the assumption of a normal gait is justified. von Meyer himself examines and analyses the "usual type of gait," which he describes as the gait "which is the most widespread because it is the most natural and most easy, without prejudice of individual variations"[8].

During the chronophotographic exposure, Marey's first experimental subject seems to have been walking as if he were progressing briskly along a country road. The length and duration of his steps suggest such a gait. In the communication cited above[9] Marey gives 1.75 m as the length of a double step and exactly 1 s as its duration. A single step thus had a length of 87.5 cm and a duration of 0.5 s so that there were 120 steps a minute. Unfortunately, Marey does not give the height of his first experimental subject. On the original of the serial picture, however, taking into account the length of the step, 87.5 cm, we can measure the greatest distance between the floor and the centre of the shoulder joint, which was about 126 cm. In our subject, this height was approximately 135 cm during walking. We can thus conclude that Marey's subject was even smaller than ours. This is also

---

[7] cf. (for example) Jendrassik BE. Klinische Beiträge zum Studium der normalen und pathologischen Gangarten. Deutsches Archiv für klinische Medizin, vol LXX, p 87

[8] Meyer H (1873) Die Statik und Mechanik des menschlichen Knochengerüstes, Leipzig, pp 332 and 330

[9] Marey (1884) Analyse cinématique de la marche. Comptes rendus, vol 98, p 1218

supported by the fact that in our subject the centre of the hip joint rose to 88 cm from the floor during walking whereas in Marey's subject it rose to only 83 cm. Consequently, Marey's subject took much longer steps than our subject, not only in absolute terms but also relative to his height. But a step of 80 cm or even more is normal when walking long distances, as shown by the experiments described below.

In 1889, before we started making chronophotographic recordings of the walking movement, I measured with W. Braune the length of a step, its duration, the velocity of the movement forward in walking, etc. in a large number of individuals of normal body build. We concentrated principally on the gait instinctively adopted for a long walk on a country road from one place to another without unnecessary stops. We described this as the *walker's step*. With this in view, we measured a distance of exactly 1 km on a little-used woodland path in the vicinity of Leipzig. We asked many young men to walk this distance and back and to count their steps. We read the time of departure and that of arrival on a chronometer and noted them so as not to bother the walker with anything but the number of his steps. We carefully avoided prescribing or even recommending the subjects a certain speed or a certain length of step. On the contrary, we made a particular point of asking each to walk quite naturally, as if he were on a walking holiday.

For this experiment, we were grateful to have a large number of soldiers from the 8th Royal Saxon Infantry Regiment Prinz Johann Georg no. 107 at our disposal. We stationed noncommissioned officers at different points along the way. They checked that the steps were correctly counted and that the subjects turned immediately to go back at the end of the 1 km. After the experiments, the height of each subject was measured indoors, as was the distance between the floor and the greater trochanter felt through the skin with the subjects standing upright and naked.

These numerous measurements have not yet been published. They will be given below in the same order as they were carried out, without selection, partly because they allow the deduction of certain constants that are characteristic for the typical walker's step. A further reason is that they provide a more convincing demonstration than the comparison of the duration of our subject's step with that of Marey's subject, and show that the objection levelled against our photographic experiments, according to which our subject had allegedly taken unnaturally long fast steps during the chronophotographic exposures, is unjustified. This objection was expressed from different sources, for instance by Jendrassik[10].

We did not restrict the experiments to the walker's step, but also asked the men to go once relatively slowly and another time as fast as possible, to find out the interdependence between the duration and length of a step and the velocity of displacement. In addition, we took the pulse of each subject at the end of his 2-km walk. The results of these experiments will not be published here, however, since they are not relevant to the purpose of the present work. The experimental results given demonstrate that the length and duration of the walker's step generally present a certain regularity, and that the length of one step in most instances exceeds 80 cm, even in small individuals. The names of the different soldiers are given to allow comparison of the data in Table 1 with those recorded in other experiments with the same subjects, should these be published later.

To avoid overloading the table with data, the number of steps and the time spent in walking the 2 km are not given. The only data entered are the average length and duration of one step, obtained by simple calculation from the mentioned data already referred to, the number of steps per minute and the average velocity of gait. We also indicate for each subject the height and the distance from the greater trochanter to the ground in the standing position, which we call "leg length". The average age of the soldiers was 21 years. The

---

[10] Jendrassik E, loc. cit, p 87. This is the journal in which the present chapter of Fischer was published!

requirements of the army for physical fitness of conscripts guaranteed that our subjects had normal body build. The experiments were carried out over several days. Each day each subject walked the 2-km distance only twice. Some soldiers came on two different days and the results of four experiments are shown for each of these.

Table 1. Experiments to determine length and duration of the typical walker's step, carried out with soldiers of the 8th Royal Saxon Infantry Regiment Prinz Johann Georg no. 107 in October 1889

No.	Name	Body height (cm)	Leg length (cm)	Length of a single step (cm)	Duration of a single step (s)	No. of steps per minute	Average velocity of gait (cm s^{-1})
1.	LC Preuß.	170,5	93,0	84,2	0,479	125	175,5
				83,8	0,477	125	175,5
2.	Körner	166,0	88,0	82,4	0,475	126	173,2
				83,7	0,448	134	186,1
				81,6	0,472	127	173,0
				80,6	0,472	127	171,0
3.	Nietsch	183,0	98,0	104,6	0,510	117	205,0
				104,3	0,506	118	206,2
4.	LC Skobel	166,5	88,5	84,4	0,487	123	173,0
				83,9	0,484	124	173,0
5.	Steisch	166,0	88,5	91,3	0,494	121	184,9
				89,6	0,500	120	179,2
6.	LC Berger	161,5	87,0	79,6	0,483	124	165,0
				79,6	0,479	125	166,6
7.	LC Hanns	169,0	91,0	84,3	0,500	120	168,5
				83,8	0,514	117	163,1
8.	LC Hannspach. .	167,0	90,5	86,9	0,471	127	184,1
				84,5	0,472	127	178,8
				85,7	0,470	127	182,2
				85,1	0,497	121	171,2
9.	Küntzer	167,0	88,5	85,9	0,497	121	173,2
				84,9	0,504	119	168,3
10.	LC Scholl.	164,0	88,0	84,1	0,475	126	177,0
				81,6	0,465	129	175,8
11.	Schmeda	177,0	94,5	91,3	0,499	120	182,5
				90,9	0,511	117	178,0
12.	Kurth	172,0	94,5	84,4	0,512	117	165,1
				84,1	0,501	120	167,8
13.	Däbernitz.	169,5	90,5	80,3	0,506	119	158,8
				80,5	0,498	120	161,2
				80,9	0,478	125	169,3
				80,7	0,471	127	171,2
14.	Richter	167,0	89,0	79,4	0,493	122	160,5
				78,2	0,485	124	161,2
15.	Keller	165,0	89,0	83,7	0,504	119	165,6
				83,2	0,501	120	166,0
16.	Krause.	161,5	88,5	84,4	0,504	119	167,3
				84,4	0,507	118	166,6
17.	Müller I	163,0	85,5	82,3	0,529	113	155,2
				83,4	0,514	117	162,1

Table 1 (continued)

18.	Schlotte	164,5	84,0	80,9	0,495	121	163,8
				79,6	0,492	122	162,0
19.	Weber	159,5	83,5	81,2	0,508	118	159,6
				80,9	0,519	116	155,8
20.	Leipnitz	159,5	80,5	80,7	0,501	120	161,2
				79,7	0,499	120	159,6
21.	Hermann	173,0	92,0	84,9	0,508	118	167,1
				84,4	0,503	119	167,6
22.	Winter	170,5	90,5	86,5	0,492	122	176,0
				85,6	0,494	121	173,3
23.	Heidel	166,0	89,0	81,8	0,471	127	173,8
				80,3	9,482	125	166,7
24.	Müller II	162,0	88,5	89,0	0,485	124	183,5
				93,8	0,568	106	165,3
25.	LC Berge......	162,0	88,0	85,2	0,507	118	167,8
				84,6	0,504	119	168,0
26.	Buch	167,0	86,0	86,4	0,475	126	182,2
				86,8	0,467	129	185,9
27.	Heidrich	167,5	85,5	85,9	0,476	126	180,6
				86,6	0,464	129	186,6
28.	LC Paul.......	170,0	92,0	86,4	0,545	110	158,5
				88,5	0,537	112	165,0
29.	Clemens.......	168,5	92,0	79,6	0,519	116	153,3
				80,9	0,528	114	153,3
30.	Friedrich	167,0	92,0	80,2	0,501	120	160,2
				80,4	0,501	120	160,3
				81,4	0,487	123	167,0
				81,3	0,472	127	172,0
31.	Thalmann	169,5	91,5	83,5	0,524	115	159,2
				82,4	0,516	116	159,7
32.	Hößler	173,5	91,0	88,3	0,518	116	170,6
				88,2	0,518	116	170,0
33.	Hofmann I	169,0	90,0	84,2	0,526	114	160,2
				83,7	0,529	113	158,2
34.	Punk	171,0	89,5	82,5	0,518	116	159,1
				83,9	0,525	114	159,6
35.	Feodoro	164,0	89,5	87,7	0,507	118	173,1
				87,5	0,506	119	173,0
36.	Graiche	167,0	88,5	86,0	0,531	113	162,2
				85,7	0,494	121	173,8
37.	Wächtler	160,0	84,0	85,6	0,520	115	164,6
				82,9	0,493	122	168,2
				82,8	0,494	121	167,3
				83,4	0,491	122	169,8
38.	Welsch	161,5	83,0	83,0	0,477	126	173,9
				82,9	0,474	126	174,7
				82,3	0,479	125	171,6
				82,4	0,478	125	172,4
39.	Veit	159,0	81,0	80,8	0,483	124	167,0
				80,0	0,485	124	164,8

Table 1 (continued)

No.	N a m e	Body height (cm)	Leg length (cm)	Length of a single step (cm)	Duration of a single step (s)	No. of steps per minute	Average velocity of gait (cm s⁻¹)
40.	Nollau	157,5	81,0	80,3 81,3	0,495 0,498	121 120	162,0 163,2
41.	Schindler	172,0	94,0	92,7 92,2	0,564 0,546	106 110	164,4 169,0
42.	Schellbach	172,0	93,0	88,6 87,4	0,524 0,507	115 118	169,1 172,3
43.	Reichstein	171,0	89,5	85,9 85,7	0,508 0,503	118 119	169,5 170,5
44.	Naumann	170,5	89,0	83,8 86,2	0,484 0,486	124 123	173,1 177,4
45.	Scheibe	167,0	88,5	85,8 85,4	0,481 0,480	125 125	178,6 178,0
46.	Tippmann	171,0	88,0	91,7 90,0	0,464 0,463	129 130	197,9 194,3
47.	Richter	165,5	88,0	82,5 82,3	0,512 0,511	117 118	161,2 161,1
48.	Kunze	163,5	87,5	95,8 94,8	0,551 0,541	109 111	174,0 175,1
49.	Lippold	168,5	87,0	86,6 87,2	0,480 0,474	125 127	180,2 184,0
50.	Landschreiber . .	159,5	84,5	83,6 84,6 85,3 85,4	0,515 0,509 0,481 0,472	117 118 125 127	162,5 166,0 177,5 181,1
51.	Wiczorek	164,5	83,5	85,8 84,4	0,471 0,467	127 129	182,0 181,0
52.	Kaiser	161,0	82,5	84,3 84,3	0,491 0,477	122 126	171,5 176,8
53.	Göllnitz	163,5	86,5	84,2 83,9	0,503 0,522	119 115	167,2 160,7
54.	Weinreich	164,5	85,0	83,8 83,6	0,493 0,501	122 120	170,3 166,8
55.	LC Schöne	171,5	91,5	93,4 93,3	0,532 0,497	113 121	175,7 187,9
56.	Herdegen	170,0	90,0	85,4 82,8	0,497 0,505	121 119	171,9 163,9
57.	Bimberg	171,0	89,0	88,7 86,6	0,522 0,520	115 115	170,1 166,7
58.	Schröder	170,0	92,0	80,4 81,2	0,507 0,516	118 116	158,8 157,3
59.	Lippmann	166,0	88,5	77,8 77,3	0,507 0,477	118 126	153,2 161,8
60.	Knorr	164,0	88,0	81,0 79,5	0,506 0,483	119 124	160,0 164,7

Table 1 (continued)

61.	Riechel	163,0	89,0	80,5	0,517	116	155,7
				79,8	0,525	114	152,0
62.	Zwick	163,0	85,5	79,2	0,493	122	160,6
				79,3	0,488	123	162,0
63.	Bauer	161,5	85,0	78,4	0,468	128	167,5
				78,5	0,463	129	169,7
64.	Hentschke	159,5	86,0	83,3	0,472	127	176,7
				82,9	0,476	126	174,2
65.	Zimmermann ...	160,5	83,5	80,7	0,458	131	176,1
				80,0	0,468	128	171,2
66.	Ibener	160,5	81,5	77,0	0,446	135	172,4
				80,8	0,478	125	168,8
67.	Steinbach......	159,0	79,0	78,4	0,461	130	170,1
				78,4	0,505	119	155,2
68.	Winzer	157,5	81,5	89,4	0,517	116	172,8
				89,3	0,525	114	170,1
69.	Rost	155,5	79,0	80,7	0,454	132	177,6
				79,8	0,451	133	177,0
70.	LC Kluge.....	171,5	90,5	90,6	0,471	127	192,3
				90,4	0,451	133	200,6
71.	Heinecke	171,5	94,0	85,9	0,473	127	181,9
.				85,9	0,480	125	178,8
72.	Kettenbeil	169,0	93,0	88,7	0,503	119	176,3
				88,4	0,491	122	180,1
73.	LC Wiedner ...	171,0	92,0	84,9	0,481	125	176,8
				85,3	0,483	124	176,3
74.	LC Lupitz	172,0	92,5	88,1	0,512	117	172,0
				84,4	0,500	120	168,8
75.	Leber	165,0	84,0	83,8	0,476	126	176,1
				84,7	0,485	124	174,6
76.	Gentsch	169,5	89,0	87,9	0,495	121	177,2
				87,4	0,483	124	181,5
77.	Förster	168,5	90,5	88,4	0,519	116	170,2
				87,7	0,532	113	165,0
78.	Beesemann.....	166,5	90,5	88,9	0,526	114	169,2
				82,6	0,479	125	172,5
79.	Wutz	167,5	88,5	93,6	0,531	113	176,2
				92,9	0,503	119	184,7
80.	Schomantzki ...	164,5	85,5	80,7	0,473	127	170,3
				81,3	0,458	131	177,1
81.	Böttcher	167,5	89,5	86,3	0,483	124	178,5
				85,6	0,497	121	172,5
82.	LC Zahlaus....	165,0	88,0	83,3	0,483	124	172,5
				89,6	0,532	113	168,3
83.	LC Wieland ...	166,5	89,5	83,6	0,463	130	180,4
				82,8	0,463	130	178,5
84.	Näbe	159,5	86,0	80,2	0,466	129	172,1
				80,3	0,465	129	172,8

Table 1 (continued)

No.	Name	Body height (cm)	Leg length (cm)	Length of a single step (cm)	Duration of a single step (s)	No. of steps per minute	Average velocity of gait (cm s⁻¹)
85.	Pippich	165,5	89,5	85,9 / 85,3	0,493 / 0,460	122 / 130	174,6 / 184,8
86.	Bowenz	167,5	87,5	88,3 / 88,4	0,473 / 0,501	127 / 120	186,3 / 176,5
87.	Grunicke	163,5	87,5	83,6 / 82,2	0,402 / 0,453	150 / 132	208,0 / 181,5
88.	Putzenhardt. . . .	163,5	85,0	82,9 / 83,0	0,477 / 0,464	126 / 129	173,7 / 178,6
89.	Geissler	159,5	84,0	83,2 / 81,3	0,460 / 0,470	130 / 128	180,8 / 173,0
90.	Drummer Graf	164,0	86,0	84,9 / 84,9	0,477 / 0,473	126 / 127	178,0 / 179,5
91.	Bässler	165,5	90,0	86,8 / 86,3	0,497 / 0,490	121 / 122	174,6 / 176,0
92.	Drummer Pika. .	164,0	87,5	86,1 / 85,2	0,470 / 0,468	128 / 128	183,2 / 181,8
93.	Heinker	166,0	89,0	79,8 / 81,8	0,539 / 0,511	111 / 117	148,2 / 160,0
94.	Schneiderheinze .	165,5	86,5	89,2 / 88,4	0,501 / 0,483	120 / 124	177,8 / 182,8
95.	Boretzki	161,5	84,0	79,8 / 79,9	0,463 / 0,471	129 / 128	172,5 / 169,6
96.	Breitling	161,0	85,0	85,7 / 84,0	0,493 / 0,477	122 / 126	174,0 / 176,3
97.	Krause.	160,0	83,5	86,5 / 87,0	0,486 / 0,472	123 / 127	178,0 / 184,5
98.	Schwarz	159,5	82,5	83,4 / 83,2	0,461 / 0,455	130 / 132	181,0 / 182,5
99.	Hoppe	160,5	82,5	83,3 / 83,4	0,482 / 0,466	124 / 129	173,0 / 178,9
100.	Hoffmann	169,5	90,0	87,2 / 87,1	0,503 / 0,499	119 / 120	173,5 / 174,8
101.	Bürckmann	162,5	85,5	85,2 / 85,2	0,452 / 0,450	133 / 133	188,5 / 189,0
102.	Albrecht	162,5	84,5	88,0 / 87,4	0,494 / 0,499	122 / 120	178,3 / 175,0
103.	Voigt.	163,0	89,5	85,9 / 86,2	0,515 / 0,515	116 / 116	166,8 / 167,2

In all, 220 experiments on gait were carried out on 103 soldiers. The average length of the walker's step was over 80 cm. Only in 22 cases was it shorter than 80 cm. In 106 cases it was between 80 cm and 84.9 cm, in 74 cases between 85 cm and 89.9 cm. It reached 90 cm or more in 18 cases. In one case (no. 3) the length of one step was over 1 m. In this case, the subject was

Table 2. Control experiments to determine length and duration of the typical walker's step, carried out with students in October 1889

No.	N a m e	Body height (cm)	Leg length (cm)	Length of a single step (cm)	Duration of a single step (s)	No. of steps per minute	Average velocity of gait (cm s⁻¹)
104.	Findeisen	168,0	90,0	85,5 83,9	0,481 0,497	125 121	177,8 168,8
105.	Garten	166,5	87,5	76,4 80,0	0,503 0,456	119 132	151,8 175,4
106.	Richter	167,0	87,0	75,5 74,0	0,503 0,495	119 121	149,8 149,1
107.	Landwehr	174,0	96,0	86,7 86,6	0,507 0,512	118 117	171,0 169,1
108.	R. His	178,5	93,0	85,8 85,5	0,533 0,533	113 113	161,0 160,4
109.	Hentschel	174,0	92,0	87,0 83,2	0,526 0,466	114 129	165,4 178,5
110.	Braune .... (Schoolboy aged 20)	169,0	87,5	83,5 82,8	0,518 0,522	116 115	161,2 158,6
111.	Welzel	167,0	86,8	84,1 85,6	0,498 0,487	120 123	168,9 175,8

very tall (183 cm) with abnormally long legs (98 cm). The length of this subject's bare foot from heel to toe was 30 cm.

The duration of a step was less than 0.5 s in 134 cases, longer in 86 cases. In no case were there less than 105 steps a minute. In 3 cases there were 105–109 per minute; in 18 cases, 110–114; in 54 cases, 115–119; in 63 cases, 120–124; in 63 cases, 125–129; in 17 cases, 130–134; and in 2 cases more than 134. It thus appears that the soldiers did not adopt a certain pace for which they might have been trained on the drill ground. On average they walked faster than on the drill ground where 110–112 steps per minute are prescribed. The average in our 220 experiments was greater than 120.

It also appears that in walking freely out of doors the number of steps per minute is greater than during walking within a restricted space, for example.

To learn whether barrack-square drill had any noticeable effect on the constants of the walkers's step we carried out a series of experiments with a control group of students of about the same age as the soldiers. The results are given in Table 2.

This shows that the length of the walker's step is more than 80 cm on average, even in subjects with no military drill training. In smaller subjects it can be a little less than 80 cm, as we observed in the soldiers. The number of steps per minute varied in the students between 113 and 132. This is in keeping with the data of Table 1. For subjects 105 and 109 the number of steps per minute was not constant whereas for the other subjects this number varied only a little in the two different experiments carried out at different times. Such discrepancies also occurred in the soldiers; see, for instance, subjects 24, 50, 66, 78 and 87 in Table 1.

The experimental subject on whose gait our chronophotographic pictures described at length in Chap. 1 are based was 167 cm tall and had a leg length of 87 cm. In the three experiments used to establish the co-ordinates, the length of a single step was 77.9 cm, 76.8 cm and 71.9 cm. The duration of a step was 0.495 s, 0.485 s and 0.4947 s. This corresponds to 121.2, 123.7 and 121.3 steps per minute. The constants of the walker's step in this subject are thus in agreement with the results of the many gait experiments we have carried out on soldiers and students. There is, therefore, no justification for claiming that the step was too long and the duration of the step too short. On the contrary, the step is a little shorter than would be expected with a height of 167 cm. The length of the step in our subject was little different, by about 1 cm, in the first two experiments. In the third experiment, however, it was 6 cm shorter than in the first two. The explanation for this lies in the heavy regulation equipment the subject had to carry in the last experiment. The duration of a step is approximately the same in the three experiments. The average velocity of gait was calculated by dividing the length of the step in centimetres by the duration of the step in seconds, and it must thus be considerably smaller in the experiment with the load than in the others. In the three experiments it attained respectively $157.3 \, \text{cm} \, \text{s}^{-1}$, $158.3 \, \text{cm} \, \text{s}^{-1}$ and $145.3 \, \text{cm} \, \text{s}^{-1}$.

The kind and sequence of movements involved in the walker's step are the same in all subjects. The individual differences are only quantitative, resulting from the different sizes of the bones, the somewhat different shapes of the joints and the different distribution of the masses in the body. This claim can only be demonstrated by the same thorough examination of the walker's step in many normal subjects with different heights and builds as we have carried out in the three experiments on our subject. However, direct observation of people walking, comparison of our chronophotographic pictures with those of Marey and the results of our experiments on the gait of 111 different individuals make its validity appear highly probable. Above all, it can be considered a basic rule of the physiology of movement that, in normal circumstances, the same effect is always obtained with the same means. Even von Meyer is forced to admit that "in the majority of individuals the same principal basic rules are observed in the realization of walking"[11] although he denies the possibility of defining a typical kind of gait. It would be in flagrant contradiction of normal anatomy to assume that, despite the essential similarity of the construction and arangement of the bones, joints and muscles of the human body, one individual uses different joints and muscles than another for a long walk on a country road. The slight differences in the shape of the joint surfaces and in the insertion of the muscles are never of such a kind that they could result in different functioning of the joints or of the muscles. When the same forces act on a body in the same conditions, the same movement must necessarily occur. The movements carried out by the limbs of a person walking at a continuous walker's step are more than probably similar in different individuals even in details which direct observation does not reliably show. The conclusions that a precise knowledge of the movements involved in human gait allows on the forces that have been involved in the process must be generally and not just individually applicable since in the walker's step of another individual the same forces are involved in the same sequence in the process of movement, even if they vary somewhat in intensity in different individuals.

If it appears in one individual that the leg swings exactly like a pendulum during the walker's step, it can thus reliably be assumed that the leg behaves in the same way in other individuals during the same sort of walking. And, if it is demonstrated that other forces than gravity must act on the different segments of the leg to move it during the swinging period, this is also true for the movements of the walker's step in other individuals.

On the Forces Involved During the Swinging Period and How They Act on the Segments of the Leg

The kind of forces that can act on the different segments of the human body and their relationships with the movement of the latter during walking have been described in Chap. 3 and, for the special case of the foot, in Chap. 4. Reference will be made to this earlier work in the present chapter.

---

[11] Die Statik und Mechanik des menschlichen Knochengerüstes, p 305

During the swinging period the circumstances are particularly simple because the ground reaction force does not act directly on the swinging leg. At a first approximation, the swinging of the leg will be considered as a plane movement parallel to the median plane through the body. In these circumstances, only the following forces applied at the different segments of the leg can be made responsible for the swinging movement.

Gravity acts on the *thigh* as a force equal to the weight G_2[12] of the thigh, pulling vertically downward and applied at the centre of gravity S_2 of the thigh, as illustrated in Fig. 2. If muscles take part in the swinging movement, those which are directly inserted into the thigh act on the latter as forces equal to the tension of the muscles and always pull towards the next point of insertion. The point of application of this force is considered to be that point in the thigh from which the muscle is spanned to the adjacent segment over the connecting joint. In many instances the point of application of the muscular force is not equivalent to the insertion of the muscle into the femur or to the midpoint of the area of insertion. This has been commented on before[13]. In Fig. 2 no representative muscular forces applied directly to the thigh have been shown to keep the illustration simple.

Besides exerting these forces, both gravity and the muscles also act on the thigh in another way, by influencing the joint pressure in the hip and in the knee.

The continuously changing compression exerted on the thigh in the hip joint must be considered as a force applied at the centre of the femoral head at all times. Its magnitude and line of action are such that, in combination with the external forces applied directly to the leg, the compressive force gives the centre of gravity of the whole leg its acceleration at the instant considered. Consequently, the compressive force exerted in the hip joint can be definitely determined if all the external forces applied directly to the leg and the acceleration (magnitude and direction) of the total centre of gravity of the leg are known. The compressive force can be considered as the resultant of a series of components. One of these components is equal to and has the same direction as the effective force of the whole leg. The others are equal but opposite to all the external forces acting directly on the leg.

The effective force of the whole leg is a force equal in magnitude to the product of the mass of the leg and the acceleration of its centre of gravity. It has the same direction as the acceleration of the centre of gravity of the leg. The mass of the leg is designated as $m_{2,4,6}$ and the acceleration of its centre of gravity as $\gamma_{2,4,6}$. The indices mean that we deal with the system formed by the segments of the body numbered 2, 4 and 6. The magnitude of the effective force of the leg then is $m_{2,4,6} \gamma_{2,4,6}$. In Fig. 2 this compression component has a positive sign, meaning that it has the same direction as the effective force of the leg.

The external forces exerted directly on the leg are of two kinds. The weight $G_{2,4,6}$ of the leg pulls vertically downwards at the centre of gravity $S_{2,4,6}$ of the leg. In addition, the muscles inserted into the leg when contracting act as forces in the manner described above. Only muscles with their origin outside the leg, actually in the trunk, are taken into account. The muscles which have their origin and insertion in the leg itself develop forces which must be considered as internal forces for the leg. Consequently, one of the components of the compression in the hip joint is vertical upwards and equal to the weight of the leg. It is designated as $-G_{2,4,6}$ in Fig. 2. The other components of compression are equal and opposite to the muscular forces. They are not shown in Fig. 2.

Similarly, the continuously varying compression exerted on the thigh at the knee is a force applied at the centre of the knee at any time when the movement of the leg is assumed to be

---

[12]   The indices used here and below indicate the numbering of the segments of the right lower limb. An index 2 indicates the right thigh, an index 4 the right lower leg, and an index 6 the right foot. The counterparts on the left are numbered 3, 5 and 7

[13]   cf., in particular, my article on: Das statische und das kinetische Maß für die Wirkung eines Muskels (The statical and kinetic measure of the effect of a muscle). Abhandlungen der math.-phys. Klasse der Königl. Sächs. Gesellschaft der Wissenschaften, vol 27, no 5 (1902)

plane. The centre of the knee is the point of intersection of the axis of the knee with the long axis of the thigh which passes through the centre of the hip joint and the centre of gravity of the thigh. This compression is equal to that exerted on the lower leg in the knee, but its direction is opposite. The compressive force exerted on the thigh in the knee is made up of (1) a component equal to the effective force $m_{4,6}\,\gamma_{4,6}$ of the lower leg + foot system but opposite in direction; (2) a component equal to the weight $G_{4,6}$ of the lower leg + foot system and also directed vertically downwards (Fig. 2); and (3) a number of components equal to the external muscular forces exerted directly on the lower leg and foot with the same direction as these forces. External muscular forces can only be exerted by muscles that have their insertions only in the lower leg and foot and their origins outside this system, i.e. in the thigh or in the trunk. Figure 2 shows only the first two components of compression in the knee joint. The components due to the muscles have again not been drawn. The weight component is designated as $+G_{4,6}$. This means that it is equal to and acts in the same direction as the weight of the lower leg + foot system. The compressive component due to the effective force of this system has a direction opposite to this effective force. It is thus a reversed effective force designated as $-m_{4,6}\,\gamma_{4,6}$.

All the forces which can act on the thigh during the swinging period have thus been listed. The thigh will move under their action as if it were completely free, separated from the

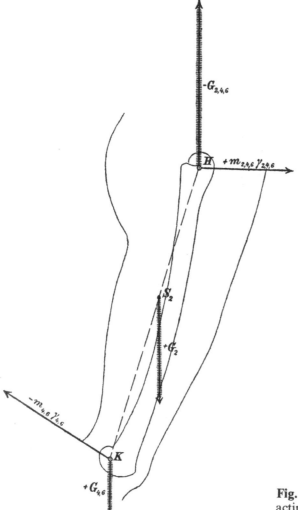

Fig. 2. Forces of gravity and effective forces acting on the thigh during the swinging period of gait

remainder of the body. The influence exerted on the moving thigh by the other parts of the body is expressed by the two compressive forces in the hip and knee joint. When the joint compression is taken into account the thigh can be considered as a free body. The forces thus mentioned represent the external forces acting on the thigh. The centre of gravity of a free body moves as if all the external forces acting on the body were applied at its centre of gravity with their magnitude and direction and as if the mass of the body were concentrated at this centre of gravity. The forces drawn in Fig. 2 combined with the muscular forces thus mentioned are displaced to the centre of gravity S_2 of the thigh and give the resultant $m_2 \gamma_2$ in which m_2 represents the mass and γ_2 the acceleration of the centre of gravity. $m_2 \gamma_2$ is the effective force of the thigh. This is easily demonstrated. The resultant of the components of compression $+ m_{2,4,6} \gamma_{2,4,6}$ in the hip H and $- m_{4,6} \gamma_{4,6}$ in the knee K is $m_2 \gamma_2$. This results from the fact that the effective force $m_{2,4,6} \gamma_{2,4,6}$ of the whole leg is the resultant of the three effective forces $m_2 \gamma_2$, $m_4 \gamma_4$ and $m_6 \gamma_6$ of the thigh, the lower leg and the foot, and that the effective force $m_{4,6} \gamma_{4,6}$ of the lower leg + foot system is the resultant of $m_4 \gamma_4$ and $m_6 \gamma_6$. Therefore, addition of $m_{2,4,6} \gamma_{2,4,6}$ and $- m_{4,6} \gamma_{4,6}$ gives the effective force $m_2 \gamma_2$. All the other forces acting on the thigh can thus exert no influence on the movement of the centre of gravity. Indeed, for each of the other forces there is one which is equal and opposite so that these forces form a series of couples.

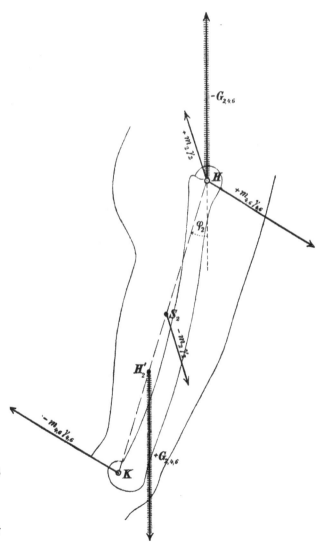

Fig. 3. Couples formed by the forces of gravity and effective forces acting on the thigh during the swinging period of gait if the forward movement of the thigh is prevented by a reversed effective force $- m_2 \gamma_2$

As far as the forces due to gravity and applied at H, S_2 and K are concerned, the last two act vertically downwards and can be combined into a resultant $G_{2,4,6}$ which is equal and opposite to the component of compression $-G_{2,4,6}$ acting on the hip H. The point of application H_2' of this resultant lies on the connecting line $S_2 K$ and divides this line in inverse proportion to the two gravity forces (Fig. 3). This point has been met before as the partial principal point of the thigh. It can also be regarded as the centre of gravity of the system of masses which is obtained by concentrating the masses of the lower leg and of the foot at the centre of the knee and by adding this mass to that of the thigh. This principal point is called partial because, in its construction, it is assumed that the thigh is separated from the trunk. If the thigh remained attached to the trunk it would be necessary to concentrate the masses of the trunk, of the opposite leg, of the two arms, and of the head at the centre of the hip and add this mass to that of the thigh. The centre of gravity of this articulated system of masses would then be the total principal point of the thigh. The force of gravity $+G_{2,4,6}$ applied at H_2' forms a couple with the component of compression $-G_{2,4,6}$ directed vertically upwards and applied at the centre H of the hip joint. This couple tends to rotate the thigh but cannot affect the acceleration of its centre of gravity. The moment of this couple represents the torque of gravity acting on the thigh during the swinging period. It will be designated as D_{S_2} and taken as positive when it tends to rotate the thigh counter-clockwise as seen from the right. It must be noted that the couple exerted by gravity on the thigh can be calculated in the manner described only during the swinging period, when the leg is not in contact with the ground. If the leg is in contact with the ground, the components of compression due to gravity acting at the knee and at the hip must be dealt with as explained in Chap. 3 (p. 214). There is another way to find the resultant torque of gravity during the swinging period, which is also explained in Chap. 3. This torque can be considered as the moment of a couple in which each force is equal to the total weight G_0 of the human body. One force is directed vertically upwards and is applied at the centre of the hip. The other is directed vertically downwards and is applied at the total principal point of the thigh. The moment of this couple is of course exactly the same as that of the couple obtained previously. The distance of the partial principal point from the centre of the hip is in the same proportion to the distance of the total principal point from the centre of the hip as the weight of the whole body is to the weight of the leg. This results from the definition of the principal points (Chap. 2, p. 135 ff.). The ratio of the horizontal moment arms of the two couples to each other is also the same as that of the distances, on the long axis of the thigh, of the two principal points from the centre of the hip to each other. Consequently, the moments of the two couples must be equal.

Similar to the gravity forces, the muscular forces directly applied to the thigh can be combined with the muscular components of compression in the hip and knee to give couples. One couple corresponds to each muscle as far as this muscle can be considered as a physiological unit. Each muscular force applied directly to the thigh has its counterpart equal and opposite to it among the components of compression in one of the two joints. This component of compression originates from the same muscle. If both the origin and insertion of a muscle lie outside the thigh so that the muscle spans the two joints, the force of the couple resulting from the proximal insertion acts at the centre of the hip; the other force is equal and opposite and acts at the centre of the knee. This has been explained and substantiated previously[14] with special reference to the muscles of the thigh[15]. However, it must be borne in mind that the nature of the couples through which the muscles tend to rotate the thigh does not depend on the temporary conditions for the movement of the thigh. Their nature is the

---

[14] Über die Drehungsmomente ein- und mehrgelenkiger Muskeln. Archiv für Anatomie und Physiologie Anat. Abt. (1894)

[15] Das statische und kinetische Maß für die Wirkung eines Muskels. pp 518 ff

same during the swinging period and during the stance period. This constitutes an essential difference from the couples by way of which the force of gravity is exerted on the thigh. If the movement is plane, only those components of the muscular couples parallel to the median plane through the body are acting. It only remains to distinguish the couples which tend to rotate the thigh counter-clockwise, as seen from the right, from those which tend to rotate clockwise. If the moment of the former is regarded as positive, that of the latter must be negative. The resultant moment of all the couples due to the muscles is equal to the algebraic sum of the moments of the different couples. It will be designated as D_{m_2}. Since all the muscular forces act in pairs, one positive, one negative, they cannot exert any influence on the movement of the centre of gravity of the thigh.

Two components of compression remain. One acts at the centre of the hip and is equal to the effective force of the whole leg. The other acts at the centre of the knee and is equal and opposite to the effective force of the lower leg + foot system. They can be combined into one resultant which, as shown above, can give the centre of gravity of the thigh its acceleration (magnitude and direction).

These two effective forces acting as compressive forces generally also exert a rotatory effect on the thigh. This is best visualized by imagining an additional force at the centre of gravity of the thigh, preventing the forward movement of the thigh without having any effect on its rotation about its centre of gravity. This additional force must be equal and opposite to the effective force of the thigh. It is drawn in Fig. 3 and labelled $-m_2\gamma_2$ in keeping with its direction. Rotation of the thigh is not disturbed by this force. It must thus be combined with the two effective forces applied at H and K to give couples so that no single force is left. This can be achieved in various ways. For instance, the force $-m_2\gamma_2$ could be resolved into two components $-m_{2,4,6}\gamma_{2,4,6}$ and $+m_{4,6}\gamma_{4,6}$. The former is equal and opposite to the effective force acting at the hip, the latter is equal and opposite to that acting at the knee (Fig. 2). Then the two couples $[+m_{2,4,6}\gamma_{2,4,6}, -m_{2,4,6}\gamma_{2,4,6}]$ and $[+m_{4,6}\gamma_{4,6}, -m_{4,6}\gamma_{4,6}]$ appear. As indicated in Fig. 3, the effective force $+m_{2,4,6}\gamma_{2,4,6}$ acting at the centre of the hip can also be resolved into its two components $+m_{4,6}\gamma_{4,6}$ and $+m_2\gamma_2$. This again gives two couples. One consists of the two forces acting at the centre of the hip and at that of the knee $+m_{4,6}\gamma_{4,6}$ and $-m_{4,6}\gamma_{4,6}$. The other is made up of the force $+m_2\gamma_2$ acting at the centre of the hip and the force $-m_2\gamma_2$ added at the centre of gravity of the thigh. This last representation of the rotatory effect of the effective force is more appropriate to the purpose of our present research which consists in determining the forces acting during the swinging period. With this approach, besides the acceleration γ_2 of the centre of gravity of the thigh, only the acceleration of the centre of gravity of the lower leg + foot system $\gamma_{4,6}$ has to be considered, and not that of the centre of gravity of the whole leg. If the movement is plane, the resultant torque of the effective forces, designated as D_{e_2} for the thigh, equals the algebraic sum of the moments of the two couples. Again, the couple tending to rotate the thigh counter-clockwise, as seen from the right, is counted as positive, and that tending to rotate the thigh clockwise as negative.

The situation in the *lower leg* is similar to that in the thigh. Gravity acts on the centre of gravity S_4 of the lower leg directly with a force G_4 directed vertically downwards and equal to the weight of the lower leg (cf. Fig. 4). Each muscle inserted directly in the lower leg, when tensed, acts with a force directed towards its other attachment. This force is applied to that point of the lower leg from which the muscle is spanned freely over the knee or over the ankle. None of these muscular forces is represented in Fig. 4. Besides these two kinds of forces applied directly to the lower leg, the movement of the lower leg is influenced only by the continuously varying compression in the knee and in the ankle joint.

The compression in the knee acts on the lower leg as a force applied to the centre of the knee joint, equal and opposite to the compression exerted on the thigh in the knee, as indicated above. This compression, according to the explanations above, results from three kinds of components. One equals the effective force $m_{4,6}\gamma_{4,6}$ of the lower leg + foot system and has the same direction. A second component equals the weight $G_{4,6}$ of this system and pulls vertically upwards. These two components are represented in Fig. 4 with their direction by $+m_{4,6}\gamma_{4,6}$ and $-G_{4,6}$. All the other components are muscular forces (not represented in Fig. 4) equal and opposite to the muscular forces acting directly on the lower leg + foot system.

The compression exerted on the lower leg at the centre of the ankle also results from three kinds of components. One is equal and opposite to the effective force $m_6\gamma_6$ of the foot. It is labelled $-m_6\gamma_6$ in Fig. 4. A second component is equal to the weight G_6 of the foot and pulls vertically downwards ($+G_6$ in Fig. 4). The other forces are muscular. They equal the muscular forces acting directly on the foot and also have the same direction.

All these facts are easy to understand and deduce if it is borne in mind that the compression exerted on the lower leg in the knee must be made up in such a way that in combination with the external forces applied directly to the lower leg + foot system, it imparts to the centre of the system its acceleration $\gamma_{4,6}$. In addition, the compression exerted on the lower leg in the ankle must be equal and opposite to the compression exerted on the foot in the ankle. The latter compression combined with the forces acting directly on the foot must impart to the centre of gravity of the foot its acceleration γ_6.

The two forces of gravity acting at the centre of gravity S_4 of the lower leg and at the centre F of the ankle can be combined into a resultant force pulling vertically downwards, equal in magnitude to the weight $G_{4,6}$ of the lower leg + foot system. Its point of application lies on the connecting line $S_4 F$ and divides this line in inverse proportion to the two weights G_4 and G_6. This point H_4'' in Fig. 5 represents the partial principal point of the lower leg. It is related to the lower leg + foot system considered as if this system were separated from the rest of the body at the knee, in contrast to the partial principal point of the thigh H_2' which is referred to the whole leg. Therefore, this partial principal point is noted with a double dash. The force of gravity acting at H_4'', shown in Fig. 5, again forms a couple with the component of compression $-G_{4,6}$ acting at K. Its moment D_{S_4} represents the torque of gravity exerted on the lower leg during the swinging period.

The muscular forces applied directly to the lower leg, like the forces of gravity, can also be combined with the muscular components of compression in the knee and in the ankle into as many couples as there are muscles acting on the lower leg during the swinging period. Some of these couples tend to rotate the lower leg counter-clockwise, as seen from the right, and

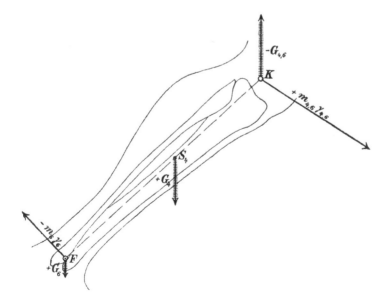

Fig. 4. Forces of gravity and effective forces acting on the lower leg during the swinging period of gait

others to rotate the lower leg clockwise. The moments of the former couples are counted as positive, those of the latter as negative. The algebraic sum of all these moments represents the resultant torque D_{m_4} of the muscles of the lower leg.

Both the forces of gravity acting directly and indirectly on the lower leg and the muscular forces applied to it directly and indirectly combine into couples. Therefore, they cannot exert any influence on the movement of the centre of gravity of the lower leg. This movement can be attributed only to the components of compression $+m_{4,6}\gamma_{4,6}$ and $-m_6\gamma_6$ corresponding to the effective forces and acting at K and F. In fact, these two forces, displaced to the centre of gravity S_4 of the lower leg, combine to give the resultant $+m_4\gamma_4$ which is necessary for the acceleration γ_4 of the centre of gravity of the lower leg. If the force $-m_4\gamma_4$ which is equal and opposite to the effective force $m_4\gamma_4$ is applied at S_4, only the movement of the centre of gravity itself is prevented whereas the rotation of the lower leg about this centre of gravity is not restricted at all. This additional force must thus be combined with the forces $+m_{4,6}\gamma_{4,6}$ and $-m_6\gamma_6$ applied at K and F to give couples. This can come about in various ways. For instance, the force $-m_4\gamma_4$ acting at S_4 can be replaced by its two components $-m_{4,6}\gamma_{4,6}$ and $+m_6\gamma_6$. Then the force $+m_{4,6}\gamma_{4,6}$ acting at K forms a couple with the component $-m_{4,6}\gamma_{4,6}$ acting at S_4, and the component $+m_6\gamma_6$ acting at S_4 forms another couple with the force $-m_6\gamma_6$ acting at F. These two couples illustrate the rotatory effect of the three forces. As in Fig. 5, however, it is also possible to replace the force at K by its components $+m_4\gamma_4$ and $+m_6\gamma_6$. Then $+m_4\gamma_4$ forms a couple with the force $-m_4\gamma_4$ acting at S_4; $+m_6\gamma_6$ forms the other couple with the force $-m_6\gamma_6$ acting at F. These two couples, of course, have exactly the same total effect as the first two. The moment is counted as positive or negative depending on whether a couple tends to rotate the lower leg counter-clockwise or clockwise, as seen from the right. The resultant torque D_{e_4} of the effective forces of the lower leg equals the algebraic sum of the moments of the two couples.

The situation is simplest at the *foot*. In view of all the data collected for all the cases in Chap. 4, only the following forces need be considered with specific reference to the swinging period. The weight G_6 of the foot pulls vertically downwards at its centre of gravity S_6 (Fig. 6). The muscles inserted in the foot can also exert forces pulling towards their origin in the lower leg or in the thigh. Apart from these there is only a compression force active at the centre of the ankle F. $-G_6$ and $+m_6\gamma_6$ are among the components of this compression. The former is directed vertically upwards and the latter acts in the direction of the acceleration of

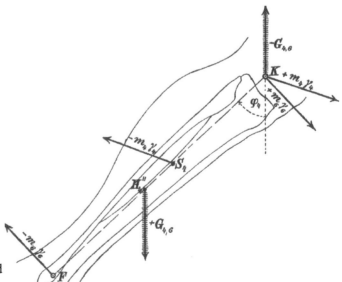

Fig. 5. Couples formed by the forces of gravity and effective forces acting on the lower leg during the swinging period of gait if the forward movement of the lower leg is prevented by a reversed effective force $-m_4\gamma_4$

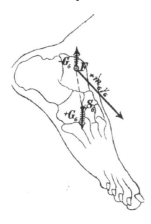

Fig. 6. Forces of gravity and effective forces acting on the foot during the swinging period of gait

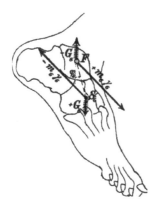

Fig. 7. Couples formed by the forces of gravity and effective forces acting on the foot during the swinging period of gait if the forwards movement of the foot is prevented by a reversed effective force $-m_6\gamma_6$

the foot γ_6. The other components of compression at F are equal and opposite to the muscular forces directly applied to the foot.

The two forces due to gravity and acting at F and at S_6 form a couple; its moment is the torque of gravity D_{s_6}. All the muscular forces acting on the foot also combine into couples, the algebraic sum of their moments being the resultant torque D_{m_6} of the muscles for the foot. The remaining component of compression $+m_6\gamma_6$ acting at F alone causes the centre of gravity of the foot S_6 to move. If this movement is prevented by a force $-m_6\gamma_6$ (Fig. 7) equal and opposite to $+m_6\gamma_6$ and applied at S_6, the two forces acting at F and S_6 form a couple depending on the effective forces. Its moment is the torque D_{e_6} of the effective forces acting on the foot.

The angular accelerations with which the thigh, lower leg and foot rotate about their centres of gravity S_2, S_4 and S_6 during the swinging period, if the movement of the leg is regarded as plane, will be designated as φ_2'', φ_4'' and φ_6''. The radii of inertia[16] of the three segments of the leg in relation to the axes through their centres of gravity and perpendicular to the median plane in the normal attitude will be called $\varkappa_2$, $\varkappa_4$ and $\varkappa_6$. As seen in a previous work[16], they represent the principal radii of inertia. The following relationships exist between these magnitudes and the torques acting on the three segments of the right leg[17] (movement equations).

---

[16] cf. Abhandlungen der mathematisch-physischen Klasse der Königlichen Sächsischen Gesellschaft der Wissenschaften, vol 18, no 8
[17] Chap. 4, p 272

$$m_2 \varkappa_2^2 \cdot \varphi_2'' = D_{m_2} + D_{s_2} + D_{e_2} \qquad \text{(right thigh)}$$

$$m_4 \varkappa_4^2 \cdot \varphi_4'' = D_{m_4} + D_{s_4} + D_{e_4} \qquad \text{(right lower leg)}$$

$$m_6 \varkappa_6^2 \cdot \varphi_6'' = D_{m_6} + D_{s_6} + D_{e_6} \qquad \text{(right foot)}$$

The three segments of the left leg are indicated by the indices 3, 5 and 7. For the swinging period, the following equations pertain:

$$m_3 \varkappa_3^2 \cdot \varphi'' = D_{m_3} + D_{s_3} + D_{e_3} \qquad \text{(left thigh)}$$

$$m_5 \varkappa_5^2 \cdot \varphi'' = D_{m_5} + D_{s_5} + D_{e_5} \qquad \text{(left lower leg)}$$

$$m_7 \varkappa_7^2 \cdot \varphi'' = D_{m_7} + D_{s_7} + D_{e_7} \qquad \text{(left foot)}$$

These are the movement equations for the three main segments of the leg in the form that seems the most appropriate to determine whether the swinging of the leg during gait is purely that of a pendulum or whether significant muscular actions are involved. Of the four variables φ'', D_m, D_s and D_e that are present in each equation, only three can be determined empirically for each instant of the swinging movement: the angular acceleration φ'', the torque D_s of gravity and the resultant torque D_e of the effective forces. The equation can thus be used to calculate the resultant torque D_m of the muscles acting on the segment of the body under consideration. This resultant torque cannot be determined empirically. From the equations it is apparent that

$$D_{m_i} = m_i \varkappa_i^2 \varphi_i'' - D_{s_i} - D_{e_i}$$

This equation applies for any segment of the right or of the left leg, regardless of the i index of the element.

To determine the angular acceleration φ'' at the different moments of the swinging of the leg, it is first necessary to find the angle φ between the segment under consideration and the vertical in the successive phases of movement. This has been done for all three segments of the leg and the results are given in Chap. 3. From these values of φ a graph can be constructed from which it is possible to derive the angular velocities φ' from which in turn the angular accelerations φ'' for the whole of the swinging period can be derived. The results can be found in the next section.

The empirical bases for determination of the torque D_s of gravity are available. The weights and lengths of the three segments of the leg are known so that the position of the partial principal points H_2' of the thigh (Fig. 3) and H_4'' of the lower leg can easily be found. The values of the φ-angle are also known for the three segments. Hence, the moment arms of the couples due to gravity can be determined. Since the weights always act vertically, these moment arms are horizontal. For the thigh they are the product of the distance HH_2' and $\sin \varphi_2$; for the lower leg the product of the distance KH_4'' and $\sin \varphi_4$; and for the foot the product of the distances FS_6 and $\sin \varphi_6$. This is clearly seen in Figs. 3, 5 and 7.

In keeping with the explanations above, determination of the torque D_e requires that the accelerations of the centres of gravity of the three segments of the leg and of the centre of gravity of the lower leg + foot system are known. However, it is sufficient to determine the accelerations of the centres of gravity of the thigh, lower leg and foot empirically since that of the centre of gravity of the lower leg + foot system can be calculated from them. The co-ordinates of the different centres of gravity given in Tables 1, 2 and 6 in Chap. 2 can be used to determine the accelerations of the centres of gravity by means of graphs, as has been explained in Chaps. 2 and 4. This yields the components of the accelerations in the directions of the axes of co-ordinates. The swinging movement has been considered as plane, as a first approximation, to determine the torques acting on the leg. Therefore, the transverse y-co-ordinates are not needed. The analysis can be restricted to the determination of the components of velocity and of acceleration in the direction of gait and in the vertical direction. Corresponding to the resolution of each acceleration into two components, each of the couples of the effective forces involved in the thigh and in the lower leg can also be resolved into two further couples. This is useful to determine the resultant torque D_e because then the moment arms of all the couples are either vertical or horizontal and can be easily calculated with the aid of the φ-angles.

It can be seen from the explanation above that determination of the angular accelerations of the rotations of the thigh, lower leg and foot about their centres of gravity and the components of the linear accelerations of these centres of gravity provides an adequate basis for the kinetic analysis of the swinging movement of the leg. The first of these determinations will be carried out in the next section and the second in the fourth section of this chapter.

Angular Velocities and Angular Accelerations of the Rotation of the Three Segments of the Leg During the Swinging Period

Before the angular velocities and accelerations of the segments of the leg can be determined, the angles φ_2, φ_4 and φ_6 (and also φ_3, φ_5 and φ_7) formed by the long axes of these segments and the vertical must be expressed in radians. The radian is the angle which defines on a circle an arc that is equal to the radius of the circle drawn with the apex of this angle as its centre. If the radius is selected equal to the unit of length, the radian is the angle which defines an arc equal to the unit of length on a circle with the radius equal to the unit of length. Then the angle of the arc coincides with the length of this arc. The length of a half circle is π. Therefore, π usually means an angle of 180° and an angle of 1 radian is equal to $180°/\pi = 57.296°$. On the other hand, an angle of 1° is equal to $\pi/180 = 0.017453$ radian and an angle $\alpha°$ equals $0.017453\,\alpha$ radian. The number of degrees in an angle must thus be multiplied by 0.017453 to express this angle in radians.

This has been done for the φ-angles occurring during the swinging period. All our chronophotographic exposures recorded the swinging of only the left leg during a double step from beginning to end. Only parts of the swinging of the right leg appeared. A series of exposures always began with the swinging of the right leg so that generally only a little more than the second half of this period was recorded. However, the exposures also extended to a part of the next right swinging period. This was long enough to complete the part of the previous swinging period recorded. This is obvious since the exposures always covered rather more than one double step. This allows examination of the whole swing period of the right leg also. It is only necessary to combine the initial exposures of the movement showing the second half of the swing, with the last exposures showing the first half of the swing: the last exposures are displaced back by the duration of a double step. The two parts thus obtained partly overlap each other, and this shows the extent to which the second swinging movement is a precise repetition of the first. Perfect superimposition of the two parts can of course hardly be expected. The smallest irregularity of the floor, for instance, is bound to provoke discrepancies in the vertical co-ordinates of the different points of the leg. These will obviously be perceptible as a consequence of the high degree of accuracy achieved in determination of the co-ordinates. However, the differences in the phases of the movement separated by the duration of a double step are small as appears in the upper graphs in Figs. 8–10.

In experiment I, swinging of the right leg began at the 25th phase (S_r)[18], i.e. 0.929 s after phase 1. Right heel strike occurred in the 10th phase (R), 0.345 s after phase 1. If the duration of a double step 0.990 s is added, heel strike occurs at 1.335 s. The right swinging period thus

---

[18] cf. Chap. 2, Tables 13–15 on p 169 and Table 7 on p 157

lasts 0.415 s. In experiment I, the left leg began to swing (S_l) between the 12th and 13th phases, at 0.426 s. Left heel strike (L) occurred in the 23th phase, at 0.843 s. The left swinging period thus lasted 0.417 s. The difference between the two swinging periods is only 0.002 s. Therefore, the duration of the right swinging period can be considered as equal to that of the left.

Table 3 gives the φ-angles related to the swing period in radians. They are calculated from Table 2 in Chap. 3 (pp. 236–237). Where the beginning or the end of this period does not exactly coincide with one of the 31 phases given by the exposures, the next phase beyond the swinging period is given. This makes it possible to find the beginning or end phase by interpolation. The times are also indicated, according to Table 13 in Chap. 2 (p. 169). For the right leg, the times corresponding to the phases 1–10 have been increased by the duration of a double step, 0.990 s, and thus displaced to the next swing period.

In experiment II, S_r occurred in the 26th phase at 0.958 s and R between the 11th and the 12th phases at 0.396 s. Addition of the duration of a double step, 0.970 s in experiment II, to the time at which R occurred gives 1.366 s. The swinging of the right leg thus lasted 0.408 s. The left swinging period started between the 13th and the 14th phases, at 0.473 s, and its end coincided with the 24th phase, at 0.882 s. It lasted 0.409 s.

Table 4 gives the φ-angles in radians for the phases occurring during the swinging periods and the times. For the phases 1–12 of the right swinging period the times have been increased by 0.970 s and the phases thus displaced to the next right swing period.

In experiment III, finally, the right swinging period began shortly before the 24th phase, at 0.878 s, and ended shortly before the 9th phase, at 0.303 s. A double step lasted 0.9895 s. Therefore, the end of the last swing period would have occurred at 1.2925 s. The duration of the right swing period thus amounted to 0.4145 s. The beginning of the left swinging period coincided with the 11th phase, at 0.383 s. Its end fell between the 21st and 22nd phases, at 0.796 s. Its duration was thus 0.413 s.

Table 5 gives the φ-angles during the swinging periods in radians and the times. The times of phases 1–9 have been increased by 0.989 s and the phases thus displaced to the next right swing period.

With a view to finding the angular velocities and accelerations, the data in Tables 3–5 have been used to construct the graphs illustrating the interdependence of the φ-angles and the time during the swinging period. The values entered along the abscissa represent time: 1 s is represented by 10 cm. Consequently, 0.01 s is represented by 1 mm. The values entered along the ordinate are the magnitudes of the φ-angle. The angle of 1 radian is represented by 10 cm. Therefore, an ordinate value of 1 mm corresponds to an angle of 0.01 radian, or about 0.573°.

All the graphs relating to the same segment of the leg but belonging to different experiments are displayed in the same figure to permit comparison of the results. Figure 8 shows the φ-angles of the thigh, Fig. 9 those of the lower leg, and Fig. 10 those of the foot. The left part of each of these figures relates to the segments of the right extremity and the right part to those of the left extremity so that the graphs relating to the same segment and to its counterpart in the three experiments are close to each other. All the graphs are bounded by vertical lines marked S_r and R or S_l and L. They indicate the beginning and the end of the swinging period, either right or left, as for the displacement curves in Chaps. 2 and 4. Underneath the marked verticals, the time is given. Time is measured from the first phase of the movement in each experiment, which is the zero point. The other figures below, 5 mm apart, indicate the number of hundredths of a second that have elapsed since the beginning of the swinging period, i.e. from S_r or S_l. This makes comparison of the six curves for the different experiments and for the two sides of the body easier. Since the duration of the swinging period is approximately the same in the six cases, two points on the ordinate with the same number in any of the six groups of graphs relate to the same phase of the swinging period.

Table 3. Values of the φ-angle in radians during the swinging period in experiment I

No.	Time (s)	Right thigh φ_2	Right lower leg φ_4	Right foot φ_6	No.	Time (s)	Left thigh φ_3	Left lower leg φ_5	Left foot φ_7
25	0,920	− 0,289	− 0,829	− 0,251	12	0,422	− 0,333	− 0,777	− 0,232
26	0,958	− 0,135	− 0,978	− 0,407	13	0,460	− 0,199	− 0,908	− 0,347
27	0,997	− 0,003	− 0,990	− 0,383	14	0,498	− 0,063	− 0,927	− 0,333
28	1,035	+ 0,117	− 0,894	− 0,236	15	0,537	+ 0,071	− 0,819	− 0,162
29	1,073	+ 0,214	− 0,744	− 0,026	16	0,575	+ 0,159	− 0,675	+ 0,017
30	1,112	+ 0,310	− 0,551	+ 0,213	17	0,613	+ 0,281	− 0,467	+ 0,270
31	1,150	+ 0,381	− 0,320	+ 0,453	18	0,652	+ 0,360	− 0,257	+ 0,512
					19	0,690	+ 0,409	− 0,008	+ 0,757
1	0,990	− 0,006	− 0,997	− 0,405	20	0,728	+ 0,407	+ 0,266	+ 1,013
2	1,028	+ 0,123	− 0,910	− 0,258	21	0,767	+ 0,376	+ 0,492	+ 1,235
3	1,067	+ 0,241	− 0,749	− 0,038	22	0,805	+ 0,388	+ 0,516	+ 1,260
4	1,105	+ 0,334	− 0,545	+ 0,193	23	0,843	+ 0,447	+ 0,358	+ 1,145
5	1,143	+ 0,408	− 0,306	+ 0,458					
6	1,182	+ 0,456	− 0,049	+ 0,695					
7	1,220	+ 0,466	+ 0,223	+ 0,927					
8	1,258	+ 0,464	+ 0,460	+ 1,141					
9	1,297	+ 0,483	+ 0,491	+ 1,178					
10	1,335	+ 0,529	+ 0,343	+ 1,082					

Table 4. Values of the φ-angle in radians during the swinging period in experiment II

No.	Time (s)	Right thigh φ_2	Right lower leg φ_4	Right foot φ_6	No.	Time (s)	Left thigh φ_3	Left lower leg φ_5	Left foot φ_7
26	0,958	− 0,243	− 0,852	− 0,287	13	0,460	− 0,326	− 0,740	− 0.146
27	0,997	− 0,090	− 0,994	− 0,401	14	0,498	− 0,202	− 0,882	− 0,313
28	1,035	+ 0,054	− 0,995	− 0,380	15	0,537	− 0,044	− 0,937	− 0,352
29	1,073	+ 0,168	− 0,895	− 0,231	16	0,575	+ 0,084	− 0,850	− 0,214
30	1,112	+ 0,266	− 0,736	− 0,015	17	0,613	+ 0,195	− 0,693	− 0,013
31	1,150	+ 0,356	− 0,520	+ 0,223	18	0,652	+ 0,295	− 0,497	+ 0,218
					19	0,690	+ 0,374	− 0,269	+ 0,485
1	0,970	− 0,213	− 0,889	− 0,352	20	0,728	+ 0,418	− 0,017	+ 0,742
2	1,008	− 0,105	− 0,987	− 0,399	21	0,767	+ 0,412	+ 0,242	+ 1,004
3	1,047	+ 0,023	− 0,952	− 0,338	22	0,805	+ 0,377	+ 0,471	+ 1,212
4	1,085	+ 0,164	− 0,825	− 0,151	23	0,843	+ 0,393	+ 0,502	+ 1,273
5	1,123	+ 0,275	− 0,651	+ 0,068	24	0,882	+ 0,450	+ 0,352	+ 1,148
6	1,162	+ 0,371	− 0,430	+ 0,312					
7	1,200	+ 0,431	− 0,171	+ 0,576					
8	1,238	+ 0,451	+ 0,079	+ 0,801					
9	1,277	+ 0,445	+ 0,374	+ 1,041					
10	1,315	+ 0,448	+ 0,509	+ 1,175					
11	1,353	+ 0,509	+ 0,388	+ 1,086					
12	1,392	+ 0,518	+ 0,278	+ 0,942					

To draw the curves the points in the system of co-ordinates which correspond to the values of the φ-angle in Tables 3–5 and to the time were first marked. Then a thin line was drawn connecting these points. This curve was smoothed to avoid irregularities. For the segments of the right extremity, where the points belong to two parts separated by a time interval of a double step and overlap to some extent, the curve was drawn as precisely as possible in the middle between points which do not exactly coincide but are very close to each other.

Table 5. Values of the φ-angle in radians during the swinging period in experiment III

No.	Time (s)	Right thigh φ_2	Right lower leg φ_4	Right foot φ_6	No.	Time (s)	Left thigh φ_3	Left lower leg φ_5	Left foot φ_7
23	0,843	−0,441	−0,544	+0,074	11	0,383	−0,251	−0,840	−0,215
24	0,882	−0,344	−0,691	−0,219	12	0,422	−0,108	−0,980	−0,312
25	0,920	−0,203	−0,818	−0,333	13	0,460	+0,034	−0,984	−0,267
26	0,958	−0,062	−0,823	−0,310	14	0,498	+0,150	−0,883	−0,137
27	0,997	+0,051	−0,723	−0,163	15	0,537	+0,260	−0,724	+0,062
28	1,035	+0,152	−0,577	+0,022	16	0,575	+0,347	−0,520	+0,308
29	1,073	+0,239	−0,381	+0,251	17	0,613	+0,403	−0,283	+0,554
30	1,112	+0,302	−0,152	+0,481	18	0,652	+0,427	−0,028	+0,819
31	1,150	+0,339	+0,094	+0,728	19	0,690	+0,412	+0,247	+1,092
					20	0,728	+0,369	+0,455	+1,268
1	0,989s	+0,028	−0,767	−0,177	21	0,767	+0,371	+0,459	+1,287
2	1,027s	+0,143	−0,599	+0,038	22	0,805	+0,390	+0,353	+1,159
3	1,066s	+0,244	−0,391	+0,290					
4	1,104s	+0,310	−0,158	+0,531					
5	1,142s	+0,337	+0,100	+0,768					
6	1,181s	+0,326	+0,358	+0,998					
7	1,219s	+0,303	+0,551	+1,191					
8	1,257s	+0,340	+0,536	+1,189					
9	1,296s	+0,391	+0,411	+1,042					

The angular velocities were determined by measuring the tangents to a sufficient number of points on the curves for the φ-angles, as described previously (p. 180). The angular velocities φ' can be derived from the curves for the φ-angles as the linear velocities were derived from the displacement curves. The velocity of displacement of the centre of gravity of a body segment is the quotient of the distance traversed in a very short (infinitely short) time, upon this time. In other words, it is the differential quotient of the distance upon the time. The angular velocity of the rotatory plane movement of a body segment is the quotient of the angle described by the long axis (variation of the φ-angle) of the body segment in a very short (infinitely short) time, upon this time. In other words, it is the differential quotient of the angle upon the time. Consequently, if the dependence of the angle φ on the time has been represented graphically, the angular velocities φ' are proportional to the trigonometric tangents of the angles formed by the tangents to the different points of the curve and the axis of the abscissa.

If the direction of the tangent has been determined for a sufficient number of time values and the corresponding angular velocities φ' have thus been found, the relationship of these velocities to time can be represented graphically by entering the velocities along the ordinate and time along the abscissa. This has been done in Figs. 8–10 for the three segments of the leg. The curves are just below those of the φ-angles from which they are derived. Here again 10 cm in the abscissa correspond to 1 s and 1 mm thus represents 0.01 s. The unit of angular velocity is represented by 1 cm in the ordinate, and 1 mm thus corresponds to an angular velocity of 0.1 radian s^{-1}. The unit of angular velocity is the angular velocity of a body segment which moves its long axis uniformly, without acceleration or deceleration, by an angle of 1 radian or 57.296° every second. A body describing a whole revolution in 1 s in a uniform rotation would thus display a velocity of 2π, and a body uniformly describing n revolutions per second would have a velocity of $2n\pi$.

From the curves of the angular velocities φ', the angular accelerations φ'' can finally be derived in the same way as the angular velocities were derived from the curves for the φ-angles. The angular acceleration is the quotient of the change in angular velocity in a very

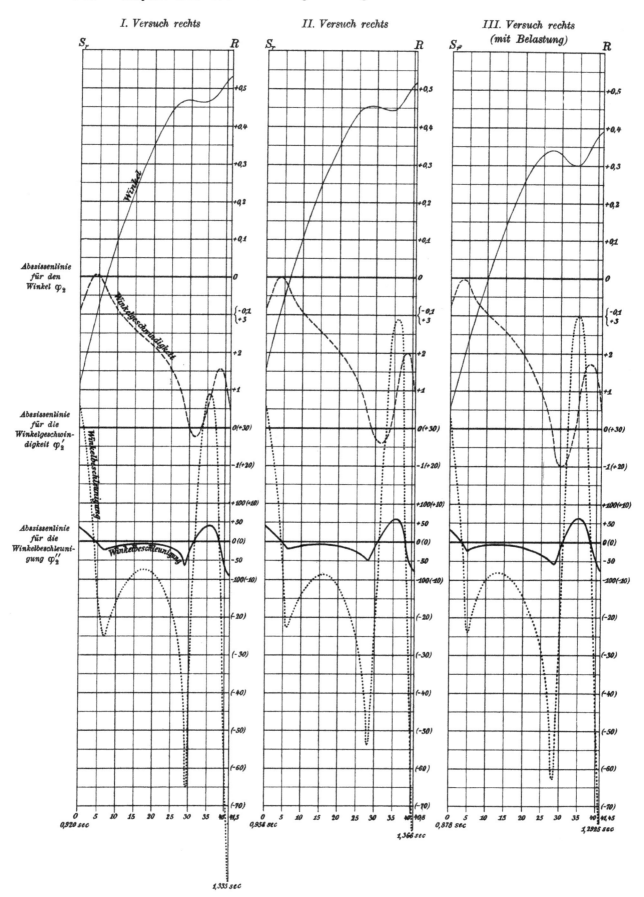

I. Versuch rechts

II. Versuch rechts

III. Versuch rechts
(mit Belastung)

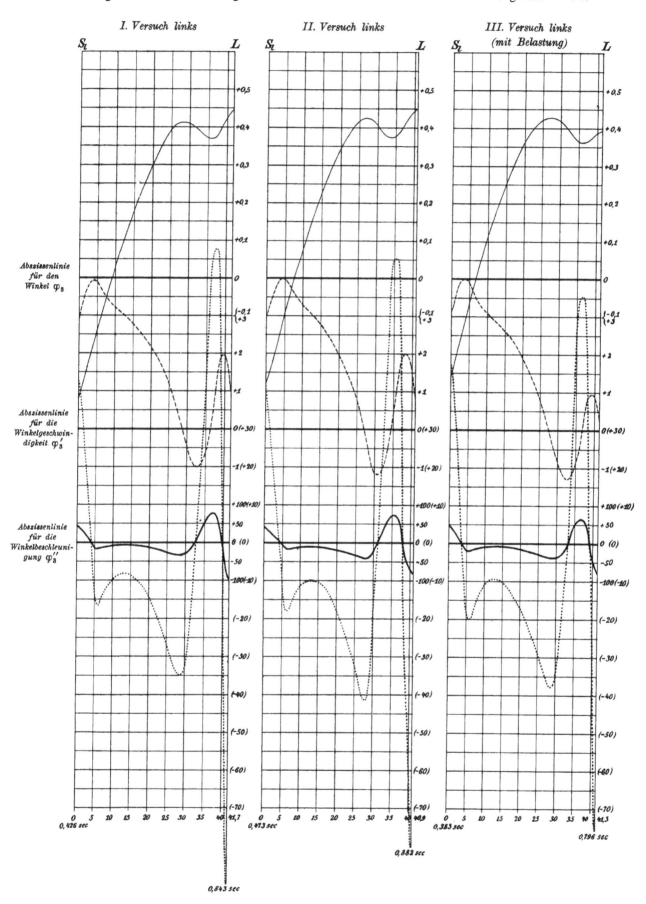

I. Versuch links

II. Versuch links

III. Versuch links
(mit Belastung)

Abszissenlinie
für den
Winkel φ₃

Abszissenlinie
für die
Winkelgeschwin-
digkeit φ₃′

Abszissenlinie
für die
Winkelbeschleuni-
gung φ₃″

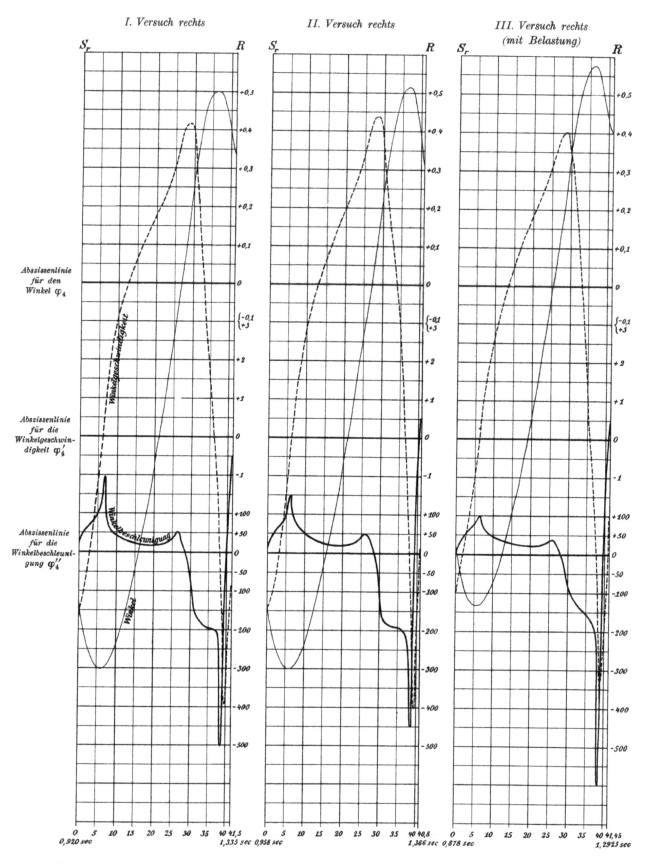

Fig. 9. Rotation of the lower leg during the swinging period of gait

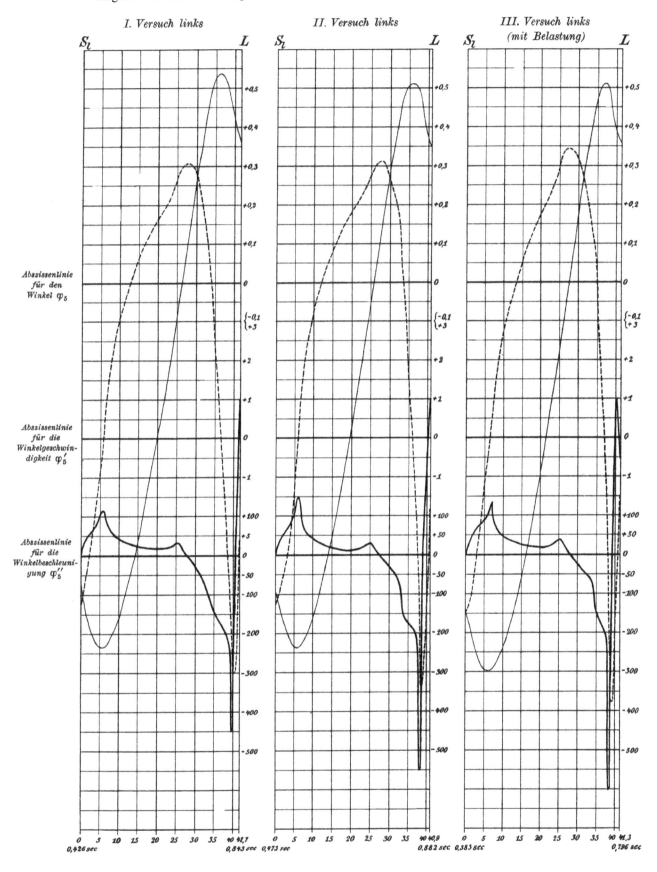

Fig. 9 (continued)

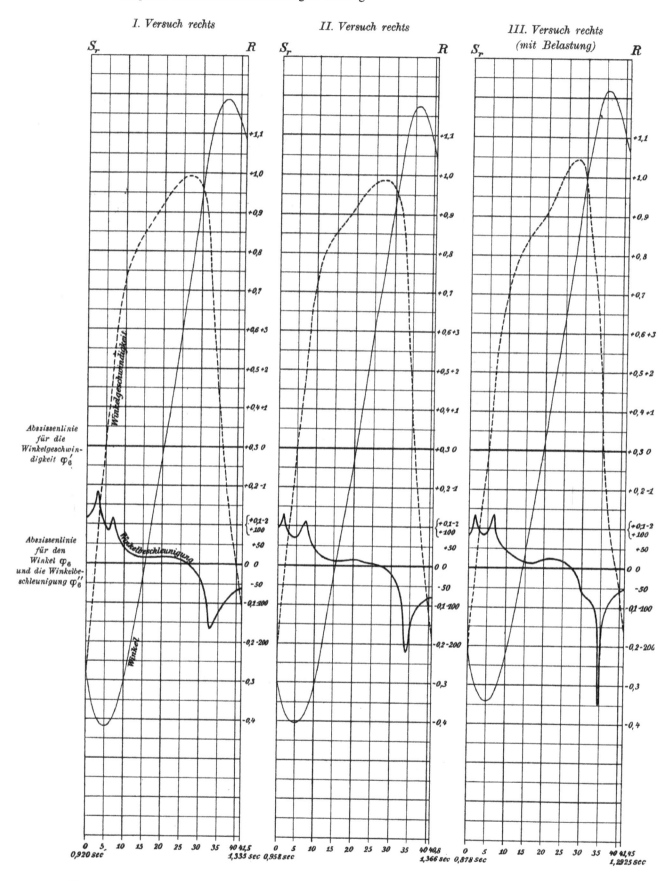

Fig. 10. Rotation of the foot during the swinging period of gait

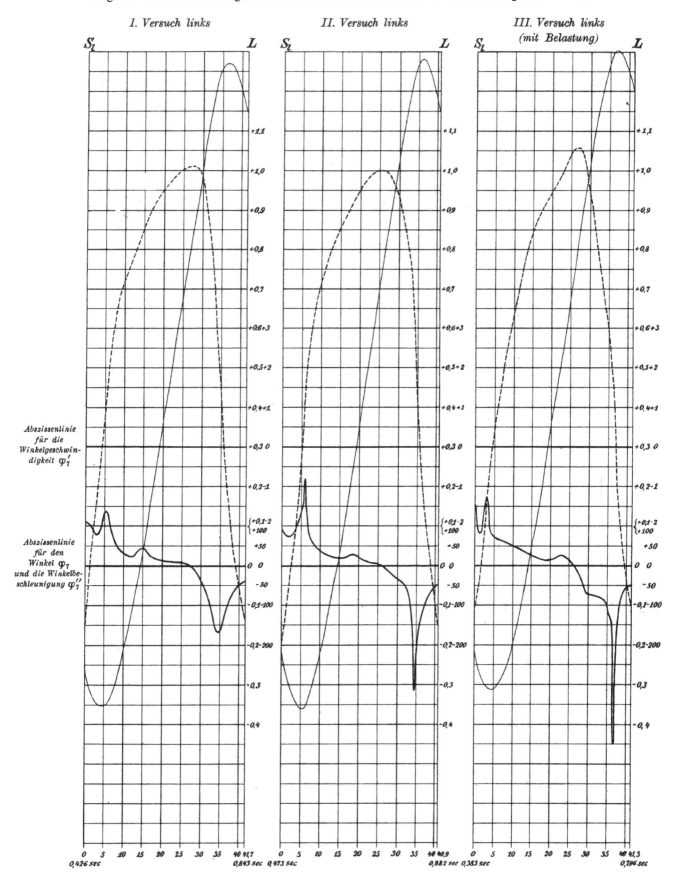

Fig. 10 (continued)

short (infinitely short) time, upon this time. In other words, it is the differential quotient of the angular velocity upon the time. The tangents to the curves of angular velocities thus have to be measured to obtain the corresponding angular accelerations. This has been done for a sufficient number of points. The result is given in the graphs below the corresponding curves of the angular velocities. A distance of 1 mm along the abscissa corresponds to 0.01 s. One unit of angular acceleration is represented in most cases by 0.1 mm in the ordinate so that 1 mm corresponds to an angular acceleration of 10 radians s^{-2}. However, in the curves related to the rotation of the thigh (Fig. 8) a distance of 1 mm also represents the unit of acceleration, because the angular acceleration of the thigh is generally not so high as that of the lower leg and foot. To allow easy comparison of the graphs for the thigh with the others, both kinds of curves have been drawn. On the smaller curves 1 mm corresponds to an angular acceleration of 10 radians s^{-2} whereas on the larger (dotted) curves 1 mm corresponds to the unit of angular acceleration (1 radian s^{-2}).

A body segment rotating with uniform acceleration displays the unit of angular acceleration if its angular velocity increases by 1 unit of velocity per second. If the angular velocity decreases, the rotation is decelerated. This results in a negative sign before the angular acceleration. This can occur even if the body segment rotates in a positive direction so that the velocity is positive.

For a proper understanding of Figs. 8–10 the following points must be emphasized. A φ-angle has a positive sign when the long axis of the body segment under consideration appears to be rotated counter-clockwise in relation to the vertical when seen from the right. The φ_2-angle is thus positive or negative depending on whether the lower end of the thigh lies in front of or behind the upper end, in the direction of gait. A positive rotation of the different body segments or a rotation with positive velocity is a rotation which increases the φ-angle. In such a rotation, the long axes of the body segments rotate counter-clockwise, as seen from the right. The positive or negative sign of the ordinate values for the upper curves thus indicates directly whether the distal or the proximal end of the long axis of the body segment under consideration is further forward. The positive or negative sign of the ordinate of the middle curves indicates whether the body segment considered rotates counter-clockwise or clockwise, as seen from the right, at the moment given by the abscissa. The sign of the ordinate in the lower curves indicates whether the rotation is accelerated or decelerated.

The φ-angles, the angular velocities φ' and the angular accelerations φ'' for each moment of the swinging period can be read off from the graphs in Figs. 8–10. To provide the following kinetic analysis of the swinging of the leg during walking with as complete a kinematic basis as possible, these data were determined for all the successive phases of the movement at time intervals of 0.01 s. On average, the swinging lasted 0.41 s. The values of the different data thus had to be measured for 42 phases, including the initial one. The results are given in Tables 6–11 (pp. 349–354).

As mentioned in the description of the curves in Figs. 8–10, for the particular analysis of the swinging movement, it appears useful to count the time, in every instance, from the beginning of the swinging period. In the same experiment, time 0 is different depending on whether the swinging of the left leg is examined or that of the right leg which begins one step later. In any case, toe-off is considered as time 0. Heel strike thus corresponds on average to 0.41 s. More precisely, in experiment I right heel strike occurs at 0.415 s and left heel strike at 0.417 s; in experiment II right heel strike occurs at 0.408 s and left heel strike at 0.409 s; and in experiment III right heel strike at 0.4145 s and left heel strike at 0.413 s.

In Tables 6–11 the number of hundredths of a second that have elapsed between the beginning of the swinging period and the phase under consideration is given in the first and last columns. These figures are completely different from the phases of movement related to the chronophotographic exposures. The first of these new phases of the movement coincides exactly with the beginning of the swinging period and is designated as 0. Therefore, the next ones directly indicate the time in hundredths of a second that has elapsed since the beginning

Table 6. Rotation of the segments of the right leg during the swinging period in experiment I

Time (0,01 s)	Right thigh			Right lower leg			Right foot			Time (0,01 s)
	Angle φ_2 (radians)	Angular velocity φ_2'	Angular acceleration φ_2''	Angle φ_4 (radians)	Angular velocity φ_4'	Angular acceleration φ_4''	Angle φ_6 (radians)	Angular velocity φ_6'	Angular acceleration φ_6''	
0	−0,289	+3,05	+ 35	−0,829	−4,67	+ 28	−0,251	−5,90	+117	0
1	−0,252	+3,33	+ 31	−0,873	−4,23	+ 48	−0,323	−5,10	+124	1
2	−0,203	+3,67	+ 24	−0,922	−3,60	+ 60	−0,366	−4,60	+138	2
3	−0,167	+3,89	+ 15	−0,956	−2,92	+ 70	−0,397	−2,20	+182	3
4	−0,128	+4,06	0	−0,977	−2,10	+ 79	−0,412	−1,00	+128	4
5	−0,091	+4,01	− 13	−0,992	−1,23	+ 90	−0,417	0	+ 96	5
6	−0,053	+3,89	− 22	−1,000	0	+107	−0,414	+1,00	+ 81	6
7	−0,019	+3,67	− 25	−0,999	+0,78	+197	−0,402	+1,90	+117	7
8	+0,017	+3,48	− 22	−0,992	+1,87	+ 63	−0,377	+2,98	+ 80	8
9	+0,053	+3,28	− 19	−0,973	+2,53	+ 49	−0,361	+3,77	+ 56	9
10	+0,085	+3,15	− 16	−0,945	+2,98	+ 43	−0,301	+4,26	+ 40	10
11	+0,118	+3,00	− 14	−0,911	+3,43	+ 36	−0,255	+4,61	+ 29	11
12	+0,143	+2,89	− 12	−0,872	+3,77	+ 34	−0,203	+4,93	+ 24	12
13	+0,170	+2,77	− 11	−0,836	+4,04	+ 30	−0,151	+5,13	+ 22	13
14	+0,197	+2,66	− 9,6	−0,792	+4,30	+ 26	−0,097	+5,31	+ 19	14
15	+0,221	+2,58	− 8,7	−0,745	+4,52	+ 23	−0,032	+5,48	+ 18	15
16	+0,250	+2,50	− 8,1	−0,700	+4,77	+ 22	+0,025	+5,61	+ 17	16
17	+0,274	+2,41	− 7,8	−0,650	+5,02	+ 20	+0,089	+5,76	+ 17	17
18	+0,297	+2,32	− 8,0	−0,601	+5,27	+ 19	+0,148	+5,93	+ 16	18
19	+0,322	+2,23	− 8,3	−0,552	+5,49	+ 18	+0,215	+6,06	+ 17	19
20	+0,343	+2,15	− 8,9	−0,498	+5,71	+ 17	+0,279	+6,23	+ 18	20
21	+0,368	+2,07	− 9,7	−0,412	+5,93	+ 19	+0,341	+6,39	+ 19	21
22	+0,387	+1,96	− 11	−0,349	+6,12	+ 22	+0,404	+6,55	+ 17	22
23	+0,406	+1,85	− 13	−0,290	+6,35	+ 25	+0,462	+6,67	+ 15	23
24	+0,422	+1,71	− 15	−0,219	+6,71	+ 28	+0,525	+6,78	+ 12	24
25	+0,439	+1,55	− 17	−0,145	+6,90	+ 33	+0,596	+6,85	+ 8	25
26	+0,452	+1,36	− 20	−0,081	+7,23	+ 47	+0,670	+6,90	+ 4	26
27	+0,461	+1,15	− 26	−0,016	+7,65	+ 54	+0,732	+6,92	0	27
28	+0,465	+0,85	− 35	+0,059	+8,03	+ 23	+0,799	+6,87	− 7	28
29	+0,467	+0,48	− 65	+0,133	+8,14	0	+0,872	+6,75	− 18	29
30	+0,468	0	− 45	+0,219	+8,00	− 48	+0,935	+6,52	− 30	30
31	+0,467	−0,21	− 15	+0,312	+7,00	− 130	+1,014	+6,16	− 52	31
32	+0,465	−0,22	0	+0,377	+5,80	− 169	+1,072	+5,30	− 95	32
33	+0,463	−0,08	+ 21	+0,426	+4,60	− 182	+1,119	+4,00	−165	33
34	+0,462	0	+ 33	+0,469	+3,00	− 189	+1,153	+2,60	−151	34
35	+0,465	+0,47	+ 38	+0,488	+1,80	− 193	+1,174	+1,40	−129	35
36	+0,470	+0,78	+ 39	+0,499	0	− 196	+1,185	0	−111	36
37	+0,477	+1,22	+ 35	+0,498	−0,80	− 198	+1,184	−0,80	− 99	37
38	+0,486	+1,50	+ 20	+0,484	−2,00	− 220	+1,175	−1,70	− 87	38
39	+0,501	+1,54	0	+0,461	−4,30	− 500	+1,155	−2,40	− 76	39
40	+0,514	+1,32	− 20	+0,420	−6,90	0	+1,130	−3,00	− 68	40
41	+0,526	+0,80	− 60	+0,367	−4,00	+ 170	+1,098	−3,70	− 64	41
41,5	+0,529	+0,55	− 90	+0,359	−3,80	+ 250	+1,086	−4,00	− 63	41,5

of the swinging period. In the different tables the phases indicated by the same number correspond with a good approximation to the same moment in the swinging process, whatever the experiment or the segment of the leg. Since the swing does not end an exact number of hundredths of a second after its beginning in all the cases, the precise time of the end is given. For instance, the last phase is numbered 41.5 for the right leg in experiment I, 41.7 for

Table 7. Rotation of the segments of the left leg during the swinging period in experiment I

Time (0,01 s)	Left thigh Angle φ_3 (radians)	Angular velocity φ_3'	Angular acceleration φ_3''	Left lower leg Angle φ_5 (radians)	Angular velocity φ_5'	Angular acceleration φ_5''	Left foot Angle φ_7 (radians)	Angular velocity φ_7'	Angular acceleration φ_7''	Time (0,01 s)
0	−0,325	+3,02	+44	−0,790	−4,27	+ 10	−0,250	−4,50	+114	0
1	−0,287	+3,30	+37	−0,824	−3,90	+ 36	−0,289	−3,50	+106	1
2	−0,251	+3,65	+26	−0,869	−3,13	+ 56	−0,320	−2,31	+ 93	2
3	−0,213	+3,87	+14	−0,898	−2,52	+ 65	−0,345	−1,42	+ 80	3
4	−0,181	+3,92	0	−0,920	−1,78	+ 78	−0,354	−0,50	+ 89	4
5	−0,137	+3,85	−13	−0,932	−1,00	+ 101	−0,357	0	+130	5
6	−0,102	+3,69	−17	−0,934	0	+ 120	−0,351	+1,50	+140	6
7	−0,072	+3,54	−14,5	−0,928	+1,19	+ 90	−0,335	+2,52	+ 90	7
8	−0,036	+3,42	−12,4	−0,912	+2,00	+ 63	−0,304	+3,28	+ 60	8
9	−0,002	+3,29	−10,7	−0,885	+2,58	+ 51	−0,263	+3,81	+ 40	9
10	+0,028	+3,19	− 9,6	−0,860	+2,96	+ 43	−0,229	+4,17	+ 31	10
11	+0,061	+3,10	− 8,9	−0,828	+3,35	+ 38	−0,179	+4,45	+ 26	11
12	+0,092	+3,02	− 8,4	−0,791	+3,69	+ 34	−0,133	+4,68	+ 23	12
13	+0,120	+2,93	− 8,2	−0,754	+3,99	+ 30	−0,085	+4,93	+ 28	13
14	+0,148	+2,84	− 8,4	−0,712	+4,30	+ 27	−0,033	+5,19	+ 41	14
15	+0,180	+2,75	− 8,8	−0,670	+4,57	+ 25	+0,024	+5,42	+ 46	15
16	+0,207	+2,64	− 9,5	−0,617	+4,81	+ 22	+0,071	+5,71	+ 40	16
17	+0,229	+2,53	−10,3	−0,571	+5,04	+ 20	+0,140	+5,98	+ 23	17
18	+0,258	+2,40	−11,5	−0,500	+5,26	+ 19	+0,212	+6,18	+ 18	18
19	+0,282	+2,27	−12,7	−0,452	+5,45	+ 18	+0,270	+6,35	+ 14	19
20	+0,304	+2,12	−14,6	−0,401	+5,59	+ 17	+0,339	+6,48	+ 12	20
21	+0,328	+1,98	−16,8	−0,345	+5,76	+ 16	+0,408	+6,59	+ 11	21
22	+0,345	+1,81	−18,8	−0,294	+5,93	+ 17	+0,468	+6,73	+ 10	22
23	+0,364	+1,64	−21,0	−0,235	+6,12	+ 21	+0,527	+6,83	+ 9	23
24	+0,381	+1,42	−24	−0,170	+6,34	+ 25	+0,592	+6,93	+ 8	24
25	+0,395	+1,13	−27	−0,108	+6,61	+ 30	+0,612	+7,00	+ 6	25
26	+0,405	+0,81	−30	−0,034	+6,89	+ 27	+0,715	+7,06	+ 4	26
27	+0,411	+0,51	−32,8	+0,025	+7,04	+ 13	+0,779	+7,10	+ 1	27
28	+0,413	0	−34,7	+0,102	+7,07	0	+0,851	+7,11	0	28
29	+0,412	−0,20	−35,2	+0,179	+7,00	− 15	+0,920	+7,03	− 18	29
30	+0,408	−0,50	−31,5	+0,240	+6,82	− 28	+0,989	+6,75	− 36	30
31	+0,403	−0,79	−23,0	+0,311	+6,50	− 43	+1,062	+6,30	− 55	31
32	+0,394	−0,98	−10	+0,372	+5,83	− 62	+1,128	+5,59	− 79	32
33	+0,385	−1,00	0	+0,438	+5,15	− 82	+1,181	+4,75	−109	33
34	+0,378	−0,84	+25	+0,487	+4,20	− 105	+1,230	+3,40	−160	34
35	+0,371	−0,52	+44	+0,523	+2,70	− 140	+1,261	+1,90	−170	35
36	+0,370	0	+70	+0,535	+1,10	− 163	+1,273	0	−140	36
37	+0,377	+0,75	+78	+0,537	0	− 176	+1,270	−1,20	−103	37
38	+0,391	+1,62	+76	+0,516	−1,90	− 193	+1,258	−2,10	− 77	38
39	+0,410	+1,95	0	+0,483	−3,80	− 450	+1,233	−2,90	− 61	39
40	+0,429	+1,88	−30	+0,438	−6,00	0	+1,199	−3,40	− 48	40
41	+0,443	+1,31	−80	+0,388	−4,80	+ 220	+1,165	−4,00	− 40	41
41,7	+0,446	+1,00	−90	+0,365	−2,60	+ 400	+1,144	−4,40	− 38	41,7

the left leg, etc. These numbers are also used in the graphs. These contain many curves, some overlapping. Therefore, in each case, the significance of the curves related to the right lower extremity in experiment I is noted over these curves. This was not necessary for the other sections because the great similarity of the corresponding curves leaves no room for doubt about what they represent.

Table 8. Rotation of the segments of the right leg during the swinging period in experiment II

Time (0,01 s)	Right thigh			Right lower leg			Right foot			Time (0,01 s)
	Angle φ_2 (radians)	Angular velocity φ_2'	Angular acceleration φ_2''	Angle φ_4 (radians)	Angular velocity φ_4'	Angular acceleration φ_4''	Angle φ_6 (radians)	Angular velocity φ_6'	Angular acceleration φ_6''	
0	−0,243	+3,14	+36	−0,852	−4,36	+ 12	−0,287	−4,70	+ 92	0
1	−0,204	+3,48	+30	−0,888	−4,09	+ 37	−0,343	−4,00	+130	1
2	−0,165	+3,71	+23	−0,928	−3,60	+ 55	−0,373	−2,70	+105	2
3	−0,127	+3,93	+16	−0,963	−3,00	+ 67	−0,391	−1,90	+ 82	3
4	−0,090	+4,00	0	−0,983	−2,10	+ 78	−0,401	−0,90	+ 73	4
5	−0,053	+3,93	− 9	−0,997	−1,20	+ 96	−0,403	0	+ 71	5
6	−0,019	+3,80	−23	−1,000	0	+150	−0,398	+0,90	+ 95	6
7	+0,015	+3,51	−21	−0,997	+0,95	+ 85	−0,385	+1,97	+117	7
8	+0,048	+3,34	−18	−0,982	+1,70	+ 67	−0,364	+3,00	+ 85	8
9	+0,080	+3,19	−15,5	−0,965	+2,30	+ 57	−0,329	+3,75	+ 59	9
10	+0,107	+3,02	−13,7	−0,939	+2,82	+ 49	−0,289	+4,25	+ 42	10
11	+0,138	+2,91	−12,2	−0,902	+2,27	+ 42	−0,242	+4,80	+ 30	11
12	+0,169	+2,80	−10,8	−0,864	+3,78	+ 36	−0,192	+5,06	+ 22	12
13	+0,195	+2,70	− 9,6	−0,823	+4,04	+ 32	−0,129	+5,27	+ 17	13
14	+0,223	+2,61	− 9,0	−0,783	+4,37	+ 28	−0,076	+5,41	+ 15	14
15	+0,249	+2,51	− 8,8	−0,736	+4,64	+ 26	−0,013	+5,54	+ 14	15
16	+0,272	+2,42	− 8,7	−0,687	+4,90	+ 22	+0,048	+5,69	+ 14	16
17	+0,295	+2,32	− 9,0	−0,638	+5,23	+ 20	+0,110	+5,82	+ 13	17
18	+0,317	+2,21	− 9,8	−0,579	+5,49	+ 19	+0,165	+5,95	+ 16	18
19	+0,338	+2,11	−10,8	−0,531	+5,78	+ 18	+0,209	+6,08	+ 17	19
20	+0,359	+2,00	−12,4	−0,459	+6,02	+ 17	+0,280	+6,22	+ 18	20
21	+0,378	+1,88	−14	−0,400	+6,25	+ 19	+0,338	+6,39	+ 19	21
22	+0,394	+1,74	−16	−0,321	+6,48	+ 22	+0,419	+6,54	+ 14	22
23	+0,410	+1,60	−19	−0,252	+6,71	+ 26	+0,479	+6,63	+ 10	23
24	+0,427	+1,43	−22	−0,182	+7,00	+ 32	+0,543	+6,75	+ 8	24
25	+0,438	+1,21	−26	−0,115	+7,38	+ 43	+0,610	+6,81	+ 6	25
26	+0,447	+0,93	−32	−0,051	+7,78	+ 50	+0,670	+6,86	+ 5	26
27	+0,451	+0,60	−40	+0,009	+8,25	+ 35	+0,729	+6,87	+ 3	27
28	+0,453	+0,24	−53	+0,088	+8,38	0	+0,798	+6,88	0	28
29	+0,454	0	−38	+0,172	+8,32	− 26	+0,869	+6,80	− 11	29
30	+0,453	−0,30	−21	+0,249	+7,60	− 120	+0,932	+6,62	− 20	30
31	+0,449	−0,39	0	+0,328	+6,55	− 175	+0,988	+6,33	− 31	31
32	+0,446	−0,36	+16	+0,390	+5,60	− 186	+1,051	+5,90	− 57	32
33	+0,442	−0,21	+34	+0,439	+4,60	− 191	+1,114	+4,70	−116	33
34	+0,440	0	+51	+0,476	+3,70	− 195	+1,153	+3,00	−220	34
35	+0,443	+0,45	+58	+0,501	+1,40	− 197	+1,172	+1,10	−157	35
36	+0,452	+0,98	+58,7	+0,513	0	− 200	+1,175	0	−118	36
37	+0,467	+1,75	+56	+0,510	−1,20	− 209	+1,165	−1,50	−102	37
38	+0,485	+1,97	0	+0,483	−3,50	− 268	+1,144	−2,49	− 92	38
39	+0,501	+1,88	−35	+0,437	−7,00	− 450	+1,113	−3,40	− 85	39
40	+0,514	+1,50	−65	+0,358	−4,70	0	+1,065	−4,05	− 79	40
40,8	+0,518	+1,00	−76	+0,333	−2,87	+ 360	+1,040	−4,70	− 78	40,8

Because of the short interval of time between two successive phases, it is now possible to determine the maxima and minima of the different data of the tables, which are important for the further analysis. In many instances these do not occur at an exact number of hundredths of a second after the beginning of the swinging period and, therefore, do not coincide precisely with any one of the movement phases in the tables. However, at worst they

Table 9. Rotation of the segments of the left leg during the swinging period in experiment II

Time (0,01 s)	Left thigh			Left lower leg			Left foot			Time (0,01 s)
	Angle φ_3 (radians)	Angular velocity φ_3'	Angular acceleration φ_3''	Angle φ_5 (radians)	Angular velocity φ_5'	Angular acceleration φ_5''	Angle φ_7 (radians)	Angular velocity φ_7'	Angular acceleration φ_7''	
0	−0,290	+2,90	+41	−0,789	*−4,20*	0	−0,223	−4,70	+ 88	0
1	−0,258	+3,20	+33	−0,827	−4,02	+ 25	−0,255	−4,10	+ 77	1
2	−0,227	+3,52	+23	−0,859	−3,55	+ 43	−0,297	−3,35	+ 70	2
3	−0,185	+3,73	+12	−0,890	−3,04	+ 63	−0,323	−2,60	+ 81	3
4	−0,149	*+3,97*	0	−0,919	−2,20	+ 81	−0,353	−1,70	+101	4
5	−0,108	+3,95	− 8	−0,934	−1,20	+107	*−0,361*	0	+132	5
6	−0,069	+3,85	*−18*	*−0,938*	0	*+150*	−0,357	+1,50	*+220*	6
7	−0,030	+3,72	−17,5	−0,928	+1,50	+ 91	−0,336	+2,50	+ 85	7
8	+0,003	+3,56	−15,3	−0,909	+2,25	+ 59	−0,308	+3,25	+ 59	8
9	+0,036	+3,40	−12,9	−0,883	+2,75	+ 48	−0,271	+3,67	+ 47	9
10	+0,070	+3,29	−11,7	−0,858	+3,30	+ 41	−0,228	+4,15	+ 38	10
11	+0,100	+3,17	−10,7	−0,822	+3,65	+ 35	−0,176	+4,50	+ 31	11
12	+0,130	+3,08	−10,1	−0,779	+4,00	+ 31	−0,131	+4,80	+ 26	12
13	+0,161	+2,97	− *9,8*	−0,732	+4,35	+ 27	−0,081	+5,05	+ 23	13
14	+0,194	+2,87	− 9,9	−0,689	+4,60	+ 23	−0,023	+5,30	+ 20	14
15	+0,221	+2,76	−10,5	−0,641	+4,82	+ 19	+0,029	+5,53	+ 19	15
16	+0,246	+2,64	−11,1	−0,591	+5,04	+ 17	+0,089	+5,75	+ *18*	16
17	+0,273	+2,47	−12,2	−0,532	+5,27	+ 15	+0,153	+5,95	+ 20	17
18	+0,298	+2,32	−13,5	−0,484	+5,48	+ 13	+0,222	+6,13	+ 23	18
19	+0,319	+2,15	−15,4	−0,432	+5,65	+ 11	+0,284	+6,35	+ *26*	19
20	+0,343	+1,98	−17,5	−0,369	+5,82	+ *10*	+0,351	+6,55	+ 21	20
21	+0,362	+1,79	−20,1	−0,308	+5,96	+ 12	+0,428	+6,70	+ 13	21
22	+0,377	+1,53	−22,5	−0,251	+6,10	+ 16	+0,499	+6,82	+ 9	22
23	+0,393	+1,30	−25	−0,182	+6,29	+ 20	+0,558	+6,92	+ 7	23
24	+0,404	+1,08	−28	−0,114	+6,50	+ 24	+0,630	+6,96	+ 5	24
25	+0,413	+0,78	−32	−0,052	+6,78	+ *30*	+0,700	+7,00	+ 2	25
26	+0,422	+0,42	−36	+0,009	+6,99	+ 20	+0,772	+7,01	0	26
27	*+0,424*	0	−40	+0,081	+7,07	+ 8	+0,831	+6,98	− 8	27
28	+0,422	−0,41	*−41,6*	+0,148	*+7,12*	0	+0,897	+6,87	− 18	28
29	+0,417	−0,78	−38,5	+0,212	+6,97	− 18	+0,967	+6,65	− 28	29
30	+0,408	−1,15	−18	+0,280	+6,67	− 30	+1,031	+6,30	− 36	30
31	+0,397	*−1,20*	0	+0,341	+6,32	− 41	+1,093	+5,85	− 41	31
32	+0,386	−1,12	+18	+0,402	+5,88	− 53	+1,141	+5,35	− 49	32
33	+0,377	−0,77	+34	+0,459	+4,95	− 90	+1,199	+4,70	− 75	33
34	*+0,372*	0	+60	+0,498	+3,40	− 160	+1,249	+3,50	−160	34
35	+0,374	+0,50	+74,5	+0,512	+1,35	− 180	+1,274	+1,30	−320	35
36	+0,383	+1,08	*+75,3*	+0,517	0	− 187	*+1,283*	0	−135	36
37	+0,394	+1,80	+60	+0,502	−1,80	− 205	+1,275	−1,50	−106	37
38	+0,412	*+1,99*	0	+0,479	−4,80	− *550*	+1,246	−2,40	− 81	38
39	+0,430	+1,84	−45	+0,425	*−6,40*	0	+1,219	−3,25	− 62	39
40	+0,444	+1,30	−70	+0,381	−4,10	+ 290	+1,182	−3,94	− 53	40
40,9	+0,450	+0,70	−79	+0,357	−1,80	+ 400	+1,146	−4,40	− 48	40,9

can only occur 0.005 s before or after one of these phases. Assigning the maximum or the minimum to the closest phase thus does not entail any significant error. The value for the appropriate phase was, therefore, replaced by the nearest maximum or minimum value. Consequently, particular tables displaying these maxima and minima such as were given in the previous parts in the same circumstances could be dispensed with. To distinguish the maxima and minima from the other data in the tables, they are printed in italic figures. A

Table 10. Rotation of the segments of the right leg during the swinging period in experiment III

Time (0,01 s)	Right thigh			Right lower leg			Right foot			Time (0,01 s)
	Angle φ_2 (radians)	Angular velocity φ_2'	Angular acceleration φ_2''	Angle φ_4 (radians)	Angular velocity φ_4'	Angular acceleration φ_4''	Angle φ_6 (radians)	Angular velocity φ_6'	Angular acceleration φ_6''	
0	−0,353	+3,10	+33	−0,670	−3,98	+0	−0,200	−5,10	+ 74	0
1	−0,324	+3,45	+27	−0,709	−3,70	+ 21	−0,247	−4,30	+ 86	1
2	−0,293	+3,75	+18	−0,753	−3,25	+ 41	−0,292	−3,60	+133	2
3	−0,255	+3,92	+10	−0,788	−2,70	+ 55	−0,319	−2,30	+102	3
4	−0,217	+3,94	0	−0,815	−2,00	+ 64	−0,337	−1,40	+ 85	4
5	−0,179	+3,85	−21	−0,827	−2,15	+ 72	−0,343	−0,60	+ 81	5
6	−0,142	+3,61	−24	−0,832	−2,40	+ 83	−0,344	0	+ 98	6
7	−0,110	+3,42	−18	−0,833	0	+103	−0,332	+1,40	+136	7
8	−0,073	+3,30	−14,5	−0,824	+1,30	+ 65	−0,308	+2,60	+ 82	8
9	−0,042	+3,18	−12,2	−0,809	+1,90	+ 55	−0,287	+3,35	+ 58	9
10	−0,011	+3,03	−10,4	−0,788	+2,35	+ 49	−0,228	+3,85	+ 47	10
11	+0,017	+2,94	− 9,1	−0,759	+2,90	+ 43	−0,193	+4,27	+ 39	11
12	+0,048	+2,86	− 8,7	−0,733	+3,35	+ 39	−0,148	+4,58	+ 31	12
13	+0,077	+2,77	− 8,2	−0,694	+3,75	+ 36	−0,100	+4,89	+ 25	13
14	+0,106	+2,70	− 8,1	−0,658	+4,10	+ 33	−0,048	+5,14	+ 20	14
15	+0,132	+2,58	− 8,8	−0,612	+4,45	+ 30	+0,012	+5,30	+ 16	15
16	+0,160	+2,48	− 9,5	−0,563	+4,78	+ 28	+0,071	+5,44	+ 13	16
17	+0,184	+2,35	−10,5	−0,521	+5,09	+ 27	+0,129	+5,58	+ 12	17
18	+0,209	+2,21	−11,7	−0,470	+5,38	+ 25	+0,185	+5,72	+ 11	18
19	+0,233	+2,05	−13,2	−0,403	+5,63	+ 22	+0,248	+5,87	+ 15	19
20	+0,253	+1,95	−14,8	−0,352	+5,89	+ 20	+0,299	+5,99	+ 19	20
21	+0,262	+1,80	−17	−0,301	+6,10	+ 19	+0,342	+6,19	+ 22	21
22	+0,288	+1,63	−19	−0,239	+6,34	+ 22	+0,417	+6,43	+ 24	22
23	+0,303	+1,43	−22	−0,158	+6,60	+ 24	+0,483	+6,68	+ 23	23
24	+0,318	+1,21	−25	−0,093	+6,88	+ 29	+0,556	+6,89	+ 21	24
25	+0,326	+0,95	−30	−0,023	+7,17	+ 32	+0,628	+7,12	+ 19	25
26	+0,334	+0,69	−35	+0,039	+7,46	+ 37	+0,689	+7,27	+ 13	26
27	+0,338	+0,35	−44	+0,104	+7,80	+ 38	+0,745	+7,38	+ 7	27
28	+0,340	0	−62	+0,187	+7,99	+ 14	+0,810	+7,42	0	28
29	+0,336	−0,65	−50	+0,255	+8,02	0	+0,896	+7,37	− 17	29
30	+0,328	−0,98	−25	+0,348	+7,70	− 67	+0,958	+7,02	− 48	30
31	+0,319	−1,02	0	+0,408	+6,70	− 98	+1,032	+6,35	− 73	31
32	+0,310	−0,97	+25	+0,469	+5,70	− 117	+1,089	+5,40	− 85	32
33	+0,302	−0,75	+45	+0,518	+4,60	− 131	+1,139	+4,65	− 98	33
34	+0,300	0	+57,5	+0,549	+3,10	− 142	+1,184	+3,70	−120	34
35	+0,304	+0,50	+60	+0,570	+1,80	− 151	+1,213	+2,00	−350	35
36	+0,312	+1,10	+58	+0,577	0	− 162	+1,216	0	−130	36
37	+0,326	+1,55	+47	+0,571	−1,30	− 177	+1,205	−1,70	− 99	37
38	+0,343	+1,73	0	+0,542	−3,00	− 205	+1,184	−2,40	− 82	38
39	+0,364	+1,70	−20	+0,500	−5,00	− 600	+1,152	−3,00	− 69	39
40	+0,379	+1,50	−45	+0,452	−6,20	0	+1,123	−3,70	− 62	40
41	+0,388	+1,00	−68	+0,424	−3,00	+ 300	+1,083	−4,25	− 56	41
41,45	+0,390	+0,70	−74	+0,418	−2,10	+ 350	+1,070	−4,50	− 55	41,45

clear view of the behaviour of the different data during the whole swinging period can thus be obtained at a glance.

Each maximum or minimum in one of the curves corresponds to a zero point in the curve just below. Therefore, these zero points which belong to the columns of the tables relating to the angular velocities and to the angular accelerations have been displaced in each such instance to the closest movement phase included in the table.

Table 11. Rotation of the segments of the left leg during the swinging period in experiment III

Time (0,01 s)	Left thigh			Left lower leg			Left foot			Time (0,01 s)
	Angle φ_8 (radians)	Angular velocity φ_8'	Angular acceleration φ_8''	Angle φ_5 (radians)	Angular velocity φ_5'	Angular acceleration φ_5''	Angle φ_7 (radians)	Angular velocity φ_7'	Angular acceleration φ_7''	
0	−0,251	+3,10	+47	−0,840	−4,50	+ 5	−0,215	−4,10	+150	0
1	−0,228	+3,39	+38	−0,871	−4,20	+ 33	−0,255	−3,55	+ 80	1
2	−0,184	+3,78	+25	−0,919	−3,60	+ 56	−0,285	−2,30	+103	2
3	−0,142	+3,95	+ 8	−0,950	−2,80	+ 71	−0,305	−1,30	+170	3
4	−0,106	+3,99	0	−0,981	−2,05	+ 82	−0,313	0	+ 87	4
5	−0,061	+3,84	−18	−0,995	−1,05	+ 91	−0,310	+0,60	+ 70	5
6	−0,028	+3,67	−20	−0,997	0	+105	−0,302	+1,15	+ 65	6
7	+0,009	+3,45	−18	−0,992	+0,80	+137	−0,283	+1,80	+ 61	7
8	+0,040	+3,28	−15	−0,979	+1,70	+ 72	−0,262	+2,50	+ 57	8
9	+0,070	+3,17	−12,5	−0,960	+2,35	+ 56	−0,236	+2,95	+ 53	9
10	+0,103	+3,07	−10,9	−0,933	+2,88	+ 47	−0,201	+3,50	+ 50	10
11	+0,136	+2,98	− 9,9	−0,907	+3,25	+ 40	−0,165	+4,00	+ 46	11
12	+0,168	+2,87	− 9,2	−0,869	+3,63	+ 37	−0,121	+4,50	+ 42	12
13	+0,194	+2,77	− 9,5	−0,833	+4,00	+ 33	−0,067	+4,90	+ 37	13
14	+0,220	+2,68	− 9,9	−0,790	+4,35	+ 31	−0,008	+5,25	+ 32	14
15	+0,245	+2,57	−10,5	−0,748	+4,66	+ 29	+0,027	+5,50	+ 26	15
16	+0,276	+2,46	−11,5	−0,707	+4,94	+ 25	+0,098	+5,78	+ 21	16
17	+0,302	+2,32	−12,6	−0,658	+5,13	+ 23	+0,160	+5,99	+ 18	17
18	+0,322	+2,18	−14	−0,599	+5,36	+ 21	+0,216	+6,13	+ 15	18
19	+0,345	+2,00	−16	−0,543	+5,60	+ 20	+0,295	+6,29	+ 16	19
20	+0,364	+1,86	−18	−0,481	+5,80	+ 19	+0,351	+6,46	+ 17	20
21	+0,380	+1,68	−19,5	−0,427	+5,96	+ 18	+0,419	+6,64	+ 19	21
22	+0,393	+1,49	−21,5	−0,353	+6,19	+ 22	+0,478	+6,83	+ 24	22
23	+0,403	+1,25	−24	−0,288	+6,43	+ 25	+0,540	+7,04	+ 27	23
24	+0,413	+0,99	−26	−0,225	+6,69	+ 33	+0,603	+7,27	+ 22	24
25	+0,420	+0,75	−29	−0,161	+7,05	+ 40	+0,678	+7,46	+ 24	25
26	+0,424	+0,38	−32,5	−0,093	+7,33	+ 28	+0,740	+7,59	0	26
27	+0,427	0	−35	−0,027	+7,41	+ 11	+0,818	+7,57	− 9	27
28	+0,426	−0,17	−37,5	+0,038	+7,43	0	+0,905	+7,40	− 29	28
29	+0,424	−0,52	−38,2	+0,124	+7,35	− 16	+0,984	+7,05	− 65	29
30	+0,419	−0,85	−34	+0,177	+7,17	− 23	+1,042	+6,25	− 72	30
31	+0,409	−1,10	−25	+0,250	+6,80	− 38	+1,095	+5,70	− 75	31
32	+0,398	−1,25	−15	+0,328	+6,30	− 51	+1,149	+5,00	− 77	32
33	+0,386	−1,30	0	+0,380	+5,75	− 58	+1,200	+4,25	− 81	33
34	+0,374	−1,13	+25	+0,429	+5,10	− 75	+1,238	+3,50	− 86	34
35	+0,365	−0,80	+57	+0,477	+3,50	−150	+1,282	+2,80	−105	35
36	+0,360	0	+65,5	+0,506	+1,30	−173	+1,304	+1,75	−135	36
37	+0,362	+0,30	+64,5	+0,511	0	−192	+1,306	0	−450	37
38	+0,367	+0,80	+35	+0,495	−2,50	−600	+1,297	−1,80	−100	38
39	+0,376	+0,93	0	+0,438	−6,70	0	+1,284	−2,70	− 62	39
40	+0,385	+0,84	−45	+0,391	−4,00	+400	+1,243	−3,30	− 54	40
41	+0,390	+0,45	−70	+0,368	−2,20	+270	+1,208	−3,80	− 49	41
41,3	+0,391	+0,39	−75	+0,366	−1,50	+230	+1,205	−4,10	− 48	41,3

These tables provide detailed information on the rotations of the three segments of the leg during the swinging period. To give a general view of the phases of the swing following each other at 0.01 s intervals and to relate the data in the tables correctly to the total swinging process, the 42 successive positions of the leg have been drawn (Fig. 11). This picture is based on the chronophotographic exposures of the swinging of the left leg in experiment II. The trajectories of the centres of the hip, knee and ankle joints and of the tip of the foot have been

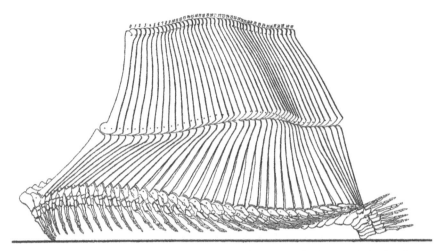

Fig. 11. Successive positions of the leg during the swinging period of gait, recorded at intervals of 0.01 s

drawn using graphical interpolation. Experiment II was selected because, in this experiment, the duration of the swinging period of the left leg was the closest to 0.41 s. Consequently, heel strike coincided sufficiently closely with the last of the 42 phases separated by time intervals of 0.01 s. The picture can also be used to study the results of the other experiments because the discrepancies during their successive phases in relation to those depicted in Fig. 11 are too small to be recognizable in a figure of this kind. This picture has thus been drawn for the right leg rather than the left, and represents the behaviour of the swinging of the leg in general. The figures above the femoral head give the time interval in hundredths of a second between each of the different phases and the first one. A description of the successive positions of the leg has been given in Chap. 3 (p. 221) and it will not be repeated here in detail. A thorough study of Fig. 11 gives quicker information than a long description can.

Figures 8–10 and Tables 6–11 do not give such a direct insight into the changes in angular velocities and accelerations during the rotation of the three segments of the leg, and an explanation, therefore, follows.

First, in each of the three figures a far-reaching similarity of the six panels can be recognized. The discrepancies are only quantitative, relating exclusively to the values of the maxima and minima. The sequence and type of the changes, however, are the same in all cases for both the angular velocities and the angular accelerations of the same body segment, except for slight irregularities in the acceleration curves which are, of course, a little less reliable.

As shown by Fig. 8, at the beginning of the swinging period the *thigh* is inclined backwards to the vertical and displays a considerable angular velocity counter-clockwise as seen from the right. It has not just acquired this angular velocity early in the swinging period, but has carried it over from the end of the stance period. This initial angular velocity at first increases further until it reaches its maximum in the fourth phase, 0.04 s after the beginning of the swinging period, in all the experiments and for both extremities. A positive angular acceleration is thus present up to this instant precisely.

From the 4th phase onward the angular velocity decreases, first quickly, then somewhat more slowly, and finally quickly again. It passes through 0 when approximately three-quarters of the swinging period has elapsed (in about the 28th phase); it then becomes negative until it reaches a minimum in about the 32nd phase, i.e., 0.32 s after the beginning of the swinging period. It comes back to 0 in the 34th phase. The negative values of the angular velocity correspond, of course, to rotation of the thigh clockwise as seen from the right. According to the previous data, this rotation lasts only about 0.06 s. When the positive angular velocity decreases the angular acceleration is negative. It is then an angular deceleration. In fact, the angular acceleration of the thigh becomes an angular deceleration from the 4th phase onward, when it is 0. This is no longer true after the angular velocity has passed through 0 and has become negative. Then a negative angular acceleration no longer means a deceleration, but an acceleration of the negative, i.e. clockwise, rotation. The deceleration of the positive rotation increases at first very quickly, reaching a first maximum in the 6th phase in nearly all the experiments. It then decreases, and passes through a minimum in about the 14th phase. It subsequently increases again until it reaches a higher maximum in the 29th phase. The values of this maximum vary perceptibly in the different experiments and, especially, between the two sides of the body. For the right extremity the maximum varies between 53 and 65, for the left extremity between 35.2 and 41.6. The values of the preceding minimum varied only between 7.8 and 9.8 for the six experiments. The values of the first, smaller, maximum were also larger for the right leg than for the left. The former varied between 23 and 25, the latter between 17 and 20. Almost exactly at the instant of the second maximum of angular deceleration the angular velocity passes through 0 and becomes negative so that the rotation is now clockwise. The angular acceleration remains negative. This negative value no longer signifies a deceleration, but rather an acceleration of the clockwise rotation of the thigh. This angular acceleration, however, decreases quickly and reaches 0 when the angular velocity reaches the minimum described above.

From this point onward the angular velocity increases again, i.e., the angular velocity, still negative at first, decreases in absolute value until it passes through 0 and progressively increases up to a maximum of 0.04 s before the end of the swinging period. Whereas the first maximum of the angular velocity was 4.0 in nearly all the experiments, the values of this second maximum vary considerably. For the right leg they are between 1.45 and 1.97. For the left leg, they are nearly the same in the first two experiments (one is 1.95, the other 1.99). In the third experiment, the very different value of 0.93 must be attributed to the influence of the heavy backpack. After the second maximum the angular velocity decreases again. However, it does not reach 0 but only about half the value of the last maximum by the end of the swinging period.

The behaviour of the angular acceleration corresponds to the variation of the angular velocity. When the latter reaches its minimum, the former is 0. Angular acceleration then is replaced by angular deceleration of the negative rotation. This deceleration increases very quickly and reaches a maximum almost at the same moment as the angular velocity becomes 0 and the rotation again changes direction. This maximum of deceleration varies between 30 and 60 for the right leg, between 65.5 and 78 for the left. The left maximum is always more pronounced than the right. Further, the thigh rotates counter-clockwise, again with an angular acceleration. The latter falls rapidly in all the experiments during the last 0.05 s of the swinging period. After about 0.03 s it passes through 0 and thus changes again into angular deceleration which amounts to between 74 and 90 at heel strike.

Figure 9 in association with Tables 6–11 gives details of the rotation of the *lower leg*. It is obvious at once that the angular velocity of the lower leg attains significantly higher values and entails much greater angular accelerations than that of the thigh. Therefore, the curves for angular acceleration had to be drawn to a smaller scale. The unit of angular acceleration is represented by 0.1 mm so that 1 mm along the ordinate corresponds to an angular acceleration of 10 radians s^{-2}.

At the beginning of the swinging period, whereas the thigh is in positive rotation, the lower leg is rotating negatively, i.e. clockwise as seen from the right, from the start. Correspondingly, its angular velocity is at first negative. It was already negative in the last part of stance, before toe-off. This negative angular velocity decreases in absolute value until it passes through 0 in the 6th phase, 0.06 s after toe-off. Then the angular velocity becomes positive and remains so for most of the swinging period. Only during the last 0.05 s does it become negative again. In the 28th phase, 0.28 s after toe-off, the positive velocity reaches a maximum, a little lower for the left lower leg than for the right. The left maximum varies between 7.07 and 7.43, the right between 8.02 and 8.38. The angular velocity falls very steeply from this maximum in all cases, becomes negative, and at less than 0.02 s before heel strike reaches a minimum varying between -6.20 and -7.00 for the right lower leg and between -6.00 and -6.70 for the left. After that, the angular velocity increases again, i.e. its absolute value decreases, but it has not yet reached 0 at heel strike. Then it is still evidently negative.

As a consequence of the decreasing negative angular velocity, the angular acceleration is positive at the beginning of the swinging period. At first, this means an angular deceleration of a negative rotation. Only after the rotation has become positive is it accelerated and this continues until the 28th phase. From this phase onward, the angular acceleration is negative, that is to say it becomes angular deceleration, until shortly before heel strike when it reverts to angular acceleration. During the first 0.28 s of the swinging period the angular acceleration reaches two maxima, the higher about 0.06 s and the lower about 0.26 s after toe-off. The higher reaches values over 100, or even nearly 200 for the right lower leg in experiment I. The lower maximum, in contrast, remains under 50 in almost all cases. Between these two maxima there is a minimum in the 20th phase. Its value varies between 10 and 19 in the different experiments. After the 28th phase the angular acceleration becomes negative. Its absolute value increases, at first quickly, then much more slowly, and finally very quickly again. Less than 0.03 s before heel strike it reaches a maximum which varies between -450 and -600. Since after the 28th phase the angular velocity initially remains positive, the negative angular acceleration means a deceleration of the positive rotation. Shortly before the absolute value of the negative angular acceleration reaches its maximum, the rotation reverses its direction. Finally, therefore, there is an acceleration of the negative rotation. Then the angular acceleration suddenly decreases almost to zero and again takes positive values so that at the end of the swing period the negative, i.e. clockwise, rotation is considerably decelerated.

Figure 10 and Tables 6–11 allow the following deductions concerning the rotation of the *foot*.

The angular velocity behaves very similarly to that of the lower leg, the only difference being that it does not present any minimum. The angular velocity of the foot is also negative at the beginning of the swinging period as it was during push-off. It reaches 0 in the 5th phase and then becomes positive and remains so until the 36th phase. From there onward it becomes negative again until the end of the swinging period so that the foot strikes the floor whilst rotating clockwise at relatively high angular velocity (between -4 and -4.7). The positive angular velocity reaches a maximum about 0.27 s after toe-off. Its value varies between 6.88 and 7.59 in the six cases.

The angular acceleration also looks essentially similar to that of the lower leg. It is at first positive for most of the swinging period, and then becomes negative until the end of the swinging period. It does not become positive again, as was the case for the lower leg. At the end of the swinging period, since the angular velocity is negative, the negative sign in front of the angular acceleration means that the clockwise rotation of the foot is finally accelerated.

This provides an overall view of the rotations of the long axes of the three segments of the leg during the swinging period. A more detailed insight will result from a thorough study of the curves in Figs. 8–10.

Velocities and Accelerations of the Centres of Gravity of the Three Segments of the Leg During the Swinging Period

To find the velocities and accelerations of the centres of gravity, we proceeded as explained at length in Chap. 2 (p. 168 ff.) for the kinematic analysis of the movement of the total centre of gravity. The displacement curves of the movements of the three centres of gravity in the direction of gait and in the vertical direction during the swinging period were drawn, and the tangent method was used to construct the velocity cuves from the displacement curves and the acceleration curves from the velocity curves. The co-ordinates in the direction of gait were related to a coronal plane moving regularly forward at the average velocity of gait, using Table 8 in Chap. 2 (p. 159). This made it possible to draw the displacement curves more conveniently and to derive the velocities and accelerations more easily. The relative

Table 12. Relative co-ordinates of the centres of gravity in the direction of gait in experiment I

No.	Time (s)	Right thigh	Right lower leg	No.	Time (s)	Left thigh	Left lower leg
25	0,920	− 5,77	− 23,46	12	0,422	− 5,37	− 23,93
26	0,958	− 2,51	− 18,33	13	0,460	− 2,55	− 19,50
27	0,997	+ 0,53	− 12,58	14	0,498	+ 0,07	− 14,13
28	1,035	+ 3,16	− 6,38	15	0,537	+ 2,42	− 7,72
29	1,073	+ 4,80	− 1,05	16	0,575	+ 3,76	− 2,92
30	1,112	+ 6,08	+ 4,85	17	0,613	+ 5,24	+ 3,77
31	1,150	+ 6,30	+ 9,68	18	0,652	+ 5,84	+ 9,12
				19	0,690	+ 6,09	+ 14,37
1	0,990	− 0,35	− 13,61	20	0,728	+ 5,64	+ 18,35
2	1,028	+ 2,31	− 7,29	21	0,767	+ 5,07	+ 20,64
3	1,067	+ 4,50	− 0,82	22	0,805	+ 5,61	+ 21,77
4	1,105	+ 6,03	+ 5,30	23	0,843	+ 7,10	+ 22,15
5	1,143	+ 6,94	+ 11,14				
6	1,182	+ 7,22	+ 16,32				
7	1,220	+ 6,74	+ 20,13				
8	1,258	+ 6,65	+ 23,22				
9	1,297	+ 7,35	+ 24,74				
10	1,335	+ 8,63	+ 25,07				

Table 13. Relative co-ordinates of the centres of gravity in the direction of gait in experiment II

No.	Time (s)	Right thigh	Right lower leg	No.	Time (s)	Left thigh	Left lower leg
26	0,958	− 4,84	− 21,77	13	0,460	− 5,72	− 23,71
27	0,997	− 1,58	− 16,58	14	0,498	− 3,25	− 20,02
28	1,035	+ 1,33	− 10,56	15	0,537	− 0,21	− 14,11
29	1,073	+ 3,59	− 4,81	16	0,575	+ 1,88	− 8,34
30	1,112	+ 5,28	+ 0,65	17	0,613	+ 3,56	− 2,57
31	1,150	+ 6,71	+ 6,66	18	0,652	+ 5,07	+ 3,43
				19	0,690	+ 6,19	+ 9,49
1	0,970	− 4,52	− 21,13	20	0,728	+ 6,66	+ 14,99
2	1,008	− 0,95	− 16,17	21	0,767	+ 5,93	+ 18,38
2	1,047	+ 1,87	− 10,21	22	0,805	+ 4,97	+ 20,24
4	1,085	+ 3,92	− 3,79	23	0,843	+ 5,94	+ 22,02
5	1,123	+ 5,22	+ 1,83	24	0,882	+ 7,30	+ 22,31
6	1,162	+ 6,15	+ 7,69				
7	1,200	+ 6,70	+ 13,40				
8	1,238	+ 6,13	+ 17,06				
9	1,277	+ 5,73	+ 20,80				
10	1,315	+ 5,72	+ 22,61				
11	1,353	+ 7,57	+ 24,17				
12	1,392	+ 8,88	+ 24,11				

Table 14. Relative co-ordinates of the centres of gravity in the direction of gait in experiment III

No.	Time (s)	Right thigh	Right lower leg	No.	Time (s)	Left thigh	Left lower leg
23	0,843	− 7,37	− 25,20	11	0,383	− 4,08	− 21,61
24	0,882	− 4,96	− 22,57	12	0,422	− 1,05	− 16,69
25	0,920	− 2,36	− 18,19	13	0,460	+ 1,84	− 10,73
26	0,958	+ 0,17	− 12,65	14	0,498	+ 3,78	− 5,07
27	0,997	+ 1,86	− 7,40	15	0,537	+ 5,43	+ 0,76
28	1,035	+ 2,98	− 2,39	16	0,575	+ 6,43	+ 6,18
29	1,073	+ 3,91	+ 2,96	17	0,613	+ 6,98	+ 11,34
30	1,112	+ 4,31	+ 8,11	18	0,652	+ 6,67	+ 15,45
31	1,150	+ 4,18	+ 12,80	19	0,690	+ 5,94	+ 18,62
				20	0,728	+ 5,23	+ 19,97
1	0,989₅	− 0,38	− 10,81	21	0,767	+ 5,78	+ 20,65
2	1,027₅	+ 1,11	− 4,82	22	0,805	+ 6,86	+ 20,65
3	1,066₅	+ 2,28	+ 1,24				
4	1,104₅	+ 2,97	+ 6,82				
5	1,142₅	+ 2,91	+ 11,56				
6	1,181₅	+ 2,54	+ 15,23				
7	1,219₅	+ 2,14	+ 17,35				
8	1,257₅	+ 3,04	+ 18,79				
9	1,296₅	+ 4,62	+ 19,44				

velocities of the three centres of gravity thus obtained again differ from the absolute velocities by the average velocity of gait in each of the three experiments. The accelerations are the same as in the immobile system of co-ordinates.

In Tables 12–14 the relative co-ordinates of the centres of gravity of the thigh and of the lower leg in the direction of gait are given in the same order as the values of the φ-angles in Tables 3–5. The relative co-ordinates were obtained by subtracting the distances given in Table 8 of Chap. 2 from the absolute co-ordinates of the centres of gravity of the thigh and of the lower leg during the swinging period, given in Tables 1, 2 and 6 of Chap. 2. The relative co-ordinates of the centre of gravity of the foot need not appear in the tables since they have been given in Tables 1–6 of Chap. 4 (pp. 286–291).

The displacement curves of the relative movement of the segments of both legs in the direction of gait, in the three experiments, were drawn in Fig. 12 in such a way that 1 mm along the abscissa represents 0.01 s and 1 mm along the ordinate represents 1 cm of co-ordinate. The ordinate thus appears at 1/10 life-size. This was necessary in order to allow the displacement curves of the relative movement in the direction of gait to be drawn in the figure. The displacement curves of the movement in the vertical direction could be drawn life-size in Fig. 13 because of the smaller magnitudes of their co-ordinates.

In both figures the velocity curves and the acceleration curves are drawn below the displacement curves. All the kinematics of the centres of gravity of the three segments of the legs during the swinging period can thus be seen at a glance in Figs. 12 and 13. For the velocity curves 1 mm along the ordinate corresponds to a velocity of $10\,\mathrm{cm\,s^{-1}}$. For the acceleration curves 1 mm along the ordinate represents an acceleration of $100\,\mathrm{cm\,s^{-2}}$. In Fig. 13 the velocity curves of the centre of gravity of the lower leg, the co-ordinates of which are relatively smaller, are also printed in dotted lines after tenfold magnification of the ordinate values. For these curves, 1 mm along the ordinate thus corresponds to a velocity of $1\,\mathrm{cm\,s^{-1}}$. Otherwise, the presentation of the curves in these two figures is exactly similar to that in Figs. 8–10, making further explanation superfluous.

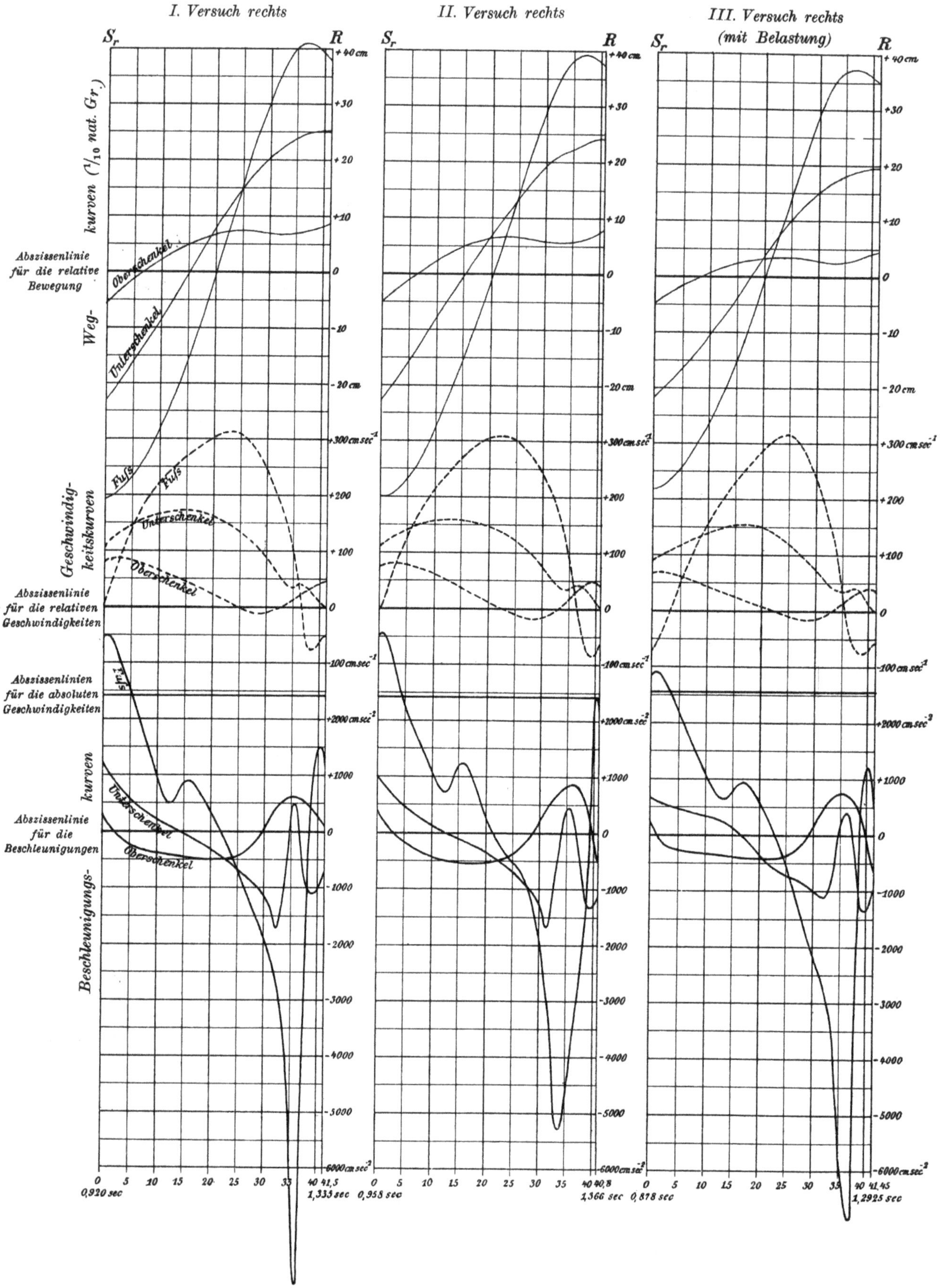

Fig. 12. Movement of the centres of gravity of the thigh, lower leg and foot in the direction of gait during the swinging period

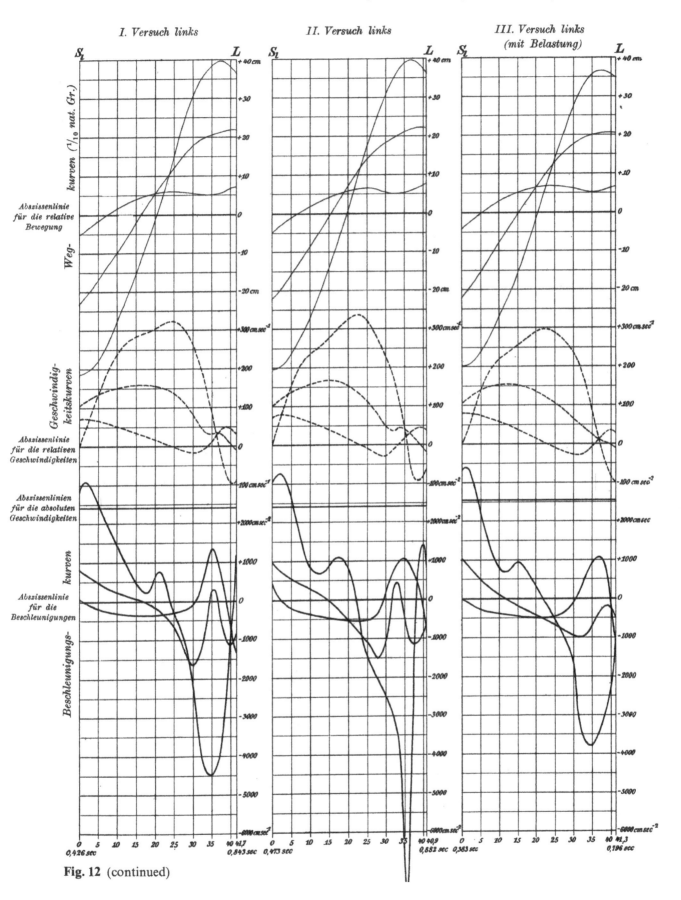

Fig. 12 (continued)

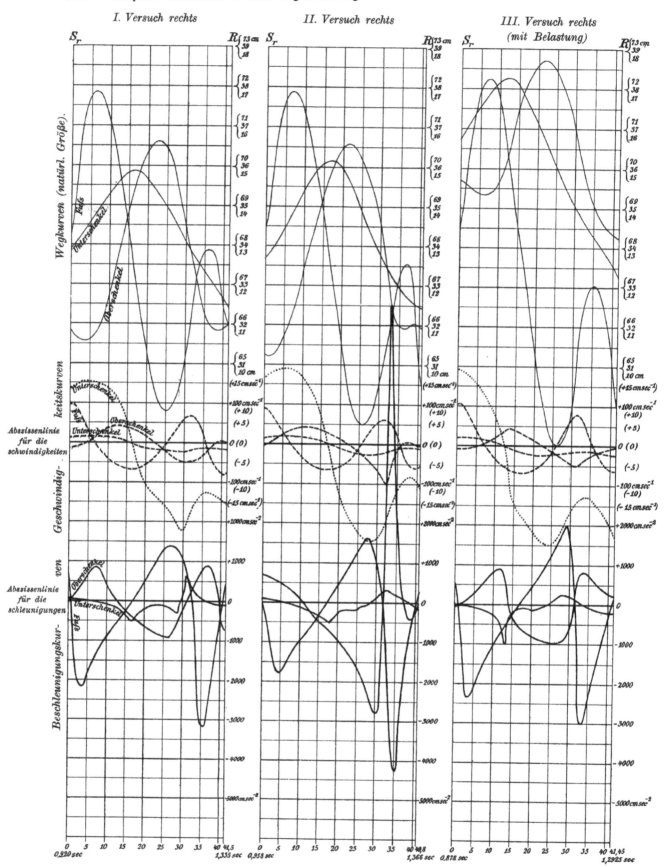

Fig. 13. Movement of the centres of gravity of the thigh, lower leg and foot in the vertical direction during the swinging period

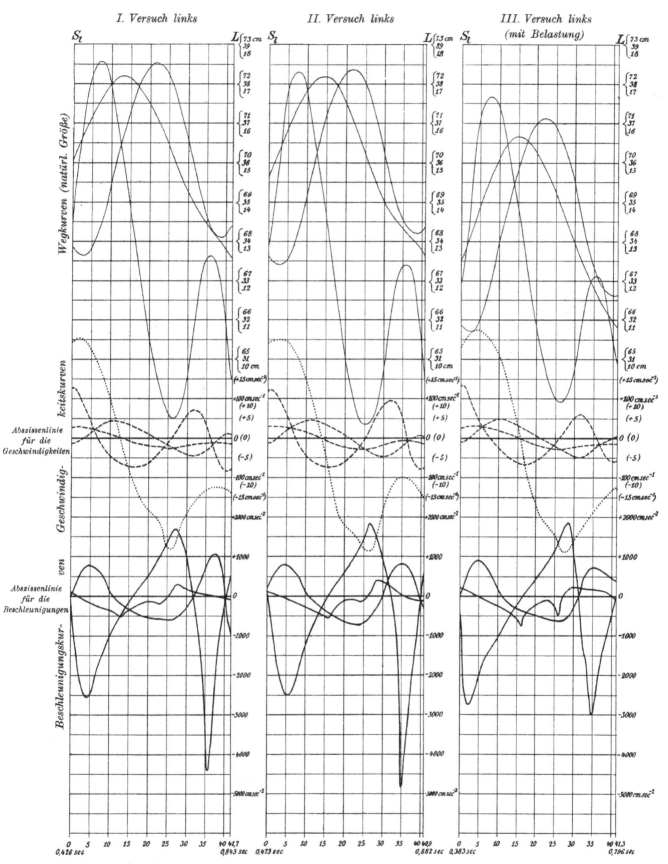

Fig. 13 (continued)

Table 15. Movement of the centre of gravity of the right thigh during the swinging period in experiment I

Time (0,01 s)	Co-ordinates (cm) ξ_2 Relative	z_2	Velocities (cm s^{-1}) Direction of gait Relative	Vertical direction	Accelerations (cm s^{-2}) Direction of gait	Vertical direction	Time (0,01 s)
0	−5,8	65,87	+78	−13	+310	0	0
1	−4,7	65,73	+81	−12	+160	+90	1
2	−3,9	65,65	+83	−11	0	+210	2
3	−3,1	65,61	+82	−7	−90	+350	3
4	−2,3	65,58	+81	0	−180	+510	4
5	−1,7	65,62	+79	+4	−230	+640	5
6	−0,8	65,68	+77	+11	−290	+790	6
7	0	65,82	+73	+19	−330	+880	7
8	+0,6	65,99	+70	+26	−370	+840	8
9	+1,3	66,30	+68	+34	−390	+650	9
10	+2,0	66,65	+64	+41	−400	+440	10
11	+2,5	67,12	+60	+43	−410	+230	11
12	+3,1	67,53	+57	+46	−420	0	12
13	+3,6	67,94	+52	+45	−430	−60	13
14	+4,0	68,39	+49	+44	−450	−140	14
15	+4,5	68,77	+45	+42	−470	−230	15
16	+4,9	69,21	+41	+39	−480	−320	16
17	+5,4	69,53	+35	+34	−500	−420	17
18	+5,8	69,85	+30	+29	−510	−520	18
19	+6,2	70,14	+26	+24	−510	−590	19
20	+6,4	70,34	+21	+18	−520	−660	20
21	+6,6	70,50	+16	+12	−510	−720	21
22	+6,8	70,62	+11	+5	−500	−780	22
23	+7,0	70,64	+8	0	−480	−820	23
24	+7,2	70,61	+3	−8	−470	−870	24
25	+7,3	70,50	0	−18	−450	−900	25
26	+7,2	70,31	−6	−26	−410	−930	26
27	+7,1	69,95	−9	−33	−340	−870	27
28	+7,0	69,54	−10	−41	−240	−710	28
29	+6,9	69,10	−10	−46	−110	−490	29
30	+6,8	68,61	−11	−48	0	−250	30
31	+6,7	68,13	−8	−49	+250	0	31
32	+6,7	67,68	−4	−47	+400	+200	32
33	+6,6	67,10	0	−43	+520	+420	33
34	+6,7	66,67	+7	−37	+580	+610	34
35	+6,8	66,36	+13	−28	+610	+740	35
36	+7,0	66,16	+18	−21	+620	+890	36
37	+7,2	65,95	+23	−13	+570	+920	37
38	+7,4	65,87	+29	0	+520	+770	38
39	+7,7	65,89	+36	+4	+450	+490	39
40	+8,0	65,96	+40	+7	+330	0	40
41	+8,3	65,99	+44	+4	+190	−630	41
41,5	+8,4	66,00	+46	0	+130	−680	41,5

These figures give the co-ordinates, velocities and accelerations for all the successive phases of the movement separated by intervals of 0.01 s. These data can also be found in Tables 15–32. The absolute co-ordinates in the direction of gait are not needed at first for the further analysis. Therefore, they are not given in the tables to avoid overloading. For the same reason, only the relative velocities are given. The absolute velocities can easily be

Table 16. Movement of the centre of gravity of the left thigh during the swinging period in experiment I

Time (0,01 s)	Co-ordinates (cm)		Velocities (cm s⁻¹)		Accelerations (cm s⁻²)		Time (0,01 s)
	ξ_3 Relative	z_3	Direction of gait Relative	Vertical direction	Direction of gait	Vertical direction	
0	−4,8	67,86	+70	−13	+ 60	0	0
1	−4,3	67,75	+71	− 9	0	+ 180	1
2	−3,6	67,67	+70	− 6	− 90	+ 450	2
3	−2,8	67,63	+69	0	− 160	+ 640	3
4	−2,3	67,68	+68	+ 6	− 190	+ 770	4
5	−1,4	67,82	+66	+14	− 230	+ 800	5
6	−0,8	68,02	+63	+20	− 260	+ 760	6
7	−0,1	68,27	+60	+27	− 280	+ 690	7
8	+0,5	68,55	+58	+34	− 290	+ 580	8
9	+1,0	68,96	+54	+40	− 300	+ 470	9
10	+1,4	69,33	+50	+43	− 310	+ 180	10
11	+1,9	69,76	+47	+44	− 320	0	11
12	+2,4	70,32	+44	+43	− 340	− 120	12
13	+3,0	70,63	+41	+41	− 350	− 170	13
14	+3,3	71,02	+38	+38	− 360	− 240	14
15	+3,8	71,37	+35	+34	− 370	− 300	15
16	+4,3	71,70	+31	+29	− 380	− 370	16
17	+4,7	71,94	+29	+25	− 390	− 420	17
18	+5,0	72,17	+24	+20	− 380	− 470	18
19	+5,3	72,24	+20	+14	− 370	− 510	19
20	+5,5	72,46	+17	+10	− 370	− 540	20
21	+5,7	72,53	+12	+ 4	− 370	− 550	21
22	+5,8	72,57	+10	0	− 360	− 550	22
23	+5,9	72,50	+ 6	− 7	− 350	− 570	23
24	+6,1	72,43	0	−11	− 330	− 590	24
25	+6,0	72,38	− 3	−17	− 310	− 610	25
26	+6,0	72,03	− 7	−22	− 290	− 620	26
27	+6,0	71,75	− 10	−28	− 280	− 570	27
28	+5,9	71,43	− 12	−33	− 260	− 510	28
29	+5,7	71,02	− 15	−37	− 210	− 420	29
30	+5,6	70,60	− 18	−39	− 160	− 300	30
31	+5,4	70,17	− 19	−42	0	− 170	31
32	+5,3	69,81	− 17	−43	+ 210	0	32
33	+5,2	69,37	− 10	−40	+ 600	+ 200	33
34	+5,0	68,99	0	−37	+ 930	+ 480	34
35	+5,1	68,65	+10	−30	+1340	+ 700	35
36	+5,2	68,40	+24	−22	+1270	+ 950	36
37	+5,3	68,22	+36	−14	+ 990	+1060	37
38	+5,8	68,10	+47	0	+ 550	+1000	38
39	+6,3	68,13	+50	+ 5	0	+ 750	39
40	+6,7	68,20	+46	+12	− 550	0	40
41	+7,2	68,36	+38	+10	−1000	− 670	41
41,7	+7,4	68,42	+29	+ 7	−1250	− 900	41,7

obtained by adding the average velocity of gait to the relative velocities. The average velocities of gait were 157.32 cm s⁻¹ in experiment I, 158.35 cm s⁻¹ in experiment II, and 145.33 cm s⁻¹ in experiment III. The relative co-ordinates in the direction of gait are designated as ζ-co-ordinates to differentiate them from the absolute x-co-ordinates.

Table 17. Movement of the centre of gravity of the right thigh during the swinging period in experiment II

Time (0,01 s)	Co-ordinates (cm)		Velocities (cm s⁻¹)		Accelerations (cm s⁻²)		Time (0,01 s)
	ξ_2 Relative	z_2	Direction of gait Relative	Vertical direction	Direction of gait	Vertical direction	
0	−4,7	65,25	+78	−5	+370	+680	0
1	−3,9	65,24	+80	0	+220	+670	1
2	−3,2	65,27	+81	+8	+100	+630	2
3	−2,5	65,36	+82	+14	0	+600	3
4	−1,7	65,51	+81	+21	−120	+550	4
5	−0,8	65,73	+79	+25	−200	+480	5
6	−0,2	65,99	+77	+31	−260	+420	6
7	+0,5	66,32	+73	+34	−300	+340	7
8	+1,1	66,67	+70	+40	−350	+280	8
9	+1,9	76,04	+67	+42	−380	+200	9
10	+2,3	67,59	+62	+43	−410	+100	10
11	+2,9	67,89	+59	+44	−430	0	11
12	+3,5	68,25	+55	+43	−470	−140	12
13	+4,1	68,64	+50	+41	−490	−270	13
14	+4,5	68,93	+46	+39	−500	−380	14
15	+5,0	69,35	+40	+35	−510	−470	15
16	+5,3	69,69	+35	+32	−520	−530	16
17	+5,6	69,96	+30	+28	−530	−620	17
18	+6,0	70,19	+24	+22	−540	−720	18
19	+6,2	70,40	+19	+17	−530	−830	19
20	+6,3	70,53	+13	+12	−520	−970	20
21	+6,4	70,57	+8	+6	−510	−1070	21
22	+6,6	70,58	+4	0	−490	−1190	22
23	+6,7	70,50	0	−6	−470	−1300	23
24	+6,6	70,34	−5	−12	−430	−1480	24
25	+6,5	70,13	−10	−20	−350	−1670	25
26	+6,4	69,91	−11	−28	−270	−1870	26
27	+6,3	69,67	−13	−38	−160	−2150	27
28	+6,1	69,32	−15	−47	−70	−2410	28
29	+6,0	68,92	−17	−56	0	−2700	29
30	+5,9	68,48	−14	−68	+220	−2770	30
31	+5,8	67,97	−8	−89	+410	−1500	31
32	+5,7	67,30	−3	−100	+580	0	32
33	+5,6	66,40	0	−62	+690	+7550	33
34	+5,7	66,10	+9	−22	+780	+5000	34
35	+5,8	65,95	+19	−6	+870	+2500	35
36	+6,0	65,92	+28	0	+890	+600	36
37	+6,2	65,97	+37	+5	+810	0	37
38	+6,6	66,01	+46	+2	+590	−410	38
39	+7,1	66,03	+50	0	0	−450	39
40	+7,7	65,99	+48	−5	−350	−180	40
40,8	+8,1	65,96	+44	−6	−550	0	40,8

These tables, in association with Tables 6–11, provide the reliable kinematic basis on which the aforementioned kinetic analysis of the forces acting on the swinging leg can be developed. Comparison of the six graphs in Figs. 12 and 13 and of Figs. 8–10 demonstrate the reliability of this basis. The six curves have the same shape except for slight irregularities and discrepancies in the acceleration curves which by their nature are subjected to greater errors than the velocity and displacement curves. There are differences only in the values of

Table 18. Movement of the centre of gravity of the left thigh during the swinging period in experiment II

Time (0,01 s)	Co-ordinates (cm)		Velocities (cm s^{-1})		Accelerations (cm s^{-2})		Time (0,01 s)
	ξ_3 Relative	z_3	Direction of gait Relative	Vertical direction	Direction of gait	Vertical direction	
0	−4,9	67,62	+71	−15	+450	+150	0
1	−4,2	67,51	+76	−11	+260	+400	1
2	−3,5	67,41	+79	−6	0	+550	2
3	−2,7	67,40	+78	0	−70	+700	3
4	−1,8	67,43	+77	+8	−130	+780	4
5	−1,2	67,55	+74	+15	−210	+810	5
6	−0,7	67,73	+72	+22	−230	+760	6
7	0	68,06	+69	+29	−280	+690	7
8	+0,7	68,38	+67	+36	−300	+580	8
9	+1,2	68,72	+64	+41	−320	+420	9
10	+1,8	69,20	+61	+44	−350	+190	10
11	+2,2	69,62	+59	+46	−380	0	11
12	+2,8	70,03	+53	+45	−400	−130	12
13	+3,1	70,48	+50	+43	−410	−230	13
14	+3,7	70,99	+46	+40	−430	−310	14
15	+4,0	71,33	+41	+36	−460	−380	15
16	+4,4	71,62	+37	+31	−480	−450	16
17	+4,9	71,85	+33	+27	−500	−510	17
18	+5,2	72,08	+28	+20	−510	−580	18
19	+5,5	72,23	+23	+13	−520	−620	19
20	+5,9	72,34	+19	+7	−530	−650	20
21	+6,1	72,40	+13	0	−540	−680	21
22	+6,3	72,39	+10	−6	−550	−710	22
23	+6,5	72,33	+5	−12	−540	−740	23
24	+6,7	72,19	0	−18	−530	−720	24
25	+6,6	72,00	−8	−25	−510	−700	25
26	+6,5	71,72	−12	−31	−480	−630	26
27	+6,4	71,43	−18	−37	−440	−510	27
28	+6,2	71,07	−22	−41	−340	−380	28
29	+5,9	70,70	−25	−46	−180	−220	29
30	+5,6	70,22	−27	−48	0	0	30
31	+5,3	69,79	−21	−46	+450	+230	31
32	+5,1	69,38	−16	−41	+700	+510	32
33	+5,0	69,02	0	−33	+910	+680	33
34	+5,2	68,72	+5	−27	+1060	+770	34
35	+5,4	68,51	+15	−19	+1040	+800	35
36	+5,7	68,37	+23	−12	+950	+770	36
37	+6,1	68,26	+33	−6	+810	+670	37
38	+6,4	68,22	+40	0	+550	+510	38
39	+6,8	68,25	+44	+4	0	+250	39
40	+7,3	68,30	+41	+7	−300	0	40
40,9	+7,5	68,36	+37	+4	−900	−240	40,9

the maxima and minima. Consequently, the swinging movements must have been qualitatively the same for the right and the left leg in the three experiments. The discrepancies revealed by the kinematic analysis are only quantitative.

The movement of the centre of gravity of the foot has been described in Chap. 4 (p. 298 ff.). Therefore, only a rough outline of the movement of the centres of gravity of the thigh and lower led during the swinging period will be given.

Table 19. Movement of the centre of gravity of the right thigh during the swinging period in experiment III

Time (0,01 s)	Co-ordinates (cm)		Velocities (cm s⁻¹)		Accelerations (cm s⁻²)		Time (0,01 s)
	ξ_2 Relative	z_2	Direction of gait Relative	Vertical direction	Direction of gait	Vertical direction	
0	− 5,2	69,68	+ 66	− 8	+ 250	− 50	0
1	− 4,4	69,62	+ 70	− 9	0	0	1
2	− 3,7	69,53	+ 68	− 7	− 90	+ 60	2
3	− 3,1	69,46	+ 66	− 6	− 200	+ 100	3
4	− 2,5	69,39	+ 63	− 4	− 270	+ 190	4
5	− 1,9	69,32	+ 60	− 2	− 290	+ 280	5
6	− 1,5	69,30	+ 58	0	− 300	+ 380	6
7	− 1,0	69,32	+ 56	+ 5	− 300	+ 480	7
8	− 0,6	69,36	+ 52	+ 11	− 310	+ 590	8
9	− 0,1	69,48	+ 49	+ 16	− 320	+ 720	9
10	+ 0,3	69,62	+ 44	+ 21	− 330	+ 800	10
11	+ 0,7	69,87	+ 40	+ 28	− 340	+ 870	11
12	+ 1,1	70,15	+ 38	+ 36	− 360	+ 900	12
13	+ 1,3	70,62	+ 33	+ 44	− 370	+ 600	13
14	+ 1,8	71,10	+ 29	+ 45	− 390	0	14
15	+ 2,0	71,49	+ 27	+ 40	− 400	− 320	15
16	+ 2,3	71,83	+ 24	+ 35	− 410	− 520	16
17	+ 2,4	72,10	+ 20	+ 30	− 420	− 610	17
18	+ 2,6	72,32	+ 17	+ 26	− 430	− 680	18
19	+ 2,8	72,48	+ 13	+ 19	− 440	− 770	19
20	+ 2,9	72,61	+ 11	+ 12	− 450	− 800	20
21	+ 3,0	72,69	+ 7	+ 7	− 440	− 870	21
22	+ 3,0	72,72	+ 4	0	− 440	− 900	22
23	+ 3,1	72,67	0	− 3	− 430	− 930	23
24	+ 3,0	72,53	− 5	− 9	− 410	− 970	24
25	+ 3,0	72,37	− 9	− 15	− 380	− 990	25
26	+ 2,9	72,20	− 11	− 22	− 330	− 1000	26
27	+ 2,8	71,95	− 13	− 29	− 260	− 980	27
28	+ 2,8	71,68	− 14	− 35	− 140	− 960	28
29	+ 2,7	71,32	− 18	− 42	0	− 860	29
30	+ 2,6	70,90	− 16	− 50	+ 150	− 550	30
31	+ 2,4	70,38	− 13	− 51	+ 270	0	31
32	+ 2,2	69,93	− 8	− 44	+ 430	+ 730	32
33	+ 2,1	69,55	− 3	− 36	+ 610	+ 820	33
34	+ 2,0	69,24	0	− 29	+ 690	+ 780	34
35	+ 2,2	69,01	+ 11	− 23	+ 740	+ 670	35
36	+ 2,4	68,75	+ 19	− 17	+ 770	+ 550	36
37	+ 2,7	68,61	+ 25	− 15	+ 680	+ 440	37
38	+ 3,1	68,47	+ 32	− 11	+ 530	+ 350	38
39	+ 3,4	68,37	+ 37	− 8	+ 330	+ 280	39
40	+ 3,9	68,29	+ 40	− 6	0	+ 240	40
41	+ 4,3	68,23	+ 36	− 4	− 550	+ 190	41
41,45	+ 4,4	68,22	+ 33	− 3	− 630	+ 170	41,45

At the beginning of the swinging period, the *centre of gravity of the thigh* already has a considerable forward relative velocity in the direction of gait. Its absolute velocity is far higher than the average velocity of gait. This initial velocity of the forward movement of the centre of gravity of the thigh is almost the greatest attained during the swinging period. In most experiments the maximum appears only a few hundredths of a second after toe-off, or

Table 20. Movement of the centre of gravity of the left thigh during the swinging period in experiment III

Time (0,01 s)	Co-ordinates (cm) ξ_3 Relative	Co-ordinates (cm) z_3	Velocities (cm s⁻¹) Direction of gait Relative	Velocities (cm s⁻¹) Vertical direction	Accelerations (cm s⁻²) Direction of gait	Accelerations (cm s⁻²) Vertical direction	Time (0,01 s)
0	−3,9	65,90	+81	−10	0	0	0
1	−3,2	65,79	+80	−7	−60	+220	1
2	−2,6	65,70	+79	0	−120	+530	2
3	−1,7	65,72	+77	+7	−190	+750	3
4	−0,7	65,78	+75	+13	−240	+840	4
5	0	65,99	+73	+21	−280	+900	5
6	+0,6	66,32	+70	+28	−300	+850	6
7	+1,2	66,63	+68	+33	−310	+760	7
8	+1,9	67,00	+64	+40	−330	+630	8
9	+2,3	67,39	+60	+45	−350	+410	9
10	+2,9	67,88	+59	+49	−370	+200	10
11	+3,3	68,38	+55	+50	−380	0	11
12	+4,0	68,77	+50	+46	−390	−120	12
13	+4,6	69,18	+45	+41	−400	−210	13
14	+5,1	69,59	+41	+37	−420	−280	14
15	+5,4	69,91	+37	+32	−450	−340	15
16	+5,7	70,21	+32	+28	−470	−380	16
17	+6,0	70,49	+28	+22	−480	−410	17
18	+6,2	70,73	+24	+18	−490	−430	18
19	+6,4	70,90	+19	+12	−500	−460	19
20	+6,7	71,05	+14	+8	−490	−490	20
21	+6,8	71,13	+10	+5	−490	−510	21
22	+6,9	71,17	+6	0	−490	−540	22
23	+7,0	71,15	0	−7	−480	−580	23
24	+6,9	71,07	−3	−12	−470	−600	24
25	+6,9	70,91	−8	−17	−450	−630	25
26	+6,8	70,72	−12	−23	−420	−640	26
27	+6,7	70,50	−16	−31	−380	−630	27
28	+6,6	70,23	−21	−37	−300	−610	28
29	+6,4	69,87	−24	−42	−200	−510	29
30	+6,2	69,42	−27	−47	0	−320	30
31	+5,9	68,97	−26	−50	+150	0	31
32	+5,8	68,51	−23	−48	+360	+350	32
33	+5,6	68,08	−18	−42	+590	+530	33
34	+5,4	67,69	−10	−37	+810	+660	34
35	+5,2	67,41	0	−30	+990	+720	35
36	+5,3	67,12	+7	−22	+1080	+710	36
37	+5,4	66,89	+18	−16	+1060	+670	37
38	+5,7	66,77	+29	−9	+950	+600	38
39	+6,0	66,68	+34	−5	+550	+510	39
40	+6,3	66,63	+37	0	0	+440	40
41	+6,7	66,64	+31	+3	−830	+380	41
41,3	+6,8	66,65	+28	+5	−1000	+350	41,3

can even coincide with toe-off, as for the left leg in experiment III. After the maximum of velocity of the movement forwards has been reached, the relative velocity decreases fairly regularly until it attains 0 in the 24th phase, or about 0.24 s after toe-off. Then, of course, the absolute velocity is equal to the average velocity of gait and, for an instant, the centre of gravity remains immobile in relation to the coronal plane moving forward at the average

Table 21. Movement of the centre of gravity of the right lower leg during the swinging period in experiment I

Time (0,01 s)	Co-ordinates (cm)		Velocities (cm s⁻¹)		Accelerations (cm s⁻²)		Time (0,01 s)
	ξ_4 Relative	z_4	Direction of gait Relative	Vertical direction	Direction of gait	Vertical direction	
0	− 23,4	33,93	+ 103	+ 14,0	+1200	+ 50	0
1	− 22,0	34,02	+ 115	+ 14,5	+1050	+ 40	1
2	− 20,8	34,17	+ 126	+ 14,7	+ 910	+ 30	2
3	− 19,7	34,30	+ 135	+ 15,1	+ 880	+ 20	3
4	− 18,2	34,46	+ 143	+ 15,3	+ 690	+ 10	4
5	− 16,7	34,59	+ 148	+ 15,5	+ 590	0	5
6	− 15,0	34,73	+ 152	+ 15,4	+ 490	− 10	6
7	− 13,6	34,87	+ 157	+ 15,0	+ 400	− 20	7
8	− 12,3	35,02	+ 160	+ 14,7	+ 330	− 30	8
9	− 10,6	35,17	+ 162	+ 14,0	+ 270	− 40	9
10	− 9,1	35,30	+ 165	+ 13,5	+ 210	− 60	10
11	− 7,2	35,43	+ 167	+ 13,0	+ 170	− 70	11
12	− 5,7	35,54	+ 168	+ 12,1	+ 120	− 110	12
13	− 4,2	35,67	+ 169	+ 11,0	+ 90	− 170	13
14	− 2,4	35,77	+ 171	+ 9,0	0	− 260	14
15	− 0,8	35,86	+ 170	+ 5,0	− 30	− 330	15
16	+ 0,7	35,88	+ 170	+ 2,0	− 90	− 450	16
17	+ 2,3	35,90	+ 169	0	− 140	− 520	17
18	+ 4,0	35,85	+ 167	− 5,0	− 190	− 380	18
19	+ 5,9	35,78	+ 164	− 8,0	− 240	− 270	19
20	+ 7,3	35,65	+ 161	− 10,5	− 310	− 170	20
21	+ 8,8	35,54	+ 157	− 11,8	− 370	− 120	21
22	+ 10,4	35,43	+ 153	− 12,7	− 450	− 90	22
23	+ 12,2	35,30	+ 148	− 13,6	− 520	− 80	23
24	+ 13,4	35,15	+ 143	− 14,4	− 590	− 90	24
25	+ 14,7	35,00	+ 135	− 15,5	− 640	− 110	25
26	+ 16,1	34,84	+ 128	− 16,7	− 720	− 140	26
27	+ 17,2	34,69	+ 119	− 18,0	− 790	− 170	27
28	+ 18,3	34,51	+ 109	− 19,5	− 890	− 270	28
29	+ 19,2	34,32	+ 99	− 21,7	− 980	− 150	29
30	+ 20,2	34,13	+ 89	− 22,1	−1100	0	30
31	+ 21,2	33,93	+ 78	− 22,0	−1250	+ 660	31
32	+ 22,0	33,78	+ 67	− 16,6	−1730	+ 400	32
33	+ 22,7	33,63	+ 49	− 15,0	−1470	+ 220	33
34	+ 23,2	33,50	+ 37	− 13,7	− 900	+ 120	34
35	+ 23,7	33,38	+ 34	− 12,6	0	+ 40	35
36	+ 24,2	33,25	+ 37	− 12,3	+ 500	0	36
37	+ 24,6	33,10	+ 38	− 12,7	0	− 40	37
38	+ 24,8	32,98	+ 32	− 13,1	− 980	− 70	38
39	+ 25,0	32,82	+ 20	− 13,8	−1100	− 80	39
40	+ 25,1	32,68	+ 11	− 14,5	−1050	− 90	40
41	+ 25,2	32,52	0	− 14,9	− 870	− 100	41
41,5	+ 25,1	32,44	− 3	− 15,4	− 730	− 110	41,5

velocity of gait. Thereafter, the relative velocity becomes negative and reaches a minimum in the 30th phase. Then it increases again, or the negative velocity decreases, until it reaches 0 in the 34th phase. After that, it becomes positive again and, except for the right thigh in experiment I, attains a maximum shortly before heel strike. This maximum, however, is smaller than the first. As long as the relative velocity is negative, of course, the centre of gravity of the thigh moves back from the coronal plane which is moving forward regularly.

Table 22. Movement of the centre of gravity of the left lower leg during the swinging period in experiment I

Time (0,01 s)	Co-ordinates (cm)		Velocities (cm s⁻¹)		Accelerations (cm s⁻²)		Time (0,01 s)
	ξ_5 Relative	z_5	Direction of gait Relative	Vertical direction	Direction of gait	Vertical direction	
0	− 23,3	36,05	+ 103	+ 24,5	+ 820	+ 130	0
1	− 22,2	36,31	+ 110	+ 25,3	+ 740	+ 70	1
2	− 21,2	36,53	+ 116	+ 25,6	+ 670	0	2
3	− 19,8	36,79	+ 124	+ 25,2	+ 600	− 70	3
4	− 18,7	36,98	+ 128	+ 24,5	+ 530	− 110	4
5	− 17,6	37,18	+ 133	+ 22,6	+ 480	− 150	5
6	− 16,0	37,39	+ 137	+ 20,6	+ 420	− 200	6
7	− 14,4	37,58	+ 141	+ 18,5	+ 380	− 230	7
8	− 12,8	37,77	+ 144	+ 16,2	+ 320	− 270	8
9	− 11,5	37,93	+ 147	+ 13,7	+ 270	− 300	9
10	− 10,0	38,06	+ 150	+ 10,8	+ 230	− 330	10
11	− 8,7	38,14	+ 152	+ 7,4	+ 190	− 380	11
12	− 7,3	38,19	+ 154	+ 4,0	+ 170	− 430	12
13	− 5,8	38,23	+ 156	0	+ 130	− 500	13
14	− 4,3	38,21	+ 157	− 4,0	+ 110	− 450	14
15	− 2,4	38,16	+ 158	− 7,2	+ 90	− 350	15
16	− 1,0	38,08	+ 159	− 9,8	+ 50	− 290	16
17	+ 0,9	37,98	+ 160	− 12,8	0	− 230	17
18	+ 2,7	37,83	+ 159	− 15,1	− 50	− 170	18
19	+ 4,1	37,64	+ 157	− 16,7	− 100	− 130	19
20	+ 5,6	37,49	+ 155	− 17,8	− 160	− 100	20
21	+ 6,8	37,32	+ 152	− 18,8	− 230	− 110	21
22	+ 8,1	37,15	+ 149	− 20,0	− 290	− 140	22
23	+ 9,6	36,94	+ 145	− 21,7	− 400	− 190	23
24	+ 10,8	36,72	+ 142	− 23,5	− 520	− 160	24
25	+ 11,9	36,49	+ 137	− 27,4	− 660	− 60	25
26	+ 13,3	36,20	+ 130	− 28,0	− 820	0	26
27	+ 14,5	35,96	+ 122	− 26,8	− 930	+ 250	27
28	+ 16,0	35,72	+ 111	− 23,5	− 1270	+ 340	28
29	+ 17,3	35,51	+ 98	− 20,7	− 1520	+ 230	29
30	+ 18,2	35,31	+ 82	− 18,8	− 1700	+ 170	30
31	+ 18,9	35,10	+ 64	− 17,5	− 1610	+ 130	31
32	+ 19,6	34,92	+ 51	− 16,1	− 1380	+ 110	32
33	+ 20,1	34,76	+ 35	− 14,9	− 1050	+ 80	33
34	+ 20,5	34,61	+ 30	− 13,8	− 470	+ 70	34
35	+ 20,7	34,48	+ 28	− 13,0	0	+ 40	35
36	+ 21,1	34,23	+ 32	− 12,6	+ 320	+ 20	36
37	+ 21,4	34,20	+ 33	− 12,2	0	0	37
38	+ 21,8	34,06	+ 27	− 12,3	− 730	− 20	38
39	+ 22,0	33,93	+ 16	− 12,5	− 1050	− 30	39
40	+ 22,1	33,79	+ 6	− 12,7	− 1130	− 60	40
41	+ 22,2	33,66	0	− 13,2	− 1000	− 80	41
41,7	+ 22,1	33,53	− 11	− 13,7	− 770	− 90	41,7

Positive acceleration corresponds to the initial small increase of the velocity. When the velocity reaches its first maximum, the acceleration passes through 0 and gives way to deceleration of the positive velocity or to acceleration of the negative relative velocity, directed backwards. In about the 20th phase, it reaches its greatest absolute value and then decreases again until it returns to 0 at the instant of minimum velocity. Then the acceleration becomes positive. At first, the relative velocity is still negative. Therefore, a positive

Table 23. Movement of the centre of gravity of the right lower leg during the swinging period in experiment II

Time (0,01 s)	Co-ordinates (cm)		Velocities (cm s⁻¹)		Accelerations (cm s⁻²)		Time (0,01 s)
	ξ_4 Relative	z_4	Direction of gait Relative	Vertical direction	Direction of gait	Vertical direction	
0	− 21,7	33,73	+ 107	+ 17,0	+ 980	+ 90	0
1	− 20,7	33,87	+ 117	+ 17,8	+ 880	+ 80	1
2	− 19,6	34,01	+ 126	+ 18,4	+ 770	+ 70	2
3	− 18,4	34,17	+ 132	+ 18,7	+ 670	+ 50	3
4	− 17,2	34,31	+ 137	+ 19,1	+ 580	+ 30	4
5	− 15,5	34,51	+ 143	+ 19,2	+ 480	+ 10	5
6	− 13,9	34,70	+ 148	+ 19,3	+ 400	0	6
7	− 12,3	34,89	+ 151	+ 19,0	+ 320	− 20	7
8	− 10,9	35,08	+ 154	+ 18,5	+ 240	− 40	8
9	− 9,2	35,27	+ 158	+ 17,8	+ 180	− 80	9
10	− 7,5	35,45	+ 159	+ 16,8	+ 120	− 110	10
11	− 6,0	35,63	+ 160	+ 15,8	+ 70	− 150	11
12	− 4,5	35,78	+ 161	+ 14,5	0	− 190	12
13	− 3,0	35,91	+ 160	+ 12,2	− 30	− 230	13
14	− 1,3	36,00	+ 159	+ 10,0	− 70	− 280	14
15	+ 0,1	36,08	+ 159	+ 7,0	− 110	− 320	15
16	+ 1,6	36,14	+ 158	+ 3,0	− 160	− 380	16
17	+ 3,0	36,17	+ 156	0	− 200	− 460	17
18	+ 4,7	36,13	+ 153	− 6,0	− 220	− 530	18
19	+ 5,9	36,05	+ 151	− 10,0	− 280	− 370	19
20	+ 7,2	35,93	+ 148	− 13,5	− 350	− 210	20
21	+ 8,7	35,76	+ 144	− 16,0	− 410	− 180	21
22	+ 9,9	35,53	+ 140	− 17,7	− 470	− 170	22
23	+ 11,4	35,31	+ 136	− 19,5	− 520	− 190	23
24	+ 12,6	35,08	+ 131	− 21,2	− 610	− 200	24
25	+ 13,9	34,83	+ 125	− 22,8	− 690	− 130	25
26	+ 15,0	34,62	+ 117	− 23,7	− 790	− 90	26
27	+ 16,1	34,42	+ 109	− 24,0	− 880	− 60	27
28	+ 17,1	34,21	+ 100	− 24,2	− 980	0	28
29	+ 18,2	34,00	+ 90	− 24,1	− 1080	+ 30	29
30	+ 19,4	33,78	+ 79	− 23,2	− 1240	+ 120	30
31	+ 20,2	33,57	+ 67	− 21,6	− 1680	+ 220	31
32	+ 20,9	33,33	+ 50	− 18,0	− 1400	+ 340	32
33	+ 21,4	33,13	+ 40	− 15,8	− 630	+ 280	33
34	+ 21,9	33,01	+ 36	− 13,5	0	+ 220	34
35	+ 22,3	32,89	+ 38	− 11,3	+ 460	+ 160	35
36	+ 22,8	32,81	+ 41	− 9,7	+ 350	+ 110	36
37	+ 23,2	32,72	+ 43	− 8,7	0	+ 60	37
38	+ 23,7	32,62	+ 36	− 8,1	− 1050	0	38
49	+ 23,9	32,55	+ 25	− 8,3	− 1300	− 80	39
40	+ 24,1	32,48	+ 12	− 9,5	− 1270	− 150	40
40,8	+ 24,2	32,44	0	− 10,6	− 1120	− 200	40,8

acceleration means a decrease of the relative velocity directed backwards. A soon as the velocity itself becomes positive, a positive acceleration means an increase of the velocity directed forwards.

The circumstances are different for the absolute velocity of the centre of gravity of the thigh, which is always positive. A positive or a negative value of the acceleration thus indicates that the absolute velocity increases or decreases, that the movement is accelerated or decelerated. This must be kept in mind to ensure correct interpretation of the curves.

Table 24. Movement of the centre of gravity of the left lower leg during the swinging period in experiment II

Time (0,01 s)	Co-ordinates (cm)		Velocities (cm s⁻¹)		Accelerations (cm s⁻²)		Time (0,01 s)
	ξ_5 Relative	z_5	Direction of gait Relative	Vertical direction	Direction of gait	Vertical direction	
0	− 22,3	35,75	+ 98	+ 24,2	+ 970	+ 80	0
1	− 21,5	35,94	+ 105	+ 24,8	+ 900	+ 70	1
2	− 20,5	36,19	+ 114	+ 25,0	+ 810	+ 30	2
3	− 19,1	36,43	+ 122	+ 25,2	+ 710	0	3
4	− 17,8	36,65	+ 130	+ 25,1	+ 630	− 40	4
5	− 16,2	36,93	+ 133	+ 24,6	+ 570	− 80	5
6	− 14,7	37,18	+ 139	+ 22,9	+ 520	− 120	6
7	− 13,3	37,39	+ 144	+ 21,5	+ 480	− 170	7
8	− 11,8	37,57	+ 148	+ 19,1	+ 420	− 210	8
9	− 10,5	37,78	+ 151	+ 17,0	+ 380	− 250	9
10	− 8,8	37,93	+ 154	+ 14,2	+ 320	− 290	10
11	− 7,5	38,03	+ 158	+ 11,2	+ 280	− 320	11
12	− 5,9	38,13	+ 161	+ 8,0	+ 220	− 380	12
13	− 4,5	38,18	+ 163	+ 5,0	+ 170	− 400	13
14	− 2,7	38,21	+ 164	0	+ 100	− 450	14
15	− 1,1	38,19	+ 167	− 5,0	0	− 510	15
16	+ 0,4	38,14	+ 166	− 9,0	− 70	− 590	16
17	+ 2,0	38,01	+ 165	− 12,5	− 140	− 470	17
18	+ 3,8	37,88	+ 162	− 17,5	− 230	− 330	18
19	+ 5,3	37,69	+ 160	− 20,5	− 350	− 210	19
20	+ 6,9	37,46	+ 156	− 22,0	− 470	− 150	20
21	+ 8,4	37,22	+ 150	− 22,8	− 600	− 110	21
22	+ 9,9	37,01	+ 143	− 23,7	− 720	− 90	22
23	+ 11,3	36,72	+ 134	− 24,9	− 830	− 110	23
24	+ 12,8	36,49	+ 126	− 26,0	− 930	− 150	24
25	+ 14,2	36,26	+ 116	− 27,6	− 1020	− 180	25
26	+ 15,2	35,99	+ 105	− 28,9	− 1150	0	26
27	+ 16,2	35,70	+ 95	− 28,3	− 1320	+ 110	27
28	+ 17,2	35,45	+ 80	− 26,0	− 1500	+ 300	28
29	+ 18,0	35,19	+ 67	− 21,5	− 1390	+ 410	29
30	+ 18,6	34,99	+ 51	− 17,8	− 1120	+ 380	30
31	+ 19,2	34,87	+ 42	− 15,0	− 580	+ 290	31
32	+ 19,7	34,69	+ 38	− 12,4	0	+ 220	32
33	+ 20,1	34,57	+ 41	− 10,6	+ 450	+ 140	33
34	+ 20,6	34,47	+ 44	− 10,1	0	+ 50	34
35	+ 21,2	34,37	+ 42	− 9,9	− 470	0	35
36	+ 21,8	34,26	+ 36	− 10,0	− 860	− 50	36
37	+ 21,9	34,15	+ 27	− 10,7	− 1070	− 80	37
38	+ 22,2	34,00	+ 16	− 11,5	− 1110	− 110	38
39	+ 22,3	33,88	0	− 12,6	− 1000	− 130	39
40	+ 22,2	33,73	− 7	− 13,8	− 810	− 150	40
40,9	+ 22,1	33,60	− 12	− 15,2	− 680	− 160	40,9

The positive acceleration reaches a maximum in the 36th phase and passes through 0 again about 0.02 s before heel strike so that it is negative at heel strike. In experiment I, however, the acceleration of the centre of gravity of the right thigh is still positive at the end of the swinging period, albeit small.

The movement of the centre of gravity of the thigh in the vertical direction is as follows.

Table 25. Movement of the centre of gravity of the right lower leg during the swinging period in experiment III

Time (0,01 s)	Co-ordinates (cm)		Velocities (cm s⁻¹)		Accelerations (cm s⁻²)		Time (0,01 s)
	ξ_4 Relative	z_4	Direction of gait Relative	Vertical direction	Direction of gait	Vertical direction	
0	− 22,3	36,84	+ 86	+ 19,3	+ 650	− 30	0
1	− 21,4	36,97	+ 92	+ 18,8	+ 610	− 40	1
2	− 20,6	37,11	+ 98	+ 18,2	+ 570	− 50	2
3	− 19,6	37,29	+ 103	+ 17,5	+ 530	− 60	3
4	− 18,4	37,46	+ 109	+ 16,8	+ 510	− 80	4
5	− 17,3	37,58	+ 113	+ 15,8	+ 480	− 110	5
6	− 16,3	37,75	+ 119	+ 14,3	+ 460	− 110	6
7	− 15,2	37,89	+ 122	+ 13,1	+ 420	− 150	7
8	− 14,0	37,99	+ 127	+ 11,2	+ 400	− 190	8
9	− 12,9	38,10	+ 131	+ 8,3	+ 380	− 210	9
10	− 11,7	38,19	+ 135	+ 7,1	+ 350	− 270	10
11	− 10,3	38,23	+ 140	+ 4,2	+ 320	− 320	11
12	− 9,2	38,29	+ 143	0	+ 290	− 480	12
13	− 7,8	38,28	+ 146	− 7,0	+ 250	−1000	13
14	− 6,4	38,22	+ 149	− 13,0	+ 180	− 400	14
15	− 5,1	38,08	+ 151	− 15,5	+ 120	− 220	15
16	− 3,4	37,92	+ 152	− 16,8	+ 70	− 150	16
17	− 1,9	37,74	+ 153	− 17,8	0	− 100	17
18	0	37,57	+ 152	− 19,3	− 110	− 110	18
19	+ 1,1	37,35	+ 151	− 20,8	− 210	− 150	19
20	+ 2,8	37,13	+ 149	− 22,0	− 340	− 180	20
21	+ 4,5	36,94	+ 145	− 23,2	− 470	− 160	21
22	+ 5,8	36,71	+ 140	− 24,1	− 540	− 90	22
23	+ 7,2	36,47	+ 133	− 24,7	− 620	− 50	23
24	+ 8,5	36,25	+ 126	− 24,9	− 680	0	24
25	+ 9,7	36,03	+ 119	− 24,6	− 710	+ 50	25
26	+ 10,8	35,77	+ 111	− 23,5	− 770	+ 100	26
27	+ 11,8	35,53	+ 103	− 21,9	− 800	+ 190	27
28	+ 12,8	35,34	+ 93	− 19,8	− 860	+ 230	28
29	+ 13,7	35,14	+ 85	− 18,5	− 900	+ 200	29
30	+ 14,7	34,98	+ 76	− 15,9	−1010	+ 160	30
31	+ 15,6	34,83	+ 65	− 14,2	−1110	+ 120	31
32	+ 16,2	34,70	+ 53	− 13,2	−1150	+ 60	32
33	+ 16,8	34,54	+ 43	− 12,7	−1000	0	33
34	+ 17,2	34,32	+ 36	− 12,9	− 550	− 60	34
35	+ 17,7	34,28	+ 33	− 13,5	0	− 110	35
36	+ 18,1	34,13	+ 36	− 14,7	+ 420	− 120	36
37	+ 18,5	33,98	+ 40	− 15,8	0	− 160	37
38	+ 18,8	33,80	+ 37	− 17,0	− 720	− 180	38
39	+ 19,1	33,64	+ 27	− 18,5	−1300	− 200	39
40	+ 19,3	33,48	+ 12	− 20,4	−1420	− 220	40
41	+ 19,4	33,30	0	− 22,7	−1040	− 240	41
41,45	+ 19,3	33,27	− 4	− 23,2	− 950	− 250	41,45

At the beginning of the swinging period, the centre of gravity of the thigh displays a negative velocity, directed downwards, so that at first it comes closer to the floor. The velocity soon increases, passes through 0 and becomes positive. From this point, the centre of gravity thus moves upwards, at first with increasing and then with decreasing velocity. The maximum velocity occurs about 0.12 s after toe-off. In the 22nd phase, the velocity is again 0.

Table 26. Movement of the centre of gravity of the left lower leg during the swinging period in experiment III

Time (0,01 s)	Co-ordinates (cm) ξ_5 Relative	z_5	Velocities (cm s⁻¹) Direction of gait Relative	Vertical direction	Accelerations (cm s⁻²) Direction of gait	Vertical direction	Time (0,01 s)
0	− 21,5	33,55	+ 107	+ 23,0	+1030	+ 180	0
1	− 20,6	33,74	+ 116	+ 24,5	+ 930	+ 160	1
2	− 19,2	34,03	+ 125	+ 25,9	+ 780	+ 110	2
3	− 17,7	34,34	+ 131	+ 27,0	+ 680	+ 70	3
4	− 16,6	34,62	+ 139	+ 27,8	+ 570	0	4
5	− 15,0	34,89	+ 142	+ 27,7	+ 480	− 50	5
6	− 13,5	35,13	+ 148	+ 26,9	+ 390	− 80	6
7	− 11,9	35,40	+ 151	+ 25,7	+ 310	− 120	7
8	− 10,5	35,68	+ 153	+ 24,5	+ 220	− 170	8
9	− 8,8	35,92	+ 154	+ 23,0	+ 160	− 210	9
10	− 7,4	36,13	+ 155	+ 20,7	+ 80	− 250	10
11	− 6,0	36,32	+ 156	+ 18,5	0	− 290	11
12	− 4,3	36,50	+ 154	+ 15,0	− 40	− 350	12
13	− 2,9	36,62	+ 153	+ 11,0	− 90	− 380	13
14	− 1,3	36,67	+ 151	+ 7,4	− 120	− 440	14
15	0	36,70	+ 150	0	− 170	− 550	15
16	+ 1,4	36,69	+ 148	− 2,0	− 210	− 770	16
17	+ 2,9	36,63	+ 146	− 7,5	− 270	− 520	17
18	+ 4,5	36,51	+ 143	− 11,0	− 320	− 380	18
19	+ 5,8	36,41	+ 140	− 14,5	− 370	− 300	19
20	+ 7,1	36,29	+ 136	− 17,0	− 420	− 240	20
21	+ 8,3	36,12	+ 131	− 19,0	− 470	− 180	21
22	+ 9,6	35,91	+ 127	− 20,9	− 530	− 150	22
23	+ 11,1	35,70	+ 120	− 22,1	− 580	− 120	23
24	+ 12,3	35,52	+ 113	− 23,5	− 630	− 110	24
25	+ 13,4	35,29	+ 107	− 24,6	− 680	− 220	25
26	+ 14,5	35,02	+ 100	− 27,0	− 730	− 470	26
27	+ 15,5	34,74	+ 93	− 28,7	− 780	0	27
28	+ 16,3	34,48	+ 85	− 28,3	− 830	+ 110	28
29	+ 17,2	34,24	+ 76	− 27,0	− 910	+ 210	29
30	+ 18,0	34,00	+ 68	− 25,5	− 960	+ 220	30
31	+ 18,7	33,76	+ 58	− 24,0	− 1000	+ 200	31
32	+ 19,2	33,48	+ 48	− 23,5	− 980	+ 190	32
33	+ 19,6	33,25	+ 39	− 21,5	− 920	+ 170	33
34	+ 19,9	33,07	+ 30	− 20,0	− 810	+ 160	34
35	+ 20,1	32,86	+ 21	− 18,5	− 630	+ 160	35
36	+ 20,3	32,67	+ 17	− 17,4	− 440	+ 160	36
37	+ 20,4	32,48	+ 9	− 16,3	− 290	+ 140	37
38	+ 20,5	32,27	+ 7	− 15,0	− 180	+ 120	38
39	+ 20,6	32,10	0	− 13,7	− 190	+ 90	39
40	+ 20,5	31,97	− 2	− 12,6	− 320	0	40
41	+ 20,5	31,83	− 7	− 12,8	− 480	− 60	41
41,3	+ 20,4	31,80	− 9	− 13,2	− 510	− 100	41,3

From then on, the centre of gravity of the thigh again moves downwards, first with increasing and then with decreasing velocity. The latter, of course, is negative. The maximum downward velocity occurs in the 31st phase. Shortly before heel strike, the centre of gravity again changes direction, except for the right thigh in experiment III. However, during this last movement upwards it does not achieve any great velocity. In one case (right leg,

Table 27. Movement of the centre of gravity of the right foot during the swinging period in experiment I

Time (0,01 s)	Co-ordinates (cm)		Velocities (cm s^{-1})		Accelerations (cm s^{-2})		Time (0,01 s)
	ξ_6 Relative	z_6	Direction of gait Relative	Vertical direction	Direction of gait	Vertical direction	
0	−40,7	13,23	0	+100	+3490	0	0
1	−40,5	14,12	+ 29	+ 96	+3470	− 500	1
2	−40,0	15,15	+ 57	+ 88	+3330	−1600	2
3	−39,3	15,96	+ 86	+ 70	+3120	−2130	3
4	−38,4	16,48	+112	+ 52	+2850	−2080	4
5	−37,2	16,73	+135	+ 35	+2540	−1770	5
6	−35,7	16,85	+158	+ 11	+2300	−1560	6
7	−34,1	16,86	+178	0	+1970	−1370	7
8	−32,5	16,80	+195	− 17	+1650	−1160	8
9	−30,7	16,47	+209	− 28	+1440	−1010	9
10	−28,5	16,10	+221	− 37	+1030	− 860	10
11	−26,0	15,75	+232	− 43	+ 720	− 730	11
12	−23,7	15,23	+242	− 50	+ 500	− 600	12
13	−21,5	14,74	+250	− 57	+ 540	− 470	13
14	−19,4	14,10	+259	− 61	+ 710	− 350	14
15	−16,5	13,40	+267	− 66	+ 860	− 230	15
16	−13,8	12,70	+273	− 67	+ 900	− 110	16
17	−10,7	12,05	+282	− 68	+ 840	0	17
18	− 7,6	11,50	+289	− 65	+ 710	+ 200	18
19	− 4,7	10,94	+295	− 59	+ 520	+ 360	19
20	− 1,2	10,33	+302	− 53	+ 330	+ 530	20
21	+ 1,8	9,84	+307	− 47	+ 170	+ 690	21
22	+ 5,5	9,38	+318	− 37	0	+ 850	22
23	+ 9,2	9,07	+312	− 28	− 210	+1010	23
24	+12,0	8,87	+311	− 19	− 400	+1150	24
25	+15,2	8,82	+307	− 8	− 670	+1270	25
26	+18,3	8,80	+302	0	− 870	+1360	26
27	+22,1	8,95	+293	+ 17	−1110	+1420	27
28	+24,7	9,20	+283	+ 31	−1300	+1370	28
29	+27,4	9,58	+267	+ 47	−1550	+1300	29
30	+30,1	10,10	+251	+ 60	−1760	+1120	30
31	+32,3	10,85	+227	+ 68	−2080	+ 850	31
32	+34,9	11,50	+204	+ 72	−2490	0	32
33	+37,0	12,14	+181	+ 66	−2950	− 600	33
34	+38,7	12,65	+152	+ 52	−3600	−2300	34
35	+39,9	12,87	+113	+ 30	−5500	−3100	35
36	+40,6	12,92	+ 68	0	−8100	−3170	36
37	+40,8	12,85	0	− 29	−6000	−2500	37
38	+40,5	12,72	− 70	− 52	−2500	−1900	38
39	+39,8	12,30	− 75	− 67	0	−1250	39
40	+39,2	11,58	− 67	− 74	+1450	− 500	40
41	+38,3	11,17	− 54	− 77	+1530	0	41
41,5	+37,8	11,00	− 52	− 75	+1200	+ 400	41,5

experiment I), the velocity is 0 at heel strike. In another case (right leg, experiment II), it even has a small negative value. It can thus be generally asserted that the velocity of the vertical movement of the centre of gravity of the thigh is 0 at the end of the swinging period.

The movement of the *centre of gravity of the lower leg* is characterized by relatively high velocity in the direction of gait but by low velocity in the vertical direction.

Table 28. Movement of the centre of gravity of the left foot during the swinging period in experiment I

Time (0,01 s)	Co-ordinates (cm) ξ_η Relative	Co-ordinates (cm) z_η	Velocities (cm s^{-1}) Direction of gait Relative	Velocities (cm s^{-1}) Vertical direction	Accelerations (cm s^{-2}) Direction of gait	Accelerations (cm s^{-2}) Vertical direction	Time (0,01 s)
0	−41,6	13,60	0	+129	+2800	0	0
1	−41,3	14,40	+ 20	+127	+3040	− 900	1
2	−40,8	15,50	+ 48	+116	+3080	−1750	2
3	−40,0	16,10	+ 77	+ 95	+2970	−2330	3
4	−39,1	16,75	+106	+ 70	+2800	−2530	4
5	−38,0	17,22	+130	+ 48	+2540	−2490	5
6	−36,3	17,46	+157	+ 24	+2350	−2200	6
7	−34,5	17,64	+178	0	+2130	−1850	7
8	−32,7	17,56	+301	− 17	+1980	−1530	8
9	−30,6	17,40	+218	− 30	+1690	−1250	9
10	−28,4	16,92	+235	− 42	+1490	−1080	10
11	−25,9	16,35	+248	− 52	+1270	− 880	11
12	−23,5	15,80	+261	− 58	+1060	− 670	12
13	−20,8	15,15	+269	− 64	+ 850	− 490	13
14	−18,1	14,50	+277	− 68	+ 670	− 330	14
15	−15,9	13,90	+282	− 71	+ 450	− 180	15
16	−13,2	13,25	+286	− 73	+ 340	0	16
17	−10,3	12,65	+289	− 71	+ 240	+ 100	17
18	− 6,5	11,85	+292	− 69	+ 220	+ 250	18
19	− 4,0	11,20	+295	− 67	+ 330	+ 390	19
20	− 1,3	10,60	+301	− 63	+ 600	+ 500	20
21	+ 1,5	10,15	+306	− 54	+ 780	+ 660	21
22	+ 4,2	9,60	+311	− 47	+ 700	+ 800	22
23	+ 8,0	9,25	+317	− 36	+ 470	+ 960	23
24	+10,8	8,90	+320	− 24	0	+1130	24
25	+14,0	8,63	+320	− 13	− 190	+1350	25
26	+17,3	8,52	+317	0	− 480	+1540	26
27	+21,5	8,56	+310	+ 13	− 750	+1720	27
28	+24,5	8,76	+298	+ 29	−1020	+1670	28
29	+28,3	9,07	+287	+ 48	−1340	+1420	29
30	+30,6	9,59	+268	+ 59	−1950	+1050	30
31	+33,0	10,30	+243	+ 71	−2980	+ 600	31
32	+34,7	11,20	+217	+ 73	−3740	0	32
33	+36,4	11,80	+182	+ 70	−4150	− 550	33
34	+37,5	12,40	+148	+ 59	−4400	−1250	34
35	+38,5	12,57	+100	+ 33	−4420	−2500	35
36	+39,3	12,63	+ 50	0	−4370	−4400	36
37	+39,8	12,58	0	− 28	−4150	−3300	37
38	+39,5	12,26	− 41	− 55	−3620	−1850	38
39	+39,0	11,60	− 71	− 72	−2700	− 850	39
40	+38,2	10,90	− 92	− 80	−1250	0	40
41	+37,2	10,10	− 96	− 79	0	+ 250	41
41,7	+36,7	9,55	− 89	− 76	+1200	+ 500	41,7

The relative velocity in the direction of gait is positive almost until the end of the swinging period and it is significantly higher than the simultaneous horizontal velocity of the centre of gravity of the thigh. Only near the end of the swinging period does the centre of gravity of the lower leg move more slowly than that of the thigh. In most cases, its relative velocity has even become negative, though not significantly so, backwards. From the beginning of the

Table 29. Movement of the centre of gravity of the right foot during the swinging period in experiment II

Time (0,01 s)	Co-ordinates (cm)		Velocities (cm s^{-1})		Accelerations (cm s^{-2})		Time (0,01 s)
	ξ_6 Relative	z_6	Direction of gait Relative	Vertical direction	Direction of gait	Vertical direction	
0	− 39,9	13,41	0	+ 89	+ 3570	0	0
1	− 39,8	14,30	+ 22	+ 87	+ 3540	− 480	1
2	− 39,4	15,15	+ 50	+ 77	+ 3270	− 1040	2
3	− 38,5	15,95	+ 81	+ 62	+ 2920	− 1570	3
4	− 37,3	16,40	+ 107	+ 45	+ 2540	− 1800	4
5	− 36,1	16,73	+ 130	+ 31	+ 2240	− 1750	5
6	− 35,0	16,87	+ 149	+ 15	+ 2000	− 1570	6
7	− 33,3	16,88	+ 168	0	+ 1780	− 1380	7
8	− 31,6	16,80	+ 186	− 11	+ 1570	− 1180	8
9	− 29,0	16,64	+ 202	− 21	+ 1350	− 1050	9
10	− 26,9	16,27	+ 218	− 30	+ 1130	− 920	10
11	− 24,7	15,95	+ 229	− 40	+ 930	− 830	11
12	− 22,3	15,44	+ 238	− 50	+ 820	− 690	12
13	− 20,0	14,90	+ 249	− 57	+ 860	− 580	13
14	− 16,9	14,43	+ 260	− 61	+ 1020	− 440	14
15	− 14,8	13,87	+ 271	− 64	+ 1210	− 310	15
16	− 12,3	13,02	+ 282	− 65	+ 1230	− 130	16
17	− 9,4	12,38	+ 290	− 66	+ 1090	0	17
18	− 6,3	11,69	+ 297	− 64	+ 870	+ 180	18
19	− 3,6	11,13	+ 301	− 61	+ 580	+ 310	19
20	0	10,39	+ 307	− 57	+ 270	+ 480	20
21	+ 2,9	9,78	+ 309	− 51	0	+ 670	21
22	+ 6,0	9,15	+ 308	− 46	− 180	+ 820	22
23	+ 8,7	8,83	+ 307	− 37	− 300	+ 1000	23
24	+ 11,9	8,62	+ 304	− 25	− 420	+ 1150	24
25	+ 15,0	8,52	+ 300	− 12	− 510	+ 1300	25
26	+ 18,1	8,50	+ 294	0	− 630	+ 1500	26
27	+ 21,0	8,58	+ 285	+ 18	− 770	+ 1650	27
28	+ 23,8	8,78	+ 275	+ 33	− 990	+ 1570	28
29	+ 26,3	9,19	+ 260	+ 48	− 1380	+ 1350	29
30	+ 29,2	9,66	+ 242	+ 57	− 1880	+ 1010	30
31	+ 32,1	10,12	+ 225	+ 63	− 2520	+ 380	31
32	+ 34,5	10,73	+ 200	+ 66	− 3350	0	32
33	+ 35,9	11,40	+ 171	+ 56	− 5260	− 2350	33
34	+ 37,0	11,85	+ 120	+ 42	− 5200	− 3700	34
35	+ 38,2	12,28	+ 78	+ 15	− 4580	− 4200	35
36	+ 38,8	12,59	+ 23	0	− 3750	− 2650	36
37	+ 39,0	12,54	0	− 38	− 2860	− 1700	37
38	+ 38,8	12,13	− 72	− 51	− 1950	− 950	38
39	+ 38,4	11,10	− 80	− 60	0	− 450	39
40	+ 37,7	10,10	− 73	− 63	+ 2410	0	40
40,8	+ 37,3	9,55	− 62	− 60	+ 2230	+ 160	40,8

swinging period onward, the positive velocity of the centre of gravity at first increases until it has reached a maximum, on average in the 15th phase. The acceleration is thus positive at first, and becomes zero 0.15 s after the beginning of the swinging period. The velocity then decreases further. The acceleration thus becomes negative, that is, it is succeeded by deceleration. The further movement of the centre of gravity of the lower leg remains

Table 30. Movement of the centre of gravity of the left foot during the swinging period in experiment II

Time (0,01 s)	Co-ordinates (cm)		Velocities (cm s⁻¹)		Accelerations (cm s⁻²)		Time (0,01 s)
	ξ_η Relative	z_η	Direction of gait Relative	Vertical direction	Direction of gait	Vertical direction	
0	− 40,5	12,75	0	+ 121	+ 3040	0	0
1	− 40,3	13,58	+ 27	+ 120	+ 3180	− 330	1
2	− 39,8	14,40	+ 58	+ 112	+ 3280	− 1150	2
3	− 39,0	15,45	+ 89	+ 95	+ 3230	− 1820	3
4	− 37,9	16,40	+ 115	+ 87	+ 3040	− 2230	4
5	− 36,7	16,93	+ 143	+ 56	+ 2750	− 2440	5
6	− 35,3	17,23	+ 169	+ 30	+ 2420	− 2480	6
7	− 33,5	17,35	+ 192	0	+ 1930	− 2330	7
8	− 31,6	17,32	+ 212	− 14	+ 1520	− 1960	8
9	− 29,6	17,10	+ 222	− 29	+ 1170	− 1640	9
10	− 27,3	16,70	+ 231	− 45	+ 930	− 1330	10
11	− 25,0	16,17	+ 240	− 54	+ 790	− 1030	11
12	− 22,4	15,55	+ 249	− 62	+ 720	− 800	12
13	− 19,2	15,00	+ 258	− 68	+ 710	− 580	13
14	− 16,9	14,30	+ 265	− 71	+ 770	− 370	14
15	− 14,2	13,54	+ 274	− 74	+ 880	− 170	15
16	− 11,3	12,82	+ 284	− 76	+ 1020	0	16
17	− 8,6	12,15	+ 295	− 73	+ 1100	+ 140	17
18	− 5,6	11,53	+ 305	− 70	+ 1110	+ 290	18
19	− 2,7	10,82	+ 313	− 68	+ 1040	+ 440	19
20	+ 0,6	10,23	+ 322	− 61	+ 900	+ 570	20
21	+ 3,5	9,61	+ 330	− 54	+ 650	+ 700	21
22	+ 7,0	9,12	+ 337	− 47	0	+ 850	22
23	+ 10,1	8,75	+ 336	− 34	− 450	+ 1070	23
24	+ 14,2	8,43	+ 333	− 22	− 930	+ 1220	24
25	+ 17,3	8,36	+ 322	− 10	− 1230	+ 1410	25
26	+ 20,2	8,33	+ 308	0	− 1510	+ 1690	26
27	+ 22,9	8,42	+ 290	+ 28	− 1770	+ 1850	27
28	+ 25,8	8,70	+ 272	+ 49	− 2020	+ 1610	28
29	+ 28,3	9,12	+ 254	+ 67	− 2200	+ 1300	29
30	+ 30,5	9,73	+ 230	+ 82	− 2380	+ 1010	30
31	+ 32,9	10,40	+ 207	+ 92	− 2610	+ 630	31
32	+ 34,8	11,22	+ 180	+ 96	− 2880	0	32
33	+ 36,5	11,82	+ 151	+ 92	− 3210	− 250	33
34	+ 37,8	12,23	+ 107	+ 77	− 3850	− 1250	34
35	+ 39,0	12,37	+ 63	+ 48	− 5500	− 4750	35
36	+ 39,8	12,40	0	0	− 7320	− 4000	36
37	+ 39,6	12,31	− 65	− 40	− 5200	− 3050	37
38	+ 39,1	11,82	− 88	− 63	0	− 1900	38
39	+ 38,4	11,12	− 87	− 76	+ 1050	− 1000	39
40	+ 37,7	10,35	− 80	− 80	+ 1410	0	40
40,9	+ 36,9	9,50	− 67	− 76	+ 1000	+ 600	40,9

decelerated, except for a short time of 0.02 s towards the end of the swinging period during which time the acceleration once again becomes temporarily positive and the velocity increases a little. The deceleration presents two maxima (minima of the acceleration), one 0.31 s, the other 0.38 s after toe-off. In the meantime, the acceleration reaches a small maximum during the brief positive period, approximately in the 36th phase. In

Table 31. Movement of the centre of gravity of the right foot during the swinging period in experiment III

Time (0,01 s)	Co-ordinates (cm) ξ_6 Relative	Co-ordinates (cm) z_6	Velocities (cm s⁻¹) Direction of gait Relative	Velocities (cm s⁻¹) Vertical direction	Accelerations (cm s⁻²) Direction of gait	Accelerations (cm s⁻²) Vertical direction	Time (0,01 s)
0	− 38,5	13,50	− 73	+105	+ 2800	+ 550	0
1	− 38,3	14,20	− 62	+107	+ 2900	0	1
2	− 37,8	15,23	− 45	+ 98	+ 2860	− 1050	2
3	− 37,1	16,00	− 20	+ 82	+ 2700	− 2220	3
4	− 36,3	16,55	0	+ 60	+ 2500	− 2350	4
5	− 35,2	16,98	+ 30	+ 39	+ 2300	− 2170	5
6	− 34,0	17,19	+ 58	+ 22	+ 2050	− 1910	6
7	− 32,3	17,26	+ 80	0	+ 1780	− 1610	7
8	− 30,9	17,24	+107	− 10	+ 1610	− 1390	8
9	− 29,3	17,11	+130	− 22	+ 1360	− 1220	9
10	− 27,4	16,80	+148	− 32	+ 1160	− 1080	10
11	− 25,5	16,25	+163	− 41	+ 950	− 920	11
12	− 23,7	15,55	+178	− 50	+ 780	− 780	12
13	− 21,6	14,73	+197	− 58	+ 650	− 590	13
14	− 19,2	14,14	+211	− 63	+ 620	− 440	14
15	− 16,5	13,45	+226	− 66	+ 760	− 290	15
16	− 14,2	12,80	+237	− 68	+ 870	− 130	16
17	− 11,2	12,10	+248	− 69	+ 930	0	17
18	− 8,1	11,42	+258	− 67	+ 900	+ 150	18
19	− 4,9	10,87	+270	− 64	+ 790	+ 300	19
20	− 1,7	10,12	+282	− 60	+ 620	+ 450	20
21	+ 1,1	9,63	+292	− 52	+ 440	+ 580	21
22	+ 4,0	9,13	+301	− 46	+ 220	+ 700	22
23	+ 7,5	8,55	+315	− 37	0	+ 870	23
24	+ 10,6	8,31	+312	− 28	− 250	+1030	24
25	+ 13,2	8,08	+312	− 12	− 510	+1220	25
26	+ 16,2	8,00	+309	0	− 800	+1410	26
27	+ 19,3	8,11	+302	+ 18	−1130	+1620	27
28	+ 22,6	8,38	+289	+ 37	−1460	+1870	28
29	+ 25,5	8,85	+267	+ 59	−1780	+1990	29
30	+ 28,3	9,40	+243	+ 73	−2080	+1600	30
31	+ 30,8	10,35	+220	+ 82	−2380	0	31
32	+ 32,8	11,27	+190	+ 76	−2700	− 1000	32
33	+ 34,4	11,77	+155	+ 52	− 305c	− 3000	33
34	+ 35,7	11,98	+124	+ 25	− 3850	− 2770	34
35	+ 36,6	12,05	+ 63	0	− 5140	− 2300	35
36	+ 37,3	11,97	0	− 16	− 6400	− 1880	36
37	+ 37,1	11,70	− 45	− 32	− 6800	− 1520	37
38	+ 36,8	11,25	− 70	− 48	− 3500	− 1040	38
39	+ 36,5	10,65	− 75	− 58	0	− 680	39
40	+ 35,9	9,90	− 70	− 64	+ 1170	− 310	40
41	+ 35,2	9,23	− 60	− 67	+ 1020	0	41
41,45	+ 34,8	9,00	− 58	− 65	+ 880	+ 110	41,45

experiment III the left lower leg behaves somewhat differently: the brief period of positive acceleration and the second maximum of the deceleration do not appear. Instead of the small maximum of acceleration, a minimum of deceleration occurs.

Whereas the velocity of the centre of gravity of the lower leg in the direction of gait reaches significantly greater values than that of the thigh, just the opposite occurs for the movement

Table 32. Movement of the centre of gravity of the left foot during the swinging period in experiment III

Time (0,01 s)	Co-ordinates (cm)		Velocities (cm s⁻¹)		Accelerations (cm s⁻²)		Time (0,01 s)
	ξ_7 Relative	z_7	Direction of gait Relative	Vertical direction	Direction of gait	Vertical direction	
0	− 39,8	11,64	+ 7	+133	+ 3400	0	0
1	− 39,7	12,72	+ 30	+131	+ 3380	− 2150	1
2	− 39,1	13,48	+ 62	+110	+ 3260	− 2690	2
3	− 38,2	14,75	+ 90	+ 82	+ 2990	− 2670	3
4	− 36,9	15,74	+ 120	+ 60	+ 2600	− 2400	4
5	− 35,6	16,23	+ 147	+ 40	+ 2170	− 2080	5
6	− 33,9	16,52	+ 166	+ 24	+ 1800	− 1790	6
7	− 32,3	16,70	+ 185	+ 10	+ 1500	− 1540	7
8	− 30,6	16,72	+ 200	0	+ 1200	− 1370	8
9	− 28,5	16,63	+ 211	− 14	+ 970	− 1240	9
10	− 26,4	16,45	+ 222	− 25	+ 770	− 1070	10
11	− 24,3	16,07	+ 229	− 35	+ 720	− 940	11
12	− 22,1	15,53	+ 237	− 45	+ 740	− 740	12
13	− 19,8	15,00	+ 245	− 53	+ 810	− 560	13
14	− 17,5	14,50	+ 252	− 58	+ 920	− 390	14
15	− 14,7	13,90	+ 260	− 62	+ 970	− 220	15
16	− 12,0	13,35	+ 267	− 64	+ 860	− 70	16
17	− 9,2	12,70	+ 273	− 66	+ 750	0	17
18	− 6,3	12,10	+ 280	− 63	+ 600	+ 170	18
19	− 3,7	11,43	+ 287	− 61	+ 450	+ 300	19
20	− 0,7	10,90	+ 293	− 58	+ 280	+ 430	20
21	+ 2,1	10,40	+ 296	− 51	+ 130	+ 540	21
22	+ 5,0	9,92	+ 297	− 45	0	+ 670	22
23	+ 8,2	9,53	+ 296	− 36	− 220	+ 800	23
24	+ 11,1	9,22	+ 292	− 26	− 380	+ 970	24
25	+ 14,1	8,99	+ 287	− 15	− 550	+ 1120	25
26	+ 16,9	8,92	+ 280	0	− 720	+ 1340	26
27	+ 19,8	8,94	+ 271	+ 9	− 910	+ 1650	27
28	+ 22,5	9,08	+ 260	+ 22	− 1090	+ 1820	28
29	+ 25,2	9,33	+ 247	+ 37	− 1400	+ 1890	29
30	+ 27,8	9,22	+ 230	+ 53	− 1850	+ 1250	30
31	+ 29,9	10,20	+ 212	+ 61	− 2850	+ 550	31
32	+ 31,8	11,00	+ 184	+ 63	− 3450	0	32
33	+ 33,7	11,50	+ 151	+ 53	− 3680	− 1470	33
34	+ 35,2	11,95	+ 113	+ 38	− 3740	− 2100	34
35	+ 36,1	12,11	+ 75	+ 17	− 3720	− 3020	35
36	+ 36,7	12,13	+ 39	0	− 3580	− 2550	36
37	+ 36,9	12,00	0	− 25	− 3300	− 1900	37
38	+ 36,8	11,64	− 35	− 42	− 2950	− 1370	38
39	+ 36,5	11,05	− 63	− 52	− 2570	− 900	39
40	+ 35,8	10,60	− 84	− 58	− 1930	− 520	40
41	+ 35,1	10,02	− 98	− 59	− 1180	− 130	41
41,3	+ 34,7	9,85	− 101	− 61	− 1000	0	41,3

in the vertical direction. In this instance, the velocity of the movement of the centre of gravity of the lower leg is so small that the velocity curves in Fig. 13 remain close to the axis of the abscissa. Therefore, they have been drawn again (dotted) in Fig. 13, after a tenfold magnification of the ordinate to give a clearer presentation of the behaviour of the velocity during the swinging period. The velocity of the centre of gravity of the lower leg is positive at the beginning of the swinging period, i.e. directed upwards. In almost all cases, it increases a little at first and then decreases until it has reached 0, on average in the 15th phase. The movement upwards is thus accelerated at first but soon becomes decelerated. After the 15th phase, the velocity becomes negative until the end of the swinging period. The centre of gravity of the lower leg thus moves downwards throughout this time. This downward movement is at first accelerated until about the 27th phase. It is then decelerated for a short time and, approximately from the 36th phase onward, again accelerated.

The movement of the centre of gravity of the foot during the swinging period has been described in Chap. 4. As shown in Figs. 12 and 13, it displays much higher velocities and accelerations than the centres of gravity of the thigh and lower leg. This is seen especially clearly in the curves for the movement in the direction of gait. Moreover, almost throughout the swinging period, the centre of gravity of the foot is moving in the opposite direction to that of the thigh, above all in the vertical direction. When the latter moves upwards, the former moves downwards, and vice versa. When the velocity of the latter in the vertical direction and also when its absolute velocity in the direction of gait reaches a minimum, that of the former is passing through a maximum. Similarly, a maximum of acceleration of one in either direction almost always coincides with a minimum of acceleration of the other, and vice-versa. Between the movements of these two centres of gravity there is an opposition similar to that observed previously for the vertical displacement curves of the centres of the hip and of the ankle (Chap. 2, p. 230). This also appears clearly in Fig. 11.

Summary

The question as to whether the swinging movement of the leg in human gait should be considered as a purely pendulum-type oscillation or not was first posed by Weber and Weber who also provided an answer of sorts, but it has not yet been possible to find a definitive and unequivocal answer to it. As explained in the introduction to this chapter, this can only be decided on the basis of a thorough and precise knowledge of the movements of the leg during the swinging period. Only this knowledge will make deduction of the forces that bring about the swinging movement of the leg possible. As soon as the type and magnitude of these forces have been determined, it becomes easy to judge whether they result only from gravity or whether they require the simultaneous contraction of muscles; and if the latter applies, to determine the torques which the muscles have exerted on the different segments of the leg.

The object of the present part of the study on human gait (Chap. 5) was to build up the kinematic bases necessary to determine the forces acting during the swinging period.

The results of such a kinematic analysis can have general significance only if it can be demonstrated that a wide variety of different individuals walking in the same kind of way essentially follow the same laws of movement. The physiologists of movement are convinced that this is so, apart from a few exceptions. However, it still has to be proved, the more so since the theory actually is contested. Absolute proof could only be provided if the movements involved in walking were analysed in a great many individuals in as much detail as in our one subject in three experiments. As long as no such comparative study exists, we shall have to content ourselves with advancing arguments that support at least the high probability that the laws of movement followed by different individuals are the same. This has been done in the first section of this chapter. On the basis of previous research carried out in more than 100 individuals, it has been shown that, during the type of gait described as the walker's step, data such as step length, number of steps per minute, velocity of gait, etc., which can be observed and measured without thorough kinematic analysis, are basically the same if related to body height and leg length. The walker's step is the kind of pace that people instinctively adopt on country roads or when they want to walk long distances without fatigue. It has also been shown that the step length and the step duration of our subject correspond to the normal walker's step in every way. Moreover, comparison of our chronophotographic exposures with the serial pictures of a walker taken by Marey shows the basic similarity of the successive positions of the upper and lower extremities of Marey's subject and those of our own subject. Consequently, it seems highly probable that the movements made in this walker's step display far-reaching typical characteristics and only quantitative discrepancies between different individuals.

In the second section an attempt was made to determine which data the kinematic analysis of the swinging movement must include to allow examination of the forces involved. With this in view, besides the different positions of the leg during the swinging period, it is necessary to know the angular accelerations of the long axes of the different segments of the leg and the components of the linear accelerations of the centres of gravity of the thigh, lower leg and foot during the swinging period. The methods used to determine these data and the results obtained are described in the third section for the angular accelerations and in the fourth for the components of acceleration of the centres of gravity. These results are illustrated by graphs.

CHAPTER 6

On the Influence of Gravity and the Muscles on the Swinging Movement of the Leg

Introduction

In the previous chapter the kinematic basis necessary for a kinetic analysis of the swinging movement of the leg was thoroughly deduced. Assuming a plane movement, the angular accelerations of the thigh, lower leg and foot during the swinging of the leg were defined for each phase of movement at intervals of 0.01 s. In addition, for the same phases the horizontal and vertical components of the accelerations imparted to the centres of gravity of the three segments of the leg by the forces acting during the swinging period as a consequence of their joint connections were calculated. The relationships between all these data for the case that not only gravity but also muscular forces exert an effect on the movement of the leg were also discussed.

All the premises allowing treatment of the problem of the forces actually exerted in the swinging period were thus collected.

As already mentioned, the equation for the rotatory movement of any of the three segments of the right or left leg during the swing period is

$$m \varkappa^2 \cdot \varphi'' = D_m + D_s + D_e$$

In this equation, m represents the mass, $\varkappa$ the radius of inertia in relation to the transverse axis through the centre of gravity, φ'' the angular acceleration of the segment under consideration and D_m, D_s, D_e the resultant torques exerted by the muscles, gravity and effective forces on this segment.

The masses and radii of inertia of the different segments of the human body were determined previously [1], and the accelerations φ'' in Chap. 5. Therefore, the products $m \varkappa^2 \cdot \varphi''$ for each of the 41 phases of movement defined in Chap. 5 can easily be calculated.

When the masses are known, the torques D_s of gravity can also be calculated for the successive positions of the three segments of the leg. According to the discussion in Chap. 5 [2], gravity acts on each of the three segments of the leg with a couple the forces of which are equal to the weight of the whole leg at thigh level, that of the lower leg and foot at lower leg level and that of the foot at foot level. One force of the couple in each case is applied at the centre of the proximal joint of the segment concerned and pulls vertically upwards. The other, equal and opposite to the first, pulls vertically downwards and is applied at a well-

[1] This information is contained in: Braune W, Fischer O (1889) Über den Schwerpunkt des menschlichen Körpers mit Rücksicht auf die Ausrüstung des Deutschen Infanteristen. Abhandlungen der mathematisch-physischen Klasse der Königlichen Sächsischen Gesellschaft der Wissenschaften, vol 15, no 7, pp 559–672. English translation (1985) On the centre of gravity of the human body. Springer, Berlin Heidelberg New York Tokyo; and (1892) Bestimmung der Trägheitsmomente des menschlichen Körpers und seiner Glieder. Abhandlungen der mathematisch-physischen Klasse der Königlichen Sächsischen Gesellschaft der Wissenschaften, vol 18, no 8, pp 409–492

[2] cf. Figs. 3, 5, and 7 in Chap. 5 with Figs. 3–5 in the present chapter

determined point on the long axis of the body segment. To find the position of this point, it is necessary to imagine that the masses of the body segments distal to the segment under consideration are concentrated at the centre of the distal joint. Then the common centre of gravity of this mass and of the segment under consideration is searched for. Since no distal segment is suspended from the foot, the point of application of the second force of the couple due to gravity coincides with the centre of gravity of the foot itself. For the lower leg, it is necessary to imagine that the mass of the foot is concentrated at the centre of the ankle. The point of application of the second force of the couple thus divides that part of the long axis of the lower leg connecting the centre of gravity of the latter and the ankle in inverse proportion to the masses of the lower leg and of the foot. To find the point of application of the second force of the couple acting on the thigh, the mass of the lower leg and foot is imagined to be concentrated at the centre of the knee. The point of application sought for divides that part of the long axis of the thigh which connects the centre of gravity of the latter and the centre of the knee in inverse proportion to the masses of the thigh and of the lower leg + foot system. The point of application of the second force of any of the three couples can thus be deduced from the masses of the three segments.

The torques of gravity D_s are the moments of these couples. The moment of a couple is the product of the magnitude of one of its two forces and its moment arm. The moment arm is the shortest distance between the lines of action of the two forces. It thus remains to determine the horizontal distance between the lines of action of the two forces for each segment of the leg and for each phase of movement. This is easy as long as the angles formed by the long axes of the three body segments and the vertical are known. These angles have been measured and are given in Tables 6–11 of Chap. 5. They are expressed in radians. A simple calculation allows conversion to degrees. The masses and positional angles indicated previously are thus sufficient to calculate the torques of gravity D_s during the swinging of the leg.

The torque of the effective forces D_e acting on a body segment generally represents the resultant moment of two couples. The forces of one couple are equal in magnitude to the effective force of the body segment itself. The forces of the other couple are equal to the total effective force of these body segments distal to the segment under consideration. The forces of the second couple acting on the thigh are determined by the total effective force of the lower leg and foot. The forces of the second couple acting on the lower leg are equal to the effective force of the foot only. For the foot there is no second couple since the foot itself is the end of the lower extremity.

One force of each couple is applied at the centre of the proximal joint of the segment considered. Its direction is the same as that of the actual effective force whereas the force of the couple due to gravity and applied at the centre of the proximal joint is opposite to gravity. The second force of each couple due to the effective forces is applied either at the centre of gravity or at the centre of the distal joint of the segment studied. The centre of gravity is the point of application of the effective force of the body segment itself. The centre of the distal joint is the point of application of the total effective force of the segments of the leg distal to the body segment under consideration. The second force of each couple, of course, has a direction opposite to that of the actual effective force.

The points of application of the two forces of all the couples due to the effective forces are known. Their moments can thus be found if the magnitude and direction of each effective force have been determined. These can be deduced from the accelerations of the centres of gravity of the three segments of the leg which are given in Tables 15–32 of Chap. 5. The effective force of a body segment is proportional to the acceleration of the centre of gravity of this body segment. It equals the product of this acceleration and the mass of the body segment. Its direction is the same as that of the acceleration of the centre of gravity. It is thus easy to determine the effective forces of the three segments of the leg. From these, the total effective force of two of the three segments or of the whole leg can be calculated. The total effective force of multiple body segments is the resultant of the effective forces of the different body segments.

In Chap. 5 the accelerations of the centres of gravity have been represented by their components in the direction of gait and in the vertical direction, assuming a plane movement. Therefore, it also seems useful to resolve the effective forces into two rectangular components along these directions. In this way, each couple gives two new couples. The forces of one are horizontal and those of the other, vertical. Consequently, the moment arm of the former is vertical and that of the latter is horizontal. Both can easily be calculated from the positional angle of the long axis of the segment under consideration and the distances of the centre of gravity and of the centre of the distal joint of the segment from the proximal joint centre.

The data obtained in Chap. 5 are thus sufficient to calculate the resultant torque of the effective forces D_e.

In the present chapter the two components of the total effective forces acting on the lower extremity during the swinging period will be calculated. The compression provoked in the hip, knee and ankle joints by the movement of the different segments of the leg will also be discussed. Then the torques due to gravity D_s and to the effective forces D_e will be calculated. In the last section the torques due to the internal forces and particularly to the muscles D_m will be deduced and, hence, conclusions concerning the involvement of the muscles in the swinging movement will be reached.

The Components of the Effective Forces

To obtain the components of the effective forces of the three segments of each leg, the components of the accelerations of the centres of gravity of the thigh, lower leg and foot must be multiplied by the masses of these segments. The components of acceleration are given in Tables 15–32 (pp. 364–381) of Chap. 5. The proportional figure for the mass of the thigh is 6.93, that for the lower leg, 3.15 and that for the foot 1.07 in our subject (Table 13 p. 301, Chap. 4). In the system of measurement thus adopted the unit of force is 1 gram, the unit of length 1 centimetre, and the unit of time 1 second. These units have also been used to measure the accelerations of the centres of gravity in Tables 15–32 in Chap. 5. Multiplication of these accelerations by the proportional figures of the masses gives the effective forces in grams. Because this unit of force is so small the figures obtained are so high as to be unwieldy. Moreover, the degree of accuracy attainable does not extend to the units, and even the tens are only very approximate. Therefore, in subsequent tables, the components of the effective forces will not be expressed in grams, but rather in kilograms[3]. This is achieved either by dividing the figures obtained by 1000 or by starting the calculation with figures for the masses that are 1/1000 those used when 1 g was the unit of force: 0.00693 for the thigh, 0.00315 for the lower leg and 0.00107 for the foot.

Multiplying of these figures by the accelerations of the centres of gravity in Tables 15–32 in Chap. 5 gives the components of the effective forces of the thigh, lower leg and foot noted in Tables 1–6. A positive sign in front of the components in the direction of gait means a forward direction whereas a negative sign means a backward direction. A positive sign in front of the components in the vertical direction means upwards and a negative sign down-

---

[3]   It was decided earlier (Chap. 4, p. 300) to use the gram as the unit of force to avoid dealing with too small figures when the masses of the different body segments are concerned. Further analysis, however, showed that this yielded figures for the effective forces and for their torques that were unmanageably high. Therefore, the forces will be expressed in kilograms in the future, as in Tables 21 and 22 of Chap. 2 for the external forces acting on the whole body and for the ground reaction force

Table 1. Components of the effective forces (kg) for the segments of the right leg during the swinging period in experiment I

Time (0,01 s)	Right thigh		Right lower leg		Right foot		Whole of right leg		Right lower leg + foot		Time (0,01 s)
	Direction of gait X_2	Vertical Z_2	Direction of gait X_4	Vertical Z_4	Direction of gait X_6	Vertical Z_6	Direction of gait $X_{2,4,6}$	Vertical $Z_{2,4,6}$	Direction of gait $X_{4,6}$	Vertical $Z_{4,6}$	
0	+2,15	0	+3,78	+0,16	+3,73	0	+9,66	+0,16	+7,51	+0,16	0
1	+1,11	+0,62	+3,31	+0,13	+3,71	−0,54	+8,13	+0,21	+7,02	−0,41	1
2	0	+1,46	+2,87	+0,09	+3,56	−1,71	+6,43	−0,16	+6,43	−1,62	2
3	−0,62	+2,43	+2,77	+0,06	+3,34	−2,28	+5,49	+0,21	+6,11	−2,22	3
4	−1,25	+3,53	+2,17	+0,03	+3,05	−2,23	+3,97	+1,33	+5,22	−2,20	4
5	−1,59	+4,44	+1,86	0	+2,72	−1,89	+2,99	+2,55	+4,58	−1,89	5
6	−2,01	+5,47	+1,54	−0,03	+2,46	−1,67	+1,99	+3,77	+4,00	−1,70	6
7	−2,29	+6,10	+1,26	−0,06	+2,11	−1,47	+1,08	+4,57	+3,37	−1,53	7
8	−2,56	+5,82	+1,04	−0,09	+1,77	−1,24	+0,25	+4,49	+2,81	−1,33	8
9	−2,70	+4,50	+0,85	−0,13	+1,43	−1,08	−0,42	+3,39	+2,28	−1,21	9
10	−2,77	+3,05	+0,66	−0,19	+1,10	−0,92	−1,01	+1,94	+1,76	−1,11	10
11	−2,84	+1,59	+0,54	−0,22	+0,77	−0,78	−1,53	+0,59	+1,31	−1,00	11
12	−2,91	0	+0,38	−0,35	+0,54	−0,64	−1,99	−0,99	+0,92	−0,99	12
13	−2,98	−0,42	+0,28	−0,54	+0,58	−0,50	−2,12	−1,46	+0,86	−1,04	13
14	−3,12	−0,97	0	−0,82	+0,76	−0,37	−2,36	−2,16	+0,76	−1,19	14
15	−3,26	−1,59	−0,09	−1,04	+0,92	−0,25	−2,43	−2,88	+0,83	−1,29	15
16	−3,33	−2,22	−0,28	−1,42	+0,96	−0,12	−2,65	−3,76	+0,68	−1,54	16
17	−3,47	−2,91	−0,44	−1,64	+0,90	0	−3,01	−4,55	+0,46	−1,64	17
18	−3,53	−3,60	−0,60	−1,20	+0,76	+0,21	−3,37	−4,59	+0,16	−0,99	18
19	−3,53	−4,09	−0,76	−0,85	+0,56	+0,39	−3,73	−4,55	−0,20	−0,46	19
20	−3,60	−4,57	−0,98	−0,54	+0,35	+0,57	−4,23	−4,54	−0,63	+0,03	20
21	−3,53	−4,99	−1,17	−0,38	+0,18	+0,74	−4,52	−4,63	−0,99	+0,36	21
22	−3,47	−5,41	−1,42	−0,28	0	+0,91	−4,89	−4,78	−1,42	+0,63	22
23	−3,33	−5,68	−1,64	−0,25	−0,22	+1,08	−5,19	−4,85	−1,86	+0,83	23
24	−3,26	−6,03	−1,86	−0,28	−0,43	+1,23	−5,55	−5,08	−2,29	+0,95	24
25	−3,12	−6,24	−2,02	−0,35	−0,72	+1,36	−5,86	−5,23	−2,74	+1,01	25
26	−2,84	−6,44	−2,27	−0,44	−0,93	+1,46	−6,04	−5,42	−3,20	+1,02	26
27	−2,36	−6,03	−2,49	−0,54	−1,19	+1,52	−6,04	−5,05	−3,68	+0,98	27
28	−1,66	−4,92	−2,80	−0,85	−1,39	+1,47	−5,85	−4,30	−4,19	+0,62	28
29	−0,76	−3,40	−3,09	−0,47	−1,66	+1,39	−5,51	−2,48	−4,75	+0,92	29
30	0	−1,73	−3,47	0	−1,88	+1,20	−5,35	−0,53	−5,35	+1,20	30
31	+1,73	0	−3,93	+2,08	−2,23	+0,91	−4,43	+2,99	−6,16	+2,99	31
32	+2,77	+1,39	−5,44	+1,26	−2,66	0	−5,33	+2,65	−8,10	+1,26	32
33	+3,60	+2,91	−4,63	+0,69	−3,16	−0,64	−4,19	+2,96	−7,79	+0,05	33
34	+4,02	+4,23	−2,84	+0,38	−3,85	−2,46	−2,67	+2,15	−6,69	−2,08	34
35	+4,23	+5,13	0	+0,13	−5,89	−3,32	−1,66	+1,94	−5,89	−3,19	35
36	+4,30	+6,17	+1,58	0	−8,67	−3,39	−2,79	+2,78	−7,09	−3,39	36
37	+3,95	+6,38	0	−0,13	−6,42	−2,68	−2,47	+3,57	−6,42	−2,81	37
38	+3,60	+5,34	−3,09	−0,22	−2,68	−2,03	−2,17	+3,09	−5,77	−2,25	38
39	+3,12	+3,40	−3,47	−0,25	0	−1,34	−0,35	+1,81	−3,47	−1,59	39
40	+2,29	0	−3,31	−0,28	+1,55	−0,54	+0,53	−0,82	−1,76	−0,82	40
41	+1,32	−4,37	−2,74	−0,32	+1,64	0	+0,22	−4,69	−1,10	−0,32	41
41,5	+0,90	−4,71	−2,30	−0,35	+1,28	+0,43	−0,12	−4,63	−1,02	+0,08	41,5

wards. Addition of the corresponding components of the effective forces of the three segments gives the components of the effective force of the whole leg. Addition of the corresponding components of the effective forces of the lower leg and foot gives the components of the effective force of the lower leg + foot system. These have also been

Table 2. Components of the effective forces (kg) for the segments of the left leg during the swinging period in experiment I

Time (0,01 s)	Left thigh		Left lower leg		Left foot		Whole of left leg		Left lower leg + foot		Time (0,01 s)
	Direction of gait X_3	Vertical Z_3	Direction of gait X_5	Vertical Z_5	Direction of gait X_7	Vertical Z_7	Direction of gait $X_{2,5,7}$	Vertical $Z_{3,5,7}$	Direction of gait $X_{5,7}$	Vertical $Z_{5,7}$	
0	+0,42	0	+2,58	+0,41	+3,00	0	+ 6,00	+0,41	+5,58	+0,41	0
1	0	+1,25	+2,33	+0,22	+3,25	−0,96	+ 5,58	+0,51	+5,58	−0,74	1
2	−0,62	+3,12	+2,11	0	+3,30	−1,87	+ 4,79	+1,25	+5,41	−1,87	2
3	−1,11	+4,44	+1,89	−0,22	+3,18	−2,49	+ 3,96	+1,73	+5,07	−2,71	3
4	−1,32	+5,34	+1,67	−0,35	+3,00	−2,71	+ 3,35	+2,28	+4,67	−3,06	4
5	−1,59	+5,54	+1,51	−0,47	+2,72	−2,66	+ 2,64	+2,41	+4,23	−3,13	5
6	−1,80	+5,27	+1,32	−0,63	+2,51	−2,35	+ 2,03	+2,29	+3,83	−2,98	6
7	−1,94	+4,78	+1,20	−0,72	+2,28	−1,98	+ 1,54	+2,08	+3,48	−2,70	7
8	−2,01	+4,02	+1,01	−0,85	+2,12	−1,64	+ 1,12	+1,53	+3,13	−2,49	8
9	−2,08	+3,26	+0,85	−0,95	+1,81	−1,34	+ 0,58	+0,97	+2,66	−2,29	9
10	−2,15	+1,25	+0,72	−1,04	+1,59	−1,16	+ 0,16	−0,95	+2,31	−2,20	10
11	−2,22	0	+0,60	−1,20	+1,36	−0,94	− 0,26	−2,14	+1,96	−2,14	11
12	−2,36	−0,83	+0,54	−1,35	+1,13	−0,72	− 0,69	−2,90	+1,67	−2,07	12
13	−2,43	−1,18	+0,41	−1,58	+0,91	−0,52	− 1,11	−3,28	+1,32	−2,10	13
14	−2,49	−1,66	+0,35	−1,42	+0,72	−0,35	− 1,42	−3,43	+1,07	−1,77	14
15	−2,56	−2,08	+0,28	−1,10	+0,48	−0,19	− 1,80	−3,37	+0,76	−1,29	15
16	−2,63	−2,56	+0,16	−0,91	+0,36	0	− 2,11	−3,47	+0,52	−0,91	16
17	−2,70	−2,91	0	−0,72	+0,26	+0,11	− 2,44	−3,52	+0,26	−0,61	17
18	−2,63	−3,26	−0,16	−0,54	+0,24	+0,27	− 2,55	−3,53	+0,08	−0,27	18
19	−2,56	−3,53	−0,32	−0,41	+0,35	+0,42	− 2,53	−3,52	+0,03	+0,01	19
20	−2,56	−3,74	−0,50	−0,32	+0,64	+0,54	− 2,42	−3,52	+0,14	+0,22	20
21	−2,56	−3,81	−0,72	−0,35	+0,83	+0,71	− 2,45	−3,45	+0,11	+0,36	21
22	−2,49	−3,81	−0,91	−0,44	+0,75	+0,86	− 2,65	−3,39	−0,16	+0,42	22
23	−2,43	−3,95	−1,26	−0,60	+0,50	+1,03	− 3,19	−3,52	−0,76	+0,43	23
24	−2,29	−4,09	−1,64	−0,50	0	+1,21	− 3,93	−3,38	−1,64	+0,71	24
25	−2,15	−4,23	−2,08	−0,19	−0,20	+1,44	− 4,43	−2,98	−2,28	+1,25	25
26	−2,01	−4,30	−2,58	0	−0,51	+1,65	− 5,10	−2,65	−3,09	+1,65	26
27	−1,94	−3,95	−2,93	+0,79	−0,80	+1,84	− 5,67	−1,32	−3,73	+2,63	27
28	−1,80	−3,53	−4,00	+1,07	−1,09	+1,79	− 6,89	−0,67	−5,09	+2,86	28
29	−1,46	−2,91	−4,79	+0,72	−1,43	+1,52	− 7,68	−0,67	−6,22	+2,24	29
30	−1,11	−2,08	−5,36	+0,54	−2,09	+1,12	− 8,56	−0,42	−7,45	+1,66	30
31	0	−1,18	−5,07	+0,41	−3,19	+0,64	− 8,26	−0,13	−8,26	+1,05	31
32	+1,46	0	−4,35	+0,35	−4,00	0	− 6,89	+0,35	−8,35	+0,35	32
33	+4,16	+1,39	−3,31	+0,25	−4,44	−0,59	− 3,59	+1,05	−7,75	−0,34	33
34	+6,44	+3,33	−1,48	+0,22	−4,71	−1,34	+ 0,25	+2,21	−6,19	−1,12	34
35	+9,29	+4,85	0	+0,13	−4,73	−2,68	+ 4,56	+2,30	−4,73	−2,55	35
36	+8,80	+6,58	+1,01	+0,06	−4,68	−4,71	+ 5,13	+1,93	−3,67	−4,65	36
37	+6,86	+7,35	0	0	−4,44	−3,53	+ 2,42	+3,82	−4,44	−3,53	37
38	+3,81	+6,93	−2,30	−0,06	−3,87	−1,98	− 2,36	+4,89	−6,17	−2,04	38
39	0	+5,20	−3,31	−0,09	−2,89	−0,91	− 6,20	+4,20	−6,20	−1,00	39
40	−3,81	0	−3,56	−0,19	−1,34	0	− 8,71	−0,19	−4,90	−0,19	40
41	−6,93	−4,64	−3,15	−0,25	0	+0,27	−10,08	−2,62	−3,15	+0,02	41
41,7	−8,66	−6,24	−2,43	−0,28	+1,28	+0,54	− 9,81	−5,98	−1,15	+0,26	41,7

calculated and the results appear in Tables 1–6. To keep the equations for the torques of the effective forces simple, all the components of the effective forces in the direction of gait will be designated as X and those in the vertical direction as Z with indices indicating the body segment considered.

Table 3. Components of the effective forces (kg) for the segments of the right leg during the swinging period in experiment II

Time (0,01 s)	Right thigh		Right lower leg		Right foot		Whole of right leg		Right lower leg + foot		Time (0,01 s)
	Direction of gait X_2	Vertical Z_2	Direction of gait X_4	Vertical Z_4	Direction of gait X_6	Vertical Z_6	Direction of gait $X_{2,4,6}$	Vertical $Z_{2,4,6}$	Direction of gait $X_{4,6}$	Vertical $Z_{4,6}$	
0	+2,56	+ 4,71	+3,09	+0,28	+3,82	0	+9,47	+ 4,99	+6,91	+0,28	0
1	+1,52	+ 4,64	+2,77	+0,25	+3,79	−0,51	+8,08	+ 4,38	+6,56	−0,26	1
2	+0,69	+ 4,37	+2,43	+0,22	+3,50	−1,11	+6,62	+ 3,48	+5,93	−0,89	2
3	0	+ 4,16	+2,11	+0,16	+3,12	−1,68	+5,23	+ 2,64	+5,23	−1,52	3
4	−0,83	+ 3,81	+1,83	+0,09	+2,72	−1,93	+3,72	+ 1,97	+4,55	−1,84	4
5	−1,39	+ 3,33	+1,51	+0,03	+2,40	−1,87	+2,52	+ 1,49	+3,91	−1,84	5
6	−1,80	+ 2,91	+1,26	0	+2,14	−1,68	+1,60	+ 1,23	+3,40	−1,68	6
7	−2,08	+ 2,36	+1,01	−0,06	+1,90	−1,48	+0,83	+ 0,82	+2,91	−1,54	7
8	−2,43	+ 1,94	+0,76	−0,13	+1,68	−1,26	+0,01	+ 0,55	+2,44	−1,39	8
9	−2,63	+ 1,39	+0,57	−0,25	+1,44	−1,12	−0,62	+ 0,02	+2,01	−1,37	9
10	−2,84	+ 0,69	+0,38	−0,35	+1,21	−0,98	−1,25	− 0,64	+1,59	−1,33	10
11	−2,98	0	+0,22	−0,47	+1,00	−0,89	−1,76	− 1,36	+1,22	−1,36	11
12	−3,26	− 0,97	0	−0,60	+0,88	−0,74	−2,38	− 2,31	+0,88	−1,34	12
13	−3,40	− 1,87	−0,09	−0,72	+0,92	−0,62	−2,57	− 3,21	+0,83	−1,34	13
14	−3,47	− 2,63	−0,22	−0,88	+1,09	−0,47	−2,60	− 3,98	+0,87	−1,35	14
15	−3,53	− 3,26	−0,35	−1,01	+1,29	−0,33	−2,59	− 4,60	+0,94	−1,34	15
16	−3,60	− 3,67	−0,50	−1,20	+1,32	−0,14	−2,78	− 5,01	+0,82	−1,34	16
17	−3,67	− 4,30	−0,63	−1,45	+1,17	0	−3,13	− 5,75	+0,54	−1,45	17
18	−3,74	− 4,99	−0,69	−1,67	+0,93	+0,19	−3,50	− 6,37	+0,24	−1,38	18
19	−3,67	− 5,75	−0,88	−1,17	+0,62	+0,33	−3,93	− 6,59	−0,26	−0,84	19
20	−3,60	− 6,72	−1,10	−0,66	+0,29	+0,51	−4,41	− 6,87	−0,81	−0,15	20
21	−3,53	− 7,42	−1,29	−0,57	0	+0,72	−4,82	− 7,27	−1,29	+0,15	21
22	−3,40	− 8,25	−1,48	−0,54	−0,19	+0,88	−5,07	− 7,91	−1,67	+0,34	22
23	−3,26	− 9,01	−1,64	−0,60	−0,32	+1,07	−5,22	− 8,54	−1,96	+0,47	23
24	−2,98	−10,26	−1,92	−0,63	−0,45	+1,23	−5,35	− 9,66	−2,37	+0,60	24
25	−2,43	−11,57	−2,17	−0,41	−0,55	+1,39	−5,15	−10,59	−2,72	+0,98	25
26	−1,87	−12,96	−2,49	−0,28	−0,67	+1,61	−5,03	−11,63	−3,16	+1,33	26
27	−1,11	−14,90	−2,77	−0,19	−0,82	+1,77	−4,70	−13,32	−3,59	+1,58	27
28	−0,49	−16,70	−3,09	0	−1,06	+1,68	−4,64	−15,02	−4,15	+1,68	28
29	0	−18,71	−3,40	+0,09	−1,48	+1,44	−4,88	−17,18	−4,88	+1,53	29
30	+1,52	−19,29	−3,91	+0,38	−2,01	+1,08	−4,40	−17,83	−5,92	+1,46	30
31	+2,84	−10,40	−5,29	+0,69	−2,70	+0,41	−5,15	− 9,30	−7,99	+1,10	31
32	+4,02	0	−4,41	+1,07	−3,58	0	−3,97	+ 1,07	−7,99	+1,07	32
33	+4,78	+52,32	−1,98	+0,88	−5,63	−2,51	−2,83	+50,69	−7,61	−1,63	33
34	+5,41	+34,65	0	+0,69	−5,56	−3,96	−0,15	+31,38	−5,56	−3,27	34
35	+6,03	+17,33	+1,45	+0,50	−4,90	−4,49	+2,58	+13,34	−3,45	−3,99	35
36	+6,17	+ 4,16	+1,10	+0,35	−4,01	−2,84	+3,26	+ 1,67	−2,91	−2,49	36
37	+5,61	0	0	+0,19	−3,06	−1,82	+2,55	− 1,63	−3,06	−1,63	37
38	+4,09	− 2,84	−3,31	0	−2,09	−1,02	−1,31	− 3,86	−5,40	−1,02	38
39	0	− 3,12	−4,10	−0,25	0	−0,48	−4,10	− 3,85	−4,10	−0,73	39
40	−2,43	− 1,25	−4,00	−0,47	+2,58	0	−3,85	− 1,72	−1,42	−0,47	40
40,8	−3,81	0	−3,53	−0,63	+2,39	+0,17	−4,95	− 0,46	−1,14	−0,46	40,8

Figure 1 illustrates the data presented in Tables 1–6 concerning the X- and Z-components of the different effective forces in a similar way to the graphs in Chap. 5. The time is entered along the abscissa with 1 s equal to 10 cm so that 0.01 s is represented by 1 mm. The components of the effective forces are entered along the ordinate: 1 mm represents a force of 1 kg. The two components of one effective force are presented in the same graph. The curve

Table 4. Components of the effective forces (kg) for the segments of the left leg during the swinging period in experiment II

Time (0,01 s)	Left thigh Direction of gait X_3	Left thigh Vertical Z_3	Left lower leg Direction of gait X_5	Left lower leg Vertical Z_5	Left foot Direction of gait X_7	Left foot Vertical Z_7	Whole of left leg Direction of gait $X_{2,5,7}$	Whole of left leg Vertical $Z_{3,5,7}$	Left lower leg + foot Direction of gait $X_{5,7}$	Left lower leg + foot Vertical $Z_{5,7}$	Time (0,01 s)
0	+3,12	+1,04	+3,06	+0,25	+3,25	0	+9,43	+1,29	+ 6,31	+0,25	0
1	+1,80	+2,77	+2,84	+0,22	+3,40	−0,35	+8,04	+2,64	+ 6,24	−0,13	1
2	0	+3,81	+2,55	+0,09	+3,51	−1,23	+6,06	+2,67	+ 6,06	−1,14	2
3	−0,49	+4,85	+2,24	0	+3,46	−1,95	+5,21	+2,90	+ 5,70	−1,95	3
4	−0,90	+5,41	+1,98	−0,13	+3,25	−2,39	+4,33	+2,89	+ 5,23	−2,52	4
5	−1,46	+5,61	+1,80	−0,25	+2,94	−2,61	+3,28	+2,75	+ 4,74	−2,86	5
6	−1,59	+5,27	+1,64	−0,38	+2,59	−2,65	+2,64	+2,24	+ 4,23	−3,03	6
7	−1,94	+4,78	+1,51	−0,54	+2,07	−2,49	+1,64	+1,75	+ 3,58	−3,03	7
8	−2,08	+4,02	+1,32	−0,66	+1,63	−2,10	+0,87	+1,26	+ 2,95	−2,76	8
9	−2,22	+2,91	+1,20	−0,79	+1,25	−1,75	+0,23	+0,37	+ 2,45	−2,54	9
10	−2,43	+1,32	+1,01	−0,91	+1,00	−1,42	−0,42	−1,01	+ 2,01	−2,33	10
11	−2,63	0	+0,88	−1,01	+0,85	−1,10	−0,90	−2,11	+ 1,73	−2,11	11
12	−2,77	−0,90	+0,69	−1,20	+0,77	−0,86	−1,31	−2,96	+ 1,46	−2,06	12
13	−2,84	−1,59	+0,54	−1,26	+0,76	−0,62	−1,54	−3,47	+ 1,30	−1,88	13
14	−2,98	−2,15	+0,32	−1,42	+0,82	−0,40	−1,84	−3,97	+ 1,14	−1,82	14
15	−3,19	−2,63	0	−1,61	+0,94	−0,18	−2,25	−4,42	+ 0,94	−1,79	15
16	−3,33	−3,12	−0,22	−1,86	+1,09	0	−2,46	−4,98	+ 0,87	−1,86	16
17	−3,47	−3,53	−0,44	−1,48	+1,18	+0,15	−2,73	−4,86	+ 0,74	−1,33	17
18	−3,53	−4,02	−0,72	−1,04	+1,19	+0,31	−3,06	−4,75	+ 0,47	−0,73	18
19	−3,60	−4,30	−1,10	−0,66	+1,11	+0,47	−3,59	−4,49	+ 0,01	−0,19	19
20	−3,67	−4,50	−1,48	−0,47	+0,96	+0,61	−4,19	−4,36	− 0,52	+0,14	20
21	−3,74	−4,71	−1,89	−0,35	+0,70	+0,75	−4,93	−4,31	− 1,19	+0,40	21
22	−3,81	−4,92	−2,27	−0,28	0	+0,91	−6,08	−4,29	− 2,27	+0,63	22
23	−3,74	−5,13	−2,61	−0,35	−0,48	+1,14	−6,83	−4,34	− 3,09	+0,79	23
24	−3,67	−4,99	−2,93	−0,47	−1,00	+1,31	−7,60	−4,15	− 3,93	+0,84	24
25	−3,53	−4,85	−3,21	−0,57	−1,32	+1,51	−8,06	−3,91	− 4,53	+0,94	25
26	−3,33	−4,37	−3,62	0	−1,62	+1,81	−8,57	−2,56	− 5,24	+1,81	26
27	−3,05	−3,53	−4,16	+0,35	−1,89	+1,98	−9,10	−1,20	− 6,05	+2,33	27
28	−2,36	−2,63	−4,73	+0,95	−2,16	+1,72	−9,25	+0,04	− 6,89	+2,67	28
29	−1,25	−1,52	−4,38	+1,29	−2,35	+1,39	−7,98	+1,16	− 6,73	+2,68	29
30	0	0	−3,53	+1,20	−2,55	+1,08	−6,08	+2,28	− 6,08	+2,28	30
31	+3,12	+1,59	−1,83	+0,91	−2,79	+0,67	−1,50	+3,17	− 4,62	+1,58	31
32	+4,85	+3,53	0	+0,69	−3,08	0	+1,77	+4,22	− 3,08	+0,69	32
33	+6,31	+4,71	+1,42	+0,44	−3,43	−0,27	+4,30	+4,88	− 2,01	+0,17	33
34	+7,34	+5,34	0	+0,16	−4,12	−1,34	+3,22	+4,16	− 4,12	−1,18	34
35	+7,21	+5,54	−1,48	0	−5,89	−5,08	−0,16	+0,46	− 7,37	−5,08	35
36	+6,58	+5,34	−2,71	−0,16	−7,83	−4,28	−3,96	+0,90	−10,54	−4,44	36
37	+5,61	+4,64	−3,37	−0,25	−5,56	−3,26	−3,32	+1,13	− 8,93	−3,51	37
38	+3,81	+3,53	−3,50	−0,35	0	−2,03	+0,31	+1,15	− 3,50	−2,38	38
39	0	+1,73	−3,15	−0,41	+1,12	−1,07	−2,03	+0,25	− 2,03	−1,48	39
40	−2,08	0	−2,55	−0,47	+1,51	0	−3,12	−0,47	− 1,04	−0,47	40
40,9	−6,24	−1,66	−2,14	−0,50	+1,07	+0,64	−7,31	−1,52	− 1,07	+0,14	40,9

relative to the Z-component (vertical direction) is dotted whereas that relative to the X-component (direction of gait) is continuous. The left half of the graph represents the effective forces of the right leg resolved into their components, the right half those of the left leg. For each leg and for each experiment five diagrams were drawn. From top to bottom they represent the effective forces of the thigh, the lower leg, the foot, the whole leg and

Table 5. Components of the effective forces (kg) for the segments of the right leg during the swinging period in experiment III

Time (0,01 s)	Right thigh		Right lower leg		Right foot		Whole of right leg		Right lower leg + foot		Time (0,01 s)
	Direction of gait X_2	Vertical Z_2	Direction of gait X_4	Vertical Z_4	Direction of gait X_6	Vertical Z_6	Direction of gait $X_{2,4,6}$	Vertical $Z_{2,4,6}$	Direction of gait $X_{4,6}$	Vertical $Z_{4,6}$	
0	+1,73	−0,35	+2,05	−0,09	+3,00	+0,59	+6,78	+0,15	+5,05	+0,50	0
1	0	0	+1,92	−0,13	+3,10	0	+5,02	−0,13	+5,02	−0,13	1
2	−0,62	+0,42	+1,80	−0,16	+3,06	−1,12	+4,24	−0,86	+4,86	−1,28	2
3	−1,39	+0,69	+1,67	−0,19	+2,89	−2,38	+3,17	−1,88	+4,56	−2,57	3
4	−1,87	+1,32	+1,61	−0,25	+2,68	−2,51	+2,42	−1,44	+4,29	−2,76	4
5	−2,01	+1,94	+1,51	−0,35	+2,46	−2,32	+1,96	−0,73	+3,97	−2,67	5
6	−2,08	+2,63	+1,45	−0,35	+2,19	−2,04	+1,56	+0,24	+3,64	−2,39	6
7	−2,08	+3,33	+1,32	−0,47	+1,90	−1,72	+1,14	+1,14	+3,22	−2,19	7
8	−2,15	+4,09	+1,26	−0,60	+1,72	−1,49	+0,83	+2,00	+2,98	−2,09	8
9	−2,22	+4,99	+1,20	−0,66	+1,46	−1,31	+0,44	+3,02	+2,66	−1,97	9
10	−2,29	+5,54	+1,10	−0,85	+1,24	−1,16	+0,05	+3,53	+2,34	−2,01	10
11	−2,36	+6,03	+1,01	−1,01	+1,02	−0,98	−0,33	+4,04	+2,03	−1,99	11
12	−2,49	+6,24	+0,91	−1,51	+0,83	−0,83	−0,75	+3,90	+1,74	−2,34	12
13	−2,56	+4,16	+0,79	−3,15	+0,70	−0,63	−1,07	+0,38	+1,49	−3,78	13
14	−2,70	0	+0,57	−1,26	+0,66	−0,47	−1,47	−1,73	+1,23	−1,73	14
15	−2,77	−2,22	+0,38	−0,69	+0,81	−0,31	−1,58	−2,22	+1,19	−1,00	15
16	−2,84	−3,60	+0,22	−0,47	+0,93	−0,14	−1,69	−4,21	+1,15	−0,61	16
17	−2,91	−4,23	0	−0,32	+1,00	0	−1,91	−4,55	+1,00	−0,32	17
18	−2,98	−4,71	−0,35	−0,32	+0,96	+0,16	−2,37	−4,87	+0,61	−0,16	18
19	−3,05	−5,34	−0,66	−0,47	+0,85	+0,32	−2,86	−5,49	+0,19	−0,15	19
20	−3,12	−5,54	−1,07	−0,57	+0,66	+0,48	−3,53	−5,63	−0,41	−0,09	20
21	−3,05	−6,03	−1,48	−0,50	+0,47	+0,62	−4,06	−5,91	−1,01	+0,12	21
22	−3,05	−6,24	−1,70	−0,28	+0,24	+0,75	−4,51	−5,77	−1,46	+0,47	22
23	−2,98	−6,44	−1,95	−0,16	0	+0,93	−4,93	−5,67	−1,95	+0,77	23
24	−2,84	−6,72	−2,14	0	−0,28	+1,10	−5,26	−5,62	−2,42	+1,10	24
25	−2,63	−6,86	−2,24	+0,16	−0,55	+1,31	−5,42	−5,39	−2,79	+1,47	25
26	−2,29	−6,93	−2,43	+0,32	−0,86	+1,51	−5,58	−5,10	−3,29	+1,83	26
27	−1,80	−6,79	−2,52	+0,60	−1,21	+1,73	−5,53	−4,46	−3,73	+2,33	27
28	−0,97	−6,65	−2,71	+0,72	−1,56	+2,00	−5,24	−3,93	−4,27	+2,72	28
29	0	−5,96	−2,84	+0,63	−1,90	+2,13	−4,74	−3,20	−4,74	+2,76	29
30	+1,04	−3,81	−3,18	+0,50	−2,23	+1,71	−4,37	−1,60	−5,41	+2,21	30
31	+1,87	0	−3,50	+0,38	−2,55	0	−4,18	+0,38	−6,05	+0,38	31
32	+2,98	+5,06	−3,62	+0,19	−2,89	−1,07	−3,53	+4,18	−6,51	−0,88	32
33	+4,23	+5,68	−3,15	0	−3,26	−3,21	−2,18	+2,47	−6,41	−3,21	33
34	+4,78	+5,41	−1,73	−0,19	−4,12	−2,96	−1,07	+2,26	−5,85	−3,15	34
35	+5,13	+4,64	0	−0,35	−5,50	−2,46	−0,37	+1,83	−5,50	−2,81	35
36	+5,34	+3,81	+1,32	−0,38	−6,85	−2,01	−0,19	+1,42	−5,53	−2,39	36
37	+4,71	+3,05	0	−0,50	−7,28	−1,63	−2,57	+0,92	−7,28	−2,13	37
38	+3,67	+2,43	−2,27	−0,57	−3,75	−1,11	−2,35	+0,75	−6,02	−1,68	38
39	+2,29	+1,94	−4,10	−0,63	0	−0,73	−1,81	+0,58	−4,10	−1,36	39
40	0	+1,66	−4,47	−0,69	+1,25	−0,33	−3,22	+0,64	−3,22	−1,02	40
41	−3,81	+1,32	−3,28	−0,76	+1,09	0	−6,00	+0,56	−2,19	−0,76	41
41,45	−4,37	+1,18	−2,99	−0,79	+0,94	+0,12	−6,42	+0,51	−2,05	−0,67	41,45

the lower leg + foot system. Consequently, all the curves relating to the same part of the body are close to each other. This allows easy comparison.

It can be seen again at a glance that the results recorded for the six instances of swinging movement of the leg completely independently of each other are qualitatively the same. Their

Table 6. Components of the effective forces (kg) for the segments of the left leg during the swinging period in experiment III

Time (0,01 s)	Left thigh		Left lower leg		Left foot		Whole of left leg		Left lower leg + foot		Time (0,01 s)
	Direction of gait X_3	Vertical Z_3	Direction of gait X_5	Vertical Z_5	Direction of gait X_7	Vertical Z_7	Direction of gait $X_{2,5,7}$	Vertical $Z_{3,5,7}$	Direction of gait $X_{5,7}$	Vertical $Z_{5,7}$	
0	0	0	+3,24	+0,57	+3,64	0	+6,88	+0,57	+6,88	+0,57	0
1	−0,42	+1,52	+2,93	+0,50	+3,62	−2,30	+6,13	−0,28	+6,55	−1,80	1
2	−0,83	+3,67	+2,46	+0,35	+3,49	−2,88	+5,12	+1,14	+5,95	−2,53	2
3	−1,32	+5,20	+2,14	+0,22	+3,20	−2,86	+4,02	+2,56	+5,34	−2,64	3
4	−1,66	+5,82	+1,80	0	+2,78	−2,57	+2,92	+3,25	+4,58	−2,57	4
5	−1,94	+6,24	+1,51	−0,16	+2,32	−2,23	+1,89	+3,85	+3,83	−2,39	5
6	−2,08	+5,89	+1,23	−0,25	+1,93	−1,92	+1,08	+3,72	+3,16	−2,17	6
7	−2,15	+5,27	+0,98	−0,38	+1,61	−1,65	+0,44	+3,24	+2,59	−2,03	7
8	−2,29	+4,37	+0,69	−0,54	+1,28	−1,47	−0,32	+2,36	+1,97	−2,01	8
9	−2,43	+2,84	+0,50	−0,66	+1,04	−1,33	−0,89	+0,85	+1,54	−1,99	9
10	−2,56	+1,39	+0,25	−0,79	+0,82	−1,14	−1,49	−0,54	+1,07	−1,93	10
11	−2,63	0	0	−0,91	+0,77	−1,01	−1,86	−1,92	+0,77	−1,92	11
12	−2,70	−0,83	−0,13	−1,10	+0,79	−0,79	−2,04	−2,72	+0,66	−1,89	12
13	−2,77	−1,46	−0,28	−1,20	+0,87	−0,60	−2,18	−3,26	+0,59	−1,80	13
14	−2,91	−1,94	−0,38	−1,39	+0,98	−0,42	−2,31	−3,75	+0,60	−1,81	14
15	−3,12	−2,36	−0,54	−1,73	+1,04	−0,24	−2,62	−4,33	+0,50	−1,97	15
16	−3,26	−2,63	−0,66	−2,43	+0,92	−0,07	−3,00	−5,13	+0,26	−2,50	16
17	−3,33	−2,84	−0,85	−1,64	+0,80	0	−3,38	−4,48	−0,05	−1,64	17
18	−3,40	−2,98	−1,01	−1,20	+0,64	+0,18	−3,77	−4,00	−0,37	−1,02	18
19	−3,47	−3,19	−1,17	−0,95	+0,48	+0,32	−4,16	−3,82	−0,69	−0,63	19
20	−3,40	−3,40	−1,32	−0,76	+0,30	+0,46	−4,42	−3,70	−1,02	−0,30	20
21	−3,40	−3,53	−1,48	−0,57	+0,14	+0,58	−4,74	−3,52	−1,34	+0,01	21
22	−3,40	−3,74	−1,67	−0,47	0	+0,72	−5,07	−3,49	−1,67	+0,25	22
23	−3,33	−4,02	−1,83	−0,38	−0,24	+0,86	−5,40	−3,54	−2,07	+0,48	23
24	−3,26	−4,16	−1,98	−0,35	−0,41	+1,04	−5,65	−3,47	−2,39	+0,69	24
25	−3,12	−4,37	−2,14	−0,69	−0,59	+1,20	−5,85	−3,86	−2,73	+0,51	25
26	−2,91	−4,44	−2,30	−1,48	−0,77	+1,43	−5,98	−4,49	−3,07	−0,05	26
27	−2,63	−4,37	−2,46	0	−0,97	+1,77	−6,06	−2,60	−3,43	+1,77	27
28	−2,08	−4,23	−2,61	+0,35	−1,17	+1,95	−5,86	−1,93	−3,78	+2,30	28
29	−1,39	−3,53	−2,87	+0,66	−1,50	+2,02	−5,76	−0,85	−4,37	+2,68	29
30	0	−2,22	−3,02	+0,69	−1,98	+1,34	−5,00	−0,19	−5,00	+2,03	30
31	+1,04	0	−3,15	+0,63	−3,05	+0,59	−5,16	+1,22	−6,20	+1,22	31
32	+2,49	+2,43	−3,09	+0,60	−3,69	0	−4,29	+3,03	−6,78	+0,60	32
33	+4,09	+3,67	−2,90	+0,54	−3,93	−1,57	−2,74	+2,64	−6,83	−1,03	33
34	+5,61	+4,57	−2,55	+0,50	−4,00	−2,25	−0,94	+2,82	−6,55	−1,75	34
35	+6,86	+4,99	−1,98	+0,50	−3,98	−3,23	+0,90	+2,26	−5,96	−2,73	35
36	+7,48	+4,92	−1,39	+0,50	−3,83	−2,73	+2,26	+2,69	−5,22	−2,23	36
37	+7,34	+4,64	−0,91	+0,44	−3,53	−2,03	+2,90	+3,05	−4,44	−1,59	37
38	+6,58	+4,16	−0,57	+0,38	−3,16	−1,47	+2,85	+3,07	−3,73	−1,09	38
39	+3,81	+3,53	−0,60	+0,28	−2,75	−0,96	+0,46	+2,85	−3,35	−0,68	39
40	0	+3,05	−1,01	0	−2,07	−0,56	−3,08	+2,49	−3,08	−0,56	40
41	−5,75	+2,63	−1,51	−0,19	−1,26	−0,14	−8,52	+2,30	−2,77	−0,33	41
41,3	−6,93	+2,43	−1,61	−0,32	−1,07	0	−9,61	+2,11	−2,68	−0,32	41,3

shape and evolution are essentially similar. The discrepancies are mainly in the vertical spread of the curves. The vertical component of the effective force of the right leg and that of the right thigh, which determines it, assumes abnormally high values downwards and especially upwards in the second half of the swinging period in experiment II. Afterwards, of

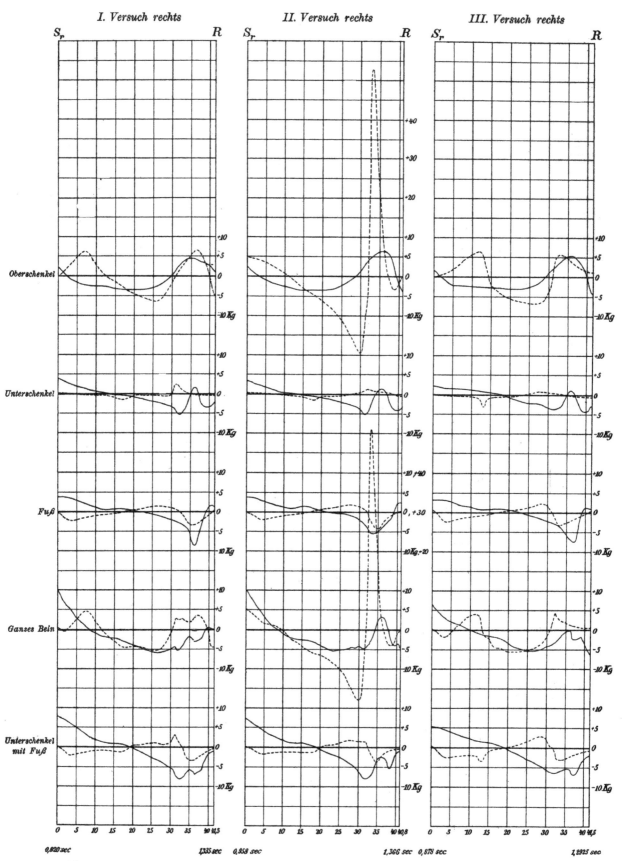

Fig. 1. Components of the effective forces in the direction of gait (———) and in the vertical direction (– – – –) during the swinging of the leg

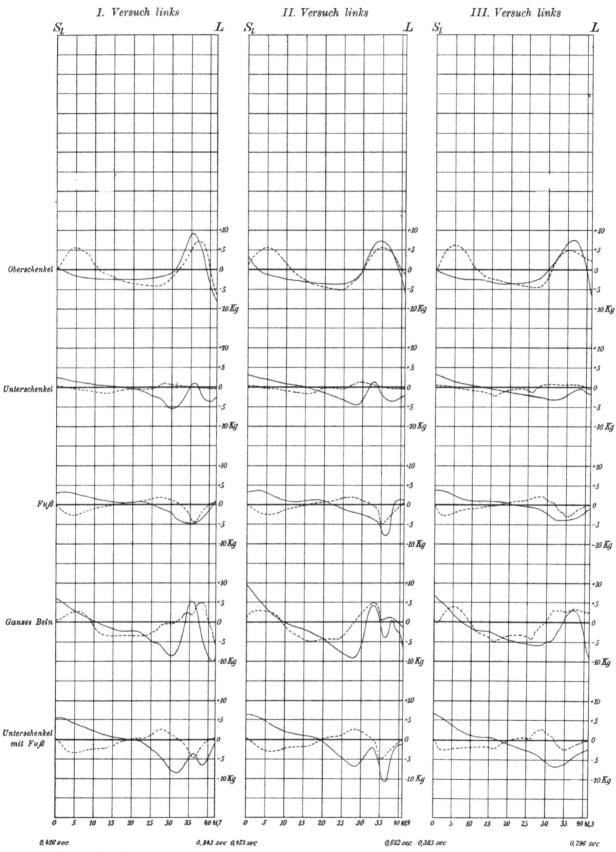

Fig. 1 (continued)

course, it is not possible to decide whether the discrepancy results from an abnormal behaviour of the right leg in experiment II or from an error in observation and measurement. The component of these effective forces in the direction of gait looks normal even quantitatively, and the effective forces of the other segments of the right leg in experiment II behave as in the other five instances. These facts rather suggest an error somewhere in the course of the analysis. The initial aspect of the two curves could, however, result from abnormal behaviour of the right leg in experiment II. Whatever the truth, the discrepancies in this case also appear to be essentially quantitative.

The qualitative and quantitative similarity of the adjacent curves for the effective forces of the thigh, lower leg and foot elsewhere in Fig. 1 are not surprising since a similarity of this kind was apparent between the curves for the accelerations of the centres of gravity of the three segments of the leg in Chap. 5. The effective forces are proportional to these accelerations. But the similarity of the two lower curves in Fig. 1 constitutes a further argument in support of the reliability of the data thus obtained. These curves result from those above which are independent of each other. Therefore, the unavoidable small errors in observation or measurement must be cumulative in the lower curves and restrict the accuracy. If, however, the curves for the effective forces of the whole leg and of the lower leg + foot system match the other curves so well, it must be concluded that the sources of error in the study do not detract from the accuracy aimed at.

To make it easy to relate the data recorded in this work to the phases of the movement of the swinging leg, Fig. 11 from Chap. 5 is reproduced here as Fig. 2.

According to Tables 1–6 and the curves in Fig. 1, the behaviour of the different effective forces during the swinging period appears as follows. At toe-off the horizontal component (in the direction of gait) of the effective force of the thigh is directed forwards. Its initial magnitude varies a little from case to case. On average it is about 2 kg. In all instances, however, it decreases very quickly and has reached 0 about 0.02 s after toe-off. From then on, the horizontal component of this effective force is directed backwards during most of the swinging movement. Its magnitude soon reaches 3 kg, on average, and it remains at about this level for some time. Towards the end of the third quarter of the swinging period, roughly 0.3 s after toe-off, the horizontal component directed backwards passes through 0, and from then on it is again directed forwards. While it is directed forwards its magnitude increases relatively quickly until it reaches a maximum, about 5 kg for the right leg and over 7 kg for the left, in all three experiments. After this maximum, which occurs approximately 0.36 s after toe-off, the horizontal component decreases again as quickly as it had increased so that

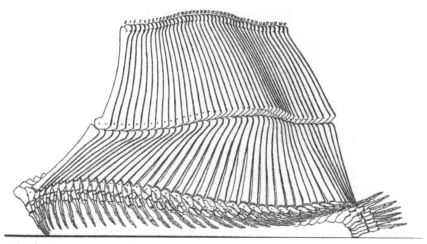

Fig. 2. Successive positions of the leg during the swinging period of gait, recorded at intervals of 0.01 s

it passes through 0 and is directed backwards again before heel strike. Only the right thigh is an exception, and only in experiment I. The horizontal component of the effective force of this thigh does not quite reach 0 at the end and thus remains directed forwards at heel strike.

The vertical component of the effective force of the thigh is equal or close to 0 at the beginning of the swinging period for both legs in experiments I and III. Only in experiment II is it significantly above 0. This means an upwards direction. It maintains this upwards direction in the other cases also, during the first quarter of the swinging period. It attains a maximum of roughly 6 kg for the left leg after 0.05 s, for the right a little later in experiments I and III, and just at the beginning in experiment II. In the second and third quarters of the swinging period the vertical component of the effective force of the thigh points downwards after having passed through 0 between the first and second quarters of the period. Its magnitude increases progressively to about 5 kg on the left, on the right to nearly 7 kg in experiments I and III, to 19 kg in experiment II if, however, importance can be attached to this last figure. This maximum of the components directed downwards occurs on average 0.26 s after toe-off. Shortly, about 0.05 s, afterwards it has decreased to 0. From then on, the vertical component is directed upwards again and reaches a maximum approximately 0.35 s after toe-off. This maximum reaches about 6 kg, on average, if the value of 50 kg obtained for the right leg in experiment II is disregarded. Finally, the magnitude of the vertical component decreases again and, for both legs in experiments I and II, it passes through 0 before the end of the swinging period, so that it is directed downwards at heel strike. In experiment III (with a load) this component keeps its positive direction till the end.

Comparison of the evolution of the two components of the effective force of the thigh shows the direction of this force at the different times and its variations during the swinging period.

At the beginning of the swinging period, the effective force of the thigh is directed quite markedly forwards. Immediately after toe-off, however, it changes and points more upwards. It thus turns counter-clockwise, as seen from the right, and points directly upwards for a while, about 0.02 s later. Its direction does not remain vertical but keeps changing in the same direction so that it is directed upwards and backwards and finally directly backwards for a time. This occurs between the first and second quarters of the swinging period. At this time the vertical component is 0. From then on, the effective force is directed backwards and downwards, at first more backwards than downwards, then as much downwards as backwards (this corresponds to the intersection of the two curves in Fig. 1), and finally more downwards than backwards. Between the third and last quarters of the swinging period the effective force is directed vertically downwards even if in some cases its magnitude is then very small. With such a small magnitude the effective force continues its counter-clockwise rotation with a relatively very high speed, passes through the forwards orientation and, finally, adopts a direction that is about as much upwards as forwards. This direction is maintained almost to the end of the swing. Except for small discrepancies, the direction of the effective force of the thigh during the swinging period carries out a little more than one revolution with very variable magnitude counter-clockwise as seen from the right.

As can be seen from Tables 1–6 and Fig. 1, the horizontal component of the effective force of the lower leg is initially directed forwards and its magnitude reaches 3 kg on average. It then decreases at a fairly constant rate until it reaches 0 between the first and second thirds of the swinging period. From then on, the horizontal component is directed backwards, and it keeps this direction almost to the end. Only 0.05 s before heel strike is it again directed forwards, and then its magnitude is very small. Whilst it is directed backwards for the first time it increases progressively until, between the third and last quarters of the swinging period, it reaches a maximum of roughly 5 kg in experiments I and II, and between 3 and 4 kg in experiment III. Thereafter, the horizontal component which is directed backwards decreases rather quickly to 0, after which, as already mentioned, it is directed forwards for about 0.02 s with a small magnitude. Later in the swinging period, it is again directed backwards, at first with increasing and finally with decreasing magnitude. The last maximum attains almost 4 kg on average, and occurs about 0.02 s before heel strike. At heel strike the magnitude of the horizontal component directed backwards is between 2 kg and 3 kg. The left leg in experiment III deviates somewhat from this description. The horizontal component of the effective force of the lower leg does not have a forwards orientation in the last quarter of the swinging period. It remains backward with a minimum of magnitude. At heel strike its magnitude is about 1.5 kg.

The vertical component of the effective force of the lower leg is very small at the beginning of the swinging period. In most cases it is directed upwards. Only for the right leg in experiment III is this component directed downwards at first, but its magnitude is then very small. Shortly after, however, the vertical component is directed downwards in the other cases also, and remains so almost until the beginning of the last third of the swinging period. At the same time its magnitude increases, slowly at first and then a little more quickly, until at the beginning of the second third of the period it attains a distinct maximum. In most cases this does not quite reach 2 kg. After this maximum, the magnitude of the vertical component falls quite quickly and then increases again a little, with a second, less distinct, maximum. Shortly afterwards, the vertical component changes direction. It pulls upwards again and soon reaches a maximum of 1–2 kg. Then it decreases to 0 and changes direction once more so that it is directed downwards at heel strike, but with an insignificant magnitude. The vertical component of the effective force of the lower leg thus remains within very close limits during the swinging period. This is because the trajectory of the centre of gravity of the lower leg does not deviate much from a straight line parallel to the floor during the swinging of the leg. This can be seen, for example, in Fig. 4 of Chap. 2.

Again, the direction of the effective force of the lower leg and its variation during the swinging period can be deduced from the behaviour of the two components.

The horizontal component is much larger than the vertical and is thus decisive for the direction of the effective force itself. The following results are, therefore, obtained.

At the beginning of the swinging period, the effective force of the lower leg is almost exactly horizontal and forwards. In the first quarter of this period it deviates only slightly downwards from this initial direction but decreases considerably in magnitude. Then its direction changes relatively quickly rotating clockwise as seen from the right. It is vertical downwards at the beginning of the second third of the period and horizontal backwards at the end of the second third. Thereafter, it deviates only slightly from this direction, at first upwards and finally downwards. But the magnitude of the effective force varies considerably. At the beginning of the last third of the period it is relatively large, falls very soon to 0 and then increases again, almost up to its initial level. In the meantime, the horizontal component not only drops to 0 but goes a little beyond. It turns forwards once again for a short while, but with a very small magnitude. If this is disregarded, the effective force of the lower leg can be said to act horizontally forwards for a time at the beginning of the swinging period and horizontally backwards for a longer time towards the end of the period. In between it rotates clockwise as seen from the right from its initial to its final position. Thus, it carries out half a revolution.

The horizontal component of the effective force of the foot behaves similarly to that of the lower leg. At first, it is directed forwards and its magnitude is 3–4 kg. Although the mass of the foot constitutes only about a third of that of the lower leg, this component of the effective force of the foot is no smaller than that of the lower leg. This applies not only at the beginning but on average throughout the whole period. If anything it is a little larger. This is because, during the swinging period, the centre of gravity of the foot undergoes horizontal accelerations much greater than that of the lower leg. During the first half of the swinging period the horizontal component is directed forwards. During the second half of this period, almost to the end, it is directed backwards. Only at the very end, about 0.02 s before heel strike, is it directed forwards again. As in the case of the effective force of the lower leg, here too the left leg in experiment III presents a slight discrepancy: at the end, the horizontal component of the effective force of the foot does not alter its direction. The magnitude of the component directed forwards in the first half of the period decreases progressively until it is reduced to 1 kg at the beginning of the second quarter. It then remains at this value with some variations up and down during the second quarter of the period. Only at the end does it fall and reach 0. In the second half of the swinging period the component is directed downwards almost to the end. Its magnitude increases, at first slowly, then more quickly, until it has attained a maximum of over 5 kg on average. This occurs approximately between the two last eighths of the swinging period. Then the magnitude drops quite quickly to 0 in such a way that, in general, the component is again directed upwards just before heel strike but without any considerable magnitude.

The vertical component of the effective force of the foot behaves in a completely different manner from that of the lower leg. It is nearly always larger than and its direction is often different from that of the vertical component of the lower leg. However, it is clear that the effective force of the foot in the vertical direction behaves by and large in a manner opposite to the effective force of the thigh, apart from its smaller magnitude. At the beginning of the swinging period its vertical component is 0. Then it adopts a downward direction which it maintains until the beginning of the second third of the swinging period. This is a little longer than the time during which the vertical component of the effective force of the thigh maintains its initial upward direction. Relatively soon, about 0.04 s after toe-off, the vertical component of the effective force of the foot reaches a maximum of a little over 2 kg. Up to the beginning of the last quarter of the swinging period, it is directed upwards after having passed through 0 at the beginning of the second third of the period. Its magnitude thereupon increases progressively up to 2 kg. This maximum of the component directed upwards occurs on average 0.28 s after toe-off. Only shortly, about 0.04 s, afterwards, the magnitude has dropped to 0. From then on, the vertical component is directed downwards again and shortly after, about 0.35 s after toe-off, it attains a maximum of about 4 kg on average. Finally its magnitude falls again and reaches 0 at heel strike or shortly before.

Comparison of the two components of the effective force of the foot results in the following conclusions on the direction of this force.

At the beginning the effective force of the foot is directed horizontally and forwards. Then it rotates clockwise as seen from the right until it approximately bisects the angle formed by the horizontal forwards and the vertical downwards. Thereafter, it rotates backwards, i.e., counter-clockwise, through the horizontal forward orientation, the vertical upward, the horizontal backward, and the vertical downward until it has returned to the horizontal forward orientation. This change in direction is accompanied by very wide variations in magnitude. The magnitude is very small whilst the component is rotating from the horizontal direction forwards into the vertical direction upwards. It becomes relatively large in the area of the direction backwards whilst it is moving further downwards from the latter. Finally, at heel strike the magnitude is again very small so that it does not make much difference in experiment III that the effective force of the left foot, exceptionally, is directed backwards. Apart from this and other slight discrepancies, in the swinging period the effective force of the foot first rotates clockwise through about 45° from the horizontal forward direction and then counter-clockwise by more than a whole revolution.

The horizontal component of the effective force of the whole leg also behaves similarly to that of the lower leg, except that its magnitude is generally even larger than those of the effective forces of the lower leg and foot. At toe-off the horizontal component is directed forwards with an average magnitude of 8 kg. Its magnitude decreases at a fairly fast but constant rate and reaches 0 before the end of the first quarter of the swinging period. Then it changes direction. The horizontal component pulls backwards and increases progressively in magnitude until it reaches a maximum of over 6 kg on average, approximately between the second and last thirds of the period. Thereafter, its magnitude decreases, reaching 0 in the first half of the last quarter of the swinging period in most cases. The component is then directed forwards again for a short while, and in this direction it reaches a maximum of 3 kg in some cases, and even of 4 and 5 kg in others. But it soon decreases quickly, and has fallen to 0 in the second half of the last quarter of the period or at the latest shortly after. During the remainder of the swinging movement, the horizontal component is again directed backwards. Its magnitude increases continuously in some cases. In others, it increases, decreases a little and then increases again. In all cases it is relatively large at heel strike, 5 kg and even up to 10 kg. The effective force of the right leg in experiment I is an exception: it is very close to 0 at heel strike. Moreover, in this case, the preceding stage during which the horizontal component is directed forwards is very short and the magnitude of the horizontal component is very small. In experiment III the effective force of the right leg, except in the first quarter of the period, does not return to the forward direction, but the component directed backwards returns briefly almost to 0 in the middle of the last quarter of the period. Except for these small discrepancies, the horizontal component of the effective force presents essentially the same behaviour in all cases, as does the vertical component which is discussed below. This is remarkable in view of the number of factors involved in arriving at the tables of the effective force of the whole leg.

The vertical component of the effective force of the whole leg behaves in essentially the same way as that of the effective force of the thigh. At the beginning, it is close to 0. Except for a few deviations it is directed upwards after a short while, and in most cases it reaches a maximum of roughly 4 kg somewhere in the middle of the first quarter of the swinging period or slightly later. Its magnitude then decreases and is 0 by the end of the first quarter. From then on, the component is directed downwards. It increases fairly quickly in magnitude at first until it reaches a little less than 5 kg. It maintains this magnitude, with slight variations up and down, for a time, until the end of the second third of the period. Thereafter, its magnitude decreases and reaches 0 at the beginning of the last quarter of the period. Then it changes direction again. The vertical component keeps pointing upwards until almost the end of the swinging period. It reaches a magnitude between 3 kg and 5 kg. In some cases, however, it finally changes direction again and is directed downwards. As a consequence of the abnormal behaviour of the right thigh in experiment II, the vertical component of the effective force of the right leg in experiment II also reaches exceptional magnitudes and displays a somewhat different behaviour at the very beginning and at the very end of the period. These extreme values of the magnitude have not been taken into account in calculation of the average magnitudes. Otherwise, the variations in direction are the same and occur approximately at the same times in this case as in the others.

Comparison of the tables or of the curves for the two components reveals the following behaviour for the effective force of the leg.

At toe-off the orientation of the effective force of the whole leg is almost exactly horizontal and forward. Then it rotates counter-clockwise as seen from the right, in some cases reaching or even passing a vertical upward direction and in other cases falling a little short of this. Thereafter it rather suddenly assumes a backward and downward direction after its magnitude has decreased almost to 0 for a short time. This occurs between the first and second quarters of the period. It remains directed backwards and downwards for a time, almost throughout the second quarter of the period, and it increases in magnitude. More precisely, it first comes closer to the vertical downwards and then returns to its middle position. During the third quarter of the swinging period, the effective force of the leg rotates clockwise to resume a horizontal direction backwards. During the last quarter of the swinging period it behaves differently in the six instances examined. By and large, during this last quarter the effective force of the leg rotates from its backward direction to an upward direction and then upwards and forwards before rotating back to its horizontal direction backwards or only as far as an upward and backward direction or through the horizontal to a backward and downward direction.

The horizontal component of the effective force of the lower leg + foot system behaves very simply. Almost up to the middle of the swinging period it is directed forwards and from then until heel strike backwards. At the beginning of the swinging period the component directed forwards is at its greatest; its magnitude is about 6 kg. Then it decreases at a fairly constant rate to about 1 kg and maintains this value for some time before falling to 0. After the change of direction the magnitude of the component which is now directed backwards increases again at a fairly constant rate up to a maximum of 7–8 kg. Then it decreases and subsequently increases again until it reaches a new maximum higher than the first in some cases and lower in others. Finally, it decreases until the end of the swinging period, but without reaching 0. At heel strike its magnitude is still between 1 and 2 kg; it is even more than 2 kg in the experiment with a load.

The vertical component of the effective force of the lower leg + foot system also displays a fairly simple behaviour which is similar in all six instances. At toe-off its magnitude is close to 0 in all experiments and its direction is upwards. After less than 0.01 s, however, its magnitude is 0 and it has changed direction. It is directed downwards and its magnitude quickly increases up to a maximum of 2–3 kg. Then it decreases somewhat and, for some time, remains almost constant at a value between 1 and 2 kg. Towards the end of the second quarter of the swinging period it decreases further, reaching 0 approximately at the beginning of the third quarter of the period. From then on, the component is directed upwards, but at first its magnitude is practically negligible. It is not until towards the end of the third quarter that it increases to almost 3 kg. Very soon after this it falls to 0 again. Thereafter it changes direction. The magnitude of the component directed downwards increases quickly so that in the middle of the last quarter it has reached a maximum of over 3 kg on average. Finally it decreases again and becomes 0 almost exactly at heel strike.

The effective force of the lower leg + foot system thus behaves as follows. At the beginning it is directed almost exactly horizontally and forwards, as are all the other effective forces considered so far. Then it rotates with decreasing magnitude clockwise as seen from the right, until it bisects the angle between the forward and downward directions. It keeps this direction for a time, except for some slight variations until it disappears completely in the middle of the swinging period. After that, it adopts a direction nearly exactly opposite to the preceding one, at first with a small magnitude and then with gradually increasing magnitude. It is directed backwards and upwards, more backwards than upwards. At the beginning of the last quarter of the swinging period it rotates counter-clockwise until it is directed horizontally backwards and thereafter its direction bisects the angle between the backward and downward directions. Finally it rotates back again, i.e. clockwise, and has just reached the horizontal and backward direction at heel strike.

On the Effect of Movement on Joint Compression

The effective forces must be known for calculation of the torques of the internal forces. This knowledge also allows an insight into the changes in joint compression resulting from the movement of the leg.

The compression in a joint represents the resultant of a series of compressive forces which can be divided into two groups (Chap. 5). The compressive forces of one group are brought about by the forces applied to the segments of the body and are present even when the body segments are not made to move by these forces. They can be called "static compressive forces". The compressive forces of the other group depend on the movement of the body segments and arise only when the body segments move. They are referred to as "kinetic compressive forces". The resultants of these kinetic forces are actually the effective forces.

For example, during the swinging period compression is exerted on the thigh in the hip joint (Chap. 5, p. 329). Its static components are equal and opposite to the weight of the whole leg and the tension of the muscles inserted in any segment of the leg and spanning the hip joint. The total effect of the kinetic compressive forces, on the other hand, is equal to the effective force of the whole leg. The same compressive components act on the pelvis with the same magnitude but in the opposite direction, according to the law of equality of action and reaction. One of these two series of opposite compressive components acts on the thigh, and the other on the pelvis. They make up most of the joint compression. Air pressure must be added if the joint cavities are closed by an airtight capsule. Air pressure was disregarded when the movement equations in Chap. 5 were devised since its motive action on the body segments is always compensated by the reaction provoked in the joint.

A thorough analysis of the joint compression must distinguish precisely where the two opposing contact surfaces of the joint are mainly pressed together by the different components of joint compression. This merely requires that a radius be drawn from the centre of the femoral head in the opposite direction to that in which the compressive force under consideration acts on the thigh. If this radius intersects the two articular contact surfaces these are pressed together at this point by the compressive component under consideration. As a consequence of the deformability of the articular cartilage this compression will of course not be limited to one point. If the radius does not intersect the articular contact surfaces the compressive component will exert a tension on some joint ligaments. It then tends to separate these areas of the articular surfaces which are on the prolongation of the radius mentioned above.

Gravity, for example, entails a compressive component equal to the weight of the whole leg which pulls vertically downwards at the pelvis and vertically upwards at the thigh. The radius to be drawn from the centre of the femoral head must be directed vertically downwards. Obviously it is only in extreme positions of the leg suspended from the pelvis that this radius passes through parts of the articular surfaces. In general, it passes through the neck of the femur. Gravity alone would separate the areas of the articular surfaces that are vertically above the centre of the femoral head and would exert tension on the joint ligaments tensed in the position under consideration or on fibres of the joint capsule directed downwards if other compressive components were not at hand to exert the opposite effect. A compressive component of this kind can be caused by the air pressure, for instance. Gravity acting on the leg pulls it downwards whereas it provokes in the hip joint a compressive component directed upwards for the thigh. Conversely, the air pressure pushes the leg upwards into the socket and this provokes a reaction downwards on the thigh. This reaction represents a compressive component for the thigh. In this case, the radius must be drawn from the centre of the femoral head upwards, more precisely upwards and medially. This radius meets points of the two articular surfaces in any position of the leg suspended from the pelvis, and air pressure tends to compress these surfaces together especially at this point.

These two examples are easily checked, and, after they have been studied to clarify the action on the articular surfaces, it is not difficult to deduce from the data concerning the effective force of the leg how the kinetic compressive component acts on the articular surfaces of the hip.

The effective force of the leg is directed horizontally forwards at the beginning of the swinging period. Therefore, as a consequence of the movement, the articular surfaces of the hip will then be pressed against each other the most in areas that are horizontally behind the centre of the joint when the thigh is in its initial position. When, some time later, the same effective force is directed backwards and downwards for a longer time, the articular surfaces of the hip joint will be pressed against each other the most at the front and the top. Since the thigh continues to rotate further forwards, different areas of the articular cartilage of the femoral head are stressed in turn. This is not the case for the socket if the pelvis does not change its position in space significantly. These examples show that knowledge of the magnitude and direction of the effective force of the whole leg gives an insight into the joint compression resulting from the movement alone.

There are components of the articular compression that in some circumstances are greater than almost all the others. At present precise determination of their direction but not of their magnitude is possible. These are the compressive components resulting from contraction of the muscles. To find the areas where the isolated contraction of a muscle spanning the hip joint presses the articular surfaces against each other, a radius is drawn from the centre of the femoral head in a direction corresponding to that of the part of the muscle which pulls unimpeded from the insertion of the muscle on the thigh, or from a bony prominence of the thigh, to the origin of the muscle or to a bony prominence of the pelvis.

The compression in the knee joint during the swinging period can be deduced in the same way as that in the hip. In this case the static compressive component due to gravity is the weight of the lower leg and foot. The kinetic compressive force is the effective force of the lower leg + foot system. Only the muscles which span the knee need be considered.

For the articular pressure in the ankle, finally, it is only necessary to consider the weight of the foot, the effective force of the foot and the muscles spanning the ankle joint.

The magnitude and direction of the effective force of the lower leg + foot system and the effective force of the foot alone throughout the swinging period have been determined. Thus all the data necessary to find the magnitude and direction of the different components of the articular pressure in the knee and in the ankle at any instant of the swinging movement are available, except for the magnitude of the static compressive components due to the muscles.

The Torques Exerted by Gravity

Calculation of the torques resulting from gravity requires that the length of the thigh and the distance between the centre of gravity of each of the three segments of the leg and the centre of its proximal joint are known. The length of the thigh is considered to be the distance between the centre of the hip and that of the knee, and the length of the lower leg the distance between the centre of the knee and that of the ankle. The co-ordinates of these joint centres have been found for the 31 phases captured by chronophotography, in each of the three experiments (Tables 14–16 in Chap. 1, pp. 75–79). The lengths of the thigh and lower leg of our subject can thus be calculated using a simple equation of tridimensional analytical geometry. I have carried out this calculation for all the phases of the first two experiments to achieve the greatest possible accuracy. The average values resulting from the 31 calculations coincide to fractions of a millimetre for the thigh and for the lower leg on the same side in the two experiments. But the lengths of the thighs and lower legs differ between left and right by over 2 cm. This is not surprising if one thinks how difficult it is to project the centre of the hip and that of the knee joint accurately on the body surface, as was required for the positioning of the Geissler tubes in the experiments. Apart from this difficulty, earlier measurements we made of the lengths of the long body segments in general showed differences of 1 cm or more. Therefore, the inaccuracies in measuring the lengths of the thigh and lower leg are probably less than 1 cm. Even if they were greater they would hardly change the data concerning the directions, angular velocities and accelerations of the axes of the segments of the leg, and the velocities and accelerations of their centres of gravity during a double step. This appears unquestionable in view of the complete concordance of all the results obtained so far and related to both sides of the body.

There is no reliable way of determining any discrepancies in the masses of symmetrical body segments in vivo. The results found in cadavers can be applied to living subjects, assuming complete symmetry. These results are used taking into account the ratio of the total mass of the subject measured on scales to the total mass of the cadaver. Equal length of corresponding body segments is assumed. The average of the measurements recorded for the right and left sides of the body is 38.8 cm for the length of the thigh of our subject and 37.7 cm for that of the lower leg.

According to a previous contribution [4], the centre of gravity of the right thigh lies on the line connecting the centre of the hip and that of the knee, dividing this line in the ratio of 0.44 : 0.56. The centre of gravity of the lower leg divides the line connecting the centre of the hip and that of the ankle in the ratio of 0.42 : 0.58. The distance from the centre of gravity of the thigh to the centre of the hip is thus 38.8 cm × 0.44 = 17.1 cm in our subject, and the distance from the centre of gravity of the lower leg to the centre of the knee is 37.7 cm × 0.42 = 15.8 cm. The distance between the centre of gravity of the foot and the centre of the ankle could be calculated directly from the co-ordinates of the centre of the ankle and of the centre of gravity of the foot, which are given in Chap. 1. Calculation yields an average value of 6.8 cm.

Finally, the position of the point of application of the second force of the couple due to gravity must be determined for the thigh and for the lower leg.

This point of application of the second force of the couple is on the long axis of the thigh, designated as H_2' in Fig. 3, as it was in Fig. 3 of Chap. 5 (p. 331). Its distance from the centre of the hip H is c_2'. Point H_2' divides that part of the long axis connecting the centre of gravity S_2 of the thigh to the centre K of the knee in inverse proportion to the ratio of the mass of the thigh m_2 to that of the lower leg + foot system $m_{4,6}$. The length of the thigh, i.e. the segment $\overline{HK}$ in Fig. 3, is designated as l_2, and the distance from the centre of gravity of the thigh to the centre of the hip, the segment $\overline{HS_2}$, is designated as r_2. To calculate the distance $\overline{HH_2'}$ or c_2' the following equation is used.

$$\frac{c_2' - r_2}{l_2 - c_2'} = \frac{m_{4,6}}{m_2}$$

The mass $m_{4,6}$ can be replaced by the sum of the masses m_4 of the lower leg and m_6 of the foot. To calculate c_2' the following equation is used:

$$c_2' = \frac{m_2 r_2 + (m_4 + m_6) l_2}{m_2 + m_4 + m_6} \tag{1}$$

It is known that $m_2 = 0.00693$; $m_4 = 0.00315$; $m_6 = 0.00107$; $r_2 = 17.1$ cm; and $l_2 = 38.8$ cm. If these values are entered in the equation it appears that $c_2' = 25.3$ cm. The same value is assigned to the segment c_3' for the left thigh.

Since the distance c_2' and the positional angle φ_2 of the thigh are known, the shortest distance between the forces $-G_{2,4,6}$ and $+G_{2,4,6}$ (equal to the weight of the leg) of the couple of gravity applied at H and H_2' can be calculated. It is $c_2' \cdot \sin \varphi_2$. The angle φ_2 formed by the long axis of the thigh and the distal part of the vertical through H is positive when the centre of the knee is further forwards than the centre of the hip. This occurs in the position represented in Fig. 3 for example. The torque D_{s_2} with which gravity acts on the thigh is positive when the couple of gravity tends to increase the angle φ_2 or to rotate the thigh counter-clockwise as seen from the right. This is not true, however, when the angle φ_2 is positive, as can be seen in Fig. 3. In this case, the couple tends to rotate the thigh clockwise so that the angle formed by the long axis and the vertical becomes smaller. Therefore, the moment of the couple, which is the product of the magnitude of the forces and the distance

---

[4] Abhandlungen der mathematisch-physischen Klasse der Königlichen Sächsischen Gesellschaft der Wissenschaften, vol 15, no 7, p 622 (1889)

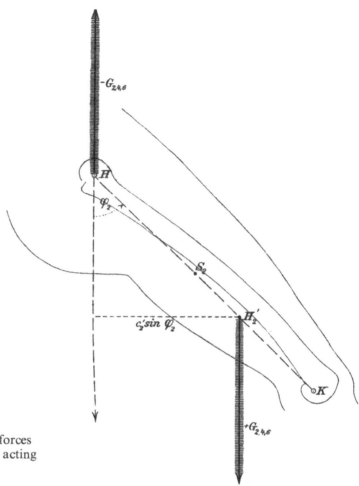

Fig. 3. Couple resulting from the forces of gravity ($-G_{2,4,6}$ and $+G_{2,4,6}$) acting on the thigh during the swinging of the leg

between their lines of action, must be negative. The total weight of the whole leg can be replaced by the sum of the weights G_2, G_4, G_6, or G_3, G_5, G_7 of the three segments of the leg. This yields the following equations for calculation of the torque of gravity.

For the right thigh: $\qquad D_{s_2} = -(G_2 + G_4 + G_6)\, c_2' \sin\varphi_2$

and for the left thigh: $\quad D_{s_3} = -(G_3 + G_5 + G_7)\, c_3' \sin\varphi_3$

$$\tag{2}$$

According to previous research on the centre of gravity of the human body[5], the weight of the thigh of our subject is 6.80 kg, that of the lower leg 3.09 kg and that of the foot 1.05 kg. If G_2, G_4 and G_6 or G_3, G_5 and G_7 are replaced by these values and if the values calculated above for c_2' and c_3' are used, the following equations are obtained:

$$D_{s_2} = -276.782 \sin\varphi_2$$
$$D_{s_3} = -276.782 \sin\varphi_3$$

$$\tag{2a}$$

[5]   Abhandlungen der mathematisch-physischen Klasse der Königlichen Sächsischen Gesellschaft der Wissenschaften, vol 15, no 7, p 629. English translation: Braune W, Fischer O (1985) On the centre of gravity of the human body. Springer, Berlin Heidelberg New York Tokyo, p 49

These equations are valid for any value of the positional angles φ_2 and φ_3. If the knee is further back than the hip the positional angle is negative and its sine is also negative. The torque is then positive. In this case, the couple of gravity tends to rotate the thigh counter-clockwise as seen from the right. According to the discussion above, this is a positive rotation.

The φ_2- and φ_3-angles for all the phases following each other at 0.01 s intervals during the swinging period are given in Tables 6–11 in Chap. 5 (pp. 349–354). They are expressed in radians. But nearly all trigonometric tables express angles in degrees. The radians must, therefore, be converted to degrees by multiplying the angles by $180/\pi$ or 57.296. The angles thus expressed in degrees and minutes are given in Tables 7–12 (pp. 408–419). The same tables also give the torques D_{s_2} and D_{s_3} calculated from Eq. (2a). Since the forces of the couples are expressed in kilograms and the moment arms of the couples in centimetres, the torques are given in kilogram-centimetres. The unit of the torque of this system is that of a couple the forces of which are 1 kg and the moment arm of which is 1 cm.

For the right lower leg, the point of application of the second force of the couple due to gravity is a point H_4'' on the long axis of the lower leg (Fig. 4 or Fig. 5, p. 335, Chap. 5). Its distance from the centre of the knee is c_4''. H_4'' divides that part of the long axis of the lower leg between the centre of gravity S_4 of the latter and the centre F of the ankle in inverse proportion to the ratio of the mass of the lower leg m_4 to that of the foot m_6. The distance $\overline{KF}$ in the lower leg is designated as l_4 and the distance $\overline{KS}_4$ between the centre of gravity of the lower leg and the centre of the knee as r_4. For calculation of c_4'' the following equations are used:

$$\frac{c_4'' - r_4}{l_4 - c_4''} = \frac{m_6}{m_4}$$

and

$$c_4'' = \frac{m_4 r_4 + m_6 l_4}{m_4 + m_6} \tag{3}$$

$r_4 = 15.8$ cm and $l_4 = 37.7$ cm. When the values given for m_4 and m_6 below Eq. (1) are used, the value 21.4 cm is obtained for c_4'', which is the same for the corresponding distance c_5'' in the left lower leg.

The moment arm of the couple formed by the forces $-G_{4,6}$ and $+G_{4,6}$ applied at K and H_4'' is $c_4'' \cdot \sin \varphi_4$, as can be seen from Fig. 4. If the positional angle of the lower leg is positive, the couple tends to rotate the lower leg clockwise, i.e., in the negative direction. The weight $G_{4,6}$ or $G_{5,7}$ can be replaced by the sum of the weights of the lower leg and of the foot. The torque due to gravity can then be calculated by using the equations below.

For the right lower leg: $D_{s_4} = -(G_4 + G_6)\, c_4'' \sin\varphi_4$

For the left lower leg: $D_{s_5} = -(G_5 + G_7)\, c_5'' \sin\varphi_5$

$$\tag{4}$$

Entering the values for the weights and for c_4'' and c_5'' in these equations gives for our subject

$$D_{s_4} = -88.596 \cdot \sin\varphi_4$$
$$D_{s_5} = -88.596 \cdot \sin\varphi_5 \tag{4a}$$

After the positional angles φ_4 and φ_5 given in radians in Tables 6–11 in Chap. 5 had been converted into degrees, the torques exerted by gravity on the lower leg were calculated. The results are given in Tables 7–12. The torques are of course expressed in kilogram-centimetres here too.

In the right foot, the point of application of the second force of the couple coincides with the centre of gravity. The symbol r_6 is used to mean the distance between the latter and the centre of the ankle. The two forces have the same magnitude as the weight of the foot. As seen in Fig. 5 (see also Fig. 7, p. 336, Chap. 5), the equations below are used to calculate the torque due to gravity.

$$\left.\begin{array}{ll} \text{For the right foot:} & D_{s_6} = -G_6 r_6 \sin\varphi_6 \\ \text{For the left foot:} & D_{s_7} = -G_7 r_7 \sin\varphi_7 \end{array}\right\} \tag{5}$$

The negative sign means that if the positional angle is positive, gravity tends to rotate the foot clockwise, that is in the negative direction.

The value of r_6 or $\overline{FS}_6$ and of r_7 or $\overline{FS}_7$ is 6.8 cm, as mentioned above. Entering this value and the weight of the foot in Eq. (5) gives:

$$\left.\begin{array}{l} D_{s_6} = -7.140 \cdot \sin\varphi_6 \\ D_{s_7} = -7.140 \cdot \sin\varphi_7 \end{array}\right\} \tag{5a}$$

The torques exerted on the foot by gravity and the positional angles φ_6 and φ_7 expressed in degrees are also given in Tables 7–12.

So that all the data relevant to the kinetics of the swinging of the leg will be together, these tables also contain the torques of the effective forces, the products $m\varkappa^2\varphi''$, and the subsequent torques of the internal forces which result mainly from the contraction of the muscles although these will not be deduced until later.

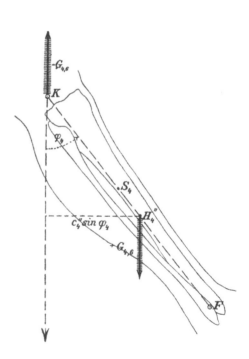

Fig. 4. Couple resulting from the forces of gravity $(-G_{4,6}$ and $+G_{4,6})$ acting on the lower leg during the swinging of the leg

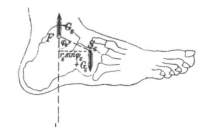

Fig. 5. Couple resulting from the forces of gravity $(-G_6$ and $+G_6)$ acting on the foot during the swinging of the leg

Table 7. Values of all the magnitudes involved in the kinetics of the swinging of the right leg in experiment I

Time (0,01 s)	Right thigh					Right lower leg					Right foot					Time (0,01 s)
	φ_2	$m_2\varkappa_2^2\varphi_2''$ (kgcm)	D_{θ_2} (kgcm)	D_{e_2} (kgcm)	D_{m_2} (kgcm)	φ_4	$m_4\varkappa_4^2\varphi_4''$ (kgcm)	D_{θ_4} (kgcm)	D_{e_4} (kgcm)	D_{m_4} (kgcm)	φ_6	$m_6\varkappa_6^2\varphi_6''$ (kgcm)	D_{s_6} (kgcm)	D_{e_6} (kgcm)	D_{m_6} (kgcm)	
0	−16°34′	+35,1	+78,9	−312,8	+269,0	−47°30′	+7,8	+65,3	−133,5	+76,0	−14°23′	+4,6	+1,8	−24,6	+27,4	0
1	−14°26′	+31,1	+69,0	−283,5	+245,6	−50°1′	+13,4	+67,9	−137,5	+83,0	−18°30′	+4,9	+2,3	−25,1	+27,7	1
2	−11°38′	+24,1	+55,8	−252,0	+220,3	−52°50′	+16,8	+70,6	−158,7	+104,9	−20°58′	+5,4	+2,6	−26,8	+29,6	2
3	−9°34′	+15,0	+46,0	−230,8	+199,8	−54°46′	+19,6	+72,4	−167,4	+114,6	−22°45′	+7,1	+2,8	−26,9	+31,2	3
4	−7°20′	0	+35,3	−182,8	+147,5	−55°59′	+22,1	+73,4	−152,8	+101,5	−23°36′	+5,0	+2,9	−25,1	+27,2	4
5	−5°13′	−13,0	+25,2	−149,7	+111,5	−56°50′	+25,2	+74,2	−131,9	+82,9	−23°54′	+3,8	+2,9	−22,1	+23,0	5
6	−3°2′	−22,1	+14,6	−119,1	+82,4	−57°18′	+30,0	+74,6	−116,6	+72,0	−23°43′	+3,2	+2,9	−19,9	+20,2	6
7	−1°5′	−25,1	+5,2	−90,8	+60,5	−57°14′	+55,2	+74,5	−101,1	+81,8	−23°2′	+4,6	+2,8	−17,1	+18,9	7
8	+0°58′	−22,1	−4,7	−65,8	+48,4	−56°50′	+17,6	+74,2	−85,6	+29,0	−21°36′	+3,1	+2,6	−14,3	+14,8	8
9	+3°2′	−19,1	−14,6	−43,8	+39,3	−55°45′	+13,7	+73,2	−73,3	+13,8	−19°32′	+2,2	+2,4	−11,5	+11,3	9
10	+4°52′	−16,0	−23,5	−21,6	+29,1	−54°9′	+12,0	+71,8	−61,0	+1,2	−17°15′	+1,6	+2,1	−9,1	+8,6	10
11	+6°46′	−14,0	−32,6	−0,8	+19,4	−52°12′	+10,1	+70,0	−49,0	−10,9	−14°37′	+1,1	+1,8	−6,3	+5,6	11
12	+8°12′	−12,0	−39,5	+19,5	+8,0	−49°58′	+9,5	+67,8	−39,5	−18,8	−11°38′	+0,9	+1,4	−4,5	+4,0	12
13	+9°44′	−11,0	−46,8	+25,3	+10,5	−47°54′	+8,4	+65,7	−37,9	−19,4	−8°39′	+0,9	+1,1	−4,4	+4,2	13
14	+11°17′	−9,6	−54,2	+35,7	+8,9	−45°23′	+7,3	+63,1	−39,3	−16,5	−5°33′	+0,7	+0,7	−5,4	+5,4	14
15	+12°40′	−8,7	−60,7	+40,2	+11,8	−42°41′	+6,4	+60,1	−42,0	−11,7	−1°50′	+0,7	+0,2	−6,4	+6,9	15
16	+14°19′	−8,1	−68,4	+53,8	+6,5	−40°6′	+6,2	+57,1	−41,6	−9,3	+1°26′	+0,7	−0,2	−6,5	+7,4	16
17	+15°42′	−7,8	−74,9	+70,7	−3,6	−37°15′	+5,6	+53,6	−37,0	−11,0	+5°6′	+0,7	−0,6	−6,1	+7,4	17

Table 7 (continued)

n															
18	+ 17° 1'	− 8,0	− 81,0	+ 81,1	**− 8,1**	− 34°26'	+ 5,3	+ 50,1	− 22,1	**− 22,7**	+ 8°29'	+ 0,6	− 1,1	− 5,3	**+ 7,0**
19	+ 18°27'	− 8,3	− 87,6	+ 92,5	**− 13,2**	− 31°38'	+ 5,0	+ 46,5	− 7,0	**− 34,5**	+ 12°19'	+ 0,7	− 1,9	− 4,3	**+ 6,9**
20	+ 19°39'	− 8,9	− 93,1	+ 106,9	**− 22,7**	− 28°32'	+ 4,8	+ 42,3	+ 8,2	**− 45,7**	+ 15°59'	+ 0,7	− 2,0	− 3,4	**+ 6,1**
21	+ 21° 5'	− 9,7	− 99,6	+ 117,8	**− 27,9**	− 23°36'	+ 5,3	+ 35,5	+ 19,5	**− 49,7**	+ 19°32'	+ 0,7	− 2,4	− 2,8	**+ 5,9**
22	+ 22°10'	− 11,0	− 104,4	+ 131,6	**− 38,2**	− 20° 0'	+ 6,2	+ 30,3	+ 31,2	**− 55,3**	+ 23° 9'	+ 0,7	− 2,8	− 2,4	**+ 5,9**
23	+ 23°16'	− 13,0	− 109,3	+ 144,2	**− 47,9**	− 16°37'	+ 7,0	+ 25,3	+ 43,3	**− 61,6**	+ 26°28'	+ 0,6	− 3,2	− 2,0	**+ 5,8**
24	+ 24°11'	− 15,0	− 113,4	+ 159,0	**− 60,6**	− 12°33'	+ 7,8	+ 19,3	+ 53,6	**− 65,1**	+ 30° 5'	+ 0,5	− 3,6	− 1,7	**+ 5,8**
25	+ 25° 9'	− 17,0	− 117,6	+ 173,3	**− 72,7**	− 8°18'	+ 9,2	+ 12,8	+ 65,0	**− 68,6**	+ 34° 9'	+ 0,3	− 4,0	− 1,1	**+ 5,4**
26	+ 25°54'	− 20,1	− 120,9	+ 186,2	**− 85,4**	− 4°38'	+ 13,2	+ 7,2	+ 74,7	**− 68,7**	+ 38°23'	+ 0,2	− 4,4	− 1,2	**+ 5,8**
27	+ 26°25'	− 26,1	− 123,1	+ 193,1	**− 96,1**	− 0°55'	+ 15,1	+ 1,4	+ 85,0	**− 71,3**	+ 41°56'	0	− 4,8	− 0,9	**+ 5,7**
28	+ 26°39'	− 35,1	− 124,1	+ 197,6	**− 108,6**	+ 3°23'	+ 6,4	− 5,2	+ 93,9	**− 82,3**	+ 45°47'	− 0,3	− 5,1	− 0,6	**+ 5,4**
29	+ 26°45'	− 65,2	− 124,6	+ 186,3	**− 126,9**	+ 7°37'	0	− 11,7	+ 104,4	**− 92,7**	+ 49°58'	− 0,7	− 5,5	0,0	**+ 4,8**
30	+ 26°49'	− 45,1	− 124,9	+ 177,6	**− 97,8**	+ 12°33'	− 13,4	− 19,3	+ 112,9	**− 107,0**	+ 53°34'	− 1,2	− 5,7	+ 1,0	**+ 3,5**
31	+ 26°45'	− 15,0	− 124,6	+ 134,8	**− 25,2**	+ 17°53'	− 36,4	− 27,2	+ 118,5	**− 127,7**	+ 58° 6'	− 2,0	− 6,1	+ 2,7	**+ 1,4**
32	+ 26°39'	0	− 124,1	+ 205,9	**− 81,8**	+ 21°36'	− 47,3	− 32,6	+ 165,9	**− 180,6**	+ 61°25'	− 3,7	− 6,3	+ 8,7	**− 6,1**
33	+ 26°32'	+ 21,1	− 123,6	+ 192,3	**− 47,6**	+ 24°24'	− 51,0	− 36,6	+ 180,6	**− 195,0**	+ 64° 7'	− 6,5	− 6,4	+ 13,4	**− 13,5**
34	+ 26°28'	+ 33,1	− 123,4	+ 174,6	**− 18,1**	+ 26°52'	− 52,9	− 40,0	+ 208,7	**− 221,6**	+ 66° 4'	− 5,9	− 6,5	+ 25,9	**− 15,3**
35	+ 26°39'	+ 38,1	− 124,1	+ 155,8	**+ 6,4**	+ 27°58'	− 54,1	− 41,5	+ 253,9	**− 266,5**	+ 67°16'	− 5,1	− 6,6	+ 36,3	**− 34,8**
36	+ 26°56'	+ 39,1	− 125,4	+ 191,5	**+ 27,0**	+ 28°35'	− 54,9	− 42,4	+ 326,2	**− 338,7**	+ 67°54'	− 4,4	− 6,6	+ 43,6	**− 41,4**
37	+ 27°20'	+ 35,1	− 127,1	+ 161,3	**+ 0,9**	+ 28°32'	− 55,5	− 42,3	+ 261,8	**− 275,0**	+ 67°50'	− 3,9	− 6,6	+ 33,4	**− 30,7**
38	+ 27°51'	+ 20,1	− 129,3	+ 141,6	**+ 7,8**	+ 27°44'	− 61,6	− 41,2	+ 169,8	**− 190,2**	+ 67°19'	− 3,4	− 6,6	+ 19,7	**− 16,5**
39	+ 28°42'	0	− 132,9	+ 72,9	**+ 60,0**	+ 26°25'	− 140,1	− 39,4	+ 73,3	**− 174,0**	+ 66°11'	− 3,0	− 6,5	+ 8,3	**− 4,8**
40	+ 29°27'	− 20,1	− 136,1	+ 40,9	**+ 75,1**	+ 24° 4'	0	− 36,1	+ 4,5	**+ 31,6**	+ 64°45'	− 2,7	− 6,5	− 1,2	**+ 5,0**
41	+ 30° 8'	− 60,2	− 139,0	+ 61,1	**+ 17,7**	+ 21° 2'	+ 47,6	− 31,8	+ 15,5	**+ 94,9**	+ 62°55'	− 2,5	− 6,4	− 5,1	**+ 9,0**
41,5	+ 30°19'	− 90,3	− 139,7	+ 60,0	**− 10,6**	+ 20°34'	+ 70,0	− 31,1	+ 15,0	**+ 116,1**	+ 62°13'	− 2,5	− 6,3	− 6,7	**+ 10,5**

Table 8. Values of all the magnitudes involved in the kinetics of the swinging of the left leg in experiment I

Time (0,01 s)	Left thigh					Left lower leg					Left foot					Time (0,01 s)
	φ_3	$m_3\kappa_3^2\varphi_3''$ (kgcm)	D_{s_3} (kgcm)	D_{e_3} (kgcm)	D_{m_3} (kgcm)	φ_5	$m_5\kappa_5^2\varphi_5''$ (kgcm)	D_{s_5} (kgcm)	D_{e_5} (kgcm)	D_{m_5} (kgcm)	φ_7	$m_7\kappa_7^2\varphi_7''$ (kgcm)	D_{s_7} (kgcm)	D_{e_7} (kgcm)	D_{m_7} (kgcm)	
0	−18°37′	+44,1	+88,4	−206,9	+162,6	−45°16′	+2,8	+62,9	−103,7	+43,6	−14°19′	+4,5	+1,8	−19,8	+22,5	0
1	−16°27′	+37,1	+78,4	−209,7	+168,4	−47°13′	+10,1	+65,0	−132,2	+77,3	−16°34′	+4,2	+2,0	−23,1	+25,3	1
2	−14°23′	+26,1	+68,8	−197,9	+155,2	−49°47′	+15,7	+67,7	−155,6	+103,6	−18°20′	+3,6	+2,2	−25,3	+26,7	2
3	−12°12′	+14,0	+58,5	−179,9	+135,4	−51°27′	+18,2	+69,3	−169,6	+118,5	−19°46′	+3,1	+2,4	−26,0	+26,7	3
4	−10°22′	0	+49,8	−160,9	+111,1	−52°43′	+21,8	+70,5	−170,2	+121,5	−20°17′	+3,5	+2,5	−25,5	+26,5	4
5	−7°51′	−13,0	+37,8	−139,2	+88,4	−53°24′	+28,3	+71,1	−161,9	+119,1	−20°27′	+5,1	+2,5	−23,6	+26,2	5
6	−5°51′	−17,0	+28,2	−121,8	+76,6	−53°31′	+33,6	+71,2	−148,0	+110,4	−20°7′	+5,5	+2,5	−21,6	+24,6	6
7	−4°8′	−14,5	+20,0	−103,2	+68,7	−53°10′	+25,2	+70,9	−131,7	+86,0	−19°12′	+3,5	+2,3	−19,0	+20,2	7
8	−2°4′	−12,4	+10,0	−87,9	+65,5	−52°15′	+17,6	+70,1	−118,2	+65,7	−17°25′	+2,4	+2,1	−17,1	+17,4	8
9	−0°7′	−10,7	+0,6	−67,7	+56,4	−50°42′	+14,3	+68,6	−102,4	+48,1	−15°4′	+1,6	+1,9	−14,3	+14,0	9
10	+1°36′	−9,6	−7,7	−51,0	+49,1	−49°16′	+12,0	+67,1	−92,0	+36,9	−13°7′	+1,2	+1,6	−12,3	+11,9	10
11	+3°30′	−8,9	−16,9	−32,8	+40,8	−47°26′	+10,6	+65,3	−81,2	+26,5	−10°15′	+1,0	+1,3	−10,2	+9,9	11
12	+5°16′	−8,4	−25,4	−15,6	+32,6	−45°19′	+9,5	+63,0	−70,3	+16,8	−7°37′	+0,9	+0,9	−8,2	+8,2	12
13	+6°53′	−8,2	−33,2	+2,7	+22,3	−43°12′	+8,4	+60,6	−60,2	+8,0	−4°52′	+1,1	+0,6	−6,5	+7,0	13
14	+8°29′	−8,4	−40,8	+15,4	+17,0	−40°48′	+7,6	+57,9	−48,0	−2,3	−1°53′	+1,6	+0,2	−5,0	+6,4	14
15	+10°19′	−8,8	−49,6	+29,4	+11,4	−38°23′	+7,0	+55,0	−32,9	−15,1	+1°23′	+1,8	−0,2	−3,3	+5,3	15
16	+11°52′	−9,5	−56,9	+40,6	+6,8	−35°21′	+6,2	+51,3	−21,4	−23,7	+4°4′	+1,6	−0,5	−2,4	+4,5	16
17	+13°7′	−10,3	−62,8	+51,9	+0,6	−32°43′	+5,6	+47,9	−12,1	−30,2	+8°1′	+0,9	−1,0	−1,9	+3,8	17

Table 8 (continued)

18	+14°47'	−11,5	−70,6	+57,4	+1,7	+28°39'	+5,3	+42,5	−4,9	32,3	+12° 9'	+0,7	−1,5	−2,0	+4,2	18
19	+16° 9'	−12,7	−77,0	+57,6	+6,7	+25°54'	+5,0	+38,7	−3,2	30,5	+15°28'	+0,5	−1,9	−3,1	+5,5	19
20	+17°25'	−14,6	−82,8	+53,2	+15,0	+22°59'	+4,8	+34,6	−8,9	20,9	+19°25'	+0,5	−2,4	−5,4	+8,3	20
21	+18°48'	−16,8	−89,2	+53,9	+18,5	+19°46'	+4,5	+30,0	−11,5	14,0	+23°23'	+0,4	−2,8	−7,0	+10,2	21
22	+19°46'	−18,9	−93,6	+62,4	+12,3	+16°51'	+4,8	+25,7	+5,9	15,0	+26°49'	+0,4	−3,2	−7,2	+10,8	22
23	+20°51'	−21,1	−98,5	+84,5	−7,1	+13°28'	+5,9	+20,6	+7,8	22,5	+30°12'	+0,4	−3,6	−6,4	+10,4	23
24	+21°50'	−24,1	−102,9	+111,2	−32,4	+9°44'	+7,0	+15,0	+32,9	40,9	+33°55'	+0,3	−4,0	−4,6	+8,9	24
25	+22°38'	−27,1	−106,5	+124,9	−45,5	+6°11'	+8,4	+9,5	+45,7	46,8	+37°21'	+0,2	−4,3	−4,8	+9,3	25
26	+23°12'	−30,1	−109,0	+145,5	−66,6	+1°57'	+7,6	+3,0	+62,1	57,5	+40°58'	+0,2	−4,7	−4,7	+9,6	26
27	+23°33'	−32,9	−110,6	+149,3	−71,6	+1°26'	+3,6	−2,2	+74,5	68,7	+44°38'	+0,0	−5,0	−5,0	+10,0	27
28	+23°40'	−34,8	−111,1	+188,8	−112,5	+5°51'	0	−9,0	+95,2	86,2	+48°45'	0	−5,4	−4,3	+9,7	28
29	+23°36'	−35,3	−110,8	+229,1	−153,6	−10°15'	−4,2	−15,8	+115,3	103,7	+52°43'	−0,7	−5,7	−2,3	+7,3	29
30	+23°23'	−31,6	−109,9	+271,4	−193,1	−13°45'	−7,8	−21,1	+146,7	133,4	+56°40'	−1,4	−6,0	−1,5	+3,1	30
31	+23° 5'	−23,1	−108,5	+286,8	−201,4	−17°49'	−12,0	−27,1	+181,4	166,3	+60°51'	−2,2	−6,2	+6,8	−2,8	31
32	+22°34'	−10,0	−106,2	+270,9	−174,7	−21°19'	−17,4	−32,2	+202,4	187,6	+64°38'	−3,1	−6,4	+11,7	−8,4	32
33	+22° 4'	0	−104,0	+208,8	−104,8	−25° 6'	−23,0	−37,6	+206,7	192,1	+67°40'	−4,3	−6,6	+15,2	−12,9	33
34	+21°39'	+25,1	−102,1	+116,0	+11,2	−27°54'	−29,4	−41,5	+199,6	187,5	+70°28'	−6,3	−6,7	+19,3	−18,9	34
35	+21°15'	+44,1	−100,3	+28,7	+115,7	−29°58'	−39,2	−44,3	+203,9	198,8	+72°15'	−6,7	−6,8	+27,1	−27,0	35
36	+21°12'	+70,2	−100,1	+17,0	+153,3	−30°39'	−45,7	−45,2	+228,1	228,6	+72°56'	−5,5	−6,8	+39,9	−38,6	36
37	+21°36'	+78,2	−101,9	+55,3	+124,8	−30°46'	−49,3	−45,3	+211,9	215,9	+72°46'	−4,0	−6,8	+31,8	−29,0	37
38	+22°24'	+76,2	−105,5	+146,1	+35,6	−29°34'	−54,1	−43,7	+195,8	206,2	+72° 5'	−3,0	−6,8	+20,9	−17,1	38
39	+23°29'	0	−110,3	+200,7	−90,4	−27°40'	−126,0	−41,1	+159,5	244,4	+70°39'	−2,4	−6,7	+12,4	−8,1	39
40	+24°35'	−30,1	−115,1	+235,3	−150,3	−25° 6'	0	−37,6	+97,9	60,3	+68°42'	−1,9	−6,7	+3,3	+1,5	40
41	+25°23'	−80,2	−118,7	+251,2	−212,7	−22°14'	+61,6	−33,5	+43,8	+51,3	+66°45'	−1,6	−6,6	+1,7	+6,7	41
41,7	+25°33'	−90,3	−119,4	+215,6	−186,5	−20°55'	+112,0	−31,6	−14,9	+158,5	+65°33'	−1,5	−6,5	+7,0	+12,0	41,7

Table 9. Values of all the magnitudes involved in the kinetics of the swinging of the right leg in experiment II

Time (0,01 s)	Right thigh					Right lower leg					Right foot					Time (0,01 s)
	φ_2	$m_2\varkappa_2^2\varphi_2''$ (kgcm)	D_{s_2} (kgcm)	D_{e_2} (kgcm)	D_{m_2} (kgcm)	φ_4	$m_4\varkappa_4^2\varphi_4''$ (kgcm)	D_{a_4} (kgcm)	D_{e_4} (kgcm)	D_{m_4} (kgcm)	φ_6	$m_6\varkappa_6^2\varphi_6''$ (kgcm)	D_{s_6} (kgcm)	D_{e_6} (kgcm)	D_{m_6} (kgcm)	
0	−13°55′	+36,1	+66,6	−280,7	+250,2	−48°49′	+3,4	+66,7	−123,7	+60,4	−16°27′	+3,6	+2,0	−24,9	+26,5	0
1	−11°41′	+30,1	+56,0	−260,7	+234,8	−50°53′	+10,4	+68,7	−129,6	+71,3	−19°39′	+5,1	+2,4	−25,5	+28,2	1
2	−9°27′	+23,1	+45,4	−232,0	+209,7	−53°10′	+15,4	+70,9	−132,9	+77,4	−21°22′	+4,1	+2,6	−24,9	+26,4	2
3	−7°17′	+16,0	+35,1	−199,8	+180,7	−55°11′	+18,8	+72,7	−136,1	+82,2	−22°24′	+3,2	+2,7	−23,9	+24,4	3
4	−5°9′	0	+24,8	−162,2	+137,4	−56°19′	+21,8	+73,7	−132,3	+80,4	−22°59′	+2,9	+2,8	−22,1	+22,2	4
5	−3°2′	−9,0	+14,6	−128,5	+104,9	−57°7′	+26,9	+74,4	−120,9	+73,4	−23°5′	+2,8	+2,8	−20,0	+20,0	5
6	−1°5′	−23,1	+5,2	−101,4	+73,1	−57°18′	+42,0	+74,6	−107,6	+75,0	−22°48′	+3,7	+2,8	−17,9	+18,8	6
7	+0°52′	−21,1	−4,2	−77,0	+60,1	−57°7′	+23,8	+74,4	−95,2	+44,6	−22°4′	+4,6	+2,7	−15,8	+17,7	7
8	+2°45′	−18,1	−13,3	−52,0	+47,2	−56°16′	+18,8	+73,7	−83,0	+28,1	−20°51′	+3,3	+2,5	−13,8	+14,6	8
9	+4°35′	−15,5	−22,1	−30,6	+37,2	−55°17′	+16,0	+72,8	−74,1	+17,3	−18°51′	+2,3	+2,3	−11,8	+11,8	9
10	+6°8′	−13,7	−29,6	−8,7	+24,6	−53°48′	+13,7	+71,5	−64,7	+6,9	−16°34′	+1,6	+2,0	−9,8	+9,4	10
11	+7°54′	−12,2	−38,0	+11,0	+14,8	−51°41′	+11,8	+69,5	−57,7	0,0	−13°52′	+1,2	+1,7	−8,1	+7,6	11
12	+9°41′	−10,8	−46;6	+32,8	+3,0	−49°30′	+10,1	+67,4	−50,0	−7,3	−11°0′	+0,9	+1,4	−6,9	+6,4	12
13	+11°10′	−9,6	−53,6	+41,7	+2,3	−47°9′	+9,0	+65,0	−48,1	−7,9	−7°23′	+0,7	+0,9	−6,7	+6,5	13
14	+12°47′	−9,0	−61,2	+46,5	+5,7	−44°52′	+7,8	+62,5	−48,9	−5,8	−4°21′	+0,6	+0,5	−7,6	+7,7	14
15	+14°16′	−8,8	−68,2	+49,7	+9,7	−42°10′	+7,3	+59,5	−51,0	−1,2	−0°45′	+0,5	+0,1	−8,8	+9,2	15
16	+15°35′	−8,7	−74,4	+59,5	+6,2	−39°22′	+6,2	+56,2	−47,8	−2,2	+2°45′	+0,5	−0,3	−9,0	+9,8	16
17	+16°54′	−9,0	−80,5	+77,7	−6,2	−36°33′	+5,6	+52,8	−41,0	−6,2	+6°18′	+0,5	−0,8	−8,0	+9,3	17

Table 9 (continued)

#														
18	+ 8,2	− 6,4	− 1,2	+ 0,6	+ 9°27′	− 12,4	− 30,8	+ 48,5	+ 5,3	− 33°10′	− 18,8	+ 95,3	+ 18°10′	− 9,8
19	+ 6,8	− 4,6	− 1,5	+ 0,7	+ 11°58′	− 28,6	− 11,3	+ 44,9	+ 5,0	− 30°25′	− 31,2	+ 112,2	+ 19°22′	− 10,8
20	+ 5,6	− 2,9	− 2,0	+ 0,7	+ 16° 3′	− 44,2	+ 9,7	+ 39,3	+ 4,8	− 26°18′	− 44,7	+ 129,5	+ 20°34′	− 12,4
21	+ 4,7	− 1,6	− 2,4	+ 0,7	+ 19°22′	− 55,0	+ 25,8	+ 34,5	+ 5,3	− 22°55′	− 59,3	+ 147,4	+ 21°39′	− 14,0
22	+ 4,6	− 1,2	− 2,9	+ 0,5	+ 24° 0′	− 58,6	+ 36,8	+ 28,0	+ 6,2	− 18°24′	− 72,4	+ 162,6	+ 22°34′	− 16,0
23	+ 5,1	− 1,4	− 3,3	+ 0,4	+ 27°27′	− 59,3	+ 44,5	+ 22,1	+ 7,3	− 14°26′	− 83,8	+ 175,0	+ 23°29′	− 19,1
24	+ 5,6	− 1,6	− 3,7	+ 0,3	+ 31° 7′	− 60,1	+ 53,1	+ 16,0	+ 9,0	− 10°26′	− 100,7	+ 193,2	+ 24°28′	− 22,1
25	+ 6,7	− 2,4	− 4,1	+ 0,2	+ 34°57′	− 58,1	+ 59,9	+ 10,2	+ 12,0	− 6°35′	− 109,7	+ 201,0	+ 25° 6′	− 26,1
26	+ 7,8	− 3,2	− 4,4	+ 0,2	+ 38°23′	− 57,9	+ 67,4	+ 4,5	+ 14,0	− 2°55′	− 125,3	+ 212,9	+ 25°37′	− 32,1
27	+ 8,7	− 3,8	− 4,8	+ 0,1	+ 41°46′	− 63,5	+ 74,1	+ 0,8	− 9,8	+ 0°31′	− 146,3	+ 226,8	+ 25°50′	− 40,1
28	+ 8,3	− 3,2	− 5,1	0	+ 45°43′	− 75,1	+ 82,9	− 7,8	0	+ 5° 3′	− 180,8	+ 248,7	+ 25°57′	− 53,2
29	+ 6,1	− 1,0	− 5,5	− 0,4	+ 49°47′	− 90,5	+ 98,4	− 15,2	− 7,3	+ 9°51′	− 201,1	+ 284,4	+ 26° 1′	− 38,1
30	+ 2,6	+ 2,3	− 5,7	− 0,8	+ 53°24′	− 133,7	+ 121,9	− 21,8	− 33,6	+ 14°16′	− 202,8	+ 302,8	+ 25°57′	− 21,1
31	− 3,0	+ 7,8	− 6,0	− 1,2	+ 56°37′	− 187,4	+ 167,0	− 28,6	− 49,0	+ 18°48′	− 174,0	+ 294,2	+ 25°44′	0
32	− 8,1	+ 12,1	− 6,2	− 2,2	+ 60°13′	− 201,3	+ 182,9	− 33,7	− 52,1	+ 22°21′	+ 64,4	+ 199,8	+ 25°33′	+ 16,0
33	− 30,3	+ 32,2	− 6,4	− 4,5	+ 63°50′	− 270,6	+ 254,8	− 37,7	− 53,5	+ 25° 9′	− 315,2	− 162,7	+ 25°20′	+ 34,1
34	− 42,0	+ 39,9	− 6,5	− 8,6	+ 66° 4′	− 263,7	+ 249,7	− 40,6	− 54,6	+ 27°16′	− 255,9	− 86,9	+ 25°13′	+ 51,1
35	− 40,6	+ 41,0	− 6,6	− 6,2	+ 67° 9′	− 232,1	+ 219,4	− 42,5	− 55,2	+ 28°42′	− 209,8	− 32,9	+ 25°23′	+ 58,2
36	− 26,3	+ 28,3	− 6,6	− 4,6	+ 67°19′	− 179,0	+ 166,5	− 43,5	− 56,0	+ 29°24′	+ 162,0	+ 17,8	+ 25°54′	+ 58,9
37	− 17,2	+ 19,7	− 6,5	− 4,0	+ 66°11′	− 148,0	+ 132,7	− 43,2	− 58,5	+ 29°13′	+ 132,0	+ 48,8	+ 26°45′	+ 56,2
38	− 9,3	+ 12,2	− 6,5	− 3,6	+ 65°33′	− 168,0	+ 134,0	− 41,1	− 75,1	+ 27°40′	− 35,6	+ 164,6	+ 27°47′	0
39	+ 0,1	+ 3,0	− 6,4	− 3,3	+ 63°46′	− 156,6	+ 68,1	− 37,5	− 126,0	+ 25° 2′	− 81,0	+ 178,8	+ 28°42′	− 35,1
40	+ 11,6	− 8,5	− 6,2	− 3,1	+ 61° 1′	+ 60,4	+ 29,3	− 31,1	0	+ 20°31′	− 32,8	+ 103,7	+ 29°27′	− 65,2
40,8	+ 12,3	− 9,2	− 6,2	− 3,1	+ 59°35′	+ 161,0	+ 31,2	− 29,0	+ 100,8	+ 19° 5′	− 42,9	+ 103,8	+ 29°41′	− 76,2

Table 10. Values of all the magnitudes involved in the kinetics of the swinging of the left leg in experiment II

Time (0,01 s)	Left thigh					Left lower leg					Left foot					Time (0,01 s)
	φ_3	$m_3\varkappa_3^2\varphi_3''$ (kgcm)	D_{s_3} (kgcm)	D_{e_3} (kgcm)	D_{m_3} (kgcm)	φ_5	$m_5\varkappa_5^2\varphi_5''$ (kgcm)	D_{s_5} (kgcm)	D_{e_5} (kgcm)	D_{m_5} (kgcm)	φ_7	$m_7\varkappa_7^2\varphi_7''$ (kgcm)	D_{s_7} (kgcm)	D_{e_7} (kgcm)	D_{m_7} (kgcm)	
0	−16°37′	+41,1	+79,2	−277,8	+239,7	−45°12′	0	+62,9	−117,6	+54,7	−12°47′	+3,4	+1,6	−21,6	+23,4	0
1	−14°47′	+33,1	+70,6	−253,1	+215,6	−47°23′	+7,0	+65,2	−124,3	+66,1	−14°37′	+3,0	+1,8	−23,0	+24,2	1
2	−13° 0′	+23,1	+62,3	−224,4	+185,2	−49°13′	+12,0	+67,1	−146,1	+91,0	−17° 1′	+2,7	+2,1	−25,4	+26,0	2
3	−10°36′	+12,0	+50,9	−207,9	+169,0	−51° 0′	+17,6	+68,9	−161,4	+110,1	−18°30′	+3,2	+2,3	−26,5	+27,4	3
4	−8°32′	0	+41,1	−186,2	+145,1	−52°39′	+22,7	+70,4	−166,6	+118,9	−20°14′	+4,0	+2,5	−26,3	+27,8	4
5	−6°11′	−8,0	+29,8	−159,6	+121,8	−53°31′	+30,0	+71,2	−165,1	+123,9	−20°41′	+5,2	+2,5	−25,0	+27,7	5
6	−3°57′	−18,1	+19,1	−138,5	+101,3	−53°45′	+42,0	+71,4	−158,4	+129,0	−20°27′	+8,6	+2,5	−22,8	+28,9	6
7	−1°43′	−17,6	+8,3	−106,8	+80,9	−53°10′	+25,5	+70,9	−143,1	+97,7	−19°15′	+3,3	+2,4	−18,9	+19,8	7
8	−0°10′	−15,3	−0,8	−78,8	+64,3	−52° 5′	+16,5	+69,9	−121,3	+67,9	−17°39′	+2,3	+2,2	−14,8	+14,9	8
9	+2° 4′	−12,9	−10,0	−55,3	+52,4	−50°36′	+13,4	+68,5	−102,7	+47,6	−15°32′	+1,8	+1,9	−11,4	+11,3	9
10	+4° 1′	−11,7	−19,4	−31,6	+39,3	−49°10′	+11,5	+67,0	−86,5	+31,0	−13° 4′	+1,5	+1,6	−8,8	+8,7	10
11	+5°44′	−10,7	−27,7	−13,8	+30,8	−47° 6′	+9,8	+64,9	−73,3	+18,2	−10° 5′	+1,2	+1,3	−7,0	+6,9	11
12	+7°27′	−10,1	−35,9	+3,3	+22,5	−44°38′	+8,7	+62,2	−64,5	+11,0	−7°30′	+1,0	+0,9	−6,0	+6,1	12
13	+9°13′	−9,8	−44,3	+14,2	+20,3	−41°56′	+7,6	+59,2	−56,6	+5,0	−4°38′	+0,9	+0,6	−5,5	+5,8	13
14	+11° 7′	−9,9	−53,4	+27,4	+16,1	−39°29′	+6,4	+56,3	−51,6	+1,7	−1°19′	+0,8	+0,2	−5,7	+6,3	14
15	+12°40′	−10,5	−60,7	+42,7	+7,5	−36°44′	+5,3	+53,0	−47,7	+0,0	+1°40′	+0,7	+0,2	−6,4	+7,3	15
16	+14° 6′	−11,1	−67,4	+53,0	+3,3	−33°52′	+4,8	+49,4	−47,6	+3,0	+5° 6′	+0,7	−0,6	−7,4	+8,7	16
17	+15°39′	−12,2	−74,7	+59,7	+2,8	−30°29′	+4,2	+44,9	−41,3	+0,6	+8°46′	+0,8	−1,1	−8,1	+10,0	17

Table 10 (continued)

n															
18	+17° 4'	—13,5	—81,2	+68,8	—1,1	—27°44'	+3,6	+41,2	—31,9	—5,7	+12°43'	+0,9	—1,6	—8,4	+10,9
19	+18°17'	—15,4	—86,8	+83,5	—12,1	—24°45'	+3,1	+37,1	—19,1	—14,9	+16°16'	+1,0	—2,0	—8,1	+11,1
20	+19°39'	—17,6	—93,1	+102,3	—26,8	—21° 9'	+2,8	+32,0	—6,3	—22,9	+20° 7'	+0,8	—2,5	—7,5	+10,8
21	+20°44'	—20,2	—98,0	+126,1	—48,3	—17°39'	+3,4	+26,9	+10,2	—33,7	+24°31'	+0,5	—3,0	—6,5	+10,0
22	+21°36'	—22,6	—101,9	+164,5	—85,2	—14°23'	+4,5	+22,0	+42,2	—59,7	+28°35'	+0,4	—3,4	—3,0	+6,8
23	+22°31'	—25,1	—106,0	+191,7	—110,8	—10°26'	+5,6	+16,0	+65,1	—75,5	+31°58'	+0,3	—3,8	—1,3	+5,4
24	+23° 9'	—28,1	—108,8	+218,7	—138,0	—6°32'	+6,7	+10,1	+88,3	—91,7	+36° 6'	+0,2	—4,2	+0,3	+4,1
25	+23°40'	—32,1	—111,1	+234,9	—155,9	—2°59'	+8,4	+4,6	+102,9	—99,1	+40° 6'	+0,1	—4,6	+0,3	+4,4
26	+24°11'	—36,1	—113,4	+239,2	—161,9	+0°31'	+5,6	—0,8	+117,7	—111,3	+44°14'	0	—5,0	+0,7	+5,7
27	+24°18'	—40,1	—113,9	+249,2	—175,4	+4°38'	+2,2	—7,2	+130,1	—120,7	+47°37'	—0,3	—5,3	+1,3	+6,3
28	+24°11'	—41,7	—113,4	+256,7	—185,0	+8°29'	0	—13,1	+142,7	—129,6	+51°24'	—0,7	—5,6	+0,1	+4,8
29	+23°54'	—38,6	—112,1	+226,7	—153,2	+12° 9'	—5,0	—18,6	+139,2	—125,6	+55°24'	—1,1	—5,9	+1,3	+3,5
30	+23°23'	—18,1	—109,9	+181,4	—89,6	+16° 3'	—8,4	—24,5	+129,5	—113,4	+59° 4'	—1,4	—6,1	+2,6	+2,1
31	+22°45'	0	—107,0	+81,9	+25,1	+19°32'	—11,5	—29,6	+113,1	—95,0	+62°37'	—1,6	—6,3	+4,6	+0,1
32	+22° 7'	+18,1	—104,2	+1,1	—121,2	+23° 2'	—14,8	—34,7	+102,5	—82,6	+65°22'	—1,9	—6,5	+8,7	+4,1
33	+21°36'	+34,1	—101,9	—59,9	—195,9	+26°18'	—25,2	—39,3	+97,2	—83,1	+68°42'	—2,9	—6,7	+10,2	+6,4
34	+21°19'	+60,2	—100,6	+15,5	+145,3	+28°32'	—44,8	—42,3	+159,3	—161,8	+71°34'	—6,3	—6,8	+17,5	+17,0
35	+21°26'	+74,7	—101,1	+188,9	—13,1	+29°20'	—50,4	—43,4	+307,8	—314,8	+73° 0'	—12,5	—6,8	+44,7	+50,4
36	+21°57'	+75,5	—103,5	+305,3	—126,3	+29°37'	—52,4	—43,8	+374,8	—383,4	+73°31'	—5,3	—6,8	+43,0	+41,5
37	+22°34'	+60,2	—106,2	+253,2	—86,8	+28°46'	—57,4	—42,6	+291,5	—306,3	+73° 3'	—4,2	—6,8	+32,2	+29,6
38	+23°36'	0	—110,8	+77,5	+33,3	+27°27'	—154,1	—40,8	+86,9	—200,2	+71°23'	—3,2	—6,8	+13,1	+9,5
39	+24°38'	—45,1	—115,4	+83,2	—12,9	+24°21'	0	—36,5	+26,2	+10,3	+69°51'	—2,4	—6,7	+4,3	+0,0
40	+25°26'	—70,2	—118,9	+76,4	—27,7	+21°50'	+81,2	—32,9	+12,6	+126,7	+67°43'	—2,1	—6,6	+3,9	+8,4
40,9	+25°47'	—79,2	—120,4	+143,4	—102,2	+20°27'	+112,0	—31,0	+11,8	+154,8	+65°40'	—1,9	—6,5	+7,0	+11,6

Table 11. Values of all the magnitudes involved in the kinetics of the swinging of the right leg in experiment III

Time (0,01 s)	Right thigh					Right lower leg					Right foot					Time (0,01 s)
	φ_2	$m_2\kappa_2^2\varphi_2''$ (kgcm)	D_{s_2} (kgcm)	D_{e_2} (kgcm)	D_{m_2} (kgcm)	φ_4	$m_4\kappa_4^2\varphi_4''$ (kgcm)	D_{s_4} (kgcm)	D_{e_4} (kgcm)	D_{m_4} (kgcm)	φ_6	$m_6\kappa_6^2\varphi_6''$ (kgcm)	D_{s_6} (kgcm)	D_{e_6} (kgcm)	D_{m_6} (kgcm)	
0	−20°14′	+33,1	+95,7	−207,0	+144,4	−38°23′	0	+55,0	−101,2	+46,2	−11°28′	+2,9	+1,4	−19,2	+20,7	0
1	−18°34′	+27,1	+88,1	−259,1	+198,1	−40°37′	+5,9	+57,7	−113,1	+61,3	−14°9′	+3,4	+1,7	−20,5	+22,2	1
2	−16°47′	+18,1	+79,9	−182,7	+120,9	−43°9′	+11,5	+60,6	−135,5	+86,4	−16°44′	+5,2	+2,1	−22,1	+25,2	2
3	−14°37′	+10,0	+69,8	−170,3	+110,5	−45°9′	+15,4	+62,8	−161,2	+113,8	−18°17′	+4,0	+2,2	−23,8	+25,6	3
4	−12°26′	0	+59,6	−149,5	+89,9	−46°42′	+17,9	+64,5	−158,5	+111,9	−19°19′	+3,3	+2,4	−22,9	+23,8	4
5	−10°15′	−21,1	+49,3	−130,2	+59,8	−47°23′	+20,2	+65,2	−147,4	+102,4	−19°39′	+3,2	+2,4	−21,0	+21,8	5
6	−8°8′	−24,1	+39,2	−110,8	+47,5	−47°40′	+23,2	+65,5	−131,9	+89,6	−19°43′	+3,8	+2,4	−18,7	+20,1	6
7	−6°18′	−18,1	+30,4	−91,9	+43,4	−47°44′	+28,9	+65,6	−115,6	+78,9	−19°1′	+5,3	+2,3	−16,0	+19,0	7
8	−4°11′	−14,5	+20,2	−79,4	+44,7	−47°13′	+18,2	+65,0	−105,7	+58,9	−17°39′	+3,2	+2,2	−14,2	+15,2	8
9	−2°24′	−12,2	+11,6	−64,7	+40,9	−46°21′	+15,4	+64,1	−94,4	+45,7	−15°52′	+2,3	+1,9	−11,9	+12,3	9
10	−0°38′	−10,4	+3,1	−51,4	+37,9	−45°9′	+13,7	+62,8	−85,7	+36,6	−13°4′	+1,8	+1,6	−10,0	+10,2	10
11	+0°58′	−9,1	−4,7	−38,8	+34,4	−43°29′	+12,0	+61,0	−75,9	+26,9	−11°3′	+1,5	+1,4	−8,1	+8,2	11
12	+2°45′	−8,7	−13,3	−25,7	+30,3	−42°0′	+10,9	+59,3	−70,9	+22,5	−8°29′	+1,2	+1,1	−6,3	+6,4	12
13	+4°25′	−8,2	−21,3	−8,2	+21,3	−39°46′	+10,1	+56,7	−77,0	+30,4	−5°44′	+1,0	+0,7	−5,2	+5,5	13
14	+6°4′	−8,1	−29,3	+5,6	+15,6	−37°42′	+9,2	+54,2	−49,8	+4,8	−2°45′	+0,8	+0,3	−4,7	+5,2	14
15	+7°34′	−8,8	−36,4	+11,3	+16,3	−35°4′	+8,4	+50,9	−42,9	+0,4	−0°41′	+0,6	−0,1	−5,5	+6,2	15
16	+9°10′	−9,5	−44,1	+17,5	+17,1	−32°15′	+7,8	+47,3	−39,4	−0,1	+4°4′	+0,5	−0,5	−6,2	+7,2	16
17	+10°33′	−10,5	−50,7	+26,3	+13,9	−29°51′	+7,6	+44,1	−35,2	+1,3	+7°23′	+0,5	−0,9	−6,7	+8,1	17

Table 11 (continued)

18	+11°58'	−11,7	−57,4	+44,7	+1,0	−26°45'	+7,0	+39,9	−27,0	−5,9	+10°36'	+0,4	−1,3	−6,6	+8,3	18
19	+13°21'	−13,2	−63,9	+66,0	−15,3	−23°5'	+6,2	+34,7	−18,1	−10,4	+14°13'	+0,6	−1,8	−6,1	+8,5	19
20	+14°30'	−14,8	−69,3	+91,7	−37,2	−20°10'	+5,6	+30,5	−4,4	−20,5	+17°8'	+0,7	−2,1	−5,3	+8,1	20
21	+15°35'	−17,0	−74,4	+114,4	−57,0	−17°15'	+5,3	+26,3	+10,0	−31,0	+20°10'	+0,9	−2,5	−4,4	+7,8	21
22	+16°30'	−19,1	−78,6	+129,4	−69,9	−13°42'	+6,2	+21,0	+23,1	−37,9	+23°54'	+0,9	−2,9	−3,6	+7,4	22
23	+17°22'	−22,1	−82,6	+144,8	−84,3	−9°3'	+6,7	+13,9	+35,5	−42,7	+27°40'	+0,9	−3,3	−2,9	+7,1	23
24	+18°13'	−25,1	−86,5	+158,0	−96,6	−5°20'	+8,1	+8,2	+48,1	−48,2	+31°51'	+0,8	−3,8	−2,4	+7,0	24
25	+18°41'	−30,1	−88,7	+164,5	−105,9	−1°19'	+9,0	+2,0	+57,3	−50,3	+35°59'	+0,7	−4,2	−2,2	+7,1	25
26	+19°8'	−35,1	−90,7	+173,3	−117,7	+2°14'	+10,4	−3,5	+68,3	−54,4	+39°29'	+0,5	−4,5	−2,1	+7,1	26
27	+19°22'	−44,1	−91,8	+174,1	−126,4	+5°58'	+10,6	−9,2	+77,1	−57,3	+42°41'	+0,3	−4,8	−2,0	+7,1	27
28	+19°29'	−62,2	−92,3	+174,6	−144,5	+10°43'	+3,9	−16,5	+83,7	−63,3	+46°25'	0	−5,2	−2,6	+7,8	28
29	+19°15'	−50,1	−91,3	+171,9	−130,7	+14°37'	0	−22,4	+89,9	−67,5	+51°20'	−0,7	−5,6	−3,2	+8,1	29
30	+18°48'	−25,1	−89,2	+175,3	−111,2	+19°56'	−18,8	−30,2	+101,6	−90,2	+54°53'	−1,9	−5,8	+0,8	+4,7	30
31	+18°17'	0	−86,8	+187,9	−101,1	+23°23'	−27,4	−35,2	+136,6	−128,8	+59°8'	−2,9	−6,1	+8,9	−5,7	31
32	+17°46'	+25,1	−84,5	+176,0	−66,4	+26°52'	−32,8	−40,0	+165,2	−158,0	+62°24'	−3,3	−6,3	+15,6	−12,6	32
33	+17°18'	+45,1	−82,3	+176,5	−49,1	+29°41'	−36,7	−43,9	+209,9	−202,7	+65°16'	−3,8	−6,5	+29,1	−26,4	33
34	+17°11'	+57,7	−81,8	+147,6	+8,1	+31°27'	−39,8	−46,2	+215,6	−209,2	+67°50'	−4,7	−6,6	+29,2	−27,3	34
35	+17°25'	+60,2	−82,8	+128,8	+14,2	+32°39'	−42,3	−47,8	+227,6	−222,1	+69°30'	−13,7	−6,7	+28,7	−35,7	35
36	+17°53'	+58,2	−85,0	+125,7	−17,5	+33°4'	−45,4	−48,3	+243,5	−240,6	+69°40'	−5,1	−6,7	+29,0	−27,4	36
37	+18°41'	+47,1	−88,7	+201,1	−65,3	+32°43'	−49,6	−47,9	+268,5	−270,2	+69°2'	−3,9	−6,7	+28,1	−25,3	37
38	+19°39'	0	−93,1	+168,8	−75,7	+31°3'	−57,4	−45,7	+278,1	−289,8	+67°50'	−3,2	−6,6	+16,5	−13,1	38
39	+20°51'	−20,1	−98,5	+119,0	−40,6	+28°39'	−168,1	−42,5	+74,9	−200,5	+66°0'	−2,7	−6,5	+4,6	−0,8	39
40	+21°43'	−45,1	−102,4	+120,1	−62,8	+25°54'	0	−38,7	+31,3	+7,4	+64°21'	−2,4	−6,4	−1,7	+5,7	40
41	+22°14'	−68,2	−104,7	+141,6	−105,1	+24°18'	+84,0	−36,5	+14,7	+105,8	+62°3'	−2,2	−6,3	+3,5	+7,6	41
41,45	+22°21'	−74,2	−105,2	+144,8	−113,8	+23°57'	+98,0	−36,0	+14,0	+120,0	+61°18'	−2,2	−6,3	−10,1	+14,2	41,45

Table 12 (continued)

18	+18°27'	−14,0	−87,6	+97,5	−23,9	−34°19'	+5,9	+49,9	−13,6	−30,4	+12°23'	+0,6	−1,5	−4,6	+6,7	18
19	+19°46'	−16,0	−93,6	+107,7	−30,1	−31° 7'	+5,6	+45,8	−1,2	−39,0	+16°54'	+0,6	−2,1	−3,8	+6,5	19
20	+20°51'	−18,1	−98,5	+116,1	−35,7	−27°34'	+5,3	+41,0	+11,0	−46,7	+20° 7'	+0,7	−2,5	−3,0	+6,2	20
21	+21°46'	−19,6	−102,6	+124,5	−41,5	−24°28'	+5,0	+36,7	+21,8	−53,5	+24° 0'	+0,7	−2,9	−2,5	+6,1	21
22	+22°31'	−21,6	−106,0	+134,3	−49,9	−20°14'	+6,2	+30,6	+31,6	−56,0	+27°23'	+0,9	−3,3	−2,3	+6,5	22
23	+23° 5'	−24,1	−108,5	+145,8	−61,4	−16°30'	+7,0	+25,2	+43,8	−62,0	+30°56'	+1,1	−3,7	−1,6	+6,4	23
24	+23°40'	−26,1	−111,1	+153,7	−68,7	−12°53'	+9,2	+19,8	+53,1	−63,7	+34°33'	+0,9	−4,0	−1,7	+6,6	24
25	+24° 4'	−29,1	−112,9	+167,9	−84,1	−9°13'	+11,2	+14,2	+60,8	−63,8	+38°51'	+0,9	−4,5	−2,0	+7,4	25
26	+24°18'	−32,6	−113,9	+185,9	−104,6	−5°20'	+7,8	+8,2	+67,8	−68,2	+42°24'	0	−4,8	−2,7	+7,5	26
27	+24°28'	−35,1	−114,6	+164,6	−85,1	−1°33'	+3,1	+2,4	+76,6	−75,9	+46°52'	−0,4	−5,2	−4,3	+9,1	27
28	+24°24'	−37,6	−114,3	+159,0	−82,3	+2°11'	0	−3,4	+82,2	−78,8	+51°51'	−1,1	−5,6	−5,6	+10,1	28
29	+24°18'	−38,3	−113,9	+158,4	−82,8	+7° 6'	−4,5	−11,0	+90,4	−83,9	+56°23'	−2,5	−5,9	−5,8	+9,2	29
30	+24° 0'	−34,1	−112,6	+160,6	−82,1	+10° 8'	−6,4	−15,6	+109,6	−100,4	+59°42'	−2,8	−6,2	−1,1	+4,5	30
31	+23°26'	−25,1	−110,1	+185,6	−100,6	+14°19'	−10,6	−21,9	+151,7	−140,4	+62°44'	−2,9	−6,3	+5,9	−2,5	31
32	+22°48'	−15,0	−107,3	+178,2	−85,9	+18°48'	−14,3	−28,6	+174,8	−160,5	+65°50'	−3,0	−6,5	+10,3	−6,8	32
33	+22° 7'	0	−104,2	+172,1	−67,9	+21°46'	−16,2	−32,9	+199,0	−182,3	+68°45'	−3,2	−6,7	+19,7	−16,2	33
34	+21°26'	+25,1	−101,1	+143,6	−17,4	+24°35'	−21,0	−36,9	+205,8	−189,9	+70°56'	−3,4	−6,7	+23,4	−20,1	34
35	+20°55'	+57,2	−98,8	+113,8	+42,2	+27°20'	−42,0	−40,7	+213,4	−214,7	+73°27'	−4,1	−6,8	+28,8	−26,1	35
36	+20°38'	+65,7	−97,5	+70,6	+92,6	+28°59'	−48,5	−42,9	+191,6	−197,2	+74°43'	−5,3	−6,9	+24,8	−23,2	36
37	+20°44'	+64,7	−98,0	+37,6	+125,1	+29°17'	−53,8	−43,3	+162,7	−173,2	+74°50'	−17,6	−6,9	+19,6	−30,3	37
38	+21° 2'	+35,1	−99,3	+19,8	+114,6	+28°22'	−168,1	−42,1	+136,2	−262,2	+74°19'	−3,9	−6,9	+15,4	−12,4	38
39	+21°33'	0	−101,7	+47,8	+53,9	+25° 6'	0	−37,6	+116,0	+78,4	+73°34'	−2,4	−6,8	+11,5	+7,1	39
40	+22° 4'	−45,0	−104,0	+99,2	−40,3	+22°24'	+112,0	−33,8	+94,9	+50,9	+71°13'	−2,1	−6,8	−8,1	−3,4	40
41	+22°21'	−70,2	−105,2	+178,1	+143,1	+21° 5'	+75,6	−31,9	+69,6	+37,9	+69°13'	−1,9	−6,7	+4,0	+0,8	41
41,3	+22°24'	−75,2	−105,5	+194,6	−164,3	+20°58'	+64,4	−31,7	+63,2	+32,9	+69° 2'	−1,9	−6,7	+2,6	+2,2	41,3

In Fig. 10 the torques D_s with which gravity acts on the three segments of the leg during the swinging period and the torques of the muscular forces are shown as solid lines, in the same way as the effective forces were illustrated in Fig. 1. In each of these graphs time is entered along the abscissa: 1 mm corresponds to 0.01 s. The torques are entered along the ordinate: 1 mm corresponds to a torque of 10 kg cm. The unit of torque is thus represented by 0.1 mm. In the curves, it would have been beneficial to show the torques related to the foot on the ordinates at a stronger magnification, as this would have enhanced the clarity. Nevertheless, the same scale was used for all the curves to allow comparison. If the ordinate values in all the curves had been magnified by about two, those for the foot would not have been much clearer and those for the thigh and lower leg would have lost considerably in clarity. Finally, the scale selected is approximately appropriate to the accuracy attainable in this analysis. The positional φ-angles are represented by dotted curves. This allows values of the torques to be assigned to the position of the leg. For these curves, 1 mm along the ordinate corresponds to an angle of $1°$.

Comparison of Tables 7–12 and the curves in Fig. 10 confirms that the sign of the torque of gravity D_s is always opposite to that of the corresponding positional angle φ. When the curve of the torque lies above the abscissa that of the positional angle lies below, and vice versa.

At the beginning of the swinging period, gravity tends to rotate the three segments of the leg counter-clockwise as seen from the right. Later it tends to rotate them clockwise. The counter-clockwise part of the period is shorter for the thigh and the foot than its clockwise part. The opposite is true for the lower leg.

For the thigh, the positive torque starts at 80 kg cm on average, decreases continuously and reaches 0 before the end of the first quarter of the swinging period. From then on it is negative and keeps increasing in absolute value fairly uniformly. Towards the end of the second third of the swinging period is has reached a value of over 120 kg cm for the right leg and over 110 kg cm for the left in the experiments without load. It remains at approximately this level for the right thigh whereas it falls progressively to some 100 kg cm for the left. In the last eighth of the swinging period, the negative torque increases in absolute value in all cases. At heel strike it has reached a maximum of about 140 kg cm for the right thigh, and 120 kg for the left. During walking with a load the left leg behaves in approximately the same way as in the other experiments, the only difference being that at the end the torque does not attain the same level. For the right leg, in contrast, the negative torque has perceptibly smaller absolute values: at the end of the period it reaches a maximum barely over 100 kg cm. However, at the beginning of the swinging period gravity acts on the right thigh with a somewhat more pronounced positive torque during walking with a load than during walking without a load.

For the lower leg, the positive torque exerted by gravity initially has a value of 60–70 kg cm. It increases a little until it has reached a maximum of over 70 kg cm some 0.06 s after toe-off. After this it decreases, at first slowly and then more quickly. It reaches 0 at the end of the second third of the swinging period, much later than for the thigh. It then becomes negative and increases in absolute value until, at the beginning of the last eighth of the swinging period, it has reached a maximum between 40 and 50 kg cm. Finally it decreases again. At heel strike it is about 30 kg cm.

The torque exerted by gravity on the foot behaves very similarly to that acting on the lower leg. But it changes sign much sooner, shortly after the beginning of the second third of the swinging period, and overall its value remains very low. In the first part of the period, when it is positive and tends to rotate the foot counter-clockwise, it is below 3 kg cm. After it has become negative its maximum value is slightly less than 7 kg cm.

The Torques Exerted by the Effective Forces

The components of the effective forces of the segments of the leg in the direction of gait and in the vertical direction were calculated in the first section of the present chapter. These components will be used to calculate the torques exerted by the effective forces. Each couple is thus replaced by two couples. The moment arms of the two component couples can be expressed by the positional angles φ much more easily than the moment arm of the initial couple.

The effective forces are referred to as E, qualified by the appropriate index or indices for the body segments under consideration and their horizontal and vertical components as X and Z with the appropriate indices. To find the torques of the effective forces D_e, each couple $[+E, -E]$ will be replaced by the couples $[+X, -X]$ and $[+Z, -Z]$ with the relevant indices.

As shown in Chap. 5, both couples $[+E_2, -E_2]$ and $[+E_{4,6}, -E_{4,6}]$ act on the right thigh. The forces $+E_2$ and $+E_{4,6}$ are applied at the centre H of the hip, the force $-E_2$ at the centre of gravity S_2 of the thigh and the force $-E_{4,6}$ at the centre of the knee. This is shown in Fig. 6 (see also Fig. 3, p. 331, Chap. 5). These two couples are replaced by the four $[+X_2, -X_2], [+Z_2, -Z_2], [+X_{4,6}, -X_{4,6}]$ and $[+Z_{4,6}, -Z_{4,6}]$.

The moment arm of the couple $[+X_2, -X_2]$ the horizontal forces of which are applied at H and S_2, is the projection of the distance $\overline{H_2S}$ or r_2 on the vertical. Its length is $r_2\cos\varphi_2$. If the positional angle φ_2 is positive, the couple tends to rotate the thigh clockwise, that is in a negative direction. The moment of this first couple is $-X_2 r_2\cos\varphi_2$.

The moment arm of the second couple $[+Z_2, -Z_2]$ the forces of which are vertical, is the projection of the distance $\overline{HS_2}$ on the horizontal. Its length is thus $r_2\sin\varphi_2$. This couple also tends to rotate the thigh in the negative direction. Its moment is $-Z_2 r_2\sin\varphi_2$.

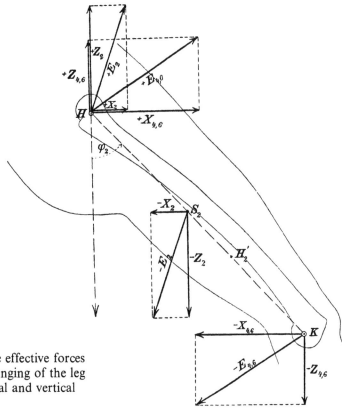

Fig. 6. Couples resulting from the effective forces acting on the thigh during the swinging of the leg and their resolution into horizontal and vertical components

The couples $[+X_{4,6}, -X_{4,6}]$ and $[+Z_{4,6}, -Z_{4,6}]$ differ from the first ones in their magnitudes and in the distance l_2 between their points of application H and K. Otherwise, the direction of their forces and their direction of rotation are the same. Their moments are $-X_{4,6}l_2\cos\varphi_2$ and $-Z_{4,6}l_2\cos\varphi_2$.

The torque D_{e_2} with which all these effective forces tend to rotate the thigh is equal to the sum of the moments of the different couples. If the four results are arranged appropriately the following equations are obtained for calculation of the torque in kilogram-centimetres:

$$\left.\begin{aligned} \text{For the right thigh:} \quad D_{e_2} &= -(X_2 r_2 + X_{4,6}l_2)\cos\varphi_2 - (Z_2 r_2 + Z_{4,6}l_2)\sin\varphi_2 \\ \text{and for the left thigh:} \quad D_{e_3} &= -(X_3 r_3 + X_{5,7}l_3)\cos\varphi_3 - (Z_3 r_3 + Z_{5,7}l_3)\sin\varphi_3 \end{aligned}\right\} \quad (6)$$

In these equations, r_2 and r_3 are replaced by the value 17.1, and l_2 and l_3 by the value 38.8. The different values of the components $X_2, Z_2, X_{4,6}, Z_{4,6}; X_3, Z_3, X_{5,7}$ and $Z_{5,7}$ of the effective forces are given in Tables 1–6 and the positional angles φ_2 and φ_3 in Tables 7–12. The torques D_{e_2} and D_{e_3} are also given in Tables 7–12 for all the successive phases of the swinging period at intervals of 0.01 s.

The effective forces act on the right lower leg with the two couples $[+E_4, -E_4]$ and $[+E_6, -E_6]$ represented in Fig. 7 (see also Fig. 5, p. 335, Chap. 5). The forces $+E_4$ and $+E_6$ are applied at the centre K of the knee, the force $-E_4$ at the centre of gravity S_4 of the lower leg, the force $-E_6$ at the centre F of the ankle. These couples are replaced by their four components $[+X_4, -X_4]$, $[+Z_4, -Z_4]$, $[+X_6, -X_6]$ and $[+Z_6, -Z_6]$, as shown in Figure 7.

The moment arms of the first two couples are the vertical and horizontal projections of the segment $\overline{KS}_4$ or r_4. Their lengths are thus $r_4\cos\varphi_4$ and $r_4\sin\varphi_4$. The moment arms of the other two couples are the vertical and horizontal projections of the segment $\overline{KF}$ or l_4. Their lengths are thus $l_4\cos\varphi_4$ and $l_4\sin\varphi_4$. When the positional angle φ_4 is positive, the four couples tend to rotate the lower leg clockwise, that is in a negative direction. Therefore, their moments are $-X_4 r_4\cos\varphi_4$, $-Z_4 r_4\sin\varphi_4$, $-X_6 l_4\cos\varphi_4$ and $-Z_6 l_4\sin\varphi_4$. The torque with which the effective forces tend to rotate the lower leg is calculated using the following equations:

$$\left.\begin{aligned} \text{For the right lower leg:} \quad D_{e_2} &= -(X_4 r_4 + X_6 l_4)\cos\varphi_4 - (Z_4 r_4 + Z_6 l_4)\sin\varphi_4 \\ \text{and for the left lower leg:} \quad D_{e_5} &= -(X_5 r_5 + X_7 l_5)\cos\varphi_5 - (Z_5 r_5 + Z_7 l_5)\sin\varphi_5 \end{aligned}\right\} \quad (7)$$

These give the torques in kilogram-centimetres if r_4 and r_5 are replaced by the value 15.8, and l_4 and l_5 by the value 37.7. The values of the components $X_4, Z_4, X_6, Z_6; X_5, Z_5, X_7$ and Z_7 of the effective forces are given in Tables 1–6 and the directional angles φ_4 and φ_5 in Tables 7–12. The torques D_{e_4} and D_{e_5} have been calculated using Eq. (7) for all the phases of the swinging period at intervals of 0.01 s. They are given in Tables 7–12.

Only the couple $[+E_6, -E_6]$ acts on the right foot. Its forces are applied at the centre F of the ankle and at the centre of gravity S_6 of the foot, as shown in Fig. 8 (see also Fig. 7, p. 336, Chap. 5). This couple can be resolved into the two couples $[+X_6, -X_6]$ and $[+Z_6, -Z_6]$, with the moment arms $r_6\cos\varphi_6$ and $r_6\sin\varphi_6$ in which r_6 is the length of segment $\overline{FS}_6$. These couples are different from the similarly designated couples for the lower leg. The magnitudes of the forces are the same but the moment arms of the couples are different for the foot and for the lower leg since the points of application of the forces are different. The couples of the effective forces acting on the foot tend to rotate the latter in the negative direction when the positional angle φ_6 is positive. The torque of the effective forces in kilogram-centimetres are calculated using simpler equations.

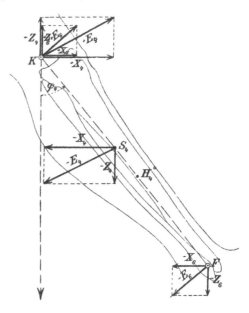

Fig. 7. Couples resulting from the effective forces acting on the lower leg during the swinging of the leg and their resolution into horizontal and vertical components

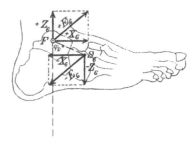

Fig. 8. Couple resulting from the effective forces acting on the foot during the swinging of the leg and its resolution into horizontal and vertical components

$$\text{For the right foot:} \quad D_{e_6} = -X_6 r_6 \cos\varphi_6 - Z_6 r_6 \sin\varphi_6$$
$$\text{and for the left foot:} \quad D_{e_7} = -X_7 r_7 \cos\varphi_7 - Z_7 r_7 \sin\varphi_7 \tag{8}$$

in which r_6 and r_7 are equal to 6.8. The values of the components X_6, Z_6; X_7 and Z_7 of the effective forces are found in Tables 1–6 and the positional angles φ_6 and φ_7 in Tables 7–12. The torques D_{e_6} and D_{e_7} have been calculated and are given in Tables 7–12 for all the phases of the swinging period at intervals of 0.01 s.

In order to provide a clear overall view of the influence of the effective forces on the swinging movement of the leg, the curves representing the torques D_e of the effective forces are drawn in Fig. 9 as broken lines. Along the abscissa 1 mm corresponds to 0.01 s, whereas along the ordinate 1 mm corresponds to a torque of 10 kg cm.

Comparison of these curves with those of the torques of gravity in Fig. 10 immediately shows that the former display greater figures on the ordinate than the latter, for almost the whole of the swinging period. Consideration of Tables 7–12 leads, of course, to the same conclusion. The effective forces thus exert a significantly greater influence on the rotation of the segments of the leg during the swinging movement than gravity. However, it must be borne in mind that the effective forces result essentially from the forces acting directly on the segments of the body. Among these forces, gravity is the most important.

The direction of the rotatory action of the effective forces on the segments of the leg is almost always opposite to that of the couples due to gravity. During the first part of the swinging period, the effective forces tend to impart to the three segments of the leg a negative rotation, a rotation clockwise. During the remainder of the swinging period, with few exceptions, they tend to rotate the segments in a positive, i.e. counter-clockwise, direction.

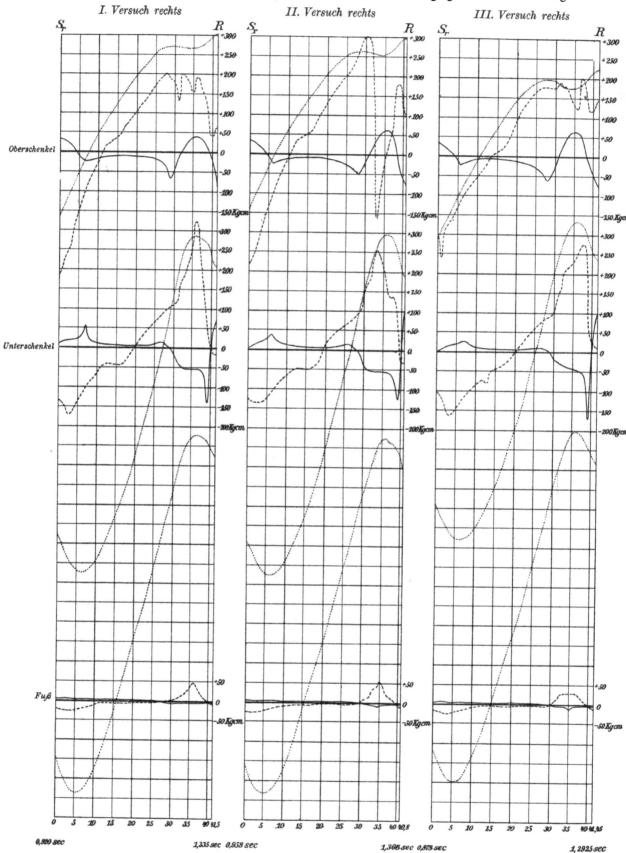

Fig. 9. Products of the moments of inertia and the angular accelerations (——) and torques of the effective forces (– – –) during the swinging of the leg. Positive ordinate values indicate a torque acting counter-clockwise. The dotted lines show the magnitude of the angle, 1 mm representing 1°

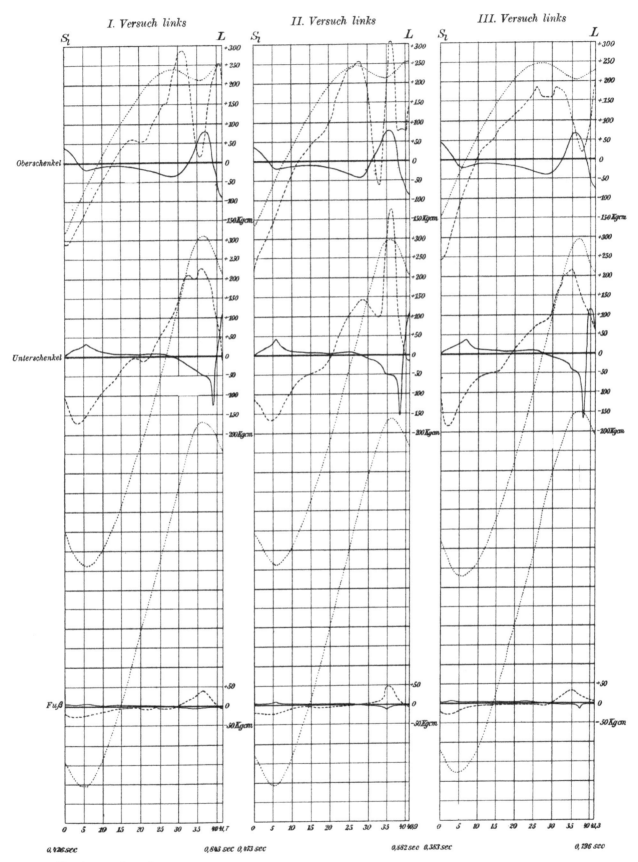

Fig. 9 (continued)

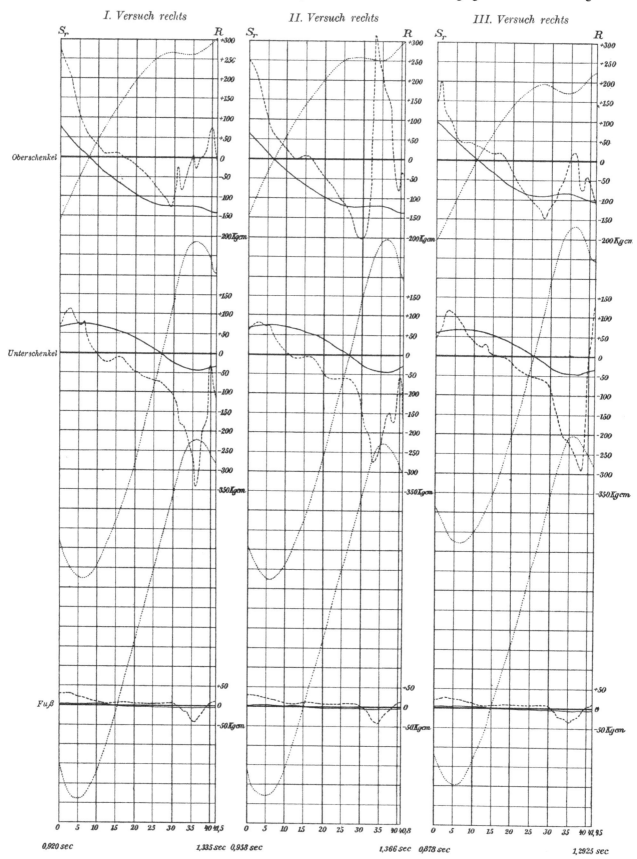

Fig. 10. Curves showing the torques with which gravity (———) and the muscles (————) act on the three segments of the leg during the swinging period. Positive ordinates indicate a torque acting counterclockwise. The dotted lines show the magnitude of the angle φ, 1 mm representing 1°

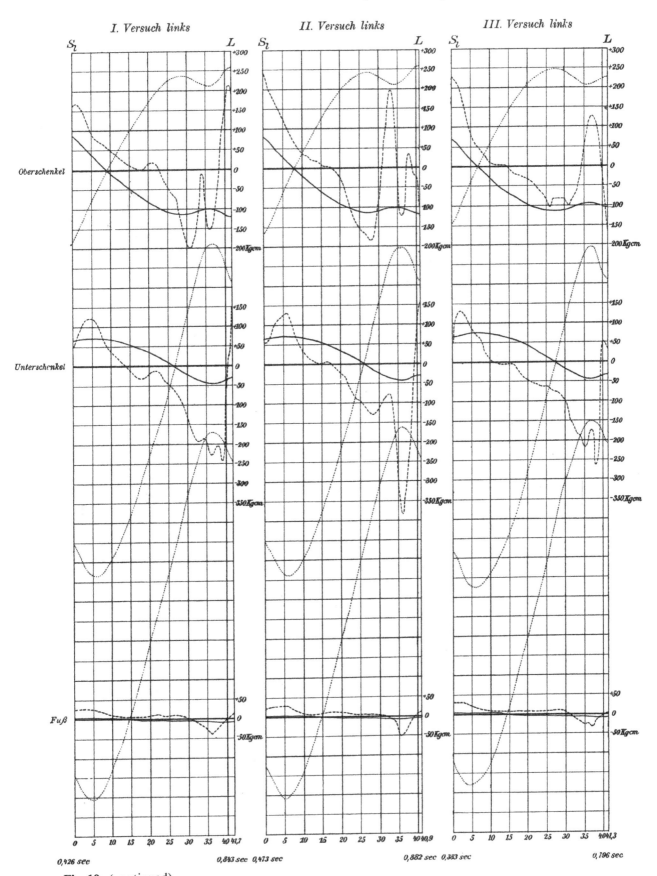

Fig. 10 (continued)

These two parts of the swinging period are fairly equal for the lower leg. For the thigh, the first is much shorter than the second. They are in a ratio of 1 : 3. For the foot, the first is much longer than the second, and they are in a ratio of 3 : 1.

For the thigh, the average negative torque at the beginning of the swinging period is 250 kg cm. It decreases very quickly so that it has reached 0 at the beginning of the second quarter of the swinging period. It becomes positive and increases fairly quickly until it has reached a maximum of 250 kg cm on average. Then it decreases, in some cases so steeply that it reaches 0 again and reverses its sign for a short while. After this it increases again and varies up and down until heel strike. Finally it has an average value hardly over 100 kg cm.

For the lower leg, the initial value of the negative torque is little more than 100 kg cm. In the first 0.03 s it increases to over 150 kg cm and then decreases, quickly at times and more slowly at others, until it reaches 0 in the middle of the swinging period. The torque then becomes positive and increases at varying speed until it reaches a maximum of 250 kg cm or even more somewhere in the middle of the last quarter of the swinging period. Then it decreases so quickly in the remaining 0.05 s that at heel strike it has dropped to 0 on average.

The torque of the effective forces acting on the foot is much smaller. At the beginning it is also negative. Its value is a little more than 20 kg cm. It remains at this level with small upward variations for 0.03 s and then decreases, at first fairly quickly and later more slowly, until it has dropped to 0 towards the end of the third quarter of the swinging period. The torque, having become positive, increases relatively quickly until it reaches its maximum of 40 kg cm on average in the middle of the last quarter of the swinging period. During the remainder of this period it decreases quickly again so that in most instances it has dropped to zero 0.02 s before heel strike. It then reverses its direction but without reaching high values. Its final value is below 10 kg cm.

The Torques Exerted by the Internal Forces

Once the torques of gravity and of the effective forces are known, only the products $m\varkappa^2\varphi''$ for the three segments of the leg are needed for calculation of the torques of the internal forces, and particularly of the muscles, during the swinging period, using the movement equations given in Chap. 5. The radii of inertia $\varkappa$ of the thigh and lower leg are deduced from the lengths of the body segments, and those of the foot from the distance between the centre of gravity of the foot and the axis of the ankle. A previous contribution[6] has shown that the radii of inertia are related to these lengths in a well-determined ratio.

The ratio of the radius of inertia $\varkappa_2$ of the right thigh to the length l_2 of the latter is 0.31. Since $l_2 = 38.8$ cm, the radius of inertia $\varkappa_2$ is 38.8 cm × 0.31 = 12.03 cm. The same value applies to the radius of inertia $\varkappa_3$ of the left thigh.

The average ratio of the radius of inertia $\varkappa_4$ of the right lower leg to the length l_4 of the latter is 0.25. The length l_4 is 37.7 cm. Therefore, the radius of inertia $\varkappa_4$ of the lower leg is 37.7 cm × 0.25 = 9.43 cm. This value applies to the right and to the left lower leg.

The ratio of the radius of inertia $\varkappa_6$ of the right foot to the distance r_6 between the axis of the ankle and the centre of gravity of the foot has been calculated from the data of the table on p. 456 of the work mentioned above and found to be 0.89. For this calculation 5.94 cm and 6.67 cm have been taken as average values of $\varkappa$ and r. Since in our subject r_6 was 6.8 cm, the value of $\varkappa_6$ is 6.8 × 0.89 = 6.05 cm. This is true for both feet.

---

6   Abhandlungen der mathematisch-physischen Klasse der Königlichen Sächsischen Gesellschaft der Wissenschaften, vol 18, no 8, pp 460 and 461 (1892)

The radii of inertia $\varkappa$ can now easily be used to calculate the moments of inertia $m\varkappa^2$, since the proportional figures are known for the masses of the thigh, 0.00693; of the lower leg, 0.00315; and of the foot, 0.00107 in our system of measurement. Calculation gives:

For the moment of inertia of the right and left thigh:

$$m_2\varkappa_2^2 = m_3\varkappa_3^2 = 1.0029$$

For the moment of inertia of the right and left lower leg:

$$m_4\varkappa_4^2 = m_5\varkappa_5^2 = 0.2801$$

For the moment of inertia of the right and left foot:

$$m_6\varkappa_6^2 = m_7\varkappa_7^2 = 0.0392$$

These values have to be multiplied by those given in Tables 6–11 of Chap. 5 (pp. 349–354) for the corresponding angular accelerations φ''. The results of this calculation are given in Tables 7–12 of the present chapter in kilogram-centimetres, the same unit as that of the torques D. Here 1 kg is not the unit of mass but rather the unit of force. These products of the moments of inertia and the angular accelerations are thus expressed in the same dimension as are the torques, and the magnitudes of both can thus be compared. To allow this comparison, the curves of these products of the moments of inertia and the angular accelerations have been drawn in Fig. 9, like those of the torques, but in solid lines. A detailed description of the curves can be omitted, especially as the behaviour of these data during the swinging period coincides exactly with that of the angular accelerations φ'' described in Chap. 5, and $m\varkappa^2\varphi''$ is directly proportional to these.

Since the torques D_s and D_e and the products $m\varkappa^2\varphi''$ are known, the torques D_m exerted on the segments of the leg by the internal forces and particularly by the muscles can be calculated. The equations to be used for this purpose result from the considerations in Chap. 5.

For the right thigh: $D_{m_2} = m_2\varkappa_2^2\varphi_2'' - D_{s_2} - D_{e_2}$
For the left thigh: $D_{m_3} = m_3\varkappa_3^2\varphi_3'' - D_{s_3} - D_{e_3}$
For the right lower leg: $D_{m_4} = m_4\varkappa_4^2\varphi_4'' - D_{s_4} - D_{e_4}$
For the left lower leg: $D_{m_5} = m_5\varkappa_5^2\varphi_5'' - D_{s_5} - D_{e_5}$ (9)
For the right foot: $D_{m_6} = m_6\varkappa_6^2\varphi_6'' - D_{s_6} - D_{e_6}$
For the left foot: $D_{m_7} = m_7\varkappa_7^2\varphi_7'' - D_{s_7} - D_{e_7}$

All the data on the right-hand side of each equation are given in Tables 7–12 for the phases of the swinging period separated by intervals of 0.01 s. The values of the torques D_m have been calculated[7] using Eq. (9) and are also given in Tables 7–12. Thus, all the data involved in the kinetics of the swinging of the leg are presented together in these tables.

The curves of the torques D_m exerted by the internal forces have been drawn in Fig. 10 in broken lines (pp. 426–427). Each millimetre along the abscissa corresponds to 0.01 s, and each millimetre on the ordinate to a torque of 10 kg cm. The curves for the torques D_m are presented together with those of the torques of gravity D_s to provide an easy view of the relations between gravity and internal forces during the swinging movement of the leg.

---

[7]   The long calculations necessary to find the different torques D_s, D_e and D_m have been deposited in the archives of the Königlich Sächsischen Gesellschaft der Wissenschaften

Throughout the study the torques of the internal forces D_m have been referred to briefly as the torques exerted by the muscles. However, the internal forces acting during the swinging period do not result solely from the voluntary contraction of muscles of the leg. In any movement of a limb some muscles are passively stretched. As a consequence of their elasticity, they thereby exert internal forces on the leg exactly as if they were actively contracted. These internal forces resulting from passive tautening of the muscles generally differ from those provoked by active contraction only in having a smaller magnitude. But they act on the segments of the leg with torques of the same kind as when the muscles are actively contracted. It can also happen that the movement of the body segments in relation to each other tightens some ligaments limiting the range of movement of the joint. For instance, at the hip joint the iliofemoral ligament (Bigelow's ligament) restricts extension and the ligamentum teres restricts adduction of the thigh. When the movement of a joint in a given direction is so far advanced that a ligament is taut, the tension in the latter can be considerable and will exert the same torques on the two articulated segments as contracting muscles. Finally, the stretching of any elastic tissue such as the skin, resulting from movement in a joint, can entail torques acting on the different body segments. However, these are usually small. All these torques are comprised in the torques of the internal forces. When present, they are thus components of the torque D_m.

The internal forces resulting from passive tightening of limiting ligaments or other tissues obviously oppose the movement of the joint that has caused them. This is also true for passively taut muscles when they are monarticular. Polyarticular muscles, however, when passively tightened can entail a rotation in one or the other of the joints which they span (of course not in all at the same time), in the same direction as the joint movement already in progress.

The question now arises as to how much such tensions in the ligaments and other tissues actually modify the swinging movement of the leg, besides the tension of the muscles. The effect of the tensions of the tissues must be very small compared with that of the other forces. It can thus be disregarded and be considered as one of the unavoidable sources of error in the study, like those resulting from simplifying assumptions, e.g., the one concerning the movement of the knee. As far as the limiting ligaments are concerned they can only exert their action in the extreme positions of the joint.

At hip level, it can hardly be assumed that, at toe-off, the iliofemoral ligament is markedly taut since at the beginning of the swinging movement the backward inclination of the long axis of the thigh to the vertical is not so pronounced as it was at the beginning of the preceding double-support period. This can be seen from the data in Chap. 3 (Tables 2–4, pp. 236–241 and the curves in Figs. 7–9, pp. 243–247). However, this cannot be definitely decided as long as the position of the pelvis is not known. At all events, at the beginning of the swinging period passive tension of this ligament could not do more than exert a counterclockwise, i.e., positive, torque on the thigh.

Approximately in the middle of the last quarter of the swinging period, exactly 0.07 s before heel strike, the knee is almost fully extended (Chap. 3). Then fibres of the ligaments limiting extension, which means the collateral ligaments according to Weber and Weber[8], are taut. These ligaments, when tightened, behave as a monarticular muscle spanning the back of the knee. They must thus act on the thigh with a positive torque, and on the lower leg with a negative torque.

The photographic exposures do not allow a reliable decision as to whether the ankle reaches extreme positions during the swinging period because the number of points on the foot from which the trajectories were recorded during walking was too small for a precise determination of the movements in the different joints of the foot. If limiting ligaments are

---

[8]   cf. Die Mechanik der menschlichen Gehwerkzeuge, §70

passively tightened during the swinging period this can only occur at toe-off. In the remainder of the swinging period, no noticeable tension of limiting ligaments could exert opposite torques on the lower leg and ankle.

On the Activity of the Muscles During the Swinging of the Leg

Except for a few moments when the limiting ligaments may be able to act on the swinging movement, the torques D_m of the internal forces result exclusively from the muscles that are actively contracting or passively stretched. They will, therefore, be referred to as torques of the muscles.

Comparison of the broken curves representing the torques of the muscles with the uninterrupted ones for the torques of gravity in Fig. 10 shows at a glance that the muscles act on the three segments of the leg with much greater torques than does gravity during most of the swinging period. The same conclusion appears in Tables 7–12.

According to Weber and Weber's concept, during walking, the leg which is not in contact with the ground is moved by its own weight and swings like a pendulum from behind forwards. The results of our research demonstrate that this concept cannot be true, even approximately. If the swinging movement were due to gravity alone the analysis would have resulted in a value 0 for all the torques D_m or, taking into account the unavoidable sources of error, in values so small that they would have been negligible compared with those of the torques of gravity. Nor is it appropriate to object that the sources of error make it impossible to answer the question with the method we have used. Despite the long and difficult calculation to find the torques of the muscles, the far-reaching concordance of their curves related to the six different swinging movements studied independently of each other indicates that the results obtained in this way are unquestionable.

It has thus been definitely demonstrated that *the swinging of the leg during gait is not purely a pendulum swing. It must be attributed much more to the action of muscles than to gravity.*

Tables 7–12 and the curves in Fig. 10 also give some insight into the muscle groups which cause the swinging movement of the leg during walking and into the magnitude of the resultant torques with which the muscles act on the three segments of the leg. However, the values of the torques D_m do not enable us reliably to decide which individual muscles are involved. But the results of the present research bring us one essential step nearer to the solution of this problem. They reveal what rotatory action the muscles exert on the thigh, lower leg and foot at each moment, even though the muscles have not yet been precisely identified. Thus, the swinging movement brought about by the muscles is no longer our main problem. But a series of static problems has yet to be solved. These consist in determining all the muscles which have to contract to ensure a particular attitude of the body at rest. The present work has thus shifted the movement itself from the centre of our endeavours and led the research from the field of kinetics into that of statics.

Tables 7–12 and Fig. 10 reveal the following results relative to the activity of the muscles during the swinging period.

At the beginning of the swinging period the muscles act on the thigh with a positive torque. They tend to rotate it counter-clockwise. At first they thus support gravity in their attempt to rotate the thigh with its lower end forwards. They develop a torque more than twice that due to gravity, on average. Gravity is thus not sufficient to bring about the rotation of the thigh which occurs at the beginning of the swinging period. If gravity alone were acting the initial rotation of the thigh might occur in the same direction but its angular velocity would be much smaller.

A model of a double pendulum can be constructed with two rulers. This model shows that, if the pendulum swings from a position corresponding to the initial attitude of the swinging leg, the segment closest to the point of suspension can be moved away from the vertical, that is against the force of gravity, at the beginning of its swing providing the second segment is sufficiently heavy. If the proximal segment moves towards the vertical initially, it tends to reach the latter with a smaller angular velocity than the distal segment. The distal segment moves from the first in such a way that its angle of inclination to the vertical becomes smaller. Comparison of the behaviour of the double pendulum with that of the leg shows that other forces must initially be acting on the leg besides gravity. At the beginning of the swinging period the lower leg rotates backwards whereas the thigh immediately moves towards the vertical with relatively higher velocity. Thigh and lower leg cannot behave otherwise at the beginning of their rotation if the foot is not to touch the ground again during the swinging.

Which muscles can exert such a positive torque on the thigh at the beginning of the swing? Four kinds of muscles must be considered: firstly muscles which span only the hip; secondly muscles which span the hip and the knee; thirdly muscles that span only the knee; and finally muscles which span both the knee and the ankle. All these muscles exert a torque on the thigh when tautened[9]. The muscles of the first group, to act on the thigh with a positive torque, must span the hip in front, like the iliopsoas muscle for instance. In the initial position of the leg the long axis of the lower leg forms an angle of about 30° with the prolongation of the long axis of the thigh[10]. In this position the gracilis is the only muscle in the second group that can exert a significant positive torque on the thigh. The rectus femoris does not do so. Muscles of the third group act on the thigh with a positive torque if they lie behind the knee, like the short head of the biceps femoris muscle. The muscles of the fourth group also need to lie behind the knee if they are to exert a positive torque on the thigh. This is actually true for only one muscle of this kind, the gastrocnemius.

These muscles must, therefore, be considered especially if we want to decide which of them contract at the beginning of the swinging movement. The values of the torques exerted simultaneously on the lower leg and the foot may perhaps give a hint. As seen in Fig. 10 and Tables 7–12, at the beginning of the swing the lower leg and the foot are subjected to positive torques brought about by the muscles. But both the gracilis and the short head of the biceps can exert only negative torques on the lower leg, whatever the position of the knee. The gastrocnemius always acts on the foot with a negative torque whereas it acts on the lower leg with a positive torque. Therefore, it is likely that flexors of the thigh such as the iliopsoas muscle have contracted at the beginning of the swing[11]. This is very probable but has not yet been proved. It may be that the negative torques exerted by the gracilis and the short head of the biceps femoris muscle on the lower leg and that exerted by the gastrocnemius on the foot are cancelled by positive torques developed by other muscles. It must be stressed that the values given here for the torques do not necessarily originate from one single muscle. They correspond to the resultant torque of all the muscles involved. The possibility cannot be excluded that muscles act with a torque that has the opposite sign to that of the resultant torque. In this case the muscles acting in the other direction would need to develop larger torques to counterbalance the effect of the former muscles. This occurs, for instance, when antagonists are passively tautened. As will be shown below, at the beginning of the swinging

---

[9] cf. my article: Das statische und das kinetische Maß für die Wirkung eines Muskels, erläutert an ein- und zweigelenkigen Muskeln des Oberschenkels. Abhandlungen der mathematisch-physischen Klasse der Königlichen Sächsischen Gesellschaft der Wissenschaften, vol 27, no 5 (1902)

[10] op. cit., Tables 16–20 and pp. 540–542; in these tables the torques are provided with opposite signs to those used in the present work. The torques acting counter-clockwise as seen from the right thus have a negative sign

[11] Duchenne's observation also supports this opinion. Subjects in whom the flexors of the thigh are paralyzed can no longer complete the swinging of the leg whereas palsy of other muscles of the thigh has no such effect. (Physiologie der Bewegungen. German by C. Wernicke, p 285)

period the rectus femoris muscle presumably contracts although when the leg is in the initial position it acts on the thigh with a negative torque, even if this is small (see Table 16 of the work cited above[12]).

After the leg has begun to swing, the torque exerted on the thigh by the muscles decreases very fast until it has reached 0 approximately at the end of the first third of the swinging period. Then either muscles are no longer acting or the muscles contracting or elastically tautened balance each other in their action on the thigh. This lasts for some time, about 0.04 s, until the torque becomes negative and again increases fairly quickly in magnitude.

Of the muscles which span the hip alone, the extensors such as the gluteus maximus exert a negative torque on the thigh. Among the muscles which span the hip and the knee, a negative torque is exerted by the rectus femoris (according to the research on the muscles of the thigh mentioned above) although it lies in front of the hip, by the long head of the biceps femoris, the semimembranosus and the semitendinosus muscles. Finally, the vastus lateralis, vastus intermedius and vastus medialis must be mentioned, which span the knee only. Since the torque simultaneously exerted on the lower leg is also negative, particular attention must be paid to the muscles which tend to rotate the thigh and the lower leg clockwise. They are the biarticular long head of the biceps femoris, the semimembranosus and semitendinosus muscles. Knowledge of the resultant torque, of course, does not make it possible to decide which of these contract at a given time.

The negative torque exerted on the thigh reaches a maximum towards the end of the third quarter of the swinging period. Up to then the muscles involved must have been contracting with increasing force. During the last quarter of the swinging period the negative torque suddenly decreases so that it very soon reaches 0 and becomes positive. It may then reach a considerable magnitude. The torques are not in complete concordance in the six cases analysed. However, there is a sudden change in the same direction in all of them. The torque exerted simultaneously on the lower leg displays the opposite behaviour. This torque is also negative and suddenly increases considerably in absolute value in the last quarter of the swinging period. This inconsistency in the variation of the two torques suggests quite a sudden involvement of a new force. One might think that new muscles become intensively involved in the movement at that moment. There is, however, a much simpler explanation.

Precisely when the sudden changes in magnitude of the two torques take place, the collateral ligaments of the knee very probably become active as limiting ligaments. The torques due to their passive tautening are such that they can result in this sudden great change in the value of D_m. Indeed, the tension of a limiting ligament passively stretched increases very quickly in magnitude and then the torque exerted by these ligaments on the thigh is positive whereas that exerted by the same ligaments on the lower leg is negative. Because of these new torques, the negative torque of the muscles acting on the thigh decreases and eventually becomes positive whereas that acting on the lower leg increases in absolute value. This is actually what is observed. The sudden involvement of the collateral ligaments would also explain the lack of coincidence at the ends of the curves. Passively induced tension of a ligament depends on more fortuitous influences and is thus less constant in magnitude than tension resulting from active contraction of a muscle.

At heel strike a negative torque is again exerted on the thigh in most cases. Its absolute magnitude, however, is not significant and is very variable.

Let us compare the action of gravity with the influence of the muscles on the movement of the thigh during the swinging period. At the beginning, muscles and gravity support each

---

[12] Das statische und das kinetische Maß für die Wirkung eines Muskels, erläutert an ein- und zweigelenkigen Muskeln des Oberschenkels. Abhandlungen der mathematisch-physischen Klasse der Königlichen Sächsischen Gesellschaft der Wissenschaft, vol 27, no 5 (1902)

other. Both tend to rotate the thigh counter-clockwise. However, the torque of gravity becomes 0 sooner than that of the muscles. Consequently, after the long axis of the thigh has passed the vertical, muscles and gravity work against each other for a time. The former still tend to rotate the thigh counter-clockwise whereas gravity tends to rotate it clockwise. After the torque of the muscles has passed through 0 and has thus changed its direction, muscles and gravity again support each other in rotating the thigh clockwise back to vertical.

The behaviour of the torque D_m exerted by the muscles on the lower leg had to be taken into account in the analysis of the muscles acting on the thigh. It appeared that at the beginning of the swinging period the muscles exert a positive torque on the lower leg. At the beginning its magnitude coincides fairly well with that of the positive torque resulting from gravity. It is now necessary to find out to which muscles this tendency to rotate can be attributed. The ones that must be considered are the quadriceps, rectus femoris, vastus lateralis, vastus intermedius and vastus medialis; the gastrocnemius; and the soleus which spans only the ankle. The three monarticular parts of the quadriceps tend to rotate the thigh in a negative direction and the gastrocnemius and soleus bring about a negative rotation of the foot. At the beginning positive torques are exerted on the thigh and on the foot. Of all the muscles above only the rectus femoris remains qualified to exert such torques. It also tends to rotate the thigh in the negative direction but the magnitude of its torque is small compared with that of the negative torques exerted on the thigh by the three other parts of the quadriceps.

The positive torque exerted on the lower leg during the swinging period increases at first until, about 0.03 s after toe-off, it reaches a maximum which is nearly twice its initial value. Consequently, at this moment the muscles, probably the rectus femoris, must be contracting much more strongly than at the beginning of the swinging period. After this, the torque decreases again relatively quickly until it reaches 0 a little sooner than the torque exerted on the thigh. At this moment, either no muscles are acting or the muscles contracting or elastically tautened counterbalance each other. For the lower leg also this lasts some time, about 0.05 s, until the torque reverses its direction. Having become negative, the torque increases in absolute magnitude, at first gradually and then, as already mentioned, very quickly up to a maximum of 300 kg cm on average after which it decreases just as quickly. At heel strike it is still negative in some cases. It has become positive in others. The average value for the six cases is close to 0. The torque of the muscles is thus uncertain at the end of the swinging period for the lower leg as well as for the thigh. The discrepancies in the final aspect of the two curves cannot be attributed to errors because the curves coincide everywhere else. Presumably, this is also an effect of the collateral ligaments of the knee.

It remains to be decided which muscles can have brought about the negative torque acting on the lower leg. This question is partly answered by considering the action on the thigh. The long head of the biceps femoris, the semimembranosus or the semitendinosus muscles may be involved. One or the other of these is probably contracted at this time, as mentioned above. Moreover, the muscles lying in front of the ankle, such as the tibialis anticus, can also be involved. These muscles simultaneously tend to rotate the foot in the positive direction. This would be consistent with the torque acting on the foot at least until the beginning of the last quarter of the swinging period. This is apparent from the lower curves in Fig. 10. During the last quarter of the swinging period, the torques acting on the lower leg and on the foot both have negative signs. An effect of this kind could result from contraction of the gastrocnemius combined with that of one or the other of the long head of the biceps femoris, the semimembranosus or the semitendinosus muscles. If the gastrocnemius tends to rotate the lower leg in the positive direction, the positive torque of this muscle will presumably be significantly smaller than the negative torque simultaneously exerted on the lower leg by the long head of the biceps femoris, the semimembranosus or the semitendinosus muscles. This is because the direction of its resultant pull causes the gastrocnemius to act on the lower leg with a smaller torque than on the foot. It is also in keeping with the fact that, because of the small torque exerted on the foot, this muscle will contract with little force at the end of the swinging period.

Let us also compare the torques of the muscles with those of gravity for the lower leg during the swinging movement. At the beginning they support each other in tending to rotate the lower leg counter-clockwise. At the beginning of the second quarter of the swinging period, the torque of the muscles becomes 0 and, some 0.05 s later, negative. Then muscles and gravity work against each other for a time. Only towards the end of the second third of the swinging period, after the torque due to gravity has also become negative, do both torques together tend to rotate the lower leg clockwise, back towards the vertical.

The resultant torque of the muscles acting on the foot remains positive longer than those exerted on the thigh and lower leg. Only after the beginning of the last quarter of the swinging period does it become negative. However, about 0.02 s before heel strike it becomes positive again and keeps this sign until the end. Whereas the torque acting on the lower leg is no smaller than that exerted on the thigh, the torque of the muscles acting on the foot is significantly smaller throughout the swinging period. Consequently, the muscles which bring about this torque contract with relatively little force during the swinging period.

The positive torque initially remains at a magnitude of less than 30 kg cm for about 0.03 s. Then it decreases progressively to an average magnitude below 10 kg cm where it remains for about 0.2 s with little variation. Finally it falls to 0. Only the muscles lying in front of the ankle, such as the tibialis anticus, can cause this positive torque. The direction of the action of these muscles on the lower leg is negative. However, the magnitude of the torque exerted on the foot is so small that, at lower leg level, this disturbing torque can easily be counterbalanced by the torque of the other muscles inserted into the lower leg. It is possible that the foot muscles already mentioned contract actively only in the first quarter of the swinging period and later act by way of their elastic tension, after the centre of gravity of the foot has passed in front of the axis of the ankle. Then, the torque of gravity and that of the effective forces are negative so that both tend to rotate the foot against the action of the tibialis anticus and may well stretch this muscle. This speculation is supported by the fact that, during the second and third quarters of the swinging period, the torque D_m of the muscles acting on the foot remains at its constant and small magnitude.

After the torque acting on the foot has become negative, it increases relatively quickly until it has reached its maximum of about 40 kg cm in the middle of the last quarter of the swinging period. Then it decreases just as quickly to 0. This negative torque can be attributed only to muscles lying behind the ankle, above all the gastrocnemius and soleus. The soleus tends to rotate the lower leg in the positive direction much more strongly than the gastrocnemius. However, the torque exerted on the lower leg by the muscles at the same time is strongly negative. It can thus hardly be assumed that the soleus is involved in the swinging movement. The gastrocnemius thus remains available for this purpose. This is a more likely candidate since it contracts towards the end of the swinging period, when it tends to rotate the thigh in the positive direction. But the torque acting on the thigh undergoes such a large positive variation precisely at that moment in most cases that it cannot be explained by the contraction of the gastrocnemius alone. The tension thus developed by this muscle cannot be very large because of the small torque exerted on the foot. The sudden positive variation of the torques acting on the thigh and lower leg have been explained by the probable involvement of the collateral ligaments of the knee. The small positive torque which the muscles exert on the foot at the very end means that before heel strike the anterior muscles, such as the tibialis anticus, contract once more or at least act by way of their elastic tension.

Let us compare again the action of the muscles with that of gravity. As can be seen from the curves in Fig. 10, during approximately the first third of the swinging period, muscles and gravity support each other in tending to rotate the foot counter-clockwise. From then on muscles and gravity work against each other until the beginning of the last quarter of the swinging period. Gravity tends to rotate the foot clockwise and the muscles to rotate it counter-clockwise. After this, both again tend to rotate the foot clockwise until 0.03 s before heel strike. From then on gravity tends to rotate the foot clockwise and the muscles counter-clockwise.

Knowledge of the torques allows conclusions concerning the muscles which probably contract at different times to bring about the swinging of the leg in combination with gravity. Of course, we could not demonstrate whether other muscles antagonistic to the former are also active. Our results must be understood as follows. Of all the muscles involved, those found in this analysis probably work on the three segments of the leg with torques the magnitude of which is greater than the algebraic sum of the torques exerted by the muscles given in Tables 7–12. So far it is not yet possible to indicate the magnitude of the torque exerted by each individual muscle. Therefore, the tensions of the different muscles cannot be calculated. This requires further research. If it could be demonstrated beyond all doubt that only one of the aforementioned muscles is active at each of the different phases of the swinging movement, its tension would be obtained by dividing the torque D_m by the moment arm of its couple.

Summary

Chapter 5 dealt with the question of how it could be decided whether the swinging movement of the leg during walking requires muscular activity. The kinematic bases necessary for this study were also derived. The study itself has been carried out and is reported in the present chapter.

In the first section, the components of the effective forces in the direction of gait and in the vertical direction have been determined for the different segments of the leg throughout the swinging movement. Their values are given in Tables 1–6 (pp. 388–393) and are illustrated by appropriate curves in Fig. 1. When these effective forces are known the effect of the movement of the three segments of the leg on the compression in the hip, knee and ankle can be found. This has been explained in the second section.

In the third section, the torques exerted by gravity on the three segments of the leg during the swinging period are calculated. They are given in Tables 7–12 (pp. 408–419) and illustrated by appropriate curves in Fig. 10. In the fourth section the torques with which the effective forces act on the three segments of the leg have been calculated. They are given in Tables 7–12. The relevant curves have been drawn and are displayed in Fig. 9. Comparison of these curves with those in Fig. 10 which illustrate the action of gravity shows that the influence of the effective forces on the rotation of the three segments of the leg is greater than that of gravity, and that by and large the two forces work against each other.

In the fifth section the moments of inertia of the three segments of the leg are discussed in relation to the axes through their centres of gravity and perpendicular to the plane of gait. These moments of inertia have been multiplied by the angular accelerations found in Chap. 5. The products thus obtained are given in Tables 7–12 and illustrated by curves in Fig. 9. Thus, all the data needed for calculation of the still unknown torques of the internal forces, which could not be obtained directly, were available. To this end, the movement equations described in Chap. 5 were used. These long calculations have been carried out and their results are given in Tables 7–12. Moreover, the relevant curves have been drawn in Fig. 10 together with the curves for the torques of gravity. Tables 7 and 12 also display, in degrees, the positional angles of the three segments of the leg. They thus contain all the data involved in the kinetics of the swinging of the leg.

As already shown, for the most part, the internal forces can be developed only by muscular contraction. The torques D_m in the tables thus relate almost exclusively to the action of the muscles. In the study they are, therefore, referred to as torques of the muscles.

Comparison of the torques exerted by the muscles with those of gravity shows that the swinging of the leg results much more from the action of the muscles than from gravity. This proves definitely that the swinging movement of the leg is not a purely pendulum swing, not even approximately. The much discussed pendulum theory of Weber and Weber is thus erroneous.

The torques of the muscles D_m by themselves do not provide any reliable conclusion as to the muscles involved in the swinging movement of the leg. However, they permit the following surmise. At the beginning of the swinging period, the flexors of the thigh, such as the iliopsoas muscle, are acting together with the rectus femoris and the tibialis anticus muscles. The tension in these muscles decreases progressively during the swing until they stop contracting altogether towards the end of the first third of the swinging period. There follows a period of about 0.04 s during which only the tibialis anticus contracts a little or is elastically tautened. After this the extensors of the thigh, such as the gluteus maximus, probably contract together with either the long head of the biceps femoris or the semimembranosus or the semitendinosus muscles whereas the tibialis anticus maintains its tension. In the last quarter of the swinging period the tibialis anticus is presumably replaced by the gastrocnemius muscle. The soleus does not seem to contract. Shortly before heel strike the tibialis anticus may perhaps become active again. However, the two groups of muscles spanning the ankle generally develop only a little tension during the swinging period.

A further specific study will show how far these conjectures on the activity of the muscles during the swinging movement of the leg are supported by other facts.

German	English	German
L. Handgelenk	left wrist	Schwerpun⟩
L. Hüftgelenk	left hip	der Fußl⟩
L. Kniegelenk	left knee	spitze
L. Oberarm	left upper arm	Schwerpun⟩
L. Oberschenkel	left thigh	Schwerpun⟩
L. Schultergelenk	left shoulder	Sec.
L. Unterarm	left forearm	Unterschen⟩
L. Unterschenkel	left lower leg	Versuch
Längsachse des linken Fußes (Schwerpunkt des linken Fußes, der vordere Punkt der Fußlängsachse liegt noch 3 cm hinter der Fußspitze)	longitudinal axis of the left foot (centre of gravity of left foot; the foremost point on the longitudinal axis of the foot is located 3 cm behind the tip of the foot)	Wegkurve / Wegkurven⟩
Längsachse des linken Oberschenkels	longitudinal axis of the left thigh	Winkel
Längsachse des linken Unterschenkels	longitudinal axis of the left lower leg	Winkelbesc⟩
Längsachse des rechten Fußes	longitudinal axis of the right foot	Winkelgesc⟩
Längsachse des rechten Oberschenkels	longitudinal axis of the right thigh	
Längsachse des rechten Unterschenkels	longitudinal axis of the right lower leg	
Mittelpunkt der Hüftlinie	midpoint of the hip line	
Mittelpunkt der Schulterline	midpoint of the shoulder line	
Mittelpunkt des linken Hüftgelenks	centre of the left hip	
Mittelpunkt des linken Schultergelenks	centre of the left shoulder	
Mittelpunkt des rechten Hüftgelenks	centre of the right hip	
Mittelpunkt des rechten Schultergelenks	centre of the right shoulder	
N. hinten	backward	
N. links	to the left	
N. oben	upward	
N. rechts	to the right	
N. unten	down	
N. vorn	forward	
Natürliche Größe	life-size	
Nummern der Bewegungsphasen	phase numbers	
Oberschenkel	right thigh	
Projektion auf	projection onto	
Projektion auf die Gangebene	projection onto the plane of gait	
Projektion auf die Horizontalebene	projection onto the horizontal plane	
Projektion auf die zur Gangrichtung senkrechten Ebene	projection onto the plane perpendicular to the direction of gait	
Projektionen des zugehörigen Hodographen	projections of the corresponding hodograph	
R. Ellenbogengelenk	right elbow	
R. Femurkopf	right head of femur	
R. Fuß	right foot	
R. Fußgelenk	right ankle	
R. Handgelenk	right wrist	
R. Hüftgelenk	right hip	
R. Kniegelenk	right knee	
R. Oberarm	right upper arm	
R. Oberschenkel	right thigh	
R. Schultergelenk	right shoulder	
R. Unterarm	right forearm	
R. Unterschenkel	right lower leg	
Rumpf Schwerpunkt	centre of gravity of trunk	
Schwerpunkt des Fußes	centre of gravity of the foot	
Schwerpunkt des Oberarms	centre of gravity of the upper arm	
Schwerpunkt des Oberschenkels	centre of gravity of the thigh	
Schwerpunkt des Unterschenkels	centre of gravity of the lower leg	
Schwerpunkt des ganzen Beines	centre of gravity of the leg	

differ-
ents

Name and Subject Index

Acceleration 212
  angular 337, 349
  curve 181
  terrestrial 198
accuracy 39
air resistance 120, 199
Alembert d' 212
angle, ankle joint 249, 252
  knee joint 249, 252
Anschütz 6
anterior complementary arc 317
axis of co-ordinates 17

Bird flight 7

Camera, arrangement of 16
  focus of optical system of 36
  nodal point of optical system of 8, 34, 36
  optical axis of 16, 34
Carlet 4
centre of gravity of leg,
  movement of 364
  of body segments 121
    calculation of 124
    device for determining 125
    trajectories of 137
  of whole body 123
    acceleration 195
    movement in the direction of gait 183
      sideways 184
      vertical 184
    relative velocity 195
    trajectory 160
    velocities and accelerations 168
chronophotography 5
compression on the ground 119, 199, 201
  in the joints 211, 329
compressive forces, kinetic 402
  static 401

co-ordinates, absolute 186
  bidimensional 43
  instrument for measurement of 40
  missing 150
  of centre of gravity of foot 74, 75, 79
  of centres of gravity 139, 154
  of joint centres 74, 76, 78
  of military regulation equipment 152
  relative 186, 358
  table of 16
    of measured 44
  of tip of foot 75, 77, 79
  tridimensional 33, 44, 60
  of vertex 75, 77, 79
core-point of trajectory 185
couple of the effective forces 272
  of forces 205, 214
  of gravity 269
  moment arm of 386
  moment of 215, 269, 386
  of the muscles 269

Degree (of arc) 341
directional cosines 65, 72, 73
displacement curve 170
displacement of system of
  co-ordinates 68
distance of centre of gravity
  from central plane moving
  forwards at the average
  velocity of gait 159
double step 157
double-support 169
  duration of 86

Effective force 212
  components of 387
  of foot 300
  reversed 265
electric spark 11
equations of the movement
  207, *216*, 257, 273, 337

external forces 118, 198
  components of 200

Filling the gaps 150
focus of optical system 36
foot-flat 167, 169
force of friction 120
forces, acting on the foot 257
  involved in the swinging of
  the leg 329
forearm hiding hip 71
friction on ground 119
frictional resistance 205

Gait of horse 5
Gassiot star experiment 14
Geissler tube, positioning of 12
Giraud Teulon 99
gravity 198
ground reaction force 119,
  199, 205

Head line 110
head, rotations of 110, 131
heel strike 86, 167, 169
heel support 169
hip line 88
  rotations of 94
hodograph 188

Instantaneous picture 11
internal forces 118

Joint angle 251

Kinematics 255
kinetics 255
  of swinging leg 408
kymograph 4

Londe 6

Marey 5, 6, 319
maxima of displacement 191
  of ground reaction force 201

maxima, of movement of centre
  of gravity of foot 292
measurement of
  co-ordinates 39
  number of 43
Meyer, von 316
microscope 40
military regulation
  equipment 152
minima of displacement 191
  of ground reaction force
  201
  of movement of centre of
  gravity of foot 292
moment arm of a couple 269
  of a couple 269
  of inertia 212
momentum of foot 300
movement, differential
  equations of 207
muscle, action of 206
  activity during swinging
  period 431
  deflection of 205
  elastic tension 205
  innervation 205
Muybridge 5

Nodal point (optics) 8, 34

Optical axis 16, 34
oscillations of the body,
  horizontal 4
  vertical 4

Partial principal point 135,
  332
path 16
pendulum, articulated 2
  double 431
  movement of the swinging leg
  315
  three-segment 315
phases of movement and time
  169, 190

phases, new – and time 220
  photographic 16
photographic projection 8
  rifle 7
photography 8
positional angle 209, 233, 236,
  242, 340
principal distance 133
principal points 128
  partial 135, 332
projection method 5
  on the horizontal floor 80,
  82
  on the plane of gait 80, 82
  on the vertical plane
  perpendicular to that of
  gait 80
proportional figures 122
push-off 169

Radian 341
radius of inertia 273, 428
reduced system 129
resultant couple of the effective
  forces 272
  gravity 271
  the muscles 269
Richer 99
rotation of system of
  co-ordinates 62
Ruhmkorff coil 11, 14

Shoulder line 88
  rotation of 103
simultaneous exposures 16
sine curve 99
single support
  duration of 86
stance period 169, 222
step, duration of 86
  length of 86
stereoscopic view 165, 166
swinging of leg 169
system of mass 212
system of segments 210

Table of co-ordinates 16
tip of the foot 67
toe-off 86, 167, 169
torque of the effective forces
  215, 386, 421
  of gravity 214, 386, 403
  of the internal forces 428
  of the muscles 214, 428
trajectories of centres of gravity
  of the feet 82
  double-curved 86
  of joint centres 82
  of long axes of body
  segments 82
  of mid-point of hip line 90,
  91
  of mid-point of shoulder line
  90, 91
  stereoscopic view of 165,
  166
  of total centre of gravity
  160
  tridimensional 81
  of model of vertex 82, 87
tridimensional co-ordinates,
  system of 16
trunk line 92
  rotations of 107
tuning fork 15, 16
turning point 180
two-sided chronophotography
  10, 32

Velocity absolute 186
  angular 341, 349
  curve 180
  relative 186
vertex 67
Vierordt 4

Walker's step 318, 322
wave line 86
Weber 1, 231, 437
Wheatstone bridge 11

Classic analysis for the first time in English

Braune/Fischer

On the Centre of Gravity of the Human Body

Translated from the German by P. G. J. Maquet, R. Furlong
1984. 33 figures, 51 tables. VII, 96 pages
ISBN 3-540-13216-3

The solution of many a problem in biomechanics requires a knowledge of the position of the centre of gravity of the human body. It varies, sometimes considerably, with the various positions of the body segments. As early as 1679, Joh. Alphonsus Borellus published a scientific work on this subject. Wilhelm Braune and Otto Fischer, using frozen cadavers, determined the centres of gravity of the human body and of its different parts and placed them in a tri-dimensional network of co-ordinates for different attitudes corresponding to life. Being subsidized by the Saxon army, they also included, in the same network, the centres of gravity of an infantry soldier, fully equipped and in soldierly attitudes. Although they date from the end of the last century, Braune and Fischer's results have hardly been questioned. They were used by the authors in their famous analysis of human gait, which is now published for the first time in English. They still remain the most accurate data on the subject and are now accessible for further research.

Springer-Verlag
Berlin
Heidelberg
New York
Tokyo

P. G. J. Maquet

Biomechanics of the Hip

As Applied to Osteoarthritis and Related Conditions
1985. 651 figures, some in color. XIII, 309 pages
ISBN 3-540-13257-0

In osteoarthritis of the hip articular pressure appears to be too high for the resistance of the joint tissues. A sensible treatment is proposed which consists of decreasing joint pressure by different surgical procedures, depending on the individual circumstances. The hanging hip procedure, varus intertrochanteric osteotomy, valgus intertrochanteric osteotomy, shortening of the opposite leg, lateral displacement of the greater trochanter are successively discussed and their results analysed.

P. G. J. Maquet

Biomechanics of the Knee

With Applications to the Pathogenesis and the Surgical Treatment of Osteoarthritis
2nd edition, expanded and revised. 1984. 243 figures.
XVIII, 306 pages
ISBN 3-540-12489-6

"... Dr. Maquet and his publishers are to be congratulated on producing a book of considerable beauty. Its general layout and the clarity of what are often complicated photographs of stress-loading in photo-elastic models, anatomical specimens and radiographs set a standard of technical excellence ..."

Rheumatology and Rehabilitation

F. Pauwels

Biomechanics of the Normal and Diseased Hip

Theoretical Foundation, Technique and Results of Treatment

An Atlas

Translated from the German by R. J. Furlong, P. Maquet
1976. 305 figures, in 853 separate illustrations.
VIII, 276 pages
ISBN 3-540-07428-7

Springer-Verlag
Berlin
Heidelberg
New York
Tokyo

Printed in the USA
CPSIA information can be obtained
at www.ICGtesting.com
LVHW081934210823
755883LV00015B/484

9 783642 703287